T0350906

Shaping Light in Nonlinear Optical Fibers

Shaping Light in Nonlinear Optical Fibers

Edited by

Sonia Boscolo
Aston University, Birmingham, UK

Christophe Finot
Université de Bourgogne - Franche-Comté, Dijon, France

This edition first published 2017

Registered Offices
John Wiley & Sons, Inc., 111 River Street, Hoboken, NJ 07030, USA
John Wiley & Sons Ltd, The Atrium, Southern Gate, Chichester, West Sussex, PO19 8SQ, UK

Editorial Office
The Atrium, Southern Gate, Chichester, West Sussex, PO19 8SQ, UK

For details of our global editorial offices, customer services, and more information about Wiley products visit us at www.wiley.com.

Wiley also publishes its books in a variety of electronic formats and by print-on-demand. Some content that appears in standard print versions of this book may not be available in other formats.

Library of Congress Cataloging-in-Publication Data

Names: Boscolo, Sonia, 1971– editor. | Finot, Christophe, 1978– editor.
Title: Shaping light in nonlinear optical fibers / edited by Sonia Boscolo, Christophe Finot.
Description: Chichester, UK ; Hoboken, NJ : John Wiley & Sons, 2017. | Includes index.
Identifiers: LCCN 2016048229 | ISBN 9781119088127 (cloth) | ISBN 9781119088141 (pdf) | ISBN 9781119088158 (epub)
Subjects: LCSH: Nonlinear optics–Materials. | Optical fibers. | Nonlinear waves.
Classification: LCC TA1800 .S46 2017 | DDC 621.382/75–dc23 LC record available at https://lccn.loc.gov/2016048229

Cover image: Julien Fatome and Stephane Pitois

Set in 10/12pt WarnockPro by Aptara Inc., New Delhi, India

Printed in Great Britain by Antony Rowe

10 9 8 7 6 5 4 3 2 1

Contents

List of Contributors

Esben R. Andresen
PhLAM, UMR 8523
CNRS-Université Lille 1
Villeneuve d'Ascq, France

Juan Diego Ania-Castañón
Instituto de Óptica
CSIC, C/Serrano 121
Madrid, Spain

Reza Ashrafi
Department of Electrical and Computing Engineering
McGill University; Energy, Materials & Telecommunications Research Centre, INRS, Montreal
Quebec, Canada

José Azaña
Energy, Materials & Telecommunications Research Centre
INRS, Montreal
Quebec, Canada

Alain Barthélémy
Institut de recherche XLIM
UMR 7252, CNRS-Université de Limoges
Limoges, France

Jean-Charles Beugnot
Department of Optics
FEMTO-ST Institute
CNRS-Université de Bourgogne - Franche-Comté
Besançon, France

Fabio Biancalana
School of Engineering and Physical Sciences
Heriot-Watt University
Edinburgh, UK

Sonia Boscolo
Aston Institute of Photonic Technologies
School of Engineering and Applied Science
Aston University
Birmingham, UK

Alejandro Carballar
Department of Electronic Engineering
University of Seville
Seville, Spain

Matteo Conforti
Univ. Lille, CNRS
UMR 8523–PhLAM–Physique des Lasers Atomes et Molécules
Lille, France

Claudio Conti
Institute for Complex Systems (ISC-CNR)
Rome, Italy

Julien Fatome
Laboratoire Interdisciplinaire CARNOT de Bourgogne
UMR 6303 CNRS-Université de Bourgogne - Franche-Comté
Dijon, France

María R. Fernández-Ruiz
Energy, Materials &
Telecommunications Research Centre
INRS, Montreal
Quebec, Canada

Christophe Finot
Laboratoire Interdisciplinaire
CARNOT de Bourgogne
UMR 6303 CNRS-Université de
Bourgogne - Franche-Comté
Dijon, France

Josselin Garnier
Centre de Mathématiques Appliquées
Ecole Polytechnique
Palaiseau, France

Silvia Gentilini
Institute for Complex Systems
(ISC-CNR)
Rome, Italy

Massimiliano Guasoni
Laboratoire Interdisciplinaire
CARNOT de Bourgogne
UMR 6303 CNRS-Université de
Bourgogne - Franche-Comté
Dijon, France

Tobias Hansson
INRS-EMT
1650 Boulevard Lionel-Boulet Varennes
Québec, Canada

Paul Harper
Aston Institute of Photonic Technologies
School of Engineering and Applied
Science
Aston University
Birmingham, UK

Bertrand Kibler
Laboratoire Interdisciplinaire
CARNOT de Bourgogne
UMR 6303 CNRS-Université de
Bourgogne - Franche-Comté
Dijon, France

Sergey M. Kobtsev
Novosibirsk State University
Novosibirsk, Russia

Alexandre Kudlinski
Univ. Lille, CNRS
UMR 8523–PhLAM–Physique des Lasers
Atomes et Molécules
Lille, France

Sophie LaRochelle
Center for Optics, Photonics and Lasers
Laval University
Quebec, Canada

Guy Millot
Laboratoire Interdisciplinaire
CARNOT de Bourgogne
UMR 6303 CNRS-Université de
Bourgogne - Franche-Comté
Dijon, France

Levon Mouradian
Ultrafast Optics Laboratory
Faculty of Physics
Yerevan State University
Yerevan, Armenia

Arnaud Mussot
Univ. Lille, CNRS
UMR 8523–PhLAM–Physique des Lasers
Atomes et Molécules
Lille, France

Francesca Parmigiani
Optoelectronics Research Centre
University of Southampton, Highfield
Southampton, United Kingdom

Antonio Picozzi
Laboratoire Interdisciplinaire
CARNOT de Bourgogne
UMR 6303 CNRS-Université de
Bourgogne - Franche-Comté
Dijon, France

Hervé Rigneault
Institut Fresnel
UMR 7249 CNRS-Aix Marseille
Université
Marseille, France

Mohammed F. Saleh
School of Engineering and Physical
Sciences
Heriot-Watt University
Edinburgh, UK

Sergey V. Smirnov
Novosibirsk State University
Novosibirsk, Russia

Thibaut Sylvestre
Department of Optics
FEMTO-ST Institute
CNRS-Université de Bourgogne -
Franche-Comté
Besançon, France

Alessandro Tonello
Institut de Recherche XLIM
UMR 7252 CNRS
Université de Limoges
Limoges, France

Stefano Trillo
Department of Engineering
University of Ferrara
Ferrara, Italy

Sergei K. Turitsyn
Aston Institute of Photonic Technologies
School of Engineering and Applied Science
Aston University
Birmingham, UK

Stefan Wabnitz
Information Engineering Department
University of Brescia
Brescia, Italy

Gang Xu
Laboratoire Interdisciplinaire
CARNOT de Bourgogne
UMR 6303 CNRS-Université de
Bourgogne - Franche-Comté
Dijon, France

Preface

The twentieth century was characterized by a tremendous growth in our capability to develop ever more sophisticated electronic devices, which have fundamentally transformed the way that society functions. We now stand on the threshold of a similar revolution due to developments in photonics, which seeks to exploit the photon in the same way that electronics is ultimately concerned with controlling the electron. Photonics plays a vital role in our daily lives, and is an imperative cross-cutting discipline of science in the twenty-first century. This effectively interdisciplinary field, at the interface of physics, material science, and engineering, is very strongly linked to the field of nonlinear science. Though nonlinear physics has a rather long history, beginning with the works of Newton and Huygens, the science and technologies of the nineteenth and most of the twentieth century have been dominated by linear mathematical models and linear physical phenomena. Over the last few decades, there has been growing recognition of physical systems in which nonlinearity introduces a rich variety of fundamentally new properties that can never be observed in linear models or implemented in linear devices. From a practical standpoint, nonlinearity adds to the difficulty of understanding and predicting the system properties, and may therefore be regarded as a spurious effect that should be avoided. However, with suitable design and control, it is possible to master and exploit nonlinear physical interactions and processes to yield tremendous benefits. Accordingly, the understanding and mastering of nonlinear optical systems have the potential to enable a new generation of engineering concepts, as well as new experimental testbeds to investigate complex nonlinear dynamics of fundamental interest, such as the formation of rogue waves, wave turbulence, and propagation in disordered media.

Nonlinear fiber optics has revolutionized photonics, providing exquisite temporal and spectral control of optical signals for telecommunications, spectroscopy, microscopy, material processing, and many other important areas of applications, as well as a myriad of specialized scientific research tools. The third-order optical nonlinearity in silica-based single-mode fibers is responsible for a wide range of phenomena, such as third-harmonic generation, nonlinear refraction (Kerr nonlinearity), and stimulated Raman and Brillouin scattering. Although the optical nonlinearity of silica is much lower than that exhibited by crystals of such materials as lithium niobate or beta barium borate, silica fibers can provide a comparatively enormous interaction length and tight confinement, which offer the long-recognized possibility of using optical fibers for nonlinear interactions. Several approaches to making highly nonlinear fibers have been used with varying success. Silica fiber manufacture is a mature technology, so the best results for many applications have been achieved by making silica fibers with a small mode area.

Other nonlinear fibers have been made with materials having a much higher nonlinearity, such as bismuth oxide, chalcogenides, and silica doped with lead or other nonlinear material. Photonic crystal fibers of various designs and materials are also used.

For more than three decades, optical fibers have been recognized as a versatile testbed for the investigation of a large variety of nonlinear concepts. The unique dispersive and nonlinear properties of optical fibers lead to various scenarios of the evolution of short pulses propagating in the fiber, which result in particular changes of the pulse temporal shape, spectrum, and phase profile. A well-known and fascinating example is the formation of optical solitons in the anomalous dispersion regime of a fiber. Solitons have been extensively studied in many and diverse branches of physics such as optics, plasmas, condensed matter physics, fluid mechanics, particle physics, and even astrophysics. Interestingly, over the past two decades, the field of solitons and related nonlinear phenomena has been substantially advanced and enriched by research and discoveries in nonlinear optics. Temporal optical solitons, as realized in fiber-optic systems, have revolutionized the field of optical communications in that they can be used as the information carrying "bits" in fibers. Temporal solitons play an important role in ultrashort pulse lasers as well. Moreover, recent analogies between optical supercontinuum generation and the occurrence of oceanic rogue waves have led to the first experimental demonstration of a peculiar hydrodynamic solution that was analytically predicted decades ago: the Peregrine soliton, later followed by the experimental observation of its counterpart in deep water hydrodynamics. The fundamental interest in optical fiber systems is not limited to pulse propagation in the anomalous group-velocity dispersion region. Indeed, recent developments in nonlinear optics have brought to the fore of intensive research an interesting class of pulses with a parabolic intensity profile and a linear instantaneous frequency shift or chirp, which propagate in optical fibers with normal group-velocity dispersion in a self-similar manner. Parabolic similaritons have opened up new avenues of fundamental and applied research in nonlinear science. The success of self-similarity analysis in nonlinear fiber optics is motivating more general studies into the dynamics of guided wave propagation as well as related self-similar evolution in physical systems such as Bose-Einstein condensates. In parallel, the unique properties of parabolic similaritons have stimulated numerous applications ranging from high-power ultrashort pulse generation to highly coherent supercontinuum sources and optical nonlinear processing of telecommunication signals.

Modern state-of-the-art communication systems are nonlinear. The only question that remains is how to respond to that fact. One possibility is simply to regard the nonlinearity as performance degradation and try to minimize its impact. A more intriguing question is whether the nonlinearity can be actively used to improve performance. Today, substantial research work is aimed at harnessing the potential of nonlinear system response. Nonlinear effects in optical fibers can be used to realize a variety of optical functions that have practical applications in the field of light wave technology. Nonlinear processes that have been exploited in demonstrations and applications include stimulated Brillouin and Raman scattering, as well as aspects of the Kerr effect variously called self-phase modulation, cross-phase modulation, four-photon (four-wave) mixing, modulation instability, Bragg scattering, phase conjugation, nonlinear polarization rotation, and parametric gain. Important examples of established and new emerging nonlinear fiber-based photonic technologies essentially relying on nonlinear phenomena include all-optical signal processing and regeneration in ultrafast telecommunications, optical gating, switching and frequency conversion, optical waveform generation and pulse shaping, optical parametric amplification, Raman amplifiers and lasers, high-power pulsed and continuous-wave lasers, broadband and supercontinuum light sources, and other applications.

This book is a contemporary overview of selected topics in fiber optics, focusing on the latest research results on light shaping using nonlinear optical fibers and exploring the very frontiers of light wave technology. Furthermore, this book provides a simple yet holistic view on the theoretical and application-oriented aspects of the various phenomena encountered by the manipulation of the fundamental properties of light, such as the intensity profile, phase, coherence and state of polarization, in optical fibers. The reader is introduced to some of the most innovative theoretical and experimental developments on fiber-optic light shaping technology, ranging from pulse propagation phenomena to pulse generation and shaping, solitons and rogue waves, novel optical fibers, supercontinuum generation, polarization management, optical signal processing, fiber lasers, optical wave turbulence, and light propagation in disordered fiber media.

Structure of the Book

In a nutshell, the contents of this book are as follows: Chapter 1 presents a comprehensive theoretical analysis of the nonlinear dynamics of the phenomenon of modulation instability, which is the main mechanism for the generation of optical solitons, supercontinuum, and rogue waves in nonlinear optical systems. The peculiar nondegenerate modulation-instability process known as Bragg scattering four-wave mixing and modulation-instability processes occurring in a coherently pumped passive fiber cavity are discussed in detail.

Chapter 2 is a review of the basic principles and recent progress in low-noise optical amplification and all-optical signal processing, mainly toward phase regeneration, electric field decomposition and quantization, for phase (and amplitude) encoded signals of various levels of coding complexity. It highlights the key underpinning technology and presents the current state of the art of such devices, mainly focusing on third-order optical nonlinearity-based platform. Gas-filled hollow-core photonic crystal fibers offer unprecedented opportunities to observe novel nonlinear phenomena. The various properties of gases that can be used to fill these fibers give additional degrees of freedom for investigating nonlinear pulse propagation in a wide range of different confined media.

Chapter 3 discusses some of the new nonlinear interactions that have been discovered in recent years, in particular, those which are based on soliton dynamics while Chapter 4 gives an overview of breakthrough research work on modulation instability in periodically modulated fibers. A new, quite unique technique for the fabrication of optical fibers with a large variety of longitudinal modulation profiles is described, along with the basic optical properties of such fibers. Experimental results highlighting the specific features of the modulation-instability process in dispersion-oscillating fibers are presented.

Chapter 5 reviews recent progress on the use of third-order nonlinear processes in optical fibers for the shaping of optical pulses in the temporal and spectral domains. It emphasizes how the normal dispersion regime of a fiber is particularly well suited for the generation of advanced temporal waveforms and stable continua with high power spectral density, as well as for spectral compression of the pulses. Furthermore, the chapter highlights some of the benefits offered by the governing soliton dynamics in the anomalous dispersion region in terms of nonlinear shaping, namely, the possibility of generating ultrashort temporal structures and ultrabroad spectra, and of tuning the central frequency of the pulses.

Chapter 6 is devoted to the demonstration of the nature and distinctive properties of similaritons generated in a fiber without gain due to the combined impacts of nonlinearity and dispersion. It is emphasized how the nonlinear-spectronic character of such similaritons, with the key specificity of linear chirping, leads to their self-spectrotemporal imaging, important for applications to signal analysis and synthesis problems in ultrafast optics and photonics. The methods of similariton-induced time lensing, aberration-free spectral focusing, similariton-based spatiotemporal imaging, and self-referencing spectral interferometry are presented.

Chapter 7 describes how soliton generation in photonic crystal fibers can be advantageously used for applications in biophotonics and microscopy. In particular, applications in conjunction with two-photon excited fluorescence, second-harmonic generation, coherent anti-Stokes Raman scattering, stimulated Raman scattering, and pump-probe microscopy modalities are discussed. Among the three independent features of a light beam propagating within an optical fiber, that is, energy, frequency and state of polarization, the latter remains the most difficult to control in many photonics applications.

Chapter 8 describes a novel all-fiber polarizing device: the omnipolarizer, in which a light beam is able to self-organize and self-trap its own state of polarization. Possible applications in all-optical signal processing, such as self-induced polarization tracking of a data signal, polarization-based optical memories, polarization switching, and polarization scramblers are discussed. Furthermore, the concept of self-organization of light is generalized to the case of the spatial and polarization modes of a multimode fiber. Fiber gratings are a stable and mature technology for the implementation of all-optical pulse shapers.

Chapter 9 presents an extensive review of recent work on the design, experimental demonstration and applications of fiber-grating filter solutions, with a focus on the recent techniques and findings that have enabled sub-picosecond processing speeds to be achieved using this key technology. The nonlinear Schrödinger equation and related modulation instability play a fundamental role in the understanding of the formation of rogue waves in almost conservative systems.

Chapter 10 focuses on the simplest breather solutions on a finite background of the nonlinear Schrödinger equation, and reviews their recent experimental evidence in nonlinear fiber optics, while Chapter 11 is dedicated to wave-breaking and dispersive shock wave phenomena in optical fibers. Recent studies that have permitted substantial progress in the understanding of these phenomena are presented. Such advances encompass the prediction and observation of dispersive shock waves in multiple four-wave mixing, the control of shock dynamics via simple wave excitation, the competition of wave-breaking and modulation instability, the complete theory of the radiation emitted by dispersive shock waves, and the role of dispersive shock waves in passive fiber cavities.

During the past decade, several remarkable phenomena inherent to the nonlinear propagation of incoherent waves in optical fibers have been reported, including supercontinuum generation, quasi-soliton turbulence, optical wave thermalization, incoherent modulation instability, and turbulent behaviors of fiber lasers. Chapter 12 reviews a theoretical formulation of wave turbulence in optical fibers based on the kinetic wave theory, which provides a non-equilibrium thermodynamic description of the system of incoherent nonlinear waves.

The interplay of disorder and nonlinearity is one of the most important topics of theoretical physics today. Starting from the work of Anderson and driven by a large variety of possible applications, the study of nonlinear waves in the presence of disorder has recently attracted a large amount of interest. Chapter 13 reviews recent theoretical and experimental investigations on the generation of optical two-dimensional solitary and shock waves driven by thermal nonlinearity in disordered optical fibers and colloidal liquid solutions. While being very practical devices, mode-locked fiber lasers exhibit quite a fascinating physics of formation and propagation of ultra-short pulses in optical fibers.

Chapter 14 explores the diversity of generation regimes observable in passively mode-locked fiber lasers, which may take place both in lasers relying on different configurations and in a single, continuously operating laser device while the parameters of its cavity are adjusted. Combining extensive numerical modeling and experimental studies, multiple, distinctly different lasing regimes are identified with a rich variety of dynamic behavior and a remarkably broad spread of key parameters of the generated pulses.

Chapter 15 is devoted to the fundamentals and applications of ultralong Raman fiber lasers, which represent a unique type of a transmission medium, offering possibilities that we are still only beginning to explore. The applications described include efficient amplification and nonlinear compensation in high-speed optical communications, secure key exchange, improving the performance of fiber-based optical sensing schemes and generating highly efficient supercontinua in conventional fibers.

Chapter 16 is devoted to the complete description of Brillouin light scattering in specialty optical fibers. It is shown in particular that light-sound interactions in these tiny waveguides are very different from those in standard optical fibers, leading to the generation of new types of acoustic waves including high-frequency shear, hybrid and surface waves. The results presented suggest that photonic crystal fibers or tapered optical fibers can be advantageously used to enhance and harness Brillouin scattering in view of potential applications to fiber-optic sensors and all-optical processing for telecommunications.

We would like to thank the Editorial Staff of Wiley for inviting us to bring a volume on fiber-based light shaping to a wider audience and for their work in coordinating the editorial process. Second, but not least, we are most grateful to all colleagues who contributed to this book for their brilliant work and continued effort in bringing this project to reality.

Sonia Boscolo
Christophe Finot

1

Modulation Instability, Four-Wave Mixing and their Applications

Tobias Hansson,[1] *Alessandro Tonello,*[2] *Stefano Trillo,*[3] *and Stefan Wabnitz*[4]

[1] *INRS-EMT, 1650 Boulevard Lionel-Boulet, Varennes, Québec, Canada*
[2] *Institut de Recherche XLIM, UMR 7252 CNRS, Université de Limoges, Limoges, France*
[3] *Department of Engineering, University of Ferrara, Ferrara, Italy*
[4] *Information Engineering Department, University of Brescia, Brescia, Italy*

1.1 Introduction

Modulation instability (MI) of a continuous wave (CW) background solution of the nonlinear Schrödinger (NLS) equation is a well-known phenomenon that occurs in a variety of fields, such as nonlinear optics, hydrodynamics, plasma physics and Bose-Einstein condensation [1, 2]. In the nonlinear optics context, MI is the main mechanism for the generation of optical solitons, supercontinuum (SC) [3,4], and rogue waves [5]. MI may be induced either by quantum noise or by a weak seed signal [6]: in the latter case, the initial stage of exponential signal amplification is followed first by the generation of higher-order sideband pairs by cascade four-photon mixing processes. Next, nonlinear gain saturation occurs, owing to pump depletion. After the maximum level of pump depletion is reached, which depends on the initial sideband detuning, the pump power and the fiber dispersion, energy flows back from the sidebands into the pump, until the initial condition is recovered, and so on. This phenomen provides a classical example of the so-called Fermi-Pasta-Ulam (FPU) recurrence [7, 8]. This process can be described in terms of exact solutions of the NLS equation [9–12], and has been experimentally observed in different physical settings, such as deep water waves [13, 14], in nonlinear optical fibers [15–20], in nematic liquid crystals [21], magnetic film strip-based active feedback rings [22], and bimodal electrical transmission lines [23]. Important qualitative physical insight into the FPU recurrence dynamics (e.g., the existence of a homoclinic structure and the associated dependence of the FPU recurrence period upon the input relative phase between pump and initial sidebands) may be obtained by means of a truncation to a finite number of Fourier modes, which may lead to simple, low-dimensional models [24–26].

In Section 1.1, we present an overview of the analysis of the nonlinear dynamics of MI by means of a simple three-mode truncation. Next we discuss how the coupling between two polarization modes in a birefringent optical fiber may extend the domain of MI to the normal dispersion regime. Also in the vector case, an important qualitative

Shaping Light in Nonlinear Optical Fibers, First Edition. Edited by Sonia Boscolo and Christophe Finot.

insight into the dynamics of the nonlinear development of MI may be obtained by using a three-mode truncation. We also briefly discuss the case where the MI is induced by two pumps and occurs on top of multiple four-wave mixing. We further present the effect of higher-order MI, which occurs whenever not only the initial modulation seed but also some of its harmonics are modulationally unstable. In this case, at some distance along the fiber one can observe the development of a full modulation of the CW pump with a frequency which is double the initial modulation. A similar situation also occurs in a random birefringence telecom fiber that may be described by means of the Manakov system. In this case, the doubling of the initial modulation frequency occurs whenever the CW pump and the signal are orthogonally polarized. We conclude the first section by examining the competition between spontaneous or noise-induced MI and induced MI, which leads to a break-up of the FPU recurrence after a small number of periods.

MI is a time domain description of a degenerate four-wave mixing (FWM) process in the frequency space. Interestingly, in optical fibers there are a variety of FWM processes: in Section 1.2, we present a peculiar non-degenerate MI process which is also known as Bragg scattering FWM. In this process, two intense pumps lead to a periodic power exchange between a seed signal and an idler, without any exponential amplification of the sidebands. Because of its conservative nature, Bragg scattering FWM has interesting potential applications to quantum optical signal processing.

Finally, in Section 1.3, we present the MI processes that occur in a coherently pumped passive fiber cavity. Again the presence of a phase-sensitive cavity feedback permits the extension of the MI domain into the normal dispersion regime. Moreover, cavity MI is a dissipative type of instability, which leads to the generation of stable trains of optical solitons also known as cavity solitons. We conclude the section by analyzing the MI of the Ikeda map which describes field recirculation in an externally driven cavity in situations (e.g., at high pump powers) that cannot be described by means of the cavity-averaged or mean-field model.

1.2 Modulation Instability

In this section, we first present the linear stability analysis of a plane wave background solution of the NLS. Next we generalize the model to include pump depletion and describe a truncated three-wave model for the pump and its immediate sidebands. A phase space description of the nonlinear MI process in terms of Stokes parameters will be introduced and applied to describing the periodic coupling between amplitude and frequency modulated signals. We also discuss the relationship between the temporal MI process, and its equivalent description in the frequency domain in terms of a degenerate four-wave mixing process.

1.2.1 Linear and Nonlinear Theory of MI

The NLS equation for the slowly varying envelope E of the electric field that propagates along the distance z in a dispersive fiber is given by

$$\frac{\partial E(z,\tau)}{\partial z} + i\frac{\beta_2}{2}\frac{\partial^2 E(z,\tau)}{\partial \tau^2} - i\gamma|E(t,\tau)|^2 E(z,\tau) = 0 \quad (1.1)$$

where β_2 is the group velocity dispersion (GVD) coefficient, $\gamma = \omega_0 n_2/(cA_{eff})$ is the nonlinear coefficient, and τ is a retarded time for a pulse moving at the group velocity at ω_0.

MI of the CW solution of Eq. (1.1) may be analyzed by considering a perturbed solution of the type $E(z,\tau) = (|E_0| + u + iv)e^{iArg\{E_0\}}$. By linearizing Eq. (1.1) around the steady-state CW solution, one obtains

$$(\tilde{u}_z + i\tilde{v}_z) - i\frac{\beta_2}{2}\omega^2(\tilde{u} + i\tilde{v}) - 2i\gamma|E_0|^2\tilde{u} = 0 \tag{1.2}$$

where tilde denotes the Fourier transform with respect to τ. The real and imaginary parts of the previous expression provide two separate linear equations: the possible growth of initial perturbations to the CW background may be investigated by studying the eigenvalues of their coefficient matrix. One finds that the eigenvalues associated with potentially unstable eigenmodes are given by the expression

$$\lambda = \pm\frac{|\beta_2|\omega}{2}\sqrt{-sgn(\beta_2)\omega_c^2 - \omega^2}, \tag{1.3}$$

where $\omega_c^2 = 4\gamma|E_0|^2/|\beta_2|$. Perturbations will experience growth, and the CW solution experiences MI, at those frequencies where the real part of these eigenvalues is positive. From Eq. (1.3) it is clear that no MI is possible in the normal GVD regime where $\beta_2 > 0$. Whereas for anomalous GVD $\beta_2 < 0$, MI occurs for $\omega < \omega_c$, with a maximum exponential growth rate 2λ of the sideband powers equal to $2\gamma|E_0|^2$ at the frequency detuning from the pump $\omega_p = \omega_c/\sqrt{2}$. Physically, this maximum exponential growth is due to the fact that, at this frequency, the four-photon interaction between the pump and the sidebands turns out to be nonlinearly phase matched. Henceforth in this section we consider the anomalous GVD case only.

For the analysis of the nonlinear development of MI past the initial stage of exponential growth of the sidebands, it is convenient to consider a dimensionless version of Eq. (1.1), namely

$$i\frac{\partial U}{\partial z} + \frac{1}{2}\frac{\partial^2 U}{\partial t^2} + |U|^2 U = 0. \tag{1.4}$$

Here z and t denote dimensionless distance and retarded time, respectively. For the study of the nonlinear development of induced MI, we set the following input condition for Eq. (1.4):

$$U(z=0,t) = 1 + \epsilon \exp\left\{i\phi^0/2\right\}\cos(\Omega t), \tag{1.5}$$

where $\epsilon \ll 1$, and Ω is the sideband detuning. In the units of Eq. (1.4), MI occurs for $\Omega < 2$, with peak gain for $\Omega = \sqrt{2}$. In real units, the pump power $|E_0|^2 = (\gamma z_0)^{-1}$, where $z_0 = \tau_0^2/|\beta_2|$, and $\tau_0 = \Omega/\omega$.

Clearly, the linear stability analysis of Eq. (1.2) cannot give any information about the saturation of MI owing to pump depletion. Exact solutions of Eq. (1.4) with the initial condition Eq. (1.5) have been found by direct substitution methods. Quite interestingly, for $\phi^0 = \pm\pi/2$, which corresponds to maximum MI gain, the exact solution of the NLS equation is periodic in time t but aperiodic in the distance coordinate z. In general, the exact solutions of the NLS equation are either periodic or quasi-periodic in both time t and space z. Thus, the aperiodic solution obtained for $\phi^0 = \pm\pi/2$ represents a separatrix solution (or homoclinic loop) in the space of time periodic solutions of the NLS equation.

A visualization of the periodic solutions of the NLS equation that are obtained inside, outside and exactly on the separatrix, is provided in Figure 1.1, where the initial frequency modulation $\Omega = \sqrt{2}$. As can be seen in Figure 1.1, the initial weak modulation

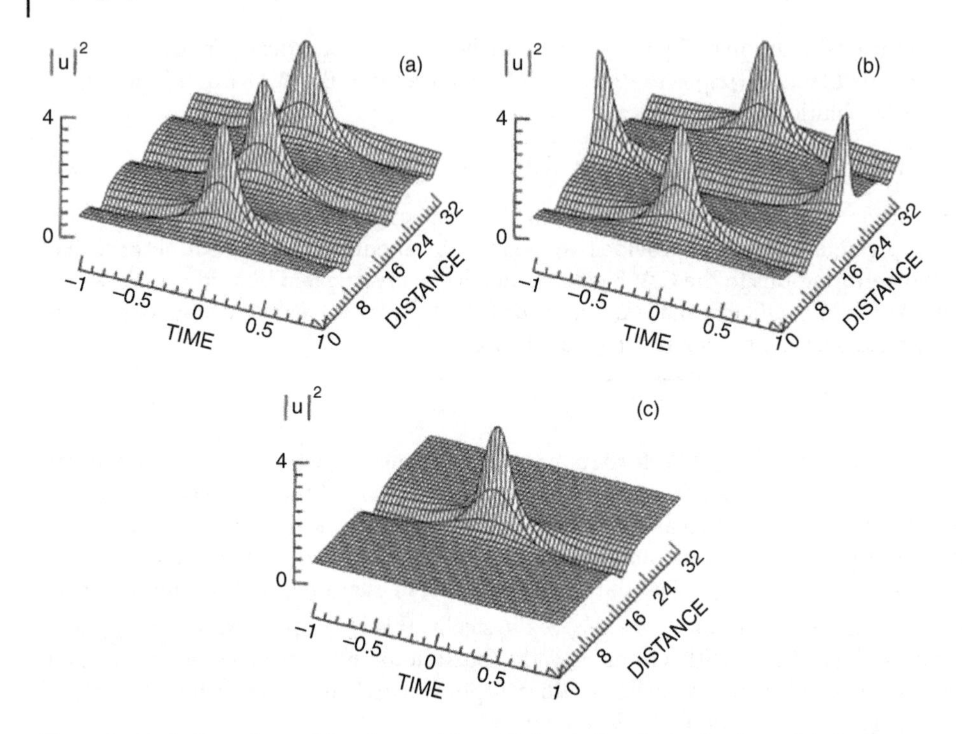

Figure 1.1 Evolutions of a modulated wave. (a) $\epsilon = 10^{-2}$ and $\phi^0 = 0$ (amplitude modulation); (b) $\epsilon = 10^{-2}$ and $\phi^0 = \pi$ (frequency modulation); (c) $\epsilon = 10^{-4}$ and $\phi^0 = -\pi/2$ (separatrix). *Source*: Trillo 1991 [26].

is amplified until most of the energy is transferred from the pump into the sidebands. Past that point, the flow of energy is reversed and the energy flows back from the sidebands into the pump, until the initial weakly modulated state is recovered, and so on. This periodic behavior of the energy flow among the pump and sidebands provides a typical example of the FPU recurrence effect. Figure 1.1 suggests that the FPU recurrence could be captured by using a finite and possibly small number of Fourier modes $A_n \exp(-in\Omega t)$, with $n = 0, \pm 1, \ldots$.

The simplest, yet accurate, model to study the nonlinear evolution of MI is provided by a three-wave truncation [25, 26]

$$U(z,t) = A_0(z,t) + A_{-1}(z,t)\exp(i\Omega t) + A_{+1}(z,t)\exp(-i\Omega t), \tag{1.6}$$

where A_0, A_{-1} and A_{+1} are the pump, Stokes and anti-Stokes wave amplitudes, respectively. By inserting Eq. (1.6) in Eq. (1.4), one obtains the three coupled NLS equations

$$\begin{aligned}
&i\frac{\partial A_0}{\partial z} + \frac{1}{2}\frac{\partial^2 A_0}{\partial t^2} + \left(|A_0|^2 + 2|A_{-1}|^2 + 2|A_{+1}|^2\right)A_0 + 2A_{-1}A_{+1}A_0^* = 0 \\
&i\frac{\partial A_{-1}}{\partial z} - \frac{\Omega^2}{2}A_{-1} + i\Omega\frac{\partial A_{-1}}{\partial t} + \frac{1}{2}\frac{\partial^2 A_{-1}}{\partial t^2} + \left(2|A_0|^2 + |A_{-1}|^2 + 2|A_{+1}|^2\right)A_{-1} + A_{+1}^*A_0^2 = 0 \\
&i\frac{\partial A_{+1}}{\partial z} - \frac{\Omega^2}{2}A_{+1} - i\Omega\frac{\partial A_{+1}}{\partial t} + \frac{1}{2}\frac{\partial^2 A_{+1}}{\partial t^2} + \left(2|A_0|^2 + 2|A_{-1}|^2 + |A_{+1}|^2\right)A_{+1} + A_{-1}^*A_0^2 = 0
\end{aligned} \tag{1.7}$$

The stationary solution $A_j(z,t) = \bar{A}_j(z)$ of Eqs. (1.7) is exactly integrable by quadratures [25]. We may set $\eta(z) = |\bar{A}_0(z)|^2/P_0$ and $\phi = \phi_{-1} + \phi_{+1} - 2\phi_0$, with $\bar{A}_j(z) = |\bar{A}_j(z)| \exp\{i\phi_j(z)\}$, where $P_0 = |\bar{A}_0|^2 + |\bar{A}_{-1}|^2 + |\bar{A}_{+1}|^2$ is the (conserved) total power of the three waves. Supposing for simplicity that the sidebands have initially equal amplitudes as in Eq. (1.5), i.e., $\bar{A}_{-1}(z=0) = \bar{A}_{+1}(z=0)$, one obtains the following equivalent particle Hamiltonian which describes the spatial evolution of (η, ϕ)

$$\frac{d\eta}{dZ} = \frac{dH}{d\phi}, \; \frac{d\phi}{dZ} = -\frac{dH}{d\eta}, \tag{1.8}$$

where $Z = P_0 z$, and H is written as

$$H = 2\eta\,(1-\eta)\cos(\phi) - (\kappa - 1)\,\eta - \frac{3}{2}\eta^2, \tag{1.9}$$

where $\kappa = -\Omega^2/P_0$ is a normalized phase mismatch between pump and sidebands. The solutions of Eqs. (1.8) and (1.9) are given in terms of Jacobian elliptic or hyperbolic functions [25]. Even though higher-order sidebands with $|n| \geq 2$ are neglected by the three mode truncation Eq. (1.6) and, accordingly, pump depletion is underestimated, Eqs. (1.8) and (1.9) provide a useful simple model to qualitatively describe the nonlinear dynamics of the induced MI process, as long as the initial sideband detuning is sufficiently large (i.e., $1 \leq \Omega \leq 2$), so that the higher-order sidebands are not MI unstable [24, 26]. The qualitative agreement between the predictions of the three-wave truncated model Eqs. (1.8) and (1.9) and the numerical solutions of the NLS equations (1.4) and (1.5) is illustrated in Figure 1.2. Here we show the solutions of Eq. (1.8) (or curves with constant Hamiltonian H) in the plane $(\eta \sin(\phi), \eta \cos(\phi))$ in the anomalous GVD regime for $\kappa = -2$ or $\Omega = \sqrt{2}$: note the presence of the separatrix solution which divides two different domains of oscillations in both the analytical and in the numerical solutions.

The validity of the three-mode truncation is based on the fact that, from the numerical solution of Eqs. (1.4) and (1.5), one finds that, at the point of maximum pump depletion which is observed for $\kappa = -1$ or $\Omega = 1$, more than 70% of the pump energy is

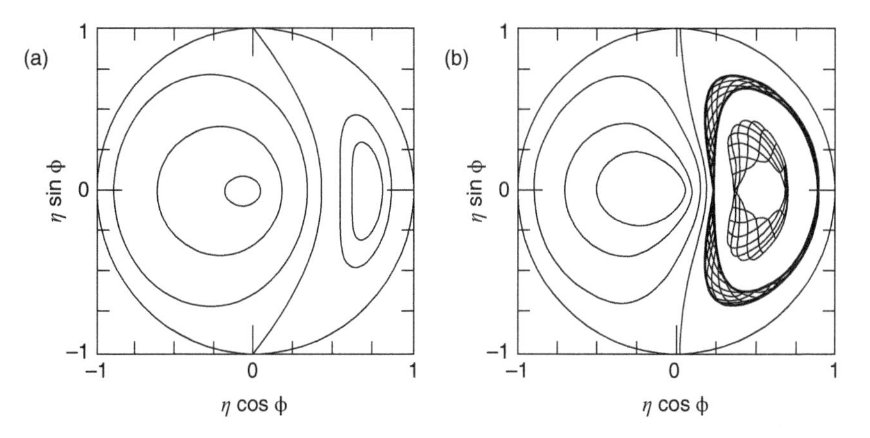

Figure 1.2 Phase space evolutions of nonlinear MI for $\kappa = -2$. (a) truncated three-wave model; (b) numerical solution of the NLS equation. *Source*: Trillo 1991 [26].

converted into the first-order sidebands with $n = \pm 1$ alone. Note that the numerical solutions reveal, in agreement with recent experiments, that maximum pump depletion occurs for a modulation frequency which is lower by a factor of $\sqrt{2}$ than the value corresponding to peak MI gain. Physically this is due to the fact that pump depletion leads to a progressive decrease of the effective optimal MI gain frequency along the fiber.

An alternative geometrical representation of the truncated dynamics of nonlinear modulations is provided by rewriting Eqs. (1.8) and (1.9) in terms of the Stokes parameters $s_i \equiv S_i/S_0$, where $S_0 \equiv P_0 = |\bar{A}_0|^2 + 2|\bar{A}_1|^2$, $S_1 \equiv |\bar{A}_0|^2 - 2|\bar{A}_1|^2$, $S_2 \equiv \sqrt{2}\bar{A}_0\bar{A}_1^* +$ c.c., and $S_3 \equiv -i\sqrt{2}\bar{A}_0\bar{A}_1^* + c.c.$ [27]. In vector notation, one obtains

$$\frac{d\mathbf{s}}{dz} = (\Omega_L + \Omega_{NL}(\mathbf{s})) \times \mathbf{s} \tag{1.10}$$

which describes the field evolution on the modulation sphere (analogous to the Poincaré sphere for polarization optics) as the motion of a rigid body subject to the sum of the fixed and the position-dependent angular velocities $\Omega_L \equiv (\kappa, 0, 0)$ and $\Omega_{NL} = (10s_1 - 1, 17s_2, 9s_3)/4$, respectively.

In the modulation sphere shown in Figure 1.3, the two points $s_1 = (\pm 1, 0, 0)$ represent the CW and the sidebands, respectively. Whereas the points on the equator with $s_3 = 0$ represent a pure amplitude modulation (AM), and points on the meridian $s_2 = 0$ represent a pure frequency modulation (FM). For $\Omega < 2$, MI is present and a separatrix trajectory emanates from the CW (see Figure 1.3 where $\Omega = 1$). At the same time, two new stable AM eigenmodulations are generated, which are analogous to the elliptically polarized eigenstates that are present in a nonlinear birefringent fiber at high powers. Note that the bifurcation of the sideband mode may also lead to the generation of new stable FM eigenmodulations.

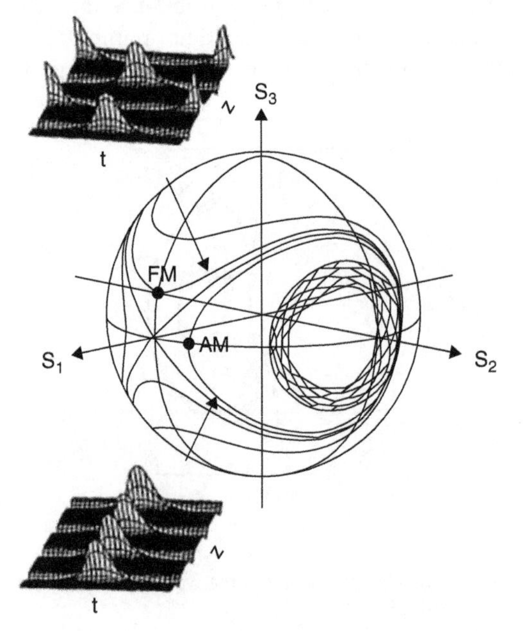

Figure 1.3 AM/FM conversion in MI from the numerical solution of the NLS equation with $\kappa = -1$. *Source:* Trillo 1991 [27].

1.2.2 Polarization MI (PMI) in Birefringent Fibers

Based on the coupled NLS equations describing linear and nonlinear coupling in birefringent optical fibers, we present the stability analysis of pump waves with different states of polarization, and show that the range of MI can be extended into the normal dispersion regime. Let us consider for simplicity the case of fibers with low birefringence (with coherent nonlinear coupling terms included). In terms of the circularly polarized components, Eq. (1.1) extends to the system of coherently coupled NLS equations [28]

$$\frac{\partial E_+}{\partial z} + i\frac{\beta_2}{2}\frac{\partial^2 E_+}{\partial \tau^2} - \frac{i\Delta\beta}{2}E_- - \frac{2i\gamma}{3}\left[|E_+|^2 + 2|E_-|^2\right]E_+ = 0$$

$$\frac{\partial E_-}{\partial z} + i\frac{\beta_2}{2}\frac{\partial^2 E_-}{\partial \tau^2} - \frac{i\Delta\beta}{2}E_+ - \frac{2i\gamma}{3}\left[|E_-|^2 + 2|E_+|^2\right]E_- = 0 \tag{1.11}$$

where $\Delta\beta = \beta_x - \beta_y > 0$ is the linear fiber birefringence. In general, the steady-state (i.e, τ-independent) solutions of Eqs. (1.11) can be expressed in terms of z-periodic Jacobian elliptic functions. The MI analysis of these solutions can be carried out by computing the spectrum of the associated Floquet exponents over one spatial period. A simpler insight can be obtained by considering the stability analysis of the spatial eigenmodes of Eqs. (1.11), for example, when the state of polarization of the input beam of power P is linear and aligned with the fast axis of the fiber. In terms of the circular polarization components and in the presence of a weak modulation, one has

$$E_\pm^f = \pm\left[i\sqrt{P/2} + u_\pm + iv_\pm\right]e^{i\left[\gamma P + \beta_y z\right]}. \tag{1.12}$$

By inserting Eqs. (1.12) into Eqs. (1.11) and keeping linear terms only, one obtains

$$\frac{\partial \tilde{u}_+}{\partial z} + i\frac{\partial \tilde{v}_+}{\partial z} - i\frac{\beta_2}{2}\omega^2(\tilde{u}_+ + i\tilde{v}_+) - \frac{i\Delta\beta}{2}\left(\tilde{u}_+ + i\tilde{v}_+ - \tilde{u}_- - i\tilde{v}_-\right) - \frac{2}{3}i\gamma P\left(\tilde{v}_+ + 2\tilde{v}_+\right) = 0 \tag{1.13}$$

along with a similar equation for $\tilde{u}_-$ and $\tilde{v}_-$ which can be obtained from Eq. (1.13) by interchanging + with −. By separating the equations for the real and imaginary parts of both circular polarization components of the perturbations, one obtains four separate linear equations. By examining the eigenvalues of the corresponding coefficient matrix, one finds that polarization coupling may lead to MI even in the normal GVD regime, provided that $\omega < \omega_{c1}$ where $\omega_{c1}^2 = \eta(p-1)$ and $p > 1$, where $p = P/P_c$, $\eta = 2\Delta\beta/|\beta_2|$ and the critical power $P_c = 3\Delta\beta/(2\gamma)$. The corresponding power growth rate reads as 2λ, where

$$\lambda = \frac{|\beta_2|}{2}\sqrt{(\eta + \omega^2)(\omega_{c1}^2 - \omega^2)}. \tag{1.14}$$

Note that, unlike the scalar MI of Eq. (1.3), in the vector case Eq. (1.14) predicts nonzero gain also for vanishing modulation frequency $\omega = 0$, which corresponds to the condition of CW polarization instability of the fast axis. On the other hand, in the anomalous GVD regime, besides experiencing the usual scalar MI as described by Eq. (1.3), the pump wave polarized along the fast axis is MI unstable for $\omega < \omega_{c2}$ whenever $p > 1$, otherwise for $\omega_{c3} < \omega < \omega_{c2}$ whenever $p \leq 1$, where $\omega_{c2}^2 = \eta$ is the low-power phase matching condition between the pump and the orthogonal sidebands, and $\omega_{c3}^2 = \eta(1-p)$. In both the normal and anomalous GVD regimes, the described MI is of vector nature, since the unstable Stokes and anti-Stokes eigenmodes are linearly polarized and orthogonal to the pump, that is they are oriented along the slow axis of the fiber [28].

The MI of a pump initially oriented along the slow axis of the fiber can be similarly studied by considering the perturbed eigenmode

$$E^s_{\pm} = \left[\sqrt{P/2} + u_{\pm} + iv_{\pm}\right] e^{i[\gamma P + \beta_x z]}. \tag{1.15}$$

By proceeding as outlined before, one finds that in the anomalous GVD regime the slow axis solution Eq. (1.15) is only subject to scalar MI. On the other hand, in the normal GVD regime there is only vector (i.e., with growing sidebands orthogonal to the pump) MI whenever $\omega_{c2} < \omega < \omega_{c5}$, with $\omega^2_{c5} = \eta(1+p)$. The corresponding MI gain is obtained from the eigenvalue

$$\lambda = \frac{|\beta_2|}{2}\sqrt{(\omega^2 - \omega^2_{c5})(\omega^2_{c2} - \omega^2)}. \tag{1.16}$$

Therefore with a pump on the slow axis, the MI always occurs at high frequencies above ω_{c2}, and there is no CW polarization instability.

As we have already done for the scalar case, to study the nonlinear stage of vector MI in a birefringent fiber, we consider the dimensionless version of Eqs. (1.11):

$$i\frac{\partial U}{\partial z} \pm \frac{1}{2}\frac{\partial^2 U}{\partial t^2} + \frac{\Delta}{2}U + (|U|^2 + \frac{2}{3}|V|^2)U + \frac{1}{3}U^2V^* = 0.$$

$$i\frac{\partial V}{\partial z} \pm \frac{1}{2}\frac{\partial^2 V}{\partial t^2} - \frac{\Delta}{2}V + (|V|^2 + \frac{2}{3}|U|^2)V + \frac{1}{3}V^2U^* = 0. \tag{1.17}$$

where U, V are the linear polarization components of the field, and Δ is the dimensionless linear birefringence [29]. We may then consider the truncation of the solutions of Eqs. (1.17) by using the following ansatz:

$$U(z,t) = A_0(z)\exp(\mp i\Omega^2 z/4),$$

$$V(z,t) = \sqrt{2}A_1(z)\exp(\mp i\Omega^2 z/4)\cos(\Omega t). \tag{1.18}$$

Accordingly, we now define $\eta(z) = |A_0(z)|^2/P_0$ and $\phi = 2\phi_{+1} - 2\phi_0$, with $A_j(z) = |A_j(z)|\exp\{i\phi_j(z)\}, j = 0, 1$, where $P_0 = |A_0|^2 + |A_1|^2$ is the total power. The spatial evolution of (η, ϕ) is again described by the motion of an equivalent particle that obeys the evolution equations [29–31]

$$\frac{d\eta}{dZ} = \frac{dH}{d\phi}, \quad \frac{d\phi}{dZ} = -\frac{dH}{d\eta}, \tag{1.19}$$

where $Z = P_0 z/3$, and the Hamiltonian H reads as

$$H = 2\eta\,(1-\eta)\cos(\phi) - (\kappa + 5)\,\eta + \frac{7}{2}\eta^2, \tag{1.20}$$

where $\kappa = -12(\Delta/2 \pm \Omega^2/4)/P_0$ is the normalized phase mismatch between pump and sidebands.

We briefly mention that in fibers with high birefringence, a different type of PMI occurs, characterized by growing sidebands which are orthogonally polarized along the birefringence axes whereas the pump is linearly polarized at 45 degrees (or more generally at any angle, though the efficiency of the process reduces) [32, 33]. This process, which also allows for extending MI to the normal dispersion regime, is described in terms of linearly polarized components U, V by incoherently coupled NLS (IC-NLS) equations where the group-velocity mismatch must be taken into account. The IC-NLS model reads in dimensionless units as

$$i\frac{\partial U}{\partial z} + i\delta\frac{\partial U}{\partial t} \pm \frac{1}{2}\frac{\partial^2 U}{\partial t^2} + (|U|^2 + X|V|^2)U = 0.$$

$$i\frac{\partial V}{\partial z} - i\delta\frac{\partial U}{\partial t} \pm \frac{1}{2}\frac{\partial^2 V}{\partial t^2} + (|V|^2 + X|U|^2)V = 0. \tag{1.21}$$

where $X = 2/3$ is the cross-phase modulation (XPM) coefficient in silica fibers. When δ is sufficiently large, the nonlinear stage of MI can be investigated by means of the two-mode truncation $U = [A_0(z) + A_1(z)\exp(i\Omega t)]/\sqrt{2}$, $V = [A_0(z) + A_1(z)\exp(-i\Omega t)]/\sqrt{2}$, where $P_0 = |A_0|^2 + |A_1|^2$ is the conserved power. Following the same approach outlined above, the dynamics is described by an equivalent integrable oscillator with Hamiltonian [34–36]

$$H = \frac{2}{3}\eta\,(1-\eta)\cos(\phi) - (\kappa - 1)\,\eta - \eta^2, \tag{1.22}$$

where η and ϕ have the same meaning as in Eqs. (1.19) and (1.20). Conversely, in this case, the normalized phase mismatch of the underlying four-photon process reads as $\kappa = (\pm\Omega^2 - \delta\Omega)/P_0$. When δ is not large enough, the nature of the MI ruled by Eq. (1.21) changes, since the sidebands possess both polarization components, and six scalar modes become effective. Under this regime the reduced Hamiltonian become two-dimensional and spatially chaotic regimes sets in [35].

1.2.3 Collective MI of Four-Wave-Mixing

On the basis of the same type of IC-NLS equations (1.21), it was predicted that MI induced by XPM can occur also by injecting two pumps with the same polarization but different frequencies $\omega_0 \pm \Omega_d$ [37], $2\Omega_d$ being the real-world frequency separation between the pumps. In this case U and V represent the field envelopes at the two carrier frequencies, whereas $X = 2$. However, the beating between the two pumps induces, via the Kerr effect, a process usually termed multiple four-wave mixing (mFWM), i.e., the generation of a cascade of multiple sideband pairs at $\omega_0 \pm n\Omega_d$, n odd integer, which is due to coherent terms neglected in the derivation of Eqs. (1.21) [38]. It was argued that, in the normal GVD regime, the most unstable MI frequencies are resonant with the generated mFWM leading-order modes, which invalidates the approach based on the IC-NLS equations [39, 40]. However, the dynamics of mFWM could be correctly captured in the framework of the single NLS equation (1.1) and a relative four-mode truncation (pump and the leading-order mFWM mode pairs) [41]. By exploiting the fact that mFWM leads, in a wide range of frequency detuning and powers, to periodic exchange of power between the pump pair and the cascaded modes, the MI linear stability analysis could be reformulated by accounting for the mFWM phenomenon. Indeed, one can linearize around the periodic orbits which describe the mFWM leading-order modes,

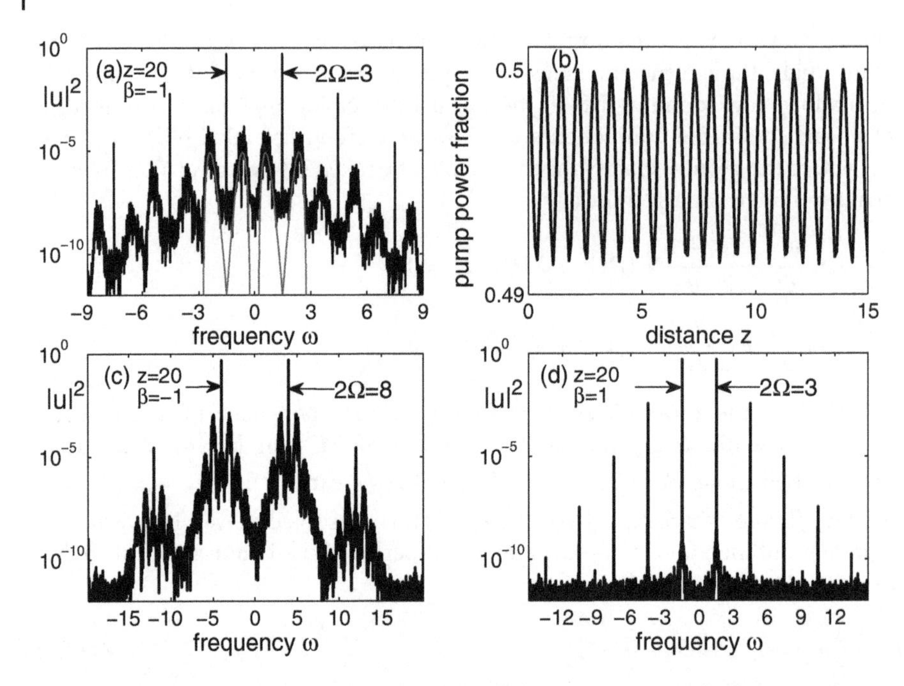

Figure 1.4 (a) Output spectra from a NLS simulation showing the onset of collective MI from noise in the anomalous GVD regime ($\beta = \beta_2/|\beta_2| = -1$). The frequencies are scaled in units of $\sqrt{\gamma P_0/\beta_2}$, e.g. $\Omega = \Omega_d\sqrt{\beta_2/(\gamma P_0)} = 1.5$ for the pump detuning. For comparison we report the calculated collective MI gain around the pumps (pale gray; the gain peaks at $\omega = \omega_p = 1$ consistently with the normalized power 0.5 for each pump (normalized total power is 1); (b) Corresponding periodic evolution of pump power fraction; (c) as in (a) for larger pump detuning $\Omega = 4$; (d) Stable FWM in normal GVD regime, $\beta = 1$. *Source*: Armaroli 2011 [42].

and study by means of Floquet techniques, the stability of the whole process against the growth of a modulation (i.e., sideband pairs) at the same frequency $\delta\omega$ around all the mFWM orders. The outcome of such analysis (for details, see [40, 42]) shows that, in the anomalous GVD regime, the mFWM pattern is unstable against the growth of MI sidebands at the frequency detuning $\delta\omega = \pm\omega_p$, where $\omega_p = \omega_c/\sqrt{2}$ is the peak frequency corresponding to the power of a single pump line, predicted by the standard scalar MI. This process has been termed *collective* MI because the modulation is transferred with the same frequency from the two pumps to all the other mFWM products. In other words the MI frequency does not scale with the power of the cascaded mFWM modes, as it would be in the event that each mFWM order would develop its own MI [see Figure 1.4 (a)].

An example of the spectrum calculated from the NLS is shown in Figure 1.4 (a). Here the MI develops on top of the spatially periodic evolution of the pumps displayed in Figure 1.4 (b). At larger frequency separations between the pumps also harmonics of the fundamental MI unstable frequency can be observed, as shown in Figure 1.4 (c). The same analysis shows that, for normal GVD, the mFWM is modulationally stable, as shown in Figure 1.4 (d) [40]. These results have been experimentally confirmed, by carefully measuring single spectra as well as the spectral behavior of mFWM, as the frequency detuning Ω_d was continuously varied [43].

1.2.4 Induced MI Dynamics, Rogue Waves, and Optimal Parametric Amplification

We briefly discuss here the nonlinear dynamics of the evolution of MI beyond the approximation intrinsic in the simple truncated models, that is, including all harmonics $\omega_0 \pm n\Omega$, $n \geq 2$ of the initial modulating signal ($n = 1$). These harmonics are indeed generated via the Kerr effect and can be important especially in the scalar MI process. In this case, thanks to the integrability of the NLS equation, the FPU recurrent evolutions that entail a periodic power exchange among the pump and the full comb of harmonics can still be described exactly in terms of doubly-periodic (in time and space) analytical solutions. Such solutions describe the homoclinic structure of the MI illustrated in Figures 1.1 and 1.2. In particular, the separatrix of Eq. (1.4) corresponds to the so-called Akhmediev breather (AB), i.e., a solution that connects the unit background to itself after a full cycle of evolution (strictly speaking, the AB is heteroclinic rather than homoclinic to the background due to the different phase at $z = \pm\infty$)

$$u_{AB}(t) = \left[1 + \frac{2(1-2a)\cosh(bz) + i\sinh(bz)}{\sqrt{2a}\cos(\omega t) - \cosh(bz)}\right] e^{iz}, \tag{1.23}$$

where the parameters $2a = \sqrt{1-(\omega/2)^2}$ and $b = \sqrt{8a(1-2a)}$ are fixed by the normalized frequency $\omega = \Omega\sqrt{|\beta_2|/(\gamma P_0)}$. Equation (1.23) describes the single cycle of conversion in the whole range of MI, namely $0 \leq \omega \leq 2$ (i.e., $1/2 \geq a \geq 0$). A remarkable limit of Eq. (1.23) is obtained for $\omega = 0$, which gives the rational soliton solution known as the Peregrine soliton [19], a prototype of (deterministic) rogue wave. Moreover for $a > 1/2$, Eq. (1.23) describes the Kuznetsov-Ma breathers [20]. The nonlinear stage of MI developing from noise usually exhibits evidence for the random excitation of AB, Peregrine and Kuznetsov-Ma structures [44]. Conversely, when the MI is induced by a sufficiently small seed, the dynamics can be accurately described in terms of ABs [4, 45], since the doubly periodic solutions lie sufficiently close to the separatrix.

Importantly, from the AB solutions one can derive a simple analytical condition for the optimum modulation frequency, that leads to maximum pump depletion. This optimum frequency does not coincide with the maximally unstable (or nonlinear phase matching) normalized frequency $\omega = \sqrt{2}$, due to the fact that the depletion tunes the underlying four-photon process out of phase-matching. Conversely, at lower modulation frequencies, even though the modulation grows initially with lower rate, it can be amplified more efficiently since pump depletion tunes the mixing process towards phase matching [25]. Quantitatively, the optimal condition for conversion can be obtained by expanding the AB at its apex [peak conversion occurring at $z = 0$ in Eq. (1.23)] in Fourier series $u_{AB}^{peak}(t) = \sum_n \tilde{u}_n \exp(in\omega t)$, where $\tilde{u}_n$ are the Fourier modal amplitudes. In this way, one can obtain simple expressions for the peak fraction of the pump ($n = 0$) and harmonic sideband modes ($\pm n$)

$$|\tilde{u}_0^{peak}|^2 = (\omega - 1)^2; \quad |\tilde{u}_n^{peak}|^2 = \omega^2 \left(\frac{2-\omega}{2+\omega}\right)^n. \tag{1.24}$$

Equation (1.24) implies that the pump is totally depleted at $\omega = 1$. As shown in Figure 1.5 (a), numerical integration of the normalized NLS Eq. (1.4) with initial condition $u = \sqrt{\eta_0} + \sqrt{1-\eta_0}\exp(i\omega t)$, confirms that, at $\omega = 1$, total pump depletion occurs in favor of multiple sideband pairs at a characteristic distance z_d, which depends on

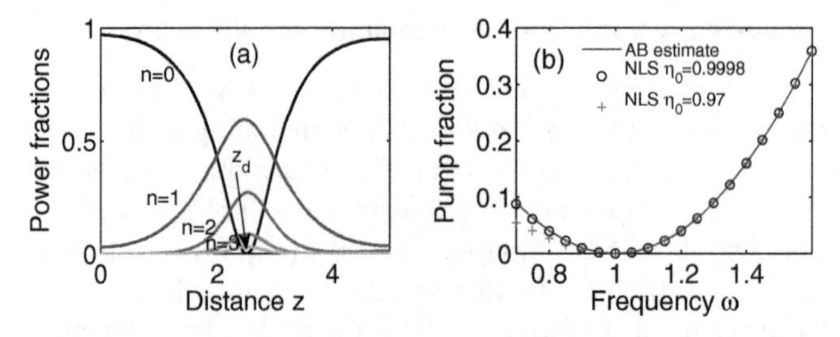

Figure 1.5 (a) Evolution of the power fraction of the pump ($n = 0$) and first four sideband pairs at optimal frequency $\omega = 1$ for pump depletion; from numerical integration of NLS equation (1.4); (b) Residual pump power fraction vs. ω: estimate from AB [Eq. (1.24)], solid line compared with NLS simulations (circles and crosses) for two values of input pump fraction $\eta_0 = 0.97$ and $\eta_0 = 0.9998$.

the power fraction of the input seed. Such simulations, repeated for different modulation frequencies ω, confirm that the parabolic law of Eq. (1.24) indeed provides a quantitatively accurate description of the maximally depleted pump in the whole range $1 \leq \omega \leq 2$, regardless of the initial power fraction of the signal. This agreement is displayed in Figure 1.5 (b), where we compare the results of the simulations, carried out for two different input pump fractions $\eta_0 = 0.9998$ and $\eta_0 = 0.97$ (signal fractions $\eta_s = 0.02\%, 3\%$), to the analytical expression [Eq. (1.24)]. As can be seen, slight discrepancies only appear for significantly high input signal fractions (see crosses for $\eta_0 = 0.97$), and in the range of modulation frequencies well below the optimum value $\omega = 1$. The validity of such arguments has been experimentally validated by measuring a 95% depletion at a frequency lower than the measured peak gain frequency, in good agreement with the normalized value $\omega = 1$ (see Figure 1.6, left panel) [47]. Noteworthy, Figure 1.6 shows that the signal probe ($n = 1$) peaks at a slightly higher frequency (yet lower than

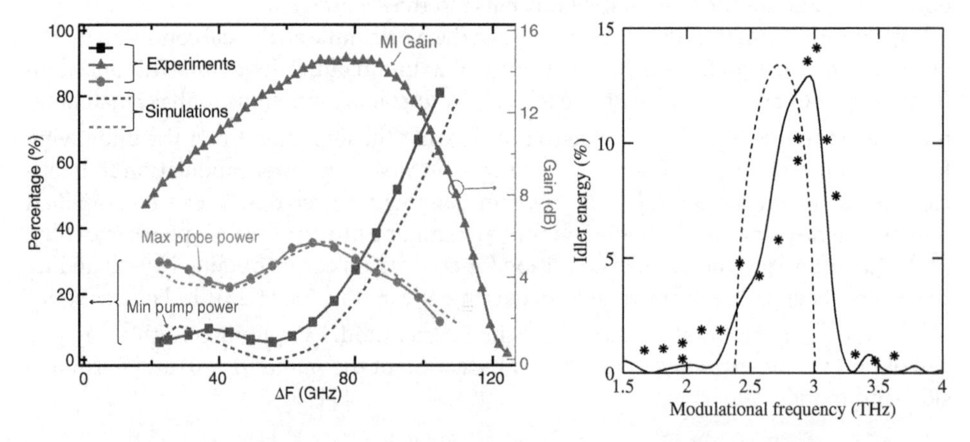

Figure 1.6 Experimental results concerning MI optimal parametric amplification. Left panel: scalar MI, output pump power and signal fractions compared with small-signal MI gain. *Source*: Bendahmane 2015 [47]. The dashed lines stand for the corresponding curves from simulations. Right panel: PMI in a highly-birefringent fiber. Output fraction of generated idler vs. modulation frequency, showing a peak conversion at frequency above the cutoff of the PMI gain (dashed), and an abrupt drop above such frequency. The solid line is the numerical result from Eqs. (1.21). The input power fraction of the signal is 0.05. *Source*: Seve 2000 [36]. Reproduced with permission of American Physical Society.

the peak gain frequency), in agreement with Eq. (1.24) that gives a maximum of $|\tilde{u}_1^{peak}|^2$ at $\omega \simeq 1.24$.

In the case of PMI processes, the deviation of optimal parametric amplification from nonlinear phase matching is even more striking. Indeed, on the basis of the Hamiltonian oscillators Eqs. (1.20)–(1.22), it was shown that the optimal conversion in the strongly depleted regime occurs outside the gain bandwidth of the MI. Despite the fact that the pump is stable, in this region the dynamics are indeed ruled by unstable phase-locked eigenmodulations, which bifurcate from the pump at the edge of the MI gain bandwidth [29–31]. As a result, a critical frequency turns out to exist, around which the conversion reaches a maximum value and then abruptly drops. In this case the three-mode truncation constitutes a good description of this regime too, since the generation of higher-order sidebands remains negligible. This has been confirmed experimentally in [46] for the low-birefringence case, as well as in [36] for the high-birefringence case. The latter case is illustrated in Figure 1.6 (right panel), which shows the measured idler fraction as a function of modulation frequency compared with the MI gain curve. Here the pump is polarized at 45 degrees and has total power $P = 56$ W, whereas the input signal is a 10% fraction of the pump.

1.2.5 High-Order Induced MI

Whenever the MI is induced in a fiber by a signal, whose harmonic is also modulationally unstable, then the so-called higher-order MI occurs. In this case, after a first stage of FPU recurrence that leads to the peak pump depletion and maximum amplification of the signal and all of its harmonics, there is a second stage of pump depletion, which is characterized by the development of the second harmonic of the initial modulation [48]. For $0 \leq \Omega \leq 1$, some higher-order sidebands (e.g., the first-harmonic of the input modulation $\pm 2\Omega$ for $0.5 \leq \Omega \leq 1$) experience exponential growth with distance [48]. Consider the solution of Eq. (1.4) with the initial condition Eq. (1.5): Figure 1.7 compares the evolution with distance z of the field amplitude with either an initial AM (that is $\phi^0 = 0$) or FM ($\phi^0 = \pi$) perturbation. As can be seen, the initial weak modulation grows

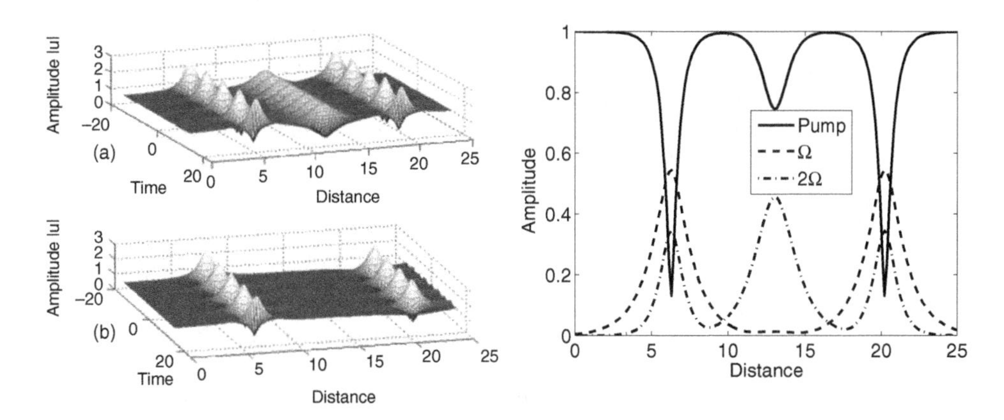

Figure 1.7 Evolution with distance of the field amplitude with $\epsilon = 0.01$, $\Omega = 0.8718$, and initial (a) AM or (b) FM. The rightmost plot shows the evolution of the energy in the pump and sidebands for the AM case. *Source*: Wabnitz 2010 [48]. Reproduced with permission of Elsevier.

until at $z \simeq 5$ the input CW pump is nearly fully depleted into the sidebands at frequency Ω and their harmonics. Next, the field energy flows back into the pump according to the usual FPU recurrence.

However, Figure 1.7 (a) also shows that in the AM case, at approximately the mid-point $z \simeq 13$ of the FPU recurrence period, a modulation develops with frequency 2Ω. Whereas Figure 1.7 (b), where a pure FM is present at the fiber input, shows that no development of the second-harmonic component occurs. We may thus conclude that the development of the frequency-doubled modulation is strongly sensitive to the relative phase between the pump and the initial sidebands. Note that the field evolution may be analytically expressed as a nonlinear superposition of all linearly unstable modes, which leads to the emergence of multiple spatial periods [49,50]. A potential application of the frequency doubling effect is the possibility of obtaining a high-extinction ratio modulation of a CW laser at frequency 2Ω by seeding its propagation in a nonlinear optical fiber with a weak modulation at frequency Ω.

An interesting extension of the MI frequency doubling can be obtained by using polarized beams in a randomly birefringent telecom fiber, where pulse propagation is described by the Manakov system

$$i\frac{\partial U}{\partial z} + \frac{1}{2}\frac{\partial^2 U}{\partial t^2} + (|U|^2 + |V|^2)U = 0,$$

$$i\frac{\partial V}{\partial z} + \frac{1}{2}\frac{\partial^2 V}{\partial t^2} + (|V|^2 + |U|^2)V = 0. \tag{1.25}$$

where the initial condition reads as

$$U(z = 0, t) = U_0 + \epsilon_U \exp\left\{i\phi^U/2\right\}\cos(\Omega t),$$

$$V(z = 0, t) = V_0 + \epsilon_V \exp\left\{i\phi^V/2\right\}\cos(\Omega t), \tag{1.26}$$

When the CW pump and its modulation are orthogonally polarized at the fiber input (e.g., with $V_0 = 0$ and $\epsilon_U = 0$ in Eq. (1.5)), no scalar MI occurs. However, MI is induced upon propagation on the CW via cross-polarization modulation (XPolM) [51]. As a result, as shown in Figure 1.8, one still observes a break-up of the pump into a pulse train which only contains even harmonics of the initial modulation. Moreover, the CW pedestal that accompanies MI-induced pulse trains in the scalar case is fully suppressed in the vector case, thus permitting in principle very large extinction ratios to be achieved. The all-optical generation of a 80-GHz high-contrast pulse train from a cross-polarized 40-GHz electro-optical weak modulation was recently experimentally demonstrated.

1.2.6 MI Recurrence Break-Up and Noise

Although the nonlinear stage of MI is characterized by the FPU recurrence, the development of supercontinuum (SC) is associated with the irreversible evolution toward a thermalization state, i.e., a nearly equal distribution of spectral energy among all frequency components [3,4]. For example, it was predicted and experimentally confirmed that third-order dispersion induced losses in Cherenkov radiation lead to the energy dissipation of the pump field, that eventually breaks the FPU recurrence [52, 53]. In fact, nonlinear fiber optics experiments typically demonstrate FPU recurrence up to a single spatial period [15–19, 47]. In addition, recent studies regarding noise-induced MI have highlighted the complex dynamics associated with the onset stage of noise

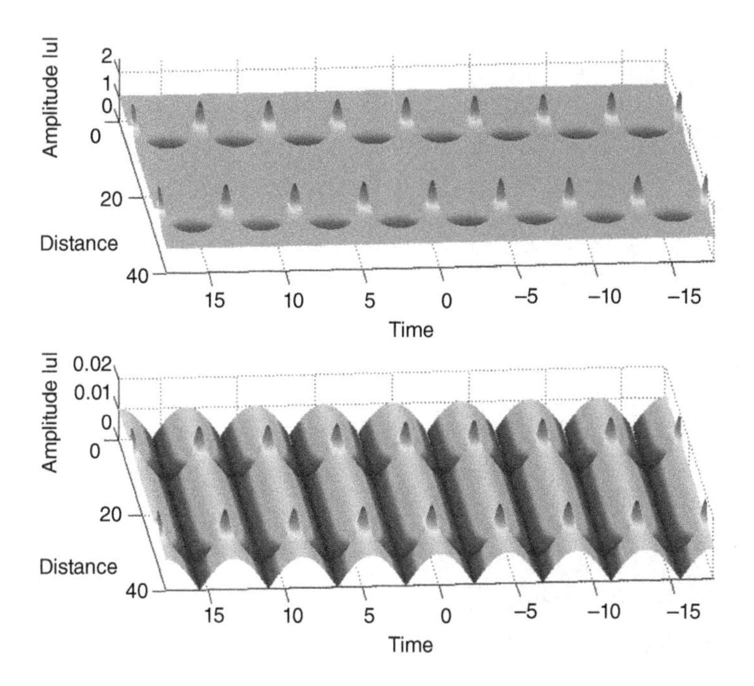

Figure 1.8 Evolution with distance of the field amplitudes $|U|$ and $|V|$ with $U_0 = 1$, $V_0 = 0$, $\epsilon_V = 0.01$, $\phi^V = 0$ and $\Omega = 1/\sqrt{2}$. *Source*: Fatome 2013 [51].

amplification [54, 55] and their links with the mechanisms of rogue wave generation [4, 56]. For example, seeding the initial stage of SC generation with a weak modulation could lead to the stabilization of the SC output [56–58], and thus reduce the impact of noise on the complex dynamics associated with SC and rogue wave generation.

We want to discuss how, in the presence of initial quantum noise, spontaneous MI competes with the induced MI process. This is a fundamental instability mechanism which breaks the FPU recurrence and leads to the irreversible evolution into statistically stationary spectra [59]. The stability of periodic nonlinear mode coupling was previously studied in the context of polarization MI in birefringent fibers [60], parametric mixing [61] or second harmonic generation in quadratic materials [62], and, more recently, in the closely related problem of dual-frequency pumped mFWM in optical fibers [40, 42, 43].

In the following, we numerically solved the NLS equation Eq. (1.4) with the initial condition Eq. (1.5), and added a broadband quantum noise floor corresponding to one photon per frequency bin with random initial phase in the spectral domain. Figure 1.9 (a) shows the evolution of the field intensity $|U(z,t)|^2$, for a particular realization of the random input noise seed (single shot case). As can be seen, after just two FPU recurrence periods, spontaneous MI leads to the field break-up into an irregular structure exhibiting frequency doubling and irregular intensity peaks formation.

Break-up of the FPU recurrence is due to the exponential growth of the initial quantum noise background, owing to MI of the periodically evolving pump and multiple FWM sidebands. This is clearly shown in the single-shot spectral domain plot of Figure 1.9 (b), showing the evolution of the log-scale spectral intensity of the field as a function of the angular frequency detuning from the pump, ω. Figure 1.9 (b) also shows

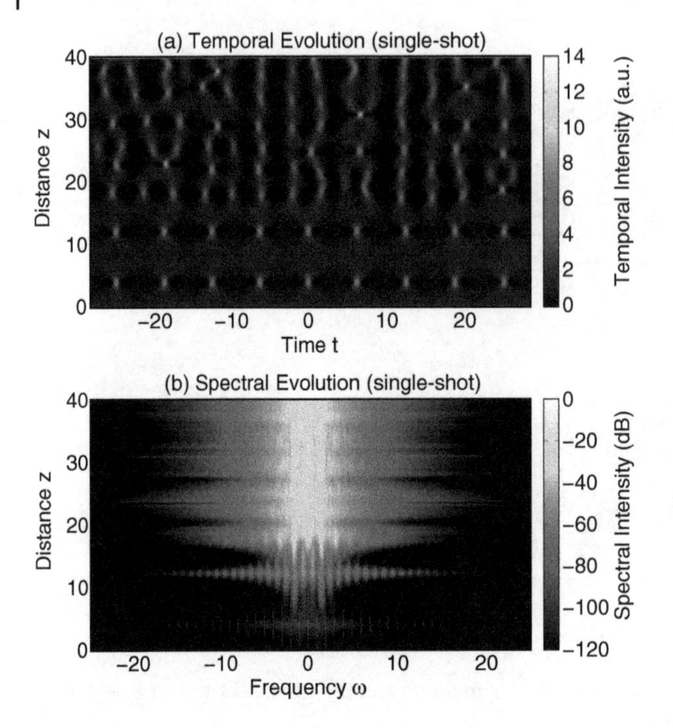

Figure 1.9 Evolution of (a) field intensity $|U|^2$; and (b) spectrum (in log scale) for $\Omega = 1$ for $\epsilon = 0.05$ and $\phi^0 = 0$. *Source*: Wabnitz 2014 [59].

that, after two periods of the FPU recurrence, the temporal field break-up is associated with the growth of a broad frequency continuum among all FWM sidebands, which leads to the irreversible equipartition of energy in frequency space [59].

The break-up of the FPU recurrence shown in Figure 1.9 is clearly displayed in terms of the spatial evolution of the power of the pump and the initial modulation sidebands. Indeed, Figure 1.10 shows that after two periods of FPU recurrence, the pump power suddenly drops down, and it exhibits an irregular evolution around a low average value.

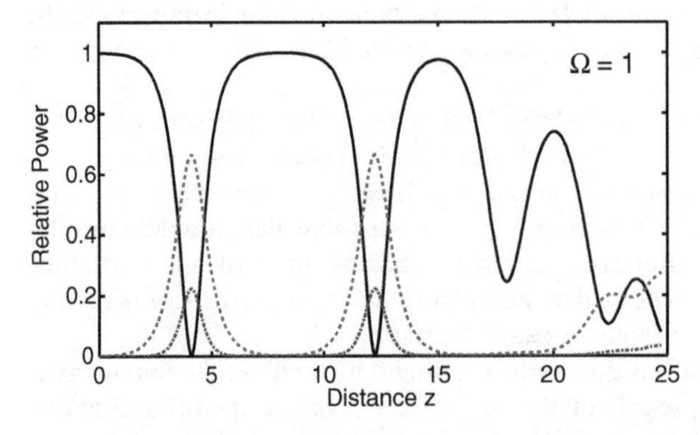

Figure 1.10 Power fraction in the pump (solid black curve) and in the sidebands with frequency shift $\pm\Omega$ (thick dashed curve) or $\pm 2\Omega$ (thin dashed curve). *Source*: Wabnitz 2014 [59]. Reproduced with permission of Elsevier.

1.3 Four-Wave Mixing Dynamics

Optical fibers permit the cubic nonlinear response of glass with long interaction lengths to be exploited. If we consider two pump wavelengths at angular frequencies ω_1 and ω_2, and a third signal at frequency ω_s, it is possible, for instance, to generate by FWM an idler at frequency $\omega_i = \omega_1 + \omega_2 - \omega_s$. When the frequencies of the two pumps coincide $\omega_1 = \omega_2$, the FWM is called degenerated. The nearly instantaneous response time of the Kerr nonlinearity makes it very interesting, for instance, for amplification or frequency conversion of optical signals with all-optical devices that are agnostic to the modulation format and directly compatible with fiber transmission systems. However, a real experiment would reveal a more complex situation: for instance, for each pump wavelength, degenerated FWM would also be possible. The efficiency of these processes is governed by different phase matching conditions.

Considering the simple case of a unique guided mode, the nonlinear dynamics of the four interacting waves with two pump wavelengths have been studied in [41] even beyond the common assumption of the undepleted pump approximation, which is instead limited to a weak signal and idler. A variety of four-photon interactions under various polarization states have been predicted and observed in optical fibers [63]; the scenario can be further expanded considering multimode waveguides, where FWM can permit energy exchange among guided modes of different temporal frequencies. The emergence of new technologies for fiber micro structuring has substantially enhanced the possibility of the application of FWM, since the dispersion profile and therefore the phase matching condition and hence the efficiency can be more easily controlled in these fibers.

The FWM process can also naturally be found as an undesired shortcoming in long fiber transmission systems with wavelength division multiplexing: the energy exchange among the interacting waves is in these cases a source of cross-talk among channels and therefore of transmission impairments.

1.3.1 FWM Processes with Two Pumps

The first type of application of FWM with two pumps is fiber optic parametric amplification. The same effect can also be obtained with a single pump with the degenerate FWM. It is possible to show that the input signal ω_s can be amplified in an optical fiber, and that at the same time an idler ω_i is created on another wavelength and with the capability of carrying the same information of the signal. The interest in using two pumps is in the gain flatness that one can obtain with the proper choice of pump wavelengths.

Another important application of FWM with two pump wavelengths is the so-called phase-sensitive amplification. The optical amplifiers commonly deployed today in fiber transmission systems are phase-insensitive, which means that they amplify an optical signal, disregarding its input phase. The introduction of more advanced modulation formats, especially phase-shift keying, has increased the interest in new types of optical amplifiers able to amplify the in-phase signal quadrature and to attenuate the out-of-phase signal quadrature. A possible implementation of a phase-sensitive amplification can be based on a FWM comprising a signal wavelength and two pump wavelengths so that $\omega_s = (\omega_1 + \omega_2)/2$. Some of the features of FWM for phase-sensitive amplification are illustrated in [64], and an example of experimental implementation is given in [65].

1.3.2 Bragg Scattering FWM

A common feature of the FWM interactions briefly listed in Section 1.3.1 is the presence of an exponential gain. The Bragg-scattering FWM (BS-FWM) is instead a special type of parametric interaction that can permit frequency translations of signals in the absence of exponential gain. The terminology has been inspired by the spatial equivalent effect [66] and this type of FWM was also previously known as wavelength exchange [67].

The absence of gain implies the absence of spontaneous noise emission: for this reason BS-FWM can permit a frequency translation that keeps preserved the quantum state of a signal and this key feature has been extensively developed by McKinstrie and coworkers [68, 69].

BS-FWM can in principle permit a unitary transformation involving a weak signal and an idler, with the same formal properties of the transformation operated by a (quantum) beam splitter involving incident reflected and transmitted waves. BS-FWM requires two pumps at two different wavelengths and the maximum conversion efficiency can be obtained at phase matching under various configurations of polarization states and wavelengths for pumps, signal and idler [68, 70, 71].

In practice, a signal at a given wavelength can be up-converted or down-converted by a frequency-shift amount equal to the beating frequency between two distinct pump wavelengths. The energy conservation requires that $\omega_i = \omega_s \pm (\omega_1 - \omega_2)$, where ω_1, ω_2 are the angular frequencies of the two pumps (1,2) and ω_s, ω_i are the angular frequencies of signal and idler respectively. The presence of the two signs reminds us that for a given signal wavelength and a pair of pump wavelengths, two distinct idler wavelengths are possible in principle.

Both up-conversion or down-conversion can be implemented in two ways, as illustrated by panels (a) and (b) of Figure 1.11. In the first implementation, shown in panel (a), the pumps are interleaved with the signal and idler, permitting broad frequency translations. In the second implementation, illustrated in panel (b), signal and idler are instead spectrally separated from the two pumps: this configuration permits a narrow-band frequency exchange, and can be of interest in quantum applications where it is important and also technically challenging to separate the pumps from the signal and idler with optical filters.

Considering for simplicity the up-conversion case shown in Figure 1.11 (a), the BS-FWM conversion efficiency is ruled by the phase matching condition

$$\beta(\omega_i) = \beta(\omega_s) + [\beta(\omega_1) - \beta(\omega_2)] + \gamma(P_2 - P_1) \tag{1.27}$$

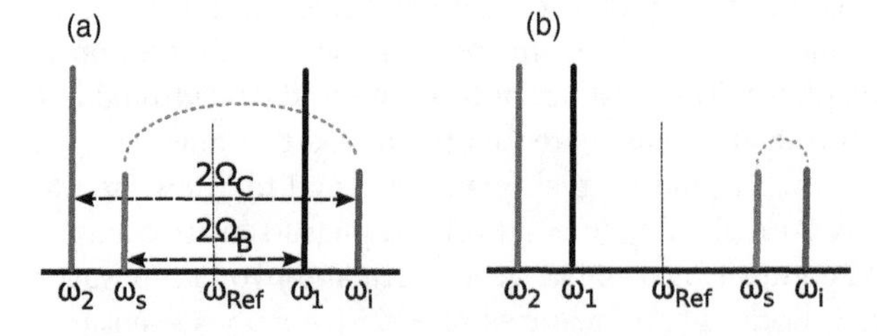

Figure 1.11 Different configurations for BS-FWM.

where $\beta(\omega) = n(\omega)\omega/c$ is the linear wave-vector at the angular frequency ω and P_1, P_2 are the powers of pumps 1 and 2. Following [68] it is convenient to identify all interacting waves by frequency detunings Ω_B, Ω_C from a common reference frequency ω_{Ref}. In doing so, it is possible to develop a Taylor series expansion of the wave-vector $\beta(\omega)$ around ω_{Ref} to highlight the role that is played by the different orders of dispersion. Similar to what happens in the case of MI, the summation of wave-vectors of opposite detuning causes the algebraic cancellation of the odd-order dispersion terms. When the expansion is truncated to the fourth order, the phase matching Eq. (1.27) imposes a condition between the second and the fourth order dispersion as follows:

$$\left(\Omega_B^2 - \Omega_C^2\right)\left[\beta_2 + \frac{\beta_4}{12}\left(\Omega_B^2 + \Omega_C^2\right)\right] = 0 \tag{1.28}$$

In particular, Eq. (1.28) shows that the phase-matching of BS-FWM requires opposite signs for β_2 and β_4: in normal dispersion ($\beta_2 > 0$) it then requires negative values of β_4; in anomalous dispersion instead β_4 should be positive. Both BS-FWM with positive and negative β_4 have been experimentally demonstrated [71].

The dynamical equations coupling the complex envelope of signal a_s and idler a_i by BS-FWM can be obtained from the standard procedure of analyzing the interplay among two pumps and two sidebands from the NLS equation [41,66]. Equations for signal and idler can be linearized in the undepleted pump approximation and the resulting coupled mode equations are

$$\begin{bmatrix} \dfrac{da_s}{dz} \\ \dfrac{da_i}{dz} \end{bmatrix} = \begin{bmatrix} i\Delta & i\Gamma \\ i\Gamma^* & -i\Delta \end{bmatrix} \begin{bmatrix} a_s \\ a_i \end{bmatrix} \tag{1.29}$$

where $2\Delta = \beta(\omega_s) + \beta(\omega_1) - \beta(\omega_i) - \beta(\omega_2) + \gamma(P_2 - P_1)$ and $\Gamma = 2\gamma A_2 A_1^*$. The corresponding Hamiltonian $H(a_s, a_s^*, a_i, a_i^*)$ is

$$H = \Delta(|a_s|^2 - |a_i|^2) + \Gamma a_s^* a_i + \Gamma^* a_s a_i^* \tag{1.30}$$

and then the coupled mode Eqs. (1.29) can be obtained directly from $da_h/dz = i\partial H/\partial a_h^*$, with $h = s, i$.

The solutions of Eqs. (1.29) at a distance z from the origin $z = 0$ are

$$\begin{aligned} a_s(z) &= \mu(z)a_s(0) + \nu(z)a_i(0) \\ a_i(z) &= -\nu^*(z)a_s(0) + \mu^*(z)a_i(0) \end{aligned} \tag{1.31}$$

with $\mu(z) = cos(\kappa z) + i\Delta sin(\kappa z)/\kappa$, $\nu(z) = i\Gamma sin(\kappa z)/\kappa$, and $\kappa^2 = \Delta^2 + |\Gamma|^2$. At phase matching, the maximum conversion efficiency is obtained when $\kappa z = \pi/2$: a slightly longer fiber length would cause a re-translation of the idler toward the signal. BS-FWM has been experimentally demonstrated in highly nonlinear fibers [72] and photonic crystal fibers (PCFs) [69] but has also been reported in other cubic nonlinear materials such as silicon nitride waveguides [73] as well as in Rb-filled PCF [74].

A similar type of noiseless parametric frequency conversion can be obtained in crystals with quadratic nonlinearities by the process of sum-frequency generation: both BS-FWM in fibers and SFG in quadratic crystals can be then represented by an equivalent geometrical interpretation through real-valued Stokes parameters and visualized as a trajectory along the surface of the Poincaré sphere [75,76].

1.3.3 Applications of BS-FWM to Quantum Frequency Conversion

Spontaneous FWM processes have been extensively studied when generating single photons in an all-fiber configuration [77]. Quantum frequency conversions requires instead a combination of a quantum source and a noiseless parametric translation. Quantum frequency translation was first observed by sum-frequency generation in quadratic crystals (see, for instance, [78]).

The first experimental observation of single photon emission and subsequent frequency translation by BS-FWM in an experimental setup entirely based on PCF was reported by McGuinness and coworkers in [69]. To reveal the non-classical nature of the converted signals, besides conversion efficiency and signal to noise ratio, it is necessary to measure correlations. For instance, the authors of [69] measured the conditional second-order degree of coherence $g^2(\tau)$, which is proportional to the probability of detecting a second photon at time $t = \tau$ given the fact that a first photon was detected at time $t = 0$. The authors measured a value of $g^2(0) < 1$ that is an indicator of a non-classical state of light for translated and untranslated light.

Although the process of BS-FWM is intrinsically noiseless, the presence of two pumps generates a series of competitive spontaneous emissions by FWM that can degrade the signal-to-noise ratio in the quantum channels. In optical fibers the spontaneous Raman effect also contributes to generate optical noise at room temperature due to the nature of glass (see, for instance, [70]). Spontanous Raman noise can be reduced by drastically lowering the temperature (for instance, with liquid nitrogen), or by using different materials with lower Raman gain. Silicon has a narrowband Raman gain, and chalcogenide glasses have a spectral window of low spontaneous Raman scattering [79].

1.4 Fiber Cavity MI and FWM

1.4.1 Dynamics of MI in a Passive Fiber Cavity

A fiber ring cavity is a simple optical device that can be made by connecting the two ends of an optical fiber into a loop configuration using a beam splitter or a fiber coupler. The fiber ring allows light to recirculate inside the cavity over multiple round trips to create an optical resonator. In the following we will focus on passive fiber cavities that are pumped using an external continuous wave (CW) laser source and that do not contain any gain medium. This is in constrast to active fiber cavities that may contain a gain medium such as an erbium doped fiber amplifier and can be used to create a fiber ring laser.

The field in a fiber cavity will experience losses, both due to propagation (intrinsic absorption) and output coupling. The condition when the absorption losses are equal to the coupling losses is known as critical coupling and allows for the complete extraction of the output field when the pump frequency is resonant. A low power sweep of the pump frequency across the resonance will then result in the pump field power showing a dip with zero transmission on resonance. The frequency separation between the pump laser and the resonant frequency of the pump mode is referred to as the pump detuning.

The intracavity field will interfere constructively with itself and be resonant with the cavity whenever the length L of the resonator circumference corresponds to an integer number m of wavelengths. This is expressed by the condition $\beta_m L = 2\pi m$, where

$\beta_m = k_m n_{eff}$ is the propagation constant of the mode with vacuum wavenumber $k_m = 2\pi/\lambda_m$ and effective refractive index n_{eff}. The resonances are in the absence of dispersion separated by an equidistant frequency spacing known as the free-spectral-range (FSR) given by $\Delta\nu_{FSR} = 1/(\beta_1 L)$, with $\beta_1 = d\beta_m/d\omega|_{\omega_m} = (1/c)(n + \omega dn/d\omega)$ being the inverse group velocity.

Resonators are commonly characterized either by their quality factor or their optical finesse. The finesse $\mathcal{F}$ is inversely proportional to the losses as $\mathcal{F} = \pi/\alpha$ and is a measure of the ratio between the FSR and the resonance linewidth $\Delta\nu$, i.e., $\mathcal{F} = \Delta\nu_{FSR}/\Delta\nu$. Meanwhile, the quality or Q-factor measures the ratio between the frequency and the linewidth, i.e., $Q = \nu/\Delta\nu$, and is related to the finesse as $\mathcal{F} = Q\lambda\Delta\nu_{FSR}/c$. The Q-factor is further related to the photon lifetime which is the characteristic decay time of the mode by $Q = \omega t_{ph}$.

We now consider the temporal evolution of the intracavity field in the so-called mean-field approximation. This approximation assumes that the field envelope changes little between each round trip so that the input pump field and the coupling losses can be taken as being distributed along the length of the cavity. The mean-field approximation is particularly convenient since it allows for the evolution of the field over multiple round trips to be modeled by a single partial differential equation in the form of a driven and damped nonlinear Schrödinger (NLS) equation. The validity of this approximation and the derivation of the evolution equation will be considered further in the next section.

The driven and damped NLS equation for the slowly varying envelope E of the intracavity electric field in a dispersive fiber ring cavity is given by [80]

$$t_R \frac{\partial E(t,\tau)}{\partial t} + i\frac{\beta_2 L}{2}\frac{\partial^2 E(t,\tau)}{\partial \tau^2} - i\gamma L|E(t,\tau)|^2 E(t,\tau) = -(\alpha + i\delta_0)E(t,\tau) + \sqrt{\theta}E_{in} \tag{1.32}$$

where $t_R = 1/\Delta\nu_{FSR}$ is the round trip time, α the total round trip loss, δ_0 the pump detuning, θ the (intensity) coupling coefficient and E_{in} the driving field. Equation (1.32) is written using two different time-scales, with a fast time τ corresponding to the ordinary retarded time for a pulse moving at the group velocity and a separate slow time t that measures the evolution of the field over multiple round trips. Equation (1.32) is also known as the Lugiato-Lefever equation (LLE) since it is formally equivalent to a model originally used for describing spatially transverse structures in diffractive and dissipative nonlinear cavities [81]. The LLE has also recently been used for modeling the formation of optical frequency combs in crystalline whispering-gallery-mode resonators and glass-based microring resonators [82, 83]. These are very similar to fiber ring resonators in most regards except for their smaller dimensions and larger FSR. Note that the LLE is often generalized by including higher-order dispersion and other effects such as self-steepening and Raman scattering when describing broadband fields.

The constant CW solution of Eq. (1.32) is found by setting the derivative terms to zero and is given by $E_0 = \sqrt{\theta}E_{in}/(\alpha + i(\delta_0 - \gamma L|E_0|^2))$, where the intracavity power $|E_0|^2$ satisfies the bistable cubic equation

$$\theta|E_{in}|^2 = |E_0|^2[(\delta_0 - \gamma L|E_0|^2)^2 + \alpha^2]. \tag{1.33}$$

This equation has either one or three simultaneous real solutions depending on the pump detuning. It is single valued for $\delta_0 \leq \sqrt{3}\alpha$ and displays bistability with three solutions for $\delta_0 > \sqrt{3}\alpha$, see Figure 1.12. The middle branch of the response

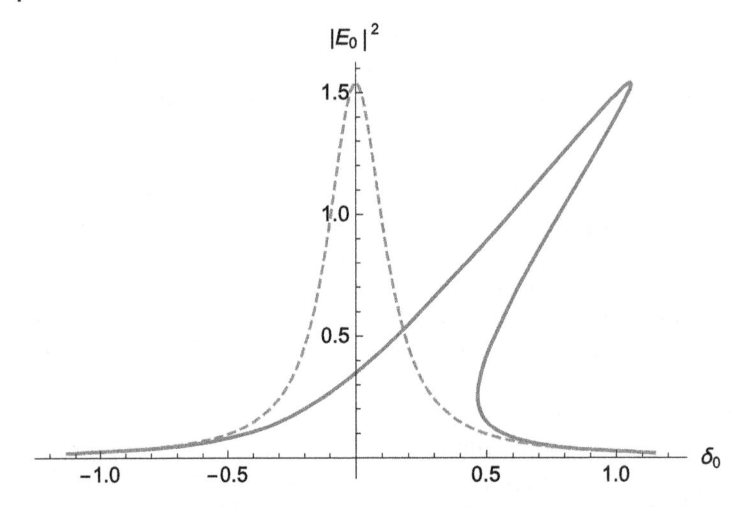

Figure 1.12 Bistability of the continuous wave solution induced by nonlinear Kerr tilt. The solid line shows the resonance bistability for the intracavity power as a function of the detuning parameter while the dashed line is the corresponding resonance in the absence of any nonlinearity ($\gamma \to 0$). Parameters $\alpha = \theta = 0.13$, $\gamma = 1.8$ W^{-1} km^{-1}, $L = 380$ m and $P_{in} = 200$ mW.

curve is, however, always unstable to CW perturbations and is not observable in practice.

In the presence of dispersion, the driven and damped NLS Eq. (1.32) can also exhibit modulational instability with qualitatively new features compared to propagation in a straight length of fiber due to the extra degree of freedom provided by the detuning. In particular, one finds that Eq. (1.32) can display modulational instability not only for anomalous dispersion, but also in the normal dispersion regime [80, 84]. To analyse the stability we look for a perturbed solution of the form $E = (|E_0| + u(t,\tau) + iv(t,\tau))e^{iArg\{E_0\}}$ and linearize Eq. (1.32) around the steady-state CW solution to obtain

$$t_R(\tilde{u}_t + i\tilde{v}_t) - i\frac{\beta_2 L}{2}\omega^2(\tilde{u} + i\tilde{v}) - i\gamma L|E_0|^2(3\tilde{u} + i\tilde{v}) = -(\alpha + i\delta_0)(\tilde{u} + i\tilde{v}) \tag{1.34}$$

where again the tilde denotes the Fourier transform with respect to τ. The real and imaginary part of this expression provides two separate linear equations and the potential growth of the perturbations can be investigated by studying the eigenvalues of their coefficient matrix. Performing the calculation, one finds that the eigenvalues are given by the expression

$$\lambda = -\alpha \pm \sqrt{(\gamma L|E_0|^2)^2 - (\delta_0 - (\beta_2 L/2)\omega^2 - 2\gamma L|E_0|^2)^2}. \tag{1.35}$$

Modulational instability is observed whenever the real part of an eigenvalue is positive. The maximum growth rate is $\lambda_{max} = \gamma L|E_0|^2 - \alpha$ and is found for frequencies satisfying the condition that the wavevector mismatch $\Delta k = \delta_0 - (\beta_2 L/2)\omega^2 - 2\gamma L|E_0|^2$ is equal to zero, i.e., $\omega^2_{max} = (2/\beta_2 L)(\delta_0 - 2\gamma L|E_0|^2)$. Contrary to the case of propagation in a straight fiber we see that this equation can have real solutions, and thus be modulational unstable, also in the normal dispersion regime when $\beta_2 > 0$, provided that the detuning $\delta_0 > 2\gamma L|E_0|^2$. From the maximum growth rate we also find that there is a minimum power threshold for the instability to occur, namely, $|E_0|^2 \geq \alpha/(\gamma L)$. This threshold can be understood to signify that the parametric gain must balance the round trip losses.

To further investigate the influence of the detuning we introduce $g_\pm = 2\gamma L|E_0|^2 \pm \sqrt{(\gamma L|E_0|^2)^2 - \alpha^2}$ and write the threshold condition $\lambda = 0$ for the instability as $(\beta_2 L/2)\omega^2 = \delta_0 - g_\pm$. For anomalous dispersion ($\beta_2 < 0$) we must then require a detuning $\delta_0 < g_+$ for MI to occur, while we should have $\delta_0 > g_-$ for normal dispersion ($\beta_2 > 0$).

It should, however, be kept in mind that the stability analysis involves the power of the intracavity field, which is related to the pump field though Eq. (1.33), rather than the experimentally accessible power of the pump field itself. This means that even though MI can occur also in the normal dispersion regime, it will not arise spontaneously unless the detuning and power are in the correct range. Beyond the initial growth of the sidebands the subsequent dynamics of the modulational instability within the dissipative cavity can in different regimes give rise not only to unstable MI, but also to the formation of stable patterns of periodic temporal structures as well as localized cavity soliton solutions, cf. Section 1.4.3.

1.4.2 Parametric Resonances and Period Doubling Phenomena

The mean-field equation of the previous section is very convenient for modeling the cavity dynamics and the formation of frequency combs when the intracavity field is changing slowly. More generally the evolution of the field inside a fiber ring resonator can be modeled using an infinite-dimensional Ikeda map [85] that consists of an ordinary NLS equation for the propagation of the field inside the fiber waveguide, together with boundary conditions that relates the fields between each round trip. The evolution of the slowly varying field envelope at round trip m is then described by [80, 86, 87]

$$E^{m+1}(\tau, 0) = \sqrt{\theta}E_{in} + \sqrt{1-\theta}e^{i\phi_0}E^m(\tau, L) \tag{1.36}$$

$$\frac{\partial E^m(\tau, z)}{\partial z} = -\frac{\alpha_i}{2}E^m(\tau, z) - i\frac{\beta_2}{2}\frac{\partial^2 E^m(\tau, z)}{\partial \tau^2} + i\gamma|E^m(\tau, z)|^2 E^m(\tau, z) \tag{1.37}$$

where z is the coordinate along the circumference of the fiber, α_i is the intrinsic fiber loss ($\alpha = (\alpha_i L + \theta)/2$) and $\phi_0 = 2\pi l - \delta_0$ is the linear phase-shift of the pump mode, which is assumed to correspond to longitudinal mode number $l = 0$. The driven and damped NLS Eq. (1.32) is obtained by averaging this map over one round trip, which is permissible if the detuning is small $\delta_0 \ll 1$ and the characteristic nonlinear length scale $L_{nl} = 1/(\gamma|E|^2)$ is much longer than the cavity length L. However, if the intracavity field changes appreciably over a single round trip, it is necessary to consider the dynamics using the full map Eqs. (1.36, 1.37). As we shall see, the presence of the boundary conditions will in fact give rise to new instabilities which leads to phenomena such as period doubling that cannot be modeled using the simple mean-field theory [86, 88].

The CW solution of the map that is periodically restored after each round trip is a fixed point of the equation $E_0 = \rho e^{\phi}E_0 + \sqrt{\theta}E_{in}$ with $\rho = \sqrt{1-\theta}e^{-\alpha_i L/2}$, $\phi = \delta_0 - \gamma L_{eff}|E_0|^2$ and $L_{eff} = (1 - e^{-\alpha_i L})/\alpha_i$. The intracavity power satisfies an equation similar to Eq. (1.33), namely

$$\theta|E_{in}|^2 = |E_0|^2[4\rho\sin^2(\phi/2) + (1-\rho)^2]. \tag{1.38}$$

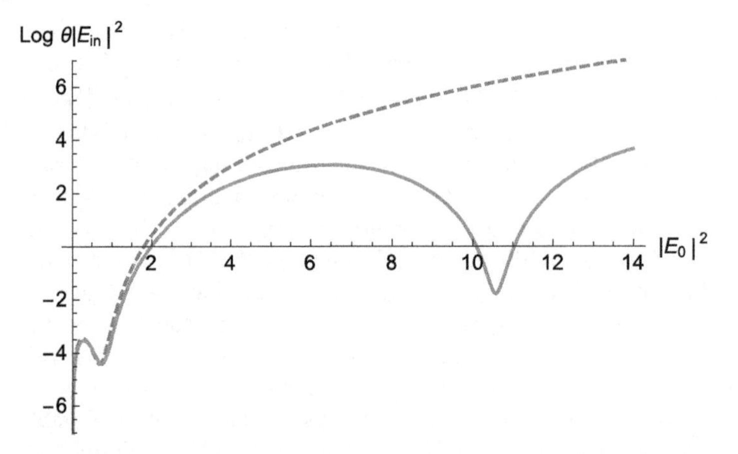

Figure 1.13 Comparison of pump dependence on the intracavity power for multi-valued stationary continuous wave solutions of the LLE (dashed) and the Ikeda map (solid). Parameters $\alpha = \theta = 0.13$, $\gamma = 1.8\ \text{W}^{-1}\ \text{km}^{-1}$, $L = 380$ m and $\delta_0 = 0.5$.

However, contrary to the mean-field case, this equation is not only bistable, but multistable and may have additional simultaneous real solutions, cf. Figure 1.13. In fact, it is easily seen that resonances occur whenever the nonlinear phase $\phi(|E_0|^2)$ is an integer multiple of 2π.

The stability of the CW solution of the map Eqs. (1.36, 1.37) can be analysed by linearizing Eq. (1.37) while assuming a perturbation of the form $E^m(\tau, z) = [E_0 + u^m(\tau, z) + iv^m(\tau, z)] \exp[-\alpha_i z/2 + i\gamma(1 - e^{-\alpha_i z})|E_0|^2/\alpha_i + iArg\{E_0\}]$. The Fourier transform of the real and imaginary part of the perturbation functions $w^m(z) = [\tilde{u}^m(\omega, z), \tilde{v}^m(\omega, z)]^T$ is then found to satisfy the linear equation system

$$\frac{dw^m}{dz} = \begin{bmatrix} 0 & -(\beta_2/2)\omega^2 \\ (\beta_2/2)\omega^2 + 2\gamma|E_0|^2 e^{-\alpha_i z} & 0 \end{bmatrix} w^m. \tag{1.39}$$

In the absence of absorption losses, i.e., $\alpha_i \to 0$, the coefficient matrix becomes independent of z which allows the eigenvalues to be calculated explicitly in order to recover the familiar result for the modulational instability gain in a lossless fiber described by the NLS equation (cf. Eq. (1.3)), namely

$$\mu = \omega\sqrt{-\beta_2\gamma|E_0|^2 - (\beta_2/2)^2\omega^2}. \tag{1.40}$$

Although it is possible to analytically calculate the instability gain of the Ikeda map for the case when the absorption losses are ignored, it is generally simpler to use numerical Floquet analysis to investigate the stability [87]. The Floquet analysis is based on investigating the eigenvalues of the fundamental matrix $W = [w_1^{m+1}(0), w_2^{m+1}(0)]$. This fundamental matrix can be obtained by first integrating Eq. (1.39) numerically over one round trip for two independent initial conditions, e.g., $w_{1,2}^m(0) = [1, 0]^T, [0, 1]^T$, in order to find $w_{1,2}^m(L)$ before applying the boundary condition Eq. (1.36) to finally get $w_{1,2}^{m+1}(0)$.

Contrary to the case of the driven and damped NLS Eq. (1.32), the Ikeda map Eqs. (1.36, 1.37) has multiple instability bands for high intracavity power that form so-called resonance tongues, see Figure 1.14 in the color plate section. These occur alternatively under both resonant and anti-resonant conditions. For the resonant case

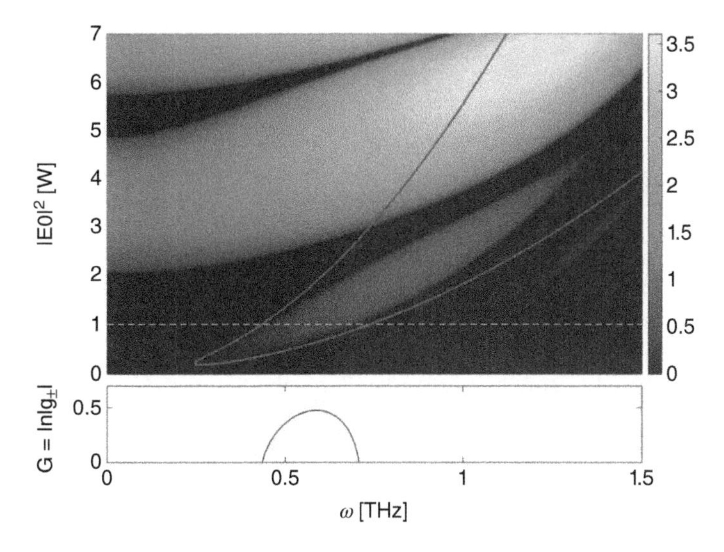

Figure 1.14 Parametric instability tongues of the Ikeda map for anomalous dispersion. The red contour shows the predicted range of modulational instability for the LLE. Below a cross-section corresponding to the dashed line is shown with the MI growth rate. Parameters $\alpha = \theta = 0.13$, $\beta_2 = -20$ ps^2 km^{-1}, $\gamma = 1.8$ W^{-1} km^{-1}, $L = 380$ m and $\delta_0 = 0$. For a color version of this figure please see color plate section.

the unstable perturbation repeats itself periodically each round trip, while for the anti-resonant case the perturbation will be π out of phase after one round trip and will not recover its original phase until two round trips have elapsed. The latter case is a form of period doubling instability which is referred to as P2-MI to distinguish it from the ordinary resonant instability or CW-MI [86]. The P2-MI is in fact usually the first instability to occur, i.e., the instability with the lowest power threshold, for normal dispersion fibers where the detuning has not been exploited in order to achieve phase matching for the lowest order CW-MI tongue that corresponds to the mean-field MI considered in the previous section.

Finally, we point out that parametric instabilities of similar origin might occur also in the driven and damped NLS Eq. (1.32) in the presence of dispersion (or nonlinearity) intracavity management. In this case the periodicity can occur on a scale $1/n, n = 1, 2, \ldots$ in units of the cavity length, and for odd n the instability is of the P2-MI type [89]. Recently, such a cavity has been implemented and employed to observe, for the first time, the steady-state excitation of sidebands on the lower branch due to the mean-field MI (temporal Turing instability) in the normal dispersion regime. This instability competes with the parametric (Faraday) instability on the upper branch of the bistable response induced by the dispersion map [90].

1.4.3 FWM in a Fiber Cavity for Optical Buffer Applications

In this section we briefly consider an application of the four-wave mixing process in passive fiber cavities. This application relies on the fact that the growth rate of the cavity MI is purely real and can give rise to stationary periodic patterns and temporal cavity solitons. The cavity solitons are a special kind of localized pulse structures that can be found in driven and damped nonlinear cavities [91]. The spectral signature of a temporal cavity soliton is a mode-locked optical frequency comb having the frequency spacing

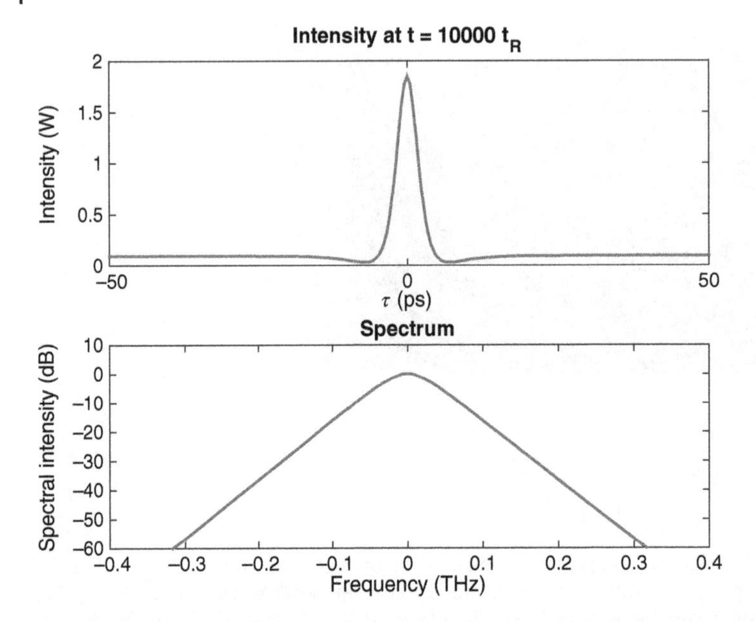

Figure 1.15 Intensity profile and normalized spectrum for a stationary temporal cavity soliton obtained from numerical solution of Eq. (1.32). The dense mode spectrum is treated as a continuum and only a portion of the round trip duration is simulated in order to simultaneously achieve sufficient temporal and spectral resolution with only 4096 sampling points. Note that the soliton is much less energetic than the continuous wave background so the latter has been filtered out from the spectrum. Parameters $\alpha = \theta = 0.13$, $\beta_2 = -20$ ps^2 km^{-1}, $\gamma = 1.8$ W^{-1} km^{-1}, $L = 380$ m, $\delta_0 = 0.5$ and $P_{in} = 150$ mW.

of a single FSR. Like other temporal solitons they represent a balance between dispersion and nonlinearity, but being dissipative structures they also simultaneously balance the gain originating from the driving beam with losses due to absorption and output coupling. An example of a cavity soliton solution of Eq. (1.32) is shown in Figure 1.15. The cavity soliton coexists with a constant low-intensity CW background corresponding to the lowest lying solution branch of Eq. (1.33) that is stable against MI, and can be considered as a compound object of a soliton plus background that is a form of low-dimensional dynamical attractor.

For a given set of pump parameters, i.e., pump power and detuning, there is a single type of cavity soliton with both amplitude and width proportional to the square root of the detuning. The solitons are phase-locked to the pump in such a way that the nonlinear phase-shift compensates for the pump field detuning. Although not integrable, these solitons are remarkably robust against different sorts of perturbations which suggests that they may be suitable candidates for use as optical bits to encode binary data [92]. By dividing the round trip time of the cavity into time slots and using cavity solitons to indicate logical one and the absence of a soliton, the logical zero, it is possible to form an optical buffer memory that in the presence of the pump beam can hold an arbitrary bit pattern for a potentially indefinite amount of time. This has possible applications, e.g., for all-optical signal processing which can benefit from the higher speeds afforded by optics in comparison to electronics.

A benefit of an all-optical bit buffer using cavity solitons in a passive fiber cavity is that it will, contrary to an active system where amplification is used, not suffer from signal degradation due to the accumulation of noise. Moreover, the soliton formation process is rather robust, which allows for pulse reshaping and wavelength conversion. The excitation of individual cavity solitons can be accomplished using an addressing beam with pulses that are close in shape and frequency to the final soliton profile. Numerical simulations have shown that such pulses can simultaneously both write and erase cavity solitons in a way that is reminiscent of XOR (eXclusive OR) gate logic.

The first experimental demonstration of temporal Kerr cavity solitons was performed by Leo et al. in 2010 [93]. In their experiment they used a passive fiber ring resonator made of 380 m of standard single-mode telecom fiber and having a round trip time of 1.85 μs to demonstrate the stable storage of individual cavity solitons and cavity soliton patterns without distortion for a time duration of over 1 s. The system was pumped with 225 mW of power using an ultra-narrow linewidth continuous wave laser operating at 1550 nm and the 4 ps cavity soliton pulses were characterized using both direct time domain oscilloscope measurements as well as frequency domain measurements using an optical spectrum analyzer. To excite the cavity solitons an additional mode-locked fiber laser was used to produce writing pulses that could address individual time slots. The writing process was confirmed to be very robust, as well as wavelength- and phase-insensitive, so that the addressing pulse did not have to precisely match the cavity soliton profile. The particular configuration studied was estimated to allow for the potential all-optical storage of up to 45000 bits at a bitrate of 25 Gbit/s.

The maximum storage time of the cavity solitons in the above experiment was limited by environmental perturbations. However, an additional limit is also set by the interactions due to the tail overlap between individual solitons that can lead to either attractive or repulsive behavior when the solitons are closely spaced in time. This relative motion of soliton pairs can result in timing-jitter and even the annihilation of solitons due to collisions. Various method for controlling these interactions have been proposed: it has, for example, been shown that they can be completely suppressed by the injection of CW light above a certain amplitude [94]. Alternatively, by introducing a shallow periodic modulation, it has been found that the solitons will be attracted to the peaks of the modulation phase profile [95]. Yet another method of suppressing the interactions between cavity solitons is using bichromatic pumping. Two pumps with equal amplitude and phase but with a frequency separation corresponding to a multiple of the FSR have also been shown to be capable of supporting cavity solitons [96]. In this case there is no continuous wave background and the solitons are superimposed on a periodic low intensity pattern with the motion of the solitons constrained by the need for them to remain on the crests of the pattern.

References

1 Bespalov, V. and Talanov, V. (1966) Filamentary structure of light beams in nonlinear liquids. *ZhETF Pisma Redaktsiiu*, **3**, 471 (JETP Lett. 3 (1966) 307).
2 Benjamin, T.B. and Feir, J. (1967) The disintegration of wave trains on deep water part 1. theory. *Journal of Fluid Mechanics*, **27**, 417–430, DOI:10.1017/S002211206700045X.

3 Dudley, J.M., Genty, G., and Coen, S. (2006) Supercontinuum generation in photonic crystal fiber. *Reviews of Modern Physics*, **78**, 1135–1184, DOI:10.1103/RevModPhys.78.1135.
4 Dudley, J.M., Genty, G., Dias, F., Kibler, B., and Akhmediev, N. (2009) Modulation instability, Akhmediev Breathers and continuous wave supercontinuum generation. *Optics Express*, **17**, 21 497–21 508, DOI:10.1364/OE.17.021497.
5 Solli, D., Ropers, C., Koonath, P., and Jalali, B. (2007) Optical rogue waves. *Nature*, **450**, 1054–1057, DOI:10.1038/nature06402.
6 Tai, K., Tomita, A., Jewell, J.L., and Hasegawa, A. (1986) Generation of subpicosecond solitonlike optical pulses at 0.3 THz repetition rate by induced modulational instability. *Applied Physics Letters*, **49**, 236–238, DOI:10.1063/1.97181.
7 Fermi, E., Pasta, J., and Ulam, S. (1955) Studies of nonlinear problems, Los Alamos Scientific Laboratory Report No. LA-1940, Los Alamos, New Mexico.
8 Zabusky, N.J. and Kruskal, M.D. (1965) Interaction of "solitons" in a collisionless plasma and the recurrence of initial states. *Physical Review Letters*, **15**, 240–243, DOI:10.1103/PhysRevLett.15.240.
9 Akhmediev, N. and Korneev, V. (1986) Modulation instability and periodic solutions of the nonlinear Schrödinger equation. *Theoretical and Mathematical Physics*, **69**, 1089–1093, DOI:10.1007/BF01037866.
10 Akhmediev, N., Korneev, V., and Mitskevich, N. (1988) N-modulation signals in a single-mode optical waveguide under nonlinear conditions. *Soviet Physics JETP*, **67**, 89.
11 Akhmediev, N., Eleonskii, V., and Kulagin, N. (1987) Exact first-order solutions of the nonlinear Schrödinger equation. *Theoretical and Mathematical Physics*, **72**, 809–818, DOI:10.1007/BF01017105.
12 Ablowitz, M.J. and Herbst, B. (1990) On homoclinic structure and numerically induced chaos for the nonlinear Schrödinger equation. *SIAM Journal of Applied Mathematics*, **50**, 339–351, DOI:10.1137/0150021.
13 Lake, B.M., Yuen, H.C., Rungaldier, H., and Ferguson, W.E. (1977) Nonlinear deep-water waves: theory and experiment. part 2. evolution of a continuous wave train. *Journal of Fluid Mechanics*, **83**, 49–74, DOI:10.1017/S0022112077001037.
14 Chabchoub, A., Hoffmann, N.P., and Akhmediev, N. (2011) Rogue wave observation in a water wave tank. *Physical Review Letters*, **106**, 204 502, DOI:10.1103/PhysRevLett.106.204502.
15 Van Simaeys, G., Emplit, P., and Haelterman, M. (2001) Experimental demonstration of the Fermi-Pasta-Ulam recurrence in a modulationally unstable optical wave. *Physical Review Letters*, **87**, 033 902, DOI:10.1103/PhysRevLett.87.033902.
16 Van Simaeys, G., Emplit, P., and Haelterman, M. (2002) Experimental study of the reversible behavior of modulational instability in optical fibers. *Journal of the Optical Society of America B*, **19**, 477–486, DOI:10.1364/JOSAB.19.000477.
17 Zhao, B., Tang, D., and Tam, H. (2003) Experimental observation of FPU recurrence in a fiber ring laser. *Optics Express*, **11**, 3304–3309, DOI:10.1364/OE.11.003304.
18 Hammani, K., Wetzel, B., Kibler, B., *et al.* (2011) Spectral dynamics of modulation instability described using Akhmediev breather theory. *Optics Letters*, **36**, 2140–2142, DOI:10.1364/OL.36.002140.
19 Kibler, B., Fatome, J., Finot, C., *et al.* (2010) The Peregrine soliton in nonlinear fibre optics. *Nature Physics*, **6**, 790–795, DOI:10.1038/nphys1740.

20 Kibler, B., Fatome, J., Finot, C., *et al.* (2012) Observation of Kuznetsov-Ma soliton dynamics in optical fibre. *Scientific Reports*, **2**, 463, DOI:10.1038/srep00463.

21 Beeckman, J., Hutsebaut, X., Haelterman, M., and Neyts, K. (2007) Induced modulation instability and recurrence in nematic liquid crystals. *Optics Express*, **15**, 11 185–11 195, DOI:10.1364/OE.15.011185.

22 Wu, M. and Patton, C.E. (2007) Experimental observation of Fermi-Pasta-Ulam recurrence in a nonlinear feedback ring system. *Physical Review Letters*, **98**, 047 202, DOI:10.1103/PhysRevLett.98.047202.

23 Farota, A.K. and Faye, M.M. (2013) Experimental study of the Fermi-Pasta-Ulam recurrence in a bi-modal electrical transmission line. *Physica Scriptai*, **88**, 55 802–55 805, DOI:10.1088/0031-8949/88/05/055802.

24 Infeld, E. (1981) Quantitive theory of the Fermi-Pasta-Ulam recurrence in the nonlinear Schrödinger equation. *Physical Review Letters*, **47**, 717–718, DOI:10.1103/PhysRevLett.47.717.

25 Cappellini, G. and Trillo, S. (1991) Third-order three-wave mixing in single-mode fibers: exact solutions and spatial instability effects. *Journal of the Optical Society of America B*, **8**, 824–838, DOI:10.1364/JOSAB.8.000824.

26 Trillo, S. and Wabnitz, S. (1991) Dynamics of the nonlinear modulational instability in optical fibers. *Optics Letters*, **16**, 986–988, DOI:10.1364/OL.16.000986.

27 Trillo, S. and Wabnitz, S. (1991) Self-injected spatial mode-locking and coherent all-optical AM/FM switching based on modulational instability. *Optics Letters*, **16**, 1566–1568, DOI:10.1364/OL.16.001566.

28 Wabnitz, S. (1988) Modulational polarization instability of light in a nonlinear birefringent dispersive medium. *Physical Review, A*, **38**, 2018–2021, DOI:10.1103/PhysRevA.38.2018.

29 Trillo, S. and Wabnitz, S. (1991) Nonlinear modulation of coupled waves in birefringent optical fibers. *Physics Letters, A*, **159**, 252–256, DOI:10.1016/0375-9601(91)90519-E.

30 Cappellini, G. and Trillo, S. (1991) Energy conversion in degenerate four-photon mixing in birefringent fibers. *Optics Letters*, **16** (12), 895–897, DOI:10.1364/OL.16.000895. URL:http://ol.osa.org/abstract.cfm?URI=ol-16-12-895.

31 Cappellini, G. and Trillo, S. (1991) Bifurcations and three-wave-mixing instabilities in nonlinear propagation in birefringent dispersive media. *Physical Review, A*, **44**, 7509–7523, DOI:10.1103/PhysRevA.44.7509. URL:http://link.aps.org/doi/10.1103/PhysRevA.44.7509.

32 Drummond, P., Kennedy, T., Dudley, J., Leonhardt, R., and Harvey, J. (1990) Cross-phase modulational instability in high-birefringence fibers. *Optics Communications*, **78** (2), 137–142. URL:http://dx.doi.org/10.1016/0030-4018(90)90110-F.

33 Rothenberg, J.E. (1990) Modulational instability for normal dispersion. *Physical Review, A*, **42**, 682–685, DOI:10.1103/PhysRevA.42.682. URL:http://link.aps.org/doi/10.1103/PhysRevA.42.682.

34 De Angelis, C., Trillo, S., and Santagiustina, M. (1994) Induced modulational instability in high-birefringence fibers: the strong conversion regime. *Optics Letters*, **19** (5), 335–337, DOI:10.1364/OL.19.000335. URL:http://ol.osa.org/abstract.cfm?URI=ol-19-5-335.

35 De Angelis, C., Santagiustina, M., and Trillo, S. (1995) Four-photon homoclinic instabilities in nonlinear highly birefringent media. *Physical Review, A*, **51**, 774–791, DOI:10.1103/PhysRevA.51.774. URL:http://link.aps.org/doi/10.1103/PhysRevA.51.774.

36 Seve, E., Millot, G., and Trillo, S. (2000) Strong four-photon conversion regime of cross-phase-modulation-induced modulational instability. *Physical Review, E*, **61**, 3139–3150, DOI:10.1103/PhysRevE.61.3139. URL:http://link.aps.org/doi/10.1103/PhysRevE.61.3139.

37 Agrawal, G.P. (1987) Modulation instability induced by cross-phase modulation. *Physical Review Letters*, **59**, 880–883, DOI:10.1103/PhysRevLett.59.880. URL:http://link.aps.org/doi/10.1103/PhysRevLett.59.880.

38 Thompson, J.R. and Roy, R. (1991) Nonlinear dynamics of multiple four-wave mixing processes in a single-mode fiber. *Physical Review, A*, **43**, 4987–4996, DOI:10.1103/PhysRevA.43.4987. URL:http://link.aps.org/doi/10.1103/PhysRevA.43.4987.

39 Rothenberg, J.E. (1990) Modulational instability of copropagating frequencies for normal dispersion. *Physical Review Letters*, **64**, 813–813, DOI:10.1103/PhysRevLett.64.813. URL:http://link.aps.org/doi/10.1103/PhysRevLett.64.813.

40 Armaroli, A. and Trillo, S. (2014) Modulational instability due to cross-phase modulation versus multiple four-wave mixing: the normal dispersion regime. *Journal of the Optical Society of America B*, **31**, 551–558, DOI:10.1364/JOSAB.31.000551.

41 Trillo, S., Wabnitz, S., and Kennedy, T.A.B. (1994) Nonlinear dynamics of dual-frequency-pumped multiwave mixing in optical fibers. *Physical Review, A*, **50** (2), 1732–1747, DOI:10.1103/PhysRevA.50.1732. URL:http://link.aps.org/doi/10.1103/PhysRevA.50.1732.

42 Armaroli, A. and Trillo, S. (2011) Collective modulation instability of multiple four-wave mixing. *Optics Letters*, **36**, 1999–2001, DOI:10.1364/OL.36.001999.

43 Fatome, J., Finot, C., Armaroli, A., and Trillo, S. (2013) Observation of modulationally unstable multi-wave mixing. *Optics Letters*, **38**, 181–183, DOI:10.1364/OL.38.000181.

44 Dudley, J.M., Dias, F., Erkintalo, M., and Genty, G. (2014) Instabilities, breathers and rogue waves in optics. *Nature Photonics*, **8**, 755–764, DOI:10.1038/nphoton.2014.220.

45 Erkintalo, M., Genty, G., Wetzel, B., and Dudley, J.M. (2011) Akhmediev breather evolution in optical fiber for realistic initial conditions. *Physics Letters, A*, **375** (19), 2029–2034. URL:http://dx.doi.org/10.1016/j.physleta.2011.04.002.

46 Millot, G., Seve, E., Wabnitz, S., and Trillo, S. (1998) Observation of a novel large-signal four-photon instability in optical wave mixing. *Physical Review Letters*, **80**, 504–507, DOI:10.1103/PhysRevLett.80.504. URL:http://link.aps.org/doi/10.1103/PhysRevLett.80.504.

47 Bendahmane, A., Mussot, A., Kudlinski, A., *et al.* (2015) Optimal frequency conversion in the nonlinear stage of modulation instability. *Optics Express*, **23** (24), 30 861–30 871, DOI:10.1364/OE.23.030861. URL:http://www.opticsexpress.org/abstract.cfm?URI=oe-23-24-30861.

48 Wabnitz, S. and Akhmediev, N. (2010) Efficient modulation frequency doubling by induced modulation instability. *Optics Communications*, **283**, 1152–1154, DOI:10.1016/j.optcom.2009.11.030.

49 Akhmediev, N., Eleonskii, V., and Kulagin, N. (1985) Generation of periodic trains of picosecond pulses in an optical fiber: exact solutions. *Soviet Physics JETP*, **62**, 894–899.

50 Erkintalo, M., Hammani, K., Kibler, B., *et al.* (2011) Higher-order modulation instability in nonlinear fiber optics. *Physical Review Letters*, **107**, 253 901, DOI:10.1103/PhysRevLett.107.253901.

51 Fatome, J., Mansouri, I.E., Blanchet, J.L., *et al.* (2013) Even harmonic pulse train generation by cross-polarization-modulation seeded instability in optical fibers. *Journal of the Optical Society of America B*, **30**, 99–106, DOI:10.1364/JOSAB.30.000099.

52 Soto-Crespo, J.M., Ankiewicz, A., Devine, N., and Akhmediev, N. (2012) Modulation instability, Cherenkov radiation, and Fermi-Pasta-Ulam recurrence. *Journal of the Optical Society of America B*, **29**, 1930–1936, DOI:10.1364/JOSAB.29.001930.

53 Mussot, A., Kudlinski, A., Droques, M., Szriftgiser, P., and Akhmediev, N. (2014) Fermi-Pasta-Ulam recurrence in nonlinear fiber optics: the role of reversible and irreversible losses. *Physical Review, X*, **4**, 011 054, DOI:10.1103/PhysRevX.4.011054.

54 Solli, D., Herink, G., Jalali, B., and Ropers, C. (2012) Fluctuations and correlations in modulation instability. *Nature Photonics*, **6**, 463–468, DOI:10.1038/nphoton.2012.126.

55 Wetzel, B., Stefani, A., Larger, L., *et al.* (2012) Real-time full bandwidth measurement of spectral noise in supercontinuum generation. *Scientific Reports*, **2**, 882, DOI:10.1038/srep00882.

56 Solli, D.R., Ropers, C., and Jalali, B. (2008) Active control of rogue waves for stimulated supercontinuum generation. *Physical Review Letters*, **101**, 233 902, DOI:10.1103/PhysRevLett.101.233902.

57 Dudley, J.M., Genty, G., and Eggleton, B.J. (2008) Harnessing and control of optical rogue waves in supercontinuum generation. *Optics Express*, **16**, 3644–3651, DOI:10.1364/OE.16.003644.

58 Nguyen, D.M., Godin, T., Toenger, S., *et al.* (2013) Incoherent resonant seeding of modulation instability in optical fiber. *Optics Letters*, **38**, 5338–5341, DOI:10.1364/OL.38.005338.

59 Wabnitz, S. and Wetzel, B. (2014) Instability and noise-induced thermalization of Fermi-Pasta-Ulam recurrence in the nonlinear Schrödinger equation. *Physics Letters, A*, **378**, 2750–2756, DOI:10.1016/j.physleta.2014.07.018.

60 Trillo, S. and Wabnitz, S. (1997) Bloch wave theory of modulational polarization instabilities in birefringent optical fibers. *Physical Review, E*, **56**, 1048–1058, DOI:10.1103/PhysRevE.56.1048.

61 Trillo, S. and Wabnitz, S. (1997) Dynamic spontaneous fluorescence in parametric wave coupling. *Physical Review, E*, **55**, R4897–R4900, DOI:10.1103/PhysRevE.55.R4897.

62 Fuerst, R.A., Baboiu, D.M., Lawrence, B., *et al.* (1997) Spatial modulational instability and multisolitonlike generation in a quadratically nonlinear optical medium. *Physical Review Letters*, **78**, 2756–2759, DOI:10.1103/PhysRevLett.78.2756.

63 Golovchenko, E.A. and Pilipetskii, A.N. (1994) Unified analysis of four-photon mixing, modulational instability, and stimulated Raman scattering under various polarization conditions in fiber. *Journal of the Optical Society of America B*, **11** (1), 92–101, DOI:10.1364/JOSAB.11.000092. URL:http://josab.osa.org/abstract.cfm?URI=josab-11-1-92.

64 McKinstrie, C. and Radic, S. (2004) Phase-sensitive amplification in a fiber. *Optics Express*, **12** (20), 4973–4979, DOI:10.1364/OPEX.12.004973. URL:http://www.opticsexpress.org/abstract.cfm?URI=oe-12-20-4973.

65 Ettabib, M.A., Jones, L., Kakande, J., *et al.* (2012) Phase sensitive amplification in a highly nonlinear lead-silicate fiber. *Optics Express*, **20** (2), 4973–4979, DOI:10.1364/OE.20.001629. URL:http://www.opticsexpress.org/abstract.cfm?URI=oe-20-2-1629.

66 McKinstrie, C.J., Radic, S., and Chraplyvy, A.R. (2002) Parametric amplifiers driven by two pump waves. *IEEE Journal of Selected Topics in Quantum Electronics*, **8** (3), 538–547, DOI:10.1109/JSTQE.2002.1016357. URL:http://ieeexplore.ieee.org/stamp/stamp.jsp?tp=&arnumber=1016357&isnumber=21871.

67 Marhic, M.E., Yang, F.S., Kazovsky, L.G., and Park, Y. (1996) Widely tunable spectrum translation and wavelength exchange by four-wave mixing in optical fibers. *Optics*

Letters, **21** (23), 1906–1908, DOI:10.1364/OL.21.001906. URL:http://ol.osa.org/abstract.cfm.?URI=ol-21-23-1906.

68 McKinstrie, C., Harvey, J., Radic, S., and Raymer, M. (2005) Translation of quantum states by four-wave mixing in fibers. *Optics Express*, **13** (22), 9131–9142, DOI:10.1364/OPEX.13.009131. URL:http://www.opticsexpress.org/abstract.cfm?URI=oe-13-22-9131.

69 McGuinness, H.J., Raymer, M.G., McKinstrie, C.J., and Radic, S. (2010) Quantum frequency translation of single-photon states in a photonic crystal fiber. *Physical Review Letters*, **105** (9), 093 604, DOI:10.1103/PhysRevLett.105.093604. URL:http://link.aps.org/doi/10.1103/PhysRevLett.105.093604.

70 Krupa, K., Tonello, A., Kozlov, V.V., *et al.* (2012) Bragg-scattering conversion at telecom wavelengths towards the photon counting regime. *Optics Express*, **20** (24), 27 220–27 225, DOI:10.1364/OE.20.027220. URL:http://www.opticsexpress.org/abstract.cfm?URI=oe-20-24-27220.

71 Provo, R., Murdoch, S., Harvey, J.D., and Méchin, D. (2010) Bragg scattering in a positive β_4 fiber. *Optics Letters*, **35** (22), 3730–3732, DOI:10.1364/OL.35.003730. URL:http://ol.osa.org/abstract.cfm?URI=ol-35-22-3730.

72 Méchin, D., Provo, R., Harvey, J.D., and McKinstrie, C.J. (2006) 180-nm wavelength conversion based on Bragg scattering in an optical fiber. *Optics Express*, **14** (20), 8995–8999, DOI:10.1364/OE.14.008995. URL:http://www.opticsexpress.org/abstract.cfm?URI=oe-14-20-8995.

73 Agha, I., Davanço, M., Thurston, B., and Srinivasan, K. (2012) Low-noise chip-based frequency conversion by four-wave-mixing Bragg scattering in SiN_x waveguides. *Optics Letters*, **37** (14), 2997–2999, DOI:10.1364/OL.37.002997. URL:https://www.osapublishing.org/ol/abstract.cfm?URI=ol-37-14-2997.

74 Donvalkar, P.S., Venkataraman, V., Clemmen, S., Saha, K., and Gaeta, A.L. (2014) Frequency translation via four-wave mixing Bragg scattering in Rb filled photonic bandgap fiber. *Optics Letters*, **39** (6), 1557–1560, DOI:110.1364/OL.39.001557. URL:http://ol.osa.org/abstract.cfm?URI=ol-39-6-1557.

75 McKinstrie, C.J. (2009) Stokes-space formalism for Bragg scattering in a fiber. *Optics Communications*, **282** (8), 1557–1562, DOI:10.1016/j.optcom.2008.12.066. URL:http://www.sciencedirect.com/science/article/pii/S0030401808013163.

76 Suchowski, H., Oron, D., Arie, A., and Silberberg, Y. (2008) Geometrical representation of sum frequency generation and adiabatic frequency conversion. *Physical Review, A*, **78** (6), 063 821, DOI:10.1103/PhysRevA.78.063821. URL:http://link.aps.org/doi/10.1103/PhysRevA.78.063821.

77 Chen, J., Li, X., and Kumar, P. (2005) Two-photon-state generation via four-wave mixing in optical fibers. *Physical Review, A*, **72** (3), 033 801, DOI:10.1103/PhysRevA.72.033801. URL:http://link.aps.org/doi/10.1103/PhysRevA.72.033801.

78 Huang, J. and Kumar, P. (1992) Observation of quantum frequency conversion. *Physical Review Letters*, **68** (14), 2153–2156, DOI:10.1103/PhysRevLett.68.2153. URL:http://link.aps.org/doi/10.1103/PhysRevLett.68.2153.

79 Collins, M.J., Clark, A.S., He, J., *et al.* (2012) Low Raman-noise correlated photon-pair generation in a dispersion-engineered chalcogenide As2S3 planar waveguide. *Optics Letters*, **37** (16), 3393–3395, DOI:10.1364/OL.37.003393. URL:http://ol.osa.org/abstract.cfm?URI=ol-37-16-3393.

80 Haelterman, M., Trillo, S., and Wabnitz, S. (1992) Dissipative modulation instability in a nonlinear dispersive ring cavity. *Optics Communications*, **91** (5-6), 401–407. URL:http://dx.doi.org/10.1016/0030-4018(92)90367-Z.

81 Lugiato, L.A. and Lefever, R. (1987) Spatial dissipative structures in passive optical systems. *Physical Review Letters*, **58**, 2209–2211, DOI:10.1103/PhysRevLett.58.2209.

82 Coen, S., Randle, H.G., Sylvestre, T., and Erkintalo, M. (2013) Modeling of octave-spanning Kerr frequency combs using a generalized mean-field Lugiato-Lefever model. *Optics Letters*, **38** (1), 37–39. URL:https://doi.org/10.1364/OL.38.000037.

83 Hansson, T., Modotto, D., and Wabnitz, S. (2013) Dynamics of the modulational instability in microresonator frequency combs. *Physical Review, A*, **88** (2), DOI:10.1103/PhysRevA.88.023819.

84 Haelterman, M., Trillo, S., and Wabnitz, S. (1992) Additive-modulation-instability ring laser in the normal dispersion regime of a fiber. *Optics Letters*, **17** (10), 745–747. URL:https://doi.org/10.1364/OL.17.000745.

85 Ikeda, K. (1979) Multiple-valued stationary state and its instability of the transmitted light by a ring cavity system. *Optics Communications*, **30** (2), 257–261, DOI:http://dx.doi.org/10.1016/0030-4018(79)90090-7.

86 Coen, S. and Haelterman, M. (1997) Modulational instability induced by cavity boundary conditions in a normally dispersive optical fiber. *Physical Review Letters*, **79**, 4139–4142, DOI:10.1103/PhysRevLett.79.4139. URL:http://link.aps.org/doi/10.1103/PhysRevLett.79.4139.

87 Hansson, T. and Wabnitz, S. (2015) Frequency comb generation beyond the Lugiato–Lefever equation: multi-stability and super cavity solitons. *Journal of the Optical Society of America B*, **32** (7), 1259, DOI:10.1364/JOSAB.32.001259.

88 McLaughlin, D.W., Moloney, J.V., and Newell, A.C. (1985) New class of instabilities in passive optical cavities. *Physical review letters*, **54** (7), 681. URL:https://doi.org/10.1103/PhysRevLett.54.681.

89 Conforti, M., Kudlinski, A., Mussot, A., and Trillo, S. (2014) Modulational instability in dispersion oscillating fiber ring cavities. *Optics Letters*, **39** (14), 4200–4203. URL:https://doi.org/10.1364/OL.39.004200.

90 Copie, F., Conforti, M., Kudlinski, A., Trillo, S., and Mussot, A. (2016) Competing Turing and Faraday instabilities in longitudinally modulated passive resonators. *Physical Review Letters*, **116**, 143901. URL:https://doi.org/10.1103/PhysRevLett.116.143901.

91 Grelu, P. and Akhmediev, N. (2012) Dissipative solitons for mode-locked lasers. *Nature Photonics*, **6** (2), 84–92, DOI:10.1038/nphoton.2011.345.

92 McDonald, G.S. and Firth, W.J. (1990) Spatial solitary-wave optical memory. *Journal of the Optical Society of America B*, **7** (7), 1328–1335. URL:https://doi.org/10.1364/JOSAB.7.001328.

93 Leo, F., Coen, S., Kockaert, P., Gorza, S.P., Emplit, P., and Haelterman, M. (2010) Temporal cavity solitons in one-dimensional Kerr media as bits in an all-optical buffer. *Nature Photonics*, **4** (7), 471–476, DOI:10.1038/nphoton.2010.120.

94 Wabnitz, S. (1993) Suppression of interactions in a phase-locked soliton optical memory. *Optics Letters*, **18** (8), 601–603. URL:https://doi.org/10.1364/OL.18.000601.

95 Luo, K., Jang, J.K., Coen, S., Murdoch, S.G., and Erkintalo, M. (2015) Spontaneous creation and annihilation of temporal cavity solitons in a coherently driven passive fiber resonator. *Optics Letters*, **40** (16), 3735, DOI:10.1364/OL.40.003735.

96 Hansson, T. and Wabnitz, S. (2014) Bichromatically pumped microresonator frequency combs. *Physical Review, A*, **90** (1), DOI:10.1103/PhysRevA.90.013811.

2

Phase-Sensitive Amplification and Regeneration

Francesca Parmigiani

Optoelectronics Research Centre, University of Southampton, Highfield, Southampton, United Kingdom

2.1 Introduction to Phase-Sensitive Amplifiers

The progress in optical communication systems over the last 50 years to improve or extend the data-carrying capacity of the transmission fibers used in our networks has been enormous, mainly as a consequence of a series of technological breakthroughs. For example, low loss single mode fibers, erbium doped fiber amplifiers (EDFAs), wavelength division multiplexing (WDM), advanced modulation techniques, coherent detection, and advanced digital signal processing (DSP) have now allowed data throughput in excess of 100 Gbit/s per signal to be reached, with many such signals covering the entire telecommunication C-band (1530–1565 nm) [1–3]. However, the complexity of, for example, the corresponding single channel transponders has increased (where the signal needs to be converted from the optical to the electrical domain, electrically processed and then converted back into the optical domain for further transmission), adding greater demand on the already power-hungry electronic routers and switches that process the data within the networks, conflicting with energy-scaling constraints [4, 5]. Thus, the development of energy-efficient, ultra-high capacity communication networks capable of connecting people and businesses seamlessly everywhere is still one of the most important challenges facing modern telecommunications. In 2012 alone, roughly 2.4 billion people worldwide used the Web, 175 million tweets per day were recorded, 4 billion hours of video was viewed via YouTube (Google Inc.'s video website) each month and 7 petabytes of photo content was added to Facebook every single month [6]. To cope with this increasing volume of Internet traffic, the bandwidth of fiber will need to be used more efficiently in future communication systems or the occupation of a new bandwidth is required to drastically increase the amount of information that can be packed and transmitted in the same fiber to approach the overall capacity of the current deployed fibers (50 THz).

Maintaining high bandwidth signals in the optical domain wherever possible, by increasing the usage of all-optical signal processing, may offer a low latency, energy-efficient solution [7], especially if higher spectral efficiency modulation formats are used.

Shaping Light in Nonlinear Optical Fibers, First Edition. Edited by Sonia Boscolo and Christophe Finot.

All optical signal processing techniques are intrinsically capable of operating at speeds far in excess of those of electronics. To ease the burden upon the electronics, high baud rate and multi-wavelength photonic solutions should be used wherever possible, bringing with them benefits such as improved energy efficiency, low latency and format flexibility. All of these factors could potentially aid in the cost effective realization of future ultrafast networks which will be required to satisfy the demand for increasing network capacity.

Over the years there has been substantial work in the field of all-optical signal processing with many interesting and impressive results achieved in the laboratory, mainly for simple intensity modulated signals. However, given the clear and necessary migration to more sophisticated modulation techniques, based on the use of phase and amplitude, the challenges in this field have now shifted to techniques that can operate on such complex, phase-based modulation format signals. The aim of this chapter is to review the impressive progress made to date, mainly toward all-optical phase regeneration and noiseless amplification of differing levels of coding complexity via phase-sensitive amplifiers (PSAs), highlighting the key underpinning technology and presenting the current state of the art.

This chapter is structured as follows. Section 2.1 gives a general description of parametric processes and the schematics of the corresponding possible devices, including a review of several practical issues that need to be addressed to achieve optimal PSA performance. We then proceed to review the history and state of the art in one-mode, two-mode and four-mode PSA-based approaches towards noiseless/improved amplification, phase regeneration and electric field decomposition of complex modulation format signals, such as (differential) binary phase keying signals ((D)BPSK) and other M-PSK signals (where M is the number of phase levels used). Within the discussion toward regeneration, we review the prospects for simultaneous amplitude and phase regeneration of higher modulation format signals and possible device integration solutions, such as semiconductors, PPLNs and silicon waveguides. Section 2.6 concludes the chapter.

2.2 Operation Principles and Realization of Phase-Sensitive Parametric Devices

In conventional optical amplifiers, usually called phase-insensitive amplifiers (PIAs), with erbium doped fiber amplifiers (EDFAs) and semiconductor optical amplifiers (SOAs) being typical examples, the signal is amplified with a certain gain, G ($G = g^2$ in Figure 2.1 (a)), regardless of its initial phase. Considering an initial signal (the input dot in Figure 2.1 (a)) with a certain phase, φ_{in}, that may vary between 0 and 2π, thus describing a circle about the origin of a Cartesian plane (constellation diagram), after amplification it will still describe a circle, centered in the same point of the plane (illustrated as the output dot in Figure 2.1 (a)), just with a bigger radius, see Figure 2.1 (a). Amplifiers are usually assessed in terms of noise and a standard metric is the noise figure (NF), defined as the ratio between the input and output signal-to-noise ratios (SNRs):

$$NF = \frac{SNR_{in}}{SNR_{out}}, \tag{2.1}$$

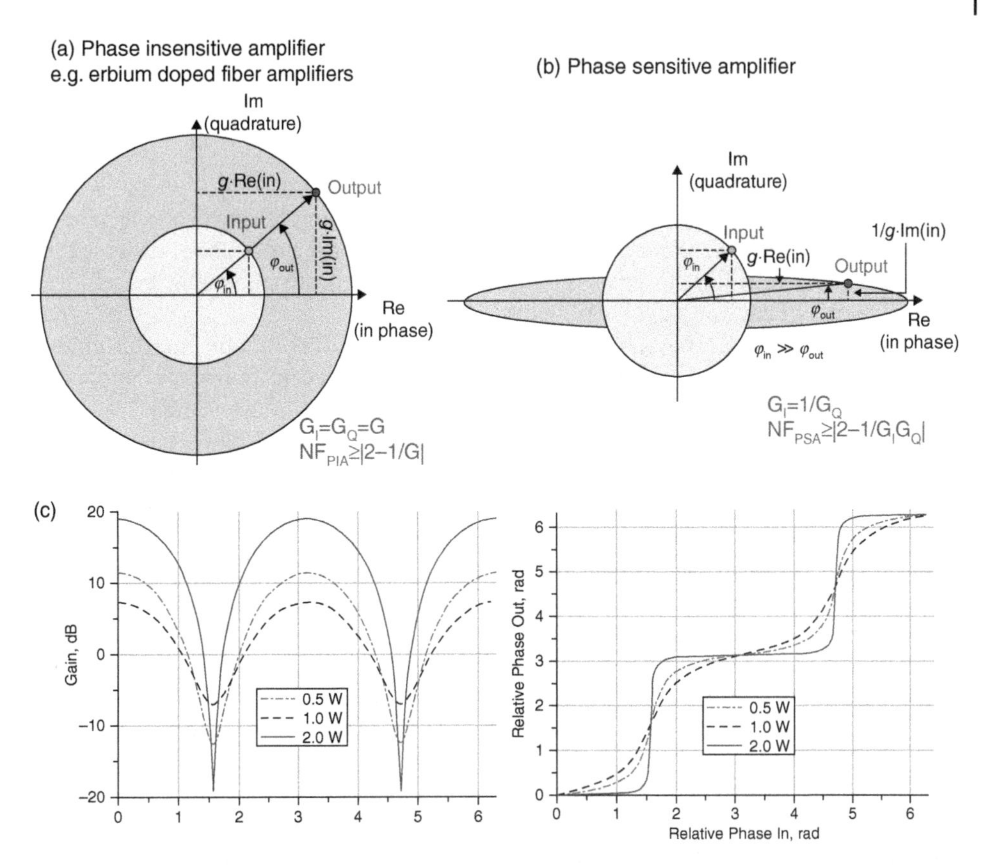

Figure 2.1 Illustration of the PIA (a) and PSA (b) operation principle; (c) Output power and phase transfer functions versus input signal phase for different PSA gains.

provided that the input light is shot-noise limited (or quantum-noise dominated) [8, 9]. In PIAs, the NF is as follows [10]:

$$NF_{PIA} \geq \left| 2 - \frac{1}{G} \right|. \tag{2.2}$$

On the other hand, in a PSA, depending on the initial signal phase relative to the PSA's gain axes (in the example shown in Figure 2.1 (b) this coincides with the real axis of the Cartesian plane, i.e., the in-phase axis), the signal will be amplified by G_I or G_Q, where $G_I = g^2$ and $G_Q = 1/g^2$ in Figure 2.1 (b). Consequently, the PSA squeezes a continuous input phase into one of two π-separated phase levels (or in general into discrete periodic step phases as we will discuss later on in this chapter), accompanied by a sinusoidal phase-to-amplitude transfer characteristic. This is clearly illustrated in Figure 2.1 (c), where the transfer functions of the gain and output phase as a function of the initial signal phase for various pump powers are shown. Clearly, the scheme is ideal for phase regeneration, phase quantization, and electric field decomposition, as will be discussed later in the chapter. Furthermore, the corresponding NF of a PSA is given by [10]:

$$NF_{PSA} \geq \left| 2 - \frac{1}{G_1 G_Q} \right| . \tag{2.3}$$

If $G_I = 1/G_Q$, $NF_{PSA} = 1$ (or $NF_{PSA} = 0$ dB), then noiseless amplification of one quadrature can be achieved (at the expense of deamplification of the other one) or, as will be seen later in this chapter, noiseless amplification of both quadratures may be achieved at the expense of half the signal bandwidth.

In general, PSAs can be achieved via interferometric [11] or non-interferometric [12, 13] means and only the second types will be discussed in the rest of the chapter, as they are nowadays the ones most commonly used. Non-interferometric schemes are based on optical parametric processes. They typically involve nonlinear interactions among several optical fields through modulating the medium's parameters, mainly the refractive index. They require conservation of both energy (i.e. the optical frequencies must be matched) and momentum (usually referred to as the phase matching condition). Depending on whether the optical material has a χ^2 (second-order) or χ^3 (third-order) susceptibility, parametric processes can be classified as three-wave mixing (TWM) or four-wave mixing (FWM), respectively. For example, in the case of FWM, four optical waves are involved: typically two pumps, one signal and one idler. In this chapter we will mainly focus on PSAs based on FWM and more specifically on the ones based on fiber optical parametric amplifiers (FOPAs), as they are the most typical examples.

If one defines the (internal) mode of parametric devices as distinct frequencies of interacting signal/idler waves, then FWM can be realized with diversified pump(s)-signal-idler schemes, which can be categorized in terms of mode number (one-mode, two-mode, and three-mode), with the pumps being degenerate or nondegenerate, see Figure 2.2. In Figure 2.2, for simplicity, only scalar processes, where the pumps, signal,

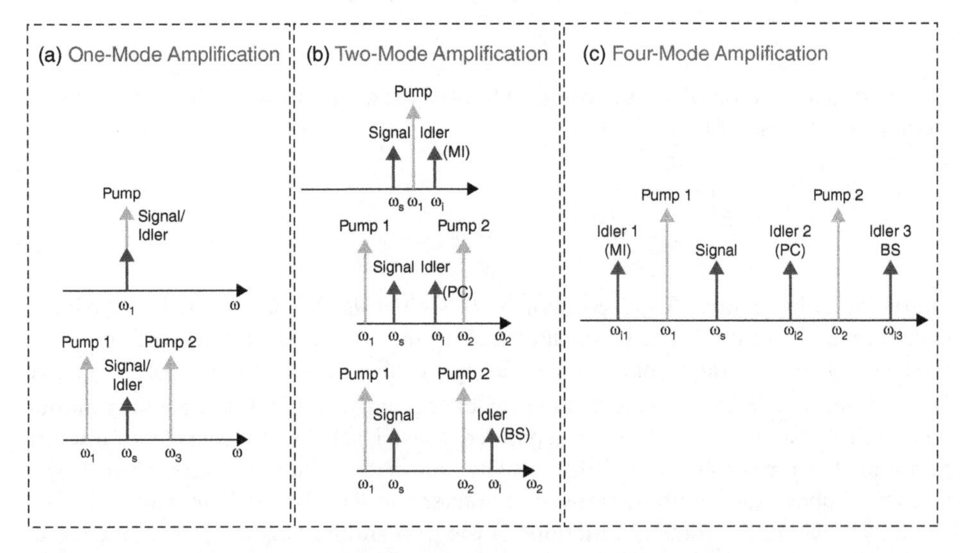

Figure 2.2 Typical frequency configurations of FWM processes for different mode numbers: (a) one-mode amplification; (b) two-mode amplification; and (c) four-mode amplification, respectively.

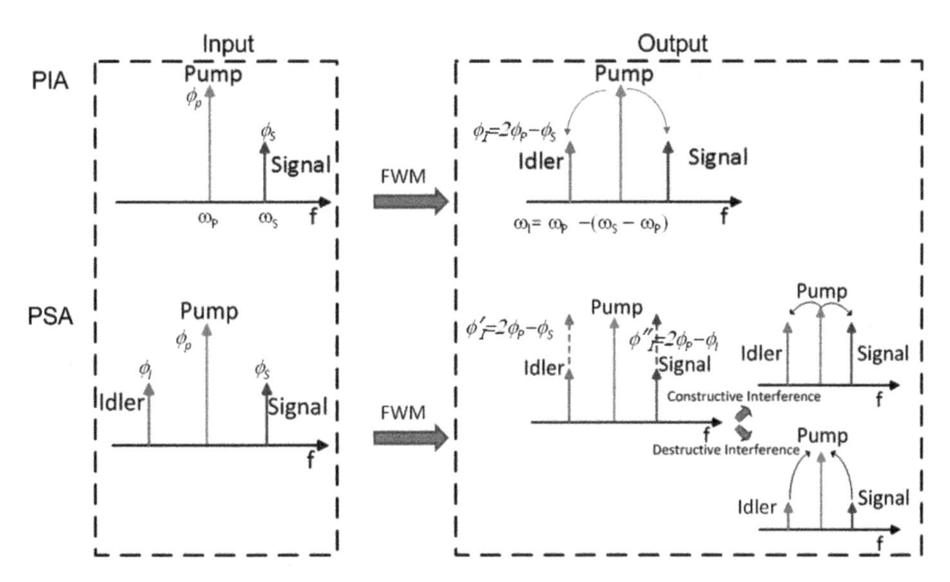

Figure 2.3 Examples of input and output frequency configurations and interactions for PIA (top row) and PSA (bottom row).

and idlers waves have the same polarization state, are considered; however, polarization states can be used to break the degeneracy of the optical modes and, thus, add freedom, as will be discussed later in the chapter. The four-mode interaction can be understood as the result of three two-mode effects happening simultaneously, which are modulation instability (MI), phase conjugation (PC), and Bragg scattering (BS) [14].

Figure 2.3 illustrates the difference in operation between the PIA-based (top) and the PSA-based (bottom) parametric process in a two-mode pump degenerate FWM. In a PIA process, the idler is generated through the FWM process only (two photons of the pump will generate one signal photon and one idler photon) and its phase will automatically follow the signal and pump phase relation $\phi_I = 2\phi_P - \phi_S$, as shown in the top of Figure 2.3. On the other hand, in a PSA process, the idler is already present at the input of the parametric interaction, together with the signal, and should be phase/frequency locked to the signal and pump. Following the previous discussion of the PIA, the PSA can be seen as the sum of two PIA processes when only the signal or the idler are considered together with the pump at the input of the system. The incoming signal interacts with the pump through FWM to generate a new idler with a certain phase relation, given by $\phi_I{}' = 2\,\phi_P - \phi_S$ (located at the same frequency of the incoming idler, causing them to interfere, see bottom of Figure 2.3. The same discussion applies for the incoming idler which produces a new FWM-generated idler with phase given by $\phi_I{}'' = 2\phi_P - \phi_I$ and located at the same frequency of the incoming signal. Thus, depending on the relative phase among the initial waves, the fields of the new FWM-generated idlers will add constructively or destructively to the fields of the original signal and idler. It is useful to introduce the phase-sensitive extinction ratio (PSER), defined as the difference between the maximum phase-sensitive (PS) gain and the maximum PS deamplification, the value of which determines the steepness of the steps in the phase response, as will appear evident in Section 2.3.

While PSAs are more than five decades old [10,15], several issues have only been tackled recently enabling the implementation of PSAs for practical network applications. First, as previously mentioned, all the incoming waves need to be phase and frequency locked and, depending on the particular application that is considered, this may be achieved via optical comb generation [11,16], modulation stripping [17,18] or a copier stage [19,20]. Second, the pump waves that are taking part in the process need to have low phase and intensity noise to avoid the transfer of their noise onto the signal/idler via the FWM process and this may be achieved via optical injection locking of the pumps to reduce their amplitude noise and the high-frequency phase noise. Third, per-wave phase and polarization control among pumps, signal and idler are needed and this requires dispersion compensation and polarization tracking. Fourth, phase-lock loops need to be implemented to deal with the slow phase variations of the signals and pumps. Finally, and critically, the nonlinear materials need to be carefully selected in order to achieve high PSA gain and high PSER. PSA-based applications have been demonstrated so far mainly using linearly strained Germanium and Aluminum doped highly nonlinear fibers (HNLFs) [17,21,22], periodically poled lithium niobate (PPLN) waveguides [23–25]; however, we will also discuss the appearance of the first PSA demonstrations in very compact devices, such as semiconductor optical amplifiers [26, 27], silicon waveguides with reversed biased pin junction [28], and more recently SiGe waveguides [29, 30].

2.3 One-Mode Parametric Processes

One-mode phase sensitive processes involve three waves in the pump non-degenerate configurations (two strong pumps and one usually weak signal), as illustrated in Figure 2.2 (a). Due to their simpler realization (there is no need to create any correlated input waves at distant frequencies), one-mode parametric devices have been extensively discussed and used for noiseless amplification, particularly during the 1990s [31–35]. More recently, one-mode PSAs were demonstrated based on both PPLN waveguides [36, 37] and HNLFs [38], showing considerable performance improvement over previous experiments.

Besides the ability for noiseless amplification, PSAs can be also used for signal phase regeneration, where only the original data signal propagates through the transmission fiber link, guaranteeing efficient bandwidth occupation. Indeed, the pumps and the idler, carrying identical or an exact multiple of the original signal phase information, depending on the specific case, are generated locally. However, for this very reason, they are modulation format-specific and, as they are usually based on dual-pump configurations, symmetrically about each signal to process, there is no easy way of scaling to wavelength division multiplexing systems without added complexity.

The first experimental demonstration of differential phase shift keying (DPSK) phase (and amplitude) regeneration using PSA was reported by Croussore *et al.* [11], where the locking among the interacting waves was guaranteed by generating an optical comb and using an amplitude modulator before printing the information on the signal carrier. Subsequently, modulation stripping techniques followed by injection locking were used to demonstrate the first black-box PSA operations of DPSK [17], QPSK and ideally any M-PSK [21] regenerators. In addition to phase locking, the ability of the device to cope with real-world impairments, such as wideband noise in phase and frequency, residual chromatic and polarization mode dispersion, and slow polarization drifts, was addressed in [39], where the DPSK regenerator was put in the middle of a deployed transmission line with an overall length of 800 km. One mode multicasting has also been demonstrated where up to four copies have shown phase regenerative properties [39].

Despite the incredible amount of work that followed these first demonstrations mainly in HNLF and PPLNs, relatively high powers (e.g., Watt-level pump powers in HNLFs) or large nonlinear interactions (defined in fiber as the γPL products, where γ is the nonlinear coefficient, P is the pump power and L is the fiber length) are required to generate phase conjugated copies at the signal wavelength, which may not always be easy to achieve. Indeed, in the case of degenerate dual pump PSA, the signal at the output can be approximated as the coherent addition of the initial signal, characterized with its phase ϕ_{IN}, with its phase conjugated version, $exp(-\phi_{IN})$, back at the signal wavelength:

$$E_{OUT}(\phi_{OUT}) \propto \exp(i\phi_{IN}) + m * \exp(-\phi_{IN}), \quad (2.4)$$

where m is the mixing ratio and describes the weight of the idler in respect to the signal. In this simple analysis and for the sake of clarity, the phases of the two pumps are set to zero, the pump noise is ignored and no pump depletion is considered.

For $m = 1$, i.e. $exp(-\phi_{IN})$ comparable to $exp(\phi_{IN})$, the phase transfer function is very sharp, with an extremely high PSER and a perfect π step, see Figure 2.4 (a), and this is usually achieved at high pump powers. However, it is worth noting in Eq. (2.4) that it is only the relative strength of the two waves that determines the PSER value (and not the absolute gain). Figures 2.4 (b) and (c) also illustrate the PSERs and the phase transfer functions versus the input phase for other two values of m (m = 0.9 and m = 0.3), where the corresponding phase quantization effects are also highlighted. If large PSERs are achieved at low pump powers, the schemes can become more energy- and cost-efficient. To this end, by carefully balancing the fiber parameters and the pump/signal powers and wavelengths, higher-order FWM components have been exploited to partially "deplete" the signal wave and hence equalize the two terms in Eq. (2.4) [22, 40]. This provided a large asymmetric PSER in a scalar configuration (achieved by obtaining a much larger PS de-amplification than PS amplification). Using this scheme, BPSK regeneration and separate in-phase and quadrature decomposition were demonstrated in [22] and [40], respectively, at nonlinear phase shifts (NPSs) (defined as the product of the total pump power, the nonlinear coefficient and the fiber length) as low as 0.8 rad, providing a PSER of ~25 dB.

A more efficient way to optimize the coherent addition of the signal/idler pair (and achieve high PSERs) would be to perform this summation in a different physical dimension, for example, at a different wavelength or polarization (to the signal) [41–43], so that it would be possible to independently control and access their relative weights without the requirement of generating too much idler in the first place. It is worth pointing out, however, that such flexibility may come at the price of reduced net PS amplification and optical signal to noise ratio (OSNR) of the final signal. In [43], an asymmetric PSER of about 20 dB was achieved using hybrid optical phase squeezers (HOPSs) to demonstrate the simultaneous phase regeneration of two coherent BPSK signals, see Figure 2.5 for the corresponding operation principle. In more detail, in this method, the coherent addition of the signal and the idler (which is generated locally in a separate nonlinear process) occurs at a different wavelength (to the signal) using a linear optical element (such as

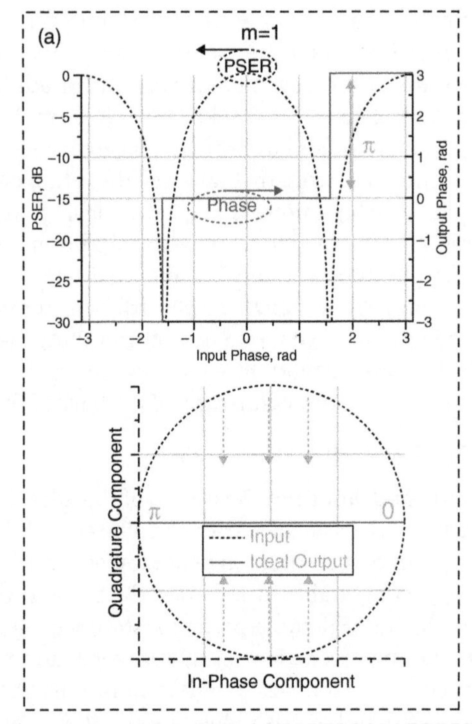

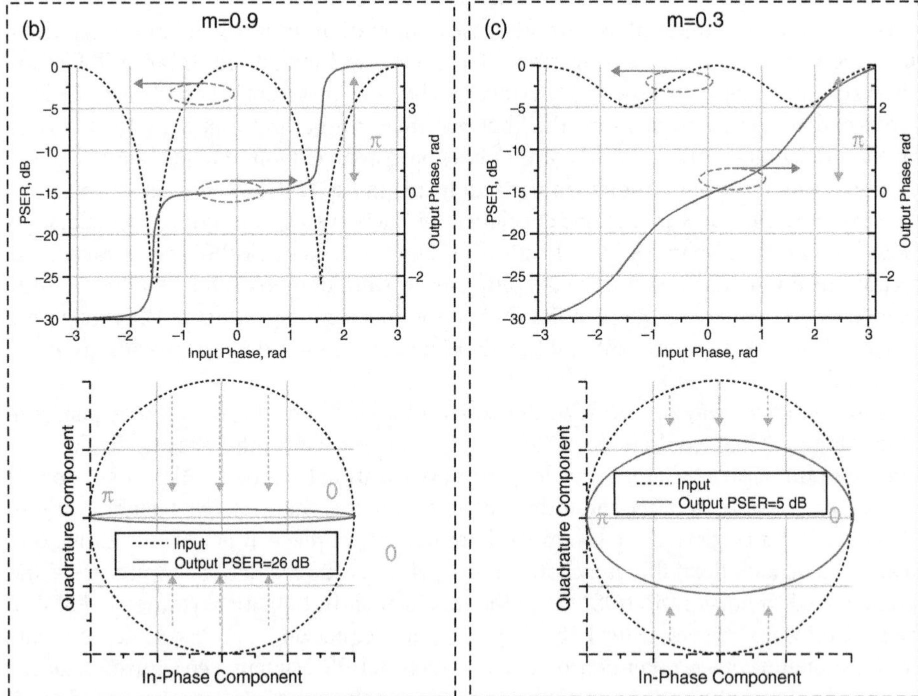

Figure 2.4 Transfer functions of the output amplitude/phase versus input phase (top) and constellation diagrams (top) derived from (4) for various values of the mixing coefficient m.

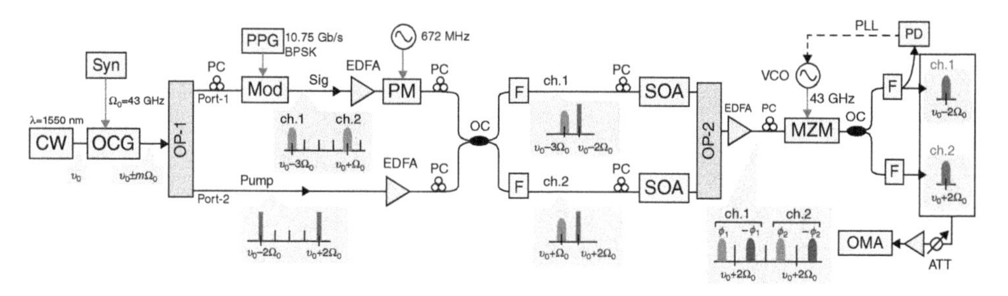

Figure 2.5 Schematic of the two-channel phase regenerator. *Source*: Kurosu 2014 [43].

an amplitude modulator), the HOPS, so that any undesired optical nonlinearity among coherent multiple channels (typically observed in parametric processes) can be avoided.

Alternatively, one can use the polarization dimension [44]. Figure 2.6 shows the schematic of this new scheme (usually referred to as polarization-assisted PSA) in comparison to more conventional schemes, such as scalar and vector ones. In more detail, the signal and pumps are still phase-locked at the input of the system as in all PSA schemes, the pumps are orthogonally polarized as in the vector PSA, but the signal is co-polarized with only one of them, as shown in Figure 2.6 (c). In this instance, the idler is generated in a phase-insensitive manner at the same frequency as the signal, but on the orthogonal polarization axis, therefore no interference (no PS operation) between signal and idler is achieved along any of the two polarization axes. However, if a polarizer is placed at the output of the nonlinear medium with its polarization angle properly rotated at an angle α with respect to the idler's axis (depending on the strength of the generated idler and, thus, on the pump power), then it is possible to exactly match the projected power of the signal and idler beams along the polarizer transmission axis. In other words, the polarizer allows the coherent addition of the projection of the signal and its phase-locked idler along its transmission axis as described in Eq. (2.4) and, thus, allows the PS operation. Even in the case when the generated idler is significantly weaker than the signal, i.e. at low pump powers, the signal/idler powers can be equalized via the optimum angle α, guaranteeing an excellent phase squeezing response. Figure 2.7 shows such a simulated angle α as a function of the total input power (the sum of signal and pump powers) for a power ratio of the signal to each of the pumps at the input of the HNLF of −14 dB, while considering the two pump powers to be equal [44]. As expected, at high powers, the optimum angle becomes $\pi/4$ (or $3/4\,\pi$), while at low power the polarizer's transmission axis will be very close to the polarization axis of the idler. Using this scheme, an asymmetric PSER of ~26 dB at NPSs as low as 0.3 rad was demonstrated [44] with excellent squeezing capabilities as shown in Figure 2.8 [45]. It is worth noting that in Figure 2.6 (c) there are two possible polarization angles of the PA-PSA, α and $\pi-\alpha$, resulting in vector summations between the signal and idler that are π-phase shifted relative to each other and Figure 2.9 attempts to explain this procedure.

Assuming, for simplicity, the high pump power case ($\alpha \sim \pi/4$), a geometrical decomposition of the PA-PSA as two orthogonal systems (as in the top row of Figure 2.9) reveals that it is constituted of two scalar PSAs (PSA 1 and PSA 2) that are π-phase shifted relative to each other. Consequently, aligning the transmission axis of a polarizer

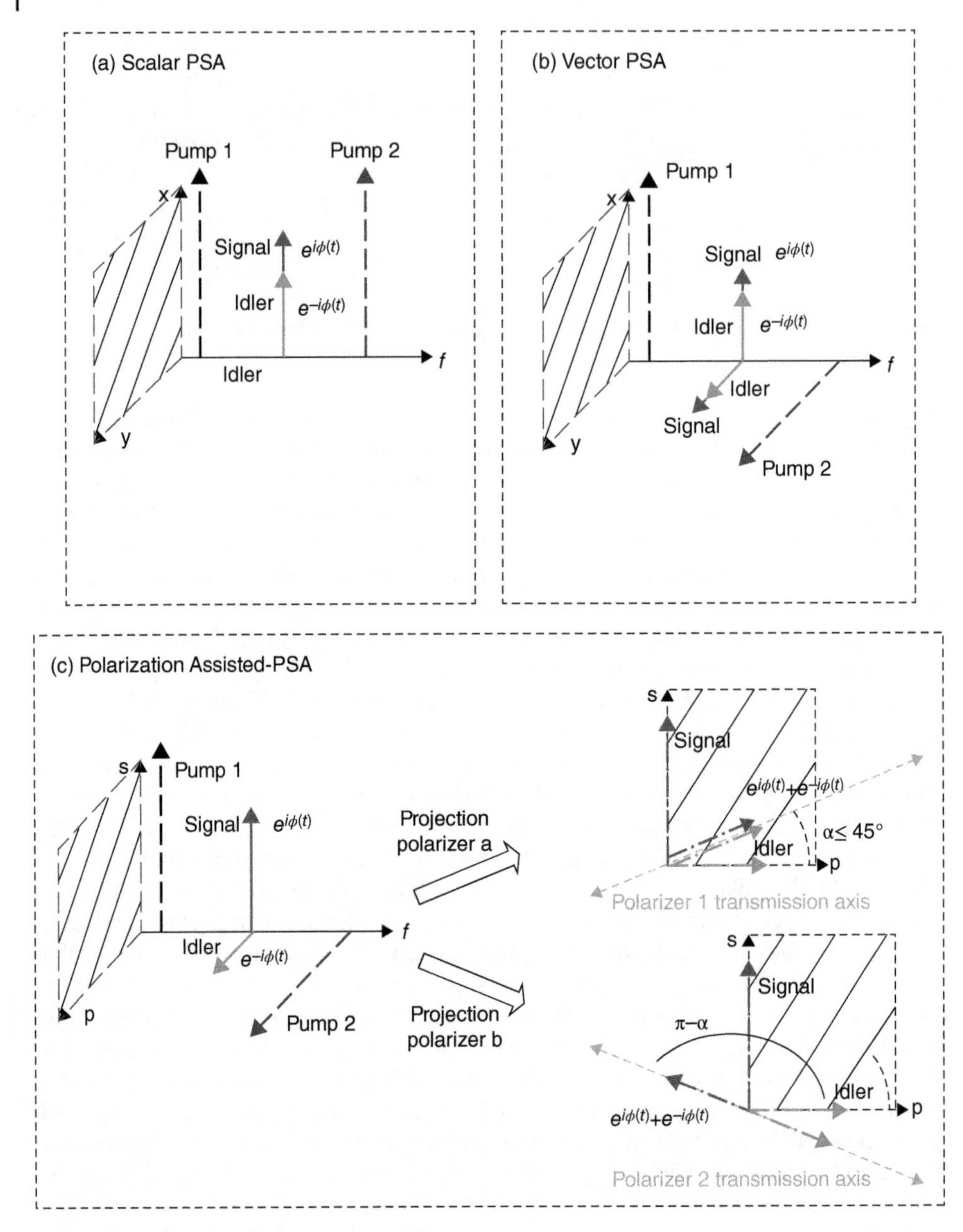

Figure 2.6 Schematic configurations of degenerate dual-pump PSAs: (a) scalar; (b) vector; and (c) polarization-assisted.

to either of the two PSAs (at $\pm\pi/4$ relative to the input signal polarization), it is possible to extract one PSA's output, which corresponds to the projection of the signal onto that specific quadrature. Figure 2.10 (a) shows how an incoming QPSK signal is decomposed into the in-phase or quadrature components at low pump power (nonlinear phase shift of about 0.3 rad) [45]. At high pump power, a polarization beam splitter (PBS)

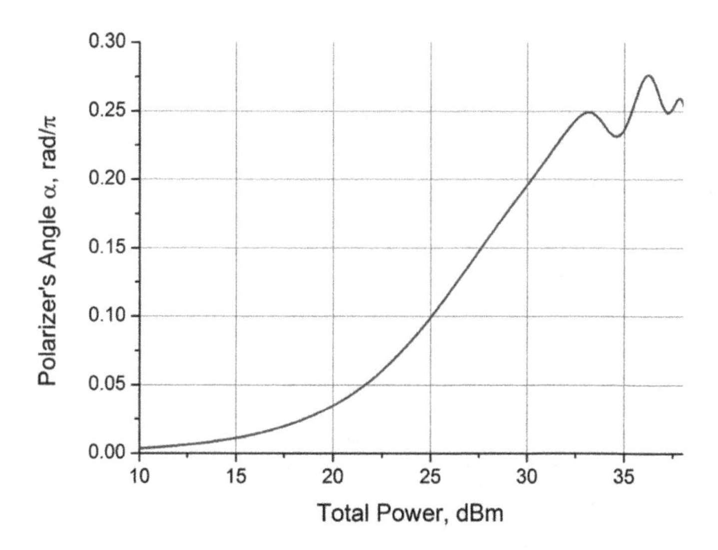

Figure 2.7 Numerical simulation of the optimum polarizer angle, α, as a function of the total input power into the nonlinear media (HNLF in the specific case) for a signal-to-pump ratio of -14 dB. *Source*: Adapted from Parmigiani 2014 [44].

in place of the polarizer can be used to simultaneously extract the two PSAs outputs and decompose the incoming signal into its quadrature components. Furthermore, it is worth noting that, for the specific case of QPSK as the incoming signal, each of the two PSA outputs represents an individually regenerated BPSK stream. Zheng *et al.* proposed and numerically investigated a scheme in [46], in which a pair of two independent PSAs (π-phase shifted relative to each other) working in saturation was used in order to suppress the phase and amplitude noise fluctuations of QPSK signals. This scheme required the recombination of the outputs of the two PSAs, which would be extremely difficult to implement in practice, due to phase coherence issues. However, by using a PA-PSA at saturation (in one fiber) any multi-path length fluctuation-related issues are mitigated and coherence is preserved with significant practical implications. Figure 2.10 (c) shows measured constellation diagrams before and after the PA-PSA, where the polarizer is aligned to the signal (idler) axis at a nonlinear phase shift of about 3.5 rad [45].

The PA-PSA scheme and more generally the one-mode PSA regenerator can be extended to allow the phase regeneration of, in principle, any M-PSK signal with an arbitrary number of phase coding levels. This requires the realization of a staircase phase response that includes the corresponding number of steps. By analogy to Eq. (2.4), this can be implemented through a coherent addition of the (M − 1)-th phase harmonic of the signal with the signal itself, which mathematically can be expressed as Eq. (2.5):

$$E_{OUT}(\phi_{OUT}) \propto \exp(i\phi_{IN}) + m * \exp(-(M-1) * \phi_{IN}), \tag{2.5}$$

where m is again the mixing ratio.

The periodicity of the phase transfer function matches M and the optimum phase quantization is achieved when m $\approx$1/(M-1) [21]. For example, four-level phase quantization steps may be achieved by coherently adding to a signal its conjugated third phase harmonic with a mixing ratio of 3:1, respectively. Many PSA-based schemes for

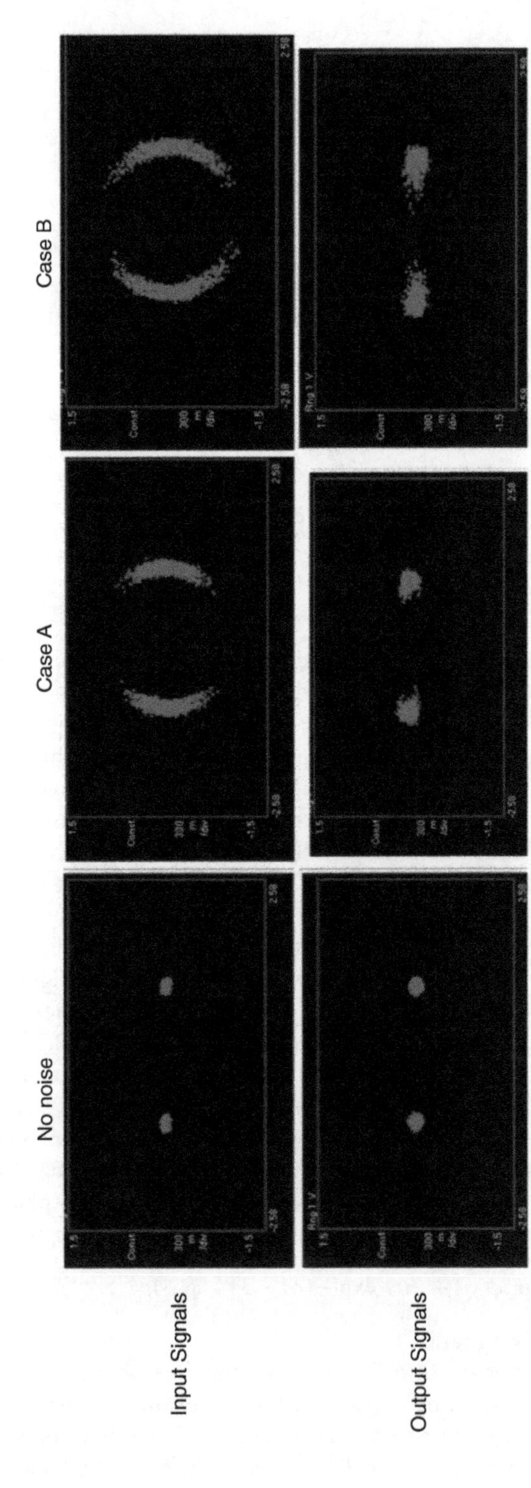

Figure 2.8 Measured constellation diagrams at the input and output of the polarization assisted PSA for different amounts of broadband phase noise added at NPS of about 0.35 rad. *Source:* Adapted from Parmigiani 2015 [45].

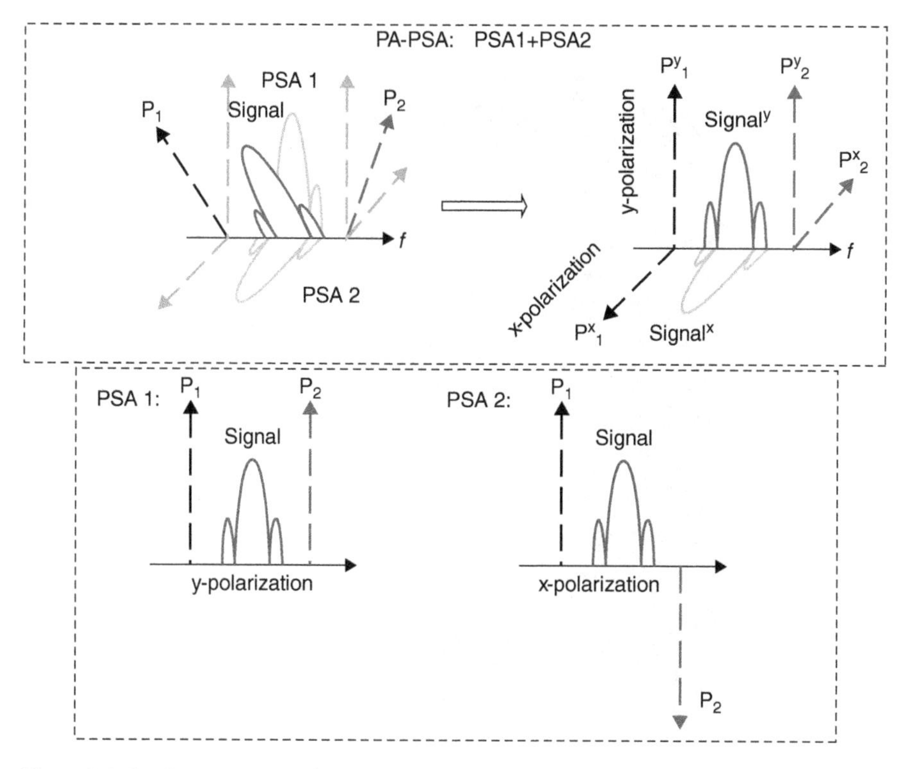

Figure 2.9 Synthesis example of the two π-phase shifted PSAs in a single medium, resulting in the PA-PSA scheme for the high pump power case ($\alpha \sim \pi/4$).

four-level phase quantization make use of two optical nonlinear processing stages [21, 47]; the first consists of a degenerate, single pump FWM scheme to produce high-order harmonics, phase locked to the signal, of which the third is chosen to be mixed back into the signal through a second stage of FWM, as illustrated in Figure 2.11 (a). A variation of this scheme (see Figure 2.11 (b)), referred to as the idler-free scheme [48], effectively combines these two stages into a single one by making use of two pumps with their power tuned so as to simultaneously produce the third phase harmonic and mix it back into the signal. Although this scheme is compact, using only a single stage, high pump powers must be used to produce enough harmonic and coherently add it back onto the signal, with the scheme only operating for very specific pump-signal power ratios. In a different idler-free configuration, the regenerated (and conjugated) signal can be produced at the location of the signal's third harmonic, as shown in Figure 2.11 (c) [49]. This last scheme is notable for its ease of tunability and flexible operating power, these advantages being paid for through its wavelength-converting nature. Indeed, phase conjugation of the output can be avoided simply by choosing the regenerated output that appears on the opposite side of the pump to the original input signal, as demonstrated in an SOA-based PSA [27]. Examples of spectra before (dashed line) and after (solid line) the FWM in SOA, where the conjugated and unconjugated regenerated output, identified by labels $\bar{S}'$ and S', are highlighted in Figure 2.12. It is worth noting that any of the idler-free schemes so far proposed cannot stand as black box regenerators.

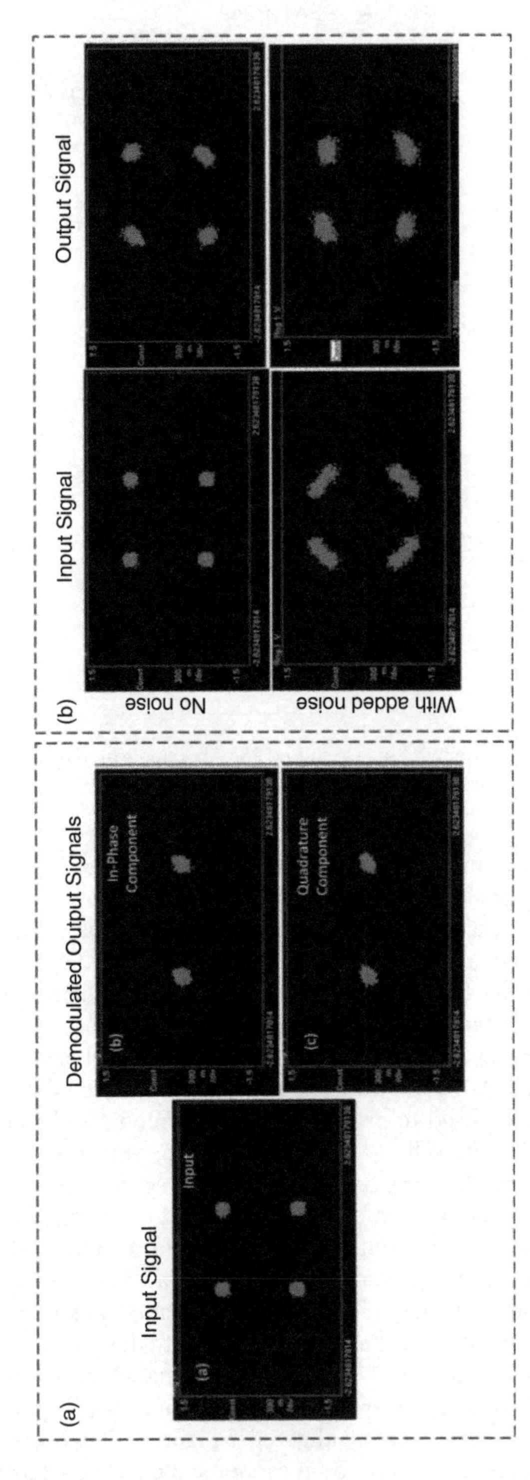

Figure 2.10 (a) Synthesis electric field decomposition of a QPSK signal at NPS of about 0.3 rad; (b) Measured constellation diagrams at the input and output of the scheme when it is implemented as a regenerator with and without noise added to the incoming QPSK signal. The PA-PSA was working at high NPS. *Source:* Adapted from Parmigiani 2015 [45].

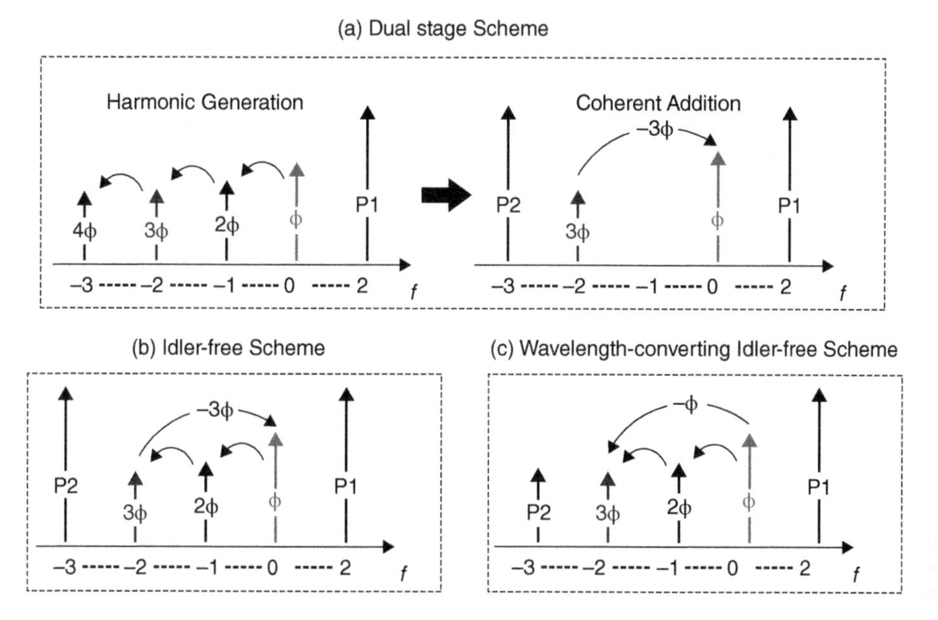

Figure 2.11 Illustration of three possible spectral configurations to achieve PSA: (a) dual stage scheme; (b) idler-free scheme; and (c) wavelength-converting idler-free scheme.

However, these schemes are compatible with any means of recovering the carrier phase, for instance, using pilot tones, direct injection locking in optical cavities [50] or 4th harmonic generation, as was achieved in [17].

Although phase regenerators alone may extend the reach of telecommunications signals, particularly PSK formats, they cannot be recursively applied to a signal ad infinitum; eventually the uncorrected amplitude noise will limit the signal-to-noise ratio of the transmission. Indeed, given that many PSA-based phase regenerators result in an

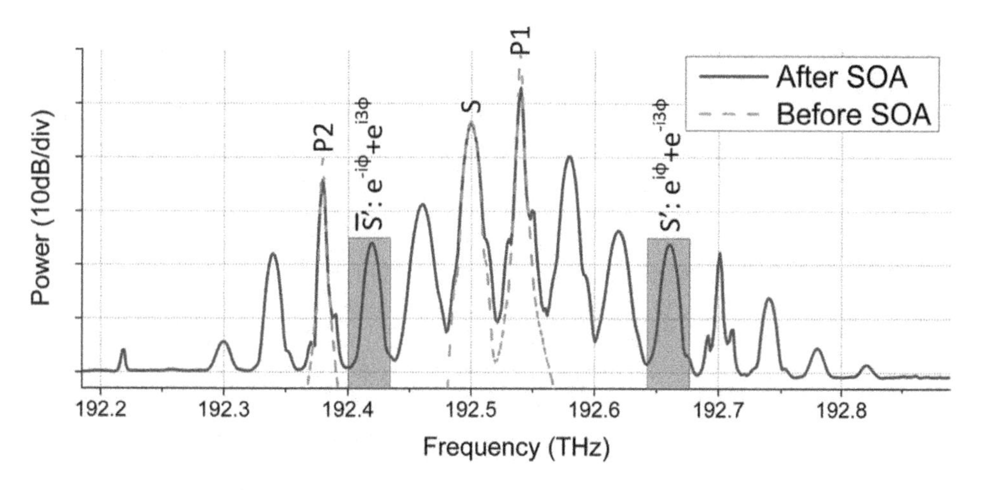

Figure 2.12 Example of spectral traces before (dashed line) and after (solid line) FWM in SOA. *Source*: Adapted from Bottrill 2015 [27].

increase in amplitude noise, relatively few consecutive phase squeezing stages may be applied before they are no longer beneficial. Ultimately, both the signal's phase and amplitude must be regenerated. This may be achieved using FWM saturation [48], operating the PSA within a nonlinear amplifying loop mirror (NALM) [51] or new nonlinear optical loop mirror-based configurations to realize two orthogonal regenerative Fourier transformations [52]. Despite successfully regenerating both phase and amplitude (and sometimes also offering multilevel functionality [51, 52]), these schemes are often complicated in their optimization and are limited in their regenerative ability, either due to intrinsic noise processes, or as a result of a compromise between the underlying regenerative operations. Instead of combining the two regenerative processes in one nonlinear stage, they may be applied sequentially in two independent processes, allowing for easier optimization, as carried out in [53, 54]. For example, in [54] the compact, highly tunable scheme for QPSK phase regeneration [48] is combined with a saturated pump-degenerate FWM-based amplitude squeezer to reduce amplitude noise to phase noise conversion [55] to achieve amplitude and phase regeneration at the original signal wavelength. Importantly for the success of the scheme, as the phase regenerator is conjugating, self-phase modulation (SPM) accrued by the signal in the phase regenerator is undone as it propagates through the amplitude regenerator. Figure 2.13 reports the corresponding spectral traces at different points in this scheme [54]. As described above, the PSA stage is wavelength converting, and so to extract the phase squeezed signal, a filter is placed at the location of the signal's third harmonic (Figures 2.13 (a), (b)). The phase squeezed, conjugated and wavelength-converted signal then enters the amplitude squeezing stage: (1) to convert the signal back to its original wavelength; (2) to restore the signal to its unconjugated state; (3) to regenerate the signal's amplitude; and (4) to compensate the SPM induced by the first nonlinear stage (Figures 2.13 (c), (d)).

Constellation plots are provided in Figure 2.13 (e) for different input noise scenarios (broadband phase noise only or amplified spontaneous emission (ASE) only added, respectively) and at different stages in the scheme [54]. When only phase noise is added to the signal, see second row of plots in Figure 2.13 (e), resulting in an input signal phase noise of 8 deg rms, the phase noise after the phase regenerator (middle column) has been greatly reduced, but at the expense of an increase in amplitude noise, which is typical behavior for this regenerator [21, 48]. Comparison of the constellation plots after phase regeneration (middle column) and after both phase and amplitude regeneration (right-hand column), clearly shows the benefit of this additional, amplitude regenerating stage. When ASE noise is added to the signal, third row in Figure 2.13 (e), the regenerator clearly reduces the noise, and the amplitude regenerator proves its worth by reducing the large amount of amplitude noise present after phase squeezing as well as undoing the SPM accrued in the PSA.

A different route toward the realization of more energy-efficient PSAs is the use of more compact and cost-effective photonic integrated circuits. For example, semiconductor optical amplifiers (SOAs) have been used in a scalar dual pump PSA configuration [26, 27]. The recent advances in and maturity of silicon photonics technology have resulted in the fabrication of highly integrated small CMOS-compatible devices with high yield and at low cost. These features, along with the ultra-high $\chi^{(3)}$ nonlinearity of silicon, have made silicon photonics a particularly attractive platform for implementing all-optical nonlinear applications [56]. So far, PSAs in silicon devices using CW pumps have been demonstrated either in a scalar scheme using a silicon waveguide with

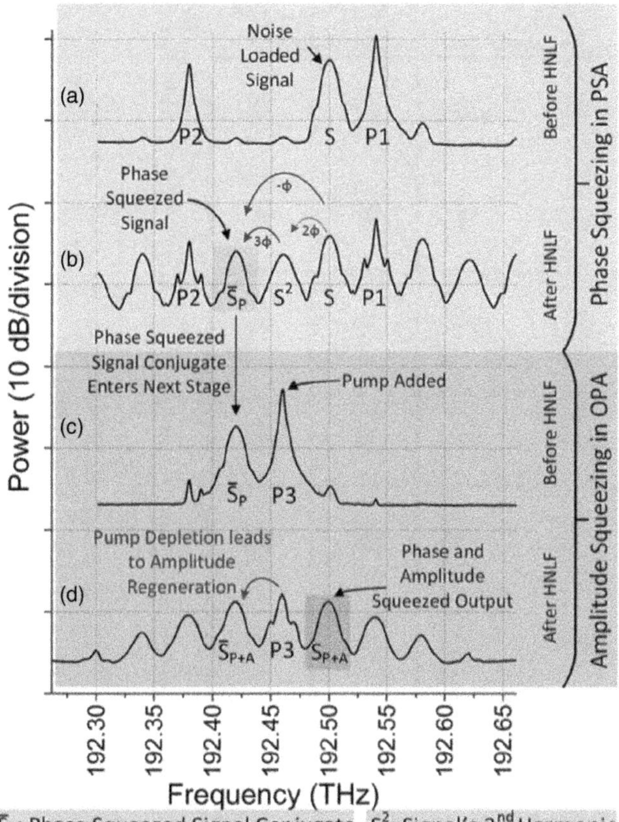

Figure 2.13 Example of spectra: (a) before PSA; (b) after PSA; (c) before amplitude regeneration; (d) After amplitude regeneration; and (e) constellation plots at different points in the system. *Source*: Adapted from Bottrill 2015 [54].

a reversed bias p-i-n junction to minimize free-carrier absorption [28] or in a PA-PSA scheme using a passive low-birefringence SiGe waveguide [29, 30].

It is worth noting that in all these QPSK schemes so far discussed [21, 48, 54], the abruptness of the corresponding phase transfer functions, given by Eq. (2.5), is less well defined (see dashed line in Figure 2.14) as compared to the one of BPSK signals, reported in Eq. (2.4), with the depth of the phase-to-amplitude conversion being reduced. However, sharper step-like phase functions could be achieved if more phase harmonics of the signal were coherently added to the original signal or if more complicated schemes, including a combination of PSA and limiting amplifier, were to be used [57], see solid line in Figure 2.14.

Most PSA experiments reported in the literature have supported single channel operation with only a few attempts to simultaneously regenerate two or more channels in a single device [43, 58–60], highlighting the challenges faced when attempting to achieve

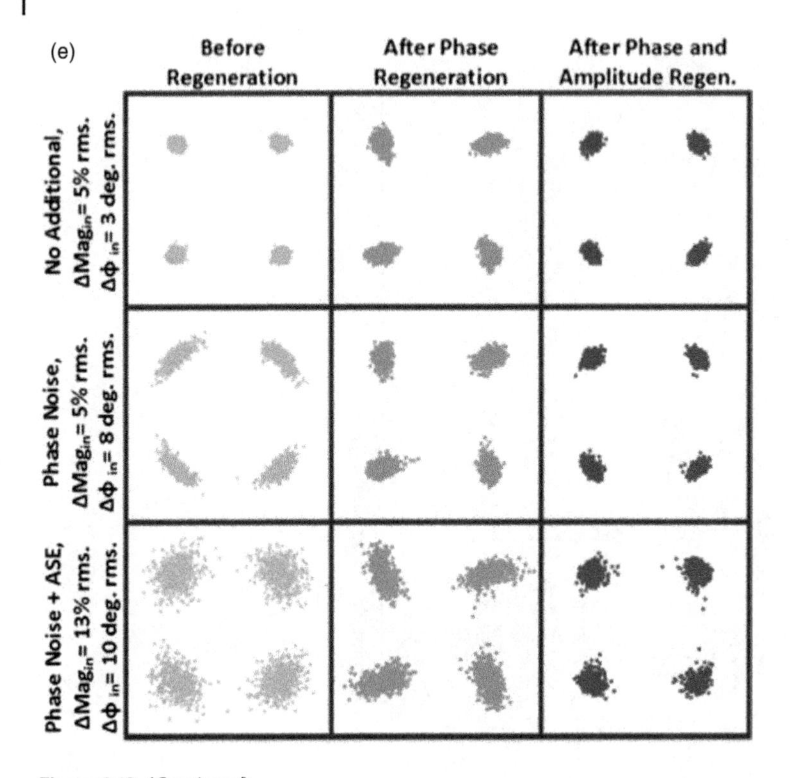

Figure 2.13 (*Continued*)

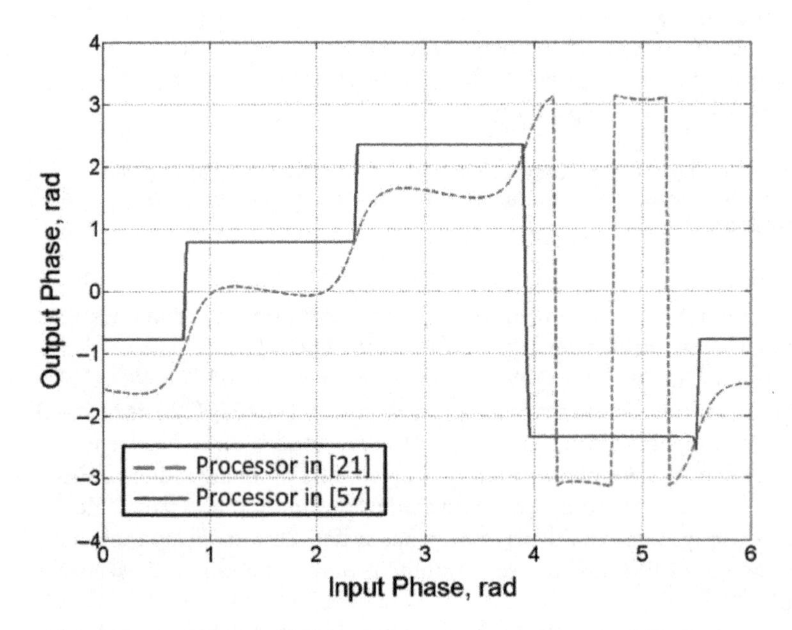

Figure 2.14 Comparison between phase transfer characteristics of [21] and [57] for M = 4 (QPSK). *Source*: From Ref. [57]; courtesy of A. Bogris.

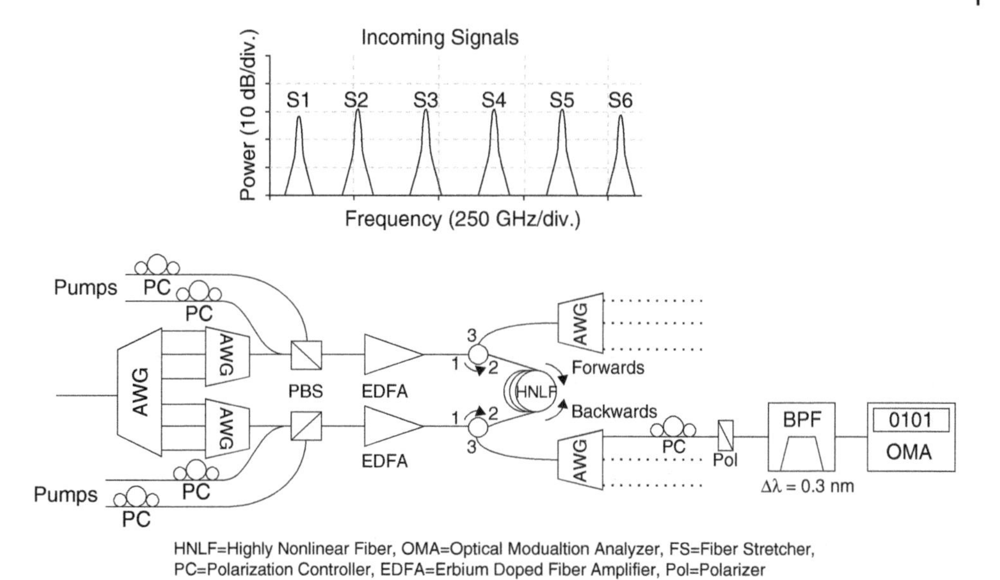

Figure 2.15 Experimental set-up of a six-channel phase regenerator based on the PA-PSA configuration and input spectrum of the six channels. *Source*: Adapted from Parmigiani 2015 [59].

multi-channel operation. For example, in [58], two scalar PSAs, sharing one pump, were used to demonstrate simultaneous regeneration of two channels only, mainly limited by the strong pump-to-pump (PtP) and pump-to-signal (PtS) FWM interaction over a large bandwidth.

In [59], the simultaneous phase regeneration of six (two times three) channels in a single nonlinear medium has been successfully achieved by exploiting the key features of the PA-PSA, which are the orthogonality of the pumps and the use of low pump powers (as previously discussed), together with bidirectional propagation and careful placement of the pump/signal wavelengths. The corresponding experimental set-up of the scheme is reported in Figure 2.15 with the spectrum of six signals at the input of the regenerator.

Figure 2.16 (a) reports the full BER curves before and after the regenerator under single- and multi-channel operation, both with and without noise, for signal S2 as an

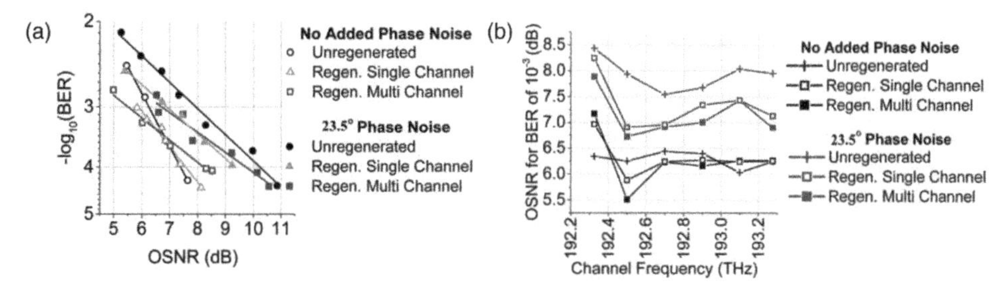

Figure 2.16 (a) Example of BER curves with and without added noise before/after the regenerator for single- and multi-channel operation (presented for signal 192.5 THz); (b) OSNRs for BER of 10^{-3} versus signal carrier frequency before/after the regenerator in single- and multi-channel operation. *Source*: Adapted from Parmigiani 2015 [59].

example [59]. A similar performance was obtained for all the channels, as shown in Figure 2.16 (b), which reports the OSNR at a BER of 10^{-3} for all of the signals. Both in single and multi-channel operation, a negligible penalty was observed in the absence of noise added to the various signals. In the presence of noise, the performance of the regeneration system was very similar in both the single- and multi-channel operation, highlighting the absence of channel cross-talk.

2.4 Two-Mode Parametric Processes

Two-mode phase-sensitive processes involve up to four waves (one/two strong pumps and four weak sidebands), as shown in Figure 2.2 (b). In principle, FWM-based Bragg scattering (BS), shown in the bottom configuration in Figure 2.2 (b), belongs to a different process compared to the other two-mode processes, modulation instability (MI) and phase conjugation (PC), illustrated in the top and middle rows of Figure 2.2 (b). Indeed, the BS process can be viewed as a photon exchange between the signal and idler waves. Its ability to achieve noiseless frequency conversion (with no associated gain) is in contrast to the amplifying PC and MI processes involving the photon transfer between the pump and the signal/idler waves [14, 61, 62], where the coupled photons between the signal and the idler are conjugated. For the latter processes, phase sensitive interaction is the only way to achieve a noiseless operation and only these schemes will be discussed in the rest of the section. In these instances, the same data needs to be carried on two signals at different wavelengths: the first signal is the original data signal and the second its complex-conjugated copy (idler). This implies occupation of twice the bandwidth as compared to a degenerate configuration, together with the requirement of perfect phase synchronization among all the waves prior to the PSA, see Figure 2.17. The PSA amplifies equally both signal quadratures of the perfect phase-matched, power-equalized signal-idler pair, experiencing 6 dB higher gain (in the high gain regime) than the uncorrelated noise, which consequently experiences phase-insensitive amplification, PIA, only, see inset cartoon in Figure 2.17 [19, 20]. Note that if the signal-idler pair is not perfectly power equalized, which can be achieved either by imbalanced noise loading (the OSNRs of the signal and idler are different) or by varying idler to signal input power ratio, the

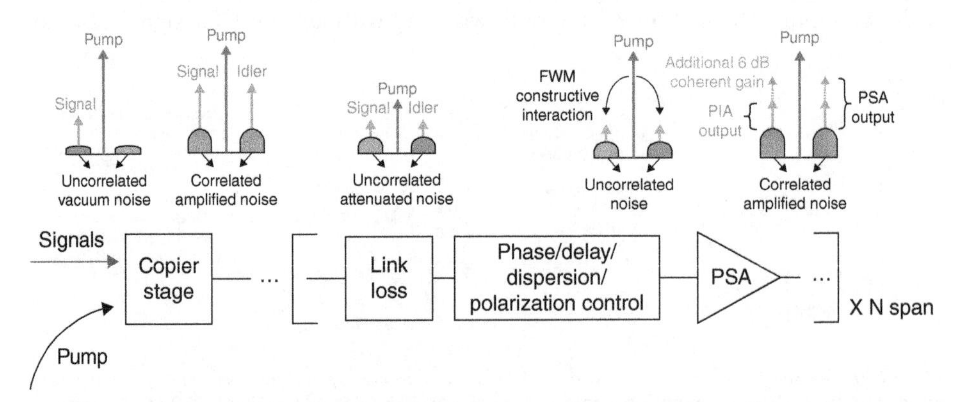

Figure 2.17 Basic principle of the two-mode parametric process and corresponding experimental schematic. *Source*: Adapted from Tong 2011 [20].

discussion above does not hold [63]. For example, an OSNR improvement (i.e., the difference in OSNR at the output and input of the PSA) for the idler wave in excess of 20 dB has been measured when the signal to idler ratio was higher than 20 dB [63]. As a consequence, the final signal OSNR degrades to about −3 dB. This can very useful in restoring low OSNR signals at the expense of other phase-locked sidebands.

Practically, the signal-idler pairs are usually generated in a parametric PIA before the transmission link (usually referred to as a "copier stage"). The copier (as well as the PSA) should operate in a linear gain regime to avoid gain-saturation-induced distortion. It is important to note that the copier-PSA scheme will only show noise performance improvement (approaching the ideal 6 dB coherent gain) if sufficient link loss is allowed between the copier and the PSA stages to decorrelate the initially strongly correlated noise components accumulated immediately after the copier stage. Key features of this type of two modes process are (1) transparency to the modulation format used (as the absolute input signal phase is tracked by that of the conjugated idler); and (2) easy extension to simultaneous multi-channel amplification without adding system complexity to the one pump configuration [20, 64]. This type of PSA has been used to demonstrate ultra-low noise optical links operating in linear and nonlinear regimes. Reference [20] reported the lowest noise figure of 1.1 dB at 26.5 dB gain, where a lump loss was simulating the corresponding transmission loss. Interestingly and importantly, the advantages of the PSA noise remained in a "real" fiber link (consisting of single mode fiber, SMF) of 80-km length, when injection locking was used for pump recovery [65] and even stronger with nonlinear transmission [66–68]. Indeed, the copier-PSA link can provide the mitigation of large accumulated nonlinear distortions through the coherent superposition function of the PSA, providing higher performance links not only through low noise figure amplification, but also by simultaneously mitigating penalties associated with high transmission power. For example, the combination of both low NF amplification and nonlinearity mitigation using an in-line PSA in a multi-span optical transmission link allowed a threefold increase in reach over an erbium-doped fiber amplifier (EDFA) using 10 GBd QPSK data, going from a transmission reach of 1050 km using an EDFA to 3465 km using an in-line fiber-based PSA [68].

In the color plate section, Figure 2.18 (a) shows the constellation diagrams at a BER $= 10^{-3}$ after linear and nonlinear transmission using either an EDFA or a PSA [68]. In the linear transmission regime (input power into the link of -1 dBm), EDFA amplification supports four round trips in a recirculating loop (with each span consisting of a fully dispersion-compensated 105-km-long standard single mode fiber) while PSA amplification supports 17 round trips. After four round trips, the constellation exhibits much less noise in the PSA case than in the EDFA case. These results are well in line with the four times (6 dB) increase in reach that is theoretically expected when using a copier-multiple-PSA link instead of a PIA-only link [69]. Note that if two waves carrying the same data were used in the EDFA link, this would only have doubled the reach at the same expense in spectral utilization. In the nonlinear transmission regime (input power into the link of 8 dBm), the reach increase is three times (going from 10 to 30 round trips) with the EDFA amplified transmission constellation clearly showing nonlinear distortions. The constellations for PSA amplification are also shown for 10 round trips, and are evidently much less distorted by nonlinear phase rotation than for EDFA amplification. The reach improvement is primarily due to the nonlinearity cancellation performed by the PSA after each loop. Figure 2.18 (b) reported the BER as a function

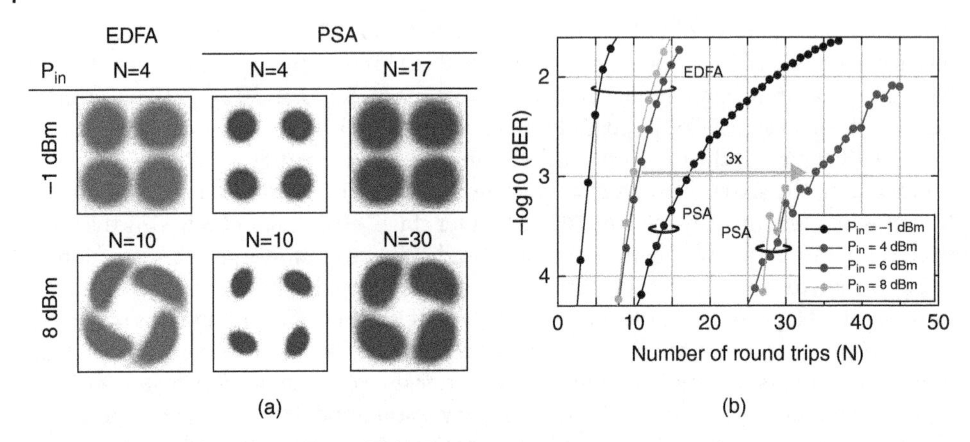

Figure 2.18 (a) Constellation diagrams in a linear and nonlinear transmission regime giving a BER of 1×10^{-3}; (b) BER versus number of round trips for EDFA and PSA in-line amplification at -1 dBm signal launch power (linear transmission), 4 dBm (optimum for PSA), 6 dBm (optimum for EDFA), and 8 dBm (strongly nonlinear transmission). *Source*: Olsson 2014 [68].

of the number of round trips in the various cases studied. The results highlight the significant increase in reach for PSA amplified transmission over EDFA amplified transmission at different launch power conditions: low, high, and optimal powers. Similar experimental demonstrations have also been carried out in PPLN waveguides [64]. By perfect analogy to the nonlinear cancellation performed by the PSA, schemes based on digital signal processing have been exploited to achieve similar improved performances in the nonlinear transmission regime. These are either based on the so-called "phase-conjugated twin waves" [70, 71] or Digital PSA [72].

Besides noiseless transmission, two-mode phase-sensitive amplification seems a promising and straightforward technique for noiseless frequency comb processing [73]. Indeed, first, all frequency lines in an optical comb are naturally frequency and phase-locked among each other, and thus ready for phase sensitive amplification. Second, by picking the central frequency line, which is usually the seed to the optical comb, as the degenerate pump in the PSA, it is possible to guarantee that the pairs of comb lines that serve as signal-idler pairs have the same power. Optical combs that have been amplified using PSAs rather than standard PIAs/EDFAs promise an improvement in: (1) spectral coverage; (2) spectral flatness/equalization; and (3) optical tone power and tone-to-noise ratio [73]. Figure 2.19 shows the corresponding initial (red curve) and PSA-amplified (black curve) spectra at the optimum pump power. A gain of up to 20 dB (at the edges of the comb), a comb flatness better than 2.5 dB over 4 THz bandwidth and an optical tone-to-noise ratio higher than 26 dB were experimentally demonstrated [73].

2.5 Four-Mode Parametric Processes

Four-mode phase-sensitive processes involve six waves (two strong pumps and four weak sidebands), which can be viewed as cascaded interactions combining three two-mode processes: MI, phase conjugation PC, and BS, respectively, as shown in Figure 2.2 (c) [14, 62, 74, 75]. Interestingly, four-mode phase-sensitive processes allow the

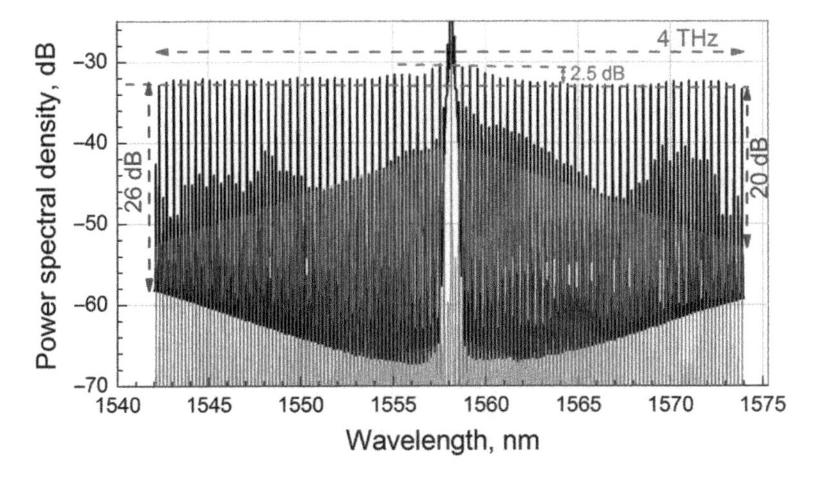

Figure 2.19 Spectra after the PSA for optimal pump power (black) and pump off (red) showing various key parameters. *Source*: Adapted from Slavik 2012 [73]. For a color version of this figure please see color plate section.

realization of both low noise amplification and noiseless signal spectral replication or frequency multicasting over broad ranges. However, as compared to two-mode ones, they have a significantly increased system complexity, hindering its practical use in some instances. For example, noiseless in-line amplification of both input signal quadratures can be realized through a phase-insensitive four-sideband parametric amplifier (or the four-mode copier) in perfect analogy to the copier stage used in the two-mode scenario if sufficient link loss is allowed to decorrelate the noise terms (see Section 2.4 and Figure 2.17). This will allow a theoretical 12 dB PS gain improvement compared to PIA and a 9 dB noise figure benefit compared to a conventional EDFA amplified link. Obviously, these performance advantages come at the expense of a fourfold bandwidth reduction as four sidebands have to co-propagate simultaneously, so this configuration may be attractive in a transmission link in which maintaining a good SNR in the link is the highest priority, rather than the information capacity. The first experimental demonstration of a four-mode copier–PSA system reported a 10.3 dB phase-sensitive gain advantage over the PIA configuration [76]. The first experimental demonstration of a parametric multicasting mixer based on a four-mode phase-sensitive process generated more than 100 copies with considerably improved optical SNR over the conventional phase-insensitive configuration, spanning a 160-nm-wide band [69]. A typical experimental set-up is shown in Figure 2.20 [69], which highlights one of the most challenging aspects of realizing a four-mode PS device: the generation of six low-noise optical waves with frequency- and phase-locked relations. In Figure 2.20, this is achieved creating an optical comb using amplitude and phase modulators, then selecting the required lines and using injection locking to recover the SNR of the pump lines. Indeed, the simpler alternative of generating the six-locked waves via a copier stage would significantly limit the output SNR improvement expected from the phase-sensitive mixer due to the noise correlation, as previously discussed. The corresponding zoomed-in output spectra under different input conditions, which are PI, two-mode PS amplification, four-mode PS amplification, four-mode PS de-amplification, and four-input with independent pump phases,

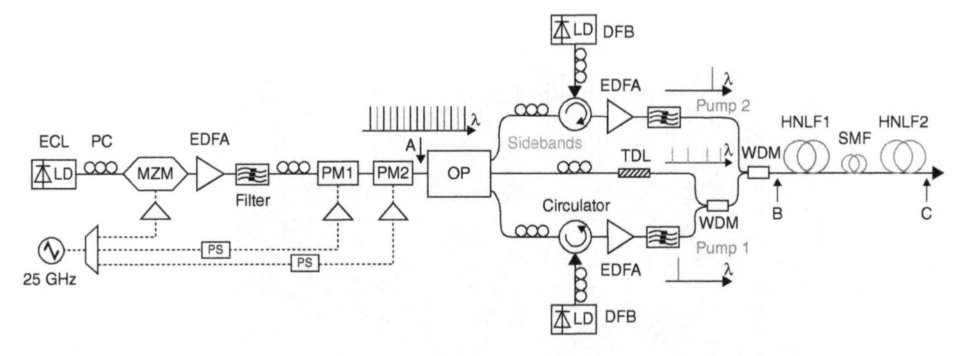

Figure 2.20 Experimental set-up of the four-mode PSA-based spectral signal replication. Key: ECL, external-cavity laser; PC, polarization controller; PS, phase shifter; MZM, Mach–Zehnder modulator; PM, phase modulator; DFB, distributed feedback laser; TDL, tunable delay line. *Source*: Tong 2012 [69]. Reproduced with permission of OSA.

respectively, are all shown in Figure 2.21 (a) [69]. The parametric multicaster for the four-input PI case will give a 6-dB power improvement over the conventional one-input PI case (as the input signals are four rather than one) and a similar conversion efficiency (CE) to the two-mode PS one. On the other hand, a four-mode PS input gives 12.6-dB and 5.1-dB (at 1579.8-nm) gain advantages over its PI and two-mode PS counterparts in good accordance with the theoretical analysis of 12-dB and 6-dB, respectively. In addition, about 25-dB gain between the optimal amplification and the de-amplification is measured and all these measurements were taken over more than the 60-nm band for all the studied cases with good uniformity, revealing perfectly preserved coherence among the interacting waves [69]. This is a very important point as it allows a practically realizable path for broadband optical signal processing with a large copy count, including signal replication and multicasting.

2.6 Conclusion

In this chapter, we have briefly reviewed the basic principles and technological approaches toward the regeneration of various signals of differing levels of complexity, noiseless amplification and noiseless frequency conversion, signal replication, and multicasting. All of them rely on ultrafast parametric processes, which we have classified depending on the number of (internal) modes. We have discussed the development of the field and have presented the current state of the art, both in terms of the demonstrations of novel device concepts that extend the range of functionality that can be achieved (e.g., range of advanced modulation format signals that can be addressed) and enhance practicality (e.g., in providing black-box operation, WDM performance, and possible routes to integration). The current technical challenges and prospects of each scheme are described, primarily to provide guidance on the trade-offs that must be made between performance and implementation difficulties.

While it should be apparent that they remain a vibrant and potentially technologically important field of research, with significant advances made in recent years, it should, however, be equally clear that many challenges remain which will need to be overcome if they are ever to make it from the laboratory to commercial deployment—not least in

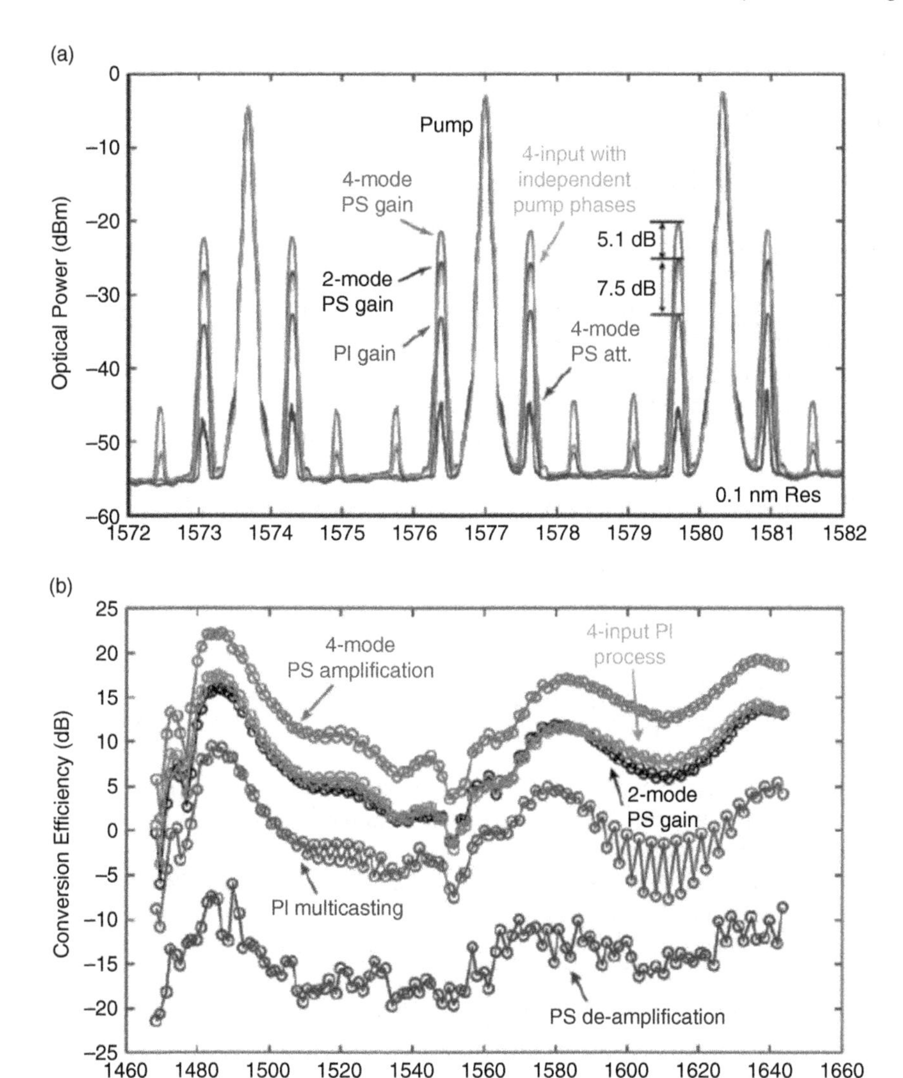

Figure 2.21 (a) High resolution spectra subset for different cases, and (b) measured conversion efficiency of each copy for different input schemes. *Source*: Tong 2012 [69]. Reproduced with permission of OSA.

terms of enabling significant levels of WDM capability and the development of compact, power-efficient variants.

Acknowledgments

The author gratefully acknowledges the contributions of Dr Kyle Bottrill, Dr Radan Slavik, Dr Graham Hesketh, Professor Periklis Petropulos, and Professor David J. Richardson.

References

1 Essiambre, R.J. (2009) Capacity limits of fiber-optic communication systems. OThL1, OFC.
2 Winzer, P. J. (2009) Modulation and multiplexing in optical communication systems. *IEEE LEOS News*, **23** (1).
3 Ellis, A.D., *et al.* (2009) Approaching the non-linear Shannon limit. *Journal of Lightwave Technology*, **28** (4), 423–433.
4 Tucker, R.S., *et al.* (2009) Evolution of WDM optical IP networks: A cost and energy perspective. *Journal of Lightwave Technology*, **27**, 243–252.
5 http://www.greentouch.org
6 Sources: Google Inc, marketingprofs.com
7 Namiki, S. *et al.* (2010) Photonics technologies for low-energy networks at AIST. Paper Mo.2.A.4, ECOC.
8 Yamamoto, Y. and Inoue, K. (2003) Noise in amplifiers. *Journal of Lightwave Technology*, **21**, 2895–2915.
9 Okoshi, T. and Kikuchi, K. (1988) *Coherent Optical Fiber Communications*. Tokyo: KTK Scientific.
10 Haus, H.A. and Mullen, J.A. (1962) Quantum noise in linear amplifiers. *Physical Review*, **128**, 2407.
11 Croussore, K., Kim, I., Kim, C., Han, Y., and Li, G.F. (2006) Phase-and amplitude regeneration of differential phase-shift keyed signals using a phase-sensitive amplifier. *Optics Express*, **14**, 2085–2094.
12 Tang, R., Lasri, J., Devgan, P.S., *et al.* (2005) Gain characteristics of a frequency nondegenerate phase sensitive fiber-optic parametric amplifier with phase self-stabilized input. *Optics Express*, **13**, 10483–10493.
13 Marhic, M.E. (2007) *Fiber Optical Parametric Amplifiers, Oscillators and Related Devices*. Cambridge University Press, Cambridge.
14 McKinstrie, C.J., Radic, S. and Chraplyvy, A.R. (2002) Parametric amplifiers driven by two pump waves. *IEEE Journal of Selected Topics in Quantum Electronics*, **8**, 538–547.
15 Caves, C.M. (1982) Quantum limits on noise in linear amplifiers. *Physical Review D*, **26**, 1817.
16 Tong, Z., *et al.* (2012) Broadband parametric multicasting via four-mode phase-sensitive interaction. *Optics Express*, **20**, 19363–19373.
17 Slavík, R., *et al.* (2010) All-optical phase and amplitude regenerator for next-generation telecommunications systems. *Nature Photonics*, **4** (10), 690–695.
18 Weerasuriya, R. *et al.* (2010) Generation of frequency symmetric signals from a BPSK input for phase sensitive amplification. Paper OWT6, OFC.
19 Vasilyev, M. (2005) Distributed phase-sensitive amplification. *Optics Express*, **13** (19), 7563–7571.
20 Tong, Z. *et al.* (2011) Towards ultrasensitive optical links enabled by low-noise phase-sensitive amplifiers. *Nature Photonics*, June 5.
21 Kakande, J., Slavik, R., Parmigiani, F. *et al.* (2011) Multilevel quantization of optical phase in a novel coherent parametric mixer architecture. *Nature Photonics*, **5**, 748–752.
22 Gao, M., Kurosu, T., Inoue, T., and Namiki, S. (2013) Efficient phase regeneration of DPSK signal by sideband-assisted dual-pump phase sensitive amplifier. *Electronics Letters*, **39** (2). 140–141.

23 Lee, K.J., Parmigiani, F., Liu, S., *et al.* (2009) Phase sensitive amplification based on quadratic cascading in a periodically poled lithium niobate waveguide. *Optics Express,* **17**, 20393–20400.

24 Puttnam, B.J., Mazroa, D., Shinada, S., and Wada, N. (2011) Phase-squeezing properties of non-degenerate PSAs using PPLN waveguides. *Optics Express,* **19**, B131–B139.

25 Umeki, T., *et al.* (2013) In-line phase sensitive amplifier based on PPLN waveguides. *Optics Express,* **21** (10), 12077.

26 Sygletos, S., Power, M.J., Garcia Gunning, F.C., *et al.* (2012) Simultaneous dual channel phase regeneration in SOAs. ECOC, Tu.1.A.2.

27 Bottrill, K., *et al.* (2015) Phase regeneration of QPSK signal in SOA using single-stage, wavelength converting PSA. accepted in *PTL.*

28 Da Ros, F., *et al.* (2014) Phase regeneration of DPSK signals in a silicon waveguide with reverse-biased p-i-n junction. *Optics Express,* **22** (5), 5029–5036.

29 Ettabib, M.A., Parmigiani, F., Kapsalis, A., *et al.* (2015) Record phase sensitive extinction ratio in a silicon germanium waveguide. Paper presented at CLEO '15 San Jose, CA, 10–15 May.

30 Ettabib, M.A., Bottrill, K., Parmigiani, F., *et al.* (2015) PSA-based phase regeneration of DPSK signals in a silicon germanium waveguide. ECOC, We_3_6_5.

31 Imajuku, W., Takada, A., and Yamabayashi, Y. (1999) Low-noise amplification under the 3 dB noise figure in high-gain phase-sensitive fibre amplifier. *Electronics Letters,* **35**, 1954–1955.

32 Imajuku, W., Takada, A., and Yamabayashi, Y. (2000) Inline coherent optical amplifier with noise figure lower than 3 dB quantum limit. *Electronics Letters,* **36**, 63–64.

33 Levandovsky, D., Vasilyev, M., and Kumar, P. (1999) Amplitude squeezing of light by means of a phase-sensitive fiber parametric amplifier. *Optics Letters,* **24**, 984–986.

34 Levenson, J.A., Abram, I., Rivera, T., and Grangier, P. (1993) Reduction of quantum noise in optical parametric amplification, *Journal of the Optical Society of Americca B,* **10**, 2233–2238.

35 Choi, S.-K., Vasilyev, M., and Kumar, P. (1999) Noiseless optical amplification of images. *Physical Review Letters,* **83**, 1938–1941.

36 Umeki, T., Tadanaga, O., Takada, A., and Asobe, M. (2011) Phase sensitive degenerate parametric amplification using directly-bonded PPLN ridge waveguides. *Optics Express,* **19**, 6326–6332.

37 Asobe, M., *et al.* (2012) Phase sensitive amplification with noise figure below the 3 dB quantum limit using CW pumped PPLN waveguide. *Optics Express,* **20** (12), 13164–13172.

38 Lorences-Riesgo, A., *et al.* (2014) Phase-sensitive amplification and regeneration of dual polarization BPSK without polarization diversity. ECOC 2014, Paper Tu.1.4.3.

39 Slavik, R., *et al.* (2012) Field-trial of an all-optical PSK regenerator/multicaster in a 40 Gbit/s, 38 channel DWDM transmission experiment. *Journal of Lightwave Technology,* **30**, 512–520.

40 Gao, M., *et al.* (2013) Low-penalty phase de-multiplexing of QPSK signal by dual-pump ase sensitive amplifiers. Paper We.3.A.5, presented at the European Conference Exhibition, Optical Communications, London.

41 Webb, R.P., *et al.* (2013) Phase-sensitive frequency conversion of quadrature modulated signals. *Optics Express,* **21**, 12713–12727.

42 Da Ros, F., *et al.* (2013) QPSK-to-2×BPSK wavelength and modulation format conversion through phase-sensitive four-wave mixing in a highly nonlinear optical fiber. *Optics Express*, **21** (23), 28743–28750.
43 Kurosu, T., *et al.* (2014) Phase regeneration of phase encoded signals by hybrid optical phase squeezer. *Optics Express*, **22** (10), 12177–12188.
44 Parmigiani, F., Hesketh, G., Slavik, R., *et al.* (2014) Optical phase quantizer based on phase sensitive four wave mixing at low nonlinear phase shifts. *IEEE Photonics Technology Letters*, **26** (21), 2146–2149.
45 Parmigiani, F., Hesketh, G., Slavík, R., *et al.* (2015) Polarization-assisted phase-sensitive processor. *Journal of Lightwave Technology*, **33** (6), 1166–1174.
46 Zheng, Z., *et al.* (2008) All-optical regeneration of DQPSK/QPSK signals based on phase-sensitive amplification. *Optics Communications*, **281**, 2755–2759.
47 Asobe, M., Umeki, T., Takenouchi, H., and Miyamoto, Y. (2013) In-line phase-sensitive amplifier for QPSK signal using multiple QPM LiNbO3 waveguide. in OECC/PS, 2013, 18 June 2013.
48 Kakande, J., *et al.* (2011) QPSK phase and amplitude regeneration at 56 Gbaud in a novel idler-free non-degenerate phase sensitive amplifier. In OFC/NFOEC, 2011.
49 Bottrill, K., *et al.* (2015) FWM-based, idler-free phase quantiser with flexible operating power. OFC 2015, Paper W4C.3.
50 Albores-Mejia, A., Kaneko, T., Banno, E., *et al.* (2015) All-optical carrier recovery using a single injection locked semiconductor laser stabilized by an incoherent optical-feedback. SF1I.4, CLEO US 2015.
51 Roethlingshoefer, T., *et al.* (2014) Cascaded phase-preserving multilevel amplitude regeneration. *Optics Express*, **22**, 31729–31734.
52 Sorokina, M., Sygletos, S., Ellis, A., and Turitsyn, S. (2015) Regenerative Fourier transformation for dual-quadrature regeneration of multilevel rectangular QAM. *Optics Letters*, **40** (13), 3117–3120.
53 Mohajerin-Ariaei, A., *et al.* (2014). Bit-rate-tunable regeneration of 30-Gbaud QPSK data using phase quantization and amplitude saturation. Paper P.3.20.ECOC 2014,
54 K. Bottrill *et al.* (2015) Phase and amplitude regeneration through sequential PSA and FWM saturation in HNLF. ECOC 2015.
55 Bottrill, K.R.H., *et al.* (2015) Investigation into the role of pump to signal power ratio in FWM-based phase preserving amplitude regeneration. Paper SM2M.1, CLEO 2015.
56 Leuthold, J., *et al.* (2010) Nonlinear silicon photonics. *Nature Photonics*, **4** (8), 535–544.
57 Bogris, A. (2015) Optical phase processor with enhanced phase quantization properties for higher order modulation formats relying on phase sensitive amplification and limiting amplifiers. Paper ci-p.13-tue, CLEO/Europe-EQEC 2015, Munich.
58 Sygletos, S., *et al.* (2011) A practical phase sensitive amplification scheme for two channel phase regeneration. *Optics Express*, **19**, B938–B945.
59 Parmigiani, F., Bottrill, K.R.H., Slavík, R., *et al.* (2015) PSA-based all-optical multi-channel phase regenerator. ECOC 2015.
60 Guan, P., Røge, K.M., Kjøller, N.K. *et al.* (2015) All-optical WDM regeneration of DPSK signals using optical Fourier transformation and phase sensitive amplification. ECOC, We_3_6_4, 2015.
61 McKinstrie, C., Harvey, J., Radic, S., and Raymer, M. (2005) Translation of quantum states by four-wave mixing in fibers. *Optics Express*, **13**, 9131–9142.

62 Tong, Z. and Radic, S. (2013) Low-noise optical amplification and signal processing in parametric devices. *Advances in Optics and Photonics*, 2013.
63 Malik, R., *et al.* (2014) Optical signal to noise ratio improvement through unbalanced noise beating in phase-sensitive parametric amplifiers. *Optics Express*, **22** (9), 10477–10486.
64 Umeki, T., *et al.* (2014) First demonstration of high-order QAM signal amplification in PPLN-based phase sensitive amplifier. *Optics Express*, **22** (3), 2473–2482.
65 Corcoran, B., *et al.* (2012) Phase-sensitive optical pre-amplifier implemented in an 80 km DQPSK link. Paper PDP5A.4.OFC 2012.
66 Olsson, S., *et al.* (2012) Mitigation of nonlinear impairments on QPSK data in phase-sensitive amplified links. ECOC 2012.
67 Olsson, S., *et al.* (2014) Phase-sensitive amplified transmission links for improved sensitivity and nonlinearity tolerance. *Journal of Lightwave Technology*, **33** (3), 710–721.
68 Olsson, S., *et al.* (2014) Long-haul (3465 km) transmission of a 10 GBd QPSK Signal with low noise phase-sensitive in-line amplification. ECOC PD 2.2.
69 Tong, Z., Wiberg, A.O.J., Myslivets, E., *et al.* (2012) Broadband parametric multicasting via four-mode phase-sensitive interaction. *Optics Express*, **20**, 19363–19373.
70 Liu, X., *et al.* (2013) Phase-conjugated twin waves for communication beyond the Kerr nonlinearity limit. *Nature Photonics*, **7**, 560.
71 Liu, X., *et al.* (2014) Generation of 1.024-Tb/s Nyquist-WDM phase conjugated twin vector waves by a polarization-insensitive optical parametric amplifier for fiber-nonlinearity-tolerant transmission. *Optics Express*, **22** (6).
72 Tian, Y., *et al.* (2013) Demonstration of digital phase-sensitive boosting to extend signal reach for long-haul WDM systems using optical phase-conjugated copy. *Optics Express*, **21** (4)
73 Slavik, R. *et al.* (2012) Processing of optical combs with fiber optic parametric amplifiers. *Optics Express*, **20**.
74 Schumaker, B.L., Perlmutter, S.H., Shelby, R.M., and Levenson, M.D. (1987) Four-mode squeezing. *Physical Review Letters*, **58**, 357–360.
75 McKinstrie, C., Radic, S., and Raymer, M. (2004) Quantum noise properties of parametric amplifiers driven by two pump waves. *Optics Express* **12**, 5037–5066.
76 Richter, T., Corcoran, B., Olsson, S.L., *et al.* (2012) Experimental characterization of a phase-sensitive four-mode fiber-optic parametric amplifier. Paper presented at European Conference and Exhibition on Optical Communication, OSA Technical Digest, Optical Society of America.

3

Novel Nonlinear Optical Phenomena in Gas-Filled Hollow-Core Photonic Crystal Fibers

Mohammed F. Saleh and Fabio Biancalana

School of Engineering and Physical Sciences, Heriot-Watt University, Edinburgh, UK

Gas-filled hollow-core photonic crystal fibers offer unprecedented opportunities to observe novel nonlinear phenomena. The various properties of gases that can be used to fill these fibers give additional degrees of freedom to investigate nonlinear pulse propagation in a wide range of different confined media. In this chapter, we will discuss some of the new nonlinear interactions that have been discovered in recent years, in particular, those which are based on soliton dynamics.

3.1 Introduction

A soliton is a nonlinear localized wave possessing a particle-like nature, which maintains its shape during propagation, even after an elastic collision with another soliton. In optics, this special wave-packet can arise due to the balance between nonlinear and dispersive effects. Based on confinement in the time or space domain, one can have either temporal or spatial solitons. Both kinds of optical solitons can occur due to the third-order Kerr nonlinearity [1] that leads to an intensity-dependent refractive index of the medium. This nonlinear dependence results in spatial self-focusing and temporal self-phase modulation [2]. A spatial soliton is formed when the self-focusing effects counteracts the natural diffraction-induced broadening of the pulse. Similarly, a temporal soliton is developed when the self-phase modulation effects compensates the usual dispersion-induced broadening.

The possibility of soliton propagation in the anomalous dispersion regime of an optical fiber was predicted by analyzing theoretically the nonlinear Schrödinger equation (NLSE) [3]. In 1980, Mollenauer *et al.* were able to excite an Nth-order soliton [4] that represents a joint state of N fundamental solitons, which propagates in a periodic way consisting of pulse splitting followed by a recovery to the original pulse after a characteristic propagation length.

Pumping in the anomalous regime, a Nth-order soliton is continuously temporally compressed and spectrally broadened during propagation. After a certain distance, the

Shaping Light in Nonlinear Optical Fibers, First Edition. Edited by Sonia Boscolo and Christophe Finot.

pulse breaks up and a series of fundamental solitons are ejected in a process known as soliton fission due to higher-order dispersion effects. Because of the intrapulse Raman scattering of silica, the central frequency of the fundamental solitons are continuously downshifted during propagation [5,6]. Also, a transfer of energy from each fundamental soliton to a weak narrowband dispersive wave in the normal dispersion regime can be stimulated via the higher-order dispersion effects [7]. This ends in a massive pulse broadening known as supercontinuum generation [8–12].

Photonic crystal fibers (PCFs) are microstructured fibers that can be engineered in various ways to tailor the fiber linear and nonlinear properties. PCF structures are fabricated using different techniques. Solid-core PCFs guides light via total internal reflection similar to step-index fibers. However, the additional degrees of freedom provided by modifying the design parameters open up different possibilities to engineer the fiber properties, such as shifting its zero dispersion wavelength (ZDW) [13], or enhancing its Kerr nonlinearity via reducing its effective-core area [14]. Hollow-core (HC) PCFs have also attracted much interest [15–17], because of their potential for lossless and distortion-free transmission, particle trapping, optical sensing, and novel applications in nonlinear optics [18–20].

Gas-filled HC-PCFs with Kagome lattice have become a powerful alternative to solid-core PCFs especially for nonlinear optical frequency conversion applications [21]. The low-loss wide transmission window, the pressure-tunability dispersion (both in the visible region), and the variety of different gases with special properties have offered several opportunities for demonstrating new nonlinear applications [22], such as Stokes generation with a drastical reduction in the Raman threshold [23], high harmonic generation [24], efficient deep ultraviolet radiation [25], ionization-induced soliton self-frequency blueshift [26–29], strong asymmetrical self-phase modulation, universal modulational instability [30], parity-time symmetric potentials [31], temporal crystals [32, 33], and tunable frequency-up/down conversion [34]. The subject of this chapter is to present in detail some of these applications, which involve mainly soliton dynamics.

This chapter is organized as follows. In Section 3.2, we give an overview of the governing equation of nonlinear pulse propagation in Kerr media. Sections 3.3 and 3.4 are dedicated to the study of ionization and Raman effects in HC-PCFs filled by atomic (Raman-inactive) and molecular (Raman-active) gases, respectively. In Section 3.5, the interplay between ionization and Raman effects in HC-PCFs filled by molecular gases is discussed. The conclusion and final remarks are presented in Section 3.6.

3.2 Nonlinear Pulse Propagation in Guided Kerr Media

Nonlinear pulse propagation in a lossless Kerr medium can be described by the scalar nonlinear Schrödinger equation (NLSE),

$$i\partial_z A + \sum_{m=2} \frac{i^m}{m!}\beta_m \partial_t^m A + \gamma |A|^2 A = 0, \tag{3.1}$$

where the slowly varying envelope approximation (SVEA) is assumed, $A(z, t)$ is the complex envelope of the electric field in units of $\mathrm{W}^{1/2}$, z is the longitudinal coordinate along the fiber, t is time in a reference frame moving with the pulse group velocity, β_m is the mth order dispersion coefficient calculated at the pulse central frequency ω_0, and γ is the nonlinear Kerr coefficient. This equation can usually be numerically integrated using the split-step Fourier method [35]. The second term is usually taken as a fit of the dispersion in the frequency domain ω as $\sum_{m=2} \beta_m(\omega - \omega_0)^m/m!$ or in an approximation-free manner as $\beta(\omega) - \beta_0 - \beta_1(\omega - \omega_0)$ in cases when the dispersion relation $\beta(\omega)$ of the waveguide under consideration is known. In a regime of deep anomalous dispersion, i.e. for $|\beta_2| \gg |\beta_{m>2}|$, the normalized solution of Eq. (3.1) is the fundamental Schrödinger soliton,

$$\psi(\xi, \tau) = N\mathrm{sech}(N\tau)\exp(iN^2\xi/2), \tag{3.2}$$

where N is an arbitrary parameter that controls the soliton amplitude and width, $\xi = z/z_0, \tau = t/t_0, \psi = A/A_0, A_0^2 = 1/(\gamma z_0), z_0 = t_0^2/|\beta_2|$ is the second-order dispersion length at ω_0, and t_0 is the input pulse duration.

There are some limitations in using the above NLSE, for instance: (i) when studying ultrashort pulses, with few optical cycles ($\omega_0 t_0 \sim 1$), since the SVEA is no longer valid; and (ii) investigating the polarization effects in birefringent waveguides requires the vector nature of the electric field to be taken into account. Alternatively, more accurate methods of simulating the propagation of electromagnetic pulses in dielectric media such as the finite difference time domain (FDTD) method [36–40] or the recent unidirectional pulse propagation equation (UPPE) [41, 42] can be used at the expense of an increased computational effort, as well as a deficiency of understanding different physical mechanisms behind the pulse dynamics. On the other hand, there are also advantages in using the NLSE: (i) suitability in generalizing the NLSE to include different phenomena such as Raman nonlinearity, self-steepening, and photo-ionization effects; and (ii) the possibility of using well-known analytical techniques such as variational perturbation theory to study new nonlinear effects [35].

3.3 Ionization Effects in Gas-Filled HC-PCFs

Photo-ionization is the physical process in which an electron is released and an ion is formed due to the interaction of a photon with an atom or a molecule. The free electron density n_e is governed by the rate equation [43]

$$\frac{\partial n_e}{\partial t} = \mathcal{W}(t)(n - n_e) - \eta\, n_e - \beta_r n_e^2, \tag{3.3}$$

where $\mathcal{W}$ is the ionization rate, n is the total density of the atoms, and η and β_r are the electron attachments, and recombination rates, respectively. For pulses with picosecond duration or less, both η and β_r are negligible. Based on the so-called Keldysh parameter p_K [44], photo-ionization can take place by multiphoton absorption $p_\mathrm{K} \gg 1$, tunneling ionization $p_\mathrm{K} \ll 1$ or both when $p_\mathrm{K} \sim 1$. Several models have been developed to determine the dependence of the ionization rate on the electric field of the optical pulse [45–48]. It has been shown experimentally that tunneling ionization is dominant over the multiphoton ionization in noble gases such as Argon for pulses with intensities $\sim 10^{14}$ W/m^2, or when $p_\mathrm{K} \sim 1$ [49, 50], which is the case under study. In the tunneling regime, the time-averaged ionization rate is given by [43]

$$\mathcal{W}(I) = d\,(I_H/I)^{1/4} \exp[-b\,(I_H/I)^{1/2}], \tag{3.4}$$

where $d = 4\,\delta_0\,[3/\pi]^{1/2}\,[U_I/U_H]^{7/4}$, $b = 2/3\,[U_I/U_H]^{3/2}$, $\delta_0 = 4.1 \times 10^{16}$ Hz is the characteristic atomic frequency, U_I is the ionization energy of the gas, $U_H \approx 13.6$ eV is the ionization energy of the atomic hydrogen, $I_H = 3.6 \times 10^{16}$ W/cm^2, and $I = |\psi|^2$ is the laser pulse intensity. This model is adequate for an analysis that is based on the evolution of the complex envelope of the pulse. Like other methods, this model is still not straightforwardly amenable to analytical manipulation, because of the complex dependence on the pulse intensity. Equation (3.4) predicts an ionization rate that is exponential-like for pulse intensities above a threshold value I_{th} [28, 51]. Any pulse with an intensity $I \gg I_{\text{th}}$ will suffer a strong ionization loss due to the absorption of photons in the plasma generation process. This limits the operating regime to near the ionization threshold I_{th} where the ionization loss is drastically reduced. A model of the ionization rate with linear dependence on the pulse intensity can then be developed using the first-order Taylor series of Eq. (3.4) [28]

$$\mathcal{W} \approx \tilde{\sigma}(I - I_{\text{th}})\,\Theta\,(I - I_{\text{th}}), \tag{3.5}$$

where $\tilde{\sigma}$ is a constant that is chosen to reproduce the physically observed value of the ionization threshold, and Θ is a Heaviside function, introduced to cut the ionization rate to zero below the value of I_{th}.

Photo-ionization decreases the refractive index of the medium by a factor proportional to the square of the plasma frequency ω_p using the Drude model, also it attenuates the pulse amplitude due to photon absorption. To include these effects Eq. (3.1) can be modified as [28]

$$i\partial_z A + \sum_{m=2} \frac{i^m}{m!}\beta_m \partial_t^m A + \gamma |A|^2 A - \frac{\omega_p^2 A}{2\omega_0 c} + i\frac{A_{\text{eff}} U_I}{2|A|^2}\partial_t n_e = 0, \tag{3.6}$$

where c is the speed of light, and A_{eff} is the effective area. Using the split-step Fourier method, the pulse amplitude can be determined at each propagation step after computing the free electron density n_e. After ignoring the electron attachments and recombination effects, Eq. (3.3) can be directly integrated analytically, and can then be substituted in Eq. (3.6) to have a single generalized NLSE that describes pulse propagation in an ionizing medium. Introducing the following rescalings and redefinitions: $\omega_{\text{T}}^2 = e^2 n/[\epsilon_0 m_{\text{e}}]$ is the maximum plasma frequency, $\phi = \frac{1}{2}k_0 z_0\,[\omega_{\text{p}}/\omega_0]^2$, $\phi_{\text{T}} = \frac{1}{2}k_0 z_0\,[\omega_{\text{T}}/\omega_0]^2$, $\sigma = \tilde{\sigma}\,t_0/[A_{\text{eff}}\gamma_{\text{K}}\,z_0]$, and $\kappa = U_I\,\tilde{\sigma}\,\epsilon_0\,m_{\text{e}}\,\omega_0^2/[k_0\,e^2]$. In this case, Eq. (3.6) becomes [51]

$$i\partial_\xi \psi + \sum_{m=2} \frac{i^m z_0}{t_0^m m!}\beta_m \partial_\tau^m \psi + |\psi|^2\psi - \phi\psi + i\alpha\psi = 0, \tag{3.7}$$

where $\phi = \phi_{\text{T}}[1 - \exp(-\sigma \int_{-\infty}^{\tau} \Delta|\psi|^2\,\Theta(\Delta|\psi|^2)\,d\tau')]$, $\Delta|\psi|^2 = |\psi|^2 - |\psi|_{\text{th}}^2$, $\alpha = \kappa\,(\phi_{\text{T}} - \phi)\Delta|\psi|^2\Theta\Delta|\psi|^2$, and $|\psi|_{\text{th}}^2 = I_{\text{th}} A_{\text{eff}}$.

3.3.1 Short Pulse Evolution

In order to extract useful analytical information from Eq. (3.7), further simplifications are necessary. First, we assume operating in the deep anomalous regime. For pulses with maximum intensities just above the ionization threshold, also called floating pulses [28], the ionization loss is not large and can be ignored as a first approximation. For such pulses, only a small portion of energy above the threshold intensity contributes to the creation of free electrons. Furthermore, the effect of the Θ-function can be approximately included via multiplying the cross-section parameter σ by a factor ε that represents the ratio between the pulse energy contributing to plasma formation (the portion above the ionization threshold) and the total energy of the pulse. In this case, ionization can be treated as a perturbation of the solution of the NLSE, which is the fundamental soliton. The solution of Eq. (3.7) can be written as $\psi(\xi,\tau) = N(\xi)\,\mathrm{sech}[N(\xi)(\tau - \bar{\tau}(\xi))]e^{-i\delta(\xi)\tau}$, where $\bar{\tau}$ is the temporal location of the soliton peak, and δ is the pulse central-frequency shift. Applying the perturbation theory [35], we found $\delta(\xi) = -g\,\xi$, and $\bar{\tau}(\xi) = g\,\xi^2/2$, where $g = -(2/3)\varepsilon\sigma\phi_T N^2$ [28]. This shows that photoionization should lead to an absolutely remarkable soliton self-frequency blueshift. This blueshift is accompanied by a constant acceleration of the pulse in the time domain—opposite to the Raman effect, which produces pulse deceleration [6].

Including the effects of both the photo-ionization loss and the Heaviside function, the perturbation theory results in two coupled differential equations that govern the spatial development of the soliton amplitude and central frequency,

$$
\begin{aligned}
\frac{\partial N}{\partial \xi} &= -2\kappa\,\phi_T\,(N\tanh\vartheta - |\psi|^2_{\mathrm{th}}\,T) \\
\frac{\partial \delta}{\partial \xi} &= \sigma\phi_T N^2\left[\frac{2}{3}\tanh^3\vartheta + \frac{|\psi|^2_{\mathrm{th}}}{N^2}(\vartheta\,\mathrm{sech}^2\vartheta - \tanh\vartheta)\right],
\end{aligned}
\tag{3.8}
$$

where $\vartheta = NT$, and $T \approx 1/N\mathrm{sech}^{-1}[|\psi|_{\mathrm{th}}/N]$ is the temporal position where the pulse amplitude exceeds the ionization threshold at $\xi = 0$ [51]. Solving these equations numerically as shown in Figure 3.1, we found that pulses with initially large intensities ($N_0^2 > |\psi|^2_{\mathrm{th}}$) will experience a boosted self-frequency blueshift. However, the ionization loss suppresses the soliton intensity after a short propagation distance to the floating-soliton regime, where the soliton can propagate for a long propagation distance with a

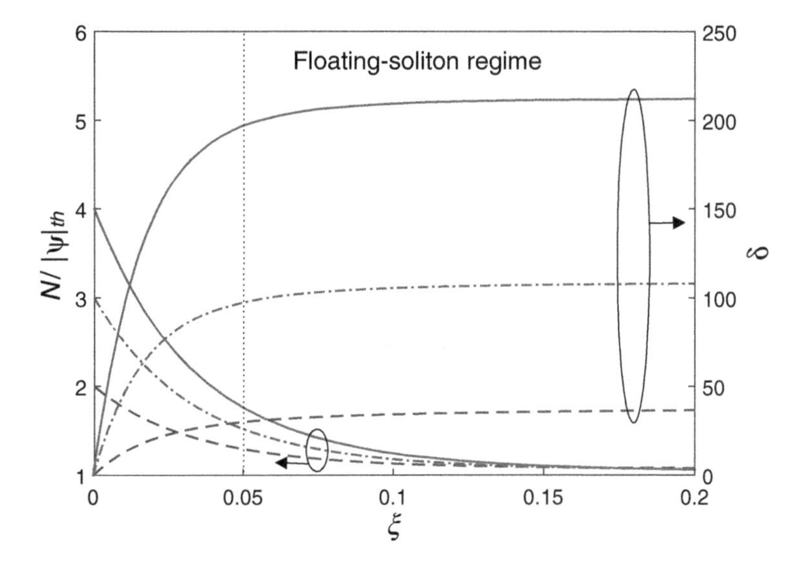

Figure 3.1 The spatial dependence of the soliton amplitude and frequency shift during a photo-ionization process for different initial pulse amplitudes. *Source*: Saleh 2011 [51]. Reproduced with permission of American Physical Society.

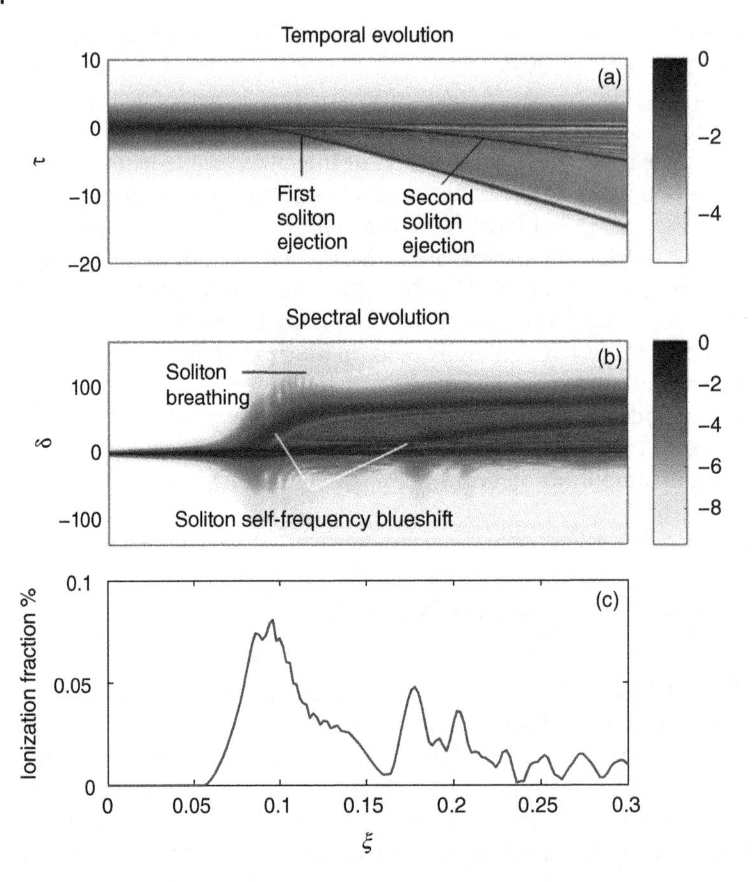

Figure 3.2 Temporal (a) and spectral (b) evolution of an ultrashort pulse in an Ar-filled HC-PCF. The temporal profile of the input pulse is $N \operatorname{sech} \tau$, with $N = 8$, $t_0 = 50$ fs. The gas pressure is 5 bar. Contour plots are given in a logarithmic scale. (c) Spatial dependence of the ionization fraction along the fiber. *Source*: Saleh 2011 [28]. Reproduced with permission of American Physical Society.

limited blueshift and negligible loss. The maximum frequency shift is achieved when the soliton intensity falls below the photo-ionization threshold.

To study the full dynamics of pulse propagation in fibers filled by an ionizing gas, Eq. (3.7) should be solved numerically via the split-step Fourier method. The temporal and spectral evolution of a pulse, with an initial temporal profile $N\text{sech}(\tau)$ and intensity less than the ionization threshold, are depicted in the panels (a,b) of Figure 3.2, respectively. Panel (c) shows the variation of the ionization fraction along the fiber. The pulse is pumped in the deep anomalous-dispersion regime of the fiber, and it undergoes self-compression. When the pulse intensity exceeds the threshold value, a certain amount of plasma is generated due to gas ionization, and a fundamental soliton is ejected from the input pulse. The soliton central frequency continues to shift towards the blue side due to the energy received from the generated plasma. However, because of the concurrent ionization loss, the soliton intensity gradually is attenuated to the regime where $|\psi|^2 \gtrapprox |\psi|^2_{\text{th}}$. Such pulses, the floating solitons, can propagate for considerably long distances with minimal attenuation and limited blueshift. When the soliton intensity goes below the ionization threshold, the blueshift process is ceased. A second

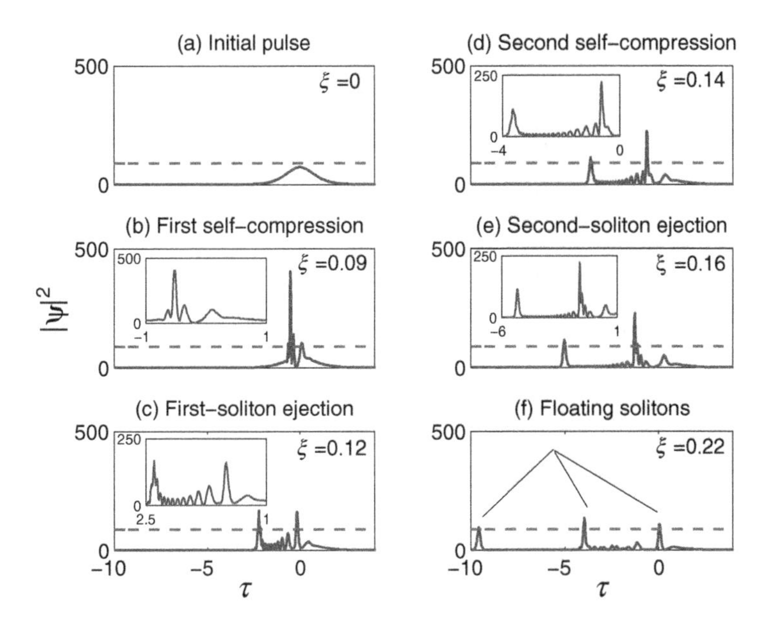

Figure 3.3 Intensity profile of an optical pulse in the time domain at different positions, ξ, inside an Ar-filled HC-PCF. The dashed line represents the threshold intensity. The simulation parameters are similar to Figure 3.2. Each panel is named after its main feature. Insets are enclosed in the panels for a better view and more detail. *Source*: Saleh 2011 [51]. Reproduced with permission of American Physical Society.

ionization event accompanied by a second-soliton emission can take place by further self-compression of the input pulse based on its initial intensity. At the end, a train of floating solitons are generated. A clear representation of the pulse dynamics in the presence of plasma is shown in Figure 3.3, where the temporal profile of the pulse intensity $|\psi|^2$ is plotted at selected positions inside the fiber.

3.3.1.1 Long-Range Non-Local Soliton Forces and Clustering

A non-local interaction between two successive solitons have been found when their temporal separation is shorter than the recombination time [28]. The leading soliton and its accompanied induced non-vanishing electron-density tail co-propagate and accelerate toward the negative-delay direction. Within the recombination time, the trailing soliton will be affected by a force in the opposite direction due to the accelerated long electron-density tail. In this case, the acceleration of the trailing soliton has an exponentially decaying dependence on the amplitude of the leading soliton. These dynamics are featured in Figure 3.4, that shows the temporal and spectral dependence on the soliton parameter N assuming that the input pulse is $N \operatorname{sech} \tau$. The scenario is as follows: As long as the intensity of the first-emitted soliton is above the threshold intensity, it prevents the ejection of a second soliton due to the presence of the non-local interaction force. When the ionization loss ends the blueshifting process of the leading soliton, the trailing soliton can be ejected and recovers its expected acceleration and blueshift. This allows the second soliton to catch up and cluster with the first soliton. In addition, the spectrum of the two solitons starts to overlap and forms spectral clustering. Similarly, when the two solitons are very close to each other, their induced electron-density tail

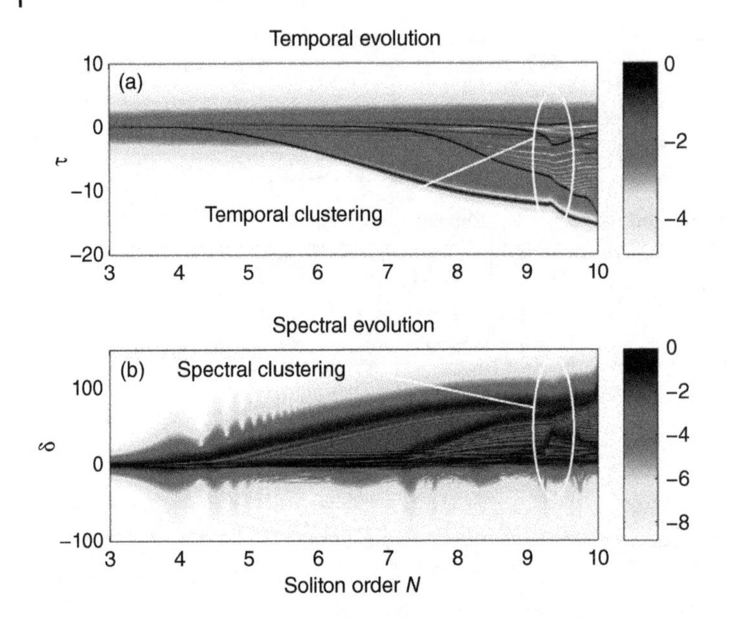

Figure 3.4 Temporal (a) and spectral (b) distribtuion of a pulse with temporal profile $N \operatorname{sech} \tau$ at the end of an Ar-filled HC-PCF with length $\xi = 1/4$ versus the soliton parameter N. Other simulation parameters are similar to Figure 3.2. *Source*: Saleh 2011 [28]. Reproduced with permission of American Physical Society.

applies a combined force on the third soliton. The development of the cross-frequency-resolved optical gating (XFROG) spectrograms for pulses with different initial intensities is depicted in the panels of Figure 3.5, where (a) represents the input pulse; (b) and (c) show the emission of the first and second solitons, respectively; and (d) depicts the temporal and spectral clustering of the first two solitons and the emission of a third soliton.

3.3.2 Long-Pulse Evolution

When the gas is excited by relatively *long* pulses with ionizing intensities, new kinds of self-phase modulation (SPM) and modulational instability (MI) can emerge during propagation. Ionization-induced SPM can be studied analytically using Eq. (3.7). In the case of small dispersion and long input pulse durations, the nonlinearity initially dominates over the group-velocity dispersion (GVD). Also by temporarily ignoring the losses (which do not change the qualitative picture, but only saturate the SPM spectrum after a certain distance), and assuming Gaussian pulse excitation, close forms of the spatial dependence of the mean frequency $\langle\Omega\rangle$ and the standard deviation $(\Delta\Omega)^2$ of the output spectrum can be derived [30],

$$\langle\Omega\rangle = \frac{1}{2}\eta\,\xi[\sqrt{2}\,\mathrm{erf}(\sqrt{2}\,\mathcal{T}) - 2\,|\psi|^2_{\mathrm{th}}\,\mathrm{erf}(\mathcal{T})], \tag{3.9}$$

$$\begin{aligned}(\Delta\Omega)^2 = {}& \frac{1}{18\,\tau_0^2}(9 + 4\sqrt{3}\,\xi^2 - 3\,\xi^2\eta^2\tau_0^2\{[\,3\,\mathrm{erf}^2(\sqrt{2}\,\mathcal{T}) \\ & -2\,\sqrt{3}\,\mathrm{erf}(\sqrt{3}\,\mathcal{T})] + 6|\psi|^2_{\mathrm{th}}[\,\mathrm{erf}(\mathcal{T}) - 1] \times [\,|\psi|^2_{\mathrm{th}}\,\mathrm{erf}(\mathcal{T}) - \sqrt{2}\,\mathrm{erf}(\sqrt{2}\,\mathcal{T})]\}),\end{aligned} \tag{3.10}$$

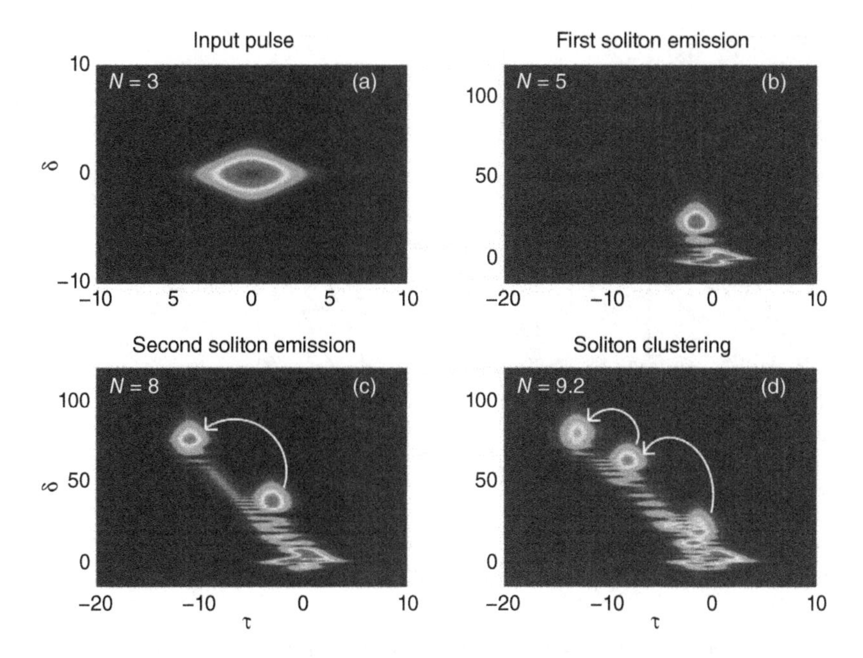

Figure 3.5 XFROG spectrograms for pulses with selected soliton parameter N in an increasing order. (a) $N = 3$, $\xi = 0$. (b) $N = 5$, $\xi = 1/4$. (c) $N = 8$, $\xi = 1/4$. (d) $N = 9.2$, $\xi = 1/4$. Each panel is named after its main feature. White arrows show the movement of the solitons. Other simulation parameters are similar to Figure 3.2. *Source*: Saleh 2011 [51]. Reproduced with permission of American Physical Society.

where erf is the error function, $\mathcal{T} = \mathrm{T}/\tau_0$, and $-\mathrm{T} \leq \tau \leq \mathrm{T}$ is the regime within which the pulse intensity exceeds the threshold intensity. For different values of $\eta = \sigma\phi_\mathrm{T}$, which measures the ionization strength, panels (a, b) in Figure 3.6 depict the spatial dependence of $\langle\Omega\rangle$ and $(\Delta\Omega)^2$ along the fiber. For $\eta = 0$, which corresponds to the absence of ionization, the mean frequency is always zero during propagation due to the well-known symmetric spectral broadening of the conventional SPM [35]. As η increases, the plasma starts to build up inside the fiber. The mean frequency changes linearly towards the blue side of the spectrum during propagation due to the ionization-induced phase-modulation, see Figure 3.6 (a). This leads to a strong and extremely asymmetric SPM, imbalanced towards the blue part of the spectrum. In addition, the pulse spectral width increases rapidly as the ionization strength increases, as shown in Figure 3.6 (b). In a real case, the spectral broadening process is certainly limited by the unavoidable ionization and fiber losses.

After the initial SPM stage is over, the interplay between nonlinear and dispersive effects can lead to an instability that modulates the temporal profile of the pulse, creating new spectral sidebands referred to as modulational instability (MI) [52–54]. MI due to the photoionization nonlinearity can be investigated by applying the standard approach described in [35,52], starting by Eq. (3.7). The Kerr-induced MI occurs only in the anomalous dispersion regime over a defined bandwidth [35]. However, the presence of the photo-ionization process induces an unusual instability that can exist in both normal and anomalous dispersion regimes, and for any frequency. The spectral dependence of the gain of these instabilities on the peak pulse power is shown in Figure 3.7 for (a)

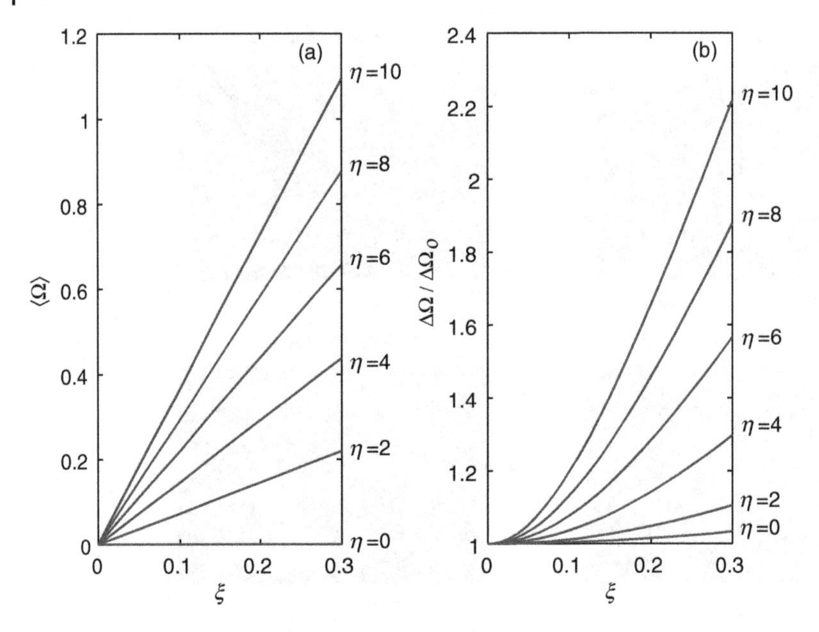

Figure 3.6 Spatial dependence of (a) the mean frequency $\langle\Omega\rangle$ and (b) the frequency standard deviation $\Delta\Omega$ of the output spectrum of an input Gaussian pulse $\exp(-\tau^2/2\,\tau_0^2)$ with $\tau_0 = 2$. The temporal position T at which the pulse intensity can initiate photoionization is assumed to be equal to τ_0. $\Delta\Omega_0$ is the spectral width at $\xi = 0$. *Source*: Saleh 2012 [30]. Reproduced with permission of American Physical Society.

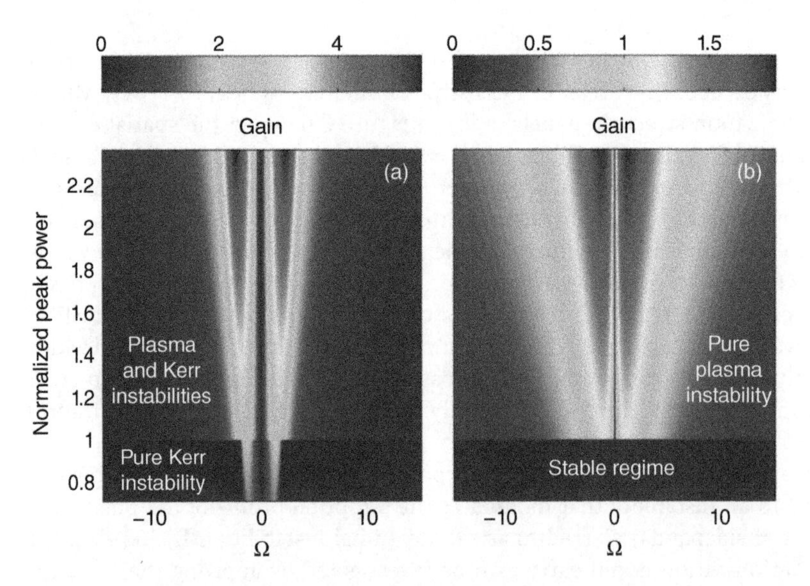

Figure 3.7 MI spectral gain profile versus normalized input peak power in (a) anomalous and (b) normal dispersion regimes with equal magnitude of β_2. For the anomalous regime, a CW with $\lambda = 1064$ nm is launched into a Kagome Ar-filled HC-PCF with core diameter 20 μm, gas pressure of 1 bar, and $\beta_2 \simeq -2.8$ ps^2/km. In this fiber, threshold ionization power is $\simeq$ 105 MW at room temperature. For the normal regime simulation, we assume hypothetically a gas-filled fiber with a β_2 of the same magnitude as in (a), but with a positive sign. *Source*: Saleh 2012 [30]. Reproduced with permission of American Physical Society.

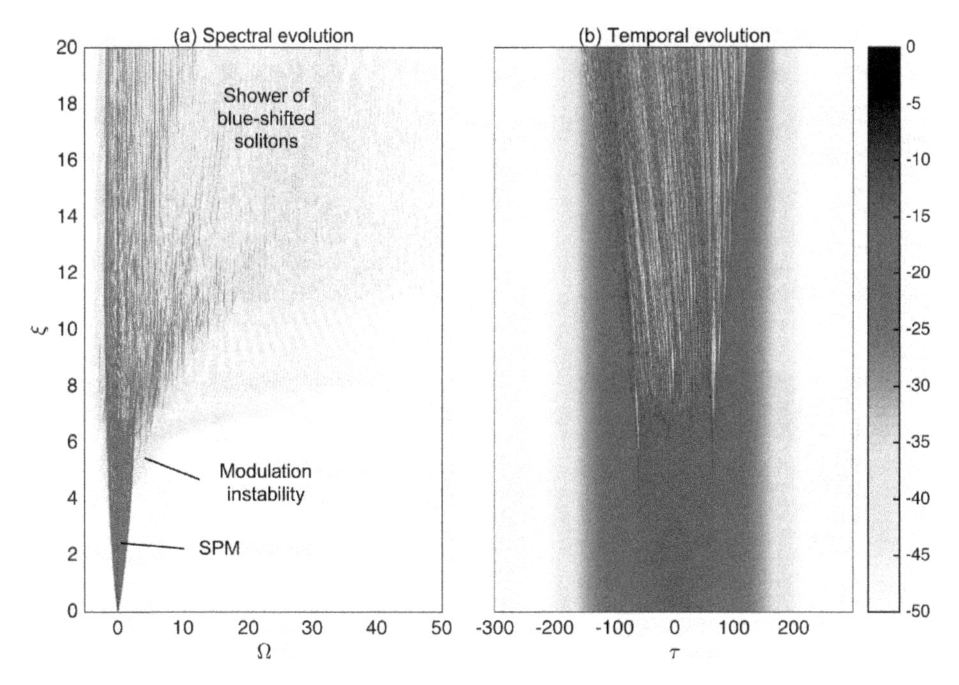

Figure 3.8 (a) Spectral and (b) temporal evolution of a long Gaussian pulse propagating in an Ar-filled HC-PCF with a gas pressure 1 bar, and a hexagonal core-diameter 20 μm. The input pulse has a peak power 200 MW and a duration 0.707 ps 1/e-intensity half-width. Contour plots are given in a logarithmic scale and truncated at −50 dB. *Source*: Saleh 2012 [30]. Reproduced with permission of American Physical Society.

anomalous and (b) normal dispersion regimes, where the physical powers are normalized to the threshold ionization power. When the normalized input power $|\psi_0|^2 \leq |\psi|^2_{\mathrm{th}}$, we have the traditional side-lobes, which exist uniquely in the anomalous dispersion regime, due to the Kerr nonlinearity. However when $|\psi_0|^2 > |\psi|^2_{\mathrm{th}}$, photo-ionization-induced instability generates unbounded side-lobes with slowly decaying tails. A similar situation occurs in the normal dispersion regime, however, the gain is slightly lower due to the absence of the Kerr contribution. In this case there are no instabilities below the threshold power since no plasma is generated.

Propagation of a long Gaussian pulse inside an anomalous dispersive gas-filled HC-PCF is portrayed in the panels (a, b) of Figure 3.8, obtained by simulating Eq. (3.7). The first stage of propagation shows asymmetric spectral broadening toward the blue due to ionization-induced SPM. Immediately after the SPM stage, dispersion starts to play a role, and due to the combined Kerr and ionization MIs, broad and slowly decaying side lobes are generated and amplified quickly. In the third and final stage in the propagation, a shower of blueshifted solitons is emitted. Further insight into the dynamics can be obtained from Figure 3.9, where the evolution of the XFROG spectrograms of the pulse at different positions along the fiber is shown. The pulse is initially asymmetrically chirped in the center of the pulse toward high frequencies, (Figure 3.9 (b)), due to the higher plasma density created at the peak intensities. At the same time two imbalanced ionization-induced MI sidebands appear in the pulse spectrum (Figure 3.9 (c)). MI facilitates the formation of many solitons. In less than half a meter of propagation the initial

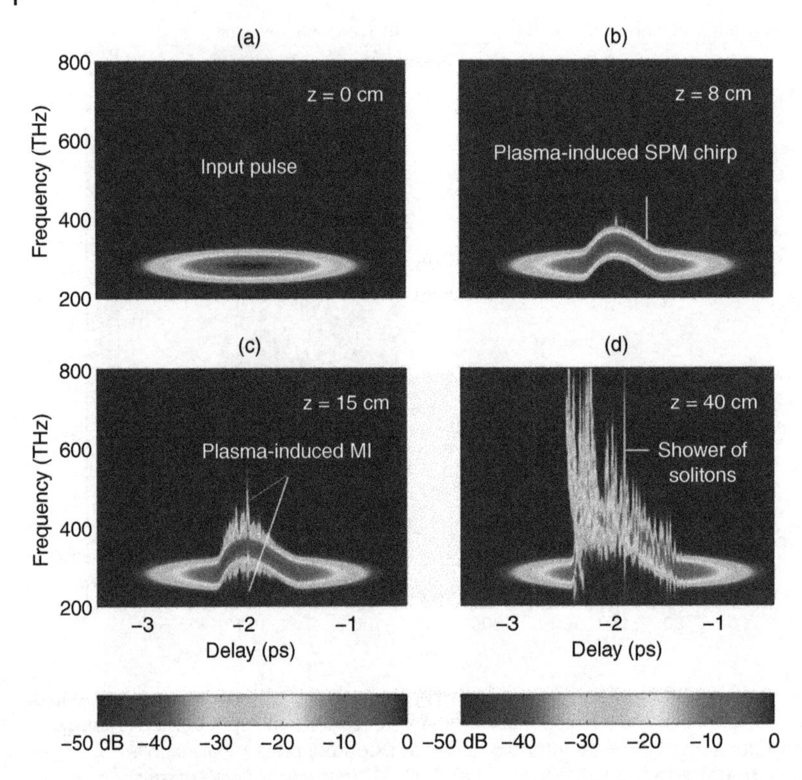

Figure 3.9 XFROG spectrograms of the propagation of a long Gaussian pulse through an Ar-filled HC-PCF. The simulation parameters are the same as in Figure 3.8. The reference pulse has a Gaussian profile with FWHM 50 fs. (a) Input pulse. (b) SPM frequency chirping. (c) Modulational instability. (d) Pulse disintegration into multiple blueshifting solitons. *Source*: Saleh 2012 [30]. Reproduced with permission of American Physical Society.

pulse disintegrates into a "shower" of solitary waves, see Figure 3.9 (d), each undergoing a strong self-frequency blueshift induced by the intrapulse photo-ionization.

3.4 Raman Effects in Gas-Filled HC-PCFs

Unlike silica glass, which has a broad Raman spectrum, stimulated Raman scattering processes in gases have a narrow Raman-gain spectrum. For this reason, gases are characterized by having a very long molecular coherence relaxation (dephasing) time, of the order of hundreds of picoseconds or even more, in comparison to the short relaxation time of phonon oscillations in silica glass (approx. 32 fs). Within this long relaxation stage, the medium can exhibit a highly non-instantaneous response to pulsed excitations. Raman responses can be manifested in either rotational or vibrational modes. Nonlinear interactions between optical pulses and Raman-active gases have usually been exploited in the synthesis of subfemtosecond pulses [55–59].

3.4.1 Density Matrix Theory

The dynamics of the Raman polarization (also called *coherence*) P_{R} due to a single mode excitation in gases can be determined by solving the Bloch equations for an effective two-level system [59,60],

$$\partial_t w + \frac{w+1}{T_1} = \frac{i\alpha_{12}}{\hbar}(\rho_{12} - \rho_{12}^*)E^2,$$
$$\left[\partial_t + \frac{1}{T_2} - i\omega_R\right]\rho_{12} = \frac{i}{2\hbar}[\alpha_{12}w + (\alpha_{11} - \alpha_{22})\rho_{12}]E^2, \quad (3.11)$$

where α_{ij} and ρ_{ij} are the elements of the 2×2 polarizability and density matrices, respectively, $E(z,t)$ is the real electric field, ω_R is the Raman frequency of the transition, $w = \rho_{22} - \rho_{11}$ is the population inversion between the excited and ground states, $\rho_{22} + \rho_{11} = 1$, $\rho_{21} = \rho_{12}^*$, $\alpha_{12} = \alpha_{21}$, N_0 is the molecular number density, T_1 and T_2 are the population and polarization relaxation times, respectively, and $\hbar$ is the reduced Planck's constant. Solving these coupled equations, the Raman polarization is then given by $P_R = [\alpha_{11}\rho_{11} + \alpha_{22}\rho_{22} + \alpha_{12}(\rho_{12} + \rho_{12}^*)]N_0 E$. For weak Raman excitation, $\rho_{11} \approx 1$ and $\rho_{22} \approx 0$, i.e. the second term in P_R can be ignored, while the first term increases the linear refractive index of the medium by a fixed amount.

Using the Maxwell and Bloch equations and applying the SVEA, one can derive the following set of normalized coupled equations that govern the pulse propagation in HC-PCFs filled by Raman-active gases,

$$\left[i\partial_\xi + \sum_{m=2} \frac{i^m z_0}{t_0^m m!}\beta_m \partial_t^m + |\psi|^2 + \frac{z_0}{z_R}\mathrm{Re}(\rho_{12})\right]\psi = 0,$$
$$\partial_\tau w + \frac{(w+1)t_0}{T_1} = -4\,\mu\, w\,\mathrm{Im}(\rho_{12})\,|\psi|^2, \quad (3.12)$$
$$\left[\partial_\tau + \frac{t_0}{T_2} - i\delta\right]\rho_{12} = i\mu\, w\,|\psi|^2,$$

where weak Raman excitation is assumed, $z_R = c\,\epsilon_0/(\alpha_{12}\,N_0\,\omega_0)$ is the nonlinear Raman length, $\mu = P_0/P_1, P_0 = A_0^2, P_1 = 2\hbar\, c\,\epsilon_0 A_{\mathrm{eff}}/(\alpha_{12}\,t_0), \delta = \omega_R t_0$, and Re and Im represent the real and imaginary parts.

For femtosecond pulses, the relaxation times of the population inversion (T_1) and the coherence (T_2) can be safely ignored, since they are of the order of hundreds of picoseconds or more. For instance, $T_1 \approx 20$ ns and $T_2 = 433$ ps for excited rotational Raman in molecular hydrogen under gas pressure 7 bar at room temperature [61,62]. Also, the population inversion is almost unchanged from its initial value for pulses with energies in the order of few μJ, i.e. $w(\tau) \approx w(-\infty) = -1$. The set of the governing equations Eq. (3.12) can be reduced to a single generalized nonlinear Schrödinger equation,

$$i\partial_\xi\psi + i\frac{\beta_1 z_0}{t_0}\partial_\tau\psi + \frac{1}{2}\partial_\tau^2\psi + |\psi|^2\psi + R(\tau)\psi = 0, \quad (3.13)$$

where $R(\tau) = \kappa\int_{-\infty}^{\tau}\sin[\delta(\tau - \tau')]|\psi(\tau')|^2 d\tau'$ is the resulting Raman convolution, and $\kappa = \mu z_0/z_R$ is the ratio between the Raman and the Kerr nonlinearities. By pumping in the deep anomalous dispersion regime ($\beta_2 < 0$), higher-order dispersion coefficients $\beta_{m>2}$ can be ignored. For ultrashort pulses with durations $t_0 \ll 1/\omega_R$ "impulsive excitation," $\sin[\delta(\tau - \tau')]$ can be expanded around the temporal location of the pulse peak by using the Taylor expansion. For instance, a fundamental soliton with amplitude N and centered at $\tau = 0$ will induce a Raman contribution in the form of $R(\tau) \approx \kappa N \sin(\delta\tau)[1 + \tanh(N\tau)]$ at the zeroth-order Taylor approximation. This

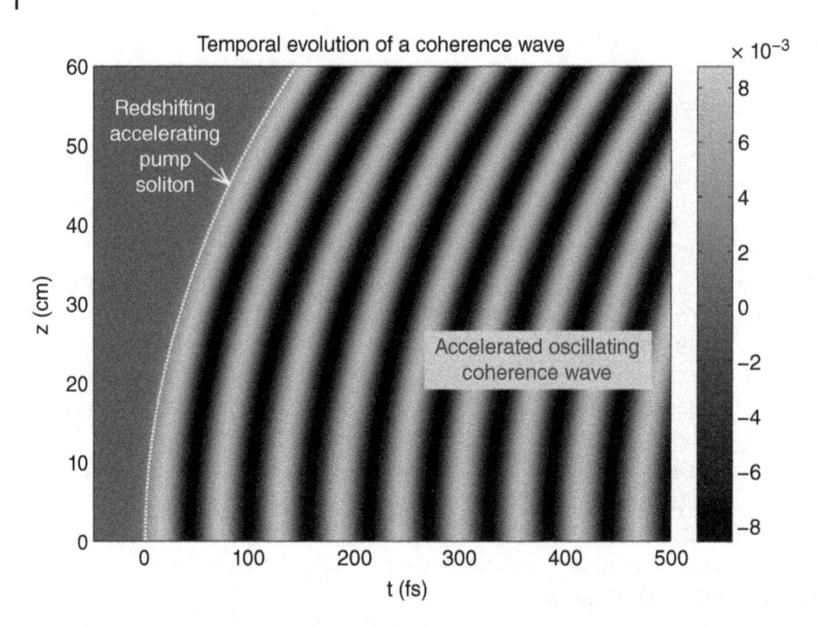

Figure 3.10 Temporal evolution of an accelerated oscillating Raman polarization with period $\Lambda = 56.7$ fs induced by a propagating fundamental soliton with an amplitude $N_1 = 1.33$, a central wavelength 1064 nm, and a FWHM 15 fs in a H_2-filled HC-PCF with a Kagome-lattice cross-section, a flat-to-flat core diameter 18 μm, a zero dispersion wavelength 413 nm, a gas pressure 7 bar, and a rotational Raman frequency $\omega_R = 17.6$ THz. The dashed line represents the temporal evolution of the soliton that excites the coherence wave. The simulation parameters are $\gamma = 7.07 \times 10^{-6}\ \mathrm{W^{-1}m^{-1}}$, $\beta_2 = -3425.5\ \mathrm{fs^2/m}$, $A_{\mathrm{eff}} = 134\ \mu\mathrm{m}^2$, $\alpha_{12} = 0.8 \times 10^{-41}\ \mathrm{C\,m^2/V}$ [61, 64], and $t_0 = 11.34$ fs. The parameter γ is calculated using the nonlinear susceptibility of H_2 [65]. For these parameters, we found that higher-order dispersion and self-steepening effects have a weak influence on the soliton dynamics. *Source*: Saleh 2015 [32].

soliton will generate a retarded sinusoidal Raman polarization that can impact the dynamics of another trailing probe pulse lagging behind it. On the other hand, for $t_0 \gg 1/\omega_R$, $R(\tau) \approx \gamma_R|\psi(\tau)|^2$ with $\gamma_R = \kappa/\delta$, i.e the Raman nonlinearity is considered to be instantaneous. Hence, the Raman contribution would induce an effective Kerr nonlinearity that is significant, and can compete directly with the intrinsic Kerr nonlinearity of the gas [61, 63].

In the following, we will exploit the long Raman coherence induced by an ultrashort pulse in controlling pulse dynamics, see Figure 3.10. We will study the propagation of two successive pulses separated by a delay $\ll T_1, T_2$ in the deep anomalous dispersion regime. The two pulses are assumed to have the same frequency, hence, they will propagate with the same group velocity, and experience the same dispersion. The leading pulse is an ultrashort strong "pump" pulse ψ_1 with $t_0 \ll 1/\omega_R$. In this case, Eq. (3.13) can be used to determine the pump solution by replacing ψ by ψ_1. For weak Raman nonlinearity, the solution of Eq. (3.13) can be assumed to be a fundamental soliton that is perturbed by the Raman polarization, i.e. $\psi_1(\xi, \tau) = N_1 \operatorname{sech}[N_1(\tau - u_1\xi - \bar{\tau}_1(\xi))] \exp[-i\Omega_1(\xi)(\tau - u_1\xi)]$, where $u_1 = \beta_{11} z_0/t_0$, β_{11} is the first-order dispersion coefficient of the pump, N_1, Ω_1, and $\bar{\tau}_1$ are the soliton amplitude, central frequency, and temporal peak, respectively. Assuming that we launch this soliton as a pump with $\bar{\tau}_1(0) = 0$, and using the variational perturbation method [35], we have found that this soliton is linearly redshifting in the frequency domain with rate $g_1 = \frac{1}{2}\kappa\pi\delta^2\text{csch}(\pi\delta/2N_1)$, and decelerating in the time domain, i.e. $\Omega_1 = -g_1\,\xi$, and $\bar{\tau}_1 = g_1\,\xi^2/2$. In the case $t_0 < 1/\omega_R$, we have found that a factor of $\approx \frac{1}{2}$ might be used to correct the overestimated value of g_1, resulting from using the zeroth-order Taylor approximation. The treatments of the dynamics of the trailing pulse 'probe' ψ_2 are presented in Sections 3.4.1.1 and 3.4.2, when it is a weak long pulse and strong ultrashort soliton, respectively [32, 33]. The study of the pump-probe dynamics has proven crucial understanding of the essential building blocks of supercontinuum generation in Raman-active gases when excited by ultrashort pulses [34, 61].

3.4.1.1 Weak Probe Evolution

When a second weak probe pulse is sent after the leading pump soliton, the probe evolution equation can be derived by substituting in Eq. (3.13),

$$i\partial_\xi\psi_2 + iu_2\partial_\tau\psi_2 + \frac{1}{2m}\partial_\tau^2\psi_2 + 2\kappa N_1\sin(\delta\tilde{\tau})\psi_2 = 0, \tag{3.14}$$

where $u_2 = \beta_{12}z_0/t_0$, $m = |\beta_{21}|/|\beta_{22}|$, β_{1j} and β_{2j} are the first- and the second-order dispersion coefficients of the jth pulse with $j = 1, 2$. Going to the reference frame of the leading decelerating soliton, $\tilde{\tau} = \tau - u_1\xi - g_1\xi^2/2$, and applying a generalized form of the Gagnon-Bélanger phase transformation [66] $\psi_2(\xi,\tilde{\tau}) = \phi(\xi,\tilde{\tau})\exp[i\tilde{\tau}(g_1\xi + u_1 - u_2) + i(g_1\xi + u_1 - u_2)^3/6g_1]$, Eq. (3.14) becomes [32]

$$i\partial_\xi\phi = -\frac{1}{2m}\partial_{\tilde{\tau}}^2\phi + [-2\kappa N_1\sin(\delta\tilde{\tau}) + g_1\tilde{\tau}]\phi. \tag{3.15}$$

This equation is the exact analogue of the time-dependent Schrödinger equation of an electron in a periodic crystal in the presence of an external electric field. In Eq. (3.15) time and space are swapped with respect to the condensed matter physics system, as is usual in optics, and we deal with a spatial-dependent Schrödinger equation of a single particle "probe" with mass m in a temporal crystal with a periodic potential $U = -2\kappa N_1\sin(\delta\tilde{\tau})$ in the presence of a constant force $-g_1$ in the positive-delay direction. The leading soliton excites a sinusoidal Raman oscillation that forms a periodic structure in the reference frame of the soliton, as shown in Figure 3.10. Due to soliton acceleration induced by the strong spectral redshift, a constant force is applied on this structure. Substituting $\phi(\xi,\tilde{\tau}) = f(\tilde{\tau})\exp(iq\xi)$, Eq. (3.15) becomes an eigenvalue problem with eigenfunctions f, and eigenvalues $-q$. The modes of this equation are the Wannier functions [67] that can exhibit Bloch oscillations [68], intrawell oscillations [69], and Zener tunneling [70] due to the applied force.

3.4.1.2 Wannier-Stark Ladder

Consider the propagation of an ultrashort soliton with FWHM 15 fs in the deep anomalous dispersion regime of a H_2-filled HC-PCF with a Kagome lattice. Exciting the rotational Raman shift frequency in the fiber via this soliton will induce a long-lived trailing temporal periodic crystal with a lattice constant $\Lambda = 56.7$ fs, see Figure 3.10, corresponding to the time required by the H_2 molecule to complete one cycle of rotation. In the absence of the applied force, the solutions are the Bloch modes, while in the presence of the applied force, the periodic potential is tilted, and the eigenstates of the system are

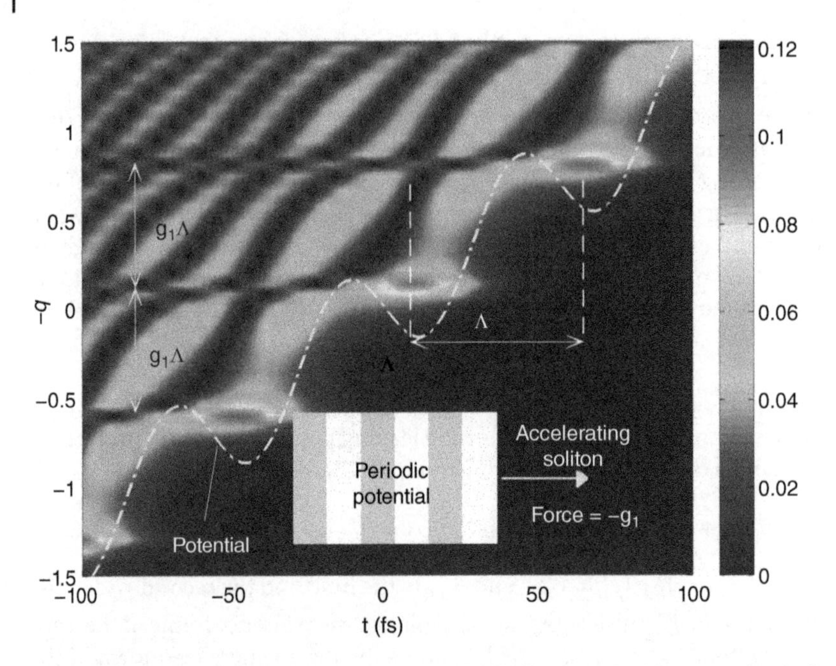

Figure 3.11 A portion of the absolute eigenstates of a Raman-induced temporal periodic crystals with a lattice constant $\Lambda = 56.7$ fs in the presence of a force with magnitude $g_1 = 0.1408$ in the positive-delay direction. The vertical axis represents the corresponding eigenvalues $-q$. The dotted-dashed line is the potential under the applied force. Other simulation parameters are similar to that used in Figure 3.10. *Source*: Saleh 2015 [32].

the Wannier functions portrayed as a 2D color plot in Figure 3.11, where the horizontal axis is the time and the vertical axis is the corresponding eigenvalue. These functions are modified Airy beams that have strong or weak oscillating decaying tails. After an eigenvalue step $g_1\Lambda$, the eigenstates are repeated, but shifted by Λ, forming the Wannier-Stark ladder, well known in condensed matter physics. As shown, each potential minimum can allow a single localized state with very weak tails. A large number of delocalized modes with long and strong tails exist between the localized states.

3.4.1.3 Bloch Oscillations and Zener Tunneling

An arbitrary weak probe following the soliton will be decomposed into the Wannier modes of the periodic temporal crystal. Due to beating between similar eigenstates in different potential wells, Bloch oscillations arise with a period δ/g_1, while beating between different eigenstates in the same potential minimum can result in intrawell oscillations. Interference between modes lying between different wells are responsible for Zener tunneling that allows transitions between different wells (or bands). In the absence of the applied force ($g_1 = 0$), the band structure of the periodic medium can be constructed by plotting the propagation constants of the Bloch modes over the first Brillouin zone $[-\delta/2, \delta/2]$, as shown in Figure 3.12 (a). Zener tunneling occurs when a particle transits from the lowest band to the next-higher band. The evolution of a delayed probe in the form of the first Bloch mode inside a H_2-filled HC-PCF under the influence of the pump-induced temporal periodic crystal, is depicted in

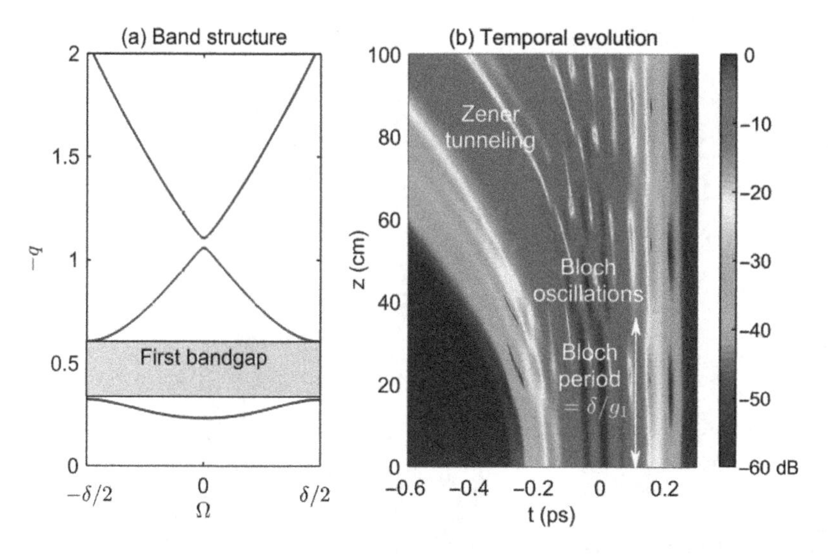

Figure 3.12 (a) Band structure of the temporal crystal induced by the leading ultrashort soliton propagating in the H_2-filled Kagome-lattice HC-PCF with $m = 1$ in the absence of the applied force. (b) Temporal evolution of a weak probe in the accelerated periodic temporal crystal. The probe initial temporal profile is a Gaussian pulse with FWHM 133.6 fs superimposed on the first Bloch mode of the periodic crystal in the absence of the applied force. The contour plot is given in a logarithmic scale and truncated at -60 dB. Other simulation parameters are similar to that used in Figure 3.10. *Source*: Saleh 2015 [32].

Figure 3.12 (b). Portions of the probe are localized in different temporal wells. Bloch oscillations are also shown at a period of 34.7 cm, which correspond to the beating between localized modes in adjacent wells. After each half of this period, an accelerated radiation to the left due to Zener tunneling is also emitted. Zener tunneling is dominant, and Bloch oscillations are weak, because the potential wells are relatively far from each other. Hence, the overlapping between the localized modes are small, consistent with the shallowness of the first band in the periodic limit (absence of the applied force). Since only a single eigenstate is allowed within each well, there is no observed intrawell oscillations in the shown simulations.

An example of this kind of dynamics occurs during supercontinuum generation in HC-PCFs filled by Raman-active gas [34, 61]. The temporal and spectral developments of the propagation of an ultrashort pulse in H_2-filled HC-PCFs are shown in Figure 3.13. Hydrogen has two Raman transition modes – a rotational mode with period 57 fs, and a vibrational mode with 8 fs. Emitted solitons from the soliton-fission process will impulsively excite the rotational Raman mode, since their durations will be shorter than 57 fs. Hence, they will experience redshift and deceleration in the spectral and time domain, and they will excite a lagging sinusoidal temporal modulation of the medium refractive index that can be thought of as the "temporal periodic crystal" [32]. The applied constant force on the temporal crystal arises from the soliton redshift. The vibrational mode of hydrogen will result in an additional Kerr nonlinearity, since it has an oscillation period less than the duration of the solitons. The long relaxation time of Raman in gases allows the lagging emitted dispersive wave to feel the induced temporal crystal. The

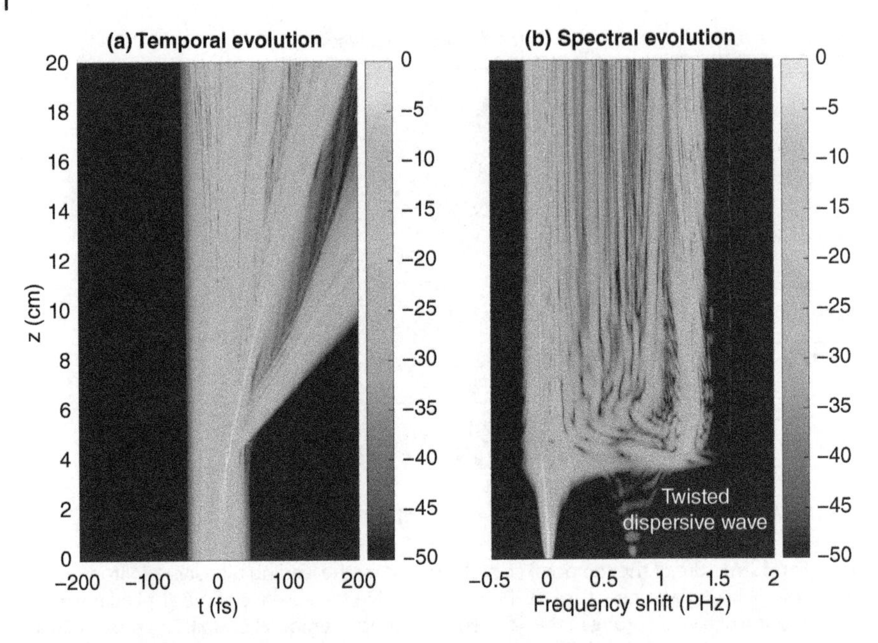

Figure 3.13 (a) Temporal and (b) spectral evolution of a Gaussian pulse propagating with central wavelength 0.805 μm, energy 3 μJ and FWHM 25 fs in a H_2-filled HC-PCF with a gas pressure 7 bar, and a hexagonal core-diameter 27 μm in absence of ionization effects. Contour plots are given in a logarithmic scale and truncated at -50 dB. The linear, nonlinear, and Raman parameters of hydrogen can be found in [61]. *Source*: Saleh 2015 [34].

dispersive wave shows periodically twist and radiation that can be interpreted as a result of combined Bloch oscillations and Zener tunneling.

3.4.2 Strong Probe Evolution

When the delayed pulse is another strong fundamental soliton, rather than a weak probe pulse as in [32], the governing equation for the strong "probe" is given by [33]

$$i\partial_\xi \psi_2 + \frac{1}{2}\partial_\tau^2 \psi_2 + |\psi_2|^2 \psi_2 + R_2(\tau)\psi_2 = 0. \tag{3.16}$$

For weak Raman nonlinearities, the solution of this equation is another perturbed fundamental soliton, $\psi_2(\xi,\tau) = N_2\,\mathrm{sech}[N_2(\tau - \bar{\tau}_2(\xi))]\exp[-i\Omega_2(\xi)\tau]$ where N_2, Ω_2, and $\bar{\tau}_2$ are the second soliton amplitude, central frequency, and temporal peak, respectively. When the soliton durations $\ll 1/\delta$, their Raman response functions can be approximated by using a Taylor expansion as [33],

$$R_2(\tau) \approx \kappa \sum_{l=1,2} N_l \sin[\delta(\tau - \bar{\tau}_l)]\{1 + \tanh[N_l(\tau - \bar{\tau}_l)]\}. \tag{3.17}$$

The superposition of the induced-Raman effects by the two solitons will affect the trailing soliton dynamics.

Adopting the variational perturbation method to understand how Raman nonlinearities can affect the pulse dynamics [35], we have derived a set of coupled governing equations that determine the evolution of each soliton parameters [33],

$$\begin{aligned}
\Omega_1 &= -g_1\,\xi, \\
\bar{\tau}_1 &= g_1\,\xi^2/2, \\
\Omega_2 &= -g_2\,\xi - g_2\frac{2N_1}{N_2}\int_0^{\xi}\cos[\delta(\bar{\tau}_2-\bar{\tau}_1)]\,d\xi, \\
\bar{\tau}_2 &= -\int_0^{\xi}\Omega_2\,d\xi,
\end{aligned} \tag{3.18}$$

where $g_j = \frac{1}{2}\kappa\pi\delta^2\text{csch}(\pi\delta/2N_j)$. The first (leading) soliton will always linearly redshift in the frequency domain with rate g_1, and decelerate in the time domain. Whereas for the second (trailing) soliton, its dynamics depends on two different components: (i) its own (self) component that will lead to a linear redshift similar to the leading soliton, with rate g_2; (ii) a cross-component representing the effect of the first soliton on the second soliton. The latter component is proportional to the ratio between their amplitudes and the cosine of the time difference between them. Since the cosine term varies between positive and negative values, the dynamics of the second soliton can switch back and forth between redshift and blueshift in the frequency domain or deceleration and acceleration in the time domain. This analytical model shows a very good agreement with the numerical model provided that the assumption of the soliton durations $\ll 1/\delta$ is satisfied. This method can also be extended easily to the case of more than two solitons.

Figure 3.14 shows four special cases of the temporal evolution of the second soliton superimposed on the total induced-Raman polarization in a reference frame moving with the leading soliton $\tilde{\tau} = \tau - g_1\xi^2/2$. N_2 is chosen smaller than N_1 so that the cross-component is comparable to the self-component in Eq. (3.18). As we are operating in the anomalous dispersion regime, the positive (negative) variation of the refractive index represents a potential well (barrier). Thus, the periodic modulation of the refractive index corresponds to a sequence of alternative wells and barriers. Based on the initial time delay between the two solitons $\Delta\bar{\tau}_i$, the dynamics of the second soliton behaves differently. Also, the uniformity of the temporal crystal will be modified along the direction of propagation, resulting in a spatiotemporal modulation of the refractive index, i.e. a spatiotemporal crystal. Looking at Figure 3.14, launching the second soliton at (a) the top of a barrier or (b) at the right edge of a well, the second soliton will be able to overcome the barriers during propagation, and transported to the left direction across the potential by the acting force induced via the leading-soliton redshift. The output spatiotemporal crystals are chirped along the direction of propagation in these cases. Interestingly, the second soliton in (b) experiences a net maximum self-frequency blueshift of 9.12 THz $\equiv$ 51.4 nm after 10 cm of propagation, before it is redshifted. Launching the second soliton at (c) a potential minimum or (d) a left edge of a well, the second soliton will not be able to overcome the barriers in these cases, so it is trapped inside the well and will oscillate indefinitely. The amplitude of oscillation in (c) is very small, since the initial velocity of the soliton in this potential is zero. The soliton will oscillate in an asymmetric manner as in (d) due to the modified potential beyond the second soliton peak as well as the acting force that is opposite to the initial velocity. The resulting spatiotemporal crystals have uniform periods along the direction of propagation in these cases.

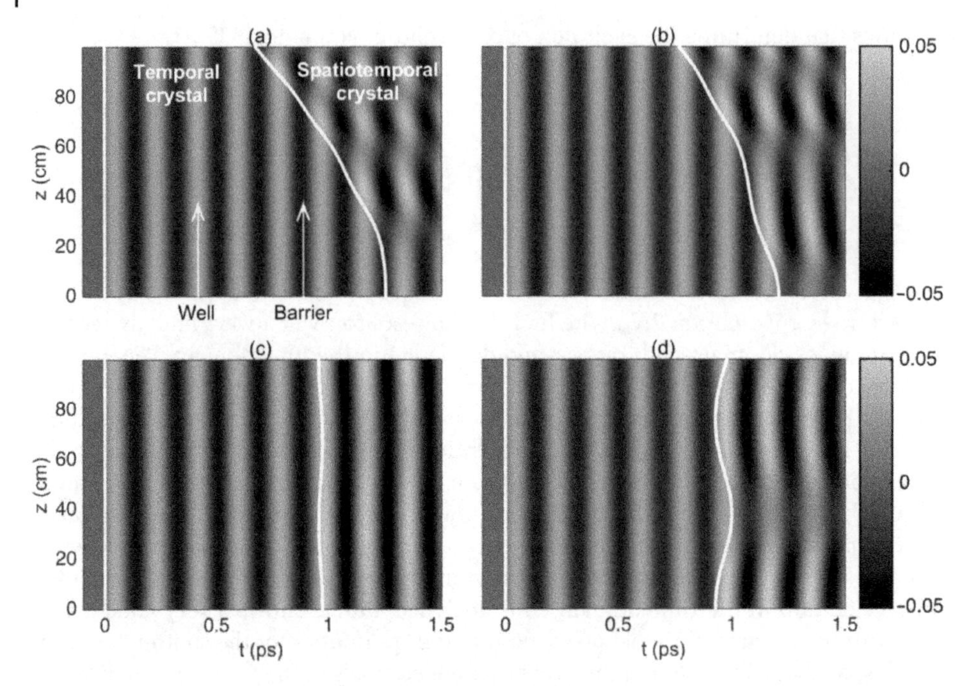

Figure 3.14 Raman polarization induced by two fundamental solitons propagating in a gas-filled HC-PCF in a reference frame moving with the leading soliton. White solid lines represent the temporal evolution of the two solitons with normalized amplitudes $N_1 = 2.5$, $N_2 = 1.25$ (corresponding to full width half maximum (FWHM) 8 fs, 16 fs, respectively), using the analytical prediction Eq. (3.18). The second soliton is launched at (a) $\bar{\tau}_2(0) = 6.75\,\bar{\Lambda}$, (b) $\bar{\tau}_2(0) = 6.5\,\bar{\Lambda}$, (c) $\bar{\tau}_2(0) = 5.25\,\bar{\Lambda}$, and (d) $\bar{\tau}_2(0) = 5\,\bar{\Lambda}$, where $\bar{\Lambda} = 2\pi/\delta$, $\delta = 0.5$ (equivalent to a Raman-mode oscillation with a period 185 fs, such as in deuterium [71]), and $\kappa = 0.16$. *Source*: Saleh 2015 [33].

The temporal evolution of two successive ultrashort Gaussian pulses rather than fundamental solitons are depicted in Figure 3.15 for different time delays. The two pulses have the same central frequencies and amplitudes, and the delay is again within the relaxation coherence time of the Raman-active gas. The two pulses will experience pulse compression and soliton fission processes. The "first" leading pulse excites the Raman polarization "potential" that will affect the "second" trailing pulse. The dynamics of the leading pulse is certainly independent of the time delay and will encounter a self-induced Raman redshift (deceleration). The trailing pulse dynamics is influenced by its self-Raman-induced effect as well as the cross-accelerated Raman polarization effect of the leading pulse. In Figure 3.15, panel (a) shows the case when the self- and cross-components are working together, resulting in a strong delay for the second pulse in comparison to the first pulse. The situation when the self- and cross-components act against each other is figured in panel (b), where initially the second pulse is nearly stopped since the cross- and self-components cancel each other. When the second pulse is launched at one minima of the potential induced by the first pulse, the pulse is well confined during propagation. Even after the pulse fission, the generated solitons are still traveling together, see panel (c). Launching at one potential-maxima, the dynamics of a tree-like behavior are obtained as shown in panel (d), where each soliton propagates in

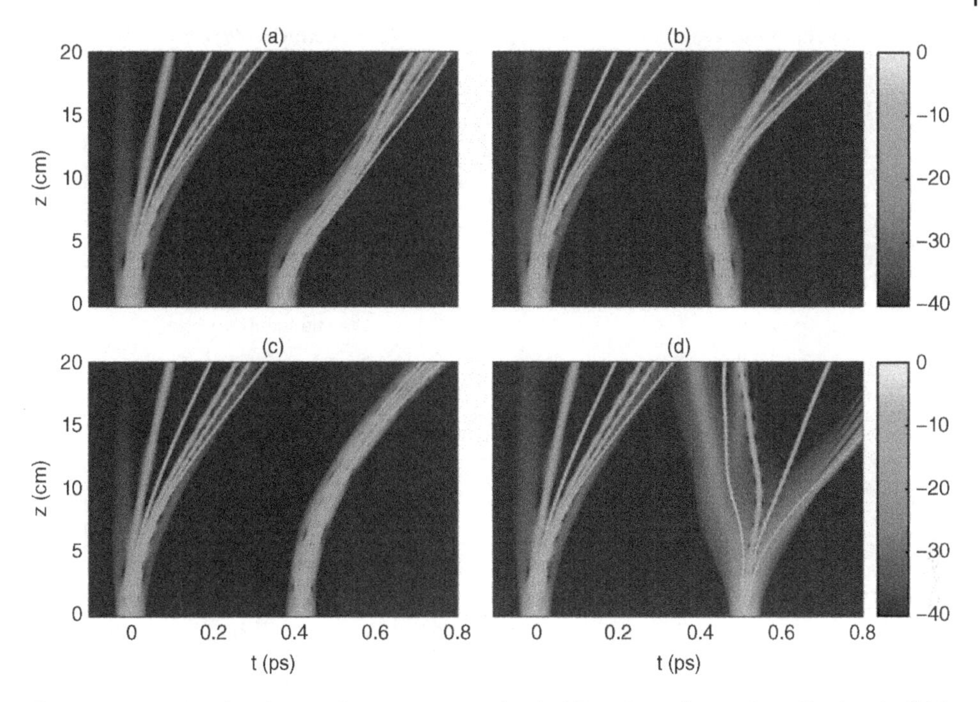

Figure 3.15 Temporal evolution of two successive identical Gaussian pulses with profile $N\exp(-\tau^2/2)$ in the gas-filled HC-PCF described in Figure 3.14, with $N = 7$, and FWHM = 25 fs. The first pulse is launched at $\bar{\tau}_1(0) = 0$. The second pulse is launched at: (a) $\bar{\tau}_2(0) = 2\,\bar{\Lambda}$, (b) $\bar{\tau}_2(0) = 2.5\,\bar{\Lambda}$, (c) $\bar{\tau}_2(0) = 2.25\,\bar{\Lambda}$, and (d) $\bar{\tau}_2(0) = 2.75\,\bar{\Lambda}$. *Source*: Saleh 2015 [33].

a different direction. In all these cases, the dynamics of each soliton depend on where exactly this soliton is born inside the total accelerated periodic potentials induced by other preceding solitons. These results are of fundamental importance in understanding the building blocks of supercontinuum generation: the multitude of solitons propagating in fibers influence each other in a well-defined way, managed by their intensities and their temporal separations.

3.5 Interplay Between Ionization and Raman Effects in Gas-Filled HC-PCFs

Raman self-frequency redshift, which continuously downshifts the central frequency of a single pulse during propagation, was first observed in solid-core fibers [6, 72, 73], and later also in molecular gases [74]. Photo-ionization-induced self-frequency blueshift due to plasma generation has been predicted and demonstrated in HC-PCFs filled by argon gas [27–29]. The core of this section is to investigate the non-trivial interplay between the nonlinear Raman and photo-ionization effects in HC-PCFs filled by Raman-active gases, and demonstrate its potential in designing novel photonic devices [34].

Working in the deep-anomalous regime, i.e., far from the zero dispersion wavelength, allows the observation of a soliton that can have a clean self-frequency shift,

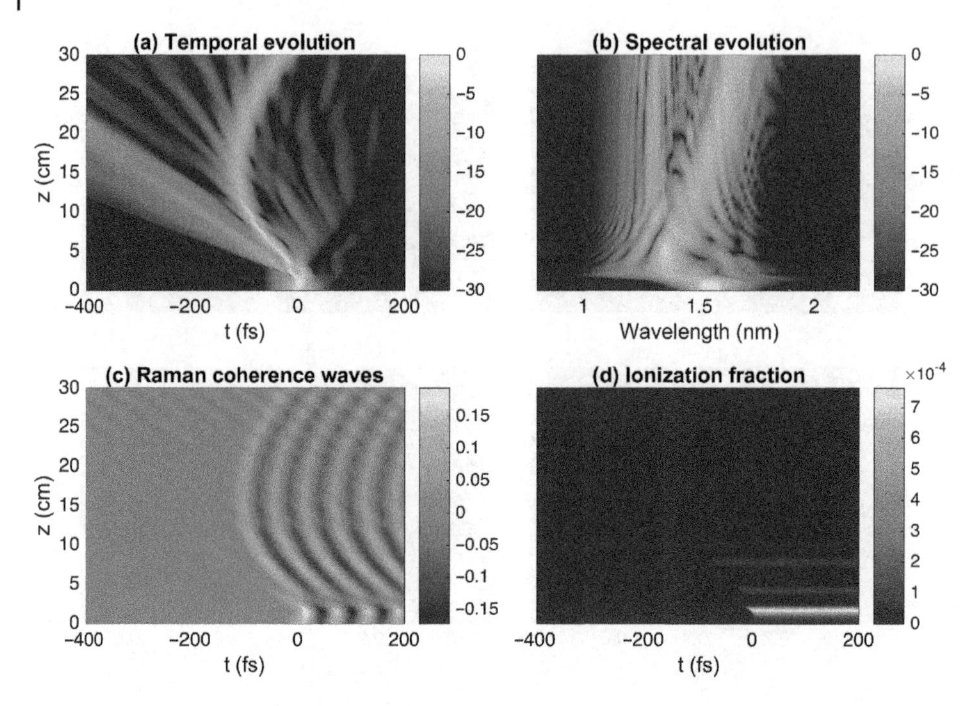

Figure 3.16 (a) Temporal and (b) spectral evolution of a sech-pulse with central wavelength 1.55 μm, energy 2.7 μJ and FWHM 30 fs in a H_2-filled HC-PCF with a gas pressure 7 bar, and an effective-core diameter 18 μm. Temporal evolution of the (c) Raman coherence waves induced by rotational and vibrational excitations, and (d) ionization fraction. Contour plots in (a) and (b) are given in a logarithmic scale and truncated at −30 dB. *Source*: Saleh 2015 [34].

without a dispersive wave emission. Hence, in this regime the interplay between photo-ionization and Raman effects can lead to interesting features. The evolution of a sech-pulse with an input energy 2.7 μJ and centered at 1.55 μm in a 30 cm-long H_2-filled PCF with a ZDW at 403 nm is depicted in Figure 3.16. Panels (a, b) show the temporal and spectral evolution of the pulse, while panels (c, d) show the temporal evolution of the Raman coherence and the ionization fraction. The net self-frequency shift depends on whether photo-ionization or Raman nonlinearity is dominant. Photo-ionizaton-induced blueshift occurs only when pulse intensity exceeds the ionization threshold [28]. However, Raman-induced redshift takes place along the whole propagation with a rate that has an approximately linear-dependence on the pulse intensity [32]. Initially, the pulse is intense enough to ionize the gas and generates enough electrons that contribute to the pulse blueshift. As the pulse shifts to shorter wavelengths, the group velocity dispersion decreases and the Kerr nonlinearity increases. This results in an adiabatic pulse compression [29] that increases the amount of ionization and blueshift. As clear from panel (b), photo-ionization is dominant during the first propagation stage along the fiber. However, due to the concurrent ionization losses, the blueshift process ceases after a certain propagation distance z_{Ion}, allowing the redshifting process to be dominant in the rest of the fiber. Panels (a, c) depict how the pulse and the Raman-induced coherence switch between acceleration and deceleration. The fiber length is a key element in determining the net frequency shift at the fiber output.

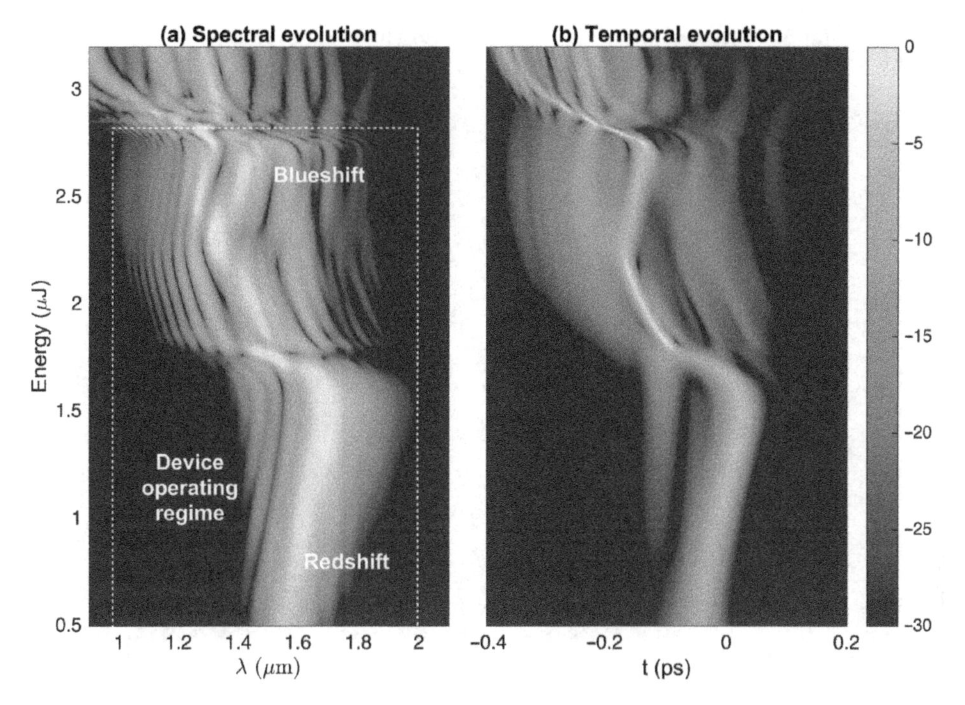

Figure 3.17 Energy dependency of the final (a) spectral and (b) temporal profiles of different sech-pulses with a central wavelength 1.55 μm, and FWHM 30 fs in a H_2-filled HC-PCF with length 10 cm, gas pressure 7 bar, and an effective-core diameter 18 μm. Contour plots are given in a logarithmic scale and truncated at −30 dB. *Source*: Saleh 2015 [34].

To explore the strong interaction between Raman and photoionization effects in this regime, the energy-dependency of the output spectral and temporal profiles of an initial sech-pulse centered at 1.55 μm is portrayed in panels (a, b) of Figure 3.17. For small input energies, the ionization threshold that allows for plasma generation is not reached. Hence, Raman nonlinearity is dominant; its vibrational mode modifies the Kerr nonlinearity, while its rotational mode introduces linear self-frequency redshift. Around 1.75 μJ photo-ionization-induced self-frequency blueshift starts to take place due to plasma formation. As the input pulse energy increases, the induced blueshift starts to compensate then take over the rotational Raman-induced redshift. Initially, the blueshift increases linearly with the input energy, then there is a slight decrease beyond about 2.4 μJ. This is because the photo-ionization process at this latter range of energies occurs very close to the fiber input, allows the Raman process to almost be dominant along the whole fiber, resulting in enhancing the redshift of the spectrum. Increasing the input energy more and more, the pulse breaks up and the clear soliton self-frequency shift is suppressed. The unambiguous interplay between photo-ionization and Raman processes for small input energies in this case suggests the proposal of a novel tunable device that can be used for either frequency-down or frequency-up conversion. The pulse central frequency can be scanned over 400 nm range from about 1.3–1.7 μm by changing its input energy over the range 0.5–2.8 μJ.

The ratio between the fiber length and z_{Ion} is a key-element in determining the frequency-up/down conversion range. Increasing the fiber length will certainly increase

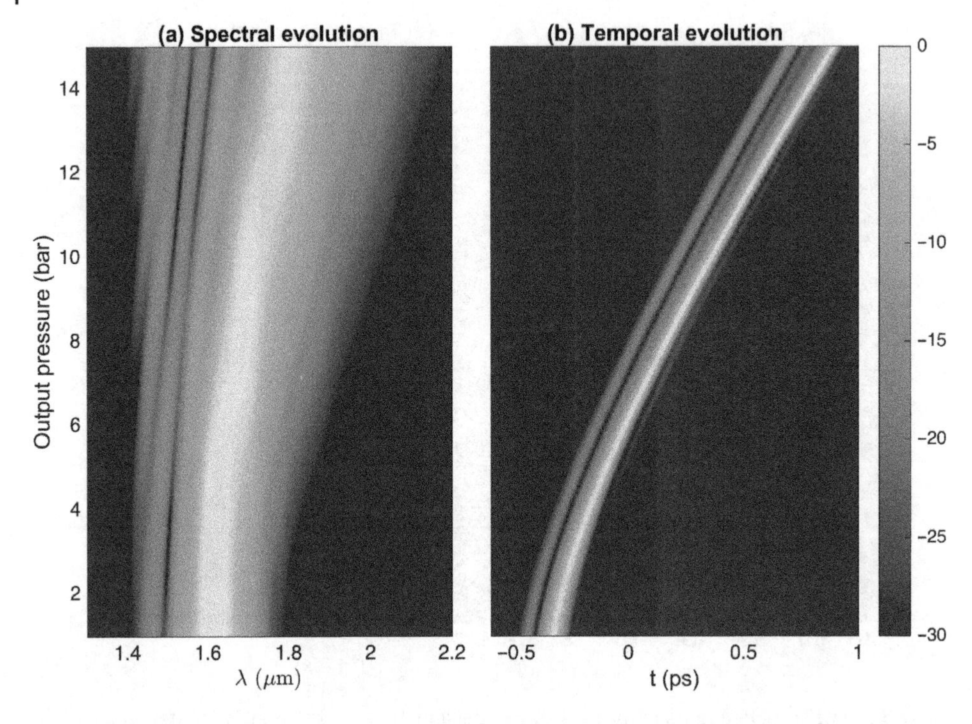

Figure 3.18 Output-pressure dependency of the final (a) spectral and (b) temporal profiles of a sech-pulse with an input pulse energy 1.3 μJ in a H_2-filled HC-PCF. The input gas pressure is fixed at 7 bar, and other simulation parameters are the same as Figure 3.17. Contour plots are given in a logarithmic scale and truncated at −30 dB. *Source*: Saleh 2015 [34].

the frequency-down conversion, since the Raman nonlinearity will become dominant eventually. The latter is because the ionization loss halts the blueshift after the first few centimeters of the fiber and allows the redshift process to rule in the remaining distance. Contrarily, decreasing the fiber length, the frequency-up conversion range is enhanced over the frequency-down conversion range. So, there is a trade-off for choosing the length of the proposed device. However, the flexibility of tuning the gas-pressure at the fiber-end offers an additional degree of freedom to overcome this problem [75]. At equilibrium, the pressure distribution across the fiber is $P(z) = [P_0^2 + \frac{z}{L}(P_L^2 - P_0^2)]^{1/2}$, where P_0 and P_L are the input and output pressures, and L is the fiber length. Panel (a) of Figure 3.18 shows the dependence of the final spectrum of an initial sech-pulse on the output pressure with a fixed input pressure and relatively-small pulse energy. By increasing the output pressure, the spectrum is linearly shifted towards the redside. For instance, by tuning the output pressure from 1 – 15 bar, the spectrum is redshifted by about 250 nm. Therefore, we could shorten the fiber length close to z_{Ion} to enhance the photo-ionization-induced blueshift. Then, the reduced redshift at lower input energies can be recovered by raising the output-fiber pressure. As shown in panel (b), the output temporal profile of the pulse is unaffected by varying the pressure at the fiber end. This enables the proposed device to have a relatively wide tunable range for both frequency-up/down conversions.

3.6 Conclusion

In this chapter, recent results concerning novel nonlinear optical phenomena that involve soliton dynamics in gas-filled HC-PCFs have been presented. Gases have been divided into two main categories: Raman-inactive (monoatomic gases) and Raman-inactive (molecular gases). Using monoatomic gases, such as argon, the effect of photoionization on the evolution of short and long pulses led to unique phenomena such as soliton self-frequency blueshift, asymmetrical self-phase modulation, universal modulation instability, and shower of blueshifted solitons. Whereas gases, such as molecular hydrogen, are characterized by long coherence in comparison to silica glass. This results in highly non-instantaneous interactions that can be detected by launching a probe pulse, delayed from the main pump pulse. For a weak probe, phenomena related to condensed matter physics such as the Wannier-stark ladder, Bloch oscillations, and Zener tunneling are predicted and deeply connected to recent observation of twisted dispersive waves [61]. However, if the probe is another ultrashort soliton, phenomena such as soliton oscillations and transport have been shown to occur. Interestingly, using the latter kind of interaction a boosted soliton self-frequency blueshift can be achieved without experiencing any additional losses. The study of the interplay between the photoionization and Raman effects has suggested the proposal of a novel device that can be used for either frequency-up/down conversion via tuning the input pulse energy and the fiber gas pressure.

These fresh theoretical results, supported by recent experiments, will pave the way for the manipulation and full control of pulse dynamics in PCFs, to demonstrate completely new optical devices. These theories predict novel nonlinear phenomena that are impossible to achieve in conventional solid-core optical fibers, and will open up new exciting venues for future discoveries.

Acknowledgments

M. Saleh would like to acknowledge the support of his research by Royal Society of Edinburgh and the Scottish Government.

References

1 Weinberger, P. (2008) John Kerr and his effects found in 1877 and 1878. *Philosophical Magazine Lett.*, **88**, 897–907.
2 Boyd, R.W. (2007) *Nonlinear Optics*, 3rd edn. Academic Press, San Diego, CA.
3 Hasegawa, A. and Tappert, F. (1973) Transmission of stationary nonlinear optical pulses in dispersive dielectric fibers. i. anomalous dispersion. *Applied Physics Letters*, **23**, 142–144.
4 Mollenauer, L.F., Stolen, R.H., and Gordon, J.P. (1980) Experimental observation of picosecond pulse narrowing and solitons in optical fibers. *Physical Review Letters*, **45**, 1095–1098.
5 Gordon, J.P. (1986) Theory of the soliton self-frequency shift. *Optics Letters*, **11**, 662–664.

6 Mitschke, F.M. and Mollenauer, L.F. (1986) Discovery of the soliton self-frequency shift. *Optics Letters*, **11**, 659–661.
7 Wai, P.K.A., Menyuk, C.R., Lee, Y.C., and Chen, H.H. (1986) Nonlinear pulse propagation in the neighborhood of the zero-dispersion wavelength of monomode optical fibers. *Optics Letters*, **11**, 464–466.
8 Hodel, P.B.W., Zysset, B., and Weber, H.P. (1987) lultrashort pulse propagation, pulse breakup, and fundamental soliton formation in a single-mode optical fiber. *IEEE Journal of Quantum Electronics*, **23**, 1938–1946.
9 Gouveia-Neto, A.S., Faldon, M.E., and Taylor, J.R. (1988) Solitons in the region of the minimum group-velocity dispersion of single-mode optical fibers. *Optics Letters*, **13**, 770–772.
10 Islam, M.N., Sucha, G., Bar-Joseph, I., *et al.* (1989) Broad bandwidths from frequency-shifting solitons in fibers. *Optics Letters*, **14**, 370–372.
11 Islam, M.N., Sucha, G., Bar-Joseph, I., *et al.* (1989) Femtosecond distributed soliton spectrum in fibers. *Journal of the Optical Society of America B*, **6**, 1149–1158.
12 Schütz, J., Hodel, W., and Weber, H.P. (1993) Nonlinear pulse distortion at the zero dispersion wavelength of an optical fibre. *Optics Communications*, **95**, 357–365.
13 Mogilevtsev, D., Birks, T.A., and P. St.J. Russell (1998) Group-velocity dispersion in photonic crystal fibers. *Optics Letters*, **23**, 1662–1664.
14 Broderick, N.G.R., Monro, T.M., Bennett, P.J., and Richardson, D.J. (1999) Nonlinearity in holey optical fibers: Measurement and future opportunities. *Optics Letters*, **24**, 1395–1397.
15 Knight, J.C., Birks, T.A., P. St.J. Russell, and Atkin, D.M. (1996) All-silica single-mode optical fiber with photonic crystal cladding. *Optics Letters*, **21**, 1547–1549.
16 Cregan, R.F., Mangan, B.J., Knight, J.C., *et al.* (1999) Single-mode photonic band gap guidance of light in air. *Science*, **285**, 1537–1539.
17 P. St.J. Russell (2003) Photonic crystal fibers. *Science*, **299**, 358–362.
18 Benabid, F., Knight, J.C., Antonopoulos, G., and P. St.J. Russell (2002) Stimulated Raman scattering in hydrogen-filled hollow-core photonic crystal fiber. *Science*, **298**, 399–402.
19 Knight, J.C. (2003) Photonic crystal fibres. *Nature*, **424**, 847–851.
20 Ouzounov, D.G., Ahmad, F.R., Müller, D., *et al.* (2003) Generation of megawatt optical solitons in hollow-core photonic band-gap fibers. *Science*, **301**, 1702–1704.
21 P. St.J. Russell, Hölzer, P., Chang, W., Abdolvand, A., and Travers, J.C. (2014) Hollow-core photonic crystal fibres for gas-based nonlinear optics. *Nature Photonics*, **8**, 278–286.
22 Travers, J.C., Chang, W., Nold, J., Joly, N.Y., and P. St.J. Russell (2011) Ultrafast nonlinear optics in gas-filled hollow-core photonic crystal fibers. *Journal of the Optical Society of America B*, **28**, A11–A26.
23 Benabid, F., Knight, J.C., Antonopoulos, G., and P. St.J. Russell (2002) Stimulated Raman scattering in hydrogen-filled hollow-core photonic crystal fiber. *Science*, **298**, 399–402.
24 Heckl, O.H., Baer, C.R.E., Kränkel, C., *et al.* (2009) High harmonic generation in a gas-filled hollow-core photonic crystal fiber. *Applied Physics B*, **97**, 369.
25 Joly, N.Y., Nold, J., Chang, W., *et al.* (2011) Bright spatially coherent wavelength-tunable deep-uv laser source in ar-filled photonic crystal fiber. *Physical Review Letters*, **106**, 203 901.
26 Chang, W., Nazarkin, A., Travers, J.C., *et al.* (2011) Influence of ionization on ultrafast gas-based nonlinear fiber optics. *Optics Express*, **19**, 21 018–21 027.

27 Hölzer, P., Chang, W., Travers, J.C., *et al.* (2011) Femtosecond nonlinear fiber optics in the ionization regime. *Physical Review Letters*, **107**, 203 901.

28 Saleh, M.F., Chang, W., Hölzer, P., *et al.* (2011) Soliton self-frequency blue-shift in gas-filled hollow-core photonic crystal fibers. *Physical Review Letters*, **107**, 203 902.

29 Chang, W., Hölzer, P., Travers, J.C., and P. St.J. Russell (2013) Combined soliton pulse compression and plasma-related frequency upconversion in gas-filled photonic crystal fiber. *Optics Letters*, **38**, 2984–2987.

30 Saleh, M.F., Chang, W., Travers, J.C., P. St.J. Russell, and Biancalana, F. (2012) Plasma-induced asymmetric self-phase modulation and modulational instability in gas-filled hollow-core photonic crystal fibers. *Physical Review Letters*, **109**, 113 902.

31 Saleh, M.F., Marini, A., and Biancalana, F. (2014) Shock-induced $\mathcal{PT}$-symmetric potentials in gas-filled photonic crystal fibers. *Physical Review A*, **89**, 023 801.

32 Saleh, M.F., Armaroli, A., Tran, T.X., *et al.* (2015) Raman-induced temporal condensed matter physics in gas-filled photonic crystal fibers. *Optics Express*, **23**, 11 879–11 886.

33 Saleh, M.F., Armaroli, A., Marini, A., Belli, F., and Biancalana, F. (2015) Raman-induced non-local soliton interactions in gas-filled photonic crystal fibers. *Optics Letters*, **40**, 4058–4061.

34 Saleh, M.F. and Biancalana, F. (2015) Tunable frequency-up/down conversion in gas-filled hollow-core photonic crystal fibers. *Optics Letters*, **40**, 4218.

35 Agrawal, G.P. (2007) *Nonlinear Fiber Optics*, 4th edn. Academic Press, San Diego, CA.

36 Goorjian, P.M., Taflove, A., Joseph, R.M., and Hagness, S.C. (1992) Computational modeling of femtosecond optical solitons from Maxwell's equations. *IEEE Journal of Quantum Electronics*, **28**, 2416–2422.

37 Joseph, R.M., Goorjian, P.M., and Taflove, A. (1993) Direct time integration of Maxwell?s equations in two-dimensional dielectric waveguides for propagation and scattering of femtosecond electromagnetic solitons. *Optics Letters*, **18**, 491.

38 Ziolkowski, R.W. and Judkins, J.B. (1993) Full-wave vector Maxwell equation modeling of the self-focusing of ultrashort optical pulses in a nonlinear Kerr medium exhibiting a finite response time. *Journal of the Optical Society of America B*, **10**, 186–198.

39 Goorjian, P.M. and Silberberg, Y. (1997) Numerical simulations of light bullets using the full-vector time-dependent nonlinear maxwell equations. *Journal of the Optical Society of America B*, **14**, 3253–3260.

40 Nakamura, S., Takasawa, N., and Koyamada, Y. (2005) Comparison between finite-difference time-domain calculation with all parameters of Sellmeier's fitting equation and experimental results for slightly chirped 12-fs laser pulse propagation in a silica fiber. *Journal of Lightwave Technology*, **23**, 855.

41 Kolesik, M. and Moloney, J.V. (2004) Nonlinear optical pulse propagation simulation: From Maxwell's to unidirectional equations. *Physical Review E*, **70**, 036 604.

42 Kinsler, P. (2010) Optical pulse propagation with minimal approximations. *Physical Review A*, **81**, 013 819.

43 Sprangle, P., Peñano, J.R., and Hafizi, B. (2002) Propagation of intense short laser pulses in the atmosphere. *Physical Review E*, **66**, 046 418.

44 Keldysh, L.V. (1965) Ionization in the field of a strong electromagnetic wave. *Soviet Physics JETP*, **20**, 1307–1314.

45 Reiss, H.R. (1990) Relativistic strong-field photoionization. *Journal of the Optical Society of America B*, **7**, 574–586.

46 Ammosov, M.V., Delone, N.B., and Krainov, V.P. (1986) Tunnel ionization of complex atoms and of atomic ions in an alternating electromagnetic field. *Soviet Physics JETP 64*, **64**, 1191–1194.

47 Perelomov, A.M., Popov, V.S., and Terent'ev, M.V. (1966) Ionization of atoms in an alternating electric field. *Soviet Physics JETP*, **23**, 924–934.

48 Yudin, G.L. and Ivanov, M.Y. (2001) Nonadiabatic tunnel ionization: Looking inside a laser cycle. *Physical Review A*, **64**, 013 409.

49 Gibson, G., Luk, T.S., and Rhodes, C.K. (1990) Tunneling ionization in the multiphoton regime. *Physical Review A*, **41**, 5049–5052.

50 Augst, S., Meyerhofer, D.D., Strickland, D., and Chint, S.L. (1991) Laser ionization of noble gases by coulomb-barrier suppression. *Journal of the Optical Society of America B*, **8**, 858–867.

51 Saleh, M.F. and Biancalana, F. (2011) Understanding the dynamics of photoionization-induced nonlinear effects and solitons in gas-filled hollow-core photonic crystal fibers. *Physical Review A*, **84**, 063 838.

52 Hasegawa, A. and Brinkman, W. (1980) Tunable coherent ir and fir sources utilizing modulational instability. *IEEE Journal of Quantum Electronics*, **16**, 694–697.

53 Tai, K., Hasegawa, A., and Tomita, A. (1986) Observation of modulational instability in optical fibers. *Physical Review Letters*, **56**, 135–138.

54 Tai, K., Tomita, A., Jewell, J.L., and Hasegawa, A. (1986) Generation of subpicosecond solitonlike optical pulses at 0.3 Thz repetition rate by induced modulational instability. *Applied Physics Letters*, **49**, 236–238.

55 Yoshikawa, S. and Imasaka, T. (1993) A new approach for the generation of ultrashort optical pulses. *Optics Communications*, **96**, 94–98.

56 Kaplan, A.E. (1994) Subfemtosecond pulses in mode-locked 2π solitons of the cascade stimulated Raman scattering. *Physical Review Letters*, **73**, 1243–1246.

57 Kawano, H., Hirakawa, Y., and Imasaka, T. (1998) Generation of high-order rotational lines in hydrogen by four-wave Raman mixing in the femtosecond regime. *IEEE Journal of Quantum Electronics*, **34**, 260–268.

58 Nazarkin, A., Korn, G., Wittmann, M., and Elsaesser, T. (1999) Generation of multiple phase-locked stokes and anti-stokes components in an impulsively excited Raman medium. *Physical Review Letters*, **83**, 2560–2563.

59 Kalosha, V.P. and Herrmann, J. (2000) Phase relations, quasicontinuous spectra and subfemtosecond pulses in high-order stimulated Raman scattering with short-pulse excitation. *Physical Review Letters*, **85**, 1226–1229.

60 Butylkin, V.S., Kaplan, A.E., Khronopulo, Y.G., and Yakubovich, E.I. (1989) *Resonant Nonlinear Interaction of Light with Matter*, Springer-Verlag, 1st edn.

61 Belli, F., Abdovaland, A., W. Chang, J.C.T., and P. St.J. Russell (2015) Vacuum-uv to ir supercontinuum in hydrogen-filled photonic crystal fiber. *Optica*, **2**, 292–300.

62 Bischel, W.K. and Dyer, M.J. (1986) Temperature dependence of the Raman linewidth and line shift for the Q(1) and Q(0) transitions in normal and para-H_2. *Physical Review A*, **33**, 3113–3123.

63 Bartels, R.A., Backus, S., Murnane, M., and Kapteyn, H. (2003) Impulsive stimulated Raman scattering excitation of molecular vibrations via nonlinear pulse shaping. *Chemical Physics Letters*, **374**, 326–333.

64 Weber, M.J. (1994) *CRC Handbook of Laser Science and Technology Supplement 2: Optical Materials*, CRC Press, Boca Raton, FL.

65 Mizrahi, V. and Shelton, D.P. (1985) Nonlinear susceptibility of h_2 and d_2 accurately measured over a wide range of wavelengths. *Physical Review A*, **32**, 3454–3460.

66 Gagnon, L. and Bélanger, P.A. (1990) Soliton self-frequency shift versus Galilean-like symmetry. *Optics Letters*, **15**, 466–468.

67 Wannier, G.H. (1960) Wave functions and effective Hamiltonian for Bloch electrons in an electric field. *Physical Review*, **117**, 432.

68 Bloch, F. (1928) Über die Quantenmechanik der Electronen in Kristallgittern. *Zeitschrift fur Physik*, **52**, 555.

69 Bouchard, A.M. and Luban, M. (1995) Bloch oscillations and other dynamical phenomena of electrons in semiconductor superlattices. *Physical Review B*, **52**, 5105.

70 Zener, C. (1934) A theory of electrical breakdown of solid dielectrics. *Royal Society of London A*, **145**, 523.

71 Burzo, A.M., Chugreev, A.V., and Sokolov, A.V. (2007) Stimulated rotational Raman generation controlled by strongly driven vibrational coherence in molecular deuterium. *Physical Review A*, **75**, 022 515.

72 Dianov, E.M., A. Ya. Karasik, Mamyshev, P.V., *et al.* (1985) Stimulated-Raman conversion of multisoliton pulses in quartz optical fibers. *JETP Letters*, **41**, 294–297.

73 Bulushev, A.G., Dianov, E.M., Okhotnikov, O.G., and Serkin, V.N. (1991) Raman self-frequency shift of the spectrum of femtosecond optical solitons and suppression of this effect in optical fibers and soliton lasers. *JETP Letters*, **54**, 619–622.

74 Korn, G., Dühr, O., and Nazarkin, A. (1998) Observation of Raman self-conversion of fs-pulse frequency due to impulsive excitation of molecular vibrations. *Physical Review Letters*, **81**, 1215–1218.

75 Suda, A., Hatayama, M., Nagasaka, K., and Midorikawa, K. (2005) Generation of sub-10-fs, 5-mj-optical pulses using a hollow fiber with a pressure gradient. *Applied Physics Letters*, **86**, 111 116.

4

Modulation Instability in Periodically Modulated Fibers

Arnaud Mussot, Matteo Conforti, and Alexandre Kudlinski

Univ. Lille, CNRS, UMR 8523–PhLAM–Physique des Lasers Atomes et Molécules, Lille, France

4.1 Introduction

Modulational instability (MI) refers to a process where a weak periodic perturbation of an intense continuous wave (CW) grows exponentially as a result of the interplay between dispersion and nonlinearity. MI constitutes one of the most basic and widespread nonlinear phenomena in physics, and it has been studied extensively in several different physical systems such as water waves, plasmas, and optical devices [1]. Despite the fact that it has been studied since the 1960s, it still attracts consistent research interest [2–4]. For cubic nonlinearities such as those of optical fibers modeled by the nonlinear Schrödinger equation (NLSE), the underlying physical mechanism can be understood in terms of four-photon mixing between the pump, signal, and idler waves. However, the scalar four-photon interactions in a homogeneous fiber can be phase matched, and hence efficient, only in the anomalous group-velocity dispersion (GVD) regime. In the normal GVD regime, on the other hand, MI can occur thanks to higher-order dispersion [5–7], or in detuned cavities [8], thanks to constructive interference between the external driving and the recirculating pulse. Alternatively, MI with normal GVD can also arise in systems with built-in periodic dispersion [9, 10], among which dispersion oscillating fibers (DOFs) have recently attracted renewed attention [11–16]. In DOFs, phase matching relies on the additional momentum carried by the periodic dispersion grating (quasi-phase-matching). The occurrence of unstable frequency bands can then be explained using the theory of parametric resonance, a well-known instability phenomenon which occurs in linearized systems for which at least one parameter is varied periodically during the evolution [11].

The additional degree of freedom brought by the periodicity has attracted a remarkable research attention, which has led to several theoretical and experimental studies in the context of optical fibers [17, 18]. This physical flexibility is of particular interest since the modulation period can be varied from the meter range up to tens of kilometers, leading to a broad range of investigation. In the 1990s, the rise of optical telecommunication networks led to the deployment of "natural" periodic optical fiber systems due to the alternation of all-optical regeneration devices and/or dispersion managed lines [10, 19]. In addition to the fundamental interest brought by these systems,

Shaping Light in Nonlinear Optical Fibers, First Edition. Edited by Sonia Boscolo and Christophe Finot.

it was then necessary to understand in depth the origin of the characteristic spurious MI sidebands [10, 19], which are highly detrimental to telecommunications since they are generated in the gigahertz range [17] due to their period in the range of kilometers. Many theoretical studies have therefore been initiated in this context ([10, 17, 19] to cite a few). Very recently, dispersion management was pushed one step further with the experimental demonstration of MI in continuously modulated waveguides [12], directly fabricated on the drawing tower, with modulation periods lying in the meter range. The direct consequence is that it was possible to report for the first time many MI side lobes in the therahertz range. This has been proved to provide another degree of freedom either for fundamental investigations or for applications where broad bandwidths are required [20].

The chapter is organized as follows. In Section 4.2, we introduce the basic theory of MI in periodically modulated fibers. In Section 4.3, we discuss the fabrication technique of such fibers and their basic optical properties. Finally, in Section 4.4, we present experimental results highlighting the specific features of the modulation instability process in dispersion oscillating fibers.

4.2 Basic Theory of Modulation Instability in Periodically Modulated Waveguides

In this section we develop a general theory that identifies the frequency bands that can be destabilized by a periodic modulation of fiber dispersion or nonlinearity. Let us consider the following NLSE

$$i\frac{\partial u}{\partial z} - \frac{\beta_2(z)}{2}\frac{\partial^2 u}{\partial t^2} + \gamma(z)|u|^2 u = 0, \tag{4.1}$$

where we assume the z-dependent GVD $\beta_2(z)$ and the nonlinear coefficient $\gamma(z)$ to be of the form

$$\beta_2(z) = \overline{\beta}_2 + \beta_m f_Z(z), \quad \gamma(z) = \overline{\gamma} + \gamma_m g_Z(z), \tag{4.2}$$

where f_Z and g_Z are periodic functions of period Z such that $\min f_Z = -1 = \min g_Z$, and their mean is vanishing, i.e. $\int_{-Z/2}^{Z/2} f_Z(z)dz = \int_{-Z/2}^{Z/2} g_Z(z)dz = 0$. A representative example of such kind of modulation function is represented by $f_Z = \sin(k_g z) = \sin(2\pi/Z \cdot z)$. Our aim is to analyse the stability of the general stationary (or CW) solution of Eq. (4.1) which reads $u_0(z) = \sqrt{P}\exp(iP\int_0^z \gamma(z')dz')$, where P is the power. We emphasize that in Eqs. (4.1 and 4.2) all the coefficients, as well as distance z and time t, are written in physical units. Hence all the formulas that we write in the following can readily be interpreted in terms of real-world quantities, thus allowing for a direct comparison with the experimental results. Nonetheless in order to simplify the graphical illustration of such results, in this theoretical section, we will assume $P = \overline{\beta}_2 = \overline{\gamma} = 1$. This is always possible, without loss of generality, by introducing the normalized distance $z/z_{nl} \to z$, time $t/t_0 \to t$, and field $u/\sqrt{P} \to u$, where $z_{nl} = (\overline{\gamma}P)^{-1}$ is the so-called nonlinear length and $t_0 = \sqrt{\overline{\beta}_2 z_{nl}}$ is a characteristic time. In this case β_m, γ_m, and Z correspond to the physical quantities measured in units of $\overline{\beta}_2, \overline{\gamma}$, and z_{nl}, respectively.

We consider a perturbation of $u_0(z)$ in the form $u(z,t) = [v(z,t) + 1]u_0(z)$, where the perturbation $v(z,t)$ satisfies $|v| \ll 1$. Inserting this expression into Eq. (4.1), and retaining only the linear terms, we find

$$i\frac{\partial v}{\partial z} - \frac{\beta_2(z)}{2}\frac{\partial^2 v}{\partial t^2} + \gamma(z)P(v + v^*) = 0. \tag{4.3}$$

Writing $v = q + ip$, with q and p real functions, we obtain the following linear system:

$$\begin{cases} \dfrac{\partial q}{\partial z} - \dfrac{\beta_2(z)}{2}\dfrac{\partial^2 p}{\partial t^2} = 0, \\ \dfrac{\partial p}{\partial z} + \dfrac{\beta_2(z)}{2}\dfrac{\partial^2 q}{\partial t^2} - 2\gamma(z)Pq = 0. \end{cases}$$

Finally, taking the Fourier transform of this system in the time variable t, leads to

$$\begin{cases} \dfrac{\partial \hat{q}}{\partial z} + \dfrac{\beta_2(z)}{2}\omega^2\hat{p} = 0, \\ \dfrac{\partial \hat{p}}{\partial z} - \dfrac{\beta_2(z)}{2}\omega^2\hat{q} - 2\gamma(z)P\hat{q} = 0. \end{cases} \tag{4.4}$$

Note that this is a Hamiltonian dynamical system in a two-dimensional phase plane with canonical coordinates $(\hat{q}, \hat{p})$. Analyzing the linear (in)stability of the stationary solution $u_0(z)$ therefore reduces to studying the solutions to Eq. (4.4) for each ω. Since the coefficients in the equation are z-periodic with period Z, Floquet theory applies [9, 21]. This amounts to studying the linearized evolution over one period Z, to obtain the Floquet map Φ which in the present situation is the two-by-two real matrix defined by $\Phi[\hat{q}(0), \hat{p}(0)]^T = [\hat{q}(Z), \hat{p}(Z)]^T$. In other words, the Floquet map is given by the fundamental matrix solution of system (4.4) evaluated at $z = Z$. As a result, $[\hat{q}(nZ), \hat{p}(nZ)]^T = \Phi^n[\hat{q}(0), \hat{p}(0)]^T$. Note that Φ necessarily has determinant one, since it is obtained by integrating a Hamiltonian dynamics, of which we know that it preserves phase space volume. As a consequence, if λ is one of its eigenvalues, then so are both its complex conjugate λ^* and its inverse λ^{-1}. This constrains the two eigenvalues of Φ considerably: they are either both real, or lie both on the unit circle. Now, the dynamics is unstable only if there is one eigenvalue λ satisfying $|\lambda| > 1$, in which case both eigenvalues are real. We will denote as $\lambda^{\pm}$ the two eigenvalues of Φ. We are interested in studying the growth rate, or gain, that is defined as

$$G(\omega) = \frac{1}{Z}\ln\left(\max\{|\lambda^+|, |\lambda^-|\}\right) \tag{4.5}$$

as a function of ω, β_m and γ_m. The gain G measures the growth of the perturbations $(\hat{q}(z), \hat{p}(z)) \propto \exp[G(\omega)z]$. It vanishes if the two eigenvalues lie on the unit circle. The regions where the gain does not vanish are commonly referred to as Arnold tongues [11]. We will explain below that, whereas their precise form depends on the choice of f_Z, g_Z, the position of their tips does not.

Since the system (4.4) is not autonomous, it cannot be solved analytically in general. Specific cases where analytical solutions can be found are piece-wise constant [9] and delta kicked [21] dispersion profiles. Nevertheless, the above observations will allow us to obtain some information about its (in)stability for small β_m, γ_m, and valid for all perturbations f_Z, g_Z, whatever their specific form.

To see this, we first consider the case $\beta_m = 0 = \gamma_m$, so that $\beta(z) = \overline{\beta}_2$, $\gamma(z) = \overline{\gamma}$. It is then straightforward to integrate the system (4.4). The Floquet map is then given by

$$\Phi_{av} = \begin{pmatrix} \cos(k_{av}Z) & -\dfrac{\overline{\beta}_2\omega^2}{2k_{av}}\sin(k_{av}Z) \\ \dfrac{2k}{\overline{\beta}_2\omega^2}\sin(k_{av}Z) & \cos(k_{av}Z) \end{pmatrix}, \tag{4.6}$$

where

$$k_{av} = \sqrt{\frac{\overline{\beta}_2}{2}\omega^2\left(\frac{\overline{\beta}_2}{2}\omega^2 + 2\overline{\gamma}P\right)}. \tag{4.7}$$

We concentrate on the normal average dispersion (defocusing) regime $\overline{\beta}_2 > 0$, since we want to characterize instabilities generated by the periodicity. The anomalous dispersion (focusing) regime is unstable even in homogeneous fibers through the standard MI. Note that the matrix Φ_{av} has determinant equal to 1, as expected. The eigenvalues of Φ_{av} can be readily computed as

$$\lambda_{av}^{\pm} = \exp(\pm ik_{av}Z). \tag{4.8}$$

What will happen if we now switch on the fiber modulation terms $f_Z(z)$ and $g_Z(z)$? It is then no longer possible, in general, to give a simple closed form expression of the solution to the system (4.4), which is no longer autonomous, and hence of the Floquet map Φ. Nevertheless, we do know that, for small β_m, γ_m, the eigenvalues of Φ must be close to the eigenvalues $\lambda_{av}^{\pm}$. We then have two cases to consider.

Case 1. $k_{av} \neq \frac{\pi m}{Z}$, $m = \pm 1, \pm 2, \ldots$. We have $\lambda_{av}^{-} = (\lambda_{av}^{+})^*$, *they are distinct*, and they both lie on the unit circle, away from the real axis. They then must remain on the unit circle under perturbation since, for the reasons explained above, they cannot move into the complex plane away from the unit circle. Consequently, in this case, the stationary solution $u_0(z)$ is linearly stable under a sufficiently small perturbation described by $\beta_m f_Z(z)$ and $\gamma_m g_Z(z)$, and this statement does not depend on the precise form of $f_Z(z)$ or of $g_Z(z)$. In fact, with growing β_m and/or γ_m, the two eigenvalues will move along the unit circle until they meet either at -1 or at $+1$ for some critical value of the perturbation parameters. Only for values of the latter above that critical value can the system become unstable. A pictorial description of this situation is shown in the left-hand side of Figure 4.1.

Case 2. $k_{av} = \frac{\pi m}{Z}$, $m = \pm 1, \pm 2, \ldots$. Now $\lambda_{av}^{+} = \lambda_{av}^{-} = \pm 1$ (upper or lower sign holds for m even or odd, respectively) is a doubly degenerate eigenvalue of Φ_{av}. Under a small perturbation, the degeneracy can be lifted and two real eigenvalues can be created, one greater than one, one less than one in absolute value. The system has then become unstable! Of course, it will now depend on the type of perturbation whether the system becomes unstable, remains marginally stable (the two eigenvalues do not move at all, but stay at 1 or -1), or becomes stable (the two eigenvalues move in opposite directions along the unit circle). A pictorial description of this situation is shown in the right-hand side of Figure 4.1.

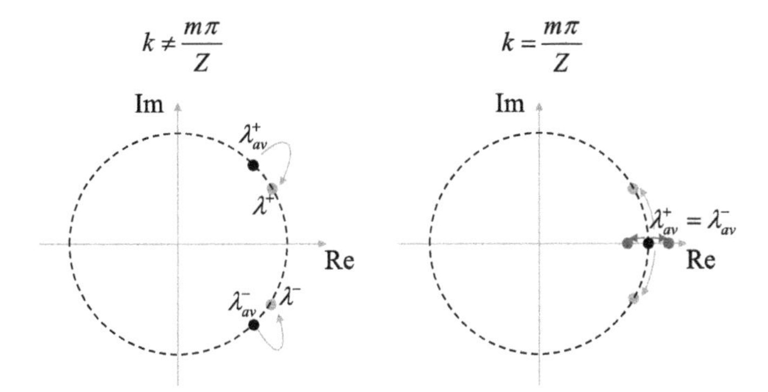

Figure 4.1 Sketch illustrating, in the complex plane, the effect of the periodic modulation terms $f_Z(z)$ and $g_Z(z)$ on the eigenvalues of the linearized Floquet map Eq. (4.6). Black dots correspond to the unperturbed eigenvalues lying on the unit circle (dashed line). Gray dots show the new position of the eigenvalues after switching on the modulations, leading to a stable regime when $k \neq \frac{\pi m}{Z}$ (left diagram) and an unstable one when $k = \frac{\pi m}{Z}$ (right diagram).

We recap by saying that, when the modulation is switched on, the instability sets in under the resonant condition $k_{av} = \frac{m\pi}{Z}$. Recalling the expression of $k_{av} = k_{av}(\omega)$ in Eq. (4.7), it is straightforward to show that the m−th order resonance is fulfilled at frequency $\omega = \omega_m$, where

$$\omega_m = \sqrt{\frac{2}{\overline{\beta}_2}\left(\sqrt{(\overline{\gamma}P)^2 + \left(\frac{m\pi}{Z}\right)^2} - \overline{\gamma}P\right)}, \tag{4.9}$$

which is therefore the frequency at which the system destabilizes for an infinitely small Hamiltonian perturbation of Φ_{av}. These values ω_m therefore correspond to the tips of the Arnold tongues, that is, to the positions of the (centers of) the unstable sidebands of the defocusing NLSE under a general periodic perturbation f_Z, g_Z.

We emphasize that the resonance condition

$$\boxed{k_{av}(\omega_m) = m\frac{\pi}{Z} = m\frac{k_g}{2}} \tag{4.10}$$

is the condition of *parametric resonance, i.e., the natural spatial frequency of the unperturbed harmonic oscillator (k_{av}) is equal to a multiple of half the forcing spatial frequency ($k_g/2 = \pi/Z$)* [11].

Additional physical insight can be obtained by expanding Eq. (4.9) for small power, i.e., assuming $\overline{\gamma}P \ll |m|\pi/Z$. This assumption is valid in a large range of experimental setups where the period of modulation lies in the meter range and with pump power up to tens of Watt. At zero order we recover the well-known quasi-phase-matching relation [10, 12, 19]

$$\overline{\beta}_2\omega_m^2 + 2\overline{\gamma}P = \frac{2\pi m}{Z}. \tag{4.11}$$

Equation (4.11) entails the conservation of the momentum (corrected for nonlinear phase-shifts), made possible thanks to the virtual momentum carried by the dispersion grating, of the four-photon mixing interaction between two photons from the pump, going into two photons in the symmetric unstable bands at lower (Stokes) and higher (antiStokes) frequencies with respect to the pump.

4.2.1 Piecewise Constant Dispersion

An example of practical interest where the Floquet analysis can be performed *analytically* is a fiber with a piecewise constant dispersion and uniform nonlinearity $\gamma(z) = \overline{\gamma} = \gamma$. This case was studied also in the context of communication system with dispersion management [9]. The Floquet map is given simply by the product of the two matrices describing each uniform segment:

$$\Psi = \Phi_a \Phi_b, \tag{4.12}$$

where $\Phi_{a,b}$ has the expression (4.6) calculated for a dispersion $\beta(z) = \beta_{a,b}$, where the two pieces of fiber have length $L_{a,b}$, such that $L_a + L_b = Z$ and the average dispersion is $\overline{\beta}_2 = (\beta_a L_a + \beta_b L_b)/Z$.

The eigenvalues of Ψ are given by

$$\lambda^{\pm} = \frac{\Delta}{2} \pm \sqrt{\frac{\Delta^2}{4} - 1}, \tag{4.13}$$

where

$$\Delta = 2\cos(k_a L_a)\cos(k_b L_b) - \sigma \sin(k_a L_a)\sin(k_b L_b), \tag{4.14}$$

is the Floquet discriminant, and $\sigma = (\beta_a \beta_b \omega^4 + 2\gamma P \omega^2(\beta_a + \beta_b))/(2k_a k_b)$. We have parametric instability if $|\Delta| > 2$, with gain $G(\omega) = \ln(\max|\lambda^{\pm}|)/Z$.

In Figure 4.2 we report an example of analytically calculated instability gain. We can clearly see the generation of several branches, the Arnold tongues, due to the periodic forcing. Their position is perfectly predicted by the vertical lines, that represent the frequency calculated by Eq. (4.9).

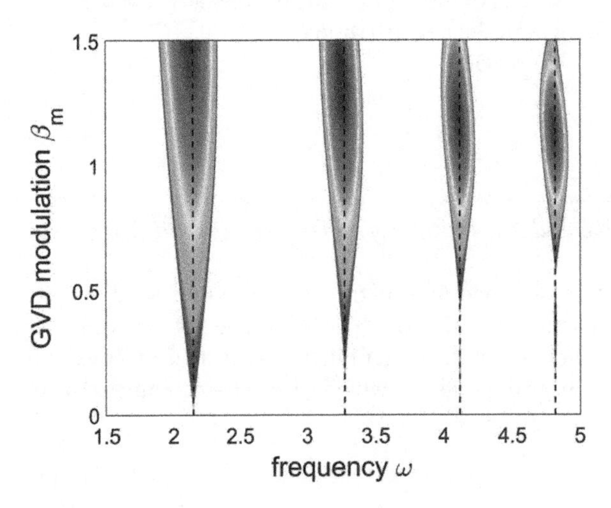

Figure 4.2 Color level plot of growth rate $G(\omega)$ for a fiber with periodic GVD modulation $\overline{\beta}_2 = 1$, $\beta_{a,b} = 1 \pm \beta_m$, $L_a = L_b = 0.5$, $\gamma(z) = \overline{\gamma} = \gamma$. The vertical lines indicate the peak gain calculated from Eq. (4.9) for $m = 1, 2, 3, 4$.

4.3 Fabrication of Periodically Modulated Photonic Crystal Fibers

4.3.1 Fabrication Principles

We use the standard stack and draw technique [22] to fabricate the photonic crystal fibers (PCFs). The stacked array of capillaries is drawn into canes of a few mm diameter, which are then jacked into pure silica tubes to form the final preform, which is drawn into fibers. The air holes of the PCF are pressurized to prevent their collapse during the drawing process. When fabricating uniform fibers with a targeted outer diameter d, the drawing speed is usually adjusted by a feedback loop in quasi real time in order to maintain the fiber diameter as close as possible to d. The typical accuracy over d is the order of $\pm 1\%$. For the fabrication of periodically modulated fibers, the longitudinal evolution of the fiber diameter is controlled by adjusting the development of the drawing speed with time (which is equivalent to the fiber length) using a servo-control system. This process is ruled by the conservation of glass volume between the preform (jacketed cane) and the fiber:

$$d_{\text{Fiber}} = d_{\text{Preform}} \sqrt{\frac{V_{\text{Preform}}}{V_{\text{Fiber}}}}, \tag{4.15}$$

where $d_{\text{Preform,Fiber}}$ are respectively the preform (jacketed cane) and the fiber outer diameter, and $V_{\text{Preform,Fiber}}$ are respectively the preform feed into the furnace and the drawing capstan speed. In our process, d_{Preform} and V_{Preform} are fixed, V_{Fiber} is adjusted with a desired $f(z)$ (where z is the longitudinal space coordinate along the fiber) function, which results in a modulation of the fiber outer diameter d_{Fiber}, and thus of the overall photonic crystal structure. By suitably adjusting the fiber tension and hole pressure during drawing, the normalized air hole diameter d/Λ (where Λ is the pitch of the microstructure) can be kept constant all along the periodically modulated PCF. This means that the pitch value (and the core diameter, for instance) follow a linear dependence with outer diameter. This results in a periodic modulation of the mode(s) propagation constant(s), and thus of all guiding properties.

4.3.2 Typical Example

The impact of the longitudinal modulations of the outer diameter on the modes propagation constants of a typical air silica PCF is illustrated in Figure 4.3 in the color plate section. The d/Λ ratio is set at 0.4 to get an endlessly single mode behavior [23]. Figure 4.3 (a) represents the development of the group velocity dispersion (GVD) versus wavelength for different pitch values. We define a reference value for the pitch, labeled Λ_{REF}, equal to 3.4 μm. In the following, modulation amplitudes will be referred to this value. For Λ_{REF}, the zero dispersion wavelength (ZDW) is located at 1074 nm (dashed black curve in Figure 4.3 (a)). By increasing the pitch value to 4.76 μm (+40%), the curve is red shifted (solid red curve) and the ZDW is located at 1153 nm. On the contrary, by decreasing the pitch value to 2.04 μm (−40%) the curve experiences a blue shift (solid blue curve) and the ZDW is located at 957 nm. Consequently, by varying the outer diameter of an air silica PCF of only a few tens of percentage, one can induce very large wavelength shift of the GVD curve (more than 200 nm on the ZDW). In order to give a clearer insight of these variations, Figures 4.3 (b)–(e) represent the development of $\beta_{2,3,4}$ and γ,

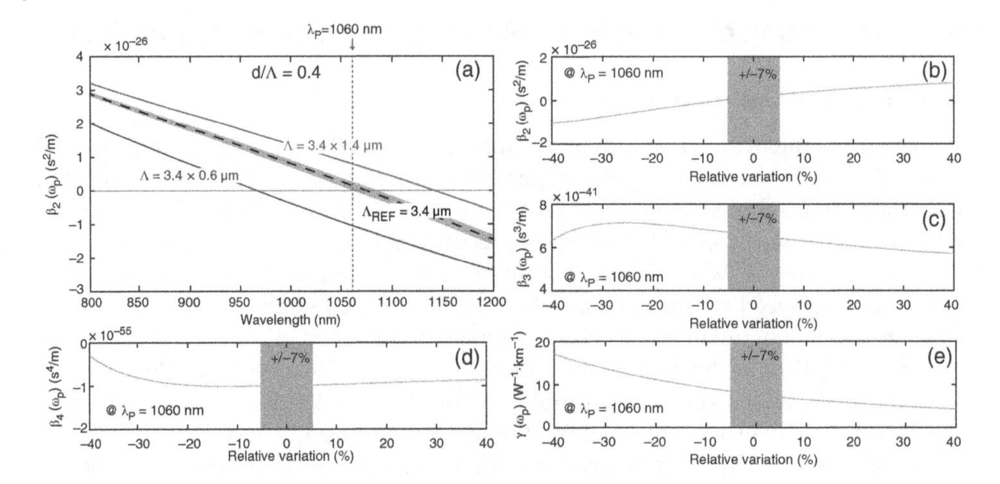

Figure 4.3 (a) GVD evolution versus wavelength for different pitch values, Λ_{REF} = 3.4 μm (dashed black lines), Λ_{MAX} = 4.76 μm (solid blue line) and Λ_{min} = 2.04 μm (solid red line). The gray area delimits a variation of ±7% of the pitch around the reference value. (b)-(e) Evolutions of $\beta_{2,3,4}$ and γ versus the normalized value of the pitch $[(\Lambda - \Lambda_{REF})/\Lambda_{REF}]$. For a color version of this figure please see color plate section.

respectively, as a function of the relative variation of the pitch $((\Lambda - \Lambda_{REF})/\Lambda_{REF})$. All these parameters are calculated at 1060 nm. The GVD experiences the largest variations compared to other parameters since it moves from negative to positive values. As can be seen in Figures 4.3 (a)–(e) these parameters are not linearly dependent on the pitch of the fiber over such large variations. However, limiting the amplitude of variation to ±7% of Λ_{REF}, we can reasonably consider that β_2, and other fiber parameters, are proportional to the pitch variations. This strongly simplifies the fabrication of dispersion oscillating fibers as the shape of the desired longitudinal variations of one physical parameter is similar to the outer diameter one (the control parameter in our drawing tower). As an example, the development of the GVD at 1060 nm of an optical fiber where the outer diameter is sinusoidally modulated with an amplitude of ±7% is represented in Figure 4.4. The period of modulation is $Z = 10$ m and we can see that the GVD also experiences a quasi sinusoidal evolution. Before performing experiments in

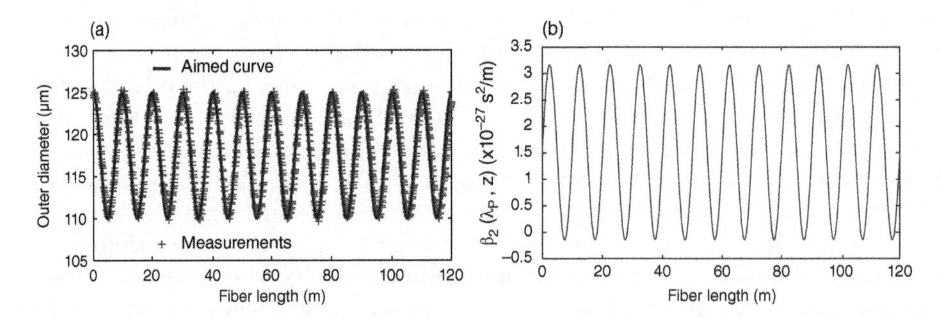

Figure 4.4 (a) Outer diameter development with ±7% variations, targeted curve (solid line), measurements (crosses). (b) Calculated GVD at 1060 nm.

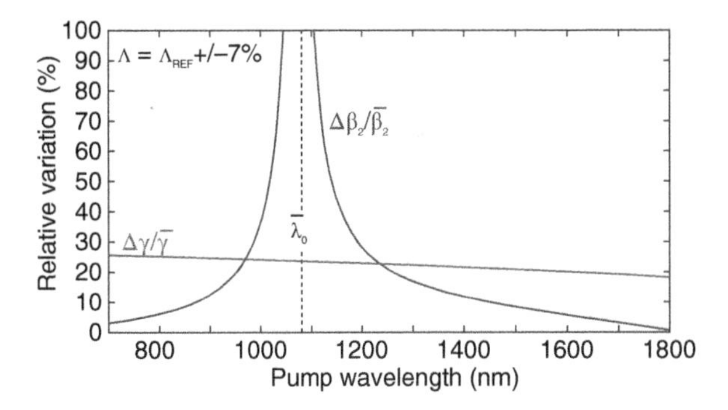

Figure 4.5 Relative variation of the GVD [$\frac{\beta_2^{Max}-\beta_2^{min}}{\overline{\beta}_2}$] and of the nonlinear coefficient [$\frac{\gamma^{Max}-\gamma^{min}}{\overline{\gamma}}$] as a function of the pump wavelength.

these fibers, one question is: what is the simplest model to use to account for nonlinear effects in such periodically modulated systems? It would be possible to account for all the longitudinal variations in the NLSE equation, however, this would lead to a complex equation unuseful to explain the dynamics of nonlinear processes. In order to identify the simplest equation, we have to establish if the modulation amplitude of a parameter is dominant with regards to other ones. To do so, the relative variations of β_2 are compared to those of γ as a function of the wavelength in Figure 4.5 for very small variations of the pitch of $\pm 7\%$ around the reference value of the pitch ($\Lambda_{REF} = 3.4$ μm). As can be seen, the relative variation of the nonlinear coefficient is almost constant while the wavelength is varied from 700 nm to 1800 nm. On the contrary, the relative variation of β_2 experiences large fluctuations on this wavelength span. It tends toward infinity close to the average ZDW ($\overline{\lambda}_0$) of the fiber and approaches zero for large wavelength shifts. This makes an important difference when modeling the behavior of a wave propagating in such fibers. We checked numerically that close to the average ZDW, the GVD variations dominate and one can consider that the nonlinear coefficient is constant. This leads to this simplified NLSE equation, which accurately reproduces the dynamics of the system:

$$i\frac{\partial u}{\partial z} - \frac{\beta_2(z)}{2}\frac{\partial^2 u}{\partial t^2} + \gamma|u|^2 u \simeq 0. \tag{4.16}$$

On the contrary, working far away from the average ZDW of the fiber is equivalent to considering that the contribution of the longitudinal variations of the nonlinear coefficient are dominant compared to those of the dispersion. The simplified NLSE equation thus reads:

$$i\frac{\partial u}{\partial z} - \frac{\beta_2}{2}\frac{\partial^2 u}{\partial t^2} + \gamma(z)|u|^2 u \simeq 0. \tag{4.17}$$

In all the experimental results presented in this chapter, we always manage to work in the vicinity of the average ZDW of the fiber. It obtains the largest amplitude variations of one of the guiding parameters while keeping relatively small outer diameter variations, that strongly relaxes fabrication constraints. The simplest governing equation to accurately model our system will be then Eq. (4.16).

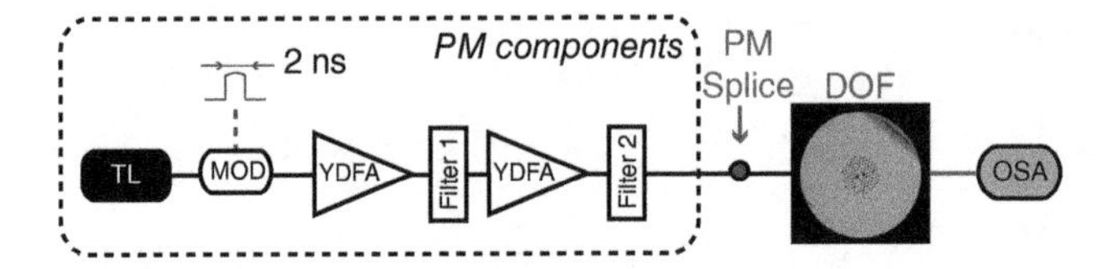

Figure 4.6 Scheme of the experimental setup.
Key: MOD: intensity modulator; YDFA: ytterbium-doped fiber amplifier; DOF: dispersion oscillating fiber; PM: polarization maintaining; TL: tunable laser; OSA: optical spectrum analyzer.

4.4 Experimental Results

4.4.1 Experimental Setup

Most of the experimental results have been obtained by using the experimental setup that is schematized in Figure 4.6. This is a typical master oscillator power amplifier (MOPA) configuration. The pump system is made of a continuous-wave tunable laser (TL) diode that is sent into an intensity modulator (MOD) in order to shape 2 ns square pulses at 1 MHz repetition rate. These pulses are both short enough to avoid the stimulated Brillouin scattering effect, and long enough to consider that the dynamics lie in the continuous regime. Typically, a thousand modulation periods are generated within a pump pulse. They are then amplified by two ytterbium-doped fiber amplifiers (YDFAs), at the output of which two successive tunable filters are inserted to remove the amplified spontaneous emission in excess around the pump. These quasi-continuous wave (QCW) laser pulses are then launched along the birefringent axis of the fiber. The neutral axis of the output fiber of the MOPA is aligned along one of the dispersion oscillating fibers (DOF) by using a polarization maintaining fusion splicer. Typical insertion losses are below 1 dB. The pump wavelength can typically be tuned between 1045 nm and 1075 nm, allowing the average group velocity dispersion during our experiments to be finely tuned.

4.4.2 First Observation of Multiple Simultaneous MI Side Bands in Periodically Modulated Fibers

The first observation of MI in periodically modulated fibers was achieved in a sinusoidally modulated optical fiber [12]. The evolution of the outer diameter as a function of fiber length measured during the drawing process is shown in Figure 4.7 (a). It has a modulation period Z of 10 m, the modulation amplitude corresponds to $\pm 7\%$ of the average fiber diameter (117 μm), and the fiber length is 120 m (12 modulation periods). The group velocity dispersion at $\lambda_P = 1072$ nm has been calculated from [24]. The amplitude of modulation of the outer diameter is weak enough ($\pm 7\%$) to consider that the GVD varies linearly as a function of the outer diameter. The average GVD dispersion at the pump wavelength is positive ($\beta_2 = 1.2 \times 10^{-3}$ ps^2/km) and the amplitude of modulation is equal to 1.5×10^{-3} ps^2/km. It is important to point out that the GVD is almost positive all along the fiber. As a consequence, no MI side lobe is expected to grow without periodic modulation of this parameter. By launching the quasi-CW pump laser with a peak power of 6.5 W in this fiber, more than 10 MI side lobes appear on

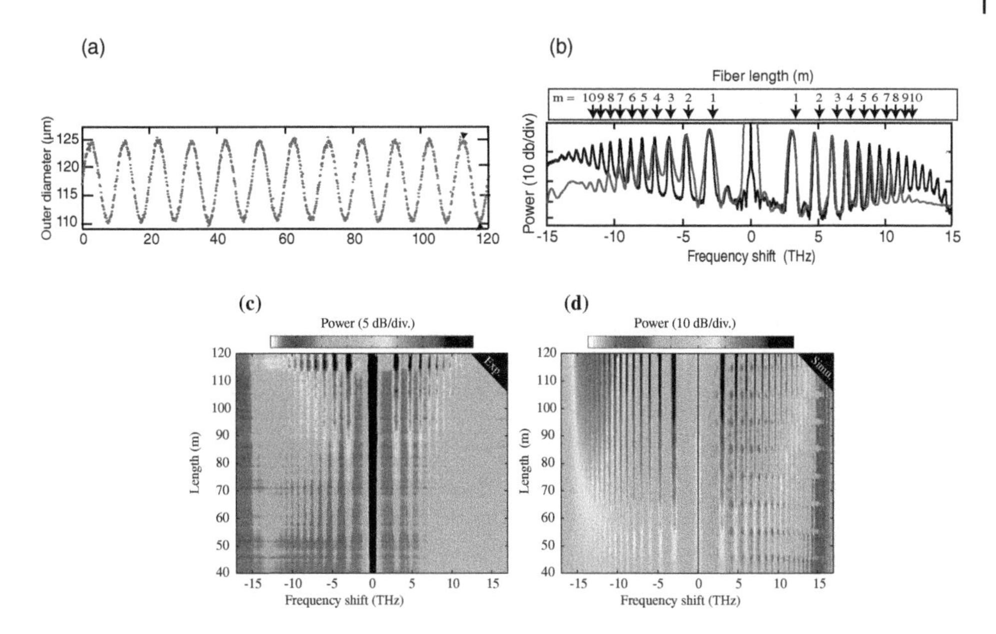

Figure 4.7 (a) Evolution of the outer diameter versus fiber length measured during drawing. (b) Output spectra from experiments (red line) and from numerical simulations (black line). (c) Experimental and (d) numerical dynamics of the structured spectrum formation in the DOF. For a color version of this figure please see color plate section.

both sides of the pump. Their position evolves as the square root of m, m being the side lobe order. A pretty good agreement is obtained with the theoretical predictions derived from the QPM theory (Eq. (4.11), indicated by arrows), and with numerical simulations (see [12] for details). The spectral dynamics of this process is illustrated in Figures 4.7 (c) and (d), where the development of the spectrum over the DOF length was experimentally recorded (by sequentially cutting back the fiber) and numerically studied. All side lobes start to grow from the beginning of the DOF, as in a standard phase-matched FWM process, but due to the experimental noise floor, the measurement was only possible from 40 m. These results constitute the first experimental observation of multiple sideband pairs over a wide spectrum through MI in an optical fiber with periodic dispersion.

4.4.3 Impact of the Curvature of the Dispersion

The impact of the curvature of the dispersion on MI process in uniform fibers was theoretically investigated a while ago [5] and confirmed experimentally a few years later [6, 7]. These works revealed that this higher-order dispersion term (β_4) must be considered when working in the low dispersion region of optical fibers. In other words, when the frequency shift of the side lobes is large, the contribution of β_4 becomes important as it scales quartically with the frequency shift. In that case, MI side lobes appear on both sides of the pump while pumping in the normal dispersion region of the fiber (provided that β_4 is negative). In DOFs, the contribution of this term is also important in the low dispersion region, as it has been shown in [15, 25].

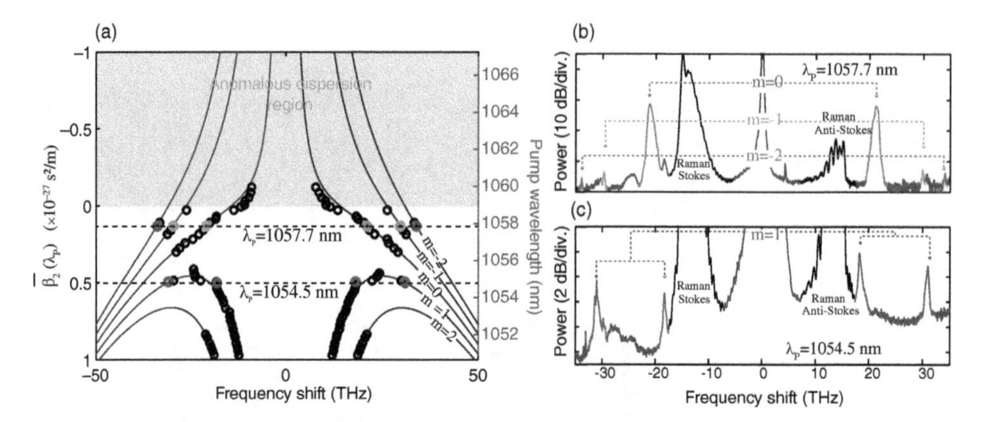

Figure 4.8 (a) QPM curves calculated from Eq. (4.18) (solid line) and measurement of MI sideband frequencies done by tuning the pump wavelength (markers). Colored markers highlight frequencies appearing in the experimental spectra shown in (b) and (c). For a color version of this figure please see color plate section.

In periodic fiber systems, QPM is achieved if the phase mismatch integrated over one period is an integer multiple of 2π (Eq. (4.11)). This equation can be modified to account for the contribution of β_4. It now reads:

$$\overline{\beta}_2\omega_m^2 + \frac{\overline{\beta}_4}{12}\omega_m^4 + 2\gamma P = \frac{2\pi m}{Z}, \tag{4.18}$$

where $\overline{\beta}_4$ is the average value of $\beta_4(z)$.

The combined effect of β_4 and of axially periodic dispersion leads to two different behaviors, one similar to the one observed in uniform fibers leading to the generation of MI sidelobes, whatever the sign of dispersion, and a second one that is specific to DOFs leading to the generation of a new family of MI frequencies. This is illustrated in Figure 4.8 in the color plate section. The fiber used has a sine modulation shape with period 1 m and amplitude $\pm 1.1 \times 10^{-3}$ ps^2/km. Other parameters can be found in [15].

The development of the QPM relation acounting for $\overline{\beta}_4$ is represented in Figure 4.8 (a). For the sake of clarity, we limit our representation to $m = \pm 2$ since they summarize the new dynamics of the process, and other m values lead to similar behaviors. Two main features related to the inclusion of β_4 in the QPM relation can be observed from Figure 4.8 (a). First, for $m \leq 0$, solutions are found in both the normal and anomalous dispersion regimes, while they only exist in anomalous dispersion when $\overline{\beta}_4$ is neglected. This phenomenon is analogous to the MI process assisted by β_4 in uniform fibers (i.e., for $m = 0$) in which the negative FOD term contributes to the linear phase mismatch to compensate for the positive nonlinear one [5–7]. Second, QPM curves obtained for $m > 0$ values exhibit two branches, i.e., two solutions are found for a fixed $\overline{\beta}_2$, while only one is expected when β_4 is neglected. The solution with the largest shift from the pump on each branch is directly linked to the presence of the $\overline{\beta}_4$ term. Typical spectra are displayed in Figure 4.8 (b) and (c). The pump component has been cut in order to clearly see the MI sidebands whose intensity is 50 dB or more below the pump. Figure 4.8 (b) focuses on $m \leq 0$ values. It shows the MI spectrum obtained for a 1057.7 nm pump, corresponding to a $\overline{\beta}_2$ value of 0.13 ps^2/km (normal average dispersion). It highlights the generation of $m = 0; -1; -2$ sideband pairs, respectively, in orange, green, and purple lines. Orange peaks correspond to $m = 0$ and therefore originate from the FOD term alone. They correspond to the same MI solutions as the one observed in uniform fiber [5–7]. MI sidebands displayed in green and purple (corresponding to k values of -1 and -2, respectively) arise from a combination of the FOD and of the periodic dispersion map. Indeed, the $\overline{\beta}_4$ term allows for these higher-order MI modes ($m \leq 0$) to exist while pumping in the normal dispersion region, similarly to the $m = 0$ case. Note that the peaks depicted in black correspond to Stokes and anti-Stokes stimulated Raman scattering (SRS) bands. Figure 4.8 (c) focuses on $m > 0$ values. It shows the simultaneous generation of two sideband pairs (in red lines, with arrows pointing to them) by increasing the $\overline{\beta}_2$ value to 0.49 ps^2/km (corresponding to 1054.5 nm pump wavelength). This feature is slightly different from the previous one in the sense that it is responsible for the generation of a second set of solutions for a given average dispersion (the ones with the largest frequency shifts from the pump, see Figure 4.8 (a)). As a consequence, this new family of unstable MI frequencies is a unique feature of DOFs when higher-order dispersion terms are accounted for. Measured MI peaks are depicted by markers in Figure 4.8 (a) and are in excellent agreement with the QPM curves obtained from Eq. (4.18). In summary, we have demonstrated that fourth-order dispersion plays an important role in the MI process in DOFs. This additional term leads to two different behaviors, one similar to the one observed in uniform fibers leading to the generation of MI side lobes whatever the sign of dispersion (provided $\overline{\beta}_4 < 0$), and a second one which is specific to DOFs. It corresponds to the generation of a new family of MI frequencies arising from a combination of fourth-order dispersion and longitudinal periodic dispersion. Our experimental results are confirmed by a relatively simple theoretical analysis based on a quasi-phase-matching process.

4.4.4 Other Modulation Formats

As illustrated in the previous section, our drawing tower capability allows us to fabricate DOFs with a large variety of longitudinal profiles. We therefore decided to investigate DOFs with more complex modulation formats than the basic sinusoidal one. Here we show two different cases: the first one consists in just adding a frequency to the basic sine modulation format and investigating how it impacts the position of the MI side lobes [14]. The simplest way to do this consists in performing an amplitude modulation of the basic sinusoidal modulation format with another sine wave having a slightly different period. In the second case, our motivation was to show that it is possible to fabricate a periodic train (or comb) of Dirac delta spikes because this is a fundamental and widespread modulation format encountered in a variety of physical systems [21]. These first experiments could thus constitute a first step before considering fiber systems as an interesting test bed to investigate kicked systems from a fundamental point of view [26].

4.4.4.1 Amplitude Modulation

Figures 4.9 (a)–(d) summarize the results obtained in an amplitude modulated DOF. For further details, refer to [14]. The 2D map shown in Figure 4.9 (a) represents the development of the MI spectrum as a function of the Z_1/Z_2 ratio from numerical simulations

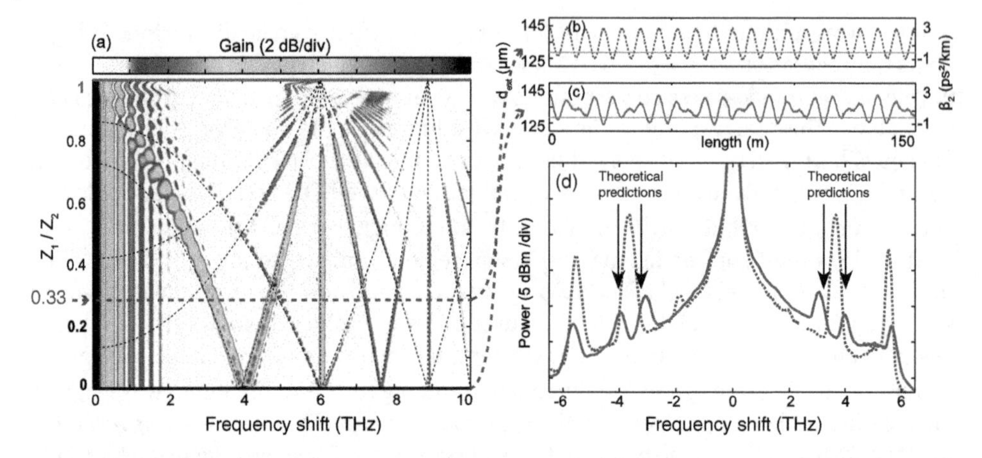

Figure 4.9 (a) Gain spectrum in amplitude modulated fibers as a function of the Z_1/Z_2 ratio (Z_1 is fixed and equal to 7.5 m, Z_2 varies from ∞ to Z1). The dotted lines represent the predicted positions of the side lobes maximum by Eq. (4.7) in [14]. (b) The reference oscillating fiber. (c) The amplitude modulated fiber (dispersion scale is calculated for $\lambda_P = 1057.5$ nm). (d) Experimental output spectra of the reference fiber (dotted line) and the amplitude modulated fiber (solid line). The black arrows point the predicted positions of the first two sidelobes in the amplitude modulated fiber (Eq. (4.7) in [14]).

[14]. Z_1 is fixed and equal to 7.5 m, Z_2 varies from ∞ to Z_1. It reveals a complex behavior with the appearance of new sidelobes compared to the simple sinusoidal modulation format (bottom of Figure 4.9 (a)). More precisely, all these original sidelobes split into multiple ones of lower gain when the modulation period decreases. In particular, Figure 4.9 clearly shows that the first [second] sidelobe splits into 2 [3] sidelobes even with a modulation period as long as 75 m ($Z_1/Z_2 = 0.1$). For $Z_2 = 45$ m (e.g., $Z_1/Z_2 = 0.33$), the gain spectrum is composed of 11 sidelobes instead of 4 without amplitude modulation of the dispersion. All these observations are in excellent agreement with the predictions of the model developed in [14] (Eq. (7)) and based on a truncated tree wave model of MI [27]. These theoretical results are superimposed in dashed lines in Figure 4.9 (a). Finally, we can state that the amplitude modulation of the dispersion dramatically modifies the modulation instability gain spectrum with the generation of new sidelobes, the positions of which depend on the ratio Z_1/Z_2. In order to experimentally validate these results, we designed an amplitude modulated DOF as well as a basic sinusoidal DOF which will serve as a reference. We optimized the parameters of the system in order to clearly observe the splitting of the first sidelobe in the MI spectrum. The modulation period of the reference fiber (pure sinusoidal modulation) is $Z_1 = 7.5$ m, and we chose $Z_2 = 45$ m ($Z_1/Z_2 = 0.33$) in order to unambiguously observe the side lobe splitting (Figure 4.9 (d)). As can be seen in Figure 4.9 (d), multiple sidelobes appear on both sides of the pump in the reference fiber (dotted line). In the amplitude modulated DOF (Figure 4.9 (c)), two new sidelobes surround the original one (solid line). They exhibit lower intensity than in the reference fiber which is consistent with our theory. To confirm their origin, we estimated their position by using the simple analytical model developed in [14] (Eq. (7)). The experimental positions of the first two sidelobes in the reference fiber allow us to us estimate the values of the average dispersion $\bar{\beta}_2$ and of the nonlinear term $2\gamma P$ thanks to the quasi-phase matching relation Eq. (4.11). Based on

these values, we were able to calculate the detunings for the split sidelobes when adding the amplitude modulation of the dispersion using our analytical model. These predicted positions (black arrows) fit very well with the experimental results, which confirms their origin. Thus, adding a single frequency in the modulation format of the DOF leads to a relatively complex behavior in the MI spectrum. Specifically, it induces a splitting of the original quasi-phase matched sidelobes into many ones whose position and gain depend on the ratio between the short and the long period of oscillation.

4.4.4.2 Dirac Delta Spikes

In order to approximate the ideal Dirac delta comb, we fabricated longitudinally modulated optical fibers by a periodic sequence of Gaussian functions. The evolution of the outer diameter is represented in Figure 4.10 (a). Their full width at half maximum is about 0.5 m. We designed three different fibers, with an increasing amplitude modulation (fibers labeled A, B and C respectively) keeping the same width as well as the same modulation period. Our motivation was to confirm experimentally that the side band gain of the first side lobe increases with the Dirac delta weight, as was predicted theoretically in [21]. Varying the outer diameter from 172 μm (fiber A) to 240 μm (fiber C) leads to a corresponding variation of GVD from 6 ps^2/km to 11 ps^2/km [21].

The spectrum recorded at the output of the periodic fiber is displayed by the curve in Figure 4.10 (b). Two MI side lobes, located at ±4.8 THz, appear on both sides of the pump. These experimental results have been compared with numerical simulations performed by integrating the generalized NLSE. As shown by the curve in Figure 4.10 (c), two symmetric MI side lobes also appear in the computed spectrum, in a very good agreement with the experiments. Their positions have been also compared with the predictions given by Eq. (4.9) [see dashed vertical lines in Figure 4.10 (c)], showing an excellent agreement. In order to show that the MI gain is larger when the weight of the Dirac delta function increases, we performed similar experiments in fibers B and C where the areas of the Gaussian pulses are larger than in fiber A. The amplitudes of the first MI

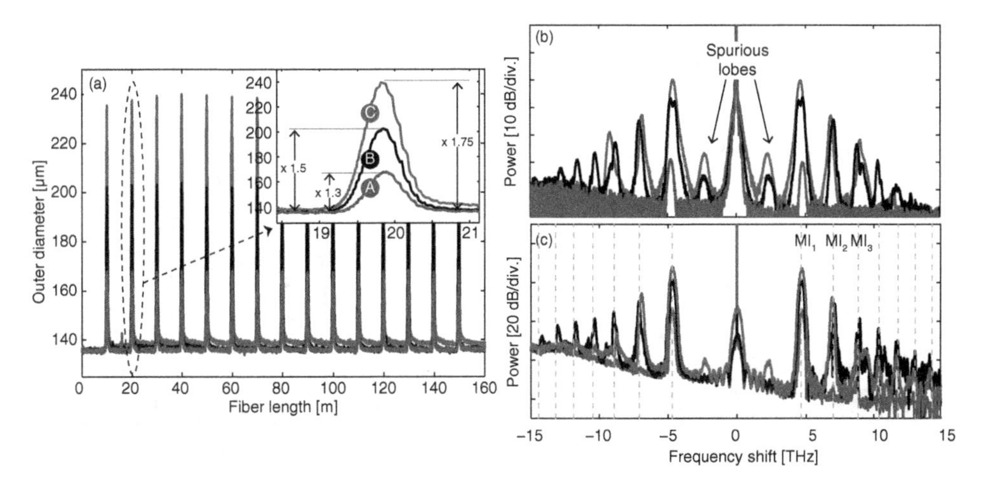

Figure 4.10 (a) Outer diameter evolution versus fiber length for fibers A, B, and C. Inset: snapshot on one leitmotif. Spectra recorded at the output of fibers A, B, and C: (b) experiments; (c) numerical simulations.

side lobes generated in fibers B and C are indeed larger compared to fiber A, as predicted by the theory. This is in pretty good agreement with numerical simulations [see Figure 4.10 (c)]. These experimental results show that MI can be observed in dispersion-kicked optical fibers. Specifically, it was shown that increasing the weights of the Dirac functions leads to larger MI gains for the first MI side lobe. We exploit the fact that the Dirac delta comb can be well approximated by a series of short Gaussian pulses in order to perform an experimental investigation using PCFs. Besides the specific interest in the engineering of parametric frequency generation in fibers, these results illustrate the fact that a fiber-based setup constitutes an interesting platform to carry out experimental investigations of fundamental phenomena such as the parametric resonance with accurate control of the shape of the perturbation.

4.4.4.3 DOFs and Fiber Optic Parametric Amplification?

MI leads to the amplification of a small perturbation provided that a phase matching or quasi-phase matching relation is fulfilled. In all the previous results, MI has been seeded by noise corresponding to random initial conditions. However, if the process is seeded by a weak coherent laser (a small signal), it will be amplified through the MI process and in that case we usually refer to fiber optical parametric amplification [28]. These amplifiers have been widely investigated in the context of telecommunication applications because of their large bandwidth and large gain of amplification [28]. Seeding a DOF by a small coherent signal will lead to the amplification of this signal without additional distortions compared to uniform fibers. We performed these experiments in a DOF, the parameters of which are listed in Figure 4.11. The experimental setup is similar to the one used for MI experiments, except that we combined a small signal with the pump at the input of the fiber with the same state of polarization. The on/off gain was measured at the output of the fiber by using an optical spectrum analyser (see pump on/off spectra in Figure 4.11 (a)).

The development of the experimental gain curve is represented in Figure 4.11 (b) in circles. We can see that important gain values around 50 dB can be achieved in DOFs on

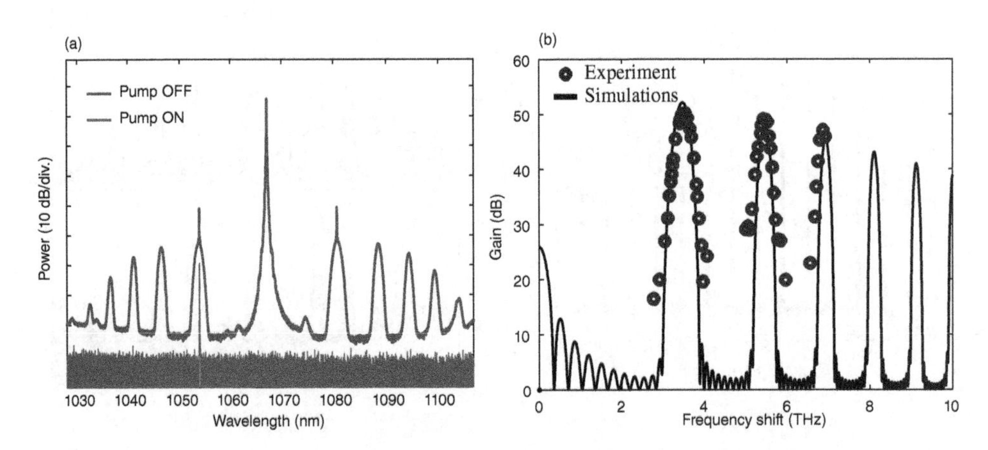

Figure 4.11 (a) Pump on/off output spectra. (b) Gain curves, experimental results (circles), numerical simulation (solid line). Parameters: pump power 26 W, signal power 10 μW, fiber length 120 m, period of modulation 10 m, amplitude of modulation 2×10^{-27} s^2/m and linear loss 8 dB/km.

the top of the first QPM MI side lobe. An excellent agreement is achieved with numerical simulations (solid black line). These basic experiments illustrate that these systems can indeed be used as amplifiers, similarly to uniform fibers. The drawback is that the gain band is neither flat nor broad, which are important characteristics for applications. To overcome these limitations, we designed a special DOF in which the period and amplitude of modulation have been optimized with a genetic algorithm. We have been able to report a nearly flat (7 dB of ripple), wide gain band (12 THz) amplifier with more than 60 dB of gain [20] that can be useful for applications.

4.5 Conclusion

This chapter gives an overview of the works related to modulation instability in dispersion oscillating fibers. We have developed a quite unique tool to fabricate optical fibers with a large variety of longitudinal profiles of modulation. Its allows us to fabricate photonic crystal fibers with basic modulation formats but also with more complex ones such as Dirac delta comb. Using a standard MOPA setup, we have been able to investigate in detail the modulation instability process in these periodically modulated waveguides. We report the first experimental evidence of MI in a sinusoidally modulated fiber with more than 10 quasi-phase matched side lobes. The impact of the higher-order dispersion term has been investigated and investigations in a more complex modulated waveguide have been achieved. Finally, we proved experimentally that these fibers can be used to develop ultra-wide and flat gain amplifiers, useful for applications. More generally, the modulation of the outer diameter of optical fibers provides a new degree of freedom in nonlinear optics. We illustrate it here in the context of modulation instability, but we also investigate the behavior of solitonic pulses propagating in these waveguides [13,29–32].

Acknowledgments

People who have contributed to this work: Andrea Armaroli, Fabio Biancalana, Géraud Bouwmans, François Copie, Stephan De Bièvre, Maxime Droques, Guillaume Dujardin, Gilbert Martinelli, Simona Rota-Nodari, and Stefano Trillo.

References

1 Zakharov, V.E. and Ostrovsky, L.A. (2009) Modulation instability: The beginning. *Physica D: Nonlinear Phenomena*, **238** (5), 540–548, DOI:10.1016/j.physd.2008.12.002.
2 Biondini, G. and Mantzavinos, D. (2016) Universal nature of the nonlinear stage of modulational instability. *Physical Review Letters*, **116** (4), 043 902, DOI:10.1103/PhysRevLett.116.043902.
3 Kibler, B., Chabchoub, A., Gelash, A., Akhmediev, N., and Zakharov, V.E. (2015) Superregular breathers in optics and hydrodynamics: Omnipresent modulation instability beyond simple periodicity. *Physical Review X*, **5** (4), 1–12, DOI:10.1103/PhysRevX.5.041026.

4 Zakharov, V.E. and Gelash, A.A. (2013) Nonlinear stage of modulation instability. *Physical Review Letters*, **111** (5), 1–5, DOI:10.1103/PhysRevLett.111.054101.

5 Cavalcanti, S.B., Cressoni, J.C., da Cruz, H.R., and Gouveia-Neto, A.S. (1991) Modulation instability in the region of minimum group-velocity dispersion of single-mode optical fibers via an extended nonlinear Schrödinger equation. *Physical Review A*, **43** (11), 6162, DOI:10.1103/PhysRevA.43.6162.

6 Pitois, S. and Millot, G. (2003) Experimental observation of a new modulational instability spectral window induced by fourth-order dispersion in a normally dispersive single-mode optical fiber. *Optics Communications*, **226** (1–6), 415–422, DOI:16/j.optcom.2003.09.001.

7 Harvey, J.D., Leonhardt, R., Coen, S., Wong, G.K.L., *et al.* (2003) Scalar modulation instability in the normal dispersion regime by use of a photonic crystal fiber. *Optics Letters*, **28** (22), 2225–2227, DOI:10.1364/OL.28.002225.

8 Haelterman, M., Trillo, S., and Wabnitz, S. (1992) Additive-modulation-instability ring laser in the normal dispersion regime of a fiber. *Optics Letters*, **17** (10), 745–747, DOI:10.1364/OL.17.000745.

9 Bronski, J.C. and Nathan Kutz, J. (1996) Modulational stability of plane waves in nonreturn-to-zero communications systems with dispersion management. *Optics Letters*, **21** (13), 937–9.

10 Smith, N.J. and Doran, N.J. (1996) Modulational instabilities in fibers with periodic dispersion management. *Optics Letters*, **21** (8), 570–572, DOI:10.1364/OL.21.000570.

11 Armaroli, A. and Biancalana, F. (2012) Tunable modulational instability sidebands via parametric resonance in periodically tapered optical fibers. *Optics Express*, **20** (22), 25 096–110, DOI:10.1364/OE.20.025096.

12 Droques, M., Kudlinski, A., Bouwmans, G., Martinelli, G., and Mussot, A. (2012) Experimental demonstration of modulation instability in an optical fiber with a periodic dispersion landscape. *Optics Letters*, **37** (23), 4832–4834, DOI:10.1364/OL.37.004832.

13 Conforti, M., Trillo, S., Mussot, A., and Kudlinski, A. (2015) Parametric excitation of multiple resonant radiations from localized wavepackets. *Scientific Reports*, **5**, DOI:10.1038/srep09433.

14 Copie, F., Kudlinski, A., Conforti, M., Martinelli, G., and Mussot, A. (2015) Modulation instability in amplitude modulated dispersion oscillating fibers. *Optics Express*, **23** (4), 3869–3875, DOI:10.1364/OE.23.003869.

15 Droques, M., Kudlinski, A., Bouwmans, G., *et al.* (2013) Fourth-order dispersion mediated modulation instability in dispersion oscillating fibers. *Optics Letters*, **38** (17), 3464–3467, DOI:10.1364/OL.38.003464.

16 Finot, C., Fatome, J., Sysoliatin, A., Kosolapov, A., and Wabnitz, S. (2013) Competing four-wave mixing processes in dispersion oscillating telecom fiber. *Optics Letters*, **38** (24), 5361, DOI:10.1364/OL.38.005361.

17 Kikuchi, K., Lorattanasane, C., Futami, F., and Kaneko, S. (1995) Observation of quasi-phase matched four-wave mixing assisted by periodic power variation in a long-distance optical amplifier chain. *IEEE Photonics Technology Letters*, **7** (11), 1378–1380, DOI:10.1109/68.473504.

18 Murdoch, S.G., Thomson, M.D., Leonhardt, R., and Harvey, J.D. (1997) Quasi-phase-matched modulation instability in birefringent fibers. *Optics Letters*, **22** (10), 682–684, DOI:10.1364/OL.22.000682.

19 Matera, F., Mecozzi, a., Romagnoli, M., and Settembre, M. (1993) Sideband instability induced by periodic power variation in long-distance fiber links. *Optics Letters*, **18** (18), 1499, DOI:10.1364/OL.18.001499.

20 Fourcade-Dutin, C., Bassery, Q., Bigourd, D., *et al.* (2015) 12 THz flat gain fiber optical parametric amplifiers with dispersion varying fibers. *Optics Express*, **23** (8), 10 103–10 110, DOI:10.1364/OE.23.010103.

21 Nodari, S.R., Conforti, M., Dujardin, G., *et al.* (2015) Modulational instability in dispersion-kicked optical fibers. *Physical Review A*, **92** (1), 013 810, DOI:10.1103/PhysRevA.92.013810.

22 Russell, P. (2003) Photonic crystal fibers. *Science*, **299** (5605), 358–362, DOI:10.1126/science.1079280.

23 Birks, T.A., Knight, J.C., and Russell, P.S.J. (1997) Endlessly single-mode photonic crystal fiber. *Optics Letters*, **22** (13), 961–963, DOI:10.1364/OL.22.000961.

24 Saitoh, K. and Koshiba, M. (2005) Empirical relations for simple design of photonic crystal fibers. *Optics Express*, **13** (1), 267–274, DOI:10.1364/OPEX.13.000267.

25 Armaroli, A. and Biancalana, F. (2014) Suppression and splitting of modulational instability sidebands in periodically tapered optical fibers because of fourth-order dispersion. *Optics Letters*, **39** (16), 4804–4807, DOI:10.1364/OL.39.004804.

26 Chirikov, B.V. (1979) A universal instability of many-dimensional oscillator systems. *Physics Reports*, **52** (5), 263–379, DOI:10.1016/0370-1573(79)90023-1.

27 Trillo, S. and Wabnitz, S. (1991) Dynamics of the nonlinear modulational instability in optical fibers. *Optics Letters*, **16** (13), 986–988, DOI:10.1364/OL.16.000986.

28 Marhic, M.E. (2007) *Fiber Optical Parametric Amplifiers, Oscillators and Related Devices*, Cambridge University Press, Cambridge.

29 Braud, F., Bendahmane, A., Mussot, A., and Kudlinski, A. (2015) Simultaneous control of the wavelength and duration of Raman-shifting solitons using topographic photonic crystal fibers. *JOSA B*, **32** (10), 2146–2152, DOI:10.1364/JOSAB.32.002146.

30 Bendahmane, A., Braud, F., Conforti, M., Barviau, B., Mussot, A., and Kudlinski, A. (2014) Dynamics of cascaded resonant radiations in a dispersion-varying optical fiber. *Optica*, **1** (4), 243–249, DOI:10.1364/OPTICA.1.000243.

31 Bendahmane, A., Mussot, A., Szriftgiser, P., *et al.* (2014) Experimental dynamics of Akhmediev breathers in a dispersion varying optical fiber. *Optics Letters*, **39** (15), 4490–4493, DOI:10.1364/OL.39.004490.

32 Wang, S.F., Mussot, A., Conforti, M., *et al.* (2015) Optical event horizons from the collision of a soliton and its own dispersive wave. *Physical Review A*, **92** (2), 023 837, DOI:10.1103/PhysRevA.92.023837.

5

Pulse Generation and Shaping Using Fiber Nonlinearities

Christophe Finot[1] and Sonia Boscolo[2]

[1] *Laboratoire Interdisciplinaire CARNOT de Bourgogne, UMR 6303 CNRS-Université de Bourgogne - Franche-Comté, Dijon, France*
[2] *Aston Institute of Photonic Technologies, School of Engineering and Applied Science, Aston University, Birmingham, UK*

5.1 Introduction

The third-order $\chi^{(3)}$ optical nonlinearity in optical fibers is responsible for a wide range of phenomena such as third-harmonic generation, nonlinear refraction (Kerr nonlinearity), and stimulated Raman and Brillouin scattering [1]. In high-speed optical communications nonlinear effects generally degrade the integrity of the transmitted signal, but the same effects, when properly managed, can be used to realize a variety of optical functions that have practical applications in the field of lightwave technology. Nonlinear processes that have been exploited in demonstrations and applications with single-mode (SM) fibers include stimulated Brillouin and Raman scattering, as well as aspects of the Kerr effect variously called self-phase modulation (SPM), cross-phase modulation (XPM), four-photon (four-wave) mixing (FWM), and parametric gain.

Techniques for generating, controlling, manipulating, and measuring ultrashort optical pulses and specialized waveforms have become increasingly strategically important in many scientific areas including, among others, ultrahigh-speed optical communications, optical signal processing, and biophotonics. Picosecond and femtosecond pulse shaping in the optical domain have been extensively implemented using the Fourier-domain approach [2], which employs spectral manipulation of the intensity and phase components of a pulse in order to create the desired field distribution. Chapter 9 by M. R. Fernández-Ruiz *et al.* gives a demonstration of the power and flexibility of such linear methods through various examples based on fiber grating devices. However, despite all its qualities, the linear pulse shaping strategy has the intrinsic drawback that the bandwidth of the output spectrum is determined by the bandwidth of the input spectrum. Indeed, a linear manipulation cannot increase the pulse bandwidth, and so in order to create shorter pulses nonlinear effects must be used. In addition, a linear pulse shaper can only subtract power from the frequency components of the signal while manipulating its intensity, thereby potentially making the whole process power-inefficient. The combination of third-order nonlinear processes and chromatic dispersion in optical fibers can provide efficient new solutions to overcome the aforementioned limitations.

Shaping Light in Nonlinear Optical Fibers, First Edition. Edited by Sonia Boscolo and Christophe Finot.

As we will see in this chapter, nonlinear fiber-based pulse shaping has become a remarkable tool to tailor both the temporal and spectral content of a light signal without the need for dedicated linear shapers.

In this chapter, we review recent progress in the research on pulse shaping using nonlinear effects in optical fibers. We would like to note that the main attention in this chapter will be focused on results obtained in our groups, and that it is not our intention here to comprehensively cover all the possible examples of fiber-based pulse shaping. The chapter is organized as follows: Section 5.2 introduces the key model governing the optical pulse propagation in a fiber. Section 5.3 emphasizes how ultrashort temporal structures can be generated by taking advantage of the fiber nonlinearity. Section 5.4 is devoted to the generation of advanced temporal waveforms in the normal dispersion regime of a fiber. Section 5.5 highlights some of the benefits offered by nonlinearity in terms of spectral shaping, namely, the possibility of tuning the central frequency of the pulses, of generating stable continua with high power spectral density, and of spectrally compressing the pulses. Finally, we conclude in Section 5.6.

5.2 Picosecond Pulse Propagation in Optical Fibers

A variety of nonlinear phenomena associated with optical pulse propagation in SM fibers is well described by the nonlinear Schrödinger (NLS) equation [3], in spite of the fact that this equation only includes two physical effects, namely, linear group velocity dispersion (GVD) and nonlinear SPM. It is important to note that though we will use mainly a silica optical fiber as a key example system to demonstrate various effects, these general nonlinear science concepts can be applied to a range of other nonlinear optical materials and waveguides. The NLS equation for the complex electric field envelope, $\psi(z,t)$, is written in its un-normalized form as

$$\psi_z = -\frac{i}{2}\beta_2\psi_{tt} + i\gamma|\psi|^2\psi, \tag{5.1}$$

where z is the propagation distance, t is the reduced time, β_2 and γ are the respective GVD and Kerr nonlinearity coefficients of the fiber. Parameters β_2 and γ can be functions of the propagation distance, as discussed in Chapter 4 by A. Mussot *et al.*

Higher-order linear effects in optical fibers, such as third- or fourth-order dispersion (FOD), can be ignored as long as the pulse wavelength is not in the vicinity of the zero-dispersion wavelength of the fiber. The effects of linear loss can be incorporated into Eq. (5.1) as a small perturbation term given the extremely low loss of silica fibers within the telecommunication wavelength window. Attenuation, however, comes into play in fibers based on other types of glass such as tellure or chalcogenide. When dealing with distributed fiber amplification, a modified NLS equation can be taken, which includes the effect of linear gain. When the polarization effects in fibers become important, Eq. (5.1) extends to a system of coupled NLS equations for the two polarization components. Examples of the richness of the resulting nonlinear dynamics are given in Chapter 1 by T. Hansson *et al.*

The NLS equation can also be readily generalized to include other nonlinear phenomena such as self-steepening, intra-pulse Raman scattering, two-photon absorption and photoionization effects. These are discussed in Chapter 3 by M. F. Saleh and

F. Biancalana. Higher-order gain effects such as gain bandwidth and saturation can also be incorporated into the model. Although all these effects can have a noticeable impact on pulses shorter than 1 ps, here we ignore them as the leading-order behavior is well approximated by Eq. (5.1). The impact of intra-pulse Raman scattering on pulse evolution will be briefly discussed in Section 5.5.2.

It is useful to normalize Eq. (5.1) by introducing the dimensionless variables: $U = \psi/\sqrt{P_0}$, $\xi = z/L_\mathrm{D}$, $\tau = t/\tau_0$, and write it in the form [3]

$$U_\xi = -\frac{i}{2}\mathrm{sgn}(\beta_2)U_{\tau\tau} + iN^2|U|^2U, \tag{5.2}$$

where the parameters τ_0 and P_0 are some characteristic temporal value and the peak power of the input pulse, respectively, $L_\mathrm{D} = \tau_0^2/|\beta_2|$ and $L_\mathrm{NL} = 1/(\gamma P_0)$ are the dispersion length and the nonlinear length, respectively, and the parameter N is introduced as $N^2 = L_\mathrm{D}/L_\mathrm{NL}$. From Eq. (5.2) it can be seen that the propagation equation depends on two characteristic lengths, each providing the length scale over which the corresponding phenomenon generates visible effects. Linear propagation regime occurs when $N^2 \ll 1$. In this case, pulse evolution can be simply determined by using the Fourier-transform method. As a result of GVD, the phase of each spectral component of the pulse is changed by an amount $\beta_2\omega^2 z/2$, where ω is the angular frequency of the pulse envelope. Even though such phase changes do not affect the pulse spectrum, they lead to temporal broadening of the pulse and a time dependence of the pulse phase, which can be characterized by the instantaneous frequency shift or chirp $\delta\omega = -\phi_t$. In the opposite case when $N^2 \gg 1$, the nonlinearity-dominant regime is applicable. As a result of the nonlinear term in Eqs. (5.1) or (5.2), upon propagation in the fiber, the pulse acquires an intensity-dependent nonlinear phase shift as $\phi_\mathrm{NL}(z,t) = \gamma|\psi(0,t)|^2 z$, namely the frequency chirp $\delta\omega = -(\phi_\mathrm{NL})_t$. In the absence of GVD and for an initially transform-limited pulse, the presence of a chirp causes a nonlinear expansion of the pulse spectrum, while the nonlinear nature of this chirp leads to spectral oscillations.

When the effects of chromatic dispersion are considered in combination with the Kerr nonlinearity, rich pulse dynamics arise from the interplay between dispersive and nonlinear effects depending on the sign of the dispersion and the relative magnitudes of the associated length scales. Sections 5.3 and 5.4 provide various examples of pulse evolution in both the anomalous ($\beta_2 < 0$) and normal ($\beta_2 > 0$) dispersion regimes of a fiber, including the generation of optical solitons and self-similar waveforms.

5.3 Pulse Compression and Ultrahigh-Repetition-Rate Pulse Train Generation

In this section, we review the generation of ultrashort pulses and high-repetition-rate pulse trains that can result from the nonlinear evolution of optical pulses in a fiber.

5.3.1 Pulse Compression

5.3.1.1 The Fundamental Soliton and its Adiabatic Evolution

A fascinating manifestation of the fiber nonlinearity occurs through optical solitons [4–6], formed in the anomalous dispersion regime of a fiber as a result of a cooperation

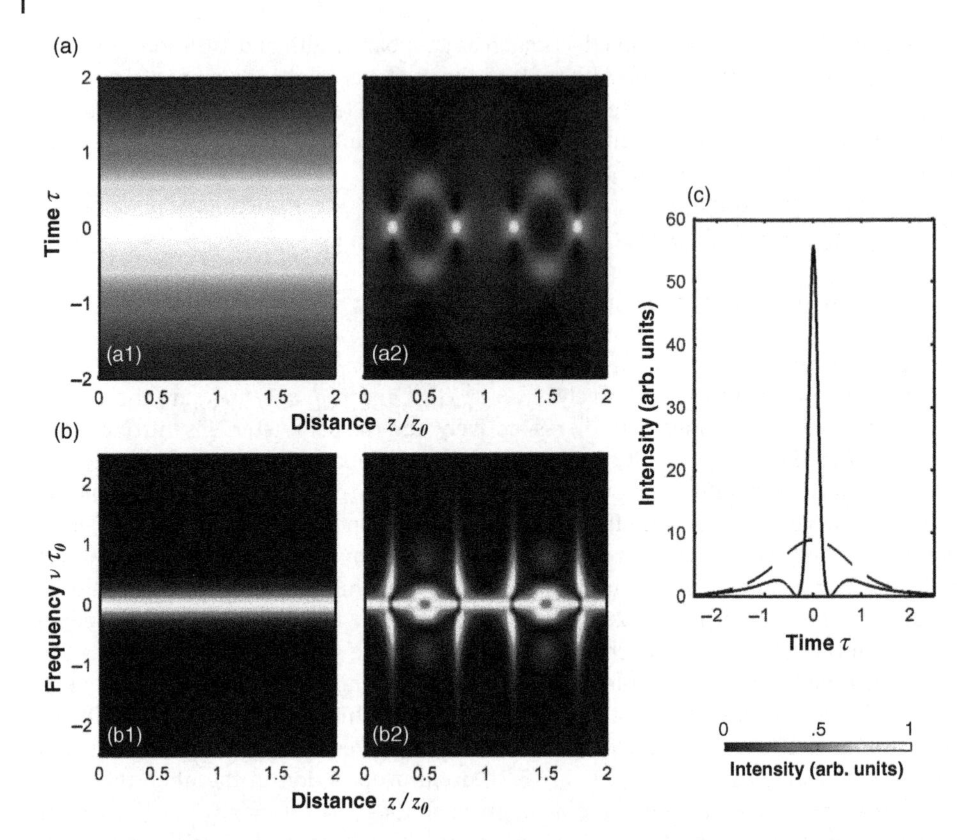

Figure 5.1 Longitudinal (a) temporal and (b) spectral evolution of fundamental (subplots 1) and third-order (subplots 2) solitons over two soliton periods, as obtained by numerical integration of the NLS equation. (c) Temporal intensity profile of the third-order soliton at the point of maximum compression ($z/z_0 = 0.32$; solid black). Also shown is the input hyperbolic-secant pulse (dashed gray).

between GVD and SPM. Solitons have been discovered and studied in many branches of physics [7, 8]. In the context of optical fibers, not only are solitons of fundamental interest but they have also found practical applications in the field of fiber-optic communications [4, 5, 9]. Optical solitons lie at the heart of other chapters of this book, incuding Chapter 3 by M. F. Saleh and F. Biancalana, and Chapter 7 by E. R. Andresen and H. Rigneault.

The NLS equation admits solutions in the form of solitons in the case where N is an integer number and β_2 is negative. The expression of the fundamental soliton (corresponding to $N = 1$) can be written as $U(\xi, \tau) = \text{sech}(\tau)\exp(i\xi/2)$. By this equation the fundamental soliton preserves its temporal and spectral shape during propagation, as highlighted by Figure 5.1 (panels a1, b1). This property derives from the interplay between GVD and the Kerr effect: the chirp induced by GVD is exactly compensated by the SPM-induced chirp, which has opposite sign. It is this feature of the fundamental solitons that makes them attractive for optical communication systems [4].

Since N must have a fixed value, to obtain a soliton (fundamental or high order), the pulse launched into the fiber must have an hyperbolic secant shape, and its power and

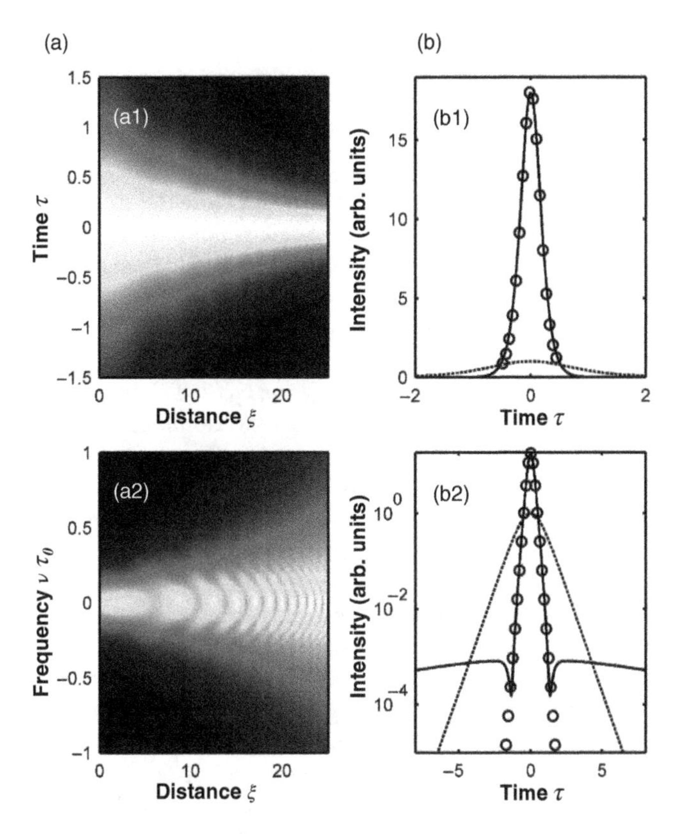

Figure 5.2 (a) Longitudinal temporal (subplot 1) and spectral (subplot 2) evolution of a fundamental soliton propagating in a distributed amplifier with a gain of 0.25 dB per dispersive length, as obtained by numerical integration of the NLS equation. The color scale is the same as that of Figure 5.1. (b) Input (dashed gray) and output (solid black) temporal intensity profiles on linear (subplot 1) and logarithmic (subplot 2) scales. Also shown is a hyperbolic-secant fit to the output profile (circles).

duration must be correlated. Using the definition of N, the following relation between the pulse energy E_s and soliton duration τ_0 is obtained for the fundamental soliton: $E_s = 2|\beta_2|/(\gamma\tau_0)$. This area theorem forms the basis for a technique to generate high-quality ultrashort pulses, which relies on the adiabatic evolution of a fundamental soliton that takes place in an anomalously dispersive fiber with longitudinally distributed gain (e.g., a Raman amplifier) (see also Figure 5.2) or decreasing dispersion (DD) [10–13]. Indeed, in the presence of gain, the temporal width of a fundamental soliton must decrease along the fiber length to compensate for the energy increase due to gain if the pulse has to preserve its soliton character. While physical distributed-amplification schemes require pumping along the fiber length, which entails an extra level of cost and complexity in the setup, soliton compression in a fiber can also be achieved by changing the dispersive properties of the fiber [10]. Indeed, the soliton area theorem indicates that for a fixed energy, a decrease of the GVD entails a shortening of the pulse duration. Step-like dispersion-profiled structures [14] using a set of short segments of conventional fibers with different dispersion values, or comb-like structures [15] where the use of a large number of fiber segments with alternate signs of dispersion approaches an average local dispersion with the desired decrease, have been successfully used to emulate continuous DD fiber profiles, whose realization requires special, not commercially available fiber-drawing technology.

5.3.1.2 Using Higher-Order Solitons

Given the soliton area theorem, the generation of high-energy fundamental solitons at a given temporal duration requires a high GVD coefficient or a low nonlinearity coefficient as provided, for instance, by hollow core fibers [16]. When N is higher than one, the exact balance between the GVD and SPM effects can no longer be maintained. Soliton theory shows that when the initial pulse shape is a hyperbolic secant, the two effects can cooperate in such a way that the pulse experiences a periodic evolution and reassumes its initial shape at multiple distances of the soliton period $z_0 = \pi L_D/2$ [17]. In the case of higher-order solitons, over a period of the evolution pattern SPM dominates initially but GVD soon catches up and leads to temporal pulse contraction. The evolution of a third-order ($N = 3$) soliton is shown in Figure 5.1 (panels a2, b2).

The stage of temporal compression occurring in the periodic evolution of higher-order solitons can be used to generate pulses with ultrashort durations. The higher the soliton number N of the input pulse, the greater the compression achievable using this approach. However, the quality of the compressed pulses is heavily compromised by non-negligible and temporally extended pedestals as seen in Figure 5.1 (c). These pedestals can be partly removed using nonlinear reshaping devices such as a nonlinear optical loop mirror [18]. Furthermore, given the scaling laws of the NLS equation, increasing the input pulse peak power leads to a decrease of the distance in the fiber at which compression takes place, while higher-order nonlinear and dispersive effects become important for pulse evolution. This may eventually make this method very sensitive to power fluctuations of the input pulse or noise. It should also be noted that as higher-order solitons can quickly reach a very large spectral extent, the validity of the simple NLS propagation model becomes questionable. Indeed, higher-order dispersive effects or intra-pulse Raman scattering may become important for pulse development and may lead to soliton fission. Feasibility of the higher-order soliton compression scheme was first shown in 1980 [9]. Various experimental demonstrations in either highly nonlinear (HNL) fibers at telecommunication wavelengths [19] or in photonic crystal (PC) fibers seeded by titanium-sapphire laser pulses [20] have been achieved since then, enabling the generation of pulses as short as a few optical cycles.

5.3.1.3 By a Combination of Normally Dispersive Fiber and Dispersive Element

A very different approach to ultrashort pulse generation relies on propagation in a HNL fiber segment that induces a large nonlinear chirp to the pulse, followed by a second element where the nonlinear chirp is compensated [21]. Typically, the HNL fiber has normal dispersion as this is beneficial to the development of a relatively linear chirp variation over the pulse. The second segment then provides the opposite anomalous chirp. This element can be strictly linear, such as a pair of diffraction gratings [22], a fiber Bragg grating, prisms [23], chirped mirror compressors [24], or an hollow core fiber [25]. It can also be nonlinear, in which case the input pulse power and length of propagation need to be carefully chosen [26]. Recent progress in the control of coherent continuum generation in normally dispersive (ND) fibers has enabled compression down to a few tens of femtoseconds [23, 24], and the use of this approach has also been demonstrated in the context of biology.

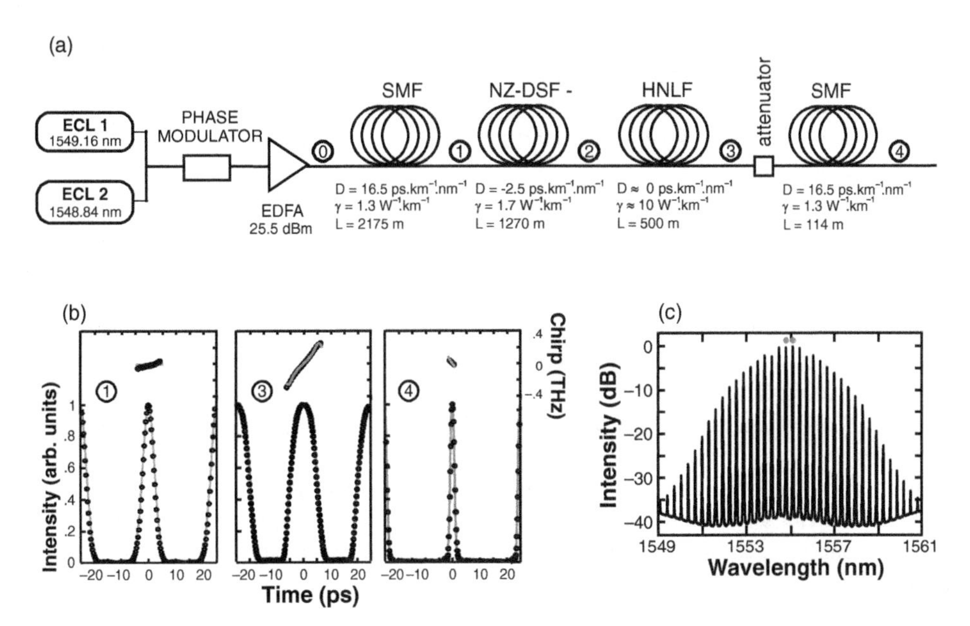

Figure 5.3 Generation of 40-GHz-repetition-rate pulse train using multiwave mixing compression of a dual-frequency beat signal and a multi-fiber-segment architecture. (a) Experimental setup. ECL: external cavity laser, EDFA: erbium-doped fiber amplifier, SMF: single-mode fiber, NZ-DSF: non-zero dispersion-shifted fiber (with negative dispersion), HNLF: highly nonlinear fiber. (b) Temporal intensity and chirp profiles at different stages obtained experimentally by FROG measurements (black circles). The experimental data are compared with the results of numerical simulations based on the NLS equation (gray). (c) Spectrum of the compressed pulse train after stage 3 or 4 (the two spectra are identical). The two input wavelengths are represented by circles. More details of the experiment can be found in [39]. *Source*: Adapted from Finot 2007 [39].

The main issue with this approach is to achieve exact compensation of the chirp induced by the nonlinear propagation stage. One way is to impart a chirp to the pulse with as much linear variation as possible in the first stage. As we will see in Section 5.4.2, parabolic pulses generated via nonlinear shaping in fiber are an attractive solution in this respect [25, 27, 28]. An example of this method is given in Figure 5.3 (b), where a parabolically shaped pulse with an enhanced linear temporal chirp obtained through propagation in a HNL fiber is linearly compressed in a segment of standard SM fiber. Another option is to use active linear shaping at the output of the nonlinear segment [24] to suppress higher-order residual chirp.

We note that several variants of this technique can be realized, including, for instance, inducing the nonlinear chirp by XPM with a co-propagating powerful additional signal [29, 30]. It is also possible to imprint a chirp on a continuous wave (CW) with a phase modulator driven with, e.g., sinusoidal modulation signals: the phase modulation will spectrally extend the pulse and linear compression may then compensate this [31, 32].

5.3.2 High-Repetition-Rate Sources

Today, the generation of pulse trains at repetition rates higher than several tens of gigahertz by optoelectronic methods represents a critical issue. Indeed, the bandwidth of optoelectronic devices usually limits their usage at repetition rates below 50 GHz.

Therefore, while it is possible to generate close-to-sinusoidal waves at such frequencies, it is not possible to generate pulse trains with low duty cycles. In this context, optical fibers can provide several technological solutions to transform a dual-frequency sinusoidal signal into a train of well-separated pulses.

5.3.2.1 The Process of Modulation Instability and Multiple Four-Wave Mixing

A nonlinear technique for generating high-repetition-rate ultrashort pulses based on modulation instability (MI) of an intense pump wave induced by a weak signal was first suggested in [33] and realized experimentally in [34]. MI in the anomalous dispersion regime of optical fibers manifests itself as breakup of the initial pump beam into a periodic pulse train whose period is inversely related to the frequency difference between the pump and signal waves, $\Omega = \omega_0 - \omega_1$, with the generation of two new spectral sidebands, and can be interpreted in terms of a FWM process that is phase-matched by SPM. Physically, the energy of two photons from the intense pump beam is used to create two different photons, one at the probe frequency ω_1 and the other at the idler frequency $2\omega_0 - \omega_1$. Since its first demonstration, this technique has been used to create optical sources capable of producing periodic trains of ultrashort pulses at high but controllable repetition rates.

When the first-order sidebands become strong, higher-order sidebands located at $\omega_0 \pm m\Omega$, $m = 2, 3, \ldots$ are created in a FWM process. An approach based on multiple FWM in a single fiber with anomalous constant dispersion has been demonstrated experimentally for the generation of compressed pulse trains from the propagation of a dual-frequency pump field, and repetition rates ranging from a few tens of gigahertz to a few terahertz have been achieved [35–39]. The initial beat signal in these experiments was obtained via temporal superposition of two CWs with slightly different frequencies delivered by external cavity lasers [36,37], which permitted higher repetition rates to be achieved than using a single CW directly modulated by an intensity modulator. Examples of phase and intensity measurements recorded with the frequency resolved optical gating (FROG) technique show that high-quality compressed pulses with a Gaussian intensity profile and nearly constant phase can be achieved at a typical duty cycle of 1:5 [39] (Figure 5.3 (c)). Lower duty cycles can still be obtained with a single fiber, but at the expense of degraded quality of the pulse train with the appearance of low-intensity pedestal components and a nonuniform phase across the pulses. Such satellites can be partly avoided by using specially designed arrangements of segments of fibers [38,39] or other comb-like structures [40]. It is worth mentioning here that similar experimental configurations can be used for the generation of rogue structures and solitons on a finite background as described in detail in Chapter 10 by B. Kibler.

5.3.2.2 Processes Based on External Gain or Phase Modulation

A different technique for transforming a sinusoidal beat signal into a train of well-separated pulses is based on the selective amplification of a weak CW signal by a sinusoidally intensity modulated pump. Fundamentally, in the presence of a strong instantaneous gain mechanism with exponential gain dependence on pump power such as FWM or stimulated Raman scattering, the temporal region of the initial signal where the intensity of the modulated pump is high will experience strong gain whereas the signal parts corresponding to low pump intensity will not be affected. The contrast induced by the gain difference between the low and high gain regions can be of several tens of

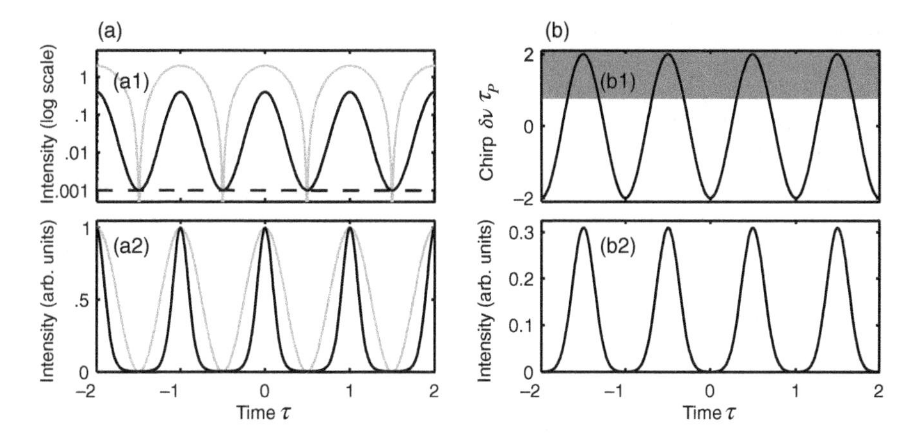

Figure 5.4 Numerical illustration of the principle of operation for high-repetition-rate pulse train generation schemes based on (a) exponential cross-gain, and (b) phase modulation and subsequent offset filtering of a CW signal. The sinusoidal pump (gray) is compared with the intensity profile of the amplified CW signal (black) in subplots a1 and a2. Subplot b1 shows the instantaneous frequency shift resulting from phase modulation of the CW signal. The transmission of the bandpass filter is represented by the gray shaded area. The resulting pulse train is shown in panel b2. The time variable is normalized to the temporal period τ_p of the input sinusoidal pump.

decibels. Therefore, after filtering the pump with an optical bandpass filter, the initially continuous seed is isolated and exhibits a pulse train structure [41,42]. Such an approach is illustrated in Figure 5.4, where the amplification of an initial CW signal by a sinusoidal pump provides short-duration pulses. Note that the position of the seed within the gain bandwidth and the saturation of the gain are crucial parameters that can affect the pulse shape, leading, for example, to Gaussian- or parabolic- or hyperbolic sine-shaped pulses [43, 44], or to pulse doubling in dispersion-oscillating fibers that are operated in the vicinity of the zero average dispersion wavelength [45]. Note also that because a gain process with a quasi-instantaneous response time as needed for ultrashort pulse generation is nonlinear and requires a strong pump, the amplification is often affected by XPM from the pump to the amplified seed signal. In order to reduce the effects of XPM, the walk-off between the pump and the amplified seed must be much shorter than the period of the pulse train generated. It is also worth mentioning here that the strong transfer of intensity fluctuations through XPM from the pump under low walk-off conditions has been identified as the main mechanism responsible for the occurrence of extreme events in a fiber amplifier: the fluctuations of a partially incoherent pump can indeed lead to the emergence of very rare but extremely intense optical spikes of light [46,47].

Another strategy to generate high-repetition-rate pulse trains relies on imparting a sinusoidal phase modulation to a CW signal. Two approaches are possible to this end. One approach is based on using a phase modulator to generate linear frequency chirp (within the time range where the sinusoidal phase modulation approximates a quadratic modulation), followed by a dispersive element to compensate the chirp. Strong compression of the modulated CW signal into pulses then occurs upon propagation in the dispersive element, and the temporal duration of the resulting pulses is directly related to the amplitude of the initial phase modulation. This is conceptually similar to the generation of short pulses described in Section 5.3.1.3, where the phase modulation is applied, not by an external linear phase modulator, but instead by SPM. The other approach relies on spectral selection of a phase-modulated signal as illustrated in Figure 5.4 (b): a phase modulator is used to introduce a time-varying frequency chirp of the CW signal, and then a spectral mask is applied to the signal, by use of an offset optical bandpass filter, to suppress all but the frequency extrema of the signal, resulting in a train of short pulses at a repetition rate set by the frequency of the sinusoid applied to the phase modulator [48, 49]. Alternatively, generation of short pulses can be achieved using nonlinear fiber self-phase spectral broadening and subsequent offset bandpass filtering [50]. The concept has many similarities with the all-optical regeneration scheme known as Mamyshev's regenerator [51].

5.4 Generation of Specialized Temporal Waveforms

We saw in Section 5.3 that the anomalous regime of dispersion of a fiber is particularly well suited for the generation of ultrashort pulses with durations well below that of the initial waveform. This section emphasizes how the interplay between normal dispersion and Kerr nonlinearity sustains the generation of advanced temporal waveforms, such as parabolic, rectangular, or triangular profiles, which are of interest for various applications in optical signal processing and manipulation [52].

5.4.1 Pulse Evolution in the Normal Regime of Dispersion

Contrary to propagation in the anomalous dispersion region on which soliton processes are dominant, in the regime of normal dispersion, pulse dynamics are highly affected by the phenomenon of optical wave breaking [21, 53–55]. In the nonlinearity-dominant regime of propagation, the combination of normal GVD and SPM makes a pulse broaden and change shape toward an almost trapezoidal pulse form with a linear frequency chirp variation over most of the pulse but with steep transition regions developed at the edges, as highlighted by Figure 5.5, which shows the temporal and spectral evolution of an initially unchirped Gaussian pulse with the normalized power $N = 6.7$. This evolution can be understood as arising from the frequency chirp, which makes the red- (blue-) shifted light near the leading (trailing) edge travel faster (slower) than the unshifted pulse parts, i.e., the pulse center and low-intensity wings. Ultimately, when the shifted light overruns the pulse tails, the wave breaks; oscillations appear in the wings of the pulse because of interference and, concomitantly, new frequencies are generated and side lobes appear in the pulse spectrum. Here we quantify the pulse evolution with the excess kurtosis parameter [56] and the parameter of misfit M_p between the pulse intensity profile $|\psi|^2$ and a compactly-supported pulse $|\psi_p|^2$ with a parabolic intensity profile of the same energy [57]:

$$M_p^2 = \int \left(|\psi(t)|^2 - |\psi_p(t)|^2 \right) \mathrm{d}t \Big/ \int |\psi(t)|^4 \,\mathrm{d}t. \tag{5.3}$$

The initial pulse form changes very fast during the first stages of propagation due to the strong action of SPM, as highlighted by the rapid variation of the excess kurtosis

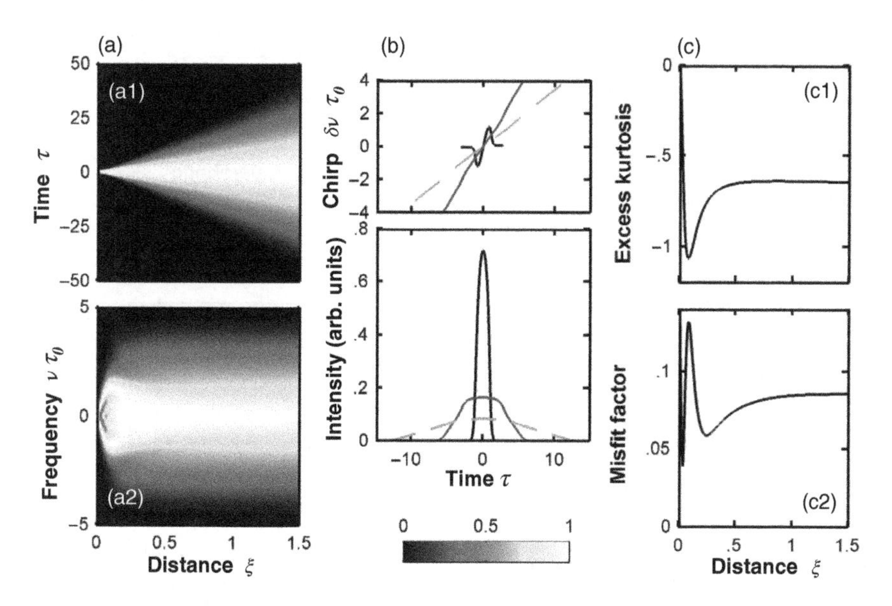

Figure 5.5 Evolution of an initially unchirped Gaussian pulse propagating in a ND fiber with $N = 6.7$, as obtained by numerical integration of the NLS equation. (a) Longitudinal evolution of the temporal and spectral intensity profiles. (b) Temporal intensity and chirp profiles at different propagation distances in the fiber: $\xi = 0.038$ (solid black), $\xi = 0.25$ (solid gray) and $\xi = 0.5$ (dashed gray). (c) Longitudinal evolution of the excess kurtosis parameter (subplot 1) and the misfit parameter to a parabolic shape (subplot 2).

(Figure 5.5 (c1)). Shortly after the distance ξ_{wb} at which wave breaking occurs [54] (here $\xi_{\text{wb}} \approx 0.04$), the spectral expansion of the pulse starts to saturate [55].

5.4.2 Generation of Parabolic Pulses

We saw in Section 5.4.1 that pulses that propagate in normal dispersion media are susceptible to distortion and breakup owing to optical wave breaking, which is a consequence of excessive nonlinear phase accumulation combined with dispersion. In 1993, Anderson *et al.* showed that a sufficient condition to avoid wave breaking is that a pulse acquires a monotonic frequency chirp as it propagates, and that wave-breaking-free solutions of the NLS equation (5.1) exist when the GVD is normal [54]. These are solutions of Eq. (5.1) in the asymptotic quasi-classical or Wentzel-Kramers-Brillouin (WKB) limit (i.e., the limit of high amplitude or small dispersion such that $\beta_2|(|\psi|)_{tt}|/(2\gamma|\psi|^3)$ can be ignored), and take the form

$$|\psi(z,t)| = a(z)\sqrt{1-\left(t/\tau_0(z)\right)^2}\theta\left(\tau_0(z)-|t|\right), \quad \arg\psi(z,t) = b(z)t^2 + \phi_0(z), \tag{5.4}$$

and thus have parabolic intensity profiles. Here, $\theta(x) = 1, x > 0; \theta(x) = 0, x < 0$ is the Heaviside function. Such a pulse maintains its shape, and is always a scaled version of itself, i.e., it evolves self-similarly. In contrast to solitons, self-similar pulses, which have been dubbed similaritons, can tolerate strong nonlinearity ($\phi_{\text{NL},0} = \int dz\,\gamma(z)P_0(z) \gg \pi$, where $\phi_{\text{NL},0}$ is the peak nonlinear phase shift) without distortion or wave breaking. The

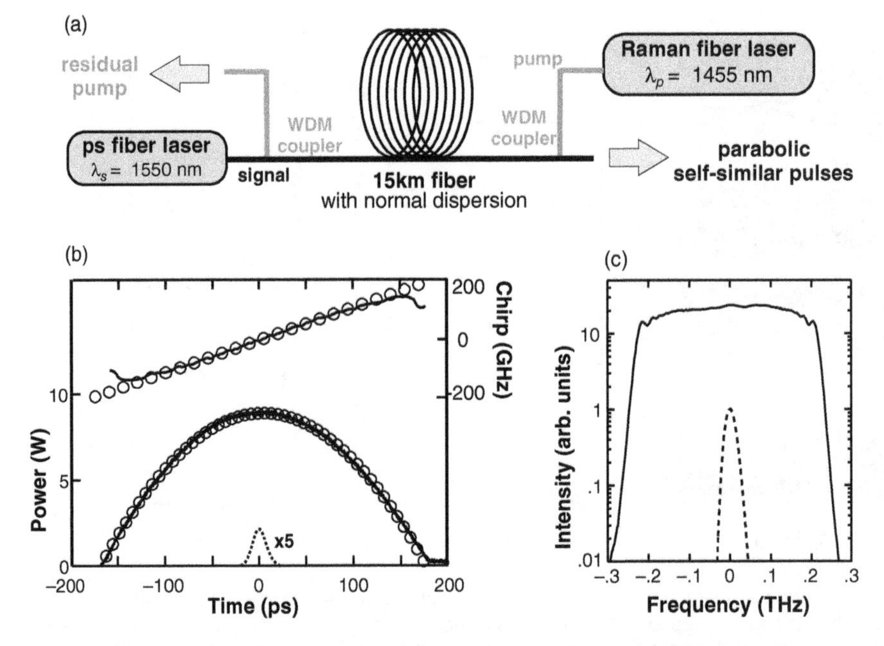

Figure 5.6 Generation of parabolic similaritons through Raman amplification at telecommunication wavelengths. (a) Experimental setup. WDM: wavelength-division multiplexing. (b) The temporal intensity profiles of the initial (dashed) and output (solid) pulses recorded in real time by a high-speed photodetector and oscilloscope are compared with a parabolic fit (circles). The chirp of the output pulse (solid) is compared with a linear fit (circles). (c) Spectra of the initial (dashed) and output (solid) pulses. More details of the experiment can be found in [67]. *Source*: Adapted from Hammani 2008 [67].

normal GVD effectively linearizes the accumulated phase of the pulse allowing the spectral bandwidth to increase without destabilizing the pulse.

Similaritons were demonstrated theoretically and experimentally in fiber amplifiers in 2000 [58], and they continue to attract much attention [59, 60]. Similaritons in fiber amplifiers are along with solitons in passive fiber the most well-known classes of nonlinear attractors for pulse propagation in optical fibers [58, 61–63], so they take on major fundamental importance. Various experimental demonstrations relying on amplification from either rare-earth doping [25,58,64,65] or Raman scattering [66,67] (Figure 5.6) have been achieved. Experimental interest has been strongly driven by the possibility of fully canceling the highly linear chirp developed during parabolic amplification and, thus, generating ultrashort high-power pulses with very low substructures [25,58,64,65]. Furthermore, recent fiber lasers that support similariton evolution in the passive [68] or gain [60, 69–72] fiber segment of the laser cavity have been demonstrated to achieve pulse energy and peak power performances much higher than those of prior approaches.

In addition to fiber amplifiers, similariton formation with nonlinear attraction can be achieved in passive fibers, provided a suitable longitudinal variation of the dispersion is introduced [73, 74, 102]. This approach is based on the observation that a longitudinal decrease of the normal dispersion is formally equivalent to linear gain, similar to the approach described in Section 5.3.1.1 for soliton adiabatic compression. A passive and simple approach to the generation of parabolic pulses that uses progressive nonlinear pulse reshaping in a ND fiber with fixed dispersion has also been demonstrated [76].

This method is illustrated in Figure 5.5 (b): the waveform that is formed at a short propagation distance in the fiber preceding the onset of wave breaking is, in fact, parabolic as indicated by low misfit value. However, as a result of the non-monotonic nature of the chirp, this parabolic waveform is not maintained with increasing propagation distance, hence, it represents a transient state of the nonlinear dynamic evolution of the pulse in the passive fiber medium [77]. As such, it has a finite life distance (characteristic length scale of the self-similar pulse dynamics) that depends on the initial conditions. This is in contrast to the asymptotic attracting solutions obtained in fiber amplifiers. Nevertheless, as shown in Figure 5.3 (b), stabilization of the parabolic features is possible by use of a second propagation stage in a fiber with specially adjusted nonlinear and dispersive characteristics relative to the first fiber [39, 77].

The evolution displayed in Figure 5.5 also reveals that after wave breaking occurs in the fiber, the accumulation of the quadratic spectral phase induced by dispersion causes the temporal and spectral contents of the pulse to become increasingly closer to each other, and the temporal and spectral profiles to evolve very little. The nearly constant kurtosis or misfit factor (panel c) clearly indicates a nearly invariant temporal profile. This corresponds to the formation of a nonlinear structure of a spectronic nature in the fiber, which describes the long-term far-field evolution in the passive medium [78–80]. The details are discussed in Chapter 6 by L. Mouradian and A. Barthélémy.

5.4.3 Generation of Triangular and Rectangular Pulses

Parabolic shapes are not the only pulse waveforms that can be generated in a passive ND fiber. Indeed, it has been shown numerically that temporal triangular intensity profiles can result from the evolution of initially rectangular pulses in the highly nonlinear (semi-classical) regime of propagation [81]. Another approach that has been more recently demonstrated theoretically relies on the progressive reshaping of initially parabolic pulses driven by the FOD of the fiber [82]. The overall temporal effect of FOD on parabolic pulse propagation is to stretch and enhance the power reduction in the pulse wings, leading to a triangular profile. In [77], the combination of pulse prechirping and nonlinear propagation in a section of ND fiber has been introduced as a method for passive nonlinear pulse shaping, which provides a simple way of generating various advanced field distributions, including flat-top- and triangular-profiled pulses with a linear frequency chirp. In this scheme, Kerr nonlinearity and GVD lead to various reshaping processes of an initial conventional field distribution (e.g., a Gaussian pulse) according to the chirping value and power level at the entrance of the fiber. In particular, triangular pulses can be generated for a negative initial chirp ((such as that imparted by an anomalously dispersive element) and sufficiently high energies. These theoretical results have been confirmed experimentally by intensity (Figure 5.7 (b)) and phase measurements of the generated pulses [83, 84]. In the experimental setup used (Figure 5.7 (a)), the control of the pulse prechirping value is realized by propagation through different lengths of standard SM fiber with anomalous GVD, which imposes a negative chirp on the pulse. The prechirped pulses are amplified to different power levels using an erbium-doped fiber amplifier and then propagated through a ND fiber to realize the pulse reshaping. Further propagation of the boosted waveform in a HNL fiber stabilizes the triangular intensity profile and leads to spectrum splitting, which can be used to realize optical signal doubling [88].

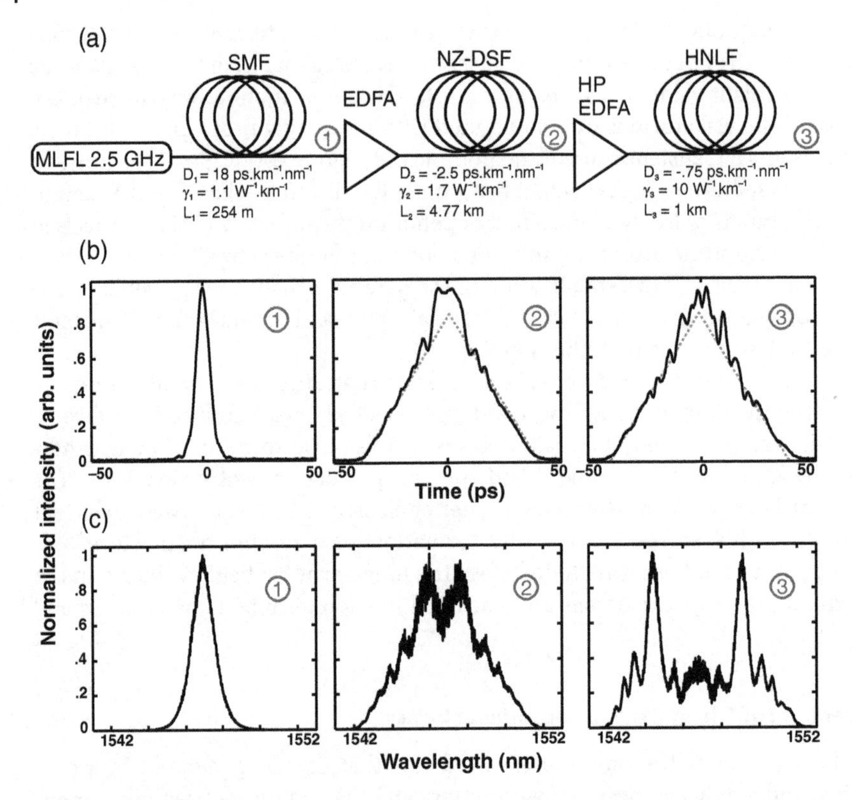

Figure 5.7 (a) Experimental setup for triangular pulse generation and spectrum splitting. MLFL: mode-locked fiber laser, SMF: single-mode fiber, EDFA: erbium-doped fiber amplifier, NZ-DSF: non-zero dispersion-shifted fiber, HP EDFA: high-power EDFA, HNLF: highly nonlinear fiber. (b) Measured temporal and spectral intensity profiles at different propagation stages in the device. Also shown are the corresponding triangular fits (dashed gray). More details can be found in [84]. *Source*: Adapted from Verscheure 2011 [84].

Furthermore, the possibility of triangular pulse shaping in mode-locked fiber lasers was first reported in [85]. It has been numerically demonstrated that for normal net dispersion, formation of two distinct steady-state solutions of stable single pulses can be obtained in a laser cavity in different regions of the system parameter space: the previously known similariton [68] and a triangular-profiled pulse with a linear chirp.

Additionally, the generation of nearly rectangular waveforms has been demonstrated both theoretically and experimentally [86, 87]. In the system used, an initially chirped pulse experiencing spectral narrowing upon nonlinear propagation in a fiber results in a near-transform-limited pulse in the region of optimum compression, as described in Section 5.5.1.1.

5.5 Spectral Shaping

Besides significant changes in the temporal domain, ultrashort pulses propagating in a fiber can also undergo substantial spectral changes as a result of the interplay between fiber dispersion and nonlinearity. In this section, we review some important examples of nonlinear pulse shaping in the spectral domain and discuss how the spectral changes induced onto a pulse can be controlled for applications that are not possible with standard linear techniques.

5.5.1 Spectral Compression

Spectral narrowing of optical pulses can be rather easily realized with standard optical bandpass filtering, but at the expense of significant optical losses. Indeed, as we mentioned in the Introduction, linear filtering introduces a power penalty that is at least proportional to the ratio of the target spectrum to the input one.

5.5.1.1 By Self-Phase Modulation

The effect of SPM on an an ultrashort pulse propagating in a nonlinear optical fiber is ordinarily associated with broadening of the pulse spectrum. However, this is not always the case with chirped pulses, where pulses with an initial negative chirp in the normal GVD region can experience spectral narrowing [89–92]. Indeed, as a result of SPM, the propagating pulse will have the frequencies of the leading edge downshifted and the frequencies of the trailing edge upshifted. Therefore, in the case of an initially negatively chirped pulse, both the trailing and leading edges of the pulse will experience a spectral shift in the direction of the central frequency. It is worth noting here that the anomalous GVD enhances the effect of SPM rather than compensating it. Figure 5.8 in the color plate section illustrates the spectral compression of an initial Gaussian pulse in a nonlinear fiber in the absence of GVD. The required negative chirp has been imprinted onto the pulse by temporal stretching in an anomalously dispersive linear element. This method of spectral compression has been demonstrated using various types of fibers [90,91,93,94], and is suitable for a very large range of wavelengths including titanium-sapphire wavelengths [90,93], the widely used 1-μm [95,96] and 1.55-μm [97] windows, and the emerging 2-μm band [98]. The process can also sustain simultaneous amplification of the pulse, and has been reported for different types of fiber amplifiers [95,98]. Therefore, spectral compression by SPM of negatively chirped pulses provides an attractive solution to convert ultrashort pulses delivered by femtosecond oscillators into powerful, near-transform-limited picosecond pulses, and to counteract the spectrum expansion that usually occurs with the direct amplification of picosecond structures. The concept can be extended to multistage architectures [99] and fiber laser cavities, where in-cavity nonlinear spectral compression in a mode-locked fiber laser has recently been demonstrated numerically [100]. A new concept of a fiber laser architecture has been presented, which supports self-similar pulse evolution in the amplifier fiber segment of the laser cavity and nonlinear spectral compression in the passive fiber, thereby enabling the generation of highly chirped parabolic pulses and nearly transform-limited spectrally compressed picosecond pulses from a single device.

However, the main limitation of SPM-driven spectral compression in the nonlinearity-dominant regime of propagation in which fiber dispersion is of little importance, is the presence of residual side lobes in the compressed spectrum, as illustrated in Figure 5.8 (b). These side lobes stem from the fact that in general an input pulse with a negative linear chirp, $|\psi(0,t)| \exp(ibt^2)$ (with $b > 0$), cannot be spectrally compressed to the Fourier transform limit because the instantaneous frequency $\phi_t = 2bt + \gamma(|\psi(0,t)|^2)_t z$ (in the absence of dispersion) cannot in general be made equal to zero for all times. A

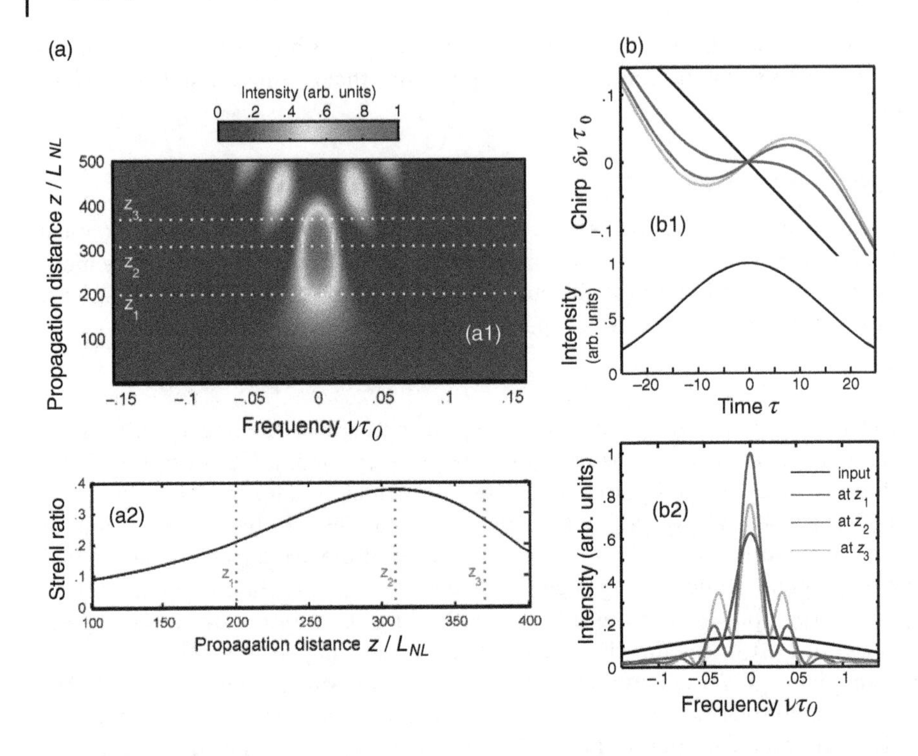

Figure 5.8 Spectral pulse compression in a purely nonlinear fiber, as obtained by numerical integration of the NLS equation. (a) Longitudinal evolution of the spectral intensity profile (subplot 1) and the Strehl ratio (suplot 2) for an initial Gaussian pulse with a stretching factor of 20. (b) Temporal intensity an chirp profiles (subplot 1) and spectral intensity profile (subplot 2) taken at different propagation distances in the fiber: the point z_1 of chirp cancellation near the pulse center (blue), the point z_2 of maximum Strehl ratio (red), and the point z_3 of maximum spectral compression factor (green). Also shown are the initial pulse profiles (black). More details can be found in [105]. *Source*: Finot 2016 [105]. For a color version of this figure please see color plate section.

convenient metric for assessing the quality of the spectral compression is the Strehl ratio S, defined as the ratio of the maximum spectral brilliance of the actual pulse to the spectral brilliance obtained assuming a flat temporal phase of the pulse [97]. Therefore, S is comprised between 0 and 1, with 1 defining an ideal compression. In the pure SPM case and with a conventional laser pulse such as a Gaussian or an hyperbolic secant pulse, the maximum achievable Strehl ratio is modest (0.4 in the example of Figure 5.8), and the propagation distance of optimum compression in terms of Strehl ratio in general differs from the the distance where the linear and nonlinear phases cancel each other in the central part of the pulse.

Several strategies have been proposed and experimentally demonstrated to enhance the performance characteristics of the spectral compression process. In [101], the balancing of the (higher-order) dispersion and chirp has enabled compression to approach the Fourier transform limit. Another approach is based on using a pre-shaped input pulse profile such as a parabolic waveform. Indeed, for an input negatively linearly chirped parabolic pulse, the instantaneous frequency $\phi_t = 2(b - \gamma a^2 z/\tau_0^2)t$ can be chosen to yield zero for all t, for a suitable combination of γ, a, z, and b [102]. Because, at the

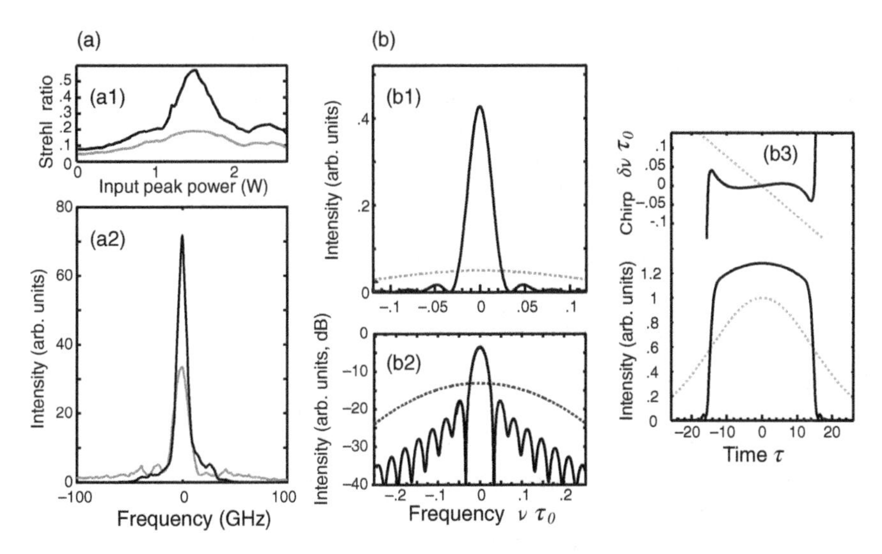

Figure 5.9 Enhanced SPM-driven spectral pulse compression. (a) Experimental results on the spectral compression of initially negatively chirped parabolic (black) and hyperbolic secant (gray) pulses in a HNL fiber. Evolution of the Strehl ratio with the input peak power—equivalent to the evolution with the propagation distance (subplot 1), and spectral intensity profiles at the point of maximum compression (subplot 2). (b) Spectral compression of an initially negatively chirped Gaussian pulse in a dispersive nonlinear fiber with normal GVD, as obtained by numerical integration of the NLS equation. Spectral intensity profiles (subplots 1 and 2) and temporal intensity and chirp profiles (subplot 3) of the input pulse (dotted gray) and the pulse at the point of optimum compression in terms of Strehl ratio (solid black). More details can be found in [105]. *Source*: Finot 2016 [105].

same time, the parabolic temporal envelope with width τ_0 remains unchanged (as long as dispersion in the fiber can be ignored), this expresses the fact that spectral compression to the transform limit takes place. Experimental demonstrations at telecommunication wavelengths using a HNL fiber [97] (Figure 5.9 (a)) or at the wavelengths around 1 μm using a PC fiber [103] have confirmed the potential of this method to achieve high-quality compressed spectra with very low substructures and enhanced brightness. In the experimental example of Figure 5.9 (a), the Strehl ratio for the spectrally compressed parabolic-shaped pulse is three times larger than that for the hyperbolic secant pulse. Another strategy to enhance the quality of the compressed pulse spectrum is to select a dispersive nonlinear regime of propagation in which the combined action of normal GVD and SPM results in a deformation of the temporal profile of the pulse tending to acquire a rectangular shape while nearly complete compensation of the pulse chirp occurs [86, 104, 105] (Figure 5.9 (b)).

5.5.1.2 Using Soliton Processes

Another way to compress an optical spectrum that takes advantage of the fiber nonlinearity relies on the adiabatic evolution of a soliton in a fiber with anomalous dispersion [106]. For example, a fundamental soliton propagating in an anomalous dispersion-increasing fiber can experience narrowing of its spectral width as a result of its adiabatic adaptation to the slowly varying fiber dispersion [107]. The working principle is the reverse operation of the well-known adiabatic soliton temporal compression in a DD fiber. The main advantage of this approach is that the spectrally compressed pulses are close to the Fourier transform limit. However, the pulse energy is limited by the soliton area theorem, as detailed in Section 5.3.1.1. Additionally, ultrashort laser pulses can be spectrally compressed in the regime of soliton self-frequency shift induced by the Raman effect in a HNL fiber [108] (details are given in Section 5.5.2). Owing to the anomalous dispersion of the HNL fiber, the laser pulses evolve toward solitons and experience a continuous frequency downshift. A lowering frequency and increasing dispersion of a red-shifting soliton dictate spectral narrowing.

5.5.2 Generation of Frequency-Tunable Pulses

One of the possibilities offered by the third-order nonlinearity of optical fibers for pulse shaping is to frequency shift the essential spectral content of ultrashort optical pulses whose width is close to or less than 1 ps. For pulses with a wide spectrum (> 1 THz), the Raman gain of the fiber (Figure 5.10 (a)) can amplify the low-frequency components of a pulse by transferring energy from the high-frequency components of the same pulse. As a result of this phenomenon, called intra-pulse Raman scattering, the pulse spectrum shifts toward the low-frequency (red) side as the pulse propagates inside the fiber, a phenomenon referred to as the self-frequency shift [109]. The physical origin of this effect is related to the delayed nature of the Raman (vibrational) response [110]. The effects

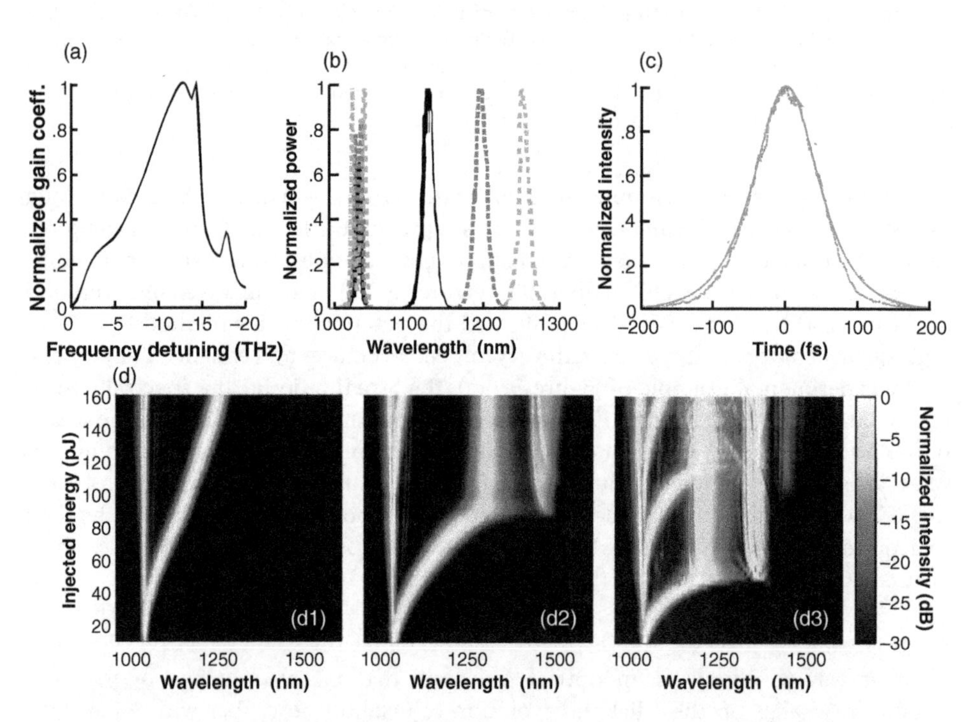

Figure 5.10 Raman-induced self-frequency shift of solitons. (a) Raman response of silica fibers. (b) Experimental spectra obtained at the output of a 2-m long PC fiber for different input pulse energies: 83 pJ (solid black), 125 pJ (dashed gray), and 173 pJ (dash-dot light gray). (c) Autocorrelation traces of the frequency-shifted pulses compared with an hyperbolic secant fit. (d) Experimentally measured dependence of the output spectral intensity profile on the input energy for 2-m long PC fibers with a single (subplot 1) or two zero-dispersion wavelengths (subplots 2 and 3).

of intra-pulse Raman scattering can be included through the NLS equation by including the term $-i\gamma\tau_R\psi(|\psi|^2)_t$ on the left-hand side of Eq. (5.1), where the parameter τ_R is related to the slope of the Raman gain spectrum [110]. For a soliton propagating inside a fiber, the amplitude is not affected by the Raman effect, but its frequency is shifted by $\delta\omega_R(z) = -8|\beta_2|\tau_R z/(15\tau_0^4)$, as illustrated experimentally in Figure 5.10 (b). The temporal intensity profile remains close to the hyperbolic secant soliton shape (Figure 5.10 (c)).

The soliton self-frequency shift (SSFS) has been exploited in the realization of fiber-based sources of spectrally tunable ultrashort optical pulses. Frequency shifts of several hundreds of nanometers have been achieved in HNL fibers. The advent of PC fibers has made it possible to observe SSFS at the operational wavelengths of titanium-sapphire lasers, and impressive shifts have been demonstrated, thereby paving the way for efficient spectrally tunable laser sources for applications in, e.g., biophotonics or spectroscopy [111, 112], as described in Chapter 7 by E. R. Andresen and H. Rigneault. In this context, the main factors that may limit SSFS generation are the absorption due to hydroxil radicals contained in the core glass of silica fibers, the fission of higher-order solitons into multiple sub-pulses, and the existence of a second zero-dispersion wavelength for some fibers, which induces a stabilization of the soliton frequency accompanied by the generation of dispersive waves [113] (Figure 5.10 (d2)). Furthermore, the combined effects of delayed nonlinear response and higher-order dispersion can result in a gradual increase of the second-order dispersion during the propagation. This leads to an adiabatic temporal broadening of the soliton, thereby progressively limiting its Raman-induced frequency shift [114]. To address this issue, optimization of the dispersion landscape "seen" by the frequency-shifted pulse has been proposed as an efficient method to enhance the performance of fiber-based tunable frequency shifters [115].

5.5.3 Supercontinuum Generation

Nonlinear pulse shaping represents the ideal way of generating spectral bandwidth that exceeds significantly that of the input seed pulse. This is obviously not possible with linear shaping techniques where one cannot shift energy outside the initial pulse bandwidth. Optical fiber-based supercontinuum (SC) sources have become a significant scientific and commercial success, with applications ranging from frequency comb production to advanced medical imaging and telecommunication applications. Several books and review articles are now available to understand the physical mechanisms driving the continuum formation and its modelling using the NLS equation in its extended form (see, e.g., [3, 116–118]), so in this section we describe only the basic temporal and spectral features of the continuum generated in the anomalous and normal dispersion regimes of a nonlinear fiber, which stem from the involved propagation dynamics. We will see that, while typically the broadest spectra are generated in anomalously dispersive fibers, where the broadening mechanism is strongly influenced by soliton dynamics, ND fibers are undoubtedly advantageous as far as the flatness and temporal coherence of the generated continuum are concerned.

5.5.3.1 Supercontinuum Stability and Extreme Events in the Anomalous Dispersion Regime

The nonlinear dynamics associated with SC generation in the anomalous dispersion regime are dominated by an initial stage of MI of the input picosecond pump pulse induced by noise, followed by a stage of higher-order soliton temporal compression,

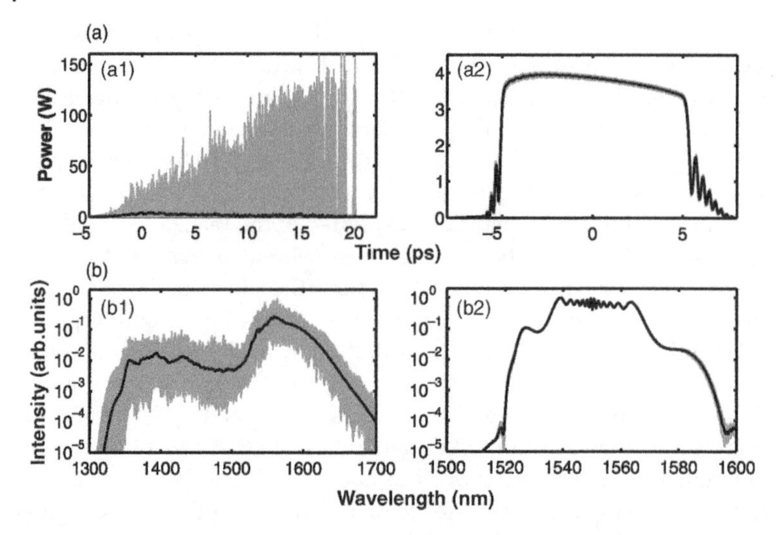

Figure 5.11 Superposition of 200 simulations of the output (a) temporal and (b) spectral intensity profiles of a hyperbolic secant pulse (3-ps full-width at half-maximum duration, 10-W peak power) propagating in a 500-m-long HNL fiber (nonlinear coefficient 10 /W/km, GVD coefficient ±0.5 ps/nm/km, dispersion slope 0.01 ps/nm/km at 1550 nm) in the anomalous (subplots 1) and normal (subplots 2) dispersion regimes. The results are obtained by numerical integration of the NLS equation with third-order dispersion and include the quantum noise as well as 1% of fluctuations on the input pump. The different shots are plotted in gray and the average curve is plotted with a solid black line. In the case of normal GVD, the gray and black curves nearly coincide. *Source*: Hammani 2012 [124]. Reproduced with permission of Elsevier.

and subsequent splitting of the ultrashort coherent structure at the point of maximum compression due to third-order dispersion or intra-pulse Raman scattering. Pulse splitting results in the emergence of several ultrashort solitons with significant shot-to-shot amplitude and timing jitters, which prevents this type of SC from being sliced in multiple wavelength channels for telecommunication applications. The spectral intensity profile is significantly broader than those typically obtained in ND fibers but also suffers from large fluctuations as shown in Figure 5.11 (panels 1), where we have plotted the results of a few hundreds of numerical simulations with distinct noise seeds [124]. Because the SC properties change significantly from shot to shot both in the temporal and spectral domains, the high level of stability of the continuum obtained in the normal regime of dispersion (Section 5.5.3.2) is generally not achieved in the anomalous dispersion region [119]. The spectral phase fluctuations were the subject of intense investigations as early as 2000 [120]. The first experimental observation of significant fluctuations in the temporal domain was reported by Solli *et al.* in 2007 [121]. Spectral filtering of the most red-shifted part of the SC spectrum generated in a PC fiber by anomalous pumping resulted into a highly-skewed distribution of the statistical fluctuations of the corresponding temporal intensity. By analogy with the typical L-shaped distribution observed for extreme events, the intense and extremely rare outliers of the distribution were called "optical rogue waves". Since then, the subject has become the focus of intense research in the international optical community. This enthusiasm has indeed largely surpassed the confines of SC generation in PC fibers, and optical rogue waves have been reported in various nonlinear optical systems. A careful analysis of various configurations has disclosed that the most red-shifted events detected in the experiment of Solli *et al.* are fundamental solitons shifted in frequency by the delayed Raman response of silica [122]. Because such giant solitons are fully coherent and persist once they form, they are now generally referred to as "rogue solitons" [123]. The emergence of such extreme solitons results from the process of a temporal collision between the substructures that are generated from the splitting of the initial pump pulse. When two ultrashort substructures collide, a temporal peak appears intermittently, thus reproducing one of the peculiar features of the infamous oceanic freak waves: their unexpected appearance and decay. Such collisions are favored by convective mechanisms such as the third-order dispersion or the Raman response of silica [125]. In the process of collision, an energy exchange also occurs, leading to the progressive growth of a giant soliton, where most of the energy of the system will eventually concentrate [126, 127].

One way to limit the degradation of the SC coherence in the anomalous dispersion regime is to use very short segments of fiber or tapered fibers [119]. An alternative solution is to reduce the spontaneous instabilities that may emerge in the early stages of propagation. By imposing an input deterministic seed located in the frequency band of the instabilities on the (coherent) pump signal, one can expect that the seed will prevail over the spontaneous effects so that enhanced stability and coherence can be ensured. Such an approach has been used in various numerical and experimental works to control the stability and spectral extent of the generated continuum [128, 129]. An accurate spectral study has confirmed that a seeded generation can be well described by the Akhmediev-Kuzentsov mathematical formalism [130]. This kind of control can be used to generate Peregrine solitons [131, 132], which represent a limiting case of Akhmediev breathers [133] for maximum temporal localization or, equivalently, of Kuznetsov-Ma breathers for maximum spatial localization. An in-depth discussion of the temporal and spectral features of these new nonlinear structures is the subject of Chapter 10 in this volume by B. Kibler.

5.5.3.2 Highly Coherent Continua in Normally Dispersive Fibers

Nonlinear pulse propagation in the normal dispersion regime of a fiber inhibits MI phenomena from occurring, which reduces significantly the spontaneous amplification of noise and thereby shot-to-shot fluctuations [119]. Consequently, a high degree of coherence is maintained. However, the main limits to spectral pulse quality in the normal dispersion regime are the spectral ripple arising from SPM of conventional laser pulses, and the effects of optical wave breaking which may lead to significant changes in the temporal pulse shape and severe energy transfer into the wings of the spectrum. These effects can in principle be avoided by using pre-shaped input pulses with a parabolic temporal intensity profile which would preserve their shape while propagating within the fiber and, thus, result in spectrally flat, highly coherent pulses (Figure 5.12 (a)). A continuum with low spectral ripple and high energy density in the central part has been indeed demonstrated by use of parabolic pulses generated in erbium-doped fibers [134] or resulting from linear shaping by fiber Bragg gratings [135] (Figure 5.12 (c)) and reconfigurable liquid-crystal-on-silicon devices [136]. More advanced techniques for achieving even flatter spectra involve the use of a feedback loop relying on a genetic algorithm to obtain a nearly rectangular output spectrum [137], as Figure 5.12 (d) shows. This demonstrates the potential of nonlinear pulse shaping assisted by a linear shaping

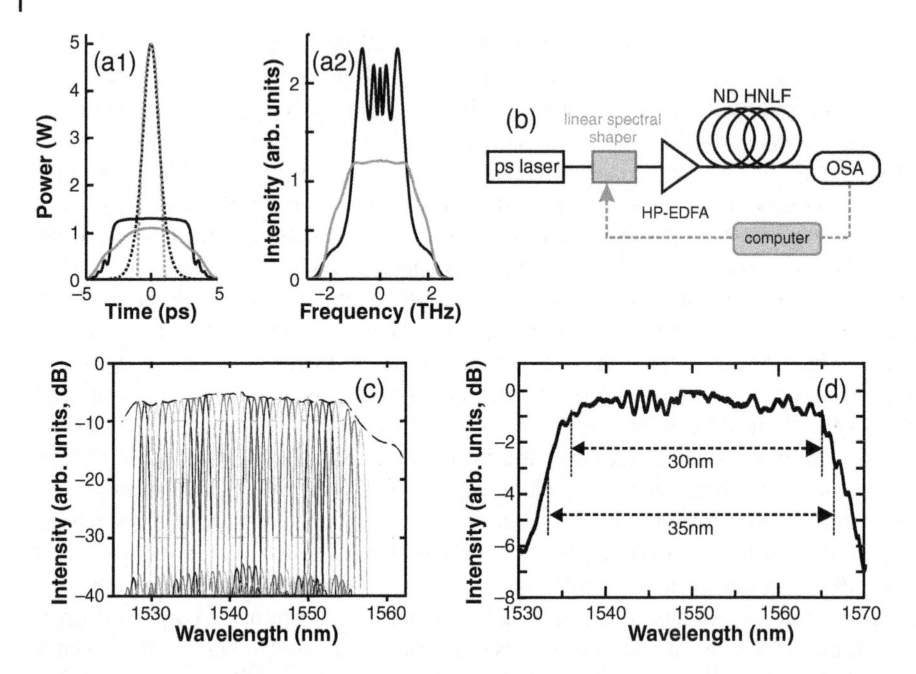

Figure 5.12 (a) Temporal (subplot 1) and spectral (subplot 2) intensity profiles at the exit of a ND-HNL fiber for initially unchirped hyperbolic secant (solid black) and parabolic (solid gray) pulses with $P_0 = 5$ W, as obtained by numerical integration of the NLS equation. Also shown are the initial pulse temporal profiles (dashed). (b) Experimental setup for the nonlinear generation of ultra-flat continua assisted by adaptive pulse shaping. HP-EDFA: high-power erbium-doped fiber amplifier, ND-HNLF: normally dispersive highly nonlinear fiber, OSA: optical spectrum analyzer. (c) Experimental spectrum of a parabolic-shaped pulse after propagation in a ND-HNL fiber (dashed black) and superposition of the measured spectra of 38 sliced channels. (d) SPM spectra generated from 2-ps Gaussian pulses after the application of the adaptive system in (b). More details can be found in [137]. *Source*: Adapted from Xin 2013 [137].

element as an additional sculpturing element to generate on-demand tailored spectra without the requirement of expensive laser sources.

The high degree of coherence of the spectra generated in the normal dispersion regime is clearly attractive for all-optical pulse processing in HNL fibers [52], the generation of optical combs for spectroscopy [138] or the generation of SC covering the entire C-band of optical telecommunications. Indeed, several studies have reported that such continua can be sliced into several tens of high-quality, high-repetition-rate picosecond channels where isolated spectral lines of the 10-GHz or 40-GHz resulting comb provide several hundreds of CWs [135, 139, 140] (Figure 5.12 (c)). Furthermore, high temporal coherence enables pulse compression with very high compression ratios as the SPM-induced nonlinear chirp can be compensated to a large extent [25, 26]. This technique has been efficiently exploited in PC fibers where the dispersion characteristics can be accurately controlled so as to obtain a fiber with normal dispersion at all wavelengths [82], which enables the generation of highly broadband coherent spectra spanning more than several hundreds of nanometers. Recently developed measurement tools have confirmed the high degree of coherence of the continua obtained in all-normal dispersion fibers [141].

5.6 Conclusion

In this chapter, we have provided an overview of several recent examples of the use of nonlinear phenomena in optical fibers for the shaping of optical pulses in the temporal and spectral domains. The propagation of short pulses in optical fibers is connected to a large variety of interesting and practically important phenomena. The unique dispersive and nonlinear properties of optical fibers lead to various scenarios of the pulse evolution which result in particular changes of the pulse temporal shape, spectrum and phase profile. Because the nonlinear dynamics of pulses propagating in fibers with normal GVD is generally sensitive to the initial pulse condition, it is possible to nonlinearly shape the propagating pulses through control of the initial pulse temporal intensity and/or phase profile. In fact, this approach can be flexibly applied to the generation of specialized temporal waveforms for applications in optical signal processing and manipulation, or the generation of highly coherent continua for optical telecommunications, or the generation of narrow-spectrum pulses for applications requiring high spectral resolution such as nonlinear vibrational microscopy. Application of nonlinear pulse shaping at normal dispersion has also provided significant progress in the field of high-power fiber amplifiers and lasers. The temporal and spectral features of pulses propagating in fibers with anomalous GVD are typically governed by soliton dynamics. The unique properties of optical solitons can be effectively used in the generation of ultrashort pulses with very high repetition rates, which are widely used in various scientific and technological areas, or the generation of frequency tunable pulses, or the generation of ultra-wide spectra with many disruptive applications and breakthroughs in fields such as optical coherence tomography, metrology and spectroscopy, biophotonics, and others. Seeding the SC generation at anomalous dispersion with a suitably chosen signal can significantly improve the SC stability. Furthermore, the same concept can be applied to stimulate the generation of Peregrine solitons, which constitute a prototype of optical rogue waves (see Chapter 10).

From a purely scientific point of view, the results presented in this chapter are a further illustration of how optical fiber systems can provide useful experimental testbeds for the study of a variety of nonlinear dynamical processes. Therefore, this research area is interesting in its own right. Despite substantial research in this field, qualitatively new phenomena are still being discovered. With respect to application requirements, integrated pulse shaping devices are desirable for applications such as in optical communications where size and robustness are important. The developments of micro-structured fibers with extremely small effective core areas and exhibiting enhanced nonlinear characteristics, and of fibres using materials with refractive indexes higher than that of the silica glass have enabled dramatic reduction of the required fiber lengths for nonlinear interactions compared to conventional fibers, thereby paving the way for integrated fiber-based pulse shaping. With the inclusion of higher-order propagation effects, nonlinear shaping can now be observed at the integrated waveguide scale. For example, higher-order soliton compression of picosecond pulses has recently been achieved in silicon [142] by using a dispersion-engineered PC waveguide. More recently, the unique dispersion properties of PC waveguides have enabled the experimental discovery of a new class of solitons originating purely from the interaction of negative FOD and SPM, which can occur even when the GVD vanishes or is normal [143]. These discoveries, combined with previous advances in the laser literature, may inspire a new wave of ultrafast laser development.

Acknowledgments

We acknowledge the important contributions of our colleagues in the original papers discussed in this chapter: J. Fatome, K. Hammani, B. Kibler, C. H. Hage, G. Millot, H. Rigneault, E. Andresen, P. Harper, A. I. Latkin, B. G. Bale, S. K. Turitsyn, F. Parmigiani, P. Petroupoulos, D. J. Richardson and J. M. Dudley.

We would also like to acknowledge the financial support of the Leverhulm Trust (grant RPG-278), the Conseil Regional de Bourgogne (PARI Photcom) and the Labex Action program (ANR-11-LABX-01-01). The experimental work has benefited from the PICASSO platform of the University of Burgundy.

References

1 Shen, Y. R. (1984) *Principles of Nonlinear Optics*. John Wiley & Sons, New York, NY.
2 Weiner, A.M. (2000) Femtosecond pulse shaping using spatial light modulators. *Review of Scientific Instruments*, **71**, 1929–1960.
3 Agrawal, G.P. (2006) *Nonlinear Fiber Optics*, 4th edn. Academic Press, San Diego, CA.
4 Hasegawa, A. and Tappert, F. (1973) Transmission of stationary nonlinear optical pulses in dispersive dielectric fibers. I. Anomalous dispersion. *Applied Physics Letters*, **23**, 142–144.
5 Mollenauer, L.F. and Gordon, G.P. (2006) *Solitons in Optical Fibers: Fundamentals and Applications*. Academic Press, San Diego, CA.
6 Kivshar, S. and Agrawal, G.P. (2003) *Optical Solitons: From Fibers to Photonic Crystals*. Academic Press, San Diego, CA.
7 Remoissenet, M. (2003) *Waves Called Solitons: Concepts and Experiments*, 3rd edn. Springer, Berlin.
8 Dauxois, T. and M. Peyrard, M. (2010) *Physics of Solitons*. Cambridge University Press, Cambridge.
9 Mollenauer, L.F., Stolen, R.H., and Gordon, J.P. (1980) Experimental observation of picosecond pulse narrowing and solitons in optical fibers. *Physical Review Letters*, **45**, 1095–1098.
10 Bogatyrev, V.A. Bubnov, M.M. Dianov, E.M., *et al.* (1991) A single-mode fiber with chromatic dispersion varying along the length. *Journal of Lightwave Technology*, **9**, 561–565.
11 Chernikov, S.V. and Mamyshev, P.V. (1991) Femtosecond soliton propagation in fibers with slowly decreasing dispersion. *Journal of the Optical Society of America B*, **8**, 1633–1641.
12 Chernikov, S.V., Richardson, D.J., Dianov, E.M., and Payne, D.N. (1992) Picosecond soliton pulse compressor based on dispersion decreasing fibre. *Electronics Letters*, **28**, 1842–1844.
13 Tse, M.L., Horak, P., Price, J.H.V. *et al.* (2006) Pulse compression at 1.06 m in dispersion-decreasing holey fibers. *Optics Letters* **31**, 3504–3506.

14 Chernikov, S.V., Taylor, J.R., and Kashyap, R. (1994) Experimental demonstration of step-like dispersion profiling in optical fibre for soliton pulse generation and compression. *Electronics Letters*, **30**, 433–435.

15 Chernikov, S.V., Taylor, J.R., and Kashyap, R. (1994) Comblike dispersion-profiled fiber for soliton pulse train generation. *Optics Letters*, **19**, 539–541.

16 Ouzounov, D.G., Ahmad, F.R., Müller, D., *et al.* (2003) Generation of Megawatt optical solitons in hollow-core photonic band-gap fibers. *Science*, **301**, 1702–1704.

17 Satsuma, J. and Yajima, N. (1974) Initial value problems of one-dimensional self-modulation of nonlinear waves in dispersive media. *Progress of Theoretical Physics Supplement*, **55**, 284–306.

18 Pelusi, M., Matsui, Y., and Suzuki, K. (1999) Pedestal suppression from compressed femtosecond pulses using a nonlinear fiber loop mirror. *IEEE Journal of Quantum Electronics*, **35**, 867–874.

19 Kibler, B., Fisher, R., Lacourt, P.A., *et al.* (2007) Optimised one-step compression of femtosecond fibre laser soliton pulses around 1550 nm to below 30 fs in highly nonlinear fibre. *Electronics Letters*, **43**.

20 Amorim, A.A., Tognetti, M.V., Oliveira, P., *et al.* (2009) Sub-two-cycle pulses by soliton self-compression in highly nonlinear photonic crystal fibers. *Optics Letters*, **34**, 3851–3853.

21 Grischkowsky, D. and Balant, A.C. (1982) Optical pulse compression based on enhanced frequency chirping. *Applied Physics Letters*, **41**, 1–3.

22 Shank, C.V., Fork, R.L. Yen, R., Stolen, R.H., and Tomlinson, W.J. (1982) Compression of femtosecond optical pulses. *Applied Physics Letters*, **40**, 761–763.

23 Hooper, L.E., Mosley, P.J., Muir, A.C., Wadsworth, W.J., and Knight, J.C. (2011) Coherent supercontinuum generation in photonic crystal fiber with all-normal group velocity dispersion. *Optics Express*, **19**, 4902–4907.

24 Demmler, S., Rothhardt, J., Heidt, A.M., *et al.* (2011) Generation of high quality, 1.3 cycle pulses by active phase control of an octave spanning supercontinuum. *Optics Express*, **19**, 20151–20158.

25 Billet, C., Dudley, J.M., Joly, N., and Knight, J.C. (2005) Intermediate asymptotic evolution and photonic bandgap fiber compression of optical similaritons around 1550 nm. *Optics Express*, **13**, 3236–3241.

26 Tomlinson, W.J., Stolen, R.H., and Shank, C.V. (1984) Compression of optical pulses chirped by self-phase modulation in fibers. *Journal of the Optical Society of America B*, **1**, 139–149.

27 Druon, F. and Georges, P. (2004) Pulse-compression down to 20 fs using a photonic crystal fiber seeded by a diode-pumped Yb:SYS laser at 1070 nm. *Optics Express*, **12**, 3383–3396.

28 Pierrot, S. and Salin, F. (2013) Amplification and compression of temporally shaped picosecond pulses in Yb-doped rod-type fibers. *Optics Express*, **21**, 20484–20496.

29 Jaskorzynska, B. and Schadt, D. (1988) All-fiber distributed compression of weak pulses in the regime of negative group-velocity dispersion. *IEEE Journal of Quantum Electronics*, **24**, 2117–2120.

30 Alfano, R.R., Agrawal, G.P., Baldeck, P.L., and Ho, P.P. (1989) Cross-phase modulation and induced focusing due to optical nonlinearities in optical fibers and bulk materials. *Journal of the Optical Society of America B*, **6**, 824–829.

31 Kobayashi, T., Yao, H., Amano, K., *et al.* (1988) Optical pulse compression using high-frequency electrooptic phase modulation. *IEEE Journal of Quantum Electronics*, **24**, 382–387.

32 Komukai, T., Yamamoto, Y., and Kawanishi, S. (2005) Optical pulse generator using phase modulator and linearly chirped fiber Bragg gratings. *IEEE Photonics Technology Letters*, **17**, 1746–1748.

33 Hasegawa, A. (1984) Generation of a train of soliton pulses by induced modulational instability in optical fibers. *Optics Letters*, **9**, 288–290.

34 Tai, K., Hasegawa, A., and Tomita, A. (1986) Observation of modulational instability in optical fibers. *Physical Review Letters*, **56**, 135–138.

35 Pitois, S., Fatome, J., and Millot, G. (2002) Generation of 160-GHz transform-limited pedestal-free pulse train trough multiwave mixing compression of a dual frequency beat signal. *Optics Letters*, **27**, 1729–1731.

36 Fatome, J., Pitois, S., and Millot, G. (2006) 20-GHz-to-1-THz repetition rate pulse sources based on multiple four-wave mixing in optical fibers. *IEEE Journal of Quantum Electronics*, **42**, 1038–1046.

37 Fatome, J., Pitois, S., Fortier, C., *et al.* (2010) Multiple four-wave mixing in optical fibers: 1.5–3.4-THz femtosecond pulse sources and real-time monitoring of a 20-GHz picosecond source. *Optics Communications*, **283**, 2425–2429.

38 El Mansouri, I., Fatome, J., Finot, C., Lintz, M., and Pitois, S. (2011) All-fibered high-quality stable 20- and 40-GHz picosecond pulse generators for 160-Gb/s OTDM applications. *IEEE Photonics Technology Letters*, **23**, 1487–1489.

39 Finot, C., Fatome, J., Pitois, S., and Millot, G. (2007) All-fibered high-quality low duty-cycle 20-GHz and 40-GHz picosecond pulse sources. *IEEE Photonics Technology Letters*, **19**, 1711–1713.

40 Inoue, T. and S. Namiki, S. (2008) Pulse compression techniques using highly nonlinear fibers. *Laser Photonics Review* **2**, 83–99.

41 Hansryd, J. and Andrekson, P.A. (2001) Wavelength tunable 40 GHz pulse source based on fibre optical parametric amplifier. *Electronics Letters*, **37**, 584–585.

42 Wiberg, A.O.G., Brès, C-S., Kuo, B.P.P., *et al.* (2009) Pedestal-free pulse source for high data rate optical time-division multiplexing based on fiber-optical parametric process. *IEEE Journal of Quantum Electronics* **45**, 1325–1329.

43 Vedadi, A.A., Shoaie, M.A., and Brès, C-S. (2012) Experimental investigation of pulse generation with one-pump fiber optical parametric amplification. *Optics Express*, **20**, 27344–27354.

44 Feng, F., Fatome, J., Sysoliatin, A., *et al.* (2014) Wavelength conversion and temporal compression of a pulse train using a dispersion oscillating fibre. *Electronics Letters*, **50**, 768–770.

45 Finot, C. and Wabnitz, S. (2015) Influence of the pump shape on the modulation instability process induced in a dispersion-oscillating fiber. *Journal of the Optical Society of America B*, **32**, 892–899.

46 Hammani, K. and Finot, C. (2013) Extreme optical fluctuations in lumped Raman fibre amplifiers. *Journal of Optics*, **15**, 064009.

47 Hammani, K., Finot, C., and Millot, G. (2009) Emergence of extreme events in fiber-based parametric processes driven by a partially incoherent wave. *Optics Letters*, **34**, 1138–1140.

48 Chapman, B.H., Doronkin, A.V., Popov, S.V., and Taylor, J.R. (2012) All-fiber integrated 10 GHz repetition rate femtosecond laser source based on Raman compression of pulses generated through spectral masking of a phase-modulated diode. *Optics Letters*, **37**, 3099–3101.

49 Chapman, B.H., Doronkin, A.V., Stone, J.M., *et al.* (2013) Femtosecond pulses at 20 GHz repetition rate through spectral masking of a phase modulated signal and nonlinear pulse compression. *Optics Express*, **21**, 5671–5676.

50 Mulvad, H.C.H., Galili, M., Oxenlowe, L.K., *et al.* (2010) Demonstration of 5.1 Tbit/s data capacity on a single-wavelength channel. *Optics Express*, **18**, 1438–1443.

51 Mamyshev, P.V. (1998) All-optical data regeneration based on self-phase modulation effect. In *Proceedings of European Conference on Optical Communication (ECOC'98)*, 475–476.

52 Boscolo, S. and Finot, C. (2012) Nonlinear pulse shaping in fibres for pulse generation and optical processing. *International Journal of Optics*, 159057.

53 Tomlinson, W.J., Stolen, R.H., and Johnson, A.M. (1985) Optical wave-breaking of pulses in nonlinear optical fibers. *Optics Letters*, **10**, 457–459.

54 Anderson, D., Desaix, M., Lisak, M, and Quiroga-Teixeiro, M.L. (1992) Wave-breaking in nonlinear optical fibers. *Journal of the Optical Society of America B*, **9**, 1358–1361.

55 Finot, C., Kibler, B., Provost, L. and Wabnitz, S. (2008) Beneficial impact of wave-breaking on coherent continuum formation in normally dispersive nonlinear fibers. *Journal of the Optical Society of America B*, **25**, 1938–1948.

56 De Carlo, L.T. (1997) On the meaning and use of kurtosis. *Psychological Methods*, **2**, 292–307.

57 Finot, C., Parmigiani, F., Petropoulos, P., and Richardson, D.J. (2006) Parabolic pulse evolution in normally dispersive fiber amplifiers preceding the similariton formation regime. *Optics Express*, **14**, 3161–3170.

58 Fermann, M.E., Kruglov, V.I., Thomsen, B.C., Dudley, J.M., and Harvey, J.D. (2000) Self-similar propagation and amplification of parabolic pulses in optical fibers. *Physical Review Letters*, **84**, 6010–6013.

59 Dudley, J.M., Finot, C., Millot, G., and Richardson, D.J. (2007) Self-similarity in ultrafast nonlinear optics. *Nature Physics*, **3**, 597–603.

60 Chong, A., Wright, L.G., and Wise, F.W. (2015) Ultrafast fiber lasers based on self-similar pulse evolution: a review of current progress. *Reports on Progress in Physics*, **78**, 113901.

61 Boscolo, S., Turitsyn, S.K., Novokshenov, V.Y., and Nijhof, J.H.B. (2002) Self-similar parabolic optical solitary waves. *Theoretical and Mathematical Physics*, **133**, 1647–1656.

62 Kruglov, V.I. and Harvey, J.D. (2006) Asymptotically exact parabolic solutions of the generalized nonlinear Schrödinger equation with varying parameters. *Journal of the Optical Society of America B*, **23**, 2541–2550.

63 Finot, C., Millot, G., and Dudley, J.M. (2004) Asymptotic characteristics of parabolic similariton pulses in optical fiber amplifiers. *Optics Letters*, **29**, 2533–2535.

64 Limpert, J.P., Schreiber, T., Clausnitzer, T., *et al.* (2002) High-power femtosecond Yb-doped fiber amplifier. *Optics Express*, **10**, 628–638.

65 Papadopoulos, D.N., Zaouter, Y., Hanna, M., *et al.* (2007) Generation of 63 fs 4.1 MW peak power pulses from a parabolic fiber amplifier operated beyond the gain bandwidth limit. *Optics Letters*, **32**, 2520–2522.

66 Finot, C., Millot, G., Pitois, S., Billet, C., and Dudley, J.M. (2004) Numerical and experimental study of parabolic pulses generated via Raman amplification in standard optical fibers. *IEEE Journal of Selected Topics in Quantum Electronics*, **10**, 1211–1218.

67 Hammani, K., Finot, C., Pitois, S., Fatome, J., and Millot, G. (2008) Real time measurement of long parabolic optical similaritons. *Electronics Letters*, **44**, 1239–1240.

68 Ilday, F. Ö., Buckley, J.R., Clark, W.G., and Wise, F.W. (2004) Self-similar evolution of parabolic pulses in a laser. *Physical Review Letters*, **92**, 213902.

69 Renninger, W.H., Chong, A., and Wise, F.W. (2010) Self-similar pulse evolution in an all-normal-dispersion laser. *Physical Review A*, 822010.

70 Bale, B.G. and Wabnitz, S. (2010) Strong spectral filtering for a mode-locked similariton fiber laser. *Optics Letters*, **35**, 2466–2468.

71 Oktem, B., Ulgudur, C., and Ilday, F. Ö. (2010) Soliton-similariton fibre laser. *Nature Photonics*, **4**, 307–311.

72 Aguergaray, C., Mechin, D., Kruglov, V.I., and Harvey, J.D. (2010) Experimental realization of a mode-locked parabolic Raman fiber oscillator. *Optics Express*, **18**, 8680–8687.

73 Hirooka, T. and Nakazawa, M. (2004) Parabolic pulse generation by use of a dispersion-decreasing fiber with normal group-velocity dispersion. *Optics Letters*, **29**, 498–500.

74 Plocky, A., Sysoliatin, A.A., Latkin, A.I., *et al.* (2007) Experiments on the generation of parabolic pulses in waveguides with length-varying normal chromatic dispersion. *JETP Letters*, **85**, 319–322.

75 Finot, C. Barviau, B., Millot, G., *et al.* (2007) Parabolic pulse generation with active or passive dispersion decreasing optical fibers. *Optics Express*, **15**, 15824–15835.

76 Finot, C., Provost, L., Petropoulos, P., and Richardson, D.J. (2007) Parabolic pulse generation through passive nonlinear pulse reshaping in a normally dispersive two segment fiber device. *Optics Express*, **15**, 852–864.

77 Boscolo, S., Latkin, A.I., and Turitsyn, S.K. (2008) Passive nonlinear pulse shaping in normally dispersive fiber systems. *IEEE Journal of Quantum Electronics*, **44**, 1196–1203.

78 Zeytunyan, A., Yesayan, G., Mouradian, L., *et al.* (2009) Nonlinear-dispersive similariton of passive fiber. *Journal of the European Optical Society: Rapid Publications*, **4**, 09009.

79 Iakushev, S. O., Shulika, O.V., and Sukhoivanov, I.A. (2012) Passive nonlinear reshaping towards parabolic pulses in the steady-state regime in optical fibers. *Optics Communications*, **285**, 4493–4499.

80 Sukhoivanov, I.A., Iakushev, S.O., Shulika, O.V., Díez, A., and Andrés, M. (2013) Femtosecond parabolic pulse shaping in normally dispersive optical fibers. *Optics Express*, **21**, 17769–17785.

81 Kodama, Y. and Wabnitz, S. (1995) Analytical theory of guiding-center nonreturn-to-zero and return-to-zero signal transmission in normally dispersive nonlinear optical fibers. *Optics Letters*, **20**, 2291–2293.

82 Bale, B.G., Boscolo, S., Hammani, K., and Finot, C. (2011) Effects of fourth-order fiber dispersion on ultrashort parabolic optical pulses in the normal dispersion regime, *Journal of the Optical Society of America B*, **28**, 2059–2065.

83 Wang, H., Latkin, A.I., Boscolo, S., Harper, P., and Turitsyn, S.K. (2010) Generation of triangular-shaped optical pulses in normally dispersive fibre. *Journal of Optics*, **12**, 035205.
84 Verscheure, N. and Finot, C. (2011) Pulse doubling and wavelength conversion through triangular nonlinear pulse reshaping. *Electronics Letters*, **47**, 1194–1196.
85 Boscolo, S. and Turitsyn, S.K. (2012) Intermediate asymptotics in nonlinear optical systems. *Physical Review A*, **85**, 043811.
86 Kutuzyan, A.A., Mansuryan, T.G., Esayan, G.L., Akobyan, R.S., and Mouradian, L.K. (2008) Dispersive regime of spectral compression. *Quantum Electronics*, **38**, 383–387.
87 Finot, C. and Boscolo, S. (2016) Design rules for nonlinear spectral compression in optical fibers. *Journal of the Optical Society of America B*, **33**, 760–767.
88 Latkin, A., Boscolo, S., Bhamber, R.S., and Turitsyn, S.K. (2009) Doubling of optical signals using triangular pulses. *Journal of the Optical Society of America B*, **26**, 1492–1496.
89 Stolen. R.H. and Lin, C. (1978) Self-phase modulation in silica optical fibers. *Physical Review A*, **17**, 1448–1453.
90 Zohrabian, A.V. and Mouradian, L.K. (1995) Compression of the spectrum of picosecond ultrashort pulses. *Quantum Electronics*, **25**, 1076.
91 Oberthaler, M. and Höpfel, R.A. (1993) Spectral narrowing of ultrashort laser pulses by self-phase modulation in optical fibers. *Applied Physics Letters*, **63**, 1017–1019.
92 Planas, S.A., Pires Mansur, N.L., Brito Cruz, C.H., and Fragnito, H.L. (1993) Spectral narrowing in the propagation of chirped pulses in single-mode fibers. *Optics Letters*, **18**, 699–701.
93 Andresen, E.R., Thogersen, J., and Keiding, S.R. (2005) Spectral compression of femtosecond pulses in photonic crystal fibers. *Optics Letters*, **30**, 2025–2027.
94 Sidorov-Biryukov, D.A., Fernandez, A., Zhu, L., *et al.* (2008) Spectral narrowing of chirp-free light pulses in anomalously dispersive, highly nonlinear photonic-crystal fibers. *Optics Express*, **16**, 2502–2507.
95 Limpert, J.P., Liem, A., Gabler, T., *et al.* (2001) High-average-power picosecond Yb-doped fiber amplifier. *Optics Letters*, **16**, 1849–1851.
96 Rusu, M. and Okhotnikov, O.G. (2006) All-fiber picosecond laser source based on nonlinear spectral compression. *Applied Physics Letters*, **89**, 091118.
97 Fatome, J., Kibler, B., Andresen, E.R., Rigneault, H., and Finot, C. (2012) All-fiber spectral compression of picosecond pulses at telecommunication wavelength enhanced by amplitude shaping. *Applied Optics*, **51**, 4547–4553.
98 Bao, C., Xiao, X., and Yang, C. (2015) Spectral compression of a dispersion-managed mode-locked Tm:fiber laser at 1.9 um. *IEEE Photonics Technology Letters*, **28**, 497–500.
99 Korobko, D.A., Okhotnikov, O.G., and Zolotovskii, I.O. (2016) Multistage fiber preamplifier comprising spectral compression for generation of high-energy laser pulses. *Journal of the Optical Society of America B*, **33**, 239–245.
100 Boscolo, S., Turitsyn, S.K., and Finot, C. (2012) Amplifier similariton fiber laser with nonlinear spectral compression. *Optics Letters*, **37**, 4531–4533.
101 Washburn, B.R., Buck, J.A., and Ralph, S.E. (2000) Transform-limited spectral compression due to self-phase modulation in fibers. *Optics Letters*, **25**, 445–447.
102 Finot, C., Provost, L., Petropoulos, P., and Richardson, D.J. (2007) Parabolic pulse generation through passive nonlinear pulse reshaping in a normally dispersive two segment fiber device. *Optics Express*, **15**, 852–864.

103 Andresen, E.R., Dudley, J.M., Finot, C., Oron, D., and Rigneault, H. (2011) Transform-limited spectral compression by self-phase modulation of amplitude shaped pulses with negative chirp. *Optics Letters*, **36**, 707–709.

104 Kalashyan, M.A., Palandzhyan, K.A., Esayan, G.L., and Muradyan, L.K. (2010) Generation of transform-limited rectangular pulses in a spectral compressor. *Quantum Electronics*, **40**, 868.

105 Finot, C. and Boscolo, S. (2016) Design rules for nonlinear spectral compression in optical fibers. *Journal of the Optical Society of America B*, **33**, 760.

106 Grigoryan, A.P., Yesayan, G.L., Kutuzyan, A.A., and Mouradian, L.K. (2016) Spectral domain soliton-effect self-compression. *Journal of Physics: Conference Series*, **672**, 012015.

107 Nishizawa, N., Takahashi, H., Ozeki, Y., and Itoh, K. (2010) Wideband spectral compression of wavelength-tunable ultrashort soliton pulse using comb-profile fibre, *Optics Express*, **18**, 11700–11706.

108 Fedotov, A. B., Voronin, A.A., Fedotov, I.V., Ivanov, A.A., and Zheltikov, A.M. (2009) Spectral compression of frequency-shifting solitons in a photonic-crystal fiber. *Optics Letters*, **34**, 662–664.

109 Mitschke, F.M. and Mollenauer, L.F. (1986) Discovery of the soliton self-frequency shift. *Optics Letters*, **11**, 659–661.

110 Gordon, J.P. (1986) Theory of the soliton self-frequency shift. *Optics Letters*, **11**, 662–664.

111 Andresen, E.R., Birkedal, V., Thogersen, J., and Keiding, S.R. (2006) Tunable light source for coherent anti-Stockes Raman scattering microspectroscopy based on the soliton self-frequency shift. *Optics Letters*, **31**, 1328–1330.

112 Johnson, T. and Diddams, S. (2012) Mid-infrared upconversion spectroscopy based on a Yb:fiber femtosecond laser. *Applied Physics B: Lasers & Optics*, **107**, 31–39.

113 Skryabin, D.V., Luan, F., Knight, J.C., and Russell, J.S. (2003) Soliton self-frequency shift cancellation in photonic crystal fibers. *Science*, **201**, 1705–1708.

114 Mamyshev, P.V., Wigley, P.G.J., Wilson, J., *et al.* (1993) Adiabatic compression of Schrödinger solitons due to the combined perturbations of higher-order dispersion and delayed nonlinear response. *Physical Review Letters*, **71**, 73–76.

115 Pant, R., Judge, A.C., Magi, E.C., *et al.* (2010) Characterization and optimization of photonic crystal fibers for enhanced soliton self-frequency shift. *Journal of the Optical Society of America B*, **27**, 1894–1901.

116 Alfano, R.R. (1989) *The Supercontinuum Laser Source*. Cambridge University Press, Cambridge.

117 Dudley, J.M. and Taylor, J.R. (2010) *Supercontinuum Generation in Optical Fibers*. Cambridge University Press, Cambridge.

118 Dudley, J.M., Genty, G., and Coen, S. (2006) Supercontinuum generation in photonic crystal fiber. *Review of Modern Physics*, **78**, 1135–1184.

119 Nakazawa, M., Tamura, K., Kubota, H., and Yoshida, E. (1998) Coherence degradation in the process of supercontinuum generation in an optical fiber. *Optical Fiber Technology*, **4**, 215–223.

120 Corwin, K.L., Newbury, N.R., Dudley, J.M., *et al.* (2003) Fundamental noise limitations to supercontinuum generation in microstructure fibers. *Physical Review Letters*, **90**, 113904.

121 Solli, D.R., Ropers, C., Koonath, P., and Jalali, B. (2007) Optical rogue waves. *Nature*, **450**, 1054.

122 Kibler, B., Hammani, K., Michel, C., Finot, C., and Picozzi, A. (2011) Rogue waves, rational solitons and wave turbulence theory. *Physics Letters A*, **375**, 3149–3155.

123 Dudley, J.M., Finot, C., Millot, G., *et al.* (2010) Extreme events in optics: Challenges of the MANUREVA project. *European Physics Journal Special Topics*, **185**, 125–133.

124 Hammani, K., Kibler, B., Fatome, J., *et al.* (2012) Nonlinear spectral shaping and optical rogue events in fiber-based systems. *Optical Fiber Technology*, **18**, 248–256.

125 Taki, M., Mussot, A., Kudlinski, A., *et al.* (2010) Third-order dispersion for generating optical rogue solitons. *Physics Letters A*, **374**, 691–695.

126 Genty, G., De Sterke, C.M., Bang, O., *et al.* (2010) Collisions and turbulence in optical rogue wave formation. *Physics Letters A*, **374**, 989–996.

127 Hammani, K., Kibler, B., Finot, C., and Picozzi, A. (2010) Emergence of rogue waves from optical turbulence. *Physics Letters A*, **374**, 3585–3589.

128 Genty, G. and Dudley, J.M. (2009) Route to coherent supercontinuum generation in the long pulse regime. *IEEE Journal of Quantum Electronics*, **45**, 1331–1335.

129 Solli, D.R., Ropers, C., and Jalali, B. (2008) Active control of rogue waves for stimulated supercontinuum generation. *Physical Review Letters*, **101**, 233902.

130 Hammani, K., Kibler, B., Finot, C., *et al.* (2011) Peregrine soliton generation and breakup in standard telecommunications fiber. *Optics Letters*, **36**, 112–114.

131 Peregrine, D.H. (1983) Water waves, nonlinear Schrödinger equations and their solutions. *Journal of Australian Mathematical Society: Series B*, **25**, 16–43.

132 Kibler, B., Fatome, J., Finot, C., *et al.* (2010) The Peregrine soliton in nonlinear fibre optics. *Nature Physics*, **6**, 790–795.

133 Akhmediev, N.N. and Korneev, V.I. (1986) Modulation instability and periodic-solutions of the nonlinear Schrödinger equation. *Theoretical and Mathematical Physics*, **69**, 1089–1093.

134 Ozeki, Y., Takushima, Y., Aiso, K., Taira, K., and Kikuchi, K. (2004) Generation of 10 GHz similariton pulse trains from 1,2 km-long erbium-doped fibre amplifier for application to multi-wavelength pulse sources. *Electronics Letters*, **40**, 1103–1104.

135 Parmigiani, F., Finot, C., Mukasa, K., *et al.* (2006) Ultra-flat SPM-broadened spectra in a highly nonlinear fiber using parabolic pulses formed in a fiber Bragg grating. *Optics Express*, **14**, 7617–7622.

136 Clarke, A.M., Williams, D.G., Roelens, M.A.F., and Eggleton, B.I. (2010) Reconfigurable optical pulse generator employing a Fourier-domain programmable optical processor. *Journal of Lightwave Technology*, **28**, 97–103.

137 Xin, Y., Richardson, D.J., and Petropoulos, P. (2013) Broadband, flat frequency comb generated using pulse shaping-assisted nonlinear spectral broadening. *IEEE Photonics Technology Letters*, **25**, 543–545.

138 Millot, G., Pitois, S., Yan, M., *et al.* (2016) Frequency-agile dual-comb spectroscopy. *Nature Photonics*, **10**, 27–30.

139 Boivin, L. and Collings, B.C. (2001) Spectrum slicing of coherent sources in optical communications. *Optical Fiber Technology*, **7**, 1–20.

140 Yousoff, Z., Petropoulos, P., Furusawa, F., Monro, T.M., and Richardson, D.J. (2003) A 36-channel x 10-GHz spectrally sliced pulse source based on supercontinuum generation in normally dispersive highly nonlinear holey fiber. *IEEE Photonics Technology Letters*, **15**, 1689–1691.

141 Klimczak, M., Sobon, G., Kasztelanic, R., Abramski, K.M., and Buczynski, R. (2016) Direct comparison of shot-to-shot noise performance of all normal dispersion and anomalous dispersion supercontinuum pumped with sub-picosecond pulse fiber-based laser. *Scientific Reports*, **6**, 19284.

142 Blanco-Redondo, A., Husko, C., Eades, D., *et al.* (2014) Observation of soliton compression in silicon photonic crystals. *Nature Communications*, **5**, 3160. DOI:10.1038/ncomms4160.

143 Blanco-Redondo, A., de Sterke, C.M., Sipe, J.E., *et al.* (2016) Pure-quartic solitons, *Nature Communications*, **7**, 10427. DOI:10.1038/ncomms10427.

6

Nonlinear-Dispersive Similaritons of Passive Fibers: Applications in Ultrafast Optics

Levon Mouradian[1] *and Alain Barthélémy*[2]

[1] *Ultrafast Optics Laboratory, Faculty of Physics, Yerevan State University, Yerevan, Armenia*
[2] *Institut de recherche XLIM, UMR 7252 CNRS-Université de Limoges, Limoges, France*

6.1 Introduction

Similariton pulses with their distinctive property of self-similar propagation, recently have attracted the attention of researchers due to fundamental interest and prospective applications in ultrafast optics and photonics, particularly for high-power pulse amplification, optical telecommunications, ultrafast all-optical signal processing, etc. [1, 2].

The self-similar propagation of high-power pulses with parabolic temporal, spectral, and phase profiles was predicted theoretically in the 1990s [3]. In practice, the generation of such parabolic similaritons is possible in active fibers, such as rare-earth-doped fiber amplifiers [4–6] and Raman fiber amplifiers [7], as well as in lasers [8]. The generation of a parabolic similariton has also been proposed in a tapered fiber with decreasing normal dispersion, using either a passive dispersion-decreasing fiber [9] or a hybrid configuration with Raman amplification [10]. Another type of similariton is generated in a conventional uniform and passive (without gain) fiber, under the sole combined impacts of Kerr nonlinearity and dispersion [11]. In contrast to the parabolic similariton with parabolic amplitude and phase profiles, this nonlinear-dispersive (NL-D) similariton has only a parabolic phase and its temporal and spectral shapes no longer follow the parabolic profiles. However, the property of self-similarity is preserved. Our spectral interferometric studies of this type of pulse [11, 12] show the linearity of their chirp (parabolic phase), with a slope given only by the fiber dispersion. This property leads, on one hand, to a spectrotemporal self-similarity, i.e., to the self-similar propagation of both the pulse amplitude and its spectrum, and on the other hand to self-spectrotemporal imaging, i.e., to a replication of the pulse profile in the spectral domain [11].

From a theoretical point of view, parabolic similariton can be observed in a waveguide with gain where the reduction in peak power due to pulse broadening can be compensated by amplification, as mentioned above. The balance can be achieved as well in passive waveguides with decreasing dispersion. In these cases parabolic similaritons are

Shaping Light in Nonlinear Optical Fibers, First Edition. Edited by Sonia Boscolo and Christophe Finot.

genuine nonlinear waves, i.e., self-similar solutions of the wave equation, nonlinearity playing a role all along the propagation. The situation is different in the case of NL-D similariton, since nonlinearity plays a major role only in the early stage of the propagation (precisely, only in the stage of shaping), and therefore, it cannot be strictly considered as a true nonlinear wave.

From a practical point of view, both the parabolic similariton of active fiber and the NL-D similariton of passive fiber are of interest for applications to ultrafast optics and photonics, namely, for femtosecond signal generation, characterization, delivery, and processing. Particularly, the similariton-based techniques of pulse compression [13, 14] and shaping [15], temporal lensing, spectrotemporal imaging, and spectral interferometry [16–18] have been demonstrated. These applications to ultrafast optics are improved by the generation of broadband similaritons. For example, the resolution of the femtosecond oscilloscope, based on the similariton-induced parabolic lens, is given by the bandwidth of the similariton, and the application range for similariton-based spectral interferometry is governed by the similariton-reference bandwidth [16, 17]. The pulse compression ratio is also fixed by the spectral broadening factor [13, 14]. Experimentally, the spectral interferometric study has led to the complete characterization of NL-D similaritons of up to 5 THz bandwidths [11,12]. For broadband similaritons (of up to 50-100-THz bandwidths), the chirp measurement has been performed by sum-frequency generation with a stretched reference for improved accuracy through spectral compression [19–22]. In this ultra-broadband regime, the application range of NL-D similaritons seems to be wider than the one of parabolic similariton because their bandwidth is restricted by the gain bandwidth.

The generation and applications of broadband NL-D similariton to nonlinear ultrafast photonics, based on our experimental and numerical studies, are the subject of this chapter. Some particular topics listed below are going to be addressed.

- *Generation of broadband NL-D similariton.* The signal synthesis and analysis needs on the femtosecond time scale can be satisfied by broadband similariton-based devices. Broadband NL-D similaritons were generated with ~80 THz-bandwidth (>150 nm at 800 nm wavelength), coupling 100-fs pulses into a short piece of standard single-mode fiber (~1 m).
- *NL-D Similariton pulse compression.* The use of similariton can improve the techniques of pulse compression and shaping, leading to accurate, aberration-free methods, since the chirp of the similariton is linear (the phase is parabolic) and its spectrotemporal profile is smooth and bell-shaped. We generated broadband NL-D similaritons which were further compressed in a conventional prism compressor down to 15–20 fs with average power of 300–500 mW, comparable with the parameters of commercial 10-fs lasers. The use of a hybrid prism-grating compressor or grism-line, free of third-order dispersion, instead of the prism compressor, would provide even shorter pulses, down to a few femtoseconds, starting from NL-D similaritons of 80 THz bandwidth.
- *Spectral focusing in similariton-induced temporal lens.* A similariton-induced parabolic temporal lens provides an effective aberration-free spectral compression method, by analogy with the beam collimation in the space domain. This method is based on the dispersive stretching of the pulse followed by the compensation of the dispersion-induced phase. It is achieved by adding a parabolic phase with the

appropriate sign and curvature through a quadratic nonlinear process instead of the traditional Kerr self-phase modulation (SPM). The method can be implemented with various frequency mixing processes, such as sum- or difference-frequency generation, CARS, etc., using for a compensation signal an NL-D similariton generated from a fraction of the input pulse. Experimentally, we have achieved an effective (up to 23 times) aberration-free spectral compression through sum-frequency generation. This type of spectral focusing is of special interest for CARS spectroscopy in view of its spectral resolution improvement.

- *Noise suppression-filtering in similariton-induced temporal lens.* Similariton regulates the optical wave parameters, and thus, by consequence the compressed pulse parameters are weakly impacted by the input noise. The noise filtering should occur as well through spectral focusing in the similariton-induced temporal lens.
- *Spectral control of signal in the similariton-induced temporal lens.* A delay between the similariton used for phase compensation and the stretched signal pulses leads in the sum frequency process to a frequency shift of the spectrally compressed radiation. The combination of spectral focusing with a fine frequency tuning in the similariton-induced parabolic temporal lens can serve for resonant spectroscopy and optical communication. This technique allows also measurement of the similariton chirp and of the dispersion of the material where the similariton is generated.
- *Similariton-based chirped CARS spectroscopy and microscopy.* The use of broadband femtosecond laser pulses in high resolution CARS microscopy was demonstrated by the use of pulse chirping and spectral focusing, providing both spectral tunability and high contrast imaging [23]. Another approach of special interest is based on the periodical amplitude-modulation of broadband radiation with high-repetition rate, in resonance with the Raman oscillations of medium [24]. The application of broadband NL-D similariton to CARS microscopy, with the two above-mentioned approaches, is promising in view of the exploitation of a single laser source associated to a substantially simplified setup.
- *Pulse spectrotemporal imaging in similariton-induced temporal lens,* i.e., conversion of the pulse temporal profile to the spectral domain for both the intensity and phase. The temporal lens serves as an optical processor which performs the operation of time to frequency Fourier Transformation (FT), by analogy with the space to wavevector Fourier Transformation achieved by a spatial lens. Direct, real-time, high-resolution temporal measurements were carried out through the spectral imaging of temporal pulse based on an aberration-free similariton-induced temporal lens, leading to the development of a femtosecond optical oscilloscope [16, 22]. The resolution of measurements is given by the bandwidth of similariton, and it is at the level of ~ 5 fs for 80-THz bandwidth similariton.
- *Similariton-based self-referencing spectral interferometry* (SI) for femtosecond pulse complete electric field characterization. The classic spectral interferometry is based on the interference of the signal and of a reference fields in a spectrometer, giving a spectral fringe pattern caused by the difference of spectral phases. The SI measurement is accurate but its application range is restricted by the need for a reference pulse with a suited bandwidth. We improved the method by using a similariton for reference and by its generation from a fraction of the signal to be measured. Thus, the method of similariton-based SI exhibits the performance of the classic SI with the benefit of self-referencing [16,17,21]. Our comparative experiments of similariton-based SI and

spectrotemporal imaging, carried out together with autocorrelation measurements, evidenced the quantitative agreement and high precision of both methods for accurate femtosecond-scale temporal measurements.

- *Reverse problem of NL-D similariton generation in view of femtosecond pulse characterization.* The NL-D similariton asymptotically has a linear chirp independent of the pulse initial parameters. The chirp slope is determined only by the fiber dispersion. Therefore the mapping of the pulse time profile in the spectral domain due to the self-spectrotemporal imaging of similariton is quantitative with a fixed coefficient. Hence, the initial pulse intensity profile can be measured by a simple spectrometer. Based on the measured spectrum module, together with the spectral phase which is known, computation of the reverse propagation provides complete information about the input field. In this way a short piece of fiber can serve as an alternative to the FROG device [26]. The solution of the reverse problem of the generation of such a similariton is also helpful for femtosecond pulse delivery [25].

Outlining the chapter, a brief analysis of the "far field" of dispersion, in spatio-temporal analogy with Fraunhofer diffraction, is presented together with "spectron" pulse shaping and other applications of such a dispersive Fourier transformation. Afterwards, we pass to the studies of the nature and peculiarities of the NL-D similaritons generated in passive fiber, and to the specificity of broadband similaritons with applications to ultrafast photonics. Then, similariton-induced time lensing, aberration-free spectral focusing, similariton-based spatiotemporal imaging, and self-referencing spectral interferometry methods are presented. Finally, the prospects of similaritons for material characterization, and for simple femtosecond signal characterization are discussed.

6.2 Spectron and Dispersive Fourier Transformation

A spectron pulse, of which the profile in time is an image of its spectrum, is known to result from propagation in the "far zone of dispersion." This is by full analogy in the time domain with what happens in space for beam, the Fraunhofer diffraction leading to far field patterns [27–29]. Such a dispersive frequency-to-time Fourier transformation is connected with the linear chirp of the pulse stretched by second-order chromatic dispersion. The phenomenon has a wealth of applications for signal synthesis and analysis at the femtosecond timescales. Obviously, the spectron pulses, because of their similar spectral and temporal profiles, evolve in a self-similar way during their (linear) propagation.

More precisely, the pulse propagation in a dispersive medium, of which the dispersion is limited to second order, is described by a parabolic equation, similar to the one for paraxial beam diffraction [29,30]. In terms of classic radiophysics, dispersion in the spectral domain acts as a spectral phase modulator with the transfer function $\tilde{D}_z(\omega) = \exp[-i\phi(\omega)]$, and a parabolic phase term under the second-order dispersion approximation: $\phi(\omega) = \beta_2\omega^2 z/2$ (ω is the frequency difference with the center frequency, $\phi(\omega)$ is the spectral phase, β_2 is the coefficient of second order dispersion supposed to be positive in the whole chapter, and z denotes the propagation distance [29,30]). Thus, for the pulse propagation trough the dispersive medium, with $A(t)$ the slowly varying temporal field amplitude and $\tilde{A}(\omega) \equiv FT^{-1}[A(t)]$ its spectral amplitude, we have

$$\tilde{A}_z(\omega) = \tilde{D}_z(\omega)\tilde{A}_0(\omega) = \tilde{A}_0(\omega)\exp(-i\tilde{C}\omega^2/2) \tag{6.1}$$

with a coefficient $\tilde{C} \equiv \ddot{\phi}(\omega = 0) = \beta_2 z$ of parabolic spectral phase. In the time domain, according to convolution theorem

$$A_z(t) \propto A_0(t) \otimes \exp(iCt^2/2) \tag{6.2}$$

with a temporal phase coefficient $C = \tilde{C}^{-1}$. In the far zone of dispersion, where the output pulse duration Δt exceeds the initial one Δt_0, $\Delta t \gg \Delta t_0$, i.e. the propagation distance significantly exceeds the characteristic dispersion length $L_D \equiv (\beta_2 \Delta\omega_0^2)^{-1}$ ($\Delta\omega_0$ stands for the pulse full bandwidth), $z \gg L_D$ [29,30], we have frequency-to-time Fourier conversion:

$$A_z(t) \propto \tilde{A}_0(\omega)\exp(-i\beta_2 z\omega^2/2)|_{\omega=Ct} = \tilde{A}_0(\omega)|_{\omega=Ct}\exp(iCt^2/2) \tag{6.3}$$

The coefficient C, which gives the scale of frequency-to-time conversion $\omega = Ct$, corresponds here to the chirp slope $C = \ddot{\varphi}(t) = \dot{\omega}(t)$, according to Eq. (6.3).

$A_z(t)$ represents a spectron pulse provided the condition of temporal Fraunhofer zone are met, which implies some sufficiently large pulse stretching $s \equiv \Delta t/\Delta t_o \approx \Delta\omega_o^2/C \gg 1$. That gives also the precision $1/s \approx C/\Delta\omega_o^2$ of the spectron's spectrotemporal similarity:

$$I(t) = |A_z(t)|^2 \propto |\tilde{A}(\omega)|^2_{\omega=Ct} = S(\omega)|_{\omega=Ct} \tag{6.4}$$

For a 100-fs pulse at 800 nm propagating in a standard single-mode fiber L_D is of ~10 cm, and at the output of 1-m fiber we will have a pulse broadening of $s \approx z/L_D \sim 10$, and a spectrotemporal similarity of spectron of $1/s$ ~10% precision.

The spectron shaping, known in its applications as dispersive Fourier transformation [31–35] or real-time Fourier transformation [36–39], is a unique measurement technique that overcomes the speed limitations of traditional optical instruments. It enables fast, continuous or single-shot measurements in optical sensing, in spectroscopy, in imaging and in the analysis of complex nonlinear pulse dynamics. Dispersive Fourier transformation, due to chromatic dispersion, maps the spectrum of an optical pulse to a temporal waveform, whose intensity can be read by a single-pixel photodetector to capture the spectrum at a scan rate significantly beyond what is possible with conventional space-domain spectrometers. This approach has brought us a new class of real-time instruments that permit the capture of rare events such as optical rogue waves and rare cancer cells in blood, which would otherwise be missed using conventional instruments. The modern implementations of dispersive Fourier transformation across a wide range of diverse applications are described in detail in the review [31].

6.3 Nonlinear-Dispersive Similariton

In this section, we present the study of NL-D similaritons of passive fiber: the shaping of such similaritons, their distinctive properties, especially their origin, nature, and relation to the spectron and flattop pulses, and their temporal, spectral and phase features in view of their potential applications. Our spectral interferometric studies for the complete characterization of the NL-D similariton generated in passive fiber are further described.

Outlining this section, a simplified analytical discussion and numerical studies are first given of the features of NL-D similariton, afterwards, experiments for the spectral interferometric characterization of NL-D similariton are presented to demonstrate the peculiarities of similariton predicted by the theory, then the technique for the measurement of similariton chirp by the use of spectrometer and autocorrelator is described, and finally, the study of the bandwidth rule of NL-D similariton is presented.

6.3.1 Spectronic Nature of NL-D Similariton: Analytical Consideration

Introducing NL-D similariton, we roughly analyze the pulse propagation in passive single-mode fiber, taking into account the "length-scales" of nonlinear and dispersive effects.

During pulse propagation in a conventional single-mode fiber, the Kerr SPM broadens the spectrum and increases the impact of dispersion, leading to the shaping of NL-D similariton with a high degree of spectrotemporal similarity. Quantitatively, for a Δt_0 $\sim$100 fs pulse ($\lambda\sim$800 nm) in a single-mode fiber with average power $P \sim 100$ mW at a 76 MHz repetition rate, the nonlinear interaction length $L_{NL} \equiv (\beta_0 n_2 I_0)^{-1}$ is much shorter than the dispersive one: L_{NL} $\sim$1 cm $\ll L_D$ $\sim$10 cm (n_2 is the coefficient of the Kerr nonlinearity, and $\beta_0 = 2\pi/\lambda_0$ is wave number). This allows us to roughly split the impact of nonlinear self-interaction and dispersion, assuming that we have first pure nonlinear SPM and spectral broadening, and afterwards pure dispersive stretching and dispersive propagation. For the dispersion-induced phase of pulse at the fiber length L_f ($\sim$1 m), we have $\varphi_D(t) = Ct^2/2$ together with an additional phase $\varphi_{NL}(t)$ that comes from the initial nonlinear propagation on a distance of $\sim L_{NL}$. Assuming $\varphi_{NL}(t)$ to be parabolic at the central energy-carrying part of the spectrum, we have at the output: $\varphi_{NL}(t) = C_{NL}^{-1}\omega^2/2|_{\omega=Ct} = (C^2/C_{NL})t^2/2$, with $C_{NL} \equiv \ddot{\varphi}_{NL}(t)$. For the overall output phase $\varphi_\Sigma = \varphi_D + \varphi_{NL}$, we have $\varphi_\Sigma(t) = Ct^2(1 + C/C_{NL})/2$. Considering the nonlinear spectral broadening and dispersive pulse stretching factors (respectively denoted by $b \equiv \Delta\omega/\Delta\omega_0$ and $s \equiv \Delta t/\Delta t_0$; $\Delta\omega$ stands for the broadened spectral bandwidth), we have for the nonlinear, dispersive, and overall chirp slopes at the output, respectively: $C_{NL} = \Delta\omega/\Delta t_0 = \Delta\omega_0^2 b$, $C = \Delta\omega_0/\Delta t = \Delta\omega_0^2/s$, and $C_\Sigma = C(1 + C/C_{NL}) = C[1 + (sb)^{-1}]$. Since $C/C_{NL} = (sb)^{-1}$, for spectral broadening b $\sim$10 and pulse stretching s $\sim$10 (with the above parameters), we will have $C_\Sigma \approx C$, with an accuracy of C/C_{NL} $\sim$1%.

Thus, for the femtosecond pulse NL-D similariton generated at the exit of a 1-m-long fiber, we have a spectrotemporal similarity of $\sim (sb)^{-1} = C/\Delta\omega^2$ $\sim$1% precision. Considering the key peculiarity of the NL-D similariton, that practically the fiber dispersion determines the chirp slope, we can describe it in the following way:

$$A_f(t) \propto |\tilde{A}_f(\omega)|_{\omega=Ct} \exp(iCt^2/2) \quad \text{with} \quad C \approx (\beta_2 L_f)^{-1} \tag{6.5}$$

Another interesting issue is the relation of NL-D similariton with flat top rectangular pulses, resulting from the propagation of quasi Gaussian pulses on the fiber length $L_f \sim 2\sqrt{L_D L_{NL}}$. For such conditions, the temporal stretching of the flat top pulses is $\Delta t \approx 2\Delta t_0$ and its spectral broadening is given by $\Delta\omega \approx 2\Delta\omega_0(L_D/L_f)$, so that the chirp slope corresponds to the value: $C = \Delta\omega/\Delta t \approx (2\Delta\omega_0 L_D/L_f)/(2\Delta t_0) = (\beta_2 L_f)^{-1}$. Therefore, during NL-D propagation in a fiber, an initial standard pulse is stretched and reshaped in a flat top profile and its chirp becomes equal to the one of a pulse under purely dispersive propagation $C \approx (\beta_2 L_f)^{-1}$, starting from the fiber length $L_f \sim 2\sqrt{L_D L_{NL}}$. Thus, NL-D flattop pulses can be considered as an earlier stage of the shaping of NL-D similaritons.

Summarizing our simplified analysis, we can state that waveguide dispersion only rules out the spectronic nature of NL-D similariton of passive fiber, as well as its spectrotemporal imaging capability which scaling parameters are fixed by the dispersion value.

6.3.2 Physical Pattern of Generation of NL-D Similariton, Its Character and Peculiarities on the Basis of Numerical Studies

To check the terms and conclusions of the qualitative analytical discussion above, a more quantitative analysis of the process was carried out through numerical modeling [11]. The mathematical description of the pulse NL-D propagation in a single-mode fiber is based on the standard nonlinear Schrödinger equation (NLSE) which is sufficient to deal with pulse durations of ≥ 100 fs in normal dispersion regime [29,30]. The split-step Fourier method was applied to solve the NLSE. In the simulations, the pulse propagation distance is expressed in units of dispersive lengths L_D; the optical intensity of the guided mode is given by the nonlinear parameter $R \equiv L_D/L_{NL} = (\beta_2 \Delta\omega_0^2)^{-1} \beta_0 n_2 I_0$ (e.g., $R = 6$ for 100 mW average power, 100-fs pulses at 76 MHz and 800 nm wavelength in a standard single-mode fiber). The dimensionless running time t and center frequency ω are normalized to the input pulse duration Δt_0 and bandwidth $\Delta\omega = 1/\Delta t_0$, respectively. Figure 6.1 shows the dynamics of similariton shaping: pulse (top row), chirp (middle row) and spectrum (bottom row) are plot after propagation on eight different distances. Figure 6.1 (a) illustrates the first step of NL-D self-interaction when, typically, an initially Gaussian-shaped pulse is stretched and reshaped into a flat top profile. This is a strongly nonlinear and weakly dispersive regime. Figure 6.1 (b) shows the following step of similariton shaping. The spectral broadening and drop of the pulse peak power make dispersion dominate over nonlinearity. The pulse gets a linear chirp (parabolic phase), and the temporal self-similarity of NL-D similariton can be observed (Figure 6.1 (b)). The simulations show that the same evolution occurs even in the case of input pulses with complex initial profile. The modulations of the temporal and spectral profiles are pushed toward the edges during the nonlinear transformation, so that pulse and spectrum become more and more similar and parabolic. One example of NL-D similariton shaping in the case where the initial pulse is modulated by a sub-structure is given in Figure 6.2 together with its dynamics.

It is important that our simulations on the generation NL-D similariton confirm the prediction of the previous analytical discussion. The output chirp slope is observed to be nearly independent of the input pulse intensity and to vary linearly with the fiber length. For a fixed fiber length, whatever the spectral broadening and the temporal stretching obtained when raising the input power, the chirp of the output signal is still the same. It extracts the full information on the NL-D similariton just from data on the spectrum and the fiber length.

We also studied the chirp of the NL-D similariton versus the chirp of the input pulse: the chirp slope of the similariton is practically constant, when the pulse intensity is high enough. In case of purely dispersive propagation, the induced chirp and the initial chirp

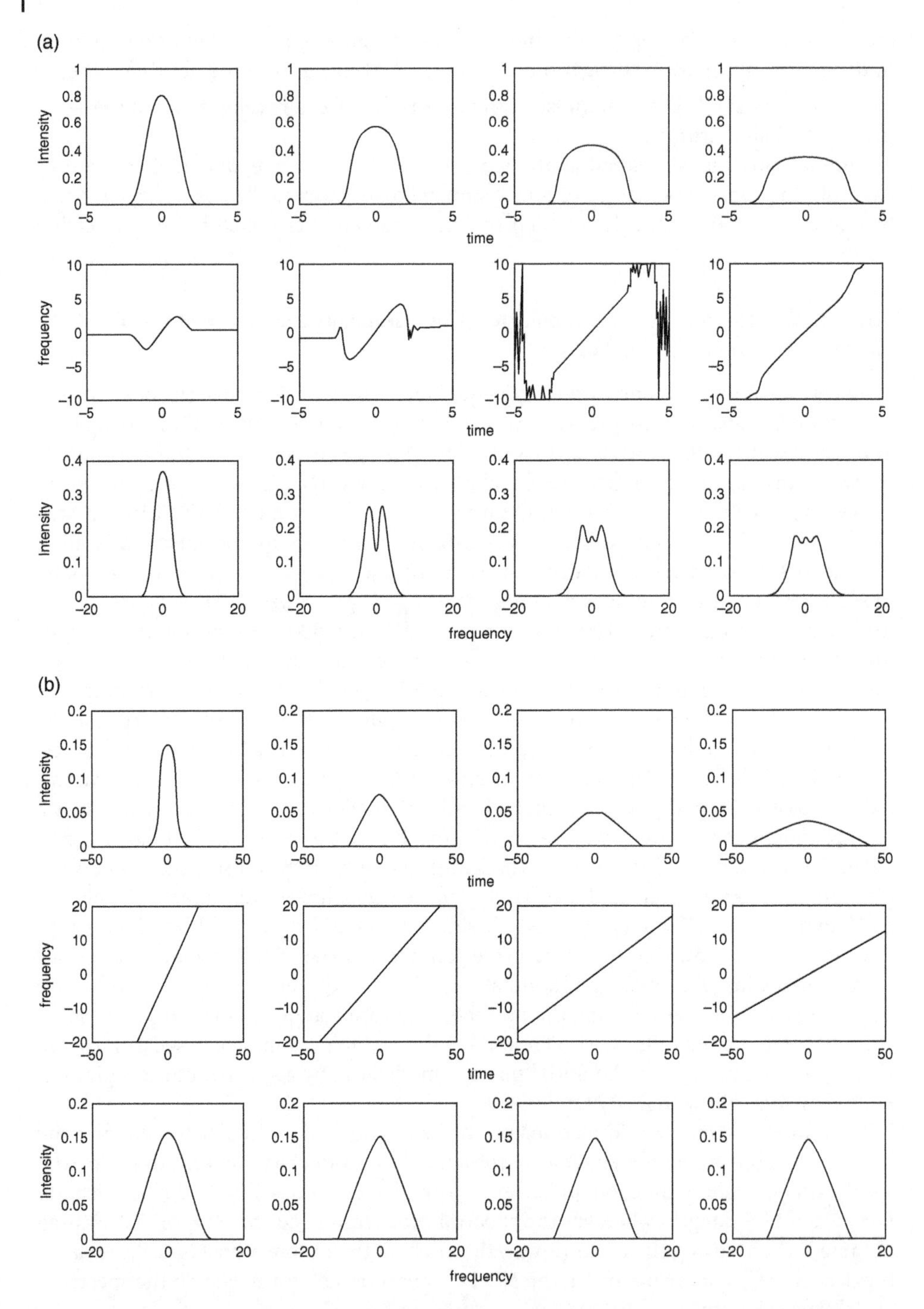

Figure 6.1 Shaping of an initial Gaussian pulse into a flattop pulses (a) and then evolution toward a NL-D similariton (b) during nonlinear dispersive propagation. From top to bottom: pulse intensity profile, chirp, and spectrum. $R = 30$. From left to right: pulse evolution in fiber for (a) $L_f/L_D = 0.1, 0.2, 0.3, 0.4$, and (b) for $L_f/L_D = 1; 2, 3, 4$.

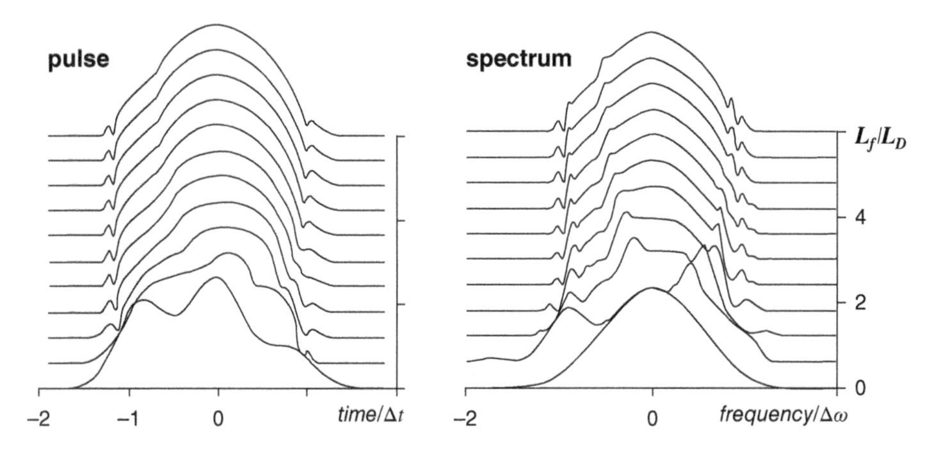

Figure 6.2 NL-D similariton shaping starting from a pulse with sub-structure: the pulse (left) and spectrum (right) dynamics. Temporal and spectral intensity profiles are normalized to their current peak values, time and frequency are normalized to current values of the pulse duration and spectral width; $R = 40$.

are obviously simply cumulated according to Eq. (6.4). In case of nonlinear dispersive propagation, the final chirp is almost independent of the initial chirp according to the above theoretical discussion so that the output pulse can be described by Eq. (6.5).

6.3.3 Experimental Study of NL-D Similariton by Spectral Interferometry (and also Chirp Measurements by Spectrometer and Autocorrelator)

In order to confirm the behaviors emphasized in the above theoretical and numerical analysis, we carried out experimental investigations. We applied the classic method of spectral interferomety [40,41] to completely characterize and study the generation process and the peculiarities of NL-D similariton of passive fiber.

Figures 6.3 and 6.4 illustrate our experiments. In all the experiments reported in this chapter the laser source was a standard Coherent Verdi V10 – Mira 900F femtosecond

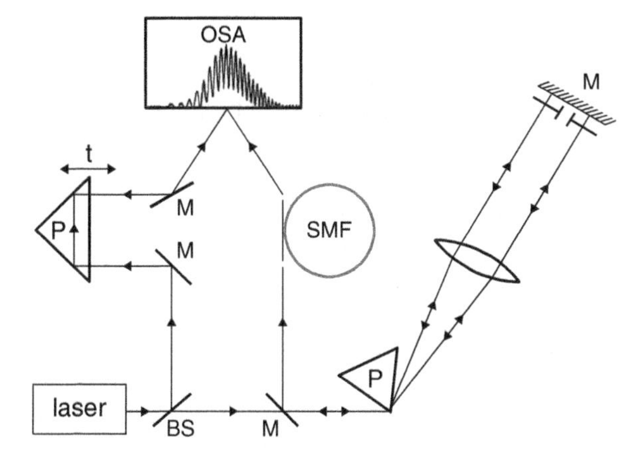

Figure 6.3 Schematic of the experiment. Key: Laser: femtosecond laser system, BS – beam splitter, M – mirrors, P – prisms, L – lens, OSA – optical spectral analyzer, and SMF – single-mode fiber.

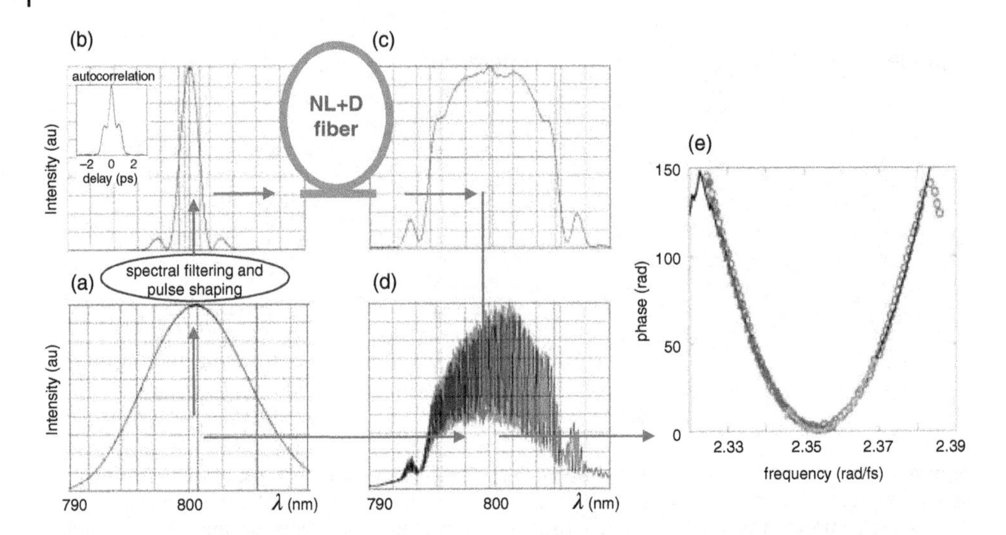

Figure 6.4 (a–d) Schematic of the experiment given by the spectrograms recorded at different locations in the set-up, and (e) spectral phases of NL-D similaritons generated from various inputs; single pulse (black), close double pulses (red o) and distant double (blue x) pulses in comparison with the one for pure dispersive propagation of a single pulse (yellow). For a color version of this figure please see color plate section.

Ti:sapphire laser system delivering at 76 MHz repetition rate ~100 fs pulses with $\Delta\lambda =$ 11 nm around 800 nm at a maximum average power of 1.5 W. The laser beam was sent in a first beam splitter giving two separate pulses carrying unequal powers. The low-power pulse served as a reference. For the high-power pulse, its spectrum was filtered out from a bandwidth $\Delta\lambda$ =11 nm down to the value $\Delta\lambda = 2$ nm. Then the light was launched into pieces of standard polarization-preserving fibers (Newport F-SPF @820 nm and ThorLabs HP @ 780 nm) of different lengths – 1 m, 9 m, and 36 m. The fiber output was combined with the reference pulse and then sent to an optical spectrum analyzer (OSA Ando 6315). The spectra of the pulses at the output of fiber were broadened, however, they still overlap and interfered with the spectrum of the reference pulse. This allowed measurement of the spectral phase of similariton within the whole range of its spectrum. The spectral interferometric fringe pattern was recorded by the OSA and the spectral phase was retrieved. Having the spectrum and retrieved spectral phase, the temporal profile of the similariton has been recovered by Fourier transformation.

The result of the experiment is given schematically in Figure 6.4 in the color plate section by means of the spectrograms of the relevant steps. Figure 6.4 (a) shows the spectrum of the laser pulse, Figure 6.4 (b) is the spectrum of the spectrally filtered pulse, Figure 6.4 (c) is the spectrum of the NL-D similariton and Figure 6.4 (d) is the spectral interference pattern. Figure 6.4 (e) shows the measured spectral phases of the similaritons generated from different input pulses. The spectral phases are parabolic ($\phi = -\tilde{C}\omega^2/2$) and their coefficients $\tilde{C}$ have nearly the same values in all cases of dispersive and NL-D propagations: $\tilde{C} = 0.32$ ps^2 for the purely dispersive propagation of a single-peak pulse, and 0.33 ps^2, 0.328 ps^2, 0.35 ps^2 for the NL-D propagations of single pulses, close double pulses and distant double pulses, respectively.

The parabolic phase (i.e., linear chirp) leads to the self-spectrotemporal imaging property of similariton. The accuracy of imaging increases with the decreasing of the chirp

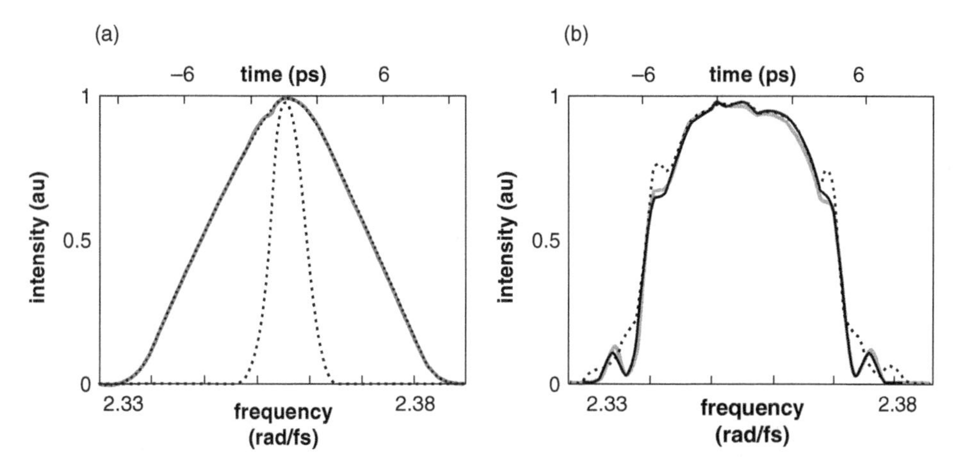

Figure 6.5 Self-spectrotemporal imaging of NL-D similaritons generated from (a) single-, and (b) double-peak pulses in a 9-m long fiber. Black solid curves show the spectra of NL-D propagated pulses (the black dotted one of (a) stands for pure dispersively propagated pulse), gray dotted lines show the temporal profiles retrieved by spectral interferometry. The gray curve of (b) shows the temporal profiles of NL-D similaritons for a 4x increased coefficient of spectral phase, corresponding to a fiber length of 36 m.

slope $C = \tilde{C}^{-1}$. Figure 6.5 shows the spectrotemporal profiles of similariton pulses in case of fiber length $L_f = 9$ m. Based on the spectral phase and spectral profile, the temporal profile of NL-D similariton is retrieved. Figure 6.5 shows the spectra, and temporal profiles of similaritons recovered using the spectral phases of Figure 6.4 (e). The black curves are the spectra and the gray-dotted curves are the reconstructed pulse. The intensity profiles are almost indistinguishable and perfectly coincide with each other, i.e., the self-spectrotemporal imaging of NL-D similariton is clearly demonstrated. A perfect spectrotemporal similarity is seen in Figure 6.5 (a) that corresponds to the case of a single-peak input pulse. For the case of a double-peak input pulse shown in Figure 6.5 (b), the matching between the spectral and temporal profiles is not complete. To obtain a better agreement, a longer fiber should be used, increasing the $\tilde{C}$ coefficient: simulations show that the spectral and temporal profiles of similaritons coincide precisely for a 36-m long fiber (the gray line in Figure 6.5 (b)).

To show the relation between similaritons and flat top pulses, spectral interferometric measurements were carried out using a short fiber piece (L_f =1 m) keeping all the other parameters similar to the ones of the previous experiments. Figure 6.6 in the color plate section reports the results of this study for flat top pulses shaped in an NL-D fiber.

Here the black curves plot the measured spectrum and spectral phase, the pink and blue curves are for parabolic and high-order polynomial fits. The measured spectral phase has a parabolic shape only at the central part of spectrum. Similarly to the case of nonlinear dispersive similaritons the chirp slope in the central part of the pulse is given only by the fiber length and dispersion ($\tilde{C} = 0.0465$ ps^2). Deviation from the parabola on the spectral wings leads to a recovered pulse closer to a flat top shape (see the blue curve in Figure 6.6 (b)). For comparison, the pink curve in Figure 6.6 (b) shows a bell-shaped pulse reconstructed by the use of a parabolic spectral phase.

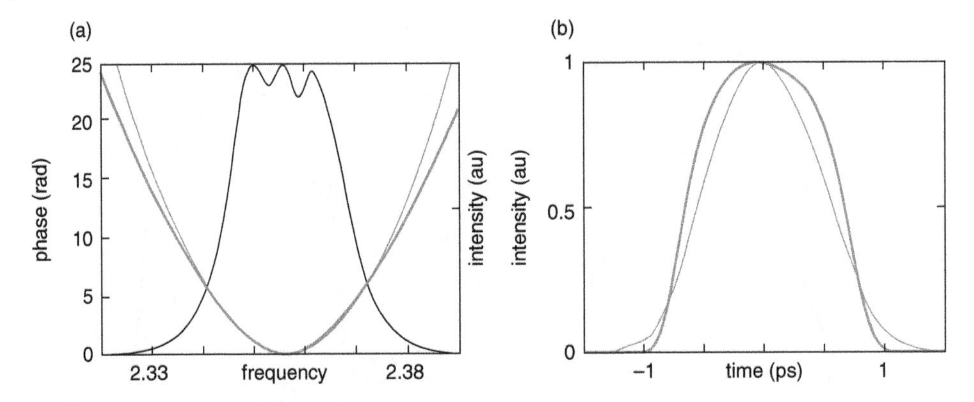

Figure 6.6 (a) Measured spectrum of NL-D flattop pulses with its retrieved spectral phase (black solid line). The pink (thin grey) and blue (thick grey) lines are for parabolic and high-order polynomial fits, respectively. (b) Reconstructed temporal profile of the flattop pulse (blue) in comparison with one reconstructed by the use of parabolic spectral phase (pink). For a color version of this figure please see color plate section.

Detailed studies of NL-D similariton, carried out through spectral interferometry together with numerical modeling, confirm its spectronic nature and description by Eq. (6.5). So its spectrotemporal similarity is demonstrated with a scaling coefficient given by the chirp slope $C = (\beta_2 L_f)^{-1}$.

This allowed us to carry out the characterization of similariton in a simpler way. We measured the spectrum and autocorrelation trace of the fiber output. We calculated the spectrum autocorrelation, and we determined the chirp coefficient C, as the frequency-to-time scaling coefficient ($\omega = Ct$) that permitted to fit the calculated autocorrelation of spectrum with the measured pulse intensity autocorrelation. The results were in good agreement with the spectral interferometric measurements. This simple method permitted us also to easily check the results of the numerical study. The experimental results, carried out in the range of the initial chirp coefficient C_0 from $-\Delta\omega_0^2/3$ to $+\Delta\omega_0^2/3$, show that the chirp of NL-D similariton is practically independent of the input pulse phase modulation in agreement with numerical studies.

6.3.4 Bandwidth and Duration of NL-D Similariton

Taking into account the discussed relation between NL-D similariton and flattop pulse, it seems reasonable to expect that the *bandwidth of similaritons* is approximately equal to the one for flattop pulses. To determine the bandwidth (and afterwards the duration) of NL-D similaritons, the relationship giving the pulse optimal compression can be used [29]. This gives the following identity for the spectral broadening of NL-D similariton: $b \equiv \Delta\omega/\Delta\omega_o \approx \sqrt{R} \equiv \sqrt{L_D/L_{NL}} = k\sqrt{P}/\Delta\omega_0$, where P is the input pulse peak power, and $k \equiv \sqrt{n_2\beta_0(\beta_2 A)^{-1}}$ is a coefficient fixed by the fiber parameters (A is the fiber mode area). The square root dependence on power of the spectral broadening (Figure 6.7, in the color plate section) was assessed numerically and experimentally (Figures 6.7 (a) and (b)). Results clearly confirm the following scaling laws for the evolution of the bandwidth $\Delta\omega$ and duration Δt of the NL-D similariton according to the input peak power P:

$$\Delta\omega = k\sqrt{P}, \quad \text{and } \Delta t = \Delta\omega/C = k\beta_2 L_f\sqrt{P}. \tag{6.6}$$

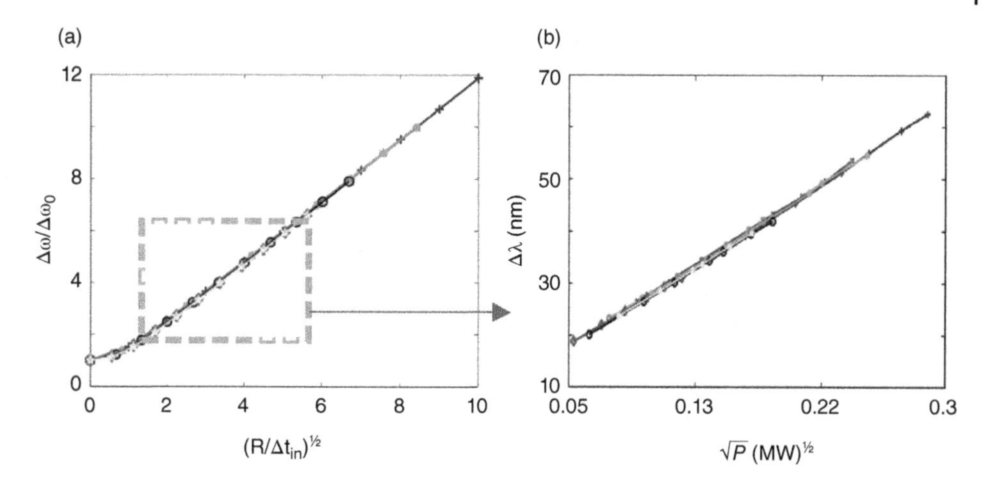

Figure 6.7 (a) Simulations: spectral broadening versus $R^{1/2}$; (b) Experiments: similariton bandwidth $\Delta\lambda$ versus $\sqrt{P}$. Blue lines correspond to transform-limited input pulses of 100 fs duration, red and cyan traces to 140 fs pulses, green and black traces to 225 fs pulses, magenta and yellow lines to 320 fs pulses. Long input pulses were obtained by broadening of the laser pulse in a dispersive line with either normal (red, green and magenta plots) or anomalous dispersion (cyan, black, and yellow plots). For a color version of this figure please see color plate section.

The property of NL-D similariton expressed in Eqs. (6.6) can be used for the measurement of femtosecond pulse duration, alternatively to the autocorrelation technique. For comparison, the spectral bandwidth of the similariton generated in a fiber amplifier is a more complex function of many parameters including the gain [6].

6.3.5 Wideband NL-D Similariton

The generation and study of *broadband similariton* are useful for ultrafast optics [18–22]. According to the above characterization of NL-D similariton with bandwidths smaller than 5 THz ($\Delta\lambda \leq 10$ nm at $\lambda \sim 800$ nm), it is actually described by Eq. (6.5), so that we have for its amplitude $A_f(t)$ and phase $\varphi_f(t)$:

$$A_f(t) \propto \tilde{A}_f^*(\omega)|_{\omega=Ct}, \quad \text{and} \quad \varphi_f(t) = -\phi_f(\omega)|_{\omega=Ct} = \beta_2 L_f \omega^2/2|_{\omega=Ct} = Ct^2/2. \tag{6.7}$$

First of all, we must check if these key peculiarities of similariton stills apply to broadband pulses with a spectral width one order of magnitude larger. For that purpose we resorted to *numerical modeling* based on the generalized NL Schrödinger equation, including high-order terms of third-order dispersion (TOD), shock wave (self-steepening) and delayed NL response (related to the Raman gain), together with the usual terms of SPM and second-order dispersion (SOD) [19, 20]. The split-step Fourier method was used to solve the propagation equation using as input conditions pulses of 100-fs duration at 800 nm wavelength with up to 70 kW peak power and considering propagation in a few meters long standard SMF (losses were ignored). Simulations in these conditions showed that the high-order nonlinearity due to the self-steepening and to the Raman delayed response do not impact on the process under study. However, the impact of TOD is visible in the generation of broadband similaritons of 50 THz bandwidth. It is conditioned by the physical pattern of the process: the NL self-interaction of powerful pulses leads to large spectral broadening (with the factor of $b \sim 10$ just at the first ~ 1 cm of propagation in fiber), substantially increasing the impact of dispersion (with the factors of $b^2 \sim 100$ for SOD, and $b^3 \sim 1000$ for TOD), resulting in the pulse stretching and peak intensity decreasing (with the factor of $b^2 \sim 100$), and thus essentially decreasing and blocking out the high-order NL effects. The impact of high-order NL effects can be significant in the case of high-power pulse propagation in low-dispersion fibers, e.g., for photonic crystal fibers with all-normal flattened dispersion, where a longer and more efficient NL self-interaction results in the generation of octave spanning supercontinuum [87].

The numerical analysis showed that the fiber TOD simply superimposed to the parabolic spectral and temporal phase profiles because it is just a perturbation of the strong SOD, and its value is the same for the NL-D similariton and D-spectron:

$$\Delta\varphi_f(t) \approx \Delta\phi_f(\omega)|_{\omega=Ct} = \beta_3 L_f \omega^3/6|_{\omega=Ct} \approx \beta_3 t^3/(6\beta_2^3 L_f^2), \tag{6.8}$$

where β_3 is the TOD coefficient. The error due to the assumption leading to Eq. (6.8) is of $\sim 4\%$ only, according to simulations made with the parameters of fused silica ($\beta_2 = 36.11\ \mathrm{fs^2/mm}$ and $\beta_3 = 27.44\ \mathrm{fs^3/mm}$). Thus, measurement of the phase (chirp) of broadband similariton becomes of interest, since it gives the TOD of fiber and permits to generalize the description of Eq. (6.7) or Eq. (6.5). The spectral interferometric study permitted the complete characterization of NL-D similariton of up to 5 THz bandwidths [11, 12]. For broadband similaritons (tens of THz bandwidth), the chirp measurement was no longer based on spectral interferometry but rather on the SFG-SC process [16–20]. It will be presented at the end of the next section.

In conclusion to this part, our spectral interferometric studies demonstrated the following properties of NL-D similaritons generated in a passive fiber:

- their linear chirp, with a slope given only by the fiber dispersion and independent of the shape, chirp and power of the input pulse;
- their link with the rectangular pulses observed in nonlinear dispersive fibers;
- their property of spectrotemporal similarity and spectral imaging, with an accuracy determined by both the spectral broadening and the pulse stretching ratios;
- their spectral bandwidth, which for a given fiber is only determined by the input pulse peak power

6.4 Time Lens and NL-D Similariton

6.4.1 Concept of Time Lens: Pulse Compression—Temporal Focusing, and Spectral Compression—"Temporal Beam" Collimation/Spectral Focusing

The optical amplitude modulation (AM) to phase modulation (PM) conversion or the reverse PM-AM conversion, as well as the soliton-shaping type processes are of great importance and of common use in ultrafast optics and laser technology [29, 30]. In the 1980s, the technique of fiber-optic pulse compression [42], based on spectral broadening of relatively long laser pulses in fiber due to Kerr nonlinearity (AM-PM) and subsequent

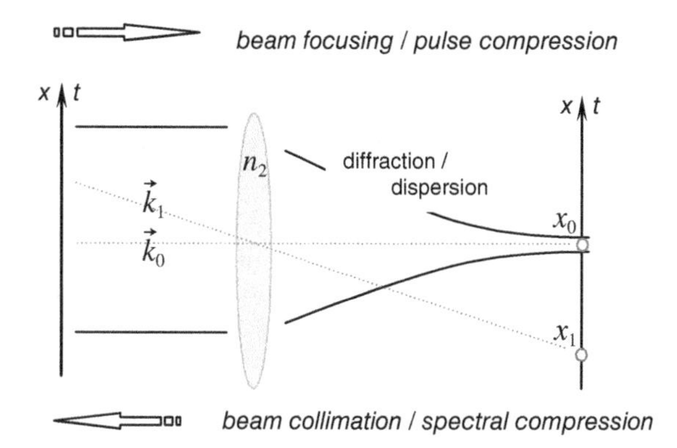

Figure 6.8 Spatiotemporal analogy and concept of time lens for pulse compression and SC. Pulse compression is the temporal focusing in SPM-induced time lens. SC is a time lensing, which "collimates" the radiation in time, and "focuses" the spectrum. For the time lens, the time t and frequency ω correspond to the spatial coordinate x and wave vector $\vec{k}$ for spatial lens, respectively.

compression in a dispersive device (PM-AM), demonstrated compression ratios up to $\sim 10^3$ [43], and shaping of pulses with a few femtosecond duration [44]. The discovery of Kerr-lens mode-locking regime of Ti:Sapphire laser in 1991, based on the temporal and spatial self-PM (SPM) of radiation in the gain medium and the management of dispersion in a prism dispersive delay line (DDL), allowed the direct generation of femtosecond pulses [45, 46]. The spectrotemporal analogy together with the analysis of reversibility of pulse compression process led to the invention of spectral compression, a NL process with prospective applications in ultrafast optics and laser physics [47–53]. The spatiotemporal analogy and concept of "time lens" are illustrative and helpful for discussing the spectral compression and its applications (Figure 6.8). While the pulse compression is a temporal analogue of beam focusing through a light-induced Kerr lens [29, 54], the compression of the spectrum [47, 48] is analogous to diffracted beam collimation by such a time lens [50]. In a system designed for spectral compression, the input pulse is stretched and chirped in a dispersive line with negative dispersion, by analogy with a beam diffraction. Further cancelation of the accumulated parabolic phase-shift by means of SPM in a nonlinear fiber leads to a nearly Fourier transform pulse with a narrowed spectrum [47–48]. The temporal phase-shift induced by the Kerr effect in the fiber, like a time lens, "collimates" the radiation in time and shrinks the spectrum.

The Kerr lens, induced by SPM from a usual bell-shape pulse, is effective only in the pulse central part, as the non-parabolic leading and trailing edges of the pulse introduce aberrations. The high-order nonlinear and dispersive effects at femtosecond time scales add supplementary "aberrations" in the temporal Kerr lens effects, limiting its performance. NL-D similariton pulses with their parabolic phase are producing aberration-free temporal lenses perfectly suited to applications of temporal lensing.

6.4.2 Femtosecond Pulse Compression

The femtosecond pulse compression to their transform limit duration demands the compensation not only of the quadratic component of the spectral phase due to the second-order dispersion (SOD) but also its higher-order terms; cubic, and sometime even quartic components. Various dispersive devices, including grating and prism pairs [55, 56], have been used for dispersion compensation. The combination of gratings and prisms was used to balance simultaneously the second and the third order dispersion (TOD), using the opposite signs of TOD for gratings and prisms [43] so that pulse compression down to 6 fs was achieved. The drawback of this configuration is the large prisms separation, when the grating and prism sequences are used separately. Mixing the prism and grating sequences was proposed to compact the setup [57]: Brewster-angle prisms were used in between the two gratings. This hybrid grating-prism compressor provides TOD tunability in a wide range, and has been applied for parabolic pulse compression to 107 fs [58] and for 33-fs pulse generation from a fiber laser [59]. A modified version with diffraction gratings coupled to prisms, called grisms [60], is more convenient for control of the TOD-to-SOD ratio.

Generally, femtosecond pulse compression is carried out in the regime of flattop "rectangular" pulse shaping [29, 30, 42], when the spectral profile is also close to a rectangular shape. In this sense the process is not optimized, as the compressed pulse for a rectangular spectrum is longer than the one for any bell-shaped spectrum of identical bandwidth (e.g., the time-bandwidth product for a rectangle pulse is twice that of a Gaussian pulse). Our approach for a better pulse compression is based on the generation of NL-D similariton pulses in a conventional passive uniform fiber [61, 62] and their subsequent compression [63]. The spectral broadening for similariton exceeds that of "rectangular" pulse, the chirp is more linear, and the spectrum is bell shaped [61, 62]. Even on a time scale of a few femtoseconds, where the impact of higher-order dispersion becomes significant, the generation and characterization of broadband similaritons [19, 64, 65] have demonstrated that their phase still is fixed only by the fiber dispersion (including high orders) [64]. Therefore, the similariton dechirping by dispersion compensation will make possible pulse compression down to the maximum ratio given by the level of spectral broadening. In [63], a 77-nm bandwidth NL-D similariton has been generated and compressed to 17 fs after two passes in a pair of 6-m separated silica prisms. The phase compensation, benefiting from the opposite signs of TOD in fiber and prism-compressor, resulted in nearly transform-limited pulses of 25-fs autocorrelation duration. However, such a phase compensation scheme is limited by the fact that the control of the second-order dispersion and TOD are not independent. Another technique of direct compression of readily-available 140-fs pulses from a Ti:sapphire laser to 10 fs [66] was demonstrated by NL spectral broadening in a noble gases-filled hollow-core waveguide at high pressure. Pulse compression was obtained in a quartz-prism pair separated by ~ 1 m. The TOD control in this case was avoided due to the small dispersion in gas, but, as compared to conventional fibers or PCFs, the use of gas introduces undesirable technical complexity.

We implemented a hybrid grating-prism compressor for similariton pulse compression [67]. Broadband NL-D similaritons of up to 75-THz FWHM bandwidth were generated in normally-dispersive standard fibers and also in photonic crystal fibers, and then launched into a hybrid grating-prism compressor (Figure 6.9). We showed that it is possible to tune the TOD-to-SOD ratio of such a compressor from −0.5 fs to 2 fs using 300 mm^{-1} and 600 mm^{-1} gratings and SF11 and fused-silica prisms.

We reached pulses of 14-fs duration (10 times pulse compression, Figure 6.9 (c)), limited by the bandwidth of our compressor transfer function and its fourth-order dispersion [19, 64].

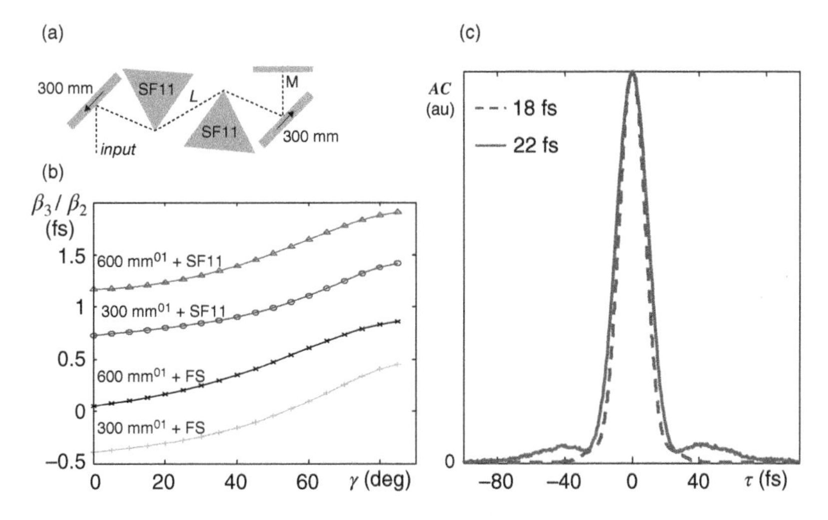

Figure 6.9 (a) Schematic of the hybrid compressor consisting of a prism pair in between a grating pair: M – mirror, $L = 10$ cm; (b) TOD-to-SOD ratio vs. grating incident angle for different combinations of gratings and prisms; (c) Measured autocorrelation trace of compressed pulse (solid) in comparison with the calculated one for transform-limited pulse (dashed).

6.4.3 Classic and "All-Fiber" Spectral Compression

Spectral compression (SC) and more generally temporal Kerr lens have demonstrated promising applications to signal analysis and synthesis problems in ultrafast optics [50]: spectrotemporal imaging for real-time femtosecond pulse measurements, fine frequency tuning along with SC for resonant spectroscopy [18, 22, 68–73], generation of dark solitons [74], nonlinear optical filtering of noise for light-wave communication [50], the measurement of Kerr nonlinearity with the D-scan technique [75] (a temporal analogue of the Z-scan method), and the fiber delivery of high peak power femtosecond pulses [76]. SC was proposed also for a similariton fiber laser architecture instead of a highly dissipative spectral filtering [53]. SC is of interest in the context of NL vibrational microscopy where a high spectral brightness is required, such as in coherent anti-Stokes Raman scattering (CARS) or stimulated Raman scattering microscopy. Through the generation of negatively chirped parabolic pulses, an aberration-free 8.7x SC was achieved, thanks to the optimal input pulse shape [77]. Further development results in a 12x SC in an all-fiber configuration for telecommunication wavelengths, utilizing a kilometer-long fiber [78].

GVD in the spectral domain, and SPM in the time domain are the two key mechanisms for SC. The comparison of the expressions for the pulse slowly varying amplitude $A(t)$ and its Fourier image $\tilde{A}(\omega)$ clearly shows that the GVD modulates the spectral phase in a parabolic way

$$\tilde{A}_z(\omega) = \tilde{A}_0(\omega)\exp[-i\omega^2(z/L_D)/2] \tag{6.9}$$

similarly to the SPM of a bell-shaped (e.g. Gaussian) pulse at its central part

$$A_z(t) = A_0(t)\exp[in_2\beta_0|A_0(t)|^2 z] \approx A_0(t)\exp(iz/L_{NL})\exp[-it^2(z/L_{NL})] \tag{6.10}$$

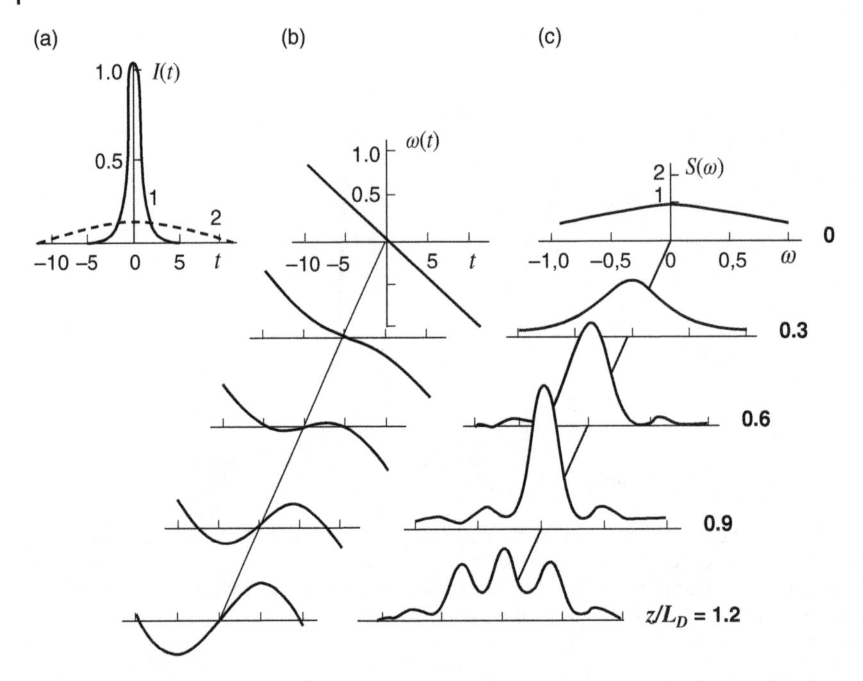

Figure 6.10 Dynamics of the SC process. The laser pulse is stretched and downchirped in a dispersive line (a) before being launched in the NL fiber where SPM leads to chirp compensation (b) and to SC (c).

The modulation "scale" for GVD is given by the L_D dispersion length of the medium, instead of the L_{NL} nonlinear length for SPM. Such a consideration points out that, since for temporal compression the pulse must experience SPM and afterwards compression in a dispersive line [47], for the spectral domain compression, first, the pulse must be stretched in a dispersive line to get a spectral phase modulated by GVD, and afterwards the pulse must undergo SPM in a nonlinear fiber.

Describing the process directly, we have $A(t) \propto \tilde{A}_0(\omega)|_{\omega=Ct} \exp\{i[\varphi_d(t) + \varphi_{nl}(t)]\}$ at the output of SC-system, in the spectron approximation for DDL. For a Gaussian pulse, an ideal chirp cancelation $\ddot{\varphi}_d(t) + \ddot{\varphi}_{nl}(t) = 0$ will result in a "s" times stretched transform-limited pulse giving a spectral width divided by "s." In reality, SPM induced by a Gaussian pulse cancels the dispersive parabolic phase partially, as it is actually parabolic in the pulse central part only, and thus, the SPM-SC is less effective.

To have a more complete pattern of the process, numerical modeling was carried out [47]. Figure 6.10 illustrates the evolution, according to the propagation distance in the fiber, of the chirp (b) and spectrum (c) of the pre-chirped input pulse because of SPM. The strongest SC occurred here for $z = 0.9\ L_D$.

With a view to achieving an efficient SC, we carried out experimental studies specifically focused on the aberrations of the temporal lensing [22, 52, 79, 80]. In the initial experiment, our SC device consisted of a 3.75-m separated SF11 prism pair as a DDL, and a 92-cm-long Newport F-SE 780 nm SMF. Starting from the 11.3 nm-bandwidth pulses, we obtained at the output 12.3x spectrally compressed pulses with 0.92 nm bandwidth (Figure 6.11 (a)), ~2 ps pulse autocorrelation duration, and 0.42 W average power. Afterwards, the prisms line was replaced by a 2 m long piece of hollow-core

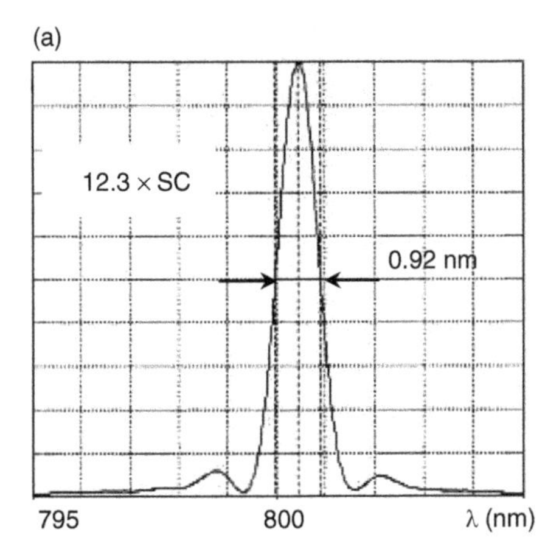

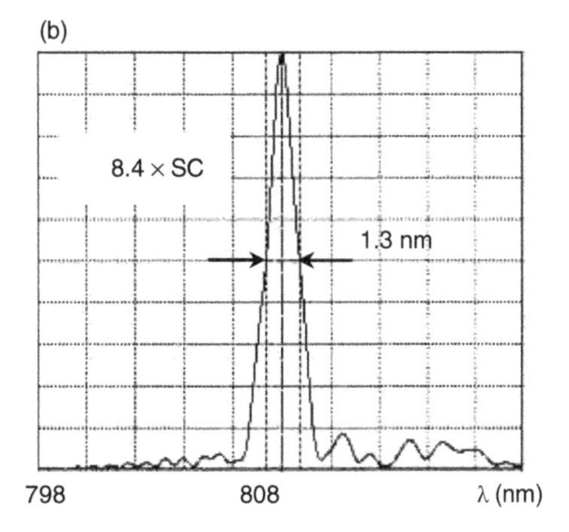

Figure 6.11 Classic (a) and all-fiber (b) SC. The measured spectra at the output of the SC system show 12.3 x (a), and 8.4 x (b) compression ratios, respectively with respect to the 11.3 nm-wide pulses at input.

fiber (HCF) (ThorLabs HCF-800) with an anomalous dispersion at the laser wavelength (800 nm).

The output pulse from the HCF was coupled into a conventional single mode fiber (Newport F-SE @780 nm, 80 cm). As a result, the initial spectrum was compressed down to 1.3 nm (Figure 6.11 (b)), corresponding to 8.4x SC, for the same 0.4 W output average power. According to our numerical simulations, the strong side lobes of the compressed spectrum of Figure 6.11 (b) are caused by the high-order dispersion of HCF that we measured via spectral interferometry. In this experiment, the HCF dispersion was tuned and the compression was optimized by setting the laser wavelength to 808 nm. Further development in this direction is expected by using HCFs with flattened dispersion, as well as by splicing the HCF and SMF.

The results of these experiments [79], comparable in efficiency with the recent achievements in this field [77, 78], clearly show that SPM-SC technique with the typical aberrations of the induced temporal Kerr-lens, demands special efforts for their compensation.

6.4.4 Spectral Self-Compression: Spectral Analogue of Soliton-Effect Compression

The NL process of pulse self-compression, benefitting from the technology of photonic crystal fibers and nanowires, recently yielded the generation of laser pulses down to a single cycle [81–84]. The so-called soliton (or solitonic) compression, exploits the behavior of high-order solitons under the combined impact of strong SPM and weak negative GVD [85, 86]. We demonstrated a spectral analogue of this process, the spectral self-compression, under reverse conditions where the impact of anomalous GVD exceeds the impact of SPM [80].

The usual scheme for SC consists in a DDL followed by an NL fiber [47,48]. For the self-SC process, GVD and NL SPM are no longer separated in two successive components but instead combined and distributed all along the propagation in a waveguide.

Experimentally, we observed the self-SC process by coupling 100 fs pulses at 800 nm wavelength in a 2-m-long hollow-core fiber (HCF) with anomalous dispersion. We

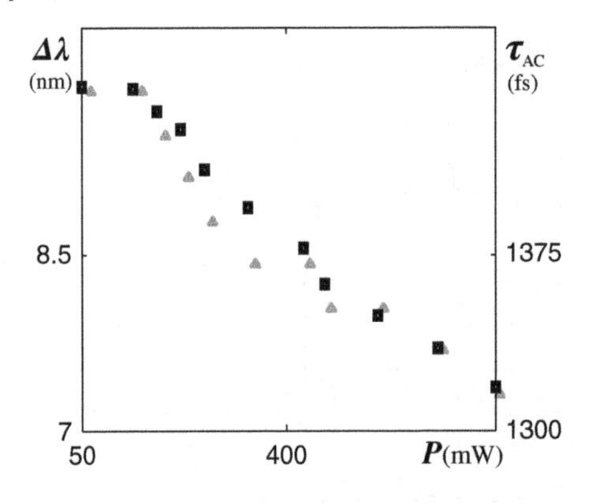

Figure 6.12 Measured spectral bandwidth (squares) and pulse autocorrelation duration (triangles) at the HCF output versus the coupled power.

measured the output spectrum (and pulse autocorrelation) versus the power coupled in the fiber. The measurements show spectral narrowing from 9.62 nm to 7.4 nm (1.3 × self-SC). Both the spectral bandwidth and pulse duration at the HCF output are decreasing monotonically with the power increase, evidencing the nonlinear character of the process (Figure 6.12). The analogy between the impacts of GVD in the spectral domain and SPM of a bell-shaped pulse in the time domain indicates that for the "spectral self-compression" to occur, GVD must dominate SPM, in contrast to the soliton time compression where nonlinearity dominates dispersion [83, 84]. In other words we have the soliton effect compression in a medium with anomalous dispersion for $L_{NL} < L_D$, and we got self-SC instead for $L_{NL} > L_D$.

To have the complete pattern of the process and to study its peculiarities, we carried out numerical simulations based on the solution of the NL Schrödinger equation considering the ratio of GVD and Kerr-nonlinearity. Figure 6.13 shows some results as a 3D evolution maps for the pulse $I_z(t)$ (a) and spectrum $\tilde{I}_z(\omega)$ (b). The strong GVD leads

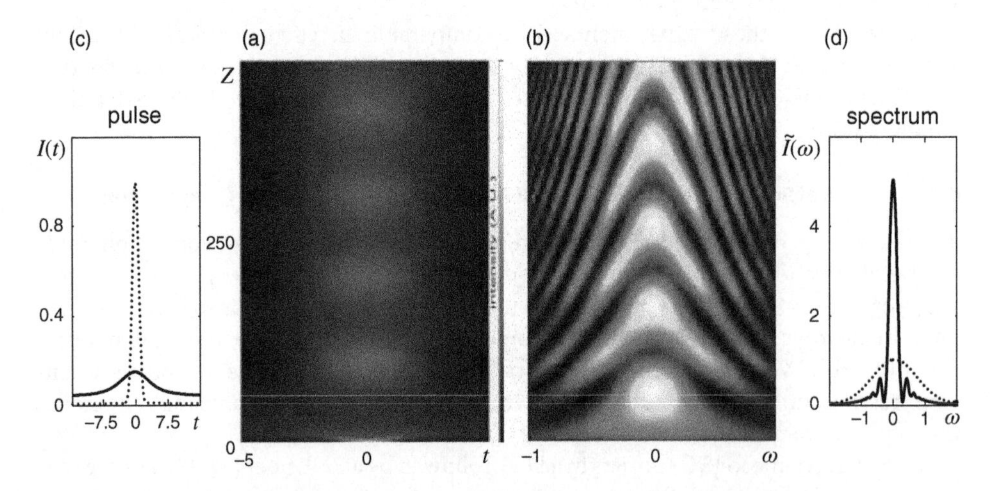

Figure 6.13 The physical pattern of the spectral analogue of the soliton effect compression for initial transform-limited Gaussian pulse ($L_D/L_{NL} = 0.6$):3D self-interaction maps for the pulse (a) and spectrum (b), with the self-compressed spectrum (d) and stretched pulse (c) at $Z \equiv z/L_D = 70$, in comparison with the initial ones at $z = 0$ (dotted curves).

to rapid pulse stretching since the initial propagation steps ($z{\sim}10L_D$), while the change in the spectrum is slower. At the distance $z{\sim}70.L_D$, we have a significant self compression of the spectrum (Figure 6.13 (d)) and a corresponding broad pulse (Figure 6.13 (c)). Beyond that distance ($z > 70L_D$), the reshaping of both the spectrum and the pulse profile exhibits a quasi-periodic behavior. The periodic character of the evolution mimics the one of high-order temporal solitons (for soliton numbers $N \equiv (L_D/L_{NL})^{1/2} \geq 2$). In the results of our numerical studies, carried out in the range of parameters $z/L_D = 0$–1100, $N^2 = 0.45$–0.9, we have demonstrated up to 33x self-SC for $L_D/L_{NL} = 0.45$, i.e. $N = 2/3$. The results of the numerical experiments are summarized in the form of an empirical expressions for the maximal self-SC ratio $F_c \equiv \Delta\omega/\Delta\omega_0$, at the given fiber length z and pulse intensity I, serving for the process optimization: $F_c(z) = (z/L_D)^{1/2} + 1$ and $F_c(I) = 0.23(L_{NL}/L_D)^6 + 3$.

Concluding, the NL process of self-SC, a spectral analogue of the soliton pulse compression was demonstrated. General rules which can serve for the process optimization, have been derived from numerical simulations. Along with the optical fibers, bulk media with strong anomalous dispersion also can be used for the process implementation, since there is no need for a strong nonlinearity.

6.4.5 Aberration-Free Spectral Compression with a Similariton-Induced Time Lens

In view of the applications of SC, it is more important to improve the process efficiency and especially through the reduction of the aberrations of the Kerr time lens. The cross-phase modulation (XPM) technique of SC, with a specially shaped reference pulse [68–72], can serve as aberration-free temporal Kerr-lensing, but it involves some technical issues connected to signal and reference walk-off, and to the signal filtering (by frequency or polarization) at the output, etc.

An aberration-free SC was achieved in [18, 22, 52, 73] using a similariton and a sum frequency generation process, instead of SPM or XPM by the Kerr effect, to get temporal lensing. The method provides an aberration-free time lens, without the requirement of separating the signal and reference radiation. It is also self-referenced, in contrast to XPM-SC. The principle of the technique is the following. We split the input pulse with a $A_0(t)$ temporal amplitude and Δt_0 duration into a low- and a high-power parts. The low-power pulse is further sent to a device with negative dispersion, resulting in a spectron with duration $\Delta t_d = s.\Delta t_0$ and $A_d(t) \propto \tilde{A}_0(\omega)|_{\omega=Ct} \exp(iC_d t^2/2)$. The high-power pulse is injected in a SMF where it is shaped into an NL-D similariton $A_f(t) \propto |\tilde{A}_f(\omega)|_{\omega=Ct} \times \exp(iC_f t^2/2)$, with the parabolic phase given by the fiber normal dispersion only [11]. Thus, in both arms of the setup, we got linearly chirped pulses, but with opposite chirp slopes. Afterwards, these two pulses are sent in a NL crystal to get sum-frequency generation (SFG). The SFG field results from the multiplication of the complex amplitudes of the two inputs: $A_{out}(t) \propto A_d(t) \times A_f(t)$. In the SFG-process, the chirps can be suppressed provided $C_d + C_f = 0$ leading to SC. Thus, at the output, we have a s x-stretched and transform-limited pulse with a compressed spectrum down to $\Delta\omega = \Delta\omega_0/s$. It is worth mentioning that the pulse carrier frequency is twice that of the input signal.

In the experimental implementation of similaritonic SC through the SFG-process [22, 52, 73], the laser beam splitting was of 80% for the high-power part, and of 20% for the low-power part (Figure 6.14). The more powerful pulse was coupled into a standard

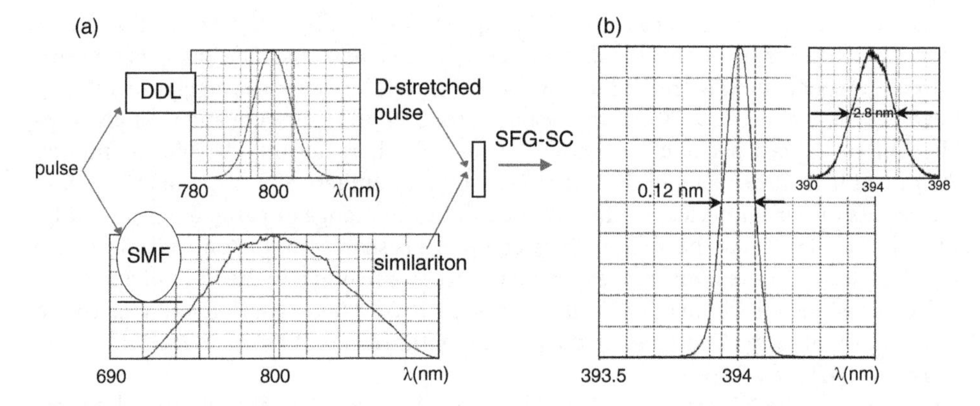

Figure 6.14 SFG-SC: (a) schematic of the set-up. The measured spectra at the output of DDL (top) and SMF (bottom) are spectrotemporal images of the D-stretched and similariton pulses respectively. (b) The spectrum at the output of the system shows 23.3x compression ratio. The inset of (b) is the SFG spectrum in the absence of DDL (control measurement).

SMF (1.65 m Newport F-SPF PP@820 nm), where an NL-D similariton with a FWHM bandwidth of ~100 nm was generated. The low-power pulse made a round-rip through a DDL consisting of a 3.5-m separated SF11 prism pair with an end mirror for feedback. The two pulses were then focused on a β-BBO crystal for SFG, resulting in 23.3x compression for the spectrum down to 0.12 nm bandwidth (Figure 6.14 (b)). The SFG-SC ratio of $\Delta\omega_{in}/\Delta\omega_{out}$ takes into account the reduced bandwidth of the SFG spectrum measured in the absence of DDL $\Delta\lambda_{in}(400\ \text{nm}) = \Delta\lambda_{in}(800\ \text{nm})/4$ (Figure 6.14 inset).

It should be mentioned that varying the delay between the pulses in interaction in the crystal allows tuning of the peak wavelength in the spectrally compressed output. This capability can be of interest, for example, for resonant spectroscopy (see Section 6.4.5 for more details).

In conclusion, various femtosecond pulse spectral compression schemes were theoretically and experimentally investigated:

- self-SC process – the spectral analogue of soliton compression;
- classic SPM-based technique with a 12.3 x SC achieved in practice;
- an all-fiber SPM-based technique with 8.4 x SC obtained in experiment;
- an aberration-free SC with an alternative method combining NL-D similariton and SFG with a 23.3 x compression demonstrated.

6.4.6 Frequency Tuning Along with Spectral Compression in Similariton-Induced Time Lens

According to the concept of temporal lens, the temporal delay Δt of an input signal should lead to the frequency shift $\Delta\omega$ at the output, in an analogy with x-to- k conversion in a spatial lens (see Figure 6.8). Such a time-to-frequency conversion was implemented through similaritonic SC [22, 52, 73]. In the system, the output is the product through SFG of the NL-D similariton with the dispersed input pulse. The NL-D similariton exhibits a linear frequency chirp. The central wavelength of the SFG product therefore depends on the overlap in time between the two pulses. Provided the D-stretched

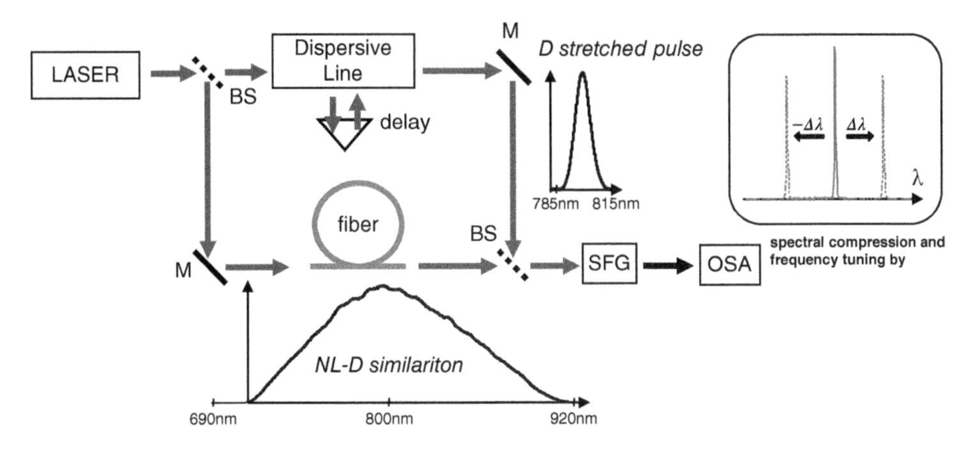

Figure 6.15 Illustration of frequency tuning of radiation in the similaritonic technique of SFG-SC. Temporal delay (Δt) between the signal and reference pulses results in wavelength shift ($\Delta\lambda$) of the SFG-signal, along with SC .

pulse is of shorter duration than the similariton, it is possible to change the output pulse carrier frequency by varying the relative delay between the two pulses sent in the SFG crystal. That was implemented using the translation of the back mirror of the dispersive line for delay settings. A schematic drawing of the setup together with some data on a preliminary proof of principle for three different delays is shown in Figure 6.15.

The frequency tuning of the spectrally compressed radiation can be applied for resonant spectroscopy, light-wave communication, etc. As shown in Section 6.5, this technique is also helpful for calibration of spectrotemporal imaging device: measurement of $\Delta\lambda$ shift, caused by a Δt delay, determines the time-to-frequency mapping parameter. Such a frequency tuning in the process of similaritonic SC can be applied for complete characterization of broadband similariton and also for the material dispersion measurement.

In the experiment, a broadband NL-D similariton of 50 THz bandwidth was generated in a piece of passive fiber and the chirp measurement was carried out through frequency tuning in the SFG-SC process [16–20]. Figure 6.16 shows the spectrum of a 107-nm (50 THz) FWHM-bandwidth similariton, generated by launching a 100 fs pulse of 7 nJ in 1.65 m piece of fiber (Newport F-SPF PP@820 nm). This spectral profile represents the spectrotemporal image of the generated similariton in the approximation of its linear chirp. The asymmetry in this spectrotemporal profile evidences the impact of TOD of the fiber. Although the initial pulse asymmetry can also cause the spectral asymmetry in the near field of dispersion, typically for the picosecond-scale experiments, the impact of the possible asymmetry of the initial laser pulse on the spectrotemporal shape of similariton becomes insignificant, according to our simulations.

For the characterization of a broadband similariton, the intensity profile can be measured also by means of SFG cross-correlation, with the laser pulse as a reference. A spectral analysis of the SFG-signal will also give information on the chirp of the similariton, similar to the X-FROG technique [88,89]. Our setup differs from X-FROG by the use of a chirped reference pulse, which provides a spectrally compressed SFG-signal in a wide spectral range. That gives a more accurate and more efficient measurement [19,22]. The

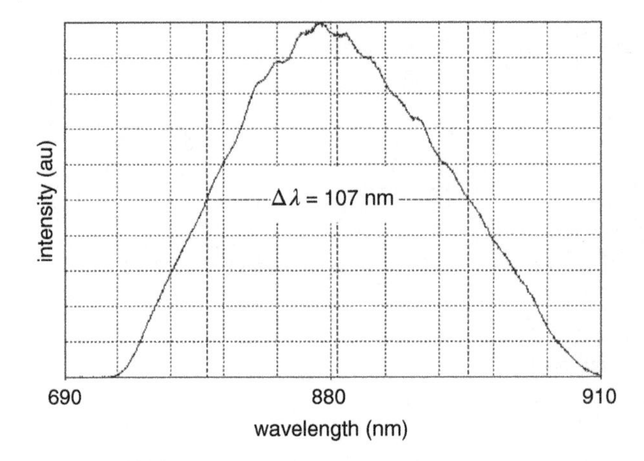

Figure 6.16 Measured spectrum of a broadband NL-D similariton.

experimental setup is the same as previously (see Figure 6.15) where the translation of the DDL feedback mirror served to vary the delay between the similariton and the reference pulse on a 32 ps range [22]. Measurements with and without the DDL were carried out, replacing the D-line with a simple temporal delay (TD), to compare the techniques of SFG-SC and XFROG. Figure 6.17 shows both measurements of the similariton chirp as 3D patterns.

The SFG-SC technique here demonstrated higher performances than XFROG, providing a sharper spectral signal in a wider spectral (and temporal) range. The temporal

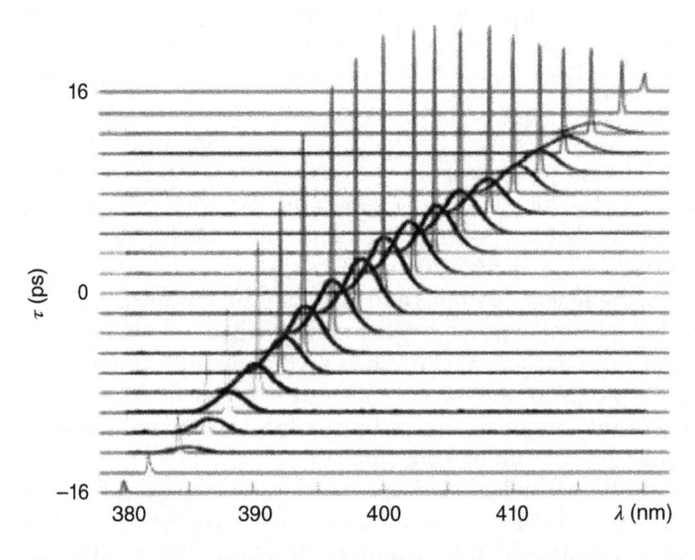

Figure 6.17 3D patterns of the similariton chirp measurement without (thick lines) and with SC (sharp peaks in line). The 40 nm (75 THz) sum frequency tuning at 400 nm agrees with the 160 nm spectral range of the similariton at 800 nm, for the 32 ps range of temporal delay between SFG-interacting pulses.

delay between the input pulses in the range of ± 16 ps resulted in a ± 20 nm wavelength shift for the spectrally compressed (23X, 0.12 nm) SFG signal, corresponding to the chirp measurement of similariton in a span of 160 nm (75 THz) around 800 nm. This 3D pattern completely characterizes the broadband similariton with the curve $\lambda(t)$, connected with the chirp $\omega(t) = \dot{\varphi}_f(t)$, as well as with the derivative of spectral phase $\dot{\phi}_f(\omega)$, according to the spectrotemporal similarity described by Eqs. (6.7) and (6.8).

Its projections on the vertical and horizontal axes respectively represent the more standard temporal and spectral intensity profiles $I(t)$ and $\tilde{I}(\lambda)$. In general, through the Fourier transformation of complex temporal amplitude, derived from the measured temporal pulse intensity and chirp, the spectral complex amplitude and phase could be retrieved. In the case of similariton, Eq. (6.8) gives the spectral phase information by a simple scaling $\omega = Ct$, and one gets the derivative of spectral phase from $\dot{\phi}_f(\omega) \approx -\dot{\varphi}_f(t)/C$.

Figure 6.18 shows the $\dot{\phi}_f(\omega)$ curve obtained in this way. The quadratic component of $\dot{\phi}_f(\omega)$ is shown separately (Figure 6.18 (b)). The circles in Figure 6.18 are the measured

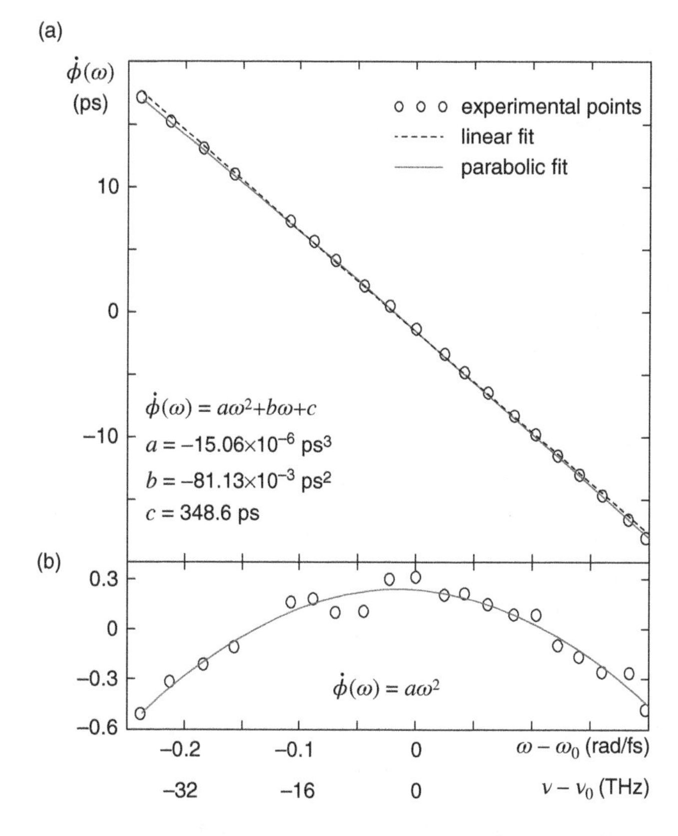

Figure 6.18 (a) Broadband NL-D similariton's chirp $\dot{\varphi}_f(t)$ measured as a derivative of spectral phase $\dot{\phi}_f(\omega)$, and (b) its isolated quadratic component. The circles are the measured experimental points; the dashed and solid curves are for the linear and parabolic fits, respectively. The ~1% difference from the linear fit in the range of ~50 THz gives the range of applications for the similariton-based methods of signal characterization (Section 6.5).

experimental points; the dashed and solid curves are for the linear and parabolic fits, respectively. The slope of the linear component gives the SOD coefficient (β_2), while the parabolic component is responsible for TOD (β_3). Taking into account the expression $\phi_f(\omega) = -\beta_3 L_f \omega^3/6 - \beta_2 L_f \omega^2/2$ for the spectral phase of similariton and fiber length $L_f = 1.65$ m, we derive the following values for SOD and TOD: $\beta_2 = 49.17$ fs^2/mm and $\beta_3 = 18.25$ fs^3/mm.

In conclusion to this part, NL-D similariton were generated with a bandwidth up to 50 THz. It was completely characterized through its precise chirp measurement, using the SFG-SC technique with delay tuning. Our studies state that only fiber dispersion determines the phase (chirp) of broadband NL-D similariton. The fiber TOD results in the same additional phase for broadband NL-D similariton and spectron. The ~1% accuracy of the linear fit for the chirp of the 50-THz bandwidth similariton gives the range of applications for similariton-based spectrotemporal imaging [22] and spectral interferometry [16]. The described approach for the generation and characterization of broadband similariton can be helpful also for its applications in pulse compression [13, 14, 67] and CARS microscopy [51]. For these applications, the value of TOD may have a significant impact and it should be considered more carefully.

6.5 Similariton for Femtosecond Pulse Imaging and Characterization

Characterization of ultrafast light signal at the femtosecond time scale employs the powerful means of modern optics, involving the methods of nonlinear optics, of Fourier optics and holography, of spectral interferometry, of adaptive optics, etc. The nonlinear-optical techniques of frequency resolved optical gating (FROG) and its derivatives [26, 91, 92] count among the most popular and commercialized tools. They provide accurate and complete determination of the temporal amplitude and phase by recording high-resolution spectrograms which are further decoded by means of iterative phase-retrieval procedures. The method of spectral interferometry (SI) has the advantage of a direct non-iterative phase retrieval, however, it requires a reference pulse of sufficiently broad bandwidth which limits its application range. To avoid this restriction, the self-referencing methods of spectral shearing interferometry, such as SPIDER [93], SPIRIT [94], and SORBETS [95], were developed at the expense of a more complicated optical arrangement. As an alternative, the method of similariton-based SI [16, 21], along with its self-referencing feature, keeps the simplicity of the principle and configuration of the classic SI [40, 41]. The MIIPS technique [96] operates with spectral phase measurement through its adaptive compensation, up to transform-limited pulse shaping, by using feedback from the SHG process. All these methods are based on the spectral phase determination together with the spectrum measurement, and on the reconstruction of the temporal pulse.

To implement a direct pulse measurement, the temporal information must be transferred to the space or frequency domain, or to the time domain with an enlarged scale in order to be measurable by contemporary electronics. The time-to-space mapping through Fourier holography, implemented in real time by using a special multiple-quantum-well photorefractive device [97] or SFG-crystal [98], provides the temporal

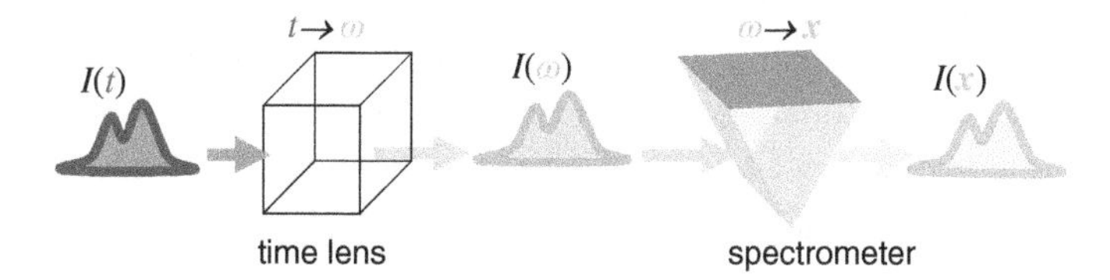

Figure 6.19 Schematic view of the principle of femtosecond scale temporal measurements based on the time-to-frequency FT and STI in a light induced time lens. Time lens transfers the temporal information to the spectral domain, reducing the problem of high-resolution temporal measurements to standard spectrometry.

resolution given by the duration of reference pulse. In the alternative approach of time stretching, the 103 × up-conversion "time microscope" demonstrated a 300-fs resolution [99]. In the time-to-frequency conversion approach, the methods of optical frequency inter-modulation by sum-frequency generation [100, 101] or by electro-optic modulation [102, 103] were limited to the picosecond domain. The technique of optical chirped-pulse gating [104] demonstrated sub-picosecond resolution. The method of SC and pulse spectrotemporal imaging (STI) in a light-induced time lens (Figure 6.19) looks more promising, the resolution being limited by the ~1 fs nonlinear time response of silica [70–73]. Using four-wave-mixing in a silicon chip, 100 ps long pulses (record length) were measured with a 220 fs resolution thanks to the time lensing approach [54, 105, 106]. The similariton-induced parabolic time lens provided self-referenced aberration-free STI, and the pulses' direct measurement in a spectrometer with few-fs resolution was demonstrated as in a femtosecond optical oscilloscope [22, 52, 107].

Thus, the progress obtained in femtosecond signal analysis has led to a wide variety of methods to perform high-resolution measurement and to adapt to the diversity of experimental situations. Nevertheless, the conventional intensity autocorrelator [108] remains the most widespread tool, because the simple measurement of pulse duration is a basic common need, for example, in the daily monitoring-adjusting procedures of ultrafast laser systems. Based on the self-shaping of NL-D similariton in a passive fiber followed by the measurement of its bandwidth, a simple alternative to the autocorrelation technique [109] for determination of femtosecond pulse duration has been developed [110]. Moreover the initial pulse shape can be recovered by solving the reverse problem of NL-D similariton generation [61, 64].

6.5.1 Fourier Conversion and Spectrotemporal Imaging in SPM/XPM-Induced Time Lens

A light-induced time lens may provide either pulse temporal compression or spectrum compression, by analogy with the space domain where a positive lens may give either beam focusing or beam collimation, respectively. Based on the more general time lens, time-to-frequency Fourier conversion, a technique for pulse spectrotemporal imaging (STI) and direct femtosecond scale measurement was developed [68–73]. The STI method with a temporal Kerr lens was experimentally demonstrated which used spectral compression (SC). The pulses were first stretched and up-chirped in a prism DDL, before the chirp was compensated by means of SPM or XPM in a single-mode fiber. The STI method reduces the problem of high-resolution temporal measurements to standard spectrometry.

For the complex temporal amplitude at the output of SC-system we have $A(t) \propto \tilde{A}_0(\omega)|_{\omega=Ct} \exp\{i[\varphi_d(t) + \varphi_{nl}(t)]\}$. Cancellation of the DDL-induced chirp by SPM in the NL-fiber ($\ddot{\varphi}_d(t) + \ddot{\varphi}_{nl}(t) = 0$) results in the mapping $A(t) \propto \tilde{A}_0(\omega)$ (Eq. (6.11')) and conversion (optical FT), with the scaling $\omega = Ct$, C being fixed by the DDL. That leads to the input pulse imaging into the output spectrum (STI):

$$\tilde{A}(\omega) \propto A_0(t) \tag{6.11}$$

$$S_{out}(\omega) = |\tilde{A}(\omega)|^2 \propto |A_0(t)|^2 = I_{in}(t) \tag{6.12}$$

It is worth mentioning that, for more visibility of the physical pattern in this consideration, Eq. (6.11) above is an approximation of the stretched pulse. It is more correct to write:

$$\tilde{A}(\omega) \propto A_0(t)\exp(iCt^2/2), \text{ with } \omega = Ct \tag{6.11'}$$

with absolutely the same STI result as Eq. (6.12). It is important that, according to Eqs. (6.11) and (6.11') we have complete optical time-to-frequency conversion for both the intensity and the phase, and special demand of phase measurements is also open to this technique, with a relevant interferometric detection.

The quality of STI strongly depends on the fulfillment of the chirp cancelation condition. For the SPM-STI, under the spectron approximation of large stretching by the DDL ($s \gg 1$), this condition requires the satisfaction of the following equality:

$$n_2\beta_0\ddot{S}_0(\omega)|_{\omega=Ct}L_f/s = -C$$

which can be implemented very approximately only in reality. Moreover, for the SPM-scheme, chirp cancelation depends on the input pulse spectrum, the shape of which may vary according to the signal. For improved accuracy, the XPM scheme has been proposed for SC and STI [68–73]. The time lens is produced here by a bell-shaped intense pulse, independent of the testing signal (Figure 6.20).

For XPM-STI with a reference pulse of Gaussian intensity profile $I_P(t)$, the chirp cancelation condition becomes: $n_2\beta_0\ddot{I}_P(t)L_f \approx -2L_f/L_{NL} = -C$. Additionally, the duration of the reference pulse must exceed that of the signal stretched in DDL: $\Delta t_P > s\Delta t_0$.

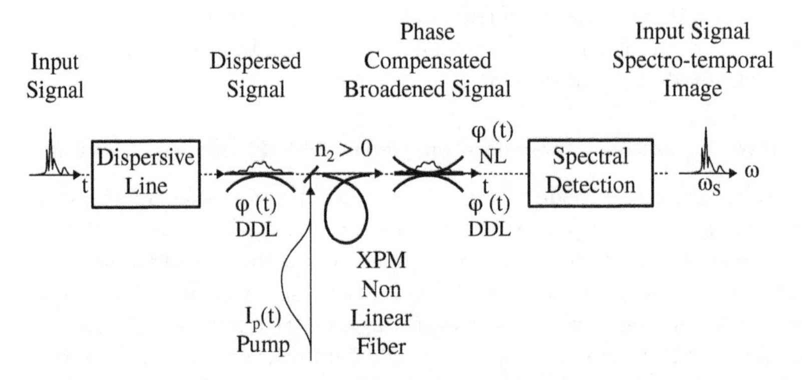

Figure 6.20 Principle of XPM-STI. In the DDL, the input signal is stretched, and negatively chirped. Then in the nonlinear fiber this chirp is canceled by means of XPM from a bell-shaped reference at a different carrier frequency (Pump). At the exit a spectrum analyzer displays the intensity image in the spectral domain of the initial pulse temporal profile with the scaling parameter $\Delta t/\Delta\omega \equiv -\tilde{C}(>0)$.

In the experiment, we used a standard KLM Ti:Sapphire laser of 120-fs pulses [70–73]. The average wavelength was 800 nm for 900 mW average power at 76 MHz repetition rate. The DDL was based on a couple of SF10 glass Brewster prisms separated by 3 m. After a double-pass in the DDL, the chirped signal pulse of approximately 1 ps duration was injected in a single-mode silica fiber. At the output, an Anritsu OSA of 0.1 nm resolution was used for spectral measurements. The signal pulse shaping, needed for the assessment of the method, was based on the use of birefringent crystals. Because of the group velocity difference between the waves with ordinary and extraordinary polarizations, a single input signal pulse could be split and shaped into a complex profile with multiple peaks. The number, the orientations and the thicknesses of crystal pieces determined the number, the intensity ratio and the temporal separation respectively in the peaks of the shaped pulses. A linear polarizer then fixed the profiling and the signal polarization.

First, we have studied SPM-STI in the classic SC scheme with a 2-m-long fiber. We have shaped the laser output into a two-peak signal with a 2-mm-long KDP crystal. The o-axis was oriented under $\theta = 40°$ angle with respect to the laser polarization. The delay between the peaks was $\Delta t_{oe} = 255$ fs and the intensity ratio was $I_e{:}I_o = 1{:}2$. On the recording of Figure 6.21 the two-peak structure is clearly visible with approximately the expected ratio and temporal separation.

However, the experimental curve as well as the numerical simulation exhibit distortions and satellite pulses. Such aberrations may have two main origins. First, SPM necessarily leads to incomplete chirp compensation on the pulse edges similar to what happens in spectral or pulse compression. Second, even in the central part of the pulse, the intensity profile of the chirped signal significantly differs from a parabola, leading also to partial chirp quenching. As the studies show, SPM-STI can be used to qualitatively verify the shape of a known signal, but it is not suited to the characterization of unknown signals with a complex shape.

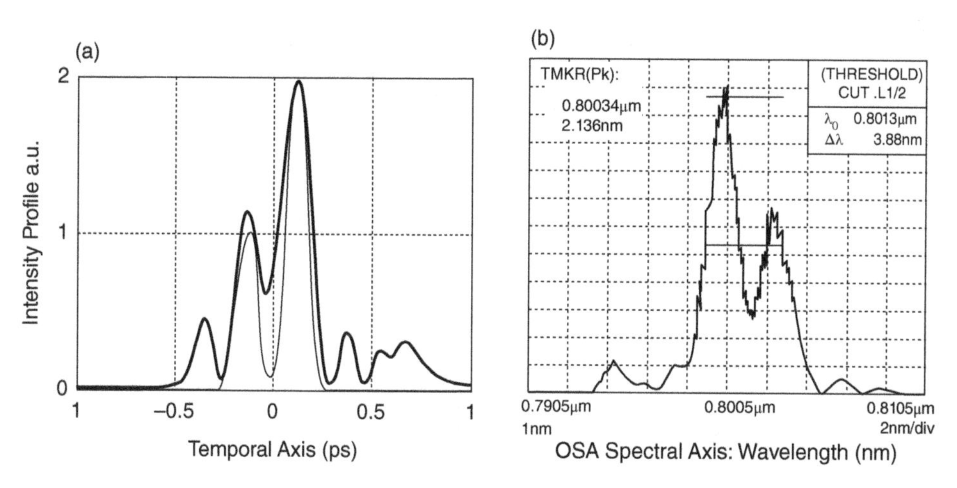

Figure 6.21 SPM-STI. (a) Simulation. Thin line: two-peak input signal with 255-fs peak separation and 1:2 intensity ratio; thick line: SPM-STI profile exhibiting aberrations due to partial SPM chirp compensation. (b) Experimental STI. The scale displayed by the OSA is −100 fs/nm, i.e., −0.2 ps/div. The expected profile shown in (a) is clearly identified experimentally in spite of spurious satellite pulses. The fiber length was 2 m.

The implementation of the XPM-STI method requires two independent pulses for the signal and pump pulses respectively (Figure 6.20). In our demonstration the two pulses were in fact derived from the same laser pulse by spectral filtering. The laser spectrum was divided into two parts of different bandwidth and central wavelength. A narrow bandwidth high-energy beam was selected to serve as a pump pulse of few picoseconds duration. The rest of the initial spectrum was used to shape a broadband signal pulse of weak energy. This configuration provided two pulses with close carrier frequency $\omega_p \approx \omega_s$well suited to an efficient pure XPM. A test signal with a periodic modulation of 255 fs period was shaped combining two KDP crystals with 2 mm and 1 mm thickness. KDP birefringent crystals o-axes were oriented at $\theta_1 = 40°$ and $\theta_2 = 50°$ with respect to the incident polarization to get four peaks with an intensity distribution approximately equal to 2:1:3:1.

Figure 6.22 illustrates the results of our experiment on XPM-STI carried out in combination with numerical studies. Experimental spectrograms recorded by the OSA with

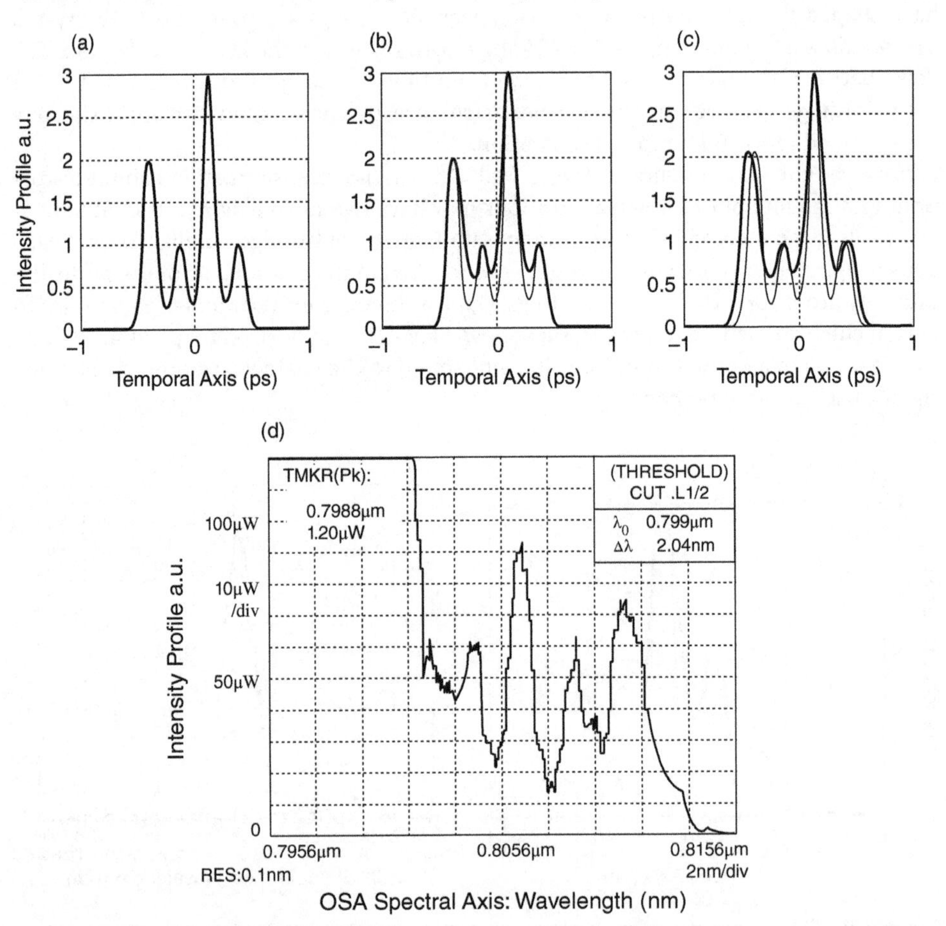

Figure 6.22 (a–c) Numerical and (d) experimental results for XPM-STI. (a) Perfect phase compensation. (b), (c) Aberrations for 1-rad (at FWHMI) phase lag with respect to the chirped signal; pump power differing by −/+10% from the ideal quenching condition. Thick line: STI profile; thin line: signal formed by four peaks with 255-fs time interval and 2:1:3:1 intensity ratio. (d) Experimental STI with the scale −100 fs/nm, i.e., −200 fs/div.

time scale of $\Delta t/\Delta\lambda \approx -100$ fs/nm reproduce the signal time intensity profiles. The four-peak structure is clearly visible in the vicinity of pump spectrum with approximately the expected intensity ratio. The temporal interval between the peaks (~220 fs) is also close to the expected value of 255 fs. The strong pump radiation with a weak separation in wavelength impaired the clarity of the image on the leading side of the pulse train.

In conclusion, the basic principles and potential of spectrotemporal imaging (STI) for direct display of femtosecond-scale optical pulse on a spectrometer has been introduced. STI is based on a time-lens induced by Kerr effect either through SPM or by XPM. Both were considered and demonstrated experimentally. The more accurate XPM-STI requires a synchronized pump pulse at a separate wavelength or polarization with an appropriate parabolic shape.

6.5.2 Aberration-Free Fourier Conversion and Spectrotemporal Imaging in Similariton-Induced Time Lens: Femtosecond Optical Oscilloscope

Aberration-free spectral focusing (SC) achieved with similariton-induced parabolic lens opens the prospect of the implementation of improved FT and STI. With the aim of designing a femtosecond optical oscilloscope, we developed a self-referencing method of aberration-free STI based on a similariton-induced time lens [16–18, 22]. The method is based on dispersive stretching of the signal pulse, and thereafter cancelation of its D-induced phase in a quadratic nonlinear process by addition of a reference parabolic phase (instead of Kerr-lensing of traditional SC). We implemented the method through sum frequency generation, using for reference an NL-D similariton generated from a fraction of the signal to measure. Direct measurements of AM and PM pulses were experimentally demonstrated at the femtosecond scale. This evolution of the STI-method offers many improvements: (i) an easy separation of the image spectrum from the reference; (ii) an enhanced accuracy because of the aberration-free time lens; and, most importantly, (iii) it makes the technique self-referenced. That latter improvement comes from the self-shaping property of the similariton [22, 52, 107].

Discussing the principle of similaritonic STI, we should remember the similariton-based SC device of Figures 6.14 and 6.15. In those schemes the signal radiation $\tilde{A}_0(\omega)$ is first split in two parts, on one branch to shape the spectron in the DDL and on the other branch to shape the NL-D similariton pulses in the fiber. The two fields are then combined in a SFG-crystal which performs their product. In the output of both the DDL and fiber arms, we have linearly chirped PM pulses with spectral amplitudes $\tilde{A}_d(\omega) = \tilde{A}_0(\omega)\exp[-i\phi_d(\omega)]$ and $\tilde{A}_f(\omega) = |\tilde{A}_f(\omega)|\exp[-i\phi_f(\omega)]$ respectively with parabolic phases $\phi_{d,f}(\omega) = \tilde{C}_{d,f}\omega^2/2$. This leads to dispersive Fourier transformation (Section 6.2), i.e., the temporal amplitudes of the spectron and of the similariton replicate their spectrum so that: $A_d(t) \propto \tilde{A}_0(\omega)\exp(iC_d t^2/2)$ and $A_f(t) \propto |\tilde{A}_f(\omega)|$ $\exp(iC_f t^2/2)$, respectively, with a scaling parameter $\omega = C_{d,f}t$. Under the condition of chirps with opposite sign and identical magnitude

$$C_f = -C_d \equiv C \tag{6.13}$$

the output temporal SFG-signal repeats the input spectral amplitude:

$$A_{SFG}(t) \propto A_d(t) \times A_f(t) \propto [\tilde{A}_0(\omega) \times |\tilde{A}_f(\omega)|]_{\omega=Ct} \propto \tilde{A}_0(\omega)|_{\omega=Ct}$$

provided the additional condition of constant similariton spectrum throughout the signal spectrum be satisfied

$$\Delta\omega_f \gg \Delta\omega_0 \tag{6.14}$$

Accordingly, the output spectral amplitude is a copy of the input temporal one

$$\tilde{A}_{SFG}(\omega) \propto A_0(t) \text{ with } \omega = Ct \tag{6.15}$$

and the output SFG-spectrum is a direct replica of the input pulse time profile:

$$S_{SFG}(\omega) = |\tilde{A}_{SFG}(\omega)|^2 \propto |A(t)|^2 = I(t) \tag{6.16}$$

with the scale $\omega = Ct$. The resolution of such a STI is determined by the similariton spectral bandwidth, since the similariton spectrum serve as a transfer function of the induced time lens [22]. In our consideration, we have used the spectron approximation for DDL (Eq. (6.3)) for more visibility, as for the Kerr-lensing STI, and in general case the Eq. (6.15) has a form

$$\tilde{A}_{SFG}(\omega) \propto A_0(t) \exp(iCt^2/2) \text{ with } \omega = Ct \tag{6.15'}$$

leading to the same STI result as Eq. (6.16). Eqs. (6.15) and (6.15') state a complete optical FT, as for the Kerr lens STI, and measure the signal temporal phase through measurements of the output spectral phase.

The experimental setup of similaritonic STI is the same as the one for SC in similariton induced time lens (Figures 6.14 and 6.15). The conditions of the experiments reported below were also similar to those of Section 6.4 dealing with aberration-free spectral compression with a similariton-induced time lens (same laser, same dispersive line, same fiber, same SFG crystal). In order to test the performance of the device, input signal pulses were shaped with various profiles of different complexity and then measured by a standard intensity autocorrelator (APE PulseCheck) for reference. The autocorrelation traces of the pulses derived from the STI measurements were computed from the recordings and compared to the experimental ones to assess the accuracy.

During experiments the spectra of the D-stretched input signal and of the NL-D similariton were measured to check if the STI conditions (such as Eq. (6.14)) were satisfied. The 107-nm bandwidth NL-D similaritons, with the spectra of Figure 6.16 serving as the transfer functions of the induced time lens, should provide ~7 fs resolution. The calibration of our similaritonic STI was achieved on the basis of the general concept of FT in a time lens, according to which the time delay between the signal and the reference pulses leads to the frequency shift of the output SFG signal. Experimentally, the shift $\Delta\lambda$ caused by the temporal delay Δt was measured and the time-to-frequency conversion coefficient $C_\lambda = \Delta\lambda/\Delta t$ was thus determined in a simple and convenient way.

To demonstrate the general character of the STI process, particularly, its independence with respect to the signal D-stretching and SC ratio according to analytic discussion above, the experiments were carried out in three different cases: (i) the spectron regime of the far field of DDL (with 2 m separation of SF11 prisms and 12 × SC), (ii) in the near field of dispersion (0.6 m prism separation, and 1.2 × SC), and (iii) for the middle case (1.2 m prism separation, and 2.5 × SC). The results confirmed that only the Eqs. (6.13) and (6.14) conditions are important for accurate and high-resolution STI. In

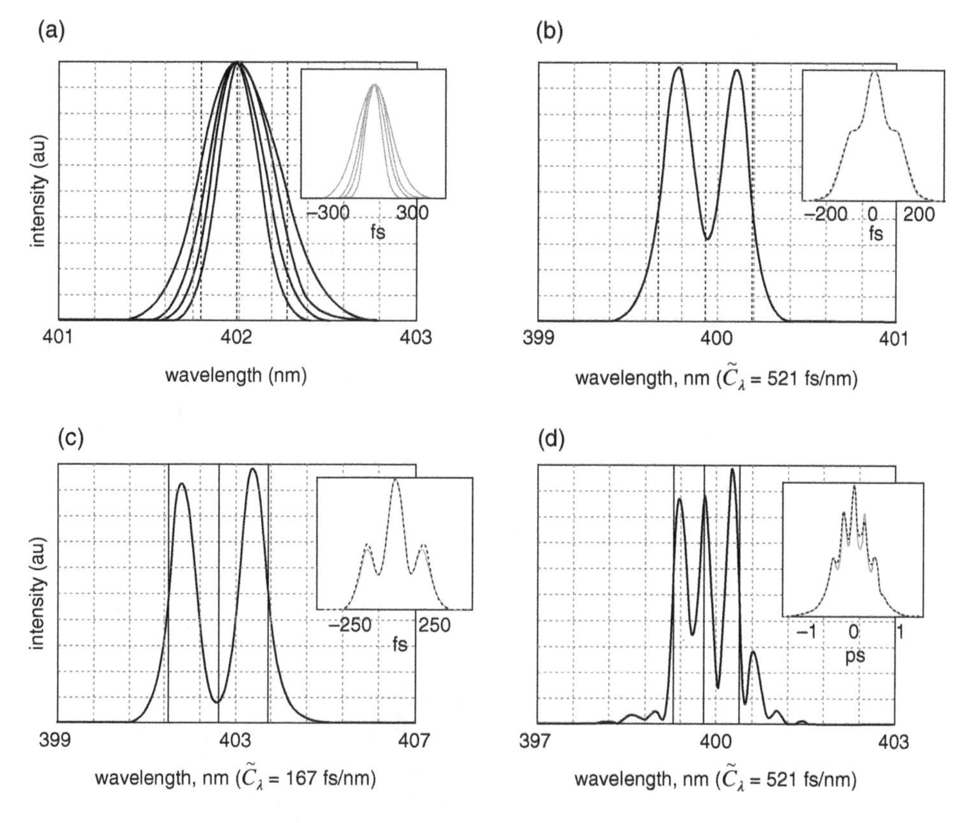

Figure 6.23 STI of different pulses: (a) pulses stretched in glasses with different thickness; (b), (c) two-peak pulses; and (d) a pulse with substructure. In insets, the autocorrelation traces of signal pulses and their images are shown for comparison. The values of scale coefficient $\tilde{C}_\lambda$ for the pulses' spectral images are the following: (b), (d) – $\tilde{C}_\lambda = 521$ fs/nm, and (c) – $\tilde{C}_\lambda = 167$ fs/nm.

practice, the weak impact of the SC ratio is important to get some freedom in the femtosecond oscilloscope implementation and to adapt to the spectrometer resolution, and to the range of signal parameters (such as spectral bandwidth).

In order to test the performance of the device, input signal pulses were shaped with various profiles of different complexity and then measured with a standard intensity autocorrelator (APE PulseCheck) for reference. The results appearing as the spectral images of the pulses of different duration and shape are shown in Figure 6.23 with the autocorrelation traces of the signal pulses given in inset (top right). The autocorrelation traces of the pulses derived from the STI measurements were computed from the recordings (dashed black line) and compared to the experimental reference traces (solid line) to assess the accuracy. Figure 6.23 (a) shows the images of the laser pulses stretched in glass with different thicknesses. Comparison with the relevant autocorrelation traces shows a good agreement: the dashed black lines for the image autocorrelation completely cover the solid lines for pulse autocorrelation traces. Figures 6.23 (b) and (c) present the spectral images of two-peak pulses with the scale coefficients $\tilde{C}_\lambda =$ 521 fs/nm (b), and $\tilde{C}_\lambda =$167 fs/nm (c): the pulse and image autocorrelation traces coincide for all regimes, practically. In Figure 6.23 (d) the spectral image of a pulse with a complex substructure is shown (with the scale coefficient $\tilde{C}_{\lambda} = 521$ fs/nm): the pulse and image autocorrelations still in good agreement.

Summarizing, the self-referenced spectrotemporal imaging of ultrafast pulses with a similariton-induced time lens was demonstrated for femtosecond pulses of different shapes. A good quantitative agreement with the intensity autocorrelation measurements attest that the method is reliable. A femtosecond optical oscilloscope (FO) providing direct temporal measurements with a 7 fs (theoretical) resolution has been designed from that result.

Figure 6.24 shows the FO based on similariton-induced time lens designed in the result of the described theoretical and experimental studies. The 3D model (a) and implemented prototype of the similariton-based time-to-frequency FT converter (b), which together with a spectrometer (OSA) serves as FO are shown. To compact the device, diffraction gratings were used instead of dispersive prisms in DDL. Further development is anticipated by the replacement of DDL by an anomalous dispersion fiber.

Actually, the method of STI with a similariton-induced time lens opens new opportunities for classic spectroscopy at femtosecond timescale, with the simultaneous measurement of temporal and spectral amplitudes of optical signal transmitted by a

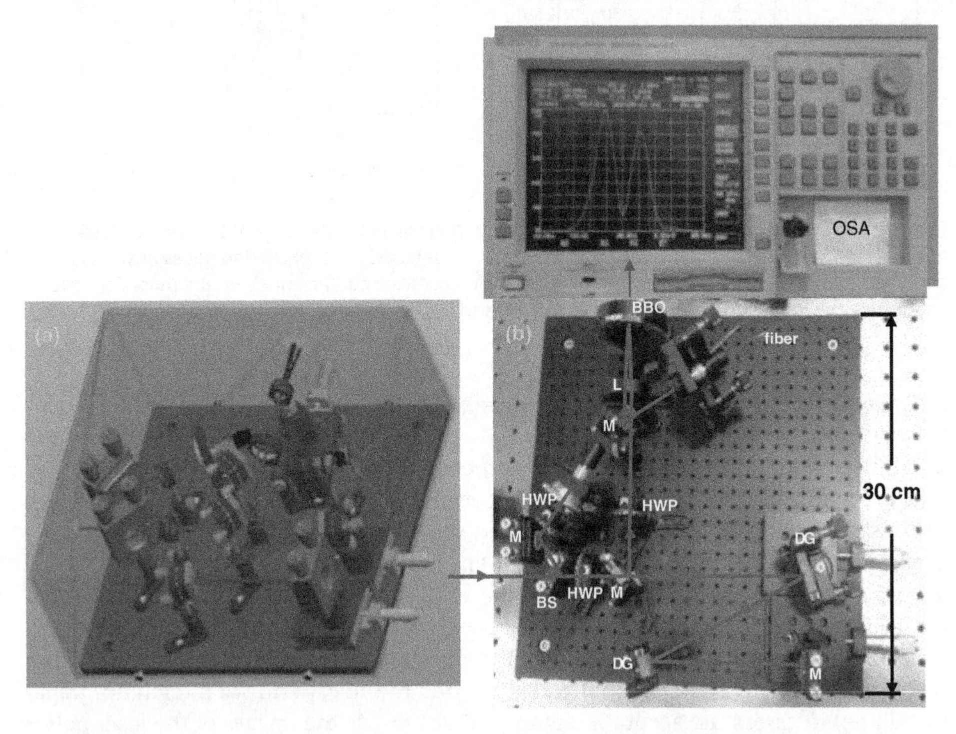

Figure 6.24 Femtosecond optical oscilloscope (FO) based on a similariton-induced time lens: 3D model (a) and an implemented prototype of similariton-based time-to-frequency FT converter (b), which together with a spectrometer (OSA) serves as a FO. Components: HWP – half-wave plate, DG – diffraction grating, M – mirror, L – lens, fiber – single-mode fiber for similariton shaping, BBO – nonlinear crystal for SFG. Spectral image of a two-peak pulse is seen on the OSA screen, as on the oscilloscope.

sample, with the possibility of full complex field recovery. Along with that, the time-to-frequency conversion in the time lens permits the direct phase measurements with an adequate interferometric detection, based on the above theoretical analysis (Eqs. (6.11), (6.11'), (6.15) and (6.15')). Technically, the phase measurement is easier to implement in the spectral domain, applying the approach of spectral interferometry (SI) [41, 42]. The method of SI, providing a complete characterization of femtosecond signal, was effectively applied for various characterization, in particular, to reveal the character and peculiarities of NL-D similariton [11, 12], to characterize the prism-lens dispersive delay line [111], to study the NL-D regime of spectral compression with the shaping of transform-limited flattop pulses [112], etc. The FO device, based on similaritonic STI, exhibits the additional benefit of self-referencing (more details, see Section 6.5.3).

6.5.3 Similariton-Based Self-Referencing Spectral Interferometry

We introduce here the self-referencing method of similariton-based SI, its motivation and principle, comparison with the alternative methods of SPIDER [93], SPIRIT [94], SORBETS [95], etc, together with a comparative experimental study with STI by means of a similariton induced time lens.

Many modern methods of femtosecond signal characterization are rooted in the approach of SI. The classic method of SI is based on the interference of the signal and reference beams spectrally dispersed in a spectrometer, with the spectral fringe pattern caused by the difference of spectral phases [41, 42]. The known spectral phase of the reference permits retrieval of the spectral phase of signal, and, together with the spectrum measurement, to recover mathematically the complex temporal amplitude of the signal through FT. The setup of classic SI is rather simple, and the measurement is accurate. However, the application range of classic SI is restricted by the spectral bandwidth of reference: SI characterization of a signal that has undergone an NL-interaction requires a special broadband reference to fully cover the broadened signal spectrum. To avoid this restriction, the self-referencing methods of spectral shearing interferometry, such as SPIDER [93], SPIRIT [94], and SORBETS [95], were developed. Self-referencing promotes the SI methods into the class of the most popular and commercialized tools for measurements at the femtosecond time scale, making them comparably competitive with the FROG devices [91,92]. The similaritonic SI method joins the advantages of both the classic SI [40, 41] and the spectral shearing interferometry [93–95], combining the simplicity of the principle and configuration with the self-referencing performance [16–18, 21].

The similariton-based SI setup is shown in Figure 6.25: the similaritonic STI-FO setup is modified for SI, simply removing the nonlinear BBO crystal and DDL.

The signal to be measured is first split in two beams of unbalanced power. The high-power part is injected into a fiber to generate the NL-D similariton, with a spectral amplitude $\tilde{A}_f(\omega) = |\tilde{A}_f(\omega)| \exp[-i\phi_f(\omega)]$, $\phi_f(\omega) = \beta_2 L_f \omega^2/2$ being fixed by the fiber parameters only. The residual part of the signal of complex spectral amplitude $\tilde{A}_0(\omega) = |\tilde{A}_0(\omega)| \exp[-i\phi_0(\omega)]$, is delayed and superimposed to the NL-D similariton before being launched in a spectrometer. The resulting spectral fringe pattern

$$S_{SI}(\omega) = 2|\tilde{A}_0(\omega)||\tilde{A}_f(\omega)| \cos[\phi_0(\omega) - \phi_f(\omega)],$$

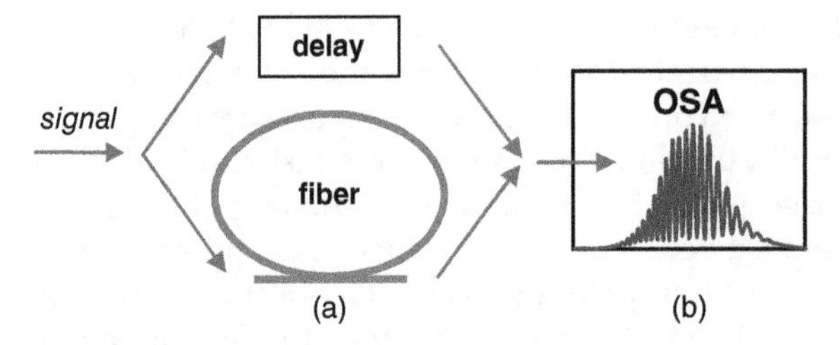

Figure 6.25 Schematic of the similaritonic SI.

in the background of the signal and similariton spectra, completely covers the signal spectrum $S_0(\omega) = |\tilde{A}_0(\omega)|^2$, and the whole phase information becomes available, for any signal. The known spectral phase of the similariton-reference allows the retrieval of the signal spectral phase $\varphi_0(\omega)$, and, by measuring also the signal spectrum, to reconstruct mathematically the complex temporal amplitude of the signal through FT: $A_0(t) = FT\{\sqrt{S_0(\omega)}\exp[-i\varphi_0(\omega)]\}$.

In the experiments, different AM and PM pulses were shaped at the setup input, and the signal radiation was split by a beam-splitter. The high-power pulse (with an average power of up to 500 mW) was injected into a standard single-mode fiber (1.65 m Newport F-SPF PP@820 nm) to generate the NL-D similariton (Figure 6.25). The low-power pulse, with an appropriate time delay (delay), was coupled with similariton in the OSA, and the displayed spectrum was recorded (Figure 6.25 (a)). To retrieve the spectral phase (Figure 6.25 (b)), the FT algorithm of the fringe-pattern analysis was used [113,114]. The similaritonic SI measurements were compared to autocorrelation traces of the same pulses delivered by a commercial autocorrelator (APE PulseCheck). Prior to measurement it was checked that the generated similariton had the expected phase curvature knowing that $L_f = 49$ cm and $\beta_2 = 40$ fs^2/mm.

First, the similaritonic SI performance was assessed with the laser pulses stretched in SF11 glasses of different thickness (Figure 6.26). Comparison of calculated intensity

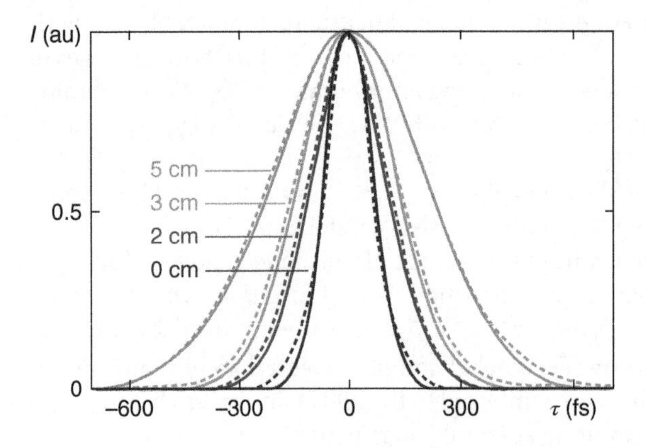

Figure 6.26 Similaritonic SI measurements for laser pulses stretched in SF11 glasses of different thickness: intensity autocorrelation functions of retrieved pulses (solid lines) in comparison with measured autocorrelation traces (dashed lines).

autocorrelation functions for retrieved pulses (solid lines) with measured autocorrelation traces (dashed lines), shows a good overlap.

Thereafter, SI-measurements of double-peak pulses were carried out. The double-peak signal pulses were shaped with separation in time T and amplitude ratio μ. The temporal amplitude $A_0(t) = a(t) + \mu a(t+T)$ corresponds to the complex spectral amplitude $\tilde{A}(\omega) = \tilde{a}(\omega)\rho(\omega)\times \exp[i\phi(\omega)]$, with the amplitude modulation $\rho(\omega) \equiv \sqrt{1+\mu^2+2\mu\cos(\omega T)}$ and the phase modulation $\phi(\omega) = \arctan[(\sin\omega T)/(\mu^{-1} + \cos\omega T)]$. To perform the shaping, the laser beam was expanded and a thin glass plate was inserted in a fraction of the beam cross-section. The beam part passed through the plate was delayed with respect to the free-propagated part. The power ratio between the peaks was adjusted by translation of the plate in the beam cross-section. Figure 6.27 in the color plate section presents the experimental results for a twin input pulse with almost balanced intensity. The measured spectrum and the SI-retrieved spectral phase are plotted respectively in black and in blue on Figure 6.27 (a) and the FT-reconstructed pulse is shown as a blue trace in Figure 6.27 (b). Validation of the similaritonic SI measurements was made by comparison of the autocorrelation profile calculated from the

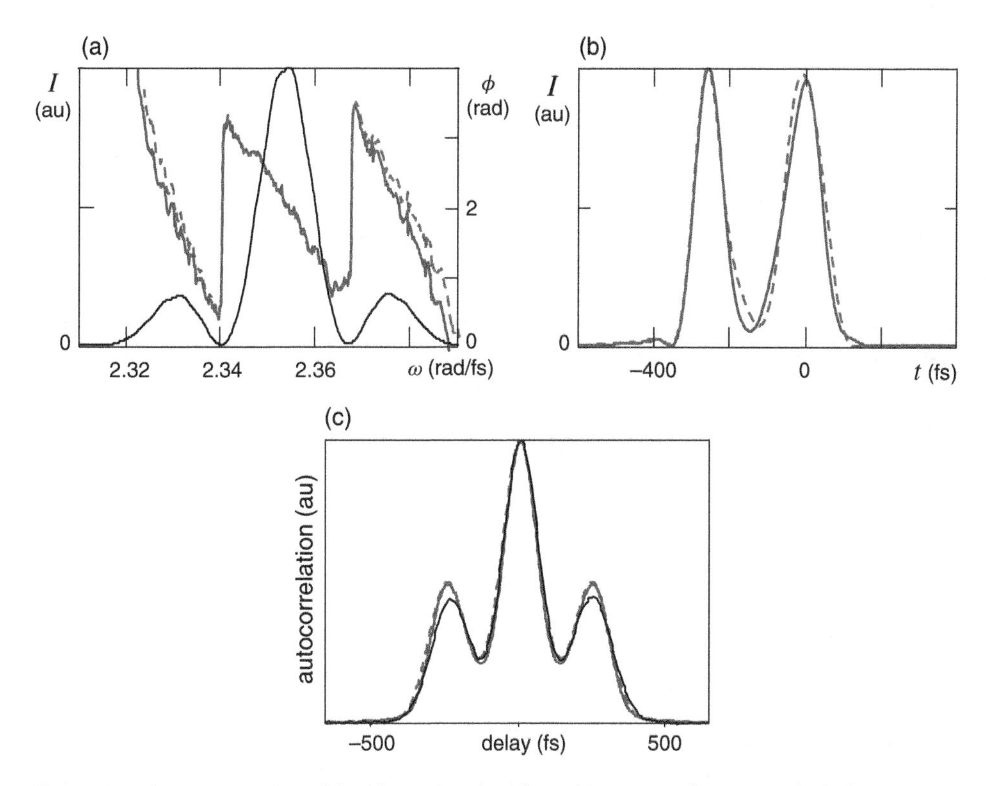

Figure 6.27 SI-reconstruction of double-peak pulse (shaped by means of a 130-μm thick glass): (a) retrieved spectral phase (color) and measured spectrum (black); (b) reconstructed pulse; and (c) measured autocorrelation of signal (black) in comparison with ones for reconstructed pulse (color). To illustrate the sensitivity-stability of retrieving, together with blue curves for $\tilde{C}_f = 2.1 \times 10^4$ fs^2, red curves for $\tilde{C}_f = 1.995 \times 10^4$ fs^2 (5% difference) is shown. For a color version of this figure please see color plate section.

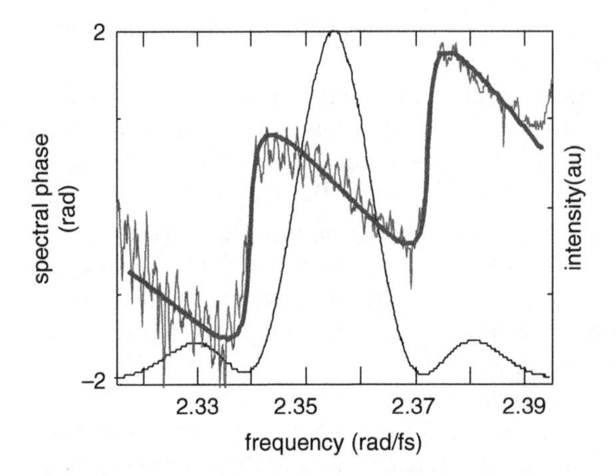

Figure 6.28 The SI-retrieved spectral phase (red/thin grey) of an "Ararat"-shaped two-peak signal in the range of its spectrum (black). The theoretical curve for spectral phase for such a two-peak pulse is shown also (blue/thick grey). For a color version of this figure please see color plate section.

retrieved pulses (in blue in Figure 6.27 (c)) with that measured directly (in black on Figure 6.27 (c)). As expected, the two traces were very similar and overlapped almost perfectly.

Finally, examining the similariton-based SI, we compared its measurements with the ones we obtained with the prototype of FO, based on the similaritonic STI. This comparative study, involving also theoretical and autocorrelation checking, serves also for a crossed evaluation of the FO prototype, the measurements of which were compared previously with the autocorrelation only [22]. The study was carried out with two-peak signal pulses clearly separated in time and with a 3:5 peak power ratio which was reminiscent of the biblical Ararat Mountains. We followed the methodology described above, and the SI-reconstructed pulses were compared with spectral images of the signal pulse. The red trace in Figure 6.28 in the color plate section shows the SI-retrieved spectral phase which perfectly agrees with the theoretical one plotted in blue, in the spectral range covered by the measured spectrum (black line). The similaritonic SI data lead to the accurate signal reconstruction shown in red on Figure 6.29 in the color plate section. A high-precision profile of the same signal obtained by use of the similaritonic STI technique is given as well in black. The similariton used for reference exhibited here a spectral bandwidth larger than 40 nm. The differences between the pulse shapes derived from these independent SI- and STI-pulse characterization and the theoretical curve (blue) are hardly seen in Figure 6.29. That evidences both the accuracy of the measurements and the potential of the similariton-based methods.

Concluding, we have developed and implemented a similariton-based self-referencing method of SI. The method is based on the similariton generation from a part of the input signal and its use as a reference for the interference with the signal in the spectrometer. Thus, the similaritonic SI provides a complete characterization of the femtosecond signal combining the advantages of the simple principle and configuration of classic SI with the self-referencing performance. Pulse measurements were achieved with the similaritonic SI and compared with the ones of STI in a similariton-induced temporal lens.

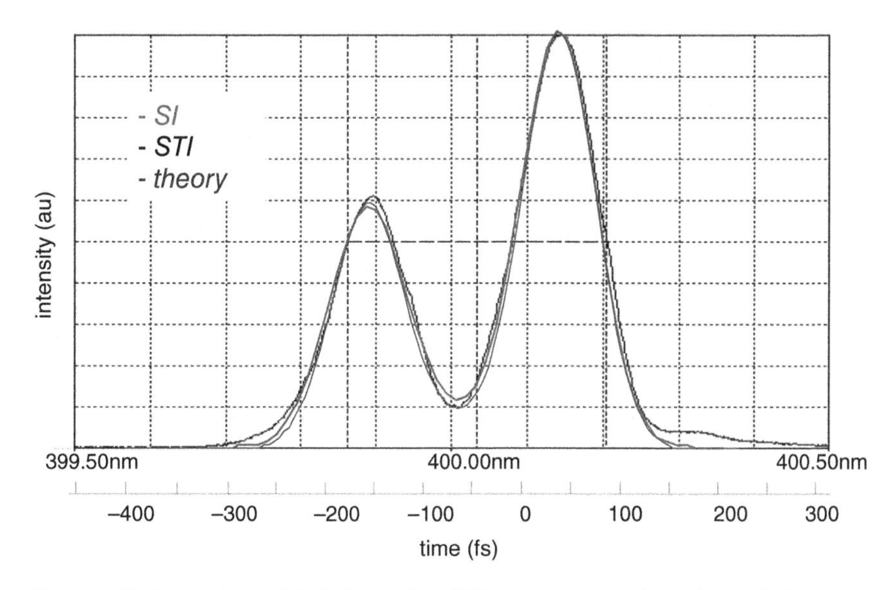

Figure 6.29 Comparison of similaritonic SI and STI measurements for an "Ararat"-shaped two-peak signal (with the 3:5 peak power ratio): SI-reconstructed pulse (red) in comparison with the spectral image of the pulse (black) obtained by STI based on similariton-induced time lens, and the theoretical profile (blue). For a color version of this figure please see color plate section.

The comparative study, carried out together with theoretical checking and autocorrelation measurements, evidenced the quantitative agreement of both the similariton-based methods of SI and STI and their potential for accurate femtosecond-scale temporal measurements. A STI device provides a direct reading of the pulse on the OSA. For complete pulse retrieval including this phase, one will choose an SI setup.

Obviously, the similariton-based STI and SI methods of femtosecond signal characterization can be implemented also by the use of "standard" parabolic similaritons generated in active or dispersion decreasing fibers. A recent paper [115] reported the generation of parabolic broadband similaritons with bandwidths of up to 11 THz (40 nm at 1050 nm central wavelength). However, the use of NL-D similaritons generated in standard passive fiber pieces is more advantageous, providing larger bandwidths and thus larger application ranges with technically simpler experimental arrangement.

6.5.4 Simple Similaritonic Technique for Measurement of Femtosecond Pulse Duration, an Alternative to the Autocorrelator

The commonest device for the femtosecond pulse diagnostics remains the conventional intensity autocorrelator [109], as a basic tool for the daily inspection or adjustment of the laser source in the sense of pulse duration. Below, we present a simple similaritonic alternative to the autocorrelation technique to determine the femtosecond laser pulse duration. The proposed technique is based on the generation of the NL-D similariton in a piece of passive fiber (without gain) and measurement of its spectral bandwidth.

As the study of the NL-D similariton (Section 6.3) shows, the initial peak power (P_0) of the pulse which is reshaped into a similariton determines its spectral bandwidth $\Delta\omega_{sim}$:

$\Delta\omega_{sim} = k\sqrt{P_0}$ with a coefficient k fixed by the fiber parameters, according to Eq. (6.6). Since the fiber losses accumulated on the distance required for NL-D similariton shaping are negligible, the energy E of the similariton is the same as that of the input pulse: $E = P_0\Delta t_0$ (Δt_0 is initial pulse duration). Thus, the initial pulse duration Δt_0 may be determined by measuring the bandwidth and energy of NL-D similariton: $\Delta t_0 = Ek^2/\Delta\omega^2_{sim}$. In practice, for high repetition rate lasers (ν), it is convenient to measure the average power $p = E\nu$ instead of the single pulse energy E. Thus, for the initial pulse duration Δt_0, we have $\Delta t_0 = (k^2/\nu) \times p/\Delta\omega^2_{sim}$, or equivalently in wavelengths:

$$\Delta t_0 = k_\lambda \times p/\Delta\lambda^2_{sim} \tag{6.17}$$

with a coefficient $k_\lambda \equiv k^2\lambda^4/\nu$.

Experiments were carried out to demonstrate the relevance of the approach [110]. Laser pulses (100 fs @800 nm) were stretched in SF11 glasses of different thickness or alternatively in a prism-DDL providing anomalous dispersion. Thereafter, a semitransparent mirror split the radiation and directed a fraction of the input toward an autocorrelator (APE PulseCheck) to measure the pulse durations at the input of the setup. The other part of the input radiation was coupled in a ~1-m long piece of single-mode fiber where NL-D similaritons were generated. Then, the bandwidths $\Delta\lambda_{sim}$ of the NL-D similariton and the radiation average power p were measured in the OSA (Ando AQ6315). Pulse durations were subsequently calculated using Eq. (6.17). These data were compared with the one given by the autocorrelation measurements. Studies were carried out for the pulse durations Δt_0 ~100–300 fs in the range of average powers p ~50–500 mW. It is worth mentioning that the power spectrum of the input signal to characterize was kept fixed since the input pulse duration was varied by dispersion. Figure 6.30 gives the results of the measurements.

The experimental points match well with the theoretical curve of $\tau_{AC} = \sqrt{2}\Delta t_0 = f(\Delta\lambda_{sim}/\sqrt{p})$ in agreement with Eq. (6.17). Therefore, the data of Figure 6.30 evidence that duration measurements performed by means of spectral analysis of similariton are similar to the ones derived from autocorrelation traces.

Note that the setup can be simplified, as the similariton bandwidths of $\Delta\lambda > 20$ nm use a diffraction grating instead of a spectrometer. Additionally, in this case, the average power measurement may be performed in the zero order of diffraction.

It important to mention also that the duration of NL-D similariton is determined by its bandwidth and fiber dispersion: $\Delta t_{sim} = \Delta\omega_{sim}/C = \beta_2 L_f k^2 (E/\Delta t_0)^{1/2}$. This allows, for the fiber lengths of ~1 km, measurement of the similariton duration in real time with a nanosecond oscilloscope instead of bandwidth measurement, and determination of the duration of the initial pulse from

$$\Delta t_0 = k_t \times p/\Delta t^2_{sim} \tag{6.17'}$$

with a coefficient $k_t = k^2\beta_2 L_f/\nu$.

Thus, a simple similaritonic technique for measurement of the duration of femtosecond laser pulses has been developed. The technique is based on the generation of NL-D similariton in a passive fiber and on measurement of its bandwidth. A short piece of fiber, a power meter and a spectrometer, three standard elements in an optics laboratory, suffice to implement the method that may be used for regular checking of femtosecond laser performance.

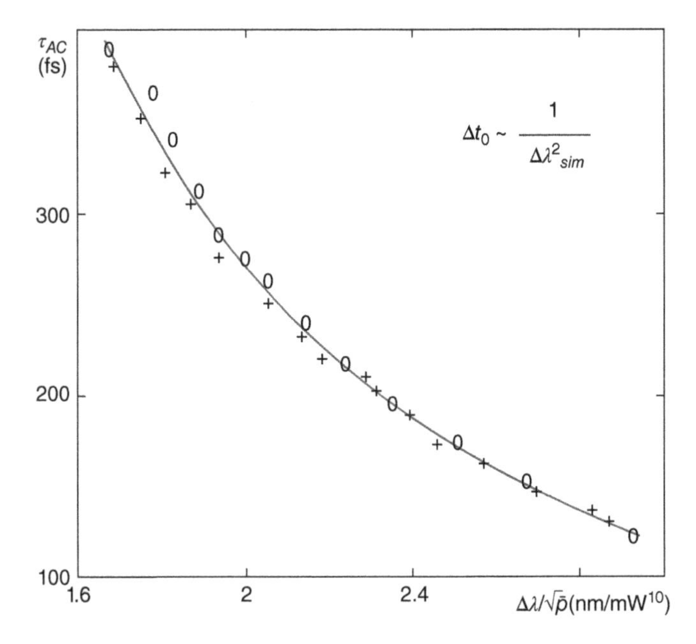

Figure 6.30 Similaritonic diagnostics of femtosecond pulses. Experimental dependence of τ_{AC} vs $\Delta\lambda_{sim}/\sqrt{p}$. Circles and crosses correspond to the pulses stretched in media with normal or anomalous dispersion, respectively; the solid line is for Eq. (6.17).

6.5.5 Reverse Problem of NL-D Similariton Generation

The process of pulse NL-D self-interaction in a fiber, in absence of losses and amplification, has a reversible character. Indeed, the process is described by NLSE, which is invariant to pair of transformations $z \to -z$, $A_z(t) \to A_z^*$. This feature determines the initial pulse parameters having the fiber output. A detailed study of this problem, for fiber delivery, has shown that reverse propagation is a feasible method of predicting the optimal input pulse that yields any desired output in a fiber, defeating all high-order NL- and D-effects [116]. Such an approach was used to design a compact and efficient arrangement for fiber delivery of sub-30 fs pulses, tunable around 800 nm. Pulses of a broadband Ti:Sapphire oscillator were pre-chirped by a grism-pair stretcher, allowing control of the SOD and TOD. At the exit of a 2.7-m-long large mode area PCF, 1-nJ pulses were compressed to 29 fs producing 30 kW of peak power [25]. Obviously, this approach can be used also for pulse characterization issue: the input pulse can be determined by solving the reverse propagation problem, starting from the characterization of the output pulse.

The reverse propagation problem is simplified by the self-shaping of NL-D similariton, with the asymptotically linear chirp, fixed by the fiber dispersion only, according to Eq. (6.5). In this case, all information on the initial pulse is passed practically to the intensity profile of the NL-D similariton, relaxing at the same time the need for a phase measurement. The mapping of the pulse time profile in the spectral domain due to the self-spectrotemporal imaging of similariton is quantitative with a fixed coefficient of chirp: $I_f(t) \propto S_f(\omega)|_{\omega=Ct}$, with $C \approx (\beta_2 L_f)^{-1}$. Hence, the initial pulse intensity profile can

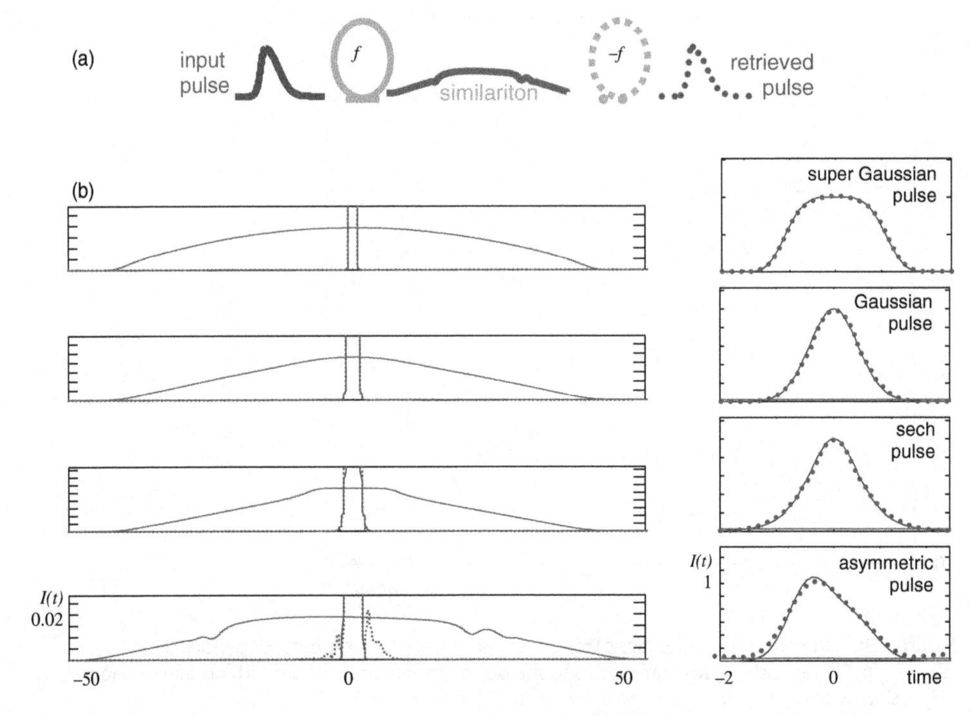

Figure 6.31 Illustration of initial studies for solution of reverse problem of NL-D similariton generation: (a) principle, and (b) results for various pulse shapes. Input (blue), similariton (green), and retrieved input (red dashed) pulses are shown. Simulation parameters: $R = L_D/L_{NL} = 50$, $L_f = 5L_D$. For a color version of this figure please see color plate section.

be simulated on the basis of a simple spectrometric measurement. Based on the measured spectrum of NL-D similariton, together with the spectral phase which is known, computation of the reverse propagation provides complete information about the input field. Figure 6.31 in the color plate section illustrates this principle (Figure 6.31 (a)) and provides some preliminary numerical results of pulse reconstruction (Figure 6.31 (b)) through the solution of the reverse NL-D similariton generation. The reconstructed pulses (red dashed) perfectly match the initial ones (blue), demonstrating the potential of this approach for pulse characterization. Further efforts should be made to get a large measurement range through an adaptation of the fiber parameters to the pulse parameters. In this way, a short piece of fiber can serve as an alternative to the FROG device [26]. The solution of the reverse problem of the generation NL-D similariton is also helpful for femtosecond pulse delivery [25].

6.5.6 Pulse Train Shaped by Similaritons' Superposition

The spectronic nature of the NL-D similariton and particularly its feature of self-spectrotemporal imaging, according to Eq. (6.5), opens the opportunity of shaping of a pulse train for applications in ultrafast optics. Superposition of such a similariton with its T-delayed replica results in a train with the $A_\Sigma(t) = A_f(t) + \mu A_f(t+T)$ temporal, and $\tilde{A}_\Sigma(\omega) = \tilde{A}_f(\omega)[1 + \mu \exp(i\omega T)]$ spectral sum amplitudes. For the spectral and temporal intensity profiles of such a train, we have $S_{\Sigma}(\omega) = |\tilde{A}_f(\omega)|^2 \times [1 + \mu^2 + 2\mu\cos(\omega T)]$, and $I_{\Sigma}(t) \propto |\tilde{A}_f(\omega)|^2 \ [1 + \mu^2 + 2\mu\cos(\omega T + CT^2/2)]_{\omega=Ct}$, respectively. Thus, as a train, we have a harmonically amplitude-modulated similariton with an adjustable amplitude and period of modulation fixed by the amplitude and delay of replica, respectively.

The periodical amplitude-modulation of spectrally broadband radiation with high-repetition rate, adjusted to the resonance of the Raman oscillations of medium, is effectively applied for CARS spectroscopy and microscopy [24, 117]. The application of broadband similariton to pulse train shaping is promising for CARS microscopy in view of the exploitation of a simplified setup with a single laser source. A similaritonic train is of interest as well for the methods based on SPM and XPM. In this case, the nonlinear PM translated into FM is proportional to the intensity derivative $\delta\omega \sim dI/dt$, and thus to the frequency of AM, fixed by the delay T. This provides the opportunity for passive adjustment of nonlinear SPM and XPM effects by simple adjustment of delay T. Figure 6.32 in the color plate section shows two examples of a pulse train after NL-D propagation in a fiber. The intensity temporal profiles, which mimic the spectra, for a train shaped by superposition of an NL-similariton with its replica (green), and the ones for a train after self-interaction in the fiber (red) are shown. The temporal intensity profile of Figure 6.32 (a), which mimics the spectrum, clearly shows the influence of NL-SPM (red), despite the decreased value of peak intensity (green). Parabolic sub-pulses of Figure 6.32 (b) (red) are of special interest for temporal Kerr-lensing issues.

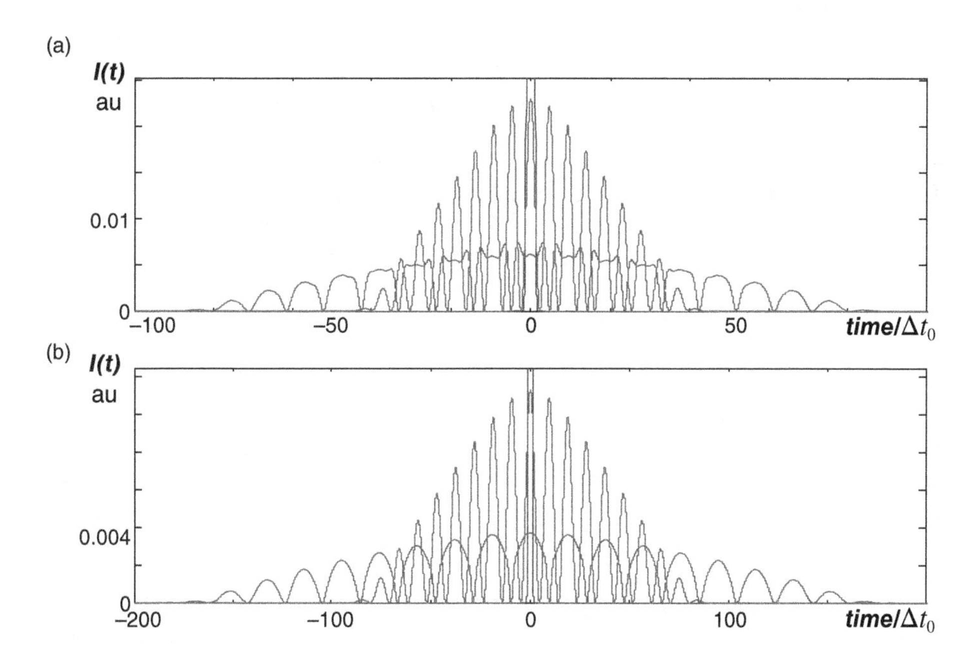

Figure 6.32 Two examples of a train self-interacted in fiber. The intensity profiles of a train shaped by superposition of NL-similariton with its replica (green), and ones for self-interacted train in the fiber (red) are shown (red). Simulation parameters: $T = 2.5$, $\mu = 1$, $R = 50$, $L_f/L_D = 5$ (a), and $L_f/L_D = 10$ (b). For a color version of this figure please see color plate section.

6.6 Conclusion

We have introduced the NL-D similariton pulses which are generated in passive optical fibers under the combined impact of Kerr-non linearity and normal dispersion. They share some features with standard similariton pulse of fibers with gain: a linear chirp and a self-similar propagation. But NL-D similaritons are more general and not restricted to parabolic pulse profile. However they lead, as well as the usual similariton, to a replication of the pulse temporal profile in the spectrum (spectrotemporal imaging). The striking features of NL-D similaritons were established through analytical, numerical and experimental studies. Fiber dispersion only determines the chirp of NL-D similariton independently from the input pulse characteristics, while their bandwidth is fixed by the input peak power. Broadband NL-D similaritons with up to 150 nm bandwidth were experimentally generated and completely characterized through precise chirp measurement technique. Their chirp were perfectly fitted by a linear law (within ~1% accuracy) as expected.

NL-D similariton can be applied to various domains of ultrafast photonics and in particular to ultrashort pulse generation, pulse processing and measurement. For example, laser pulses of 140 fs duration were compressed down to 14 fs ($P_{ave} = 500$ mW) after generation of 75 THz wide NL-D similaritons in a short piece of fiber. The performances were limited by the bandwidth and fourth order dispersion of our grating-prism-based compressor.

Regarding pulse processing, the NL-D similariton can serve for spectral compression, a particular lossless frequency filtering.

Based on the concept of time lens, we have developed an effective method of spectral compression (SC) which provides aberration-free spectral focusing, by analogy with beam collimation in the space domain. In a comparative experimental study of various femtosecond pulse SC schemes, we have achieved various compression ratio. The highest 1:23 ratio has been obtained in a scheme where an NL-D similariton generated from a fraction of the input pulse was mixed in a nonlinear crystal with the rest of the input pulse after stretching. The technique can be implemented with various frequency mixing processes, such as sum- or difference-frequency generation, four wave mixing. This type of spectral focusing is of special interest for CARS spectroscopy in view of its spectral resolution improvement. Most of the applications of NL-D similaritons regard ultrashort pulse characterization.

The basic principles and potential of spectrotemporal imaging (STI) for direct and instantaneous display of femtosecond scale optical pulse on a spectrometer were introduced. Spectral imaging of temporal light signals, STI, is based on a time-lens induced by the Kerr effect, either through SPM or XPM. Both were considered and demonstrated experimentally. The more accurate XPM-STI requires a synchronized pump pulse at a separate wavelength or polarization with an appropriate parabolic shape. The spectral imaging of light pulses can be self-referenced thanks to the use of the NL-D similariton generation. The STI of ultrafast pulses with an NL-D similariton-induced parabolic time lens was demonstrated for femtosecond pulses of different shapes. Agreement with the intensity autocorrelation measurements attested that the method is reliable. A femtosecond optical oscilloscope providing direct temporal measurements with a 7 fs (theoretical) resolution has been designed from those results.

For complete and direct characterization of femtosecond electric fields, similariton generation and spectral interferometry can be associated. The proposed method of similaritonic SI combines the performance and configuration of classic spectral interferometry with the self-referencing provided by the NL-D similariton. Our comparative experiments of similaritonic SI and STI, carried out together with theoretical checking and autocorrelation measurements, evidenced the quantitative agreement and high precision of both similariton-based methods and their potential for accurate femtosecond-scale temporal measurements.

Each time pulse characterization can be restricted to their duration measurement, NL-D similaritons offer a simple alternative to intensity autocorrelation. It is based on the generation of the NL-D similariton in a passive fiber and on subsequent measurement of its bandwidth. A short piece of fiber, a power meter and a spectrometer, three standard elements in an optics laboratory, suffice to implement the technique that may be used for regular checking of femtosecond laser performance.

Some perspectives for new uses of broadband similariton in ultrafast optics have been quickly addressed. Reverse propagation modeling of the NL-D similariton generation can provide another option for complete characterization of ultrashort pulses. Based on the measured spectrum of the NL-D similariton, together with the spectral phase which is known, computation of the reverse propagation retrieves the complete input field, as we have shown.

Analysis of a pulse train created by superposition of NL-D similariton suggest it could be relevant for applications to CARS spectroscopy, as well as for the SPM- XPM-based nonlinear methods of ultrafast signal control and characterization.

References

1 Dudley, J.M., Finot, C., Richardson, D.J., and Millot, G. (2007) Self-similarity and scaling phenomena in nonlinear ultrafast optics. *Nature Physics*, **3**, 597–603.

2 Finot, C., Dudley, J.M., Kibler, B., Richardson, D.J., and Millot, G. (2009) Optical parabolic pulse generation and applications. *IEEE Journal of Quantum Electronics*, **45**, 1482–1489.

3 Anderson, D., Desaix, M., Karlson, M., Lisak, M., and Quiroga-Teixeiro, M.L. (1993) Wave-breaking-free pulses in nonlinear optical fibers. *Journal of Optical Society of America B*, **10**, 1185–1190.

4 Fermann, M.E., Kruglov, V.I., Thomsen, B.C., Dudley, J.M., and Harvey, J.D. (2000) Self-similar propagation and amplification of parabolic pulses in optical fibers. *Physical Review Letters*, **84**, 6010–6013.

5 Kruglov, V.I., Peacock, A.C., Dudley, J.M., and Harvey, J.D. (2000) Self-similar propagation of high-power parabolic pulses in optical fiber amplifiers. *Optics Letters*, **25**, 1753–1755.

6 Kruglov, V.I., Peacock, A.C., Harvey, J.D., and Dudley J.M. (2002) Self-similar propagation of parabolic pulses in normal-dispersion fiber amplifiers. *Journal of Optical Society of America B*, **19**, 461–469.

7 Finot, C., Millot, G., Billet, C., and Dudley, J.M. (2003) Experimental generation of parabolic pulses via Raman amplification in optical fiber. *Optics Express*, **11**, 1547–1552.

8 Ilday, F.Ö., Buckley, J.R., Clark, W.G., and Wise, F.W. (2004) Self-similar evolution of parabolic pulses in a laser. *Physical Review Letters*, **92**, 213902.
9 Hirooka, T. and Nakazawa, M. (2004) Parabolic pulse generation by use of a dispersion-decreasing fiber with normal group-velocity dispersion. *Optics Letters*, **29**, 498–500.
10 Finot, C., Barviau, B., Millot, G., *et al.* (2007) Parabolic pulse generation with active or passive dispersion decreasing optical fibers. *Optics Express*, **15**, 15824–15835.
11 Zeytunyan, A., Yesayan, G., Mouradian, L., *et al.* (2009) Nonlinear-dispersive similariton of passive fiber. *Journal of European Optical Society, Rapid Publications*, **4**, 09009; Kockaert, P., Mouradian, L., Yesayan, G., and Emplit, P. (2007) Generation of similaritons in a nonlinear dispersive optical fibre without gain. Proceedings of Symposium IEEE/LEOS Benelux Chapter, p. 39; Yesayan, G., Palanjyan, K., Mansuryan, T., *et al.* (2007) Nonlinear-spectronic similariton of single-mode fiber without gain. Paper JWA18, in *Nonlinear Photonics*, OSA Technical Digest,
12 Zeytunyan, A.S., Palandjyan, K.A., Yesayan, G.L., and Mouradian, L.K. (2010) Nonlinear dispersive similariton: spectral interferometric study. *Quantum Electronics*, **40**, 327–328.
13 Kruglov, V.I., Méchin, D., and Harvey, J.D. (2007) High compression of similariton pulses under the influence of higher-order effects. *Journal of Optical Society of America B*, **24**, 833–838.
14 Palanjyan, K., Muradyan, A., Zeytunyan, A., Yesayan, G., and Mouradian, L. (2010) Pulse compression down to 17 femtoseconds by generating broadband similariton. *Proceedings of SPIE*, **7998**, 79980N.
15 Finot, C. and Millot, G. (2004) Synthesis of optical pulses by use of similaritons. *Optics Express*, **12**, 5104–5109.
16 Zeytunyan, A., Muradyan, A., Yesayan, G. *et al.* (2011) Generation of broadband similaritons for complete characterization of femtosecond pulses. *Optics Communications*, **284**, 3742–3747.
17 Zeytunyan, A., Yesayan, G., Mouradian, L., Louradour, F., and Barthélémy, A. (2009) Applications of similariton in ultrafast optics: spectral interferometry and spectrotemporal imaging. Paper FWI5, in Frontiers in Optics, OSA Tech. Digest.
18 Mouradian, L.K., Zeytunyan, A.S., Muradyan, A., *et al.* (2010) Similariton for Femtosecond Optics. Paper Mo.2.E.5, in *Proceedings of ECOC 2010* (19–23 September, 2010), Torino, Italy.
19 Zeytunyan, A., Muradyan, A., Yesayan, G., and Mouradian, L. (2010) Broadband similariton. *Laser Physics*, **20**, 1729–1732.
20 Zeytunyan, A., Muradyan, A., Yesayan, G., *et al.* (2010) Measuring of broadband similariton chirp. Paper NME46, in *Nonlinear Photonics*, OSA Tech. Digest.
21 Mouradian, L., Zeytunyan, A., and Yesayan, G. (2012) Similariton-based spectral interferometry for signal analysis on femtosecond time scale. In *Interferometry: Research and Applications in Science and Technology* (ed. I. Padron). InTech, Moscow, pp. 99–124.
22 Mansuryan, T., Zeytunyan, A., Kalashyan, M., *et al.* (2008) Parabolic temporal lensing and spectrotemporal imaging: a femtosecond optical oscilloscope. *Journal of Optical Society of America B*, **25**, A101–A110.
23 Pegoraro, A.F., Ridsdale, A., Moffatt, D.J., *et al.* (2009) Optimally chirped multimodal CARS microscopy based on a single Ti:sapphire oscillator. *Optics Express*, **17**, 2984.

24 Gershgoren, E., Bartels, R.A., Fourkas, J.T., *et al.* (2003) Simplified setup for high-resolution spectroscopy that uses ultrashort pulses. *Optics Letters*, **28**, 361.

25 Kalashyan, M., Lefort, C., Martínez-León, L., *et al.* (2012) Ultrashort pulse fiber delivery with optimized dispersion control by reflection grisms at 800 nm. *Optics Express*, **20** (23), 25624–25635.

26 Kane, D.J. and Trebino, R. (1993) Single-shot measurement of the intensity and phase of an arbitrary ultrashort pulse by using frequency-resolved optical gating. *Optics Letters*, **18**, 823–825.

27 Akhmanov, S.A. *et al.* (1968) Nonstationary nonlinear optical effects and ultrafast light pulse formation. *IEEE Journal of Quantum Electronics*, **4**, 598–605.

28 Caputi, W.J. (1971) Stretch: A time-transformation technique. *IEEE Transactions of Aerospace Electronic Systems*, **7**, 269–278.

29 Akhmanov, S.A., Vysloukh, V.A., and Chirkin, A.S. (1992) *Optics of Femtosecond Laser Pulses* AIP, New York.

30 Agrawal, G.P. (1995) *Nonlinear Fiber Optics*. Academic Press, New York,

31 Goda, K. and Jalali, B. (2013) Dispersive Fourier transformation for fast continuous single-shot measurements. *Nature Photonics*, **7**, 102–112.

32 Tong, Y.C., Chan, L.Y., and Tsang, H.K. (1997) Fiber dispersion or pulse spectrum measurement using a sampling oscilloscope. *Electronics Letters*, **33**, 983–985.

33 Kelkar, P.V., Coppinger, F., Bhushan, A.S., and Jalali, B. (1999) Time-domain optical sensing. *Electronics Letters*, **35**, 1661–1662.

34 Goda, K., Solli, D.R., Tsia, K.K., and Jalali, B. (2009) Theory of amplified dispersive Fourier transformation. *Physical Review A* **80**, 043821.

35 Solli, D.R., Chou, J., and Jalali, B. (2008) Amplified wavelength-time transformation for real-time spectroscopy. *Nature Photonics*, **2**, 48–51.

36 Jannson, T. (1983) Real-time Fourier transformation in dispersive optical fibers. *Optics Letters*, **8**, 232–234.

37 Anaza, J., Chen, L.R., Muriel, M.A., and Smith, P.W E. (1999) Experimental demonstration of real-time Fourier transformation using linearly chirped fiber Bragg gratings. *Electronics Letters*, **35**, 2223–2224.

38 Muriel, M.A., Azana, J. & Carballar, A. (1999) Real-time Fourier transformer based on fiber gratings. *Optics Letters*, **24**, 1–3.

39 Azana, J. and Muriel, M.A. (2000) Real-time optical spectrum analysis based on the time-space duality in chirped fiber gratings. *IEEE Journal of Quantum Electronics*, **36**, 517–526.

40 Piasecki, J., Colombeau, B., Vampouille, M., Froehly, C., and Arnaud, J.A. (1980) Nouvelle méthode de mesure de la réponse impulsionnelle des fibres optiques. *Applied Optics*, **19**, 3749.

41 Reynaud, F., Salin, F., and Barthélémy, A. (1989) Measurement of phase shifts introduced by nonlinear optical phenomena on subpicosecond pulses. *Optics Letters*, **14**, 275–277.

42 Tomlinson, W.J., Stolen, R.H., and Shank, C.V. (1984) Compression of optical pulses chirped by self-phase modulation in fibers. *Journal of Optical Society of America B*, **1**, 139–149.

43 Tay, K. and Tomita, A. (1986) *Applied Physics Letters*, **48**, 1033.

44 Fork, R.L., Brito Cruz, C.H., Becker, P.C., and Shank, C.H. (1987) Compression of optical pulses to six femtoseconds by using cubic phase compensation. *Optics Letters*, **12**, 483–485.

45 Spence, D.E., Kean, P.N., and Sibbett. W. (1991) 60-fsec pulse generation from a self-mode-locked Ti:sapphire laser. *Optics Letters*, **16**(1), 42–44.

46 Salin, F., Squier, J., and Piche. M. (1991) Mode-locking of Ti:Al2O3 lasers and self-focusing—a Gaussian approximation" *Optics Letters*, **16**(21):1674–1676.

47 Muradyan, L.Kh., Markaryan, N.L., Papazyan, T.A., and Ohanyan, A.A. Self-action of chirped pulses: spectral compression. In *Conference on Lasers and Electro-Optics*, Vol. 7, OSA Tech. Digest. Series (1990), paper CTUH32; Markaryan, N.L., Muradyan, L.K., and Papazyan, T.A. (1991) Spectral compression of ultrashort laser pulses" *Soviet Journal of Quantum Electronics*, **21**, 783.

48 Oberthaler, M. and Hopfel, R.A. (1993) Special narrowing of ultrashort laser pulses by self-phase modulation in optical fibers. *Applied Physics Letter*, **63**, 1017–1019.

49 Washburn, B.R., Buck, J.A., and Ralph, S.E. (2000) Transform limited spectral compression due to self-phase modulation in fibers. *Optics Letters*, **25**, 445–447.

50 Mouradian, L.Kh. *et al.* (2000) Applications of temporal Kerr lensing to signal manipulation and analysis. *CLEO-Europe'2000*, Conf. Digest **39**, CTuH6.

51 Clark, S.W., Ilday, F.Ö., and Wise, F.W. (2001) Fiber delivery of femtosecond pulses from a Ti:sapphire lase. *Optics Letters*, **26**, 1320–1322.

52 Mouradian, L.Kh., Mansuryan, T.G., Zeytunyan, A.S., *et al.* (2007) Spectro-temporal imaging through aberration-free temporal lensing: an ultrafast optical oscilloscope. Paper NThB3 in OSA Nonlinear Photonics-2007 Topical Meeting, Quebec Canada; OSA Tech. Digest.

53 Boscolo, S., Turitsyn, S.K., and Finot, C. (2012) Amplifier similariton fiber laser with nonlinear spectral compression. *Optics Letters*, **37**, 4531.

54 Salem, R., Foster, M.A., and Gaeta, A.L. (2013) Application of space-time duality to ultrahigh-speed optical signal processing. *Advances in Optics Photoics*, **5**, 274–317.

55 Treacy, E.B. (1969) Optical pulse compression with diffraction gratings. *IEEE Journal of Quantum Electronics*, **5**, 454–458.

56 Fork, R.L., Martinez, O.E., and Gordon, J.P. (1984) Negative dispersion using pairs of prisms. *Optics Letters*, **9**, 150–152.

57 Kane, S., Squier, J., Rudd, J.V., and Mourou, G. (1994) Hybrid grating-prism stretcher-compressor system with cubic phase and wavelength tunability and decreased alignment sensitivity. *Optics Letters*, **19**, 1876–1878.

58 Zaouter, Y., Papadopoulos, D.N., Hanna, M., *et al.* (2007) Third-order spectral phase compensation in parabolic pulse compression. *Optics Express* **15**, 9372–9377.

59 Buckley, J.R., Clark, S.W., and Wise, F.W. (2006) Generation of ten-cycle pulses from an ytterbium fiber laser with cubic phase compensation. *Optics Letters*, **31**, 1340–1342.

60 Kane, S. and Squier, J. (1997) Grism-pair stretcher-compressor system for simultaneous second- and third-order dispersion compensation in chirped-pulse amplification. *Journal of Optical Society of America B*, **14**, 661–665.

61 Mouradian, L.K. *et al.* (2013) Broadband similariton: Applications to ultrafast optics and photonics. Paper presented at VI Int. Conference on Advanced Optoelectronics and Lasers (CAOL 2013), Sep. 9–13, 2013, Sudak, Crimea, Ukraine (CAOL-077).

62 Zeytunyan, A. *et al.* (2012) Nonlinear-dispersive similariton for ultrafast photonics. Paper presented at 5th EPS-QEOD Europhoton Conference, "Solid State, Fibre, and Waveguide Coherent Light Sources," Stockholm, 26–31 Aug., 2012, TuP.32.

63 Palanjyan, K., Muradyan, A., Zeytunyan, A., Yesayan, G., and Mouradian, L. (2010) Pulse compression down to 17 femtoseconds by generating broadband similariton. *Proceedings of the SPIE,* **7998**, 79980N.

64 Mouradian, L.K., Zeytunyan, A.S., Yesayan, G.L., *et al.* (2012) Nonlinear-dispersive similariton for ultrafast photonics. Paper presented at Annual Meeting of European Optical Society (EOSAM 2012), TOM 6, Nonlinear Photonics, Aberdeen, Scotland, UK, 25–28 September.

65 Chong, A., Liu, H., Nie, B., *et al.* (2012) Pulse generation without gain-bandwidth limitation in a laser with self-similar evolution. *Optics Express,* **20**, 14213–14220.

66 Nisoli, M., De Silvestri, S., and Svelto, O. (1996) Generation of high energy 10 fs pulses by a new pulse compression technique. *Applied Physics Letters,* **68**, 2793–2795.

67 Zeytunyan, A., Yesayan, G., and Mouradian, L. (2013) Pulse compression to 14 fs by third-order dispersion control in a hybrid grating-prism compressor. *Applied Optics,* **52** (32), 7755–7758.

68 Markaryan, N.L. and Muradyan, L.Kh. (1995) Determination of the temporal profiles of ultrashort pulses by a fibre-optic compression technique. *Quantum Electronics,* **25**, 668–670.

69 Zohrabyan, A.V., Kutuzian, A.A., Ninoyan, V.Z., and Mouradian, L.K. (1997) Spectral compression of picosecond pulses by means of cross phase modulation. *AIP Conference Proceedings,* **406**, 395–401.

70 Mouradian, L.K., Louradour, F., Froehly, C., and Barthélémy, A. (1998) Self- and cross-phase modulation of chirped pulses: Spectral imaging of femtosecond pulses. Paper NFC4 in *Nonlinear Guided Waves and Their Applications,* Vol. 5 of OSA Technical Digest Series (Optical Society of America, 1998),

71 Mouradian, L.K., Zohrabian, A.V., Froehly, C., Louradour, F., and Barthélémy, A. (1998) Spectral imaging of pulses temoral profile. Paper CMA5 in *Conference on Lasers and Electro-Optics (CLEO/Europe),* OSA Tech. Digest Series (OSA, 1998).

72 Mouradian, L.K., Zohrabyan, A.V., Ninoyan, V.J., *et al.* (1998) Characterization of optical signals in fiber-optic Fourier converter. *Proceedings of SPIE,* **3418**, 78–85.

73 Mouradian, L.K., Louradour, F., Messager, V., Barthélémy, A., and Froehly, C. (2000) Spectro-temporal imaging of femtosecond events" *IEEE Journal of Quantum Electronics,* **36**, 795–801.

74 Kutuzyan, A.A., Mansuryan, T.G., Kirakosyan, A.A., and Mouradian, L.K. (2003) Self-forming of temporal dark soliton in spectral compressor. *Proceedings of SPIE,* **5135**, 156–160.

75 Louradour, F., Lopez-Lago, E., Couderc, V., Messager, V., and Barthélémy, A. (1999) Dispersive-scan measurement of the fast component of the third-order nonlinearity of bulk materials and waveguides. *Optics Letters,* **24**, 1361–1363.

76 Clark, S.W., Ilday, F.Ö., and Wise, F.W. (2001) Fiber delivery of femtosecond pulses from a Ti:sapphire laser. *Optics Letters,* **26**, 1320–1322.

77 Andresen, E.R., Dudley, J.M., Oron, D., Finot, C., and Rigneault, H. (2011) Transform-limited spectral compression by self-phase modulation of amplitude-shaped pulses with negative chirp. *Optics Letters,* **36**, 707–709.

78 Fatome, J., Kibler, B., Andresen, E.R., Rigneault, H., and Finot, C. (2012) All-fiber spectral compression of picosecond pulses at telecommunication wavelength enhanced by amplitude shaping. *Applied Optics*, **51**, 4547–4553.
79 Toneyan, H., Zeytunyan, A., Mouradian, L., *et al.* (2014) 8x, 12x, and 23x spectral compression by all-fiber, classic, and similaritonic techniques. Paper FW4D.5 in Frontiers in Optics, 2014, OSA Tech. Digest, 1,
80 Mouradian, L.K., Grigoryan, A., Kutuzyan, A., *et al.* (2015) Spectral analogue of the soliton effect compression: spectral self-compression. Paper FW3F.3 in Frontiers in Optics/Laser Science Conference (FiO/LS 2015), San Jose, CA, October 18–22,].
81 Foster, M.A. *et al.* (2005) Soliton-effect compression of supercontinuum to few-cycle durations in photonic nanowires. *Optics Express*, **13**, 6848.
82 Balciunas, T. *et al.* (2015) A strong-field driver in the single-cycle regime based on self-compression in a Kagome fibre. *Nature Communications*, **6**:6117 doi: 10.1038 / ncomms 7117.
83 Salem, A.B. *et al.* (2011) Soliton-self compression in highly nonlinear chalcogenide photonic nanowires with ultralow pulse energy. *Optics Express*, **19**, 1995510.
84 Amorim, A.A. *et al.* (2009) Sub-two-cycle pulses by soliton self-compression in highly nonlinear photonic crystal fibers. *Optics Letters*, **34**, 3851.
85 Mollenauer, L.F. *et al.* (1980) Experimental-observation of picosecond pulse narrowing and solitons in optical fibers. *Physical Review Letters*, **45**, 1095.
86 Mollenauer, L.F. *et al.* (1983) Extreme picosecond pulse narrowing by means of solton effect in single mode fibers. *Optics Letters*, **8**, 289.
87 Heidt, A.M. (2010) Pulse preserving flat-top supercontinuum generation in all-normal dispersion photonic crystal fibers. *Journal of Optical Society of America B*, **27**, 550–559.
88 Reid, D.T., Loza-Alvarez, P., Brown, C.T.A., Beddard, T., and Sibbett, W. (2000) Amplitude and phase measurement of mid-infrared femtosecond pulses by using cross-correlation frequency-resolved optical gating. *Optics Letters*, **25**, 1478–1480.
89 Dudley, J., Gu, X., Xu, L. *et al.* (2002) Cross-correlation frequency resolved optical gating analysis of broadband continuum generation in photonic crystal fiber: simulations and experiments. *Optics Express*, **10**, 1215–1221.
90 Pegoraro, A.F., Ridsdale, A., Moffatt, D.J. *et al.* (2009) Optimally chirped multimodal CARS microscopy based on a single Ti:sapphire oscillator. *Optics Express*, **17**, 2984–2996.
91 Trebino, R. (2002) *Frequency-Resolved Optical Gating: The Measurement of Ultrashort Laser Pulses*. Kluwer Academic Publishers, Boston.
92 Akturk, S., Kimmel, M., O'Shea, P., and Trebino, R. (2003) Measuring spatial chirp in ultrashort pulses using singleshot frequency-resolved optical gating. *Optics Express*, **11**, 68–78.
93 Iaconis, C. and Walmsley, I.A. (1998) Spectral phase interferometry for direct electric-field reconstruction of ultrashort optical pulses. *Optics Letters*, **23**, 792–794.
94 Messager, V., Louradour, F., Froehly, C., and Barthélémy, A. (2003) Coherent measurement of short laser pulses based on spectral interferometry resolved in time. *Optics Letters*, **28**, 743–745.
95 Lelek, M., Louradour, F., Barthélémy, A., *et al.* (2008) Two-dimensional spectral shearing interferometry resolved in time for ultrashort optical pulse characterization. *Journal of Optical Society of America B*, **25**, A17–A24.

96 Kockaert, P., Haelterman, M., Emplit, P., and Froehly, C. (2004) Complete characterization of (ultra)short optical pulses using fast linear detectors. *IEEE Journal of Selected Topics in Quantum Electronics*, **10**, 206–212.
97 Lozovoy, V.V., Pastirk, I., and Dantus, M. (2004) Multiphoton intrapulse interference. IV. Ultrashort laser pulse spectral phase characterization and compensation. *Optics Letters*, **29**, 775 (2004); Xu, B., Gunn, J.M., Dela Cruz, M., Lozovoy, V.V., and Dantus, M. (2004) Quantitative investigation of the multiphoton intrapulse interference phase scan method for simultaneous phase measurement and compensation of femtosecond laser pulses. *Journal of Optical Society of America B*, **23**, 750–759.
98 Nuss, M.C., Li, M., Chiu, T.H., Weiner, A.M., and Partovi, A. (1994) Time-to-space mapping of femtosecond pulses. *Optics Letters*, **19**, 664–666.
99 Sun, P.C., Mazurenko, Y.T., and Fainman, Y. (1997) Femtosecond pulse imaging: Ultrafast optical oscilloscope. *Journal of Optical Society of America A*, **14**, 1159–1170.
100 Bennett, C.V. and Kolner, B.H. (1999) Upconversion time microscope demonstrating 103X magnification of femtosecond waveforms. *Optics Letters*, **24**, 783–785.
101 Vampouille, M., Barthélémy, A., Colombeau, B., and Froehly, C. (1984) Observation et applications des modulations de fréquence dans les fibers unimodales. *Journal Optique* (Paris) **15**, 385–390.
102 Vampouille, M., Marty, J., and Froehly, C. (1986) Optical frequency intermodulation between two picosecond laser pulses. *IEEE Journal of Quantum Electronics*, **22**, 192–194.
103 Kauffman, M.T., Banyai, W.C., Godil, A.A., and Bloom, D.M. (1994) Time-to-frequency converter for measuring picosecond optical pulses. *Applied Physics Letters*, **64**, 270–272.
104 Azana, J., Berger, N.K., Levit, B. and Fischer, B. (2003) Time-to-frequency conversion of optical waveforms using a single time lens system. *Physica Scripta*, **T118**, 115–117 (2005); Azana, J., Berger, N.K., Levit, B., and Fischer, B. (2004) Spectrotemporal imaging of optical pulses with a single time lens. *IEEE Photonics Technology Letters*, **16**, 882–884; Azana, J., (2004) Time-to-frequency conversion using a single time lens. *Optics Communications*, **217**, 205–209.
105 Arons, E., Leith, E.N., Tien, A., and Wagner, R. (1997) High resolution optical chirped pulse gating. *Applied Optics*, **36**, 2603–2608.
106 Foster, M.A., Salem, R., Geraghty, D.F., *et al.* (2008) Silicon-chip based ultrafast optical oscilloscope. *Nature*, **456**, 81–84.
107 Foster, M.A., Salem, R., Geraghty, D.F., *et al.* (2008) Silicon-chip-based single-shot ultrafast optical oscilloscope. Paper FTuU1 in Frontiers in Optics, OSA Tech. Digest.
108 Zeytunyan, A., Mansuryan, T., Kalashyan, M., *et al.* (2007) Spectrotemporal imaging by sum frequency generation: ultrafast optical oscilloscope. Paper FThD3 in *Frontiers in Optics*, OSA Tech. Digest, OSA, 2007.
109 Ippen, E.P. and Shank, C.V. (1984) Techniques for measurements. In *Ultrashort Light Pulses* (ed. S. L. Shapiro). Springer-Verlag, New York, pp. 83–122.
110 Zeytunyan, A.S., Madatyan, H.R., Yeayan, G.L., and Mouradian, L.K. (2010) Diagnostics of femtosecond laser pulses based on the generation of nonlinear-dispersive similariton. *Journal of Contemporary Physics*, **45** (4), 265–273.
111 Kalashyan, M.A., Palanjyan, K.H., Khachikyan, T.J., *et al.* (2009) Prism-lens dispersive delay line. *Technical Physics Letters*, **35** (3), 211–214.

112 Kalashyan, M.A., Palandzhyan, K.A., Esayan, G.L., and Muradyan, L.K. (2010) Generation of transform-limited rectangular pulses in a spectral compressor. *Quantum Electronics*, **40** (10), 868–872.
113 Takeda, M., Ina, H., and Kobayashi, S. (1982) Fourier-transform method of fringe-pattern analysis for computer-based topography and interferometry. *Journal of the Optical Society of America*, **72** (1), 156–160.
114 Lepetit, L., Chériaux, G., and Joffre, M. (1995) Linear techniques of phase measurement by femtosecond spectral interferometry for applications in spectroscopy. *Journal of Optical Society of America B*, **12** (12), 2467–2474.
115 Renninger, W.H., Chong, A., and Wise, F.W. (2010) Self-similar pulse evolution in an all-normal-dispersion laser. *Physical Review A*, **82**, 021805(R).
116 Tsang, M., Psaltis, D., and Omenetto, F.G. (2003) Reverse propagation of femtosecond pulses in optical fibers. *Optics Letters*, **28** (20), 1873–1875.
117 Weiner, A.M., Leaird, D.E., Wiederrecht, G.P., and Nelson, K.A. (1990) Femtosecond pulse sequences used for optical manipulation of molecular motion. *Science*, **247**, 1317–1319.

7

Applications of Nonlinear Optical Fibers and Solitons in Biophotonics and Microscopy

Esben R. Andresen[1] *and Hervé Rigneault*[2]

[1] PhLAM, UMR 8523, CNRS-Université Lille 1, Villeneuve d'Ascq, France
[2] Institut Fresnel, UMR 7249 CNRS- Aix Marseille Université, Marseille, France

7.1 Introduction

The many design degrees of freedom of photonic-crystal fibers (PCFs) [1–3] have resulted in a range of PCFs with dispersion and nonlinear properties vastly different from those of standard fibers. The design degrees of freedom in effect allow producers to tailor a PCF to a certain application, e.g., frequency conversion which has already been treated to great extent in the present book in Chapters 4 and 5.

This chapter will deal with applications of nonlinear PCF in biophotonics and microscopy. In particular, it will deal with applications in conjunction with the following microscopy modalities, all of which are now commonly used in many biological communities: Two-photon excited fluorescence (TPEF) microscopy [4]; Second-harmonic generation (SHG) microscopy [5]; two modalities based on coherent anti-Stokes Raman scattering (CARS), picosecond CARS (ps-CARS) microscopy [6] and multiplex CARS (MCARS) microscopy [7,8]; stimulated-Raman scattering (SRS) microscopy [9–12]; and pump-probe microscopy [13]. The first two, TPEF and SHG microscopy, are what we might call single-beam techniques, i.e., they require a light source which provides only one pulsed excitation beam. The remaining, ps-CARS microscopy, MCARS microscopy, SRS microscopy, and pump-probe microscopy, are all two-beam techniques (apart from a few special cases which we will not treat here), i.e., they require a light source which provides two, synchronized, pulsed excitation beams which are, generally, of different color.

Although new types of light sources are being developed at a rapid pace, the most commonly used light sources for the microscopy modalities mentioned above are predominantly high repetition-rate systems (around 80 MHz). In order to name a few and associate them with some microscopy modalities we might mention a Ti:Sapphire femtosecond (fs-) laser (TPEF, SHG); a pair of electronically synchronized Ti:Sapphire picosecond (ps-) lasers (ps-CARS, SRS); a Ti:Sapphire ps-laser electronically synchronized to Ti:Sapphire fs-laser (MCARS); a ps-laser pumping a

Shaping Light in Nonlinear Optical Fibers, First Edition. Edited by Sonia Boscolo and Christophe Finot.

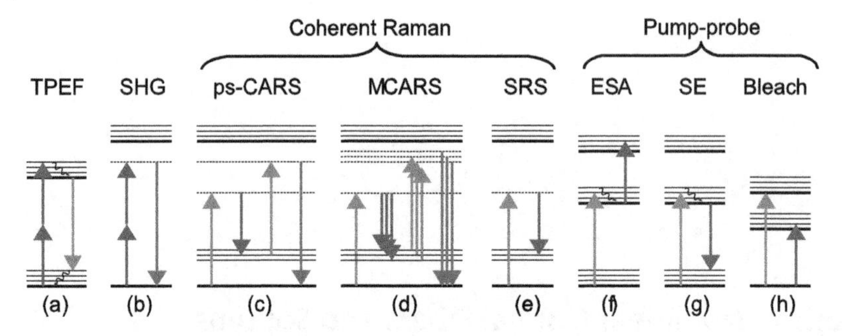

Figure 7.1 Energy level diagrams of the nonlinear light-matter interactions which underlie the considered microscopy modalities. (a) Two-photon excited fluorescence; (b) Second-harmonic generation; (c–e) Coherent Raman scattering comprising (c) picosecond coherent anti-Stokes Raman scattering; (d) multiplex coherent anti-Stokes Raman scattering; (e) stimulated Raman scattering; (f–h) Pump-probe processes comprising (f) excited-state absorption; (g) stimulated emission; (h) bleach or ground-state depletion.

synchronously-pumped optical parametric oscillator (OPO) (ps-CARS, SRS); and a fs-laser pumping a synchronously-pumped OPO (Pump-probe).

For an initial overview, Figure 7.1 presents a schematic diagram of the different light-matter interactions which underlie the microscopy modalities that are the subject of this chapter.

When selecting a light source for a certain type of microscopy there is always a common denominator which guides the selection: the maximization of signal. Indeed, the detected signal level is what puts the upper bound on the achievable pixel scan rate so, quite unequivocally, one is always interested in maximizing it. In microscopy the detected signal level is of course a function of many factors, like filters, optical aberration, sample absorption, etc. each of which can be expressed as a multiplicative factor between 0 and 1 that is multiplied onto the total signal generated in the sample. For brevity we will not treat these further and instead focus on the total signal generated in the sample which depends almost exclusively on the parameters of the light source. And the total signal generated of course gives the upper bound on the detected signal level. So therefore, in this chapter we are going to take the following approach: Each microscopy modality is treated in its own section, and at the beginning of each section we give the main equations and considerations that assess the signal generated by the underlying nonlinear interaction (Figure 7.1) as a function of the light source parameters. Based upon these considerations, we establish how to most suitably design an adapted light source based on nonlinear PCF. And in doing so we will restrict ourselves to considering only PCF operating in a parameter regime where soliton dynamics with fundamental solitons is the primary nonlinear effect in the PCF.

7.2 Soliton Generation

7.2.1 Fundamental Solitons

Even though the theory behind soliton dynamics was treated in depth in Chapter 3 of this book, we will still give a brief overview of the soliton dynamics of fundamental solitons which highlights some practical aspects which will become relevant in the coming sections.

A fundamental soliton in a PCF is characterized by having soliton order N equal to 1 where

$$N^2 = \gamma P_0 T_0^2 / |\beta_2|. \tag{7.1}$$

P_0 is the soliton's peak power, T_0 the natural half-width of its sech^2-shaped temporal intensity profile, γ the nonlinearity parameter of the fiber, and β_2 the group-velocity dispersion of the PCF [14]. A soliton can be efficiently generated in a PCF by injecting a laser pulse at a wavelength where the PCF has anomalous dispersion, i.e., $\beta_2 < 0$. This will normally result in the generation of one (or more) fundamental solitons in the PCF each of which fulfills Eq. (7.1). The soliton energy as a ratio of the input pulse energy can be optimized by adapting the input pulse parameters to also fulfill Eq. (7.1). Indeed, if the input pulse has a sech^2 profile and fulfills Eq. (7.1) (after taking fiber coupling losses into account), it should generate a fundamental soliton in the PCF with 100% efficiency. This can be thought of as temporal mode-matching, although the nonlinearities complicate somewhat the analogy to the more widely known spatial mode matching. The word "soliton" is usually taken to imply a solution to the propagation equation that is invariant upon propagation. However, for a soliton shorter than about 100 fs, its bandwidth is so large that the frequency differences between its frequency components cover a significant portion of the low-frequency Raman-active modes of silica. And so, efficient intrapulse SRS can occur which transfers energy from the blue to the red components and which, when happening in a PCF in an interplay with soliton dynamics, results in the adiabatic, continous redshift of the soliton as it propagates along the fiber. This effect is known as the soliton self-frequency shift (SSFS) [15, 16]. What is very interesting is that SSFS provides a redshifted pulse while at the same time ensuring that this redshifted pulse is always transform-limited.

In [17] a 190-cm-long piece of PCF was employed as a basis for a "soliton light source." Injecting a near-transform-limited pulse with duration FWHM 60 fs, a soliton of comparable duration and energy around 10 pJ was generated. A characterization of this soliton light source and its dependence upon PCF input power is given in Figure 7.2. The redshift of the soliton is dependent not only on the fiber length but also on the energy of the incident pulse. Thus, for input pulse energies between 0 and 66 pJ, the wavelength of the generated soliton spans as much as 300 nm between 800 and 1100 nm. This span is comparable to that of commercial, wavelength-tunable Ti:Sapphire lasers which are typically used as light sources for TPEF microscopes that need to be compatible with many different fluorophore excitation spectra. Alternatively, if we consider as well the redshifted soliton and the laser at 800 nm, their frequency difference spans 0-4000 cm^{-1} when the soliton spans its 300 nm range. This is enough to cover the entire vibrational Raman spectrum which contains bands at frequencies up to 3400 cm^{-1}. Conversion efficiencies (soliton energy as a ratio of input pulse energy) from 80% at small redshifts to 13.4% at large redshifts were observed [17]. In terms of conversion efficiency, the SSFS, which frequency-converts the initial pulse to a relatively narrow spectral region, is ahead of supercontinuum generation, which disperses the initial energy over the entire spectrum. The energy of the soliton is fundamentally limited by the constraint $N = 1$.

Typically the soliton energy is less than 100 pJ although in Section 7.10.1 we will describe a novel class of fibers that exceeds this limit.

Though the soliton self-frequency shift (SSFS) has been known since 1986 [15, 16], it was not until 2004 that it found an application in TPEF microscopy [18] and not until 2006 that it found an application in a CRS experiment [17].

We will refer to as the "one-beam soliton light source" the light source that provides just the redshifted soliton generated in a nonlinear PCF. Further on, we will show how the one-beam soliton light source can be put to use in TPEF microscopy (Section 7.3) and SHG microscopy (Section 7.4).

With "two-beam soliton light source" we will refer to the light source that provides two pulses; the redshifted soliton generated in a nonlinear PCF as well as the femtosecond laser pulse. It is understood that the two are synchronized so that they can induce nonlinear mixing processes. We will show how the two-beam soliton light source can be put to use in MCARS microscopy (Section 7.6), ps-CARS microscopy (Section 7.7), SRS microscopy (Section 7.8), and pump-probe microscopy (Section 7.9).

7.2.2 A Sidenote on Dispersive Wave Generation

A related generation scheme which also relies on soliton dynamics is to exploit the behavior of the redshifting soliton when it arrives at the second zero-dispersion wavelength (ZDW) of the PCF. This behavior was already seen in Figure 7.2 (gray line). Since the soliton is not a solution to the propation equation in the normal dispersion region,

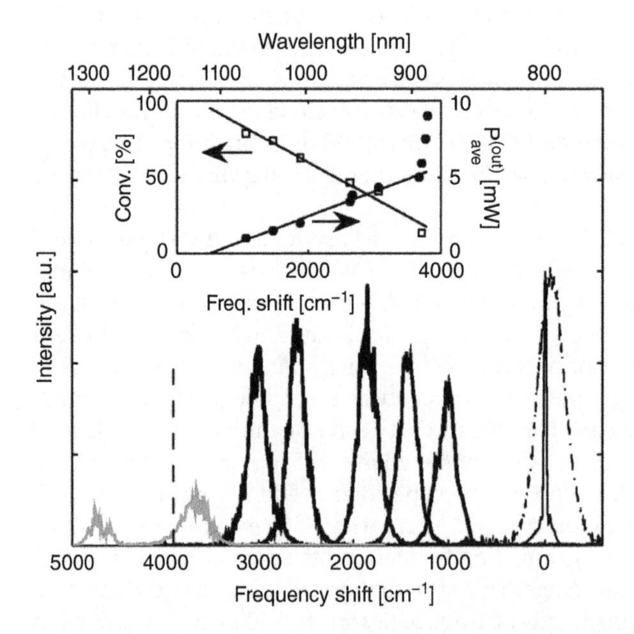

Figure 7.2 Redshift of fundamental soliton by SSFS. (Dash-dot) The laser spectrum. (Thick black) Spectra of the fundamental solitons generated for different input pulse energies. (Thick gray) spectrum of a fundamental soliton at the second ZDW marked by the dashed line and the dispersive wave shed by same. (Thin black) Typical spectrum of a spectrally compressed Gaussian pulse. Inset, (dots) total average output power and (squares) the percentage of the input energy contained in the soliton. *Source*: Andresen 2006 [17]. Reproduced with permission of OSA.

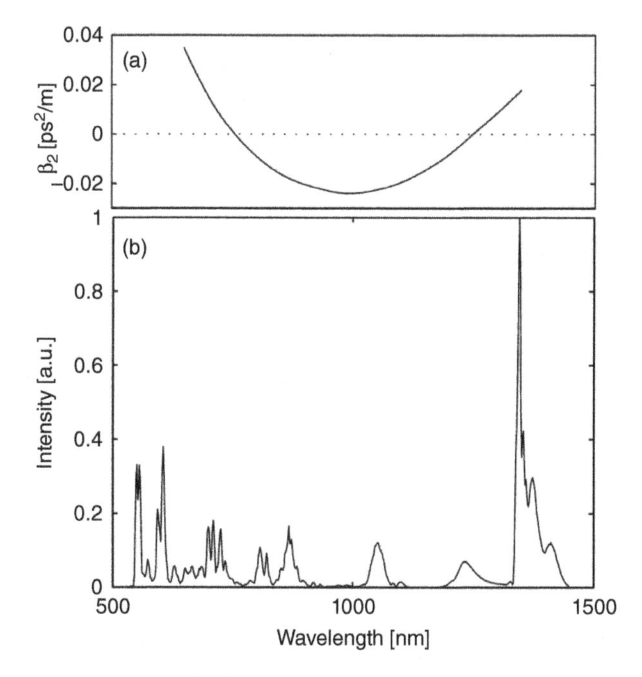

Figure 7.3 Example of dispersive wave generation at the second zero dispersion wavelength of a nonlinear PCF. (a) Dispersion curve of the PCF. (b) Output spectrum for an input pulse with the following parameters: $\lambda = 800$ nm; $\tau = 31$ fs; and pulse energy 2.8 nJ.

i.e., $\beta_2 > 0$ it is unable to redshift further than the ZDW. What happens instead is that the soliton sheds its energy as a dispersive wave. This can be exploited in order to increase spectral brightness at the ZDW as compared to the spectral brightness of a fundamental soliton, the input pulse energy can then be increased. If $N > 2$, fundamental solitons with different energies will form by soliton fission [20–22]. As each soliton redshifts and reaches the second ZDW, energy will "pile up" in a narrow wavelength range generating a dispersive wave with spectral brightness that can surpass that of a fundamental soliton. An example of this piling up is clearly seen in Figure 7.3 where the GVD is plotted for reference (Figure 7.3 (a)) along with a typical output spectrum (Figure 7.3 (b)). The PCF used was the same as in Figure 7.2 and the input pulse had similar duration but the energy was much higher so many solitons were generated ($N \gg 1$). A soliton can be clearly distinguished at $\lambda \approx 1050$ nm; while the feature at $\lambda \approx 1200$ nm is a soliton that has been blocked in its redshift by the second ZDW, hence its assymetric shape; the dominant feature at $\lambda \approx 1300$ nm is the dispersive wave shed by the >10 solitons. We see that this generation scheme allows one to direct the frequency conversion toward a certain spectral region and create pulses of spectral brightness much higher than the spectral brighness of a fundamental soliton. Unfortunately the wavelength at which the dispersive wave appears is dictated by the PCF parameters, namely by the location of the second ZDW. However, PCFs can be designed with a second ZDW anywhere within a quite wide wavelength range. A potential issue with this approach is the temporal broadening of the dispersive wave; the output pulse is not transform-limited since dispersion is no longer kept in check by the soliton dynamics. To minimize this effect the PCF length must be chosen properly. It must be long enough to allow several solitons to reach the second ZDW, but not so long that the generated dispersive wave experiences any significant dispersion. A number of papers have reported the use of such a dispersive wave for ps-CARS and MCARS microscopy and spectroscopy [23–25].

7.2.3 Spatial Properties of PCF Output

All the microscopy modalities are nonlinear modalities which, as we will see in the following sections, means that the generated signal scales with the intensity of excitation light to a potency of 2 or higher. In the coming sections we will omit the spatial dependence of signal generation and we will instead treat it in this section and draw the necessary connections to the properties of nonlinear PCF, because the considerations are applicable to all the modalities.

Let us consider a simple case representative of a biological sample, a sample consisting of a point source located in the focus of the excitation beam. In this simplified case we can simply consider the sample a δ-function in space, and so the generated signal becomes proportional to a potency 2 or higher of the local excitation intensity on the sample. The lesson to draw from this example is that the light source utilized for nonlinear microscopy must be amenable to being focused to a very small and hence very intense, diffraction-limited spot in order to maximize the total signal generated. In other words the light source must be spatially coherent. The soliton light source based on SSFS in PCF is expected to fulfill this criterion. Indeed, it was demonstrated very soon after the realization of the first functioning PCFs that this class of fibers support single-mode operation at all wavelengths [26,27]. This effect ensures the diffraction-limited output of the PCF since there are no contributions from higher-order spatial modes. This has been confirmed in direct measurements of the spatial profile, e.g., [28] which found that the spatial profile of the supercontinuum output of a nonlinear PCF was diffraction limited with M^2 close to 1 over a large wavelength range.

7.3 TPEF Microscopy

TPEF microscopy was invented in 1990 [4] and has since become one of the most essential tools in biological imaging, enabling three-dimensional imaging of biological specimens with sub-micrometer resolution [29], and especially its ability to image deep in scattering tissue is being heavily exploited [30]. TPEF has become omnipresent and a true workhorse of biological imaging, so it is commonly used for a plethora of applications, of which a few important examples are *ex vivo* and *in vivo* functional imaging, developmental biology, and 3D photochemistry [31–33].

TPEF is the process involving the absorption of two photons whose combined energy is sufficient to induce a molecular transition to an electronically excited state. The fluorescence subsequently emitted by the molecule as it decays to the ground state is the TPEF signal that is detected and used to generate image contrast in TPEF microscopy (Figure 7.1 (a)). The efficiency of two-photon excitation and the averaged fluorescence yield S_{TPEF} follows the relation [4]

$$S_{\mathrm{TPEF}} \propto P_{\mathrm{ave}}^2 \alpha / (T_0 f_{\mathrm{rep}}) \mathrm{NA}^4 = 4 P_0^2 \alpha T_0 f_{\mathrm{rep}} \mathrm{NA}^4 \tag{7.2}$$

where P_{ave} is the average power, α the molecular two-photon absorption coefficient, T_0 the pulse duration, f_{rep} the repetition frequency, NA the numerical aperture, and P_0 the peak power. From Eq. (7.2) we see particularly that it is inversely proportional to T_0 if all other excitation beam parameters are kept constant which means that a shorter excitation pulse is unequivocally beneficial. When we relate this to the description of the one-beam soliton light source (Section 7.2) we see that this light source complies with the pulse duration requirement because the soliton is transform-limited by definition and typically has a duration less than 100 fs. In addition to this, the soliton is wavelength-tunable thanks to the SSFS, permitting the generation of a soliton with a wavelength coinciding with the absorption maximum of any two-photon fluorophore. Interestingly, TPEF microscopy was the first nonlinear microscopy modality that saw the application of a one-beam soliton light source [18].

As we saw in Section (7.2) the soliton is bounded in energy and hence in peak power P_0 by the PCF parameters. This is a limitation of the one-beam soliton light source, but examining Eq. (7.2) we see that a way to circumvent this limitiation to a certain degree is to increase the repetition rate f_{rep}. The repetition rate can be increased up to the inverse excited-state lifetime of the fluorophore where saturation might occur. Since the peak power of a soliton shifted to a given wavelength is independent of the repetition rate, a 1 GHz train of solitons should yield 12.5 times higher fluorescent yield than an 80 MHz soliton train. This was the approach taken in [34] where a one-beam soliton light source using a 1 GHz Ti:Sapphire laser was shown to provide the expected signal increase compared to the case where a standard 80 MHz Ti:Sapphire laser was used (Figure 7.4).

The limitation on the soliton energy can also be alleviated in another way which is by generating the soliton not in a nonlinear PCF but in a novel kind of fiber called solid-core photonic-bandgap fiber, a subject that we will reserve for Section 7.10.1.

7.4 SHG Microscopy

SHG microscopy dates back from the 1970s [5] but modern equipment has contributed to the method becoming much more widespread in recent years. Some examples of modern applications of SHG microscopy are the imaging of membrane probes and membrane potential [35, 36] and microtubule polarity [37–39] in living cells, as well as collagen structures in mouse models of disease [39, 40].

Unlike TPEF, SHG does not involve any absorption processes. Rather, the excitation beam induces a nonlinear, second-order polarization at exactly half the frequency of the

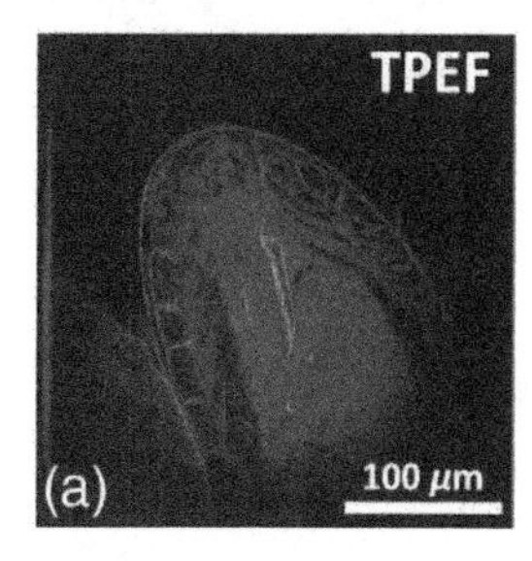

Figure 7.4 TPEF image of a fixed *Drosophilia* embryo obtained with a single-pulse soliton light source at 870 nm and a repetition rate of 1 GHz. *Source*: Saint-Jalm 2014 [34]. Reproduced with permission of SPIE.

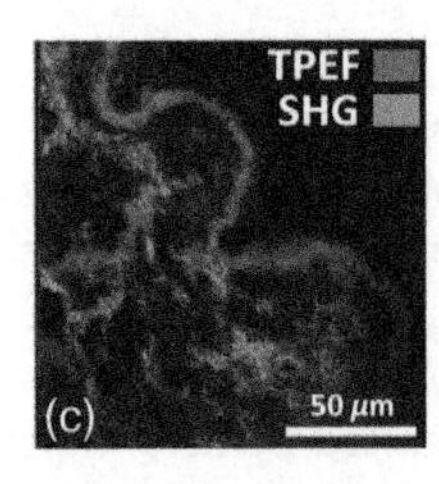

Figure 7.5 Composite TPEF and SHG image of fixed mouse tumor skin tissue obtained with a single-pulse soliton light source at 870 nm and a repetition rate of 1 GHz. *Source*: Saint-Jalm 2014 [34]. Reproduced with permission of SPIE.

excitation beam (Figure 7.1 (b)). The SHG signal is governed by an equation very similar to Eq. (7.2) with α replaced by the second-order susceptibility $\chi^{(2)}$ of the sample. There are, however, a few important differences which will only be mentioned briefly here: The susceptibility is in general a tensor; only non-centrosymmetric objects give off a SHG signal; and SHG is a coherent process so phase-matching gives rise to a highly directional SHG signal, typically the emission in the forward direction is by far the strongest. We will use the same arguments as we did in Section 7.3 to assert that the one-beam soliton light source is a suitable light source for SHG microscopy: The short pulse duration of the soliton ensures a high signal level and the wavelength-tunability assures adaptability.

Figure 7.5, shows a typical SHG image obtained using this approach. The image is actually an overlay of a TPEF image with a SHG image of the same sample; since the TPEF signal appears at longer wavelength than does the SHG signal, the two can be separated by a dichroic filter which allows for simultaneous (multi-modal) TPEF and SHG imaging.

7.5 Coherent Raman Scattering

The signal generation in the microscopy modalities MCARS, ps-CARS, and SRS to be treated in Sections 7.6, 7.7, and 7.8, respectively can all be described by very similar formalisms. They are sometimes commonly denoted as "coherent Raman scattering" (CRS) (Figures 7.1 (c)–7.1 (e)) [41,42]. They will therefore be described here in a manner that can accomodate all three [43].

CRS has the characteristic that it involves a two-photon resonance with a vibrational level. Normally, two pulses "pump" and "Stokes" at different wavelengths are used in CARS and SRS. The total complex driving field is

$$E(t) = \sum_{m} \epsilon_m(t) e^{-i\omega t_{m0}} + c.c. \tag{7.3}$$

where $m = p$ (pump) or S (Stokes), ω_{m0} the central angular frequency, ϵ_m the slowly-varying complex temporal field envelope, and *c.c.* denotes the complex conjugate. In the frequency domain the total driving field is

$$\tilde{E}(\omega) = \sum_{m} \left[\tilde{\epsilon}_m(\omega + \omega_{m0}) + \tilde{\epsilon}^*_m(-\omega - \omega_{m0})\right] \tag{7.4}$$

where $\tilde{\epsilon}_m$ is the complex spectral field envelope. The macroscopic Raman response of any sample can be expressed by the third-order nonlinear optical susceptibility tensor $\chi^{(3)}(-\omega_1 - \omega_2 - \omega_3; \omega_1, \omega_2, \omega_3)$ or $\chi^{(3)}$ for short (we ignore the tensorial nature, effectively assuming that all implicated fields have the same polarization). In the Bloch approximation and far from electronic resonance, the susceptibility takes the form

$$\chi^{(3)} = \sum_{i,j} \frac{A_{ij}}{\omega_1 + \omega_2 - \Omega_{ij} + i\Gamma_{ij}} \tag{7.5}$$

where Ω_{ij} is the angular frequency of a Raman-active mode, Γ_{ij} its inverse decoherence time, and A_{ij} a number related to the intensity of the mode. The spontaneous Raman spectrum S_{Raman} is proportional to the negative imaginary part of $\chi^{(3)}$,

$$S_{\mathrm{Raman}}(\omega_S) \propto -\mathrm{Im}[\chi^{(3)}(-\omega_S; -\omega_{p0}, \omega_S, \omega_{p0})] \tag{7.6}$$

where ω_{p0} is the frequency of a continuous-wave pump. Normally, in CRS measurements one strives to recover the Raman spectrum since it is linear in the concentration of each species in the sample.

The total nonlinear third-order polarization induced by the pump and Stokes fields is

$$\tilde{P}^{(3)}(\omega_1 + \omega_2 + \omega_3) \propto \iiint \mathrm{d}\omega_1 \mathrm{d}\omega_2 \mathrm{d}\omega_3 \chi^{(3)} \tilde{E}(\omega_1)\tilde{E}(\omega_2)\tilde{E}(\omega_3). \tag{7.7}$$

$P^{(3)}$ thus contains many terms oscillating at all the possible combinations of pump and Stokes frequencies. In the present context, CARS and SRS, the terms of interest are (i) the term oscillating at the anti-Stokes frequency $\omega_{p0} - \omega_{S0} + \omega_{p0}$, known as the CARS signal; (ii) the term oscillating at the Stokes frequency $\omega_{p0} + \omega_{S0} - \omega_{p0}$, known as the stimulated-Raman gain (SRG) signal; (iii) the term oscillating at the pump frequency $\omega_{p0} - \omega_{S0} + \omega_{S0}$, known as the stimulated-Raman loss (SRL) signal. One makes use of (i) in ps-CARS as well as in MCARS, while one makes use of either (ii) or (iii), or both, in SRS. In the context of maximization of generated signal we may make one important, general comment to Eq. (7.7) by asking the question: Which spectral width of pump and Stokes pulses maximize the generated signal? The equation finds the signal at a given frequency by summing the contributions from all the three-photon paths for which $\omega_1 + \omega_2 + \omega_3$ equals that given frequency. And the presence of $\chi^{(3)}$ in the overlap integral gives significant weight to the three-photon paths within a frequency range of approximately Γ_{ij}. From this brief reflection we can therefore convince ourselves that the optimal spectral width of the driving pulses for CARS and SRS will be close to Γ_{ij}. In practice, Γ_{ij} is on the order of 10 cm^{-1} in condensed-phase samples which includes most biological samples. This optimal spectral width translates into an optimal transform-limited pulse duration of a few picoseconds.

7.6 MCARS Microscopy

The third-order nonlinear polarization at the anti-Stokes frequency gives rise to a radiated field with a frequency higher than the pump and Stokes driving fields. This field can thus be spectrally separated from them by optical filters and homodyne-detected on a detector to yield the (frequency-dependent) CARS signal

$$S_{\mathrm{CARS}}(\omega_{\mathrm{CARS}}) \propto |i\tilde{P}_{\mathrm{CARS}}(\omega_{\mathrm{CARS}})|^2. \tag{7.8}$$

In this section we will be dealing with the modality called MCARS microscopy. Examples of applications of MCARS microscopy include polymer blend imaging [44], label-free flow cytometry [45], and imaging of lipid droplets [46] and lipophilic bioactive molecules in cells [47].

The MCARS measurement scheme comes about when the following conditions are met: (i) The spectral width of the pump pulse is much lower than the inverse decoherence time of the sample Γ_{ij}; (ii) The spectral width of the Stokes pulse is much larger than Γ_{ij}. These conditions can also be expressed in the time domain: (i) The pump pulse is much longer than the decoherence time Γ_{ij}^{-1} of the sample; and (ii) the Stokes pulse is much shorter than Γ_{ij}^{-1}.

In this case the CARS signal simplifies to (since we may assume $\tilde{\epsilon}_p(\omega) \to \tilde{\epsilon}_{p0}\delta(\omega - \omega_{p0})$)

$$S_{\mathrm{MCARS}}(\omega_{\mathrm{CARS}}) \propto |\chi^{(3)}(-\omega_{\mathrm{CARS}}; \omega_{p0}, -\omega_S, \omega_{p0})|^2 |\tilde{\epsilon}_{p0}|^4 |\tilde{\epsilon}_S(\omega_S - \omega_{S0})|^2. \tag{7.9}$$

Here multichannel detection is desired because the generated signal $S_{\mathrm{MCARS}}(\omega_{\mathrm{CARS}})$ contains spectral information within a frequency range given by the spectral width of the Stokes. Normally a spectrometer is used to detect $S_{\mathrm{MCARS}}(\omega_{\mathrm{CARS}})$ directly which—when done in a microscope—directly gives hyperspectral images.

We will now detail how the two-beam soliton source (Section 7.2.1) can be adapted as a suitable light source for MCARS microscopy. Following our above discussion of the SSFS, it is clear that the soliton provided by the two-beam soliton source corresponds to the conditions required for the Stokes for the MCARS scheme; it has a spectral width of typically 150 cm^{-1}, some 15 times larger than the typical Γ_{ij} of biological samples, and additionally it is frequency-tunable over a frequency range of 4000 cm^{-1} or more covering the entire interesting region of the vibrational Raman spectrum which contains information in the 0-3400 cm^{-1} frequency range. But when it comes to the pump pulse, the two-beam soliton light source is not immediately suitable because it can only offer as pump pulse the laser pulse used to generate the soliton. This pulse is a femtosecond pulse and so also has spectral width equal to many times Γ_{ij}—which is seemingly in conflict with the other requirement on the pump pulse for the MCARS scheme, a spectral width close to Γ_{ij}. Fortunately, this conundrum can be solved by resorting to another phenomenon readily available in nonlinear PCF called spectral compression. Spectral compression is realized when self-phase modulation in a nonlinear PCF acts on a negatively pre-chirped pulse, as it was described in Chapter 5 of this book. In spectral compression, the auxiliary components of a broadband pulse are converted to the central frequency. While spectral compression has been known for a long time [48–50] it was initially performed in standard fibers, and later results have shown that nonlinear PCF with optimized dispersion properties allow for unprecedented compression factors, as was demonstrated in [51]: Starting out with a 40 fs long transform-limited Gaussian pulse which was chirped to 2400 fs before it was sent through a 60-cm-long PCF with small, normal dispersion, a compression factor of 21 was achieved and the time-bandwidth product of the resulting pulse was three times that of a transform-limited Gaussian pulse. Figure 7.2 shows an example of a spectrum of a spectrally compressed Gaussian pulse (thin line) along with its initial spectrum (dash-dotted line). Spectral compression thus provides an almost transform-limited pump pulse which

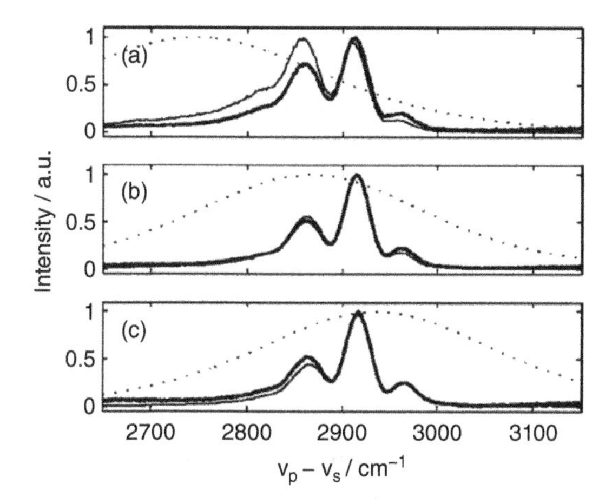

Figure 7.6 Typical examples of spectra obtained in MCARS microspectroscopy using a two-pulse soliton light source which provides a spectrally compressed Gaussian pump pulse and a redshifted soliton Stokes pulse. (Dotted) Stokes spectra. (Thin solid) raw CARS spectra. (Thick solid) CARS spectra deconvolved with the Stokes spectrum. Spectra acquired at three different Stokes frequencies: (a) 9710 cm^{-1}; (b) 9615 cm^{-1}; (c) 9520 cm^{-1}. *Source*: Andresen 2006 [17]. Reproduced with permission of OSA.

under most circumstances fits within the criteria for the MCARS scheme. In [17,52,53] a spectrally compressed pump pulse was used for CARS microspectroscopy.

Figure 7.6 show MCARS spectra acquired from a microscopic volume of a sample of neat ethanol and showcases that CARS microspectroscopy can be used to obtain quantitative spectral information, as evidenced by the fact that the three spectra Figures 7.6 (a)–7.6 (c) can be deconvolved to yield the same result regardless of the frequency of the Stokes pulse.

Spectral compression does not in general result in transform-limited pulses, whether it does is highly dependent on the exact shape of the pulse temporal envelope. Often one observes little wings on the spectrally compressed spectrum which stem from residual chirp and which in CARS microspectroscopy can lead to the kind of artefacts that have been described in for example, [25]. Attempts have been made to devise a spectral compression scheme that gives transform-limited spectrally compressed pulses. With a parabolic temporal pulse envolope, in essence, a similariton pulse, this turns out to be possible. Spectral compression of similariton pulses was treated in Chapter 6 of this book. An example can be found in [54]: A 160 fs sech^2-shaped laser pulse was shaped into a 2500-fs-long parabolic pulse with linear, negative chirp and sent through a 1-m-long nonlinear PCF with small, normal dispersion. Thus, a spectrally compressed pulse with time-bandwidth product 1.2 times the transform limit was produced and therefore the spectrum exhibited no wings on the high- and low-frequency side of the main peak. Spectral compression of similariton pulses thus in principle gives optimal spectral compression. However, if a laser with a similariton output is not available, the method in practice requires amplitude shaping which necessitates lossy optics. This is likely why spectrally compressed similariton pulses have not yet been demonstrated as pump pulses in MCARS microscopy.

7.7 ps-CARS Microscopy

In this section we will be dealing with the microscopy modality called ps-CARS microscopy. As in Section 7.6, we are dealing with the third-order nonlinear polarization at the anti-Stokes frequency which induces a field with frequency higher than the pump and Stokes driving fields which can be spectrally separated from them and homodyne-detected on a detector to yield the CARS signal

$$S_{\mathrm{CARS}} \propto \int \mathrm{d}\omega_{\mathrm{CARS}} |i\tilde{P}_{\mathrm{CARS}}(\omega_{\mathrm{CARS}})|^2. \tag{7.10}$$

ps-CARS has in particular been applied to study demyelinating diseases [55–57], cellular biology [58], and lipid droplet tracking [59].

The ps-CARS scheme comes about when the following condition is fulfilled: The spectral width of the pump as well as the Stokes pulse is much lower than the inverse decoherence time of the sample Γ_{ij}; formulated in the time domain, this condition becomes: The pump and Stokes pulses are much longer than the decoherence time Γ_{ij}^{-1} of the sample. In this case the generated signal simplifies to ($\tilde{\epsilon}_p(\omega) \rightarrow \tilde{\epsilon}_{p0}\delta(\omega - \omega_{p0})$ and $\tilde{\epsilon}_S(\omega) \rightarrow \tilde{\epsilon}_{S0}\delta(\omega - \omega_{S0})$)

$$S_{\mathrm{CARS}} \propto |\chi^{(3)}(-\omega_{\mathrm{CARS}}; \omega_{p0}, -\omega_{S0}, \omega_{p0})|^2 |\tilde{\epsilon}_{p0}|^4 |\tilde{\epsilon}_{S0}|^2 \tag{7.11}$$

In this case a two-beam soliton light source is not immediately adapted. Both pulses provided by the source will be spectrally broad, with spectral widths around 15 times larger than Γ_{ij}. There exists a quite elegant way of reconciling the parameters of the light source with the requirements of the ps-CARS scheme which is known as spectral focusing [60]. In spectral focusing, a common, linear chirp is imposed on both pump and Stokes pulses (the chirp does not strictly have to be linear, the concept works with any, common, chirp). In doing so, the initial femtosecond pulses are stretched to durations of a few ps and their frequency components are distributed over the stretched pulse envelopes. The frequency difference between pump and Stokes then remains constant at all times and, thanks to the stretched profile, the sample is probed for a few ps rather than a few hundred fs if the transform-limited pulses were used. As discussed in [23, 61, 62] the generated signal obtained in a spectral focusing scheme is the same as would be obtained in a ps-CARS scheme using the same, though unchirped, pump and Stokes envelopes. The two-beam soliton light source which provides two femtosecond pulses can thus be made to fulfill the requirements for the ps-CARS modality by using the spectral focusing scheme; it is immediately comparable to the traditional light source which provides two transform-limited picosecond pulses, and detection of the signal can be done with the same single-point detector as one uses in traditional ps-CARS.

Typical examples of ps-CARS images obtained using the two-beam soliton light source and the spectral focusing measurement scheme are presented in Figures 7.7 (a)–7.7 (c). The images, taken at different delays between the driving pump and Stokes pulses show markedly different contrast because in the spectral focusing scheme, the delay is proportional to the instantaneous frequency difference between pump and Stokes which again equals the Raman frequency that is probed. As such, Figure 7.7 (b) derives image contrast from a very strong Raman line at 1000 cm^{-1} while Figures 7.7 (a)–7.7 (c) derive contrast from regions of the Raman spectrum where no lines are present. This clearly

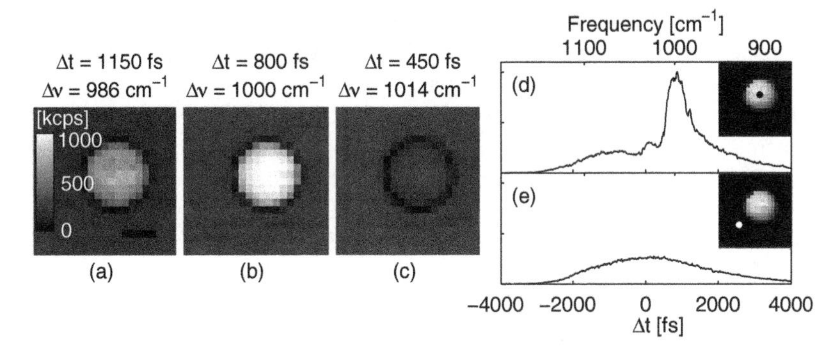

Figure 7.7 Examples of CARS images and spectra obtained with the spectral focusing measurement scheme and a soliton light source. (a)–(c) CARS images of a polystyrene bead in index-matching oil for the different pump-Stokes delays indicated; following the concept of spectral focusing, at each delay is probed the indicated frequency of the Raman spectrum. (d) CARS spectrum of the polystyrene bead. (e) CARS spectrum acquired in a point in the immersion oil outside of the bead. Scale bar, 10 μm. *Source*: Andresen 2011 [62]. Reproduced with permission of OSA.

showcases the chemical selectivity of ps-CARS microscopy in general and the spectral focusing scheme in particular.

The spectral focusing scheme can also be put to use to acquire ps-CARS spectra from within the focal region of the excitation lasers, giving a result very similar to MCARS microspectroscopy (Section 7.6) but by different means. This is showcased in Figures 7.7 (d) and 7.7 (e) where the excitation beams are parked on one spot on the sample and the signal acquired as a function of the pump-Stokes delay in order to produce ps-CARS spectra of a polystyrene bead and the surrounding immersion oil respectively.

The spectral focusing scheme has also been demonstrated in ps-CARS microscopy together with a two-beam light source that provided a dispersive wave (Section 7.2.2) in the place of a soliton [23, 24].

7.8 SRS Microscopy

The third-order nonlinear polarization at the Stokes frequency gives rise to a radiated field at the same frequency as the driving Stokes field. The radiated field is always in phase with the driving field and therefore is observed as a gain on the Stokes pulse, which is detected as a differential measurement:

$$S_{\mathrm{SRG}} \propto \int \mathrm{d}\omega_S |\tilde{E}_S(\omega_S) + i\tilde{P}^{(3)}_{\mathrm{SRG}}(\omega_S)|^2 - \int \mathrm{d}\omega_S |\tilde{E}_S(\omega_S)|^2. \tag{7.12}$$

Similarly, the third-order nonlinear polarization at the pump frequency gives rise to a radiated field at the same frequency as the driving pump pulse but it is always π out of phase with the driving field which is observed as a loss on the pump pulse, which is detected as a differential measurement:

$$S_{\mathrm{SRL}} \propto \int \mathrm{d}\omega_p |\tilde{E}_p(\omega_p) + i\tilde{P}_{\mathrm{SRL}}(\omega_p)|^2 - \int \mathrm{d}\omega_p |\tilde{E}_p(\omega_p)|^2 \tag{7.13}$$

These two modalities, SRG and SRL, are the sub-modalities that constitute SRS microscopy. Recent examples of applications of SRS microscopy are monitoring drug penetration in skin [63] and surgical guidance [64].

In practice, the requirement that S_{SRG} and S_{SRL} have to be measured by differential detection precludes the possibility of measuring them in the multiplex scheme, the reason being that only single-point detectors can reach the short read-out times that are required for the rate of the differential measurement to match the desired pixel scan rates of point-scanning microscopes. This thus confines SRS microscopy to the ps-scheme in which case the signals simplify to

$$\begin{aligned} S_{\mathrm{SRG}} &\propto -\mathrm{Im}[\chi^{(3)}(-\omega_{S0};-\omega_{p0},\omega_{S0},\omega_{p0})]|\tilde{e}_{p0}|^2|\tilde{e}_{S0}|^2 \\ S_{\mathrm{SRL}} &\propto -\mathrm{Im}[\chi^{(3)}(-\omega_{p0};-\omega_{S0},\omega_{p0},\omega_{S0})]|\tilde{e}_{p0}|^2|\tilde{e}_{S0}|^2. \end{aligned} \tag{7.14}$$

And with this we see that the requirements for the light source for SRS microscopy are identical to those for CARS microscopy, Section 7.7 so here as well the measurement scheme that can accommodate the soliton light source is spectral focusing. So similarly, and as discussed in [62] the generated signals S_{SRG} and S_{SRL} obtained in a spectral focusing scheme is the same as would be obtained in a ps-scheme using the same, though unchirped, pump and Stokes temporal envelopes.

The example SRS imaging and microspectroscopy results obtained with the two-beam soliton light source and the spectral focusing measurement scheme presented in Figure 7.8 are taken under the same conditions as Figure 7.7. Typical examples of SRS images are presented in Figures 7.8 (a)–7.8 (c). As previously the images taken at different delays between the driving pump and Stokes pulses derive contrast from different Raman frequencies due to the concept of spectral focusing. As such, Figure 7.8 (b) derives image contrast from a very strong Raman line at 1000 cm^{-1} while Figures 7.8 (a) and 7.8 (c) derive contrast from regions of the Raman spectrum where no lines are present. This showcases the chemical selectivity of SRS microscopy in general and the

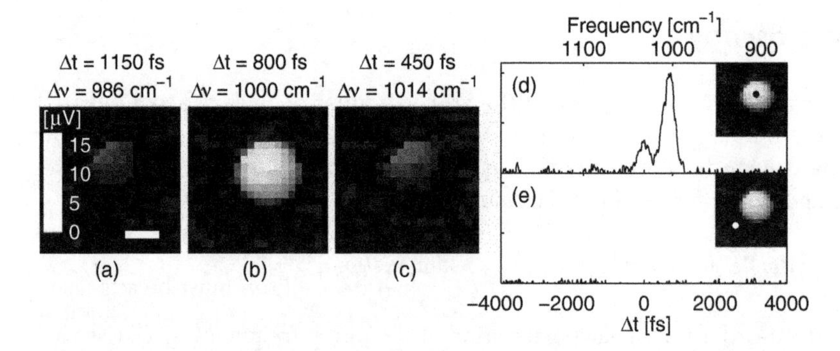

Figure 7.8 Examples of SRS images and spectra obtained with the spectral focusing measurement scheme and a soliton light source. (a)–(c) SRG images of a polystyrene bead in index-matching oil for the different pump-Stokes delays indicated; following the concept of spectral focusing, at each delay is probed the indicated frequency of the Raman spectrum. (d) SRG spectrum of the polystyrene bead. (e) SRG spectrum acquired in a point in the immersion oil outside of the bead. Scale bar, 10 μm. *Source*: Andresen 2011 [62]. Reproduced with permission of OSA.

spectral focusing scheme in particular. And, indeed, it shows that simultaneous (multimodal) ps-CARS and SRS microscopy is possible since the two signals are generated at different wavelengths and can thus be detected in separate detection paths.

Figures 7.8 (d) and 7.8 (e) showcase the ability to perform SRS microspectroscopy. When the excitation beams are parked in a bead, the SRS spectrum in Figure 7.8 (d) results, and when the excitation beams are parked in the surrounding immersion oil, the SRS spectrum in Figure 7.8 (e) results.

An added advantage of performing microspectroscopy in the spectral focusing scheme is the ability to acquire spectra rapidly, by simply scanning a mechanical delay. This ability to acquire spectra rapidly was put to use in [65] to monitor a chemical equilibrium.

7.9 Pump-Probe Microscopy

Two-color pump-probe microscopy was developed in 2007 [13]. Unlike the microscopy methods that we have mentioned in the previous sections, pump-probe microscopy was developed specifically for absorbing samples.

The principle is as follows [66]. A pump pulse arriving on a spot on the sample at time zero is partially absorbed by the sample. A probe pulse arriving on the same spot a time τ later probes the pump-induced changes in the sample. The transient absorption of the probe is found as the difference in probe transmission with the pump pulse on and off. The transient trace is obtained by measuring the transient absorption at a sequence of τ and plotting the transient absorption vs. τ. Generating a pump-probe microscopy image is thus the action of acquiring a transient trace in every pixel of the image.

These properties make two-color pump-probe microscopy a promising tool for analysis and diagnosis of absorbing samples, such as hair samples [67], imaging cutaneous pigmentation [68], imaging red blood cells [69], with endogenous contrast through principal component analysis [70] or phasor analysis [71].

In the pump-probe scheme, temporal resolution better than the excited-state lifetimes of the sample is required in order to get intelligible transient traces. Since lifetimes of the above-cited samples are in the 1 ps range, the temporal resolution must be in the 100 fs range. In the pump-probe scheme the temporal resolution is given by the duration of the pump and probe pulses. A suitable light source for two-color pump-probe microscopy must therefore have the following properties: (i) two pulses, pump and probe, of different colors; (ii) both pulses of 100 fs duration; (iii) controllable delay between the pump and probe. As we have seen above (in Section 7.2), (i) is fulfilled by the two-beam soliton light source with the laser taking on the role of pump pulse and the soliton generated through SSFS taking on the role of probe pulse, tunable over 100s of nm. In the same way (ii) is also fulfilled since the laser needed to generate a soliton must be a fs-laser, and the soliton generated through SSFS is always transform-limited. (iii) can be fulfilled by very standard means, such as a letting one of the two pulses go through a mechanical delay stage before recombining it with the other. So the two-beam soliton light source can be readily adapted for two-color pump-probe microscopy fairly trivially. Which is why, in this section, we are going to spend the bulk of the time examining an alternative way of obtaining the controllable pump-probe delay that uses properties of the soliton. We thus finish off our exposé with a nonlinear imaging application that—unlike the previous applications—employs both the temporal and spectral characteristics of the SSFS.

The group velocity v_g of a soliton of frequency ω is given by $v_g = 1/\beta_1(\omega)$. We remember that the frequency of a soliton undergoing SSFS is a function of distance z, so $\omega \rightarrow \omega(z)$. To get the group delay Δt incurred by the soliton upon propagating through a fiber of length L, we would thus have to calculate

$$\Delta t(P_0, L) = \int_0^L \beta_1[\omega(P_0, z)]\mathrm{d}z. \tag{7.15}$$

To get some intuition, we can imagine the simplest fiber in which a soliton can exist, a fiber with constant β_2 and no higher-order dispersion.

$$\Delta t(P_0, L) = \int_0^L \beta_2\omega(P_0, z)\mathrm{d}z. \tag{7.16}$$

And from this we see that Δt increases faster with L than does ω, in this simple case Δt increases by one more potency in z than ω. And this bodes well for a soliton light source for two-color pump-probe microscopy; we would need a source where $\Delta t(P_0, L) - \Delta t(P_0 + \delta P, L)$ is greater than the sample dynamics, and where $\omega(P_0, L) - \omega(P_0 + \delta P, L)$ is smaller than the sample absorption linewidth. In view of the above discussion, these criteria will be ever more fulfilled, the longer the fiber becomes.

For a more rigourous approach $\omega(z)$ can, for example, be calculated by numerical propagation methods if the fiber dispersion and nonlinear parameters are known. $\omega(z)$ can also be measured by cutback measurements.

Figure 7.9 shows calculated as well as measured values of Δt for four different fiber lengths. It is apparent that the slope increases with fiber length which is in agreement with our simple estimation above. The idea is that for sufficient long fiber, the redshift and Δt are almost decoupled, i.e., a negligibly small differential redshift is accompanied by a very large differential Δt. In this case, Δt can be scanned simply by scanning the fiber input power within the relevant range.

In [72] this "τ-scan two-beam soliton light source" was introduced for pump-probe microscopy. In effect, a laser pulse at 800 nm was used as pump pulse and a soliton shifted to 870 nm was used as probe pulse. In order to obtain transient absorption traces, the setup permitted scanning the pump-probe delay by a scanning the attenuation of an element located just in front of the soliton-generating fiber, i.e., following the approach outlined above. The setup additionally permitted scanning the pump-probe delay by traditional means, i.e., a mechanical delay stage. In Figure 7.10 is shown a comparison of the two approaches. It is seen that they give identical results.

The τ-scan two-beam soliton light source is thus an alternative to traditional light sources for pump-probe microscopy. The important difference between the two lies in the speed and flexibility of the τ-scan two-beam soliton light source. Since it contains no moving elements, it is not restricted to low scan speeds by inertia and the Pockels cell which can operate at MHz frequencies could therefore allow to acquire an entire transient trace in μs which is evidently of interest in microscopy where high pixel scan rates are normally required, or simply in samples that evolve on a fast time scale.

In conclusion, this section has shown that the delay dependence of a soliton which is redshifting under the SSFS can to a large degree be mastered and put to use in

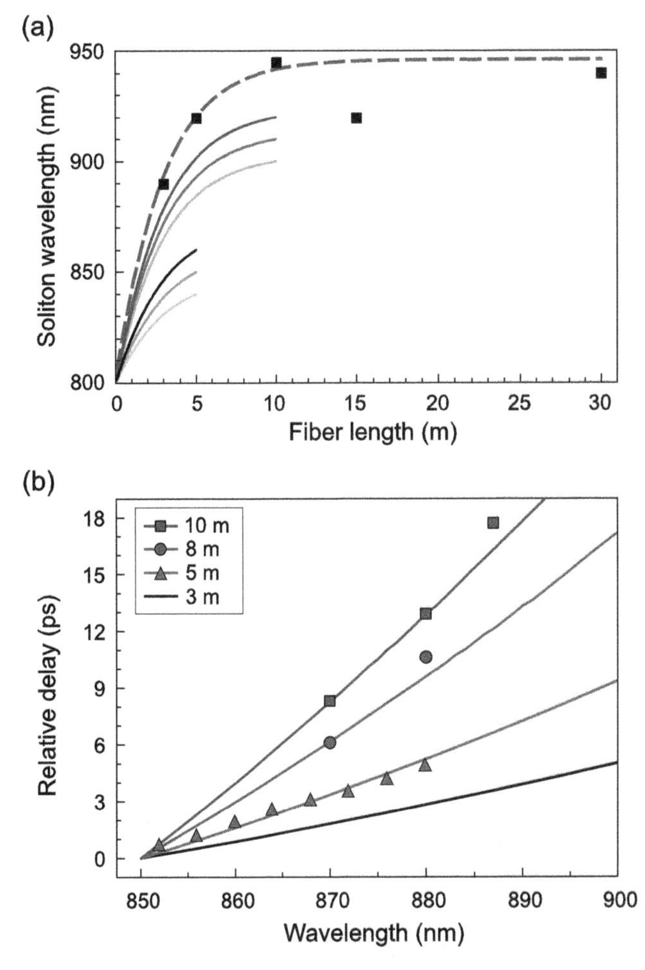

Figure 7.9 Calculation and measurement of the delay difference between solitons induced by SSFS. (a) Black squares: maximum soliton wavelength achieved in SC-PBG fibers of different lengths. Dashed line: guide to the eye. Full lines: estimated soliton trajectories in 5 m and 10 m long fibers at different input powers. (b) Calculated (lines) and experimental (markers) relative delay of redshifted solitons for four fiber lengths. The solitons exiting at 850 nm are taken as an arbitrary delay reference. *Source*: Adapted from Saint-Jalm [72].

nonlinear imaging applications. In addition, the example we provided actually made use of both the redshift and the delay dependence upon input power. As yet, there are no similar reports in the literature but this is definitely a subject to be explored in future studies.

7.10 Increasing the Soliton Energy

In this section we discuss two remedies for the inherently low soliton energy which have begun to be explored recently.

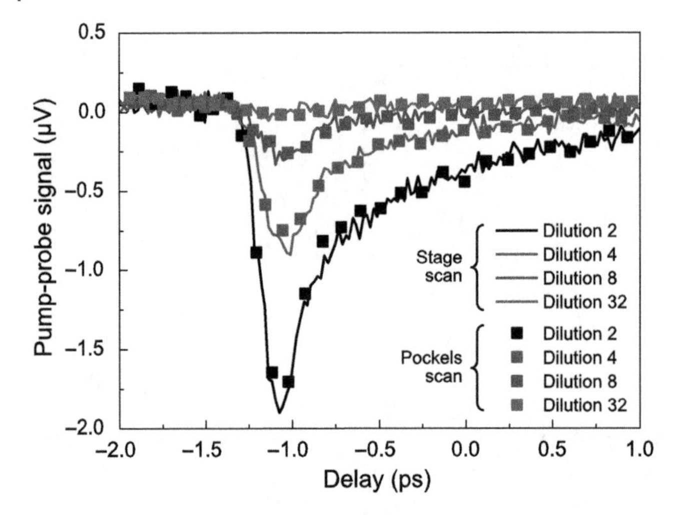

Figure 7.10 Comparison between translation stage scan (lines) and Pockels cell scan (squares) for four different concentrations of black ink. *Source*: Adapted from Saint-Jalm [72].

7.10.1 SC-PBG Fibers

Rearranging Eq. (7.1) we get

$$E_{\text{sol}} = \frac{2|\beta_0|}{T_0\gamma} \tag{7.17}$$

from which it is apparent that the achievable soliton energy depends on two fiber parameters β_2 and γ and more specifically their ratio $|\beta_2/\gamma|$. The higher this ratio, the higher the soliton energies allowed in the fiber. In nonlinear PCF, at the wavelengths typically employed in nonlinear microscopy 800–1000 nm, the two are often anti-correlated. If one wishes to increase the ratio by increasing the numerator $|\beta_2|$, this must be done by increasing waveguide dispersion, which can be done by decreasing the core size which, however, unfortunately increases the numerator γ. And vice versa if one wishes to increase the ratio by decreasing the numerator γ, this can be done by increasing the core size which has the consequence of decreasing $|\beta_2|$. So, it has been acknowledged that nonlinear PCFs have some intrinsic limitation in the soliton energies that they can allow.

Other kinds of microstructured fiber which guide light by a different mechanism than PCFs do offer $|\beta_2/\gamma|$ that are vastly out of reach of nonlinear PCFs. For example, a hollow-core photonic-bandgap fiber (HC-PBG) was shown to support fundamental solitons with energies of several hundreds of nJ, at least three orders of magnitude more than in nonlinear PCFs [73]. This kind of energy level is, however, incompatible with biological imaging where pulse energies are usually kept below 1 nJ to avoid damage and, additionally, the high-repetition-rate lasers commonly used in biological imaging (around, 80 MHz, around 10 nJ pulse energy) do not even possess sufficient pulse energy to generate solitons in a HC-PBG. A compromise that would bridge the gap between the two extremes, nonlinear PCFs and HC-PBGs would be desirable.

Such a compromise could be solid-core photonic bandgap fibers (SC-PBGs). This class of fiber guide light by the photonic bandgap effect, similarly to the HC-PBGs, and thus they exhibit the large, anomalous dispersion (large $|\beta_2|$) that is characteristic of this effect. However, by virtue of having a solid rather than a hollow core, the SC-PBG brings the $|\beta_2/\gamma|$ down significantly compared to the HC-PBGs. Importantly, though, the ratio is up to an order of magnitude greater than what can be found in nonlinear PCFs. It has been shown that such a fiber can support solitons with energies 3 to 5 times higher than PCFs which can redshift over hundreds of nm over the entire photonic bandgap [74,75]. And indeed the fiber used in the soliton light source for TPEF and SHG microscopy in [34] employed such a SC-PBG.

7.10.2 Multiple Soliton Generation

An alternative remedy to the generally low soliton energy was first hinted at in [76] which is coherent addition of multiple solitons. In that paper the possibility of phase-compensation of n-tuples of solitons generated from a single input pulse was discussed and numerically investigated. Since only a single pulse was launched into the PCF there was only one input parameter to tune, the input power. For certain values of the input power, the output spectrum displayed high-visibility spectral interference fringes hinting at temporal separations of few hundreds of fs.

Refs. [77, 78] both embraced the principle of generating multiple solitons in a PCF, in both papers however, more input parameters were available since pulse trains were launched into the PCF; each pulse of the pulse train led to one soliton generated in the PCF. Ref. [77] employed multiple solitons addition with the aim of generating a spectrum spanning the entire spectral range relevant for CRS microspectroscopy, 800 nm $< \lambda_S <$ 1050 nm for $\lambda_p = 800$ nm. This approach is potentially more efficient than supercontinuum generation because a fundamental soliton can be generated with an efficiency approaching unity and only about five solitons at equally-spaced intervals are required to cover the required spectral range; supercontinuum generation would generate a large amount of light at frequencies that are not of interest for CRS. Using a 4-f pulse shaper and a genetic algorithm to optimize the pulse trains launched into the PCF, the authors arrived at two different transfer functions that resulted in two and three solitons superposed in time. The pulse shaper alternated between the two masks every second, so that the 2.4 s average of the PCF output was five evenly spaced solitons all having the same delay and spanning 850–1080 nm. When combined with the pump pulse (the fs-laser pulse, filtered down to 3 nm), broadband CARS microspectroscopy over the entire Raman spectrum was demonstrated.

While [77] was able to temporally superpose up to three solitons of different frequency, temporal superposition of solitons of the same frequency cannot be temporally superposed without them interacting. This is unwanted if one is interested not in a broadband Stokes pulse but rather a relatively narrow Stokes pulse with a high spectral density. Ref. [78] aimed at generating pairs of spectrally overlapping solitons and to bring them as close in time as possible. The task was complicated by various phase-dependent interactions between the two pulses, primarily interpulse SRS, which is significant for solitons whose frequencies differ by less than the Raman frequencies of silica. Nevertheless, parameter regimes were found where two redshifted solitons separated by only a few 100 fs and having a large spectral overlap could be generated; due to spectral interference, the total spectrum of the pair consisted of bright and dark fringes, the brightest being almost four times as bright as a single, fundamental soliton. If more than two solitons could be simultaneously generated in this way, it could result is soliton trains of ps-duration having a spectrally narrow, very bright fringe that might be useful in the ps-CARS measurement scheme.

7.11 Conclusion

In this chapter we have reviewed how soliton generation in nonlinear PCF can be advantageously used to perform nonlinear imaging. The examples presented have all been enabled by some unique characteristics of solitons. The first is the inherently Fourier transform limited time-bandwidth product of fiber-generated solitons. Indeed, ensuring that a (non-solitonic) fiber-delivered ultra-short pulse of light is emitted as a transform-limited pulse is a difficult challenge which can, however, be overcome using complex compensation schemes [79] or specially designed hollow-core fibers [80]. The second unique characteristic is the ability of the soliton to redshift as it propagates, resulting immediately in a fs light source that is tunable over hundreds of nanometers. In terms of wavelength tunability and pulse quality, such a soliton light source is a direct competitor to the fs-OPOs that are extensively used for biophotonics applications [81], and the soliton has the advantage of being compact, simple, and uncostly. The major drawback of soliton light sources is their inherently limited peak power levels, this limitation also applies, however, to lesser extent, to fiber laser-based light sources [82] which are also being actively developed for biophotonic applications. Nevertheless several possible methods exist for increasing soliton peak power, including multiple soliton generation and specially-designed SC-PBG fibers.

Soliton light sources are presently functional for in non-exigent biological imaging applications. Looking a bit further ahead, it is to be expected that soliton light sources could soon constitute the heart of what we might call "active" endoscopes, the word "active" signifying that the endoscope fiber not only delivers and collects light but also generates the excitation light with the required parameters. Although this has not yet been demonstrated, a few encouraging works deserve to be mentioned. In [52] a 50 fs laser pulse at 797 nm was shaped into two pulses delayed relative to one another, one of which was given a negative chirp while the other was given no chirp. These were sent through a 11.9-cm-long PCF which gave rise to a spectrally compressed pulse overlapping temporally with a redshifted soliton, thus providing pump and Stokes pulses which were used in a MCARS measurement. While this example only demonstrates fiber delivery and not the collection of signal through the same fiber, it could pave the way to the aforementioned active endoscope. Looking even further ahead, light generation in active endoscopes could be combined with recently developed endoscope imaging techniques that rely on coherent wave front control to acquire spatial information through multi-core [83–85]. The road leading to these achievements is still long but it is clear that there is a lot more to come from fiber solitons in the context of biophotonics and microscopy.

References

1 Knight, J.C. (2003) Photonic crystal fibers. *Nature*, **424**, 847–851.
2 Russell, P.S.J. (2003) Photonic Crystal Fibers. *Science*, **299**, 358–362.

3 Dudley, J.M. and Taylor, J.R. (2009) Ten years of nonlinear optics in photonic crystal fibre. *Nature Photonics*, **3**, 85–90.
4 Denk, W., Strickler, J.H., and Webb, W.W. (1990) 2-photon laser scanning fluorescence microscopy. *Science*, **248** (4951), 73–76.
5 Gannaway, J.N. and Sheppard, C.J.R. (1978) Second-harmonic imaging in the scanning optical microscope. *Optical and Quantum Electronics*, **10**, 435–439.
6 Zumbusch, A., Holtom, G.R., and Xie, X.S. (1999) Three-dimensional vibrational imaging by coherent anti-Stokes Raman scattering. *Physical Review Letters*, **82** (20), 4142–4145.
7 Cheng, J.X., Volkmer, A., Book, L.D., and Xie, X.S. (2002) Multiplex coherent anti-stokes raman scattering microspectroscopy and study of lipid vesicles. *Journal of Physical Chemistry B*, **106** (34), 8493–8498.
8 Müller, M., Sciens, J.M., Nastase, N., Wurpel, S.G.W.H., and Brakenhoff, F.G.J. (2002) Imaging the thermodynamic state of lipid membranes with multiplex cars microscopy. *Journal of Physical Chemistry B*, **106** (14), 3715–3723.
9 Ploetz, E., Laimgruber, S., Berner, S., Zinth, W., and Gilch, P. (2007) Femtosecond stimulated Raman microscopy. *Applied Physics B*, **87**, 389–393.
10 Freudiger, C.W., Min, W., Saar, B.G., *et al.* (2008) Label-Free Biomedical Imaging with High Sensitivity by Stimulated Raman Scattering Microscopy. *Science*, **322**, 1857–1861.
11 Nandakumar, P., Kovalev, A., and Volkmer, A. (2009) Vibrational imaging based on stimulated Raman scattering microscopy. *New Journal of Physics*, **11**, 033 026.
12 Ozeki, Y., Dake, F., Kajiyama, S., Fukui, K., and Itoh, K. (2009) Analysis and experimental assessment of the sensitivity of stimulated Raman scattering microscopy. *Optics Express*, **17** (5), 3651–3658.
13 Fu, D., Ye, T., Matthews, T.E., Yurtsever, G., and Warren, W.S. (2007) Two-color, two-photon, and excited-state absorption microscopy. *Journal of Biomedical Optics*, **12**, 054 004.
14 Agrawal, G.P. (2007) *Nonlinear Fiber Optics*, Elsevier, Oxford.
15 Mitschke, F.M. and Mollenauer, L.F. (1986) Discovery of the soliton self-frequency shift. *Optics Letters*, **11** (10), 659–661.
16 Gordon, J.P. (1986) Theory of the soliton self-frequency shift. *Optics Letters*, **11** (10), 662–664.
17 Andresen, E.R., Birkedal, V., Thogersen, J., and Keiding, S.R. (2006) Tunable light source for coherent anti-stokes raman scattering microspectroscopy based on the soliton self-frequency shift. *Optics Letters*, **31** (9), 1328–1330.
18 McConnell, G. and Riis, E. (2004) Photonic crystal fibre enables short-wavelength two-photon laser scanning fluorescence microscopy with fura-2. *Physics in Medicine and Biology*, **49** (20), 4757–4763.
19 Andresen, E.R. (2007) *Fiber-based Implementations of Coherent anti-Stokes Raman Scattering Microspectroscopy and Microscopy*, Ph.D. thesis, University of Aarhus.
20 Tai, K., Hasegawa, A., and Bekki, N. (1988) Fission of optical solitons induced by stimulated Raman effect. *Optics Letters*, **13**, 392–395.
21 Kodama, Y. and Hasegawa, A. (1987) Nonlinear pulse propagation in a monomode dielectric guide. *IEEE Journal of Quantum Electronics*, **23**, 510–524.
22 Tran, T.X., Podlipensky, A., Russell, P.S.J., and Biancalana, F. (2010) Theory of Raman multipeak states in solid-core photonic crystal fibers. *Journal of the Optical Society of America B*, **27** (9), 1785–1791.

23 Pegoraro, A.F., Risdale, A., Moffatt, D.J., *et al.* (2009) Optimally chirped multimodal CARS microscopy based on a single Ti:sapphire oscillator. *Optics Express*, **17** (4), 2984–2996.

24 Pegoraro, A.F., Risdale, A., Moffatt, D.J., *et al.* (2009) All-fiber CARS microscopy of live cells. *Optics Express*, **17** (23), 20 700–20 706.

25 Andresen, E.R., Paulsen, H.N., Birkedal, V., Thøgersen, J., and Keiding, S.R. (2005) Broadband multiplex coherent anti-Stokes Raman scattering microscopy employing photonic-crystal fibers. *Journal of the Optical Society of America B*, **22** (9), 1934–1938.

26 Birks, T.A., Knight, J.C., and Russell, P.S.J. (1997) Endlessly single-mode photonic crystal fiber. *Optics Letters*, **22** (13), 961–963.

27 Gander, M.J., McBride, R., Jones, J.D.C., *et al.* (1999) Measurement of the wavelength dependence of beam divergence for photonic crystal fiber. *Optics Letters*, **24** (15), 1017–1019.

28 Esposito, E., Harris, J., Burns, D., and McConnell, G. (2007) Measurement of white-light supercontinuum beam properties from a photonic crystal fibre using a laser scanning confocal microscope. *Measurement Science and Technology*, **18**, 2609–2615.

29 So, P.T., Dong, C.Y., and Berland, K.M. (2000) Two-photon excitation fluorescence microscopy. *Annual Review of Biomedical Engineering*, **2**, 399–429.

30 Helmchen, F. and Denk, W. (2005) Deep tissue two-photon microscopy. *Nature Methods*, **2** (12), 932–940.

31 Denk, W. and Svoboda, K. (1997) Photon upmanship: why multiphoton imaging is more than a gimmick. *Neuron*, **18**, 351–357.

32 Helmchen, F. and Denk, W. (2002) New developments in multiphoton microscopy. *Current Opinions in Neurobiology*, **12**, 593–601.

33 Mertz, J. (2004) Nonlinear microscopy: new techniques and applications. *Current Opinions in Neurobiology*, **14**, 610–616.

34 Saint-Jalm, S., Andresen, E.R., Ferrand, P., *et al.* (2014) Fiber-based ultrashort pulse delivery for nonlinear imaging using high-energy solitons. *Journal of Biomedical Optics*, **19** (8), 086 021.

35 Campagnola, P.J., Clark, H.A., Mohler, W.A., Lewis, A., and Loew, L.M. (2001) Second-harmonic imaging microscopy of living cells. *Journal of Biomedical Optics*, **6** (3), 277–286.

36 Moreaux, L., Sandre, O., Blanchard-Desce, M., and Mertz, J. (2000) Membrane imaging by simultaneous second-harmonic generation and two-photon microscopy. *Optics Letters*, **25** (5), 320–322.

37 Campagnola, P.J., Millard, A.C., Terasaki, M., *et al.* (2002) Three-dimensional High-resolution Second-harmonic generation imaging of endogenous structural proteins in biological tissues. *Biophysical Journal*, **82** (1), 493–508.

38 Dombeck, D.A., Kasischke, K.A., Vishwasrao, H.D., *et al.* (2003) Uniform polarity microtubule assemblies imaged in native brain tissue by second-harmonic generation microscopy. *Proceedings of the National Academy of Sciences of the USA*, **100** (12), 7081–7086.

39 Mohler, W., Millard, A.C., and Campagnola, P.J. (2003) Second harmonic generation imaging of endogenous structural proteins. *Methods*, **29** (1), 97–109.

40 Zipfel, W.R., Williams, R.M., Christie, R., *et al.* (2003) Live tissue intrinsic emission microscopy using multiphoton-excited native fluorescence and second harmonic generation. *Proceedings of the National Academy of Sciences of the USA*, **100** (12), 7075–7080.

41 Volkmer, A. (2005) Vibrational imaging and microspectroscopies based on coherent anti-Stokes Raman scattering microscopy. *Journal of Physics D*, **38** (5), R59–R81.

42 Jr., C.H.C. and Cicerone, M.T. (2015) Chemically sensitive bioimaging with coherent raman scattering. *Nature Photonics*, **9**, 295–305.

43 Mukamel, S. (1995) *Principles of Nonlinear Spectroscopy*, Oxford University Press, Oxford.

44 von Vacano, B., Meyer, L., and Motzkus, M. (2007) Rapid polymer blend imaging with quantitative broadband multiplex CARS microscopy. *Journal of Raman Spectroscopy*, **38** (7), 916–926.

45 Camp, C.H., Eftekhar, S.Y.A.A., Sridhar, H., and Adibi, A. (2009) Multiplex coherent anti-Stokes Raman scattering (MCARS) for chemically sensitive, label-free flow cytometry. *Optics Express*, **17** (25), 22 879–22 889.

46 Rinia, H.A., Burger, K.N.J., Bonn, M., and Muller, M. (2008) Quantitative label-free imaging of lipid composition and packing of individual cellular lipid droplets using multiplex CARS microscopy. *Biophysical Journal*, **95** (10), 4908–4914.

47 Day, J.P.R., Rago, G., Domke, K.F., Velikov, K.P., and Bonn, M. (2010) Label-free imaging of lipophilic bioactive molecules during lipid digestion by multiplex coherent anti-stokes raman scattering microspectroscopy. *Journal of the American Chemical Society*, **132** (24), 8433–8439.

48 Stolen, R. and Lin, C. (1978) Self-phase-modulation in silica optical fibers. *Physical Review A*, **17** (4), 1448–1453.

49 Oberthaler, M. and Hopfel, R. (1993) Special narrowing of ultrashort laser-pulses by self-phase modulation in optical fibers. *Applied Physics Letters*, **63** (8), 1017–1019.

50 Planas, S., Mansur, N., Cruz, C., and Fragnito, H. (1993) Spectral narrowing in the propagation of chirped pulses in single-mode fibers. *Optics Letters*, **18** (9), 699–701.

51 Andresen, E.R., Thogersen, J., and Keiding, S.R. (2005) Spectral compression of femtosecond pulses in photonic crystal fibers. *Optics Letters*, **30** (15), 2025–2027.

52 Tada, K. and Karasawa, N. (2011) Single-Beam Coherent Anti-Stokes Raman Scattering Spectroscopy Using Both Pump and Soliton Pulses from a Photonic Crystal Fiber. *Applied Physics Express*, **4**, 092 701.

53 Hage, C.H., Kibler, B., Andresen, E.R., *et al.* (2011) Optimization and characterization of a femtosecond tunable light source based on the soliton self-frequency shift in photonic crystal fiber, in *Proceedings of SPIE*, vol. 8071, vol. 8071, p. 80710I.

54 Andresen, E.R., Dudley, J.M., Oron, D., Finot, C., and Rigneault, H. (2011) Transform-limited spectral compression by self-phase modulation of amplitude-shaped pulses with negative chirp. *Optics Letters*, **36** (5), 707–709.

55 Fu, Y., Wang, H., Huff, T.B., Shi, T., and Cheng, J.X. (2007) Coherent anti-Stokes Raman scattering imaging of myelin degradation reveals a calcium-dependent pathway in lyso-PtdCho-induced demyelination. *Journal of Neuroscience Research*, **85**, 2870–2881.

56 Imitola, J., Côté, D., Rasmussen, S., *et al.* (2011) Multimodal coherent anti-Stokes Raman scattering microscopy reveals microglia-associated myelin and axonal dysfunction in multiple sclerosis-like lesions in mice. *Journal of Biomedical Optics*, **16** (2), 021 109.

57 Bélanger, E., Henry, F.P., Vallée, R., *et al.* (2011) In vivo evaluation of demyelination and remyelination in a nerve crush injury model. *Biomedical Optics Express*, **2** (9), 2698–2708.

58 Konorov, S.O., Glover, C.H., Piret, J.M., *et al.* (2007) In situ analysis of living embryonic stem cells by coherent anti-stokes raman microscopy. *Analytical Chemistry*, **79** (18), 7221–7225.

59 Jüngst, C., Winterhalder, M.H., and Zumbusch, A. (2011) Fast and long term lipid droplet tracking with CARS microscopy. *Journal of Biophotonics*, **4** (6), 425–441.

60 Hellerer, T., Enejder, A.M.K., and Zumbusch, A. (2004) Spectral focusing: High spectral resolution spectroscopy with broad-bandwidth laser pulses. *Applied Physics Letters*, **85** (1), 25–27.

61 Langbein, W., Rocha-Mendoza, I., and Borri, P. (2009) Coherent anti-stokes raman micro-spectroscopy using spectral focusing: theory and experiment. *Journal of Raman Spectroscopy*, **40** (7), 800–808.

62 Andresen, E.R., Berto, P., and Rigneault, H. (2011) Stimulated raman scattering microscopy by spectral focusing and fiber-generated soliton as stokes pulse. *Optics Letters*, **36** (13), 2387–2389.

63 Saar, B.G., Contreras-Rojas, L.R., Xie, X.S., and Guy, R.H. (2011) Imaging drug delivery to skin with stimulated raman scattering microscopy. *Molecular Pharmacology*, **8** (3), 969–975.

64 Ji, M.B., Orringer, D.A., Freudiger, C.W., *et al.* (2013) Rapid, label-free detection of brain tumors with stimulated raman scattering microscopy. *Science Translational Medicine*, **5** (201), 201ra119.

65 Saint-Jalm, S., Berto, P., Jullien, L., Andresen, E.R., and Rigneault, H. (2014) Rapidly tunable and compact coherent raman scattering light source for molecular spectroscopy. *Journal of Raman Spectroscopy*, **45** (7), 515–520.

66 Ye, T., Fu, D., and Warren, W.S. (2009) Nonlinear absorption microscopy. *Photochemistry and Photobiology*, **85**, 631–645.

67 Piletic, I.R., Matthews, T.E., and Warren, W.S. (2010) Probing near-infrared photorelaxation pathways in eumelanins and pheomelanins. *Journal of Physical Chemistry A*, **114**, 11 483–11 491.

68 Fu, D., Ye, T., Matthews, T.E., *et al.* (2008) Probing skin pigmentation changes with transient absorption imaging of eumelanin and phenomelanin. *Journal of Biomedical Optics*, **13**, 054 036.

69 Fu, D., Ye, T., Matthews, T.E., *et al.* (2007) High-resolution in vivo imaging of blood vessels without labeling. *Optics Letters*, **32** (18), 2641–2643.

70 Matthews, T.E., Piletic, I.R., Selim, M.A., Simpson, M.J., and Warren, W.S. (2011) Pump-probe imaging differentiates melanoma from melanocytic nevi. *Science Translational Medicine*, **3**, 71ra15.

71 Robles, F.E., Wilson, J.W., Fischer, M.C., and Warren, W.S. (2012) Phasor analysis for nonlinear pump-probe microscopy. *Optics Express*, **20**, 17 082–17 092.

72 Saint-Jalm, S., Andresen, E.R., Bendahmane, A., Kudlinski, A., and Rigneault, H. In preparation.

73 Ouzounov, D.G., Ahmad, F.R., Müller, D., *et al.* (2003) Generation of megawatt optical solitons in hollow-core photonic band-gap fibers. *Science*, **301** (5640), 1702–1704.

74 Bétourné, A., Kudlinski, A., Bouwmans, G., *et al.* (2009) Control of supercontinuum generation and soliton self-frequency shift in solid-core photonic bandgap fibers. *Optics Letters*, **34** (20).

75 Bendahmane, A., Mussot, A., Vanvincq, O., *et al.* (2013) Solid-core photonic bandgap fiber for the generation of tunable high-energy solitons, in *Workshop on Specialty Optical Fibers and their Applications*, Optical Society of America.

76 Voronin, A.A., Fedotov, I.V., Fedotov, A.B., and Zheltikov, A.M. (2009) Spectral interference of frequency-shifted solitons in a photonic-crystal fiber. *Optics Letters*, **34** (5), 569–571.

77 Tada, K. and Karasawa, N. (2009) Broadband coherent anti-Stokes Raman scattering spectroscopy using soliton pulse trains from a photonic crystal fiber. *Optics Communications*, **282**, 3948–3952.

78 Andresen, E.R., Dudley, J.M., Oron, D., Finot, C., and Rigneault, H. (2011) Nonlinear pulse shaping by coherent addition of multiple redshifted solitons. *Journal of the Optical Society of America B-Optical Physics*, **28** (7), 1716–1723.

79 Lefort, C., Mansuryan, T., Louradour, F., and Barthelemy, A. (2011) Pulse compression and fiber delivery of 45 fs fourier transform limited pulses at 830 nm. *Optics Letters*, **36**, 292–294.

80 Benabid, F. Hollow-core photonic bandgap fibre: new light guidance for new science and technology. *Philosophical Transactions A*, **364** (1849), 3439–3462.

81 Brustlein, S., Ferrand, P., Walther, N., *et al.* (2011) Optical parametric oscillator-based light source for coherent raman scattering microscopy: practical overview. *Journal of Biomedical Optics*, **16** (2), 021 106.

82 Xu, C. and Wise, F. (2013) Recent advances in fibre lasers for nonlinear microscopy. *Nature Photonics*, **7**, 875–882.

83 Thompsons, A.J., Paterson, C., Neil, M.A.A., Dunsby, C., and French, P.M.W. (2011) Adaptive phase compensation for ultracompact laser scanning endomicroscopy. *Optics Letters*, **36** (9), 1707–1709.

84 Andresen, E.R., Bouwmans, G., Monneret, S., and Rigneault, H. (2013) Toward endoscopes with no distal optics: video-rate scanning microscopy through a fiber bundle. *Optics Letters*, **38** (5), 609–611.

85 Andresen, E.R., Bouwmans, G., Monneret, S., and Rigneault, H. (2013) Two-photon lensless endoscope. *Optics Express*, **21** (18), 20 713–20 721.

8

Self-Organization of Polarization State in Optical Fibers

Julien Fatome and Massimiliano Guasoni

Laboratoire Interdisciplinaire CARNOT de Bourgogne, UMR 6303 CNRS-Université de Bourgogne - Franche-Comté, Dijon, France

8.1 Introduction

Nowadays, in optical networks, the major part of signal processing is performed in higher electronic layers, i.e., after an opto-electronic conversion, rather than in the physical optical domain. This property is then expressed by defining the network as "opaque." However, as the amount of transmitted data is increasing, the necessary trend is to progressively move towards transparent networks and confine opto-electronic conversion to network boundaries, directly at transmitter and receiver side. To pursue this target of a transparent all-optical network, it then becomes mandatory to master all the parameters that characterize the transmitted light. Among the three independent properties that define a light beam propagating in a single-mode optical fiber, i.e., the wavelength, power, and state-of-polarization (SOP), the SOP remains the most elusive variable which is still challenging to predict and control. In fact, it is noteworthy that in the past decade spectacular advances in the manufacturing of optical fibers have been realized. Especially, by implementing a well-optimized spinning process during the drawing stage, fiber suppliers are now able to deliver standard telecom fibers with outstanding weak levels of polarization-mode dispersion [1–4]. Nevertheless, the remaining residual birefringence induced by random variations of the core geometry along the fiber length combined with more or less fast mechanical stress imposed by the fiber surroundings, such as bending, squeezing, vibrations or temperature variations make the polarization of light unpredictable after a few tens of meters of propagation [5–10].

On the other hand, from a general point of view, despite the recent tremendous technological developments in waveguide and fiber-based systems to mitigate polarization impairments, the basic principle of operation of these systems basically rests upon a combative strategy rather than on a preventive strategy. For instance, in high-capacity coherent transmissions, polarization impairments such as polarization randomness, polarization-mode dispersion [11–15], polarization depending loss [16] or cross-polarization interactions [17, 18] are efficiently compensated at the receiver side thanks to digital signal processing based on complex algorithms [19–21].

Shaping Light in Nonlinear Optical Fibers, First Edition. Edited by Sonia Boscolo and Christophe Finot.

Regarding highly polarization dependent systems such as on-chip integrated optical circuits or fiber-based nonlinear processing devices, special designs and more or less complex polarization-diverse schemes (polarization diversity, bi-directional loop or polarization splitting/recombination) may ensure the mitigation of polarization-dependent performances [22–25].

Another possible scenario consists in controlling the polarization of light in order to prevent or mitigate some of these impairments. Basically, in a first approach, two kinds of devices can be considered in order to control the light SOP in optical fibers, i.e., dissipative elements and non-dissipative ones. The first type of element is typically made of a Glan polarizer, thus inducing a non-negligible amount of polarization depending losses. Indeed, the output polarization state is then totally fixed and independent of the input SOP but in that case, all the input polarization fluctuations are transferred into intensity fluctuations at the device output, which is unacceptable for many practical applications in photonics. The second one is typically Lefebvre loops, wave-plates or electro-optic devices, which seem to be more attractive since they do not induce any polarization depending loss. These kinds of devices transform any input state of polarization into another state but the drawback is that resulting polarization fluctuations are the same as the input.

In fact, the most common and commercially available way to efficiently control the light SOP in fiber systems is to implement an opto-electronic polarization tracking solution [26–30]. These devices generally consist in linear polarization transformations followed by partial diagnostic associated with an active feedback loop control driven by complex algorithms. Thanks to these techniques, records of polarization tracking or diversity speeds have been achieved, reaching several Mrad/s. Nevertheless, these devices are essentially based on opto-electronic technologies, which could be seen as a limitation for the development of future transparent networks and are thus limited by the electronic response time of their feedback loop.

Beyond its fundamental interest, the light-by-light polarization control represents a complementary and alternative approach in this context. Indeed, the ability to all-optically master the state-of-polarization of a light beam without polarization depending loss could encounter numerous applications in photonics. To this aim, several techniques have emerged in the literature during the past decade so as to develop a nonlinear "ideal polarizer" that can repolarize an incident signal with 100% efficiency, while preserving the quality of the temporal intensity profile. More precisely, whatever its initial SOP, all the energy of an input signal has to be trapped in a single output SOP without transferring polarization fluctuations onto the intensity profile in the temporal domain. This phenomenon of polarization attraction in optical fibers or polarization pulling effect, has been the subject of numerous studies in the literature involving the Raman effect [31–35], the stimulated Brillouin backscattering [36, 37], the parametric amplification [38, 39] as well as a counter-propagating four-wave mixing process, also called nonlinear cross-polarization interaction [40–53].

For this last particular case, it has been shown that an arbitrarily polarized incident signal can be attracted toward a particular SOP, which is fixed by the polarization reference imposed by the counter-propagating pump wave injected at the opposite end of the fiber [42]. Efficient repolarization of telecom signals at 10 and 40 Gbit/s has been reported, in combination with several types of optical functionalities in a single span of fiber. For example, 2R regeneration processing for On/Off keying (OOK) telecom signals [49], noise cleaning [50], data packet processing [51], as well as spatial mode attraction have been demonstrated [53]. Nevertheless, the common feature of all of these systems is that the injection of an external reference pump wave is a prerequisite for the existence of the polarization attraction process.

As the opposite of this general rule, recent experimental observations have shown that a spontaneous organization of the light SOP can also occur in the absence of any polarization reference beam in a device called the Omnipolarizer [54, 55]. In this novel solution, the signal beam interacts with its own counter-propagating replica generated at the fiber end thanks to a single reflecting component, e.g., Fiber Bragg-Mirror (FBG), coating or amplified reflective fiber loop setup [54–56]. The signal itself evolves in time toward a stationary state imposed by the self-organization process. Furthermore, the counter-propagating geometry could also give rise to a chaotic polarization dynamics [57–63], which in turn could give rise to a tunable all-optical scrambling at the fiber output [62, 63].

The aim of this chapter is to provide a global overview of this phenomenon, as well as highlight some new results and discuss future developments. Our contribution is organized as follows. In Section 8.2 we introduce the principle of operation of the Omnipolarizer as well as three distinct working regimes of the device, called the bistability regime, the alignment regime and the scrambling regime. In Section 8.3 we describe the experimental implementation; in Section 8.4 we provide a general theoretical overview of the system dynamics. In Sections 8.5, 8.6 and 8.7, we discuss in greater detail the aforementioned regimes as well as some possible applications in the framework of all-optical signal processing, among which self-induced polarization tracking of a 40-Gbit/s OOK signal, a polarization-based optical memory, a polarization switching for 10 Gbit/s data, as well as the implementation of a polarization scrambler whose performances are directly tunable by means of the back-reflected power. Finally, in Section 8.9, we discuss new perspectives for the generalization of the idea of self-organization to the spatial modes of multimode or multicore fibers, which may open up new amazing scenarios in several application fields.

8.2 Principle of Operation

As shown in Figure 8.1, the Omnipolarizer basically consists of a few-km-long standard optical fibers enclosed between an optical circulator and a reflective element. In this configuration a forward signal interacts through a Kerr nonlinear cross-polarization process with its own backward replica.

The nonlinear length of the fiber is here defined as $L_{nl} = (\gamma P)^{-1}$, being P the input forward power and γ the nonlinear Kerr coefficient, which is typically around 2 W^{-1} km^{-1} in standard fibers. The total number of nonlinear lengths read as $N = L/L_{nl}$, being L the fiber length. The length L should be equivalent to at least some few nonlinear lengths, let us say indicatively $N \geq 4$, which leads to an efficient cross-polarization interaction among the counter-propagating beams. Consequently, in a typical configuration involving a 5-km-long fiber, a relatively high level of power close to 500 mW is necessary, which makes $L_{nl} \approx 1$ km and thus $N \approx 5$.

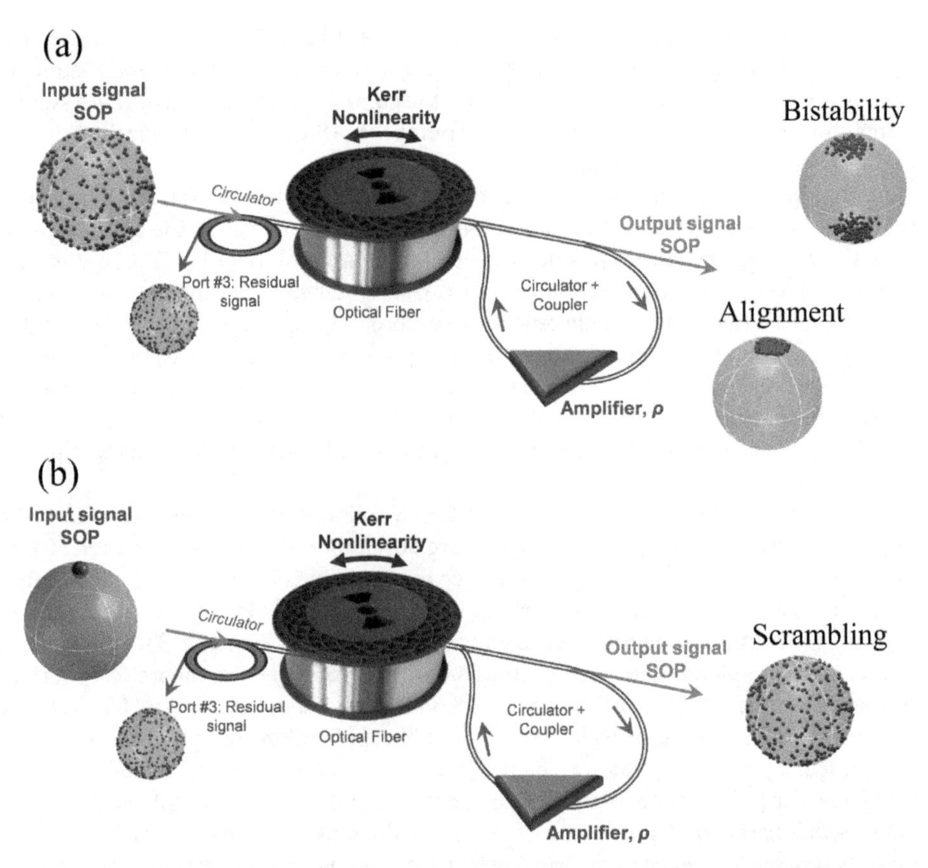

Figure 8.1 Principle of operation of the Omnipolarizer illustrating the three operational regimes. (a) In the bistable regime a depolarized input signal (input signal SOP uniformly distributed over the surface of the Poincaré sphere) is injected into the fiber. Thanks to the Kerr-nonlinear interaction with its backward replica in the fiber, the signal SOP undergoes a self-organization process so that at the fiber output it is strongly localized around the north and south pole. In the attraction regime there is only one pole of attraction, instead (here the north pole of the sphere), which is tunable by means of system parameters. (b) In the scrambling regime a fully polarized input signal (input signal SOP is a unique and fixed point over the Poincaré sphere) is injected, and the output signal turns out to be completely depolarized, i.e., its SOP is uniformly distributed all over the sphere.

The Omnipolarizer is characterized by its reflection coefficient ρ, which is defined as the ratio of power between backward and forward signals at the fiber end. Depending on the value of ρ three main operating regimes can be identified.

The first operating regime is the bistable regime, typically reached for $0.8 \leq \rho \leq 1$. A reflection coefficient below unity corresponds to a passive reflection at the fiber end, and can be thus achieved by means of a mirror, a FBG or a special coating. In this regime, whatever the polarization of the input forward signal is, two opposite poles of attraction for the output SOP are found on the Poincaré sphere. More precisely, the sign of the input signal ellipticity defines which of the two SOPs is obtained in output. Consequently, if an initially depolarized signal is injected into the Omnipolarizer, (namely, a signal quickly scrambled in time all over the Poincaré sphere), then the output signal

SOP distribution turns to be localized around the north and south poles of the Poincaré sphere at the fiber output (see Figure 8.1 (a)) [56].

The second operating regime is characterized by an amplification of the backward signal, corresponding to a reflective coefficient in the range of $1.2 \leq \rho \leq 2$. In this case, any arbitrarily polarized input signal is attracted towards a single output SOP, whose position over the Poincaré sphere can be controlled by means of a polarization controller implemented into the reflective loop [54, 55]. Therefore, an initially depolarized signal injected into the Omnipolarizer is aligned to a unique SOP at the device output: we thus define this functionality as the alignment regime.

Finally, for larger reflection coefficients, namely $\rho \gg 1$, a chaotic dynamics can be reached which leads to an all-optical scrambling of the output polarization [63]. This working principle is the opposite of the alignment regime: indeed in this case if a fully polarized signal is injected into the system, namely a unique SOP constant in time, then the corresponding output signal turns out to be scrambled all over the Poincaré sphere (see Figure 8.1 (b)).

8.3 Experimental Setup

In order to study the wide range of dynamics of the Omnipolarizer and especially the three main working regimes, the experimental setup displayed in Figure 8.2 has been implemented. The optical fiber inserted into the Omnipolarizer generally consists in a few km-long Non-Zero Dispersion-Shifted Fiber (NZDSF), typically between 5 and 6 kilometers of fiber. For most of our demonstrations, especially for telecom applications, the normal dispersion regime has been selected so as to avoid any modulational instability and soliton compression effect in the fiber under-test. At the input of the fiber, an optical circulator is inserted to inject the incident signal into the fiber as well as to reject the counter-propagating replica by means of the port-3. At the output end of the fiber, the backward signal is generated by means of an amplified reflective loop setup. This feedback loop is made of a second circulator, a 90:10 tap coupler to extract the output signal and a second EDFA whose gain can adjust the reflection coefficient of the system ρ and thus select in which regime the Omnipolarizer is operating. A fiber-based polarization controller is also implemented in order to adjust the rotation matrix of the loop and select the orthogonal SOP basis of attractors onto the Poincaré sphere. Depending on the proof-of-principle under study, the initial signal can be switched between a partially coherent source, a 10-Gbit/s return-to-zero (RZ) signal as well as a high bit rate 40-Gbit/s RZ signal. All the experimental studies have been realized in the telecom C band, around 1550 nm but the concept of self-organization of light SOP can be extended to other wavelengths.

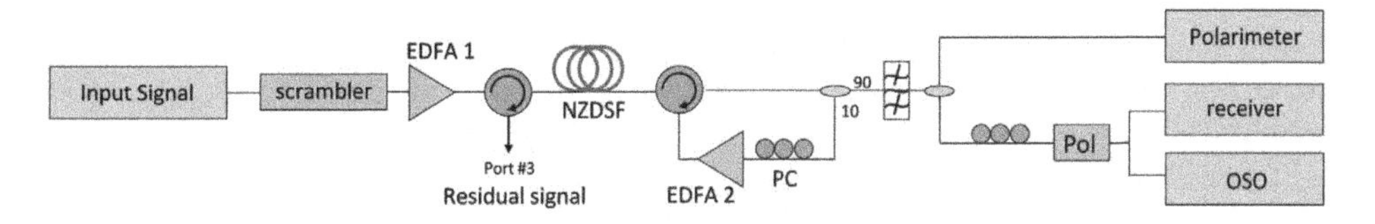

Figure 8.2 Experimental setup involved in our experiments.

For basic studies, the initial signal consists of a 100-GHz bandwidth, partially incoherent wave centered around 1550 nm. This partially incoherent signal is generated by spectral slicing of the spontaneous noise emission of an Erbium-based fiber source. An inline fiber polarizer is inserted in order to polarize this noise-based input signal. It is important to note that this spectral bandwidth, rather than a pure continuous wave is here implemented only to avoid any impairment due to the stimulated Brillouin backscattering in the optical fiber of the Omnipolarizer.

For telecom demonstrations, this partially incoherent source can be replaced by a return-to-zero (RZ) signal. Two different bit-rates have been exploited in our experiments. First of all, a 10-Gbit/s RZ signal is generated from a 10-GHz mode-locked fiber laser (MLFL) delivering 2.5-ps pulses at 1555 nm. A programmable liquid-crystal-based optical filter is used to temporarily broaden the initial pulses to 25-ps Gaussian pulses through a spectral slicing operation. This operation is required in order to limit the impact of chromatic dispersion and self-phase modulation in the fiber under test. A LiNbO3 intensity modulator driven by a $2^{31} - 1$ pseudo-random bit sequence finally modulates this resulting pulse train. For 40-Gbit/s experiments, the 10-GHz mode-locked fiber laser then operates at a central wavelength of 1562.4 nm. As in the previous case, to limit the impact of chromatic dispersion and self-phase modulation in the fiber, the pulses are also broadened to 7.5-ps. After encoding the data through the LiNbO3 intensity modulator, the 10-Gbit/s data stream is then multiplexed in the time domain in order to reach a global bit rate of 40-Gbit/s. At the input of the Omnipolarizer, a commercially available opto-electronic polarization controller/analyzer enables us to alternatively scramble or create a specific polarization trajectory onto the Poincaré sphere as well as to analyze the resulting SOP at the output of the system. For instance, a fast polarization fluctuation can be imposed on the incident signal at a repetition rate of 0.625 kHz so as to inject a fully depolarized signal into the system. For the optical memory demonstration reported in the next sections, an arbitrary sequence of 2-ms polarization spikes can be imprinted on the incident signal or alternatively an adiabatic variation of the input SOP so to catch the bistability properties of the system. Finally, whatever the resulting initial signal, the incident beam is amplified by means of an Erbium doped fiber amplifier before injection into the Omnipolarizer.

At the output of the system, for telecom purposes, the 10 and 40-Gbit/s signals are first filtered and characterized over the Poincaré sphere by means of a commercially available polarimeter. The intensity profile impairments as well as residual polarization fluctuations are then characterized in the temporal domain thanks to a systematic monitoring of the eye-diagram beyond an inline polarizer. Signal integrity is also tested by means of bit-error-rate (BER) measurements.

8.4 Theoretical Description

Let us indicate with $\mathbf{S} = [S_1, S_2, S_3]$ and $\mathbf{J} = [J_1, J_2, J_3]$ the Stokes vectors related to the forward and backward beams, respectively. Components of $\mathbf{S}$ and $\mathbf{J}$ are defined as follows: $S_1 = |\bar{E}_x|^2 - |\bar{E}_y|^2$; $S_2 = \bar{E}_x^* \bar{E}_y + \bar{E}_x \bar{E}_y^*$; $S_3 = -i\bar{E}_x \bar{E}_y^* + i\bar{E}_x^* \bar{E}_y$; $J_1 = |\bar{F}_x|^2 - |\bar{F}_y|^2$; $J_2 = \bar{F}_x^* \bar{F}_y + \bar{F}_x \bar{F}_y^*$; $J_3 = -i\bar{F}_x \bar{F}_y^* + i\bar{F}_x^* \bar{F}_y$, being $\bar{E}_x$ and $\bar{E}_y$ ($\bar{F}_x$ and $\bar{F}_y$) the fields obtained from a proper unitary transformation of the beams E_x and E_y (F_x and F_y), that are the

$x-$ and $y-$polarized envelopes of the forward (backward) electric field. Furthermore, we indicate with $\mathbf{s} = \mathbf{S}/|\mathbf{S}|$ and $\mathbf{j} = \mathbf{J}/|\mathbf{J}|$ the corresponding unitary Stokes vectors, which define the SOPs over the unitary Poincaré sphere. According to this notation the components s_3 and j_3 represent the forward and backward signal ellipticity, respectively, whereas magnitudes $|\mathbf{S}| = |\bar{E}_x|^2 + |\bar{E}_y|^2$ and $|\mathbf{J}| = |\bar{F}_x|^2 + |\bar{F}_y|^2$ indicate the total forward and backward power intensities, respectively.

The following coupled nonlinear partial differential equations describe the spatiotemporal dynamics of $\mathbf{S}$ and $\mathbf{J}$ in the fiber [54]:

$$\begin{aligned} c^{-1}\partial_t\mathbf{S} + \partial_z\mathbf{S} &= \mathbf{S} \times \gamma D\mathbf{J} - \alpha\mathbf{S} \\ c^{-1}\partial_t\mathbf{J} - \partial_z\mathbf{J} &= \mathbf{J} \times \gamma D\mathbf{S} - \alpha\mathbf{J} \end{aligned} \tag{8.1}$$

where $D = diag(-8/9, -8/9, 8/9)$ is a diagonal matrix, α represents the linear propagation loss coefficient of the fiber and c is the speed of light in the fiber. We point out that Eqs. (8.1) are valid in a fiber where birefringence varies randomly along the propagation direction, namely a telecommunication fiber. At this purpose, one should take in mind that the length scale over which a random variation of the birefringence axes occurs is typically few decade meters [42]: therefore, for fibers of 1 kilometer or more, that are those usually employed in our setup and for telecom applications, Eqs. (8.1) allow precise modeling of the polarization dynamics.

A boundary condition applies at the fiber end, which is $\mathbf{J}(z = L, t) = \rho R\mathbf{S}(z = L, t)$ [54], where R is a 3×3 rotation matrix modeling the polarization rotation induced by the reflective-loop, and $\rho = |\mathbf{J}(z = L, t)|/|\mathbf{S}(z = L, t)|$ is the power-ratio previously introduced. This boundary condition, along with Eqs. (8.1), univocally determines $\mathbf{S}(z, t)$ and $\mathbf{J}(z, t)$ once the input fields $\mathbf{S}(z = 0, t)$ and $\mathbf{J}(z = L, t)$ are fixed. Note that chromatic dispersion is neglected in our model: numerical simulations show indeed that it does not play a relevant role in the polarization dynamics under analysis.

From Eqs. (8.1) we immediately see that the intensity-shapes of $|\mathbf{S}|$ and $|\mathbf{J}|$ are preserved. Therefore $|\mathbf{S}|$ and $|\mathbf{J}|$ propagate unaltered except for linear propagation losses and a temporal shift, that is to say, $|\mathbf{S}(z, t)| = |\mathbf{S}(0, t - z/c)|exp(-\alpha z)$ and $|\mathbf{J}(z, t)| = |\mathbf{J}(0, t + z/c)|exp(\alpha z - \alpha L)$. Intensity-shape preservation is an important feature of the device under analysis, as it prevents input temporal polarization fluctuations to translate into large output intensity variations, which are referred as relative intensity noise (RIN) in literature. From this point of view the Omnipolarizer can be defined as a lossless polarizer, differently from typical linear polarizers which suffer of RIN.

The dynamics of the Omnipolarizer is related to the stability of the stationary solutions of Eqs. (8.1), which are found by dropping the time derivatives. Massive numerical simulations have shown that the stable stationary states are characterized by non-oscillatory behavior, namely, they are monotonic along the fiber length, while unstable states are oscillating [47]. Moreover, we have found that in the bistable and alignment regimes only these stable stationary states, here indicated with $\mathbf{s}_{stat}$ and $\mathbf{j}_{stat}$, play the role of natural attractors for $\mathbf{s}$ and $\mathbf{j}$.

This stability criterion is displayed in Figure 8.3. When we solve Eqs. (8.1) employing a stable-stationary state $\mathbf{s}_{stat}$ as input longitudinal field, namely, $\mathbf{s}(z, t = 0) = \mathbf{s}_{stat}(z)$, then this state is preserved in time due to the stability. Even if some additional noise $\mathbf{n}(z)$ is added to the input longitudinal field, that is $\mathbf{s}(z, t = 0) = \mathbf{s}_{stat}(z) + \mathbf{n}(z)$, however, the stable stationary state plays the role of attractor which "cancels" the noise component,

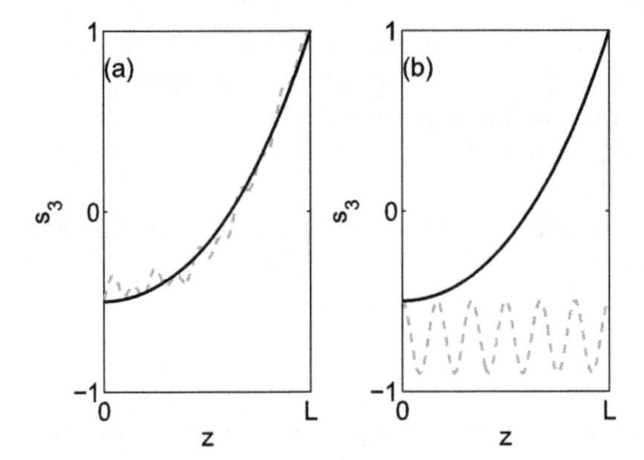

Figure 8.3 Illustration of the stability criterion. Here the s_3 component is represented, however, s_1 and s_2 follow a similar behavior. (a) A stable stationary state (non-oscillatory behavior along the fiber length) is employed as input longitudinal field with the addition of a weak noise component (dotted line). Despite the initial presence of noise, the system converges in time toward the original stationary-state (black solid line). (b) An unstable stationary state (oscillatory behavior along the fiber length) is employed as input longitudinal field (dotted line). The system leaves this unstable state and converges in time toward a stable state (black solid line).

so that the field relaxes in time toward the original stable state, that is $\mathbf{s}(z,t) = \mathbf{s}_{stat}(z)$ for $t \gg 1$ (Figure 8.3 (a)). On the contrary, if an unstable stationary state is used as input longitudinal field, then it is not preserved in time: the system gradually leaves this state and converges toward a stable stationary state with a non-oscillatory behavior (Figure 8.3 (b)).

The behavior of the Omnipolarizer in the aforementioned regimes is illustrated in Figure 8.4. Panels (a–c) show the spatiotemporal dynamics in the bistable regime ($0.8 \leq \rho \leq 1$). The evolution along the fiber of the component s_3 is depicted at three consecutive times, $t_A < L/c$, $t_B > L/c$ and $t_C \gg L/c$, where L/c defines the end-to-end propagation time in the fiber. For the sake of simplicity we consider here an input signal constant in time, i.e., $\mathbf{s}(z=0,t) = \mathbf{s}(z=0)$. At $t_A < L/c$ (panel a) the backward replica has not been generated, yet: the forward beam propagates thus unchanged into the fiber. On the other hand, at $t_B > L/c$ (panel b) the backward replica $\mathbf{J}$ has been generated and it nonlinearly interacts with $\mathbf{S}$, which makes $\mathbf{s}$ to gradually converge towards the stable state $\mathbf{s}_{stat}$: consequently $\mathbf{s}(z, t = t_C)$ and $\mathbf{s}_{stat}(z)$ practically coincide for a large time $t_C \gg L/c$ (see panel c).

Note that there is a unique stable stationary state $\mathbf{s}_{stat}(z)$ associated with a given input value $\mathbf{s}_{stat}(z=0)$: $\mathbf{s}$ converges therefore towards the stable stationary state such that $\mathbf{s}_{stat}(z=0) = \mathbf{s}(z=0)$. Moreover, the stable stationary states $\mathbf{s}_{stat}(z)$ strictly depends on ρ and N. It is interesting that in the bistable regime, that is $0.8 \leq \rho \leq 1$, and in the case of a strong system nonlinearity, that is $N \gg 1$, the output component $s_{3,stat}(z=L)$ is related to the input $s_{3,stat}(z=0)$ by the relation $s_{3,stat}(z=L) \approx sign[s_{3,stat}(z=0)]$. Consequently, being that $\mathbf{s}$ converges towards $\mathbf{s}_{stat}$, we find that $s_3(z=L,t) \approx sign[s_3(z=0)]$ for $t \gg 1$. That is to say: the output value of s_3 approaches $+1$ or -1, simply depending on its input value $s_3(z=0)$. This means that a weak variation on the input $s_3(z=0)$, causing a change of its sign, would lead to an output SOP which is switched to the orthogonal

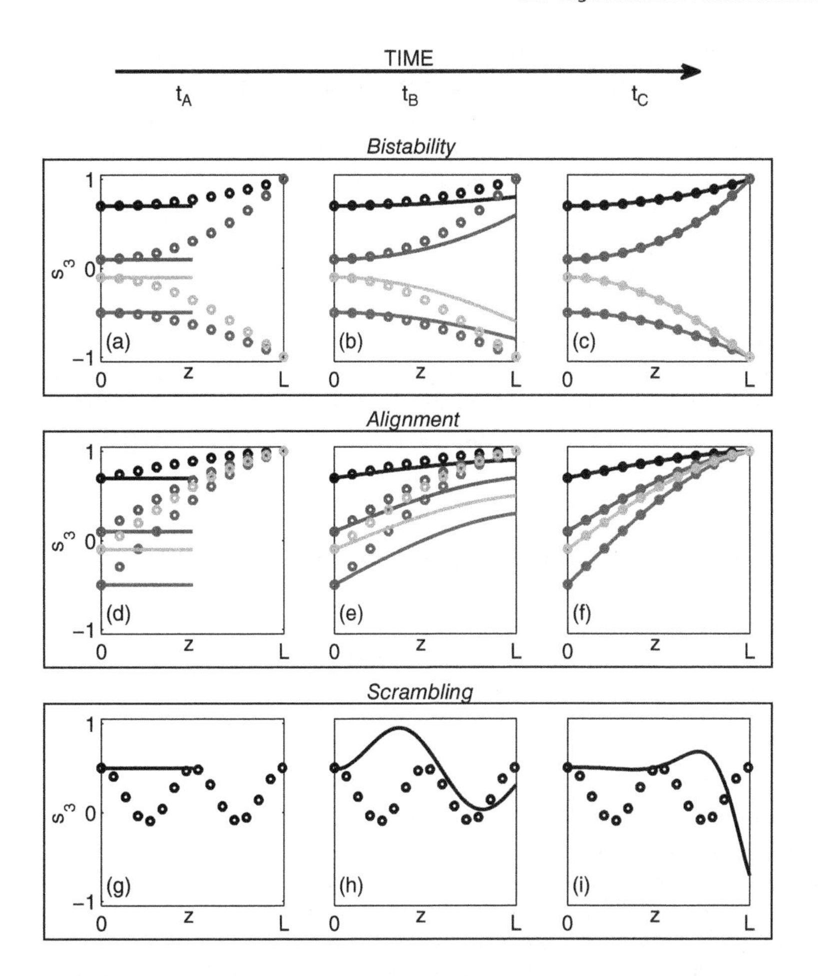

Figure 8.4 Spatial evolution along the fiber length of the normalized Stokes component s_3 (solid lines) in the 3 regimes of the Omnipolarizer. Three consecutive instants are represented: $t_A < L/c$ (left panels a,d,g); $t_B > L/c$ (central panels b,e,h); $t_C \gg L/c$ (right panels c,f,i). The corresponding stationary solutions ($s_{3,stat}$) are represented with circles. Panels a–c: bistable regime. Four different cases are represented that correspond to the input polarizations $s_3(z = 0) = +0.7$ (black lines and circles), $s_3(z = 0) = +0.1$ (blue lines and circles), $s_3(z = 0) = -0.1$ (green lines and circles), $s_3(z = 0) = -0.5$ (red lines and circles). In all cases s_3 asymptotically converges in time towards the corresponding stable stationary solution $s_{3,stat}$. Panels d–f: same as in panels a-c but in the case of the attraction regime. Here the attraction SOP at the fiber output is $s_{3,stat}(z = L) = +1$. However, it is freely tunable by means of R and ρ. Panels g–i: scrambling regime. The stationary solution is unstable and cannot thus play the role of attractor. Therefore, **s** does not converge towards $\mathbf{s}_{stat}$ but fluctuates in time without reaching a fixed state. Consequently, the output SOP $\mathbf{s}(z = L, t)$ turns out to be temporally scrambled. For a color version of this figure please see color plate section.

polarization. This phenomenon is illustrated in panels (a-c), where if $s_3(z = 0) = +0.1$ then $s_3(z = L, t = t_C) \approx +1$, whereas if $s_3(z = 0) = -0.1$ then $s_3(z = L, t = t_C) \approx -1$.

Panels (d–f) show the spatiotemporal dynamics in the alignment regime ($1.2 \leq \rho \leq 2$). By analogy with the bistable regime, we still observe an asymptotic convergence of **s** toward the related stable stationary state $\mathbf{s}_{stat}$ such that $\mathbf{s}_{stat}(z = 0) = \mathbf{s}(z = 0)$. However, due to a symmetry breaking induced by the rotation matrix R, one of the two poles of attraction becomes unstable. Consequently, whatever the input SOP is, we observe a sole attraction SOP at the fiber output; moreover, interestingly enough, such a SOP is freely tunable all over the Poincaré sphere by simply adjusting R and ρ [54], that is to say, the polarization controller and the backward power in our setup.

Panels (d–f) show the spatiotemporal dynamics in the third regime, which is the chaotic regime ($\rho \gg 1$). Here, contrary to the previous regimes, the stationary states turn to be always oscillating along the fiber length, and thus unstable, which prevents them from playing the role of attractors. Therefore, $\mathbf{s}$ does not converge toward a stable stationary solution, but it varies endlessly and chaotically in every point of the fiber without reaching a fixed state. As a result, the output signal SOP $\mathbf{s}(z = L, t)$ is scrambled and chaotic in time: as we will see in Section 8.5, this paves the way to a scrambler device where scrambling performances can be carefully adjusted by means of the backward power.

8.5 Bistability Regime and Related Applications

The stationary system associated with Eqs. (8.1) when losses are neglected admits as a constant of motion the vector $\mathbf{K} = \mathbf{S} - \gamma D\mathbf{J}$. If now we insert $\gamma D\mathbf{J} = \mathbf{S} - \mathbf{K}$ in the stationary system, we obtain the following equality:

$$\partial_z \mathbf{S} = \mathbf{K} \times \mathbf{S} \tag{8.2}$$

The normalized stationary state $\mathbf{s}_{stat}$ is computed as $\mathbf{s}_{stat} = \mathbf{S}/|\mathbf{S}|$, being $\mathbf{S}$ the solution of Eq. (8.2). We point out that this solution represents a rotation around the vector $\mathbf{K}$ corresponding to circles of spatial period $2\pi/|\mathbf{K}|$ [54].

In order to analyse the bistability regime, we initially assume that $\rho = 1$ and take into account that there is no rotation imposed on $\mathbf{J}$ at the fiber end, i.e., $\mathbf{R} = diag(1, 1, 1)$, then the boundary condition related to Eq. (8.2) in $z = L$ reads as $\mathbf{J}(z = L) = \mathbf{S}(z = L)$. The component K_3 is thus null and therefore the system Eq. (8.2) is symmetric with respect to rotations around the S_3 direction. This allows restricting the study to find the relation between the input $S_3(z = 0)$ and the output $S_3(z = L)$, and thus to trace the curve of $s_{3,stat}(z = L)$ as a function of $s_{3,stat}(z = 0)$. This curve is shown in Figure 8.5 (a), (b) in the color plate section when $N = 0.4$ (weak nonlinear regime) and $N = 4$ (strong nonlinear regime). In the first case we clearly get a quasi-straight line, namely, $S_3(z = L) \approx S_3(z = 0)$: indeed, the weak nonlinearity prevents a strong coupling between $\mathbf{S}$ and $\mathbf{J}$, which thus remain nearly unaltered along z. On the other hand, the more N increases, the more the curve is deformed until it exhibits a vertical tangent when $N = \pi/2$. For $N > \pi/2$ the curve becomes multivalued, so that for a given input $s_{3,stat}(z = 0)$ several outputs $s_{3,stat}(z = L)$ are found, each one corresponding to a particular solution $S_3(z)$ of Eq. (8.2) associated with the initial condition $S_3(z = 0)$. However, only one of these stationary solutions is stable. More precisely, if $-1 \leq s_{3,stat}(z = 0) \leq 0$ then only the lower-valued solution $s_{3,stat}(z = L)$ (black solid line in Figure 8.5 (b) in the color plate section) is the one related to a stationary state which is not oscillating along the fiber length, and is thus stable [47]. On the contrary, when $0 \leq s_{3,stat}(z = 0) \leq 1$ then only the higher-valued solution $s_{3,stat}(z = L)$ (red solid line in Figure 8.5 (b)) is the one related to a stable stationary state.

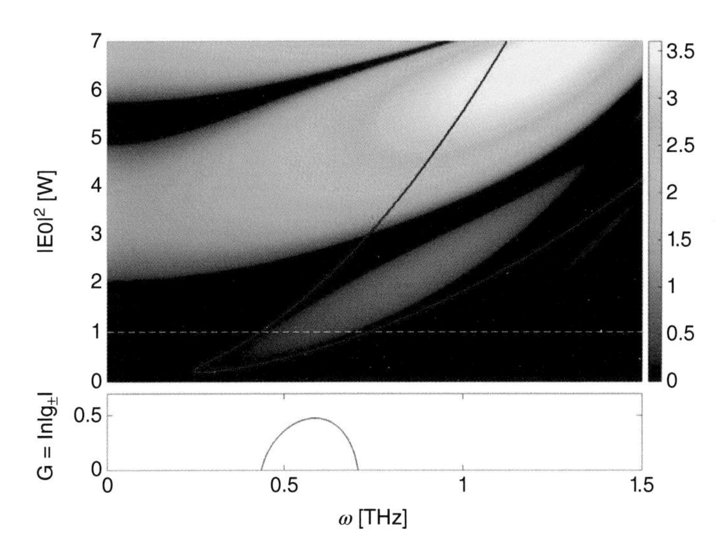

Figure 1.14 Parametric instability tongues of the Ikeda map for anomalous dispersion. The red contour shows the predicted range of modulational instability for the LLE. Below a cross-section corresponding to the dashed line is shown with the MI growth rate. Parameters $\alpha = \theta = 0.13$, $\beta_2 = -20$ ps^2 km^{-1}, $\gamma = 1.8$ W^{-1} km^{-1}, $L = 380$ m and $\delta_0 = 0$.

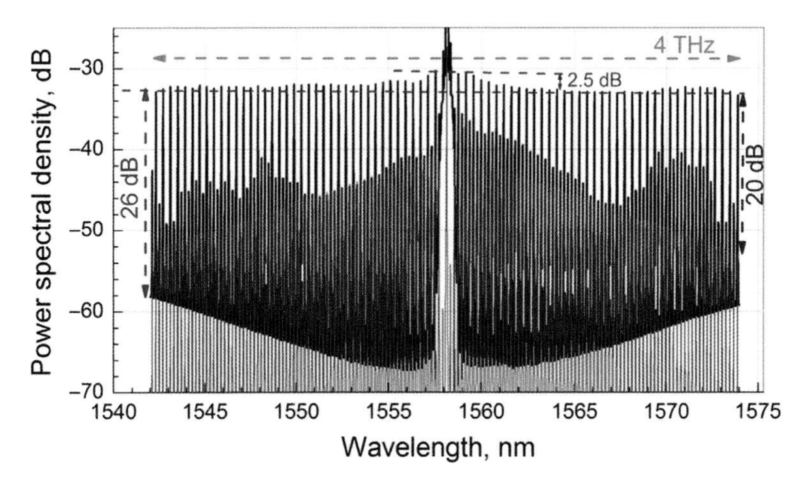

Figure 2.19 Spectra after the PSA for optimal pump power (black) and pump off (red) showing various key parameters. *Source*: Adapted from Slavik 2012 [73].

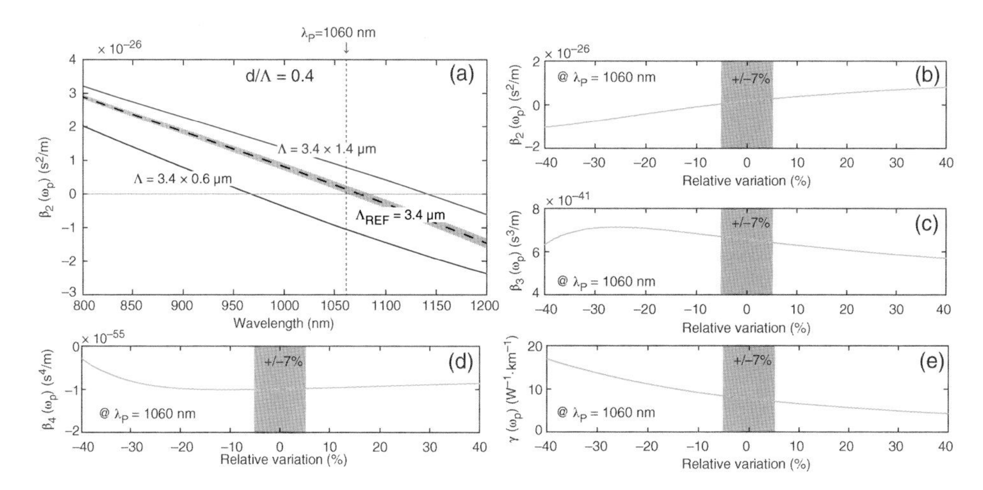

Figure 4.3 (a) GVD evolution versus wavelength for different pitch values, $\Lambda_{REF} = 3.4$ μm (dashed black lines), $\Lambda_{MAX} = 4.76$ μm (solid blue line) and $\Lambda_{min} = 2.04$ μm (solid red line). The gray area delimits a variation of ±7% of the pitch around the reference value. (b)-(e) Evolutions of $\beta_{2,3,4}$ and γ versus the normalized value of the pitch $[(\Lambda - \Lambda_{REF})/\Lambda_{REF}]$.

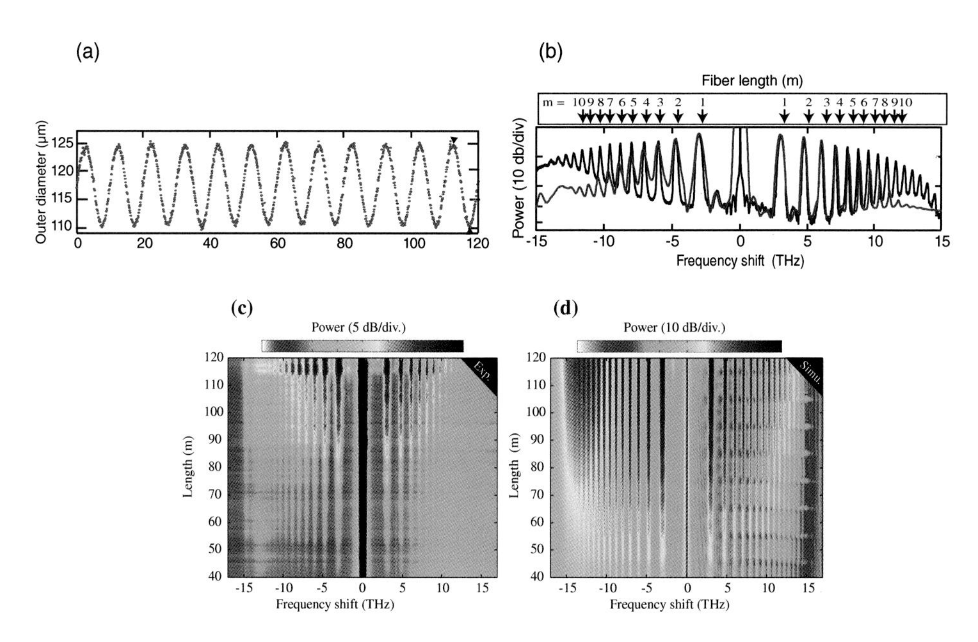

Figure 4.7 (a) Evolution of the outer diameter versus fiber length measured during drawing. (b) Output spectra from experiments (red line) and from numerical simulations (black line). (c) Experimental and (d) numerical dynamics of the structured spectrum formation in the DOF.

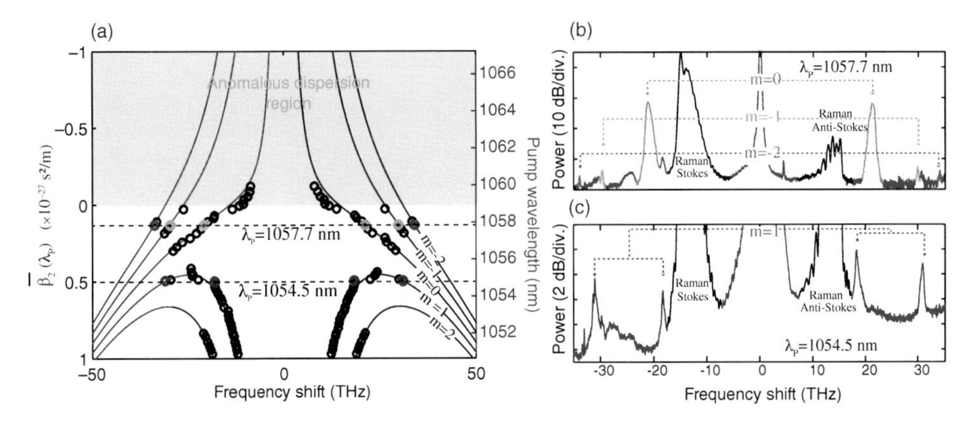

Figure 4.8 (a) QPM curves calculated from Eq. (4.18) (solid line) and measurement of MI sideband frequencies done by tuning the pump wavelength (markers). Colored markers highlight frequencies appearing in the experimental spectra shown in (b) and (c).

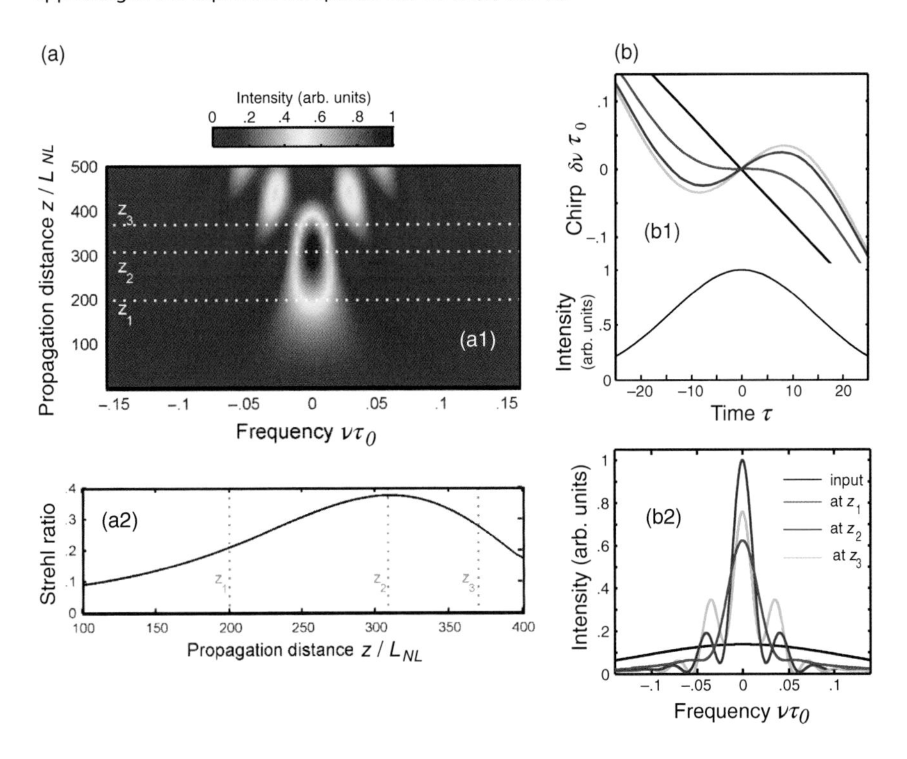

Figure 5.8 Spectral pulse compression in a purely nonlinear fiber, as obtained by numerical integration of the NLS equation. (a) Longitudinal evolution of the spectral intensity profile (subplot 1) and the Strehl ratio (suplot 2) for an initial Gaussian pulse with a stretching factor of 20. (b) Temporal intensity an chirp profiles (subplot 1) and spectral intensity profile (subplot 2) taken at different propagation distances in the fiber: the point z_1 of chirp cancellation near the pulse center (blue), the point z_2 of maximum Strehl ratio (red), and the point z_3 of maximum spectral compression factor (green). Also shown are the initial pulse profiles (black). More details can be found in [105].
Source: Finot 2016 [105].

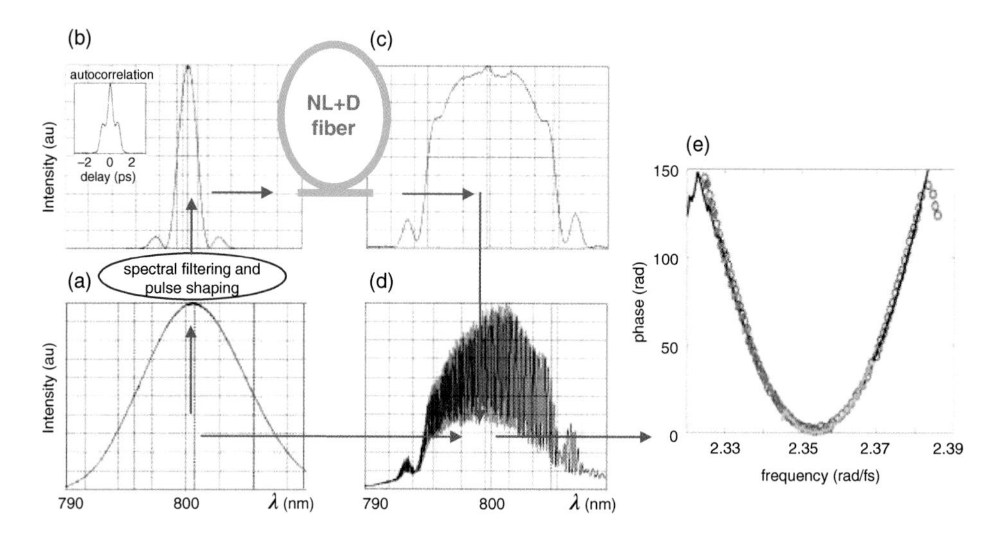

Figure 6.4 (a–d) Schematic of the experiment given by the spectrograms recorded at different locations in the set-up, and (e) spectral phases of NL-D similaritons generated from various inputs; single pulse (black), close double pulses (red o) and distant double (blue x) pulses in comparison with the one for pure dispersive propagation of a single pulse (yellow).

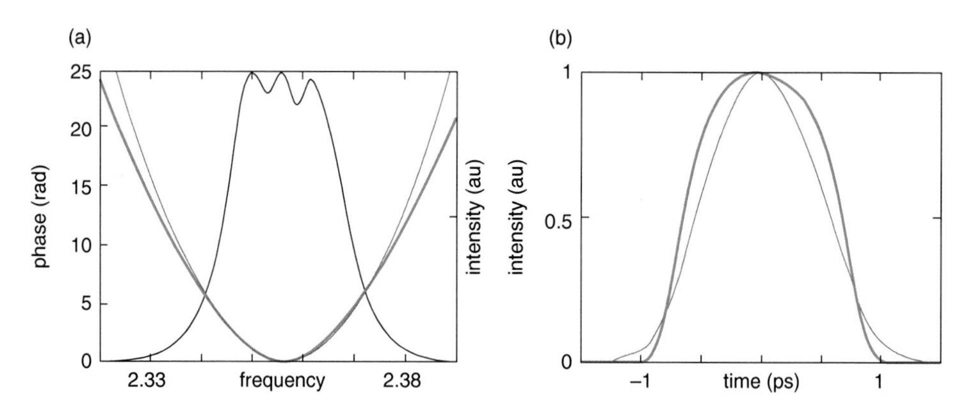

Figure 6.6 (a) Measured spectrum of NL-D flattop pulses with its retrieved spectral phase (black solid line). The pink and blue lines are for parabolic and high-order polynomial fits, respectively. (b) Reconstructed temporal profile of the flattop pulse (blue) in comparison with one reconstructed by the use of parabolic spectral phase (pink).

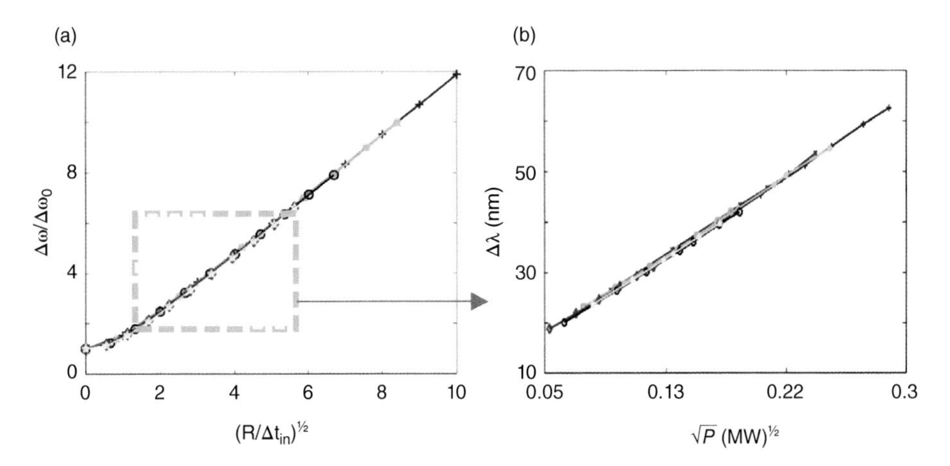

Figure 6.7 (a) Simulations: spectral broadening versus $R^{1/2}$; (b) Experiments: similariton bandwidth $\Delta\lambda$ versus $\sqrt{P}$. Blue lines correspond to transform-limited input pulses of 100 fs duration, red and cyan traces to 140 fs pulses, green and black traces to 225 fs pulses, magenta and yellow lines to 320 fs pulses. Long input pulses were obtained by broadening of the laser pulse in a dispersive line with either normal (red, green and magenta plots) or anomalous dispersion (cyan, black, and yellow plots).

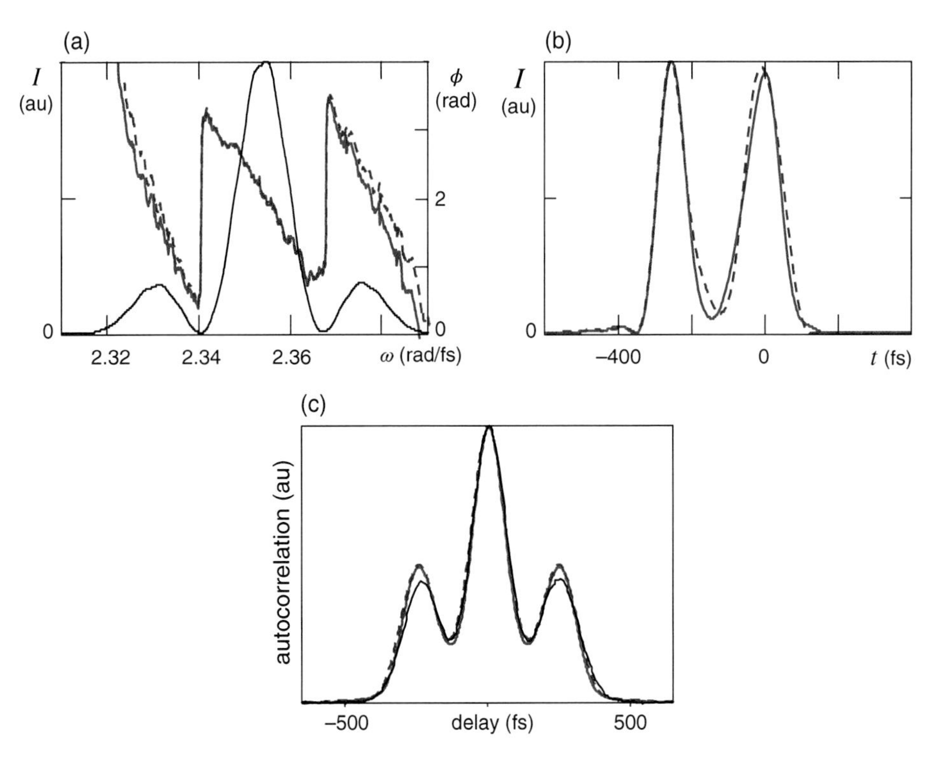

Figure 6.27 SI-reconstruction of double-peak pulse (shaped by means of a 130-μm thick glass): (a) retrieved spectral phase (color) and measured spectrum (black); (b) reconstructed pulse; and (c) measured autocorrelation of signal (black) in comparison with ones for reconstructed pulse (color). To illustrate the sensitivity-stability of retrieving, together with blue curves for $\tilde{C}_f = 2.1 \times 10^4$ fs^2, red curves for $\tilde{C}_f = 1.995 \times 10^4$ fs^2 (5% difference) is shown.

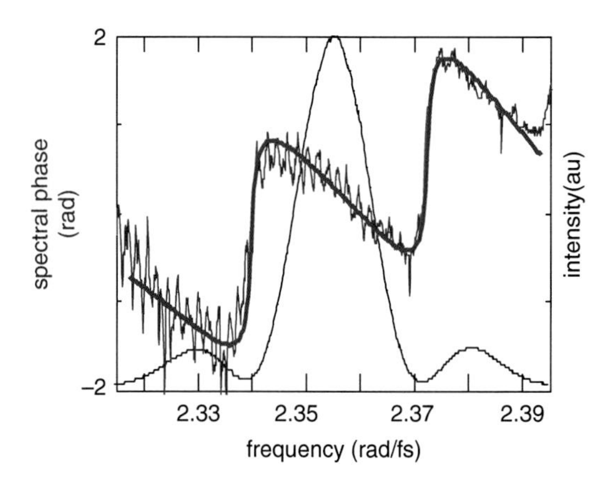

Figure 6.28 The SI-retrieved spectral phase (red) of an "Ararat"-shaped two-peak signal in the range of its spectrum (black). The theoretical curve for spectral phase for such a two-peak pulse is shown also (blue).

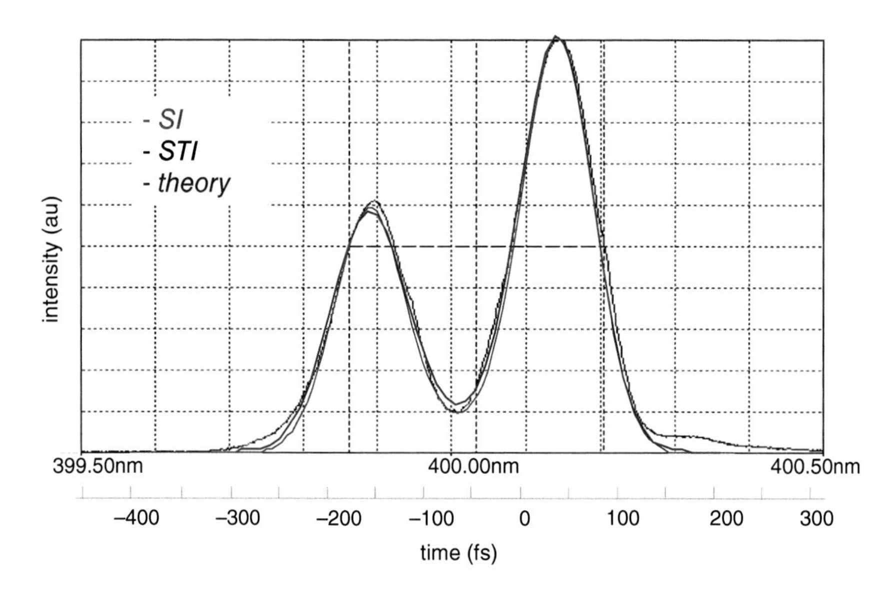

Figure 6.29 Comparison of similaritonic SI and STI measurements for an "Ararat"-shaped two-peak signal (with the 3:5 peak power ratio): SI-reconstructed pulse (red) in comparison with the spectral image of the pulse (black) obtained by STI based on similariton-induced time lens, and the theoretical profile (blue).

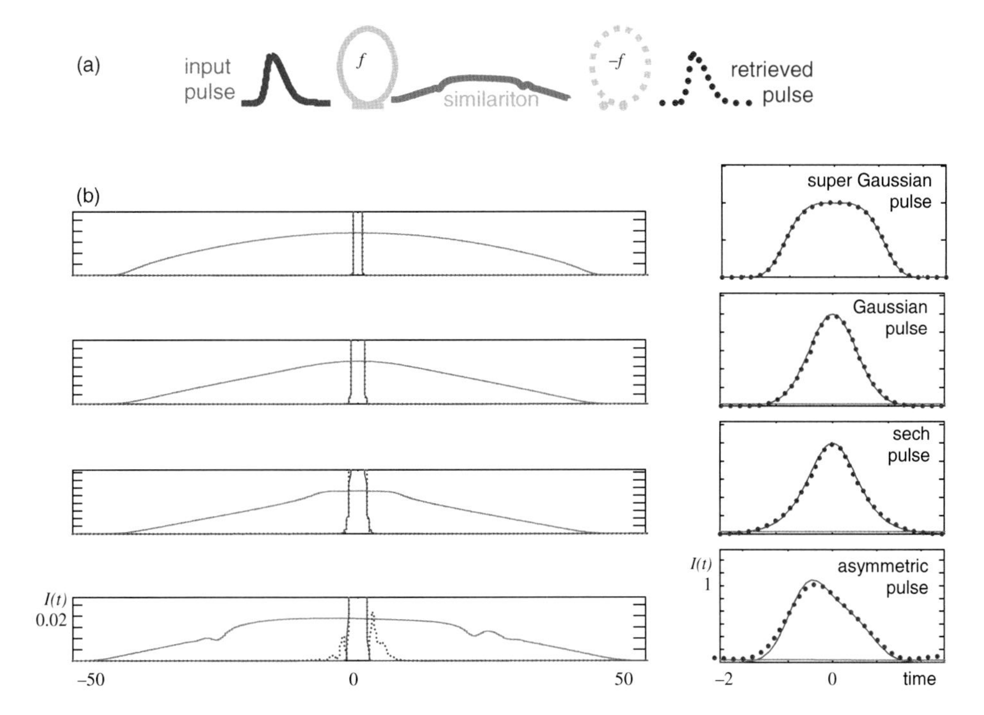

Figure 6.31 Illustration of initial studies for solution of reverse problem of NL-D similariton generation: (a) principle, and (b) results for various pulse shapes. Input (blue), similariton (green), and retrieved input (red dashed) pulses are shown. Simulation parameters: $R = L_D/L_{NL} = 50$, $L_f = 5L_D$.

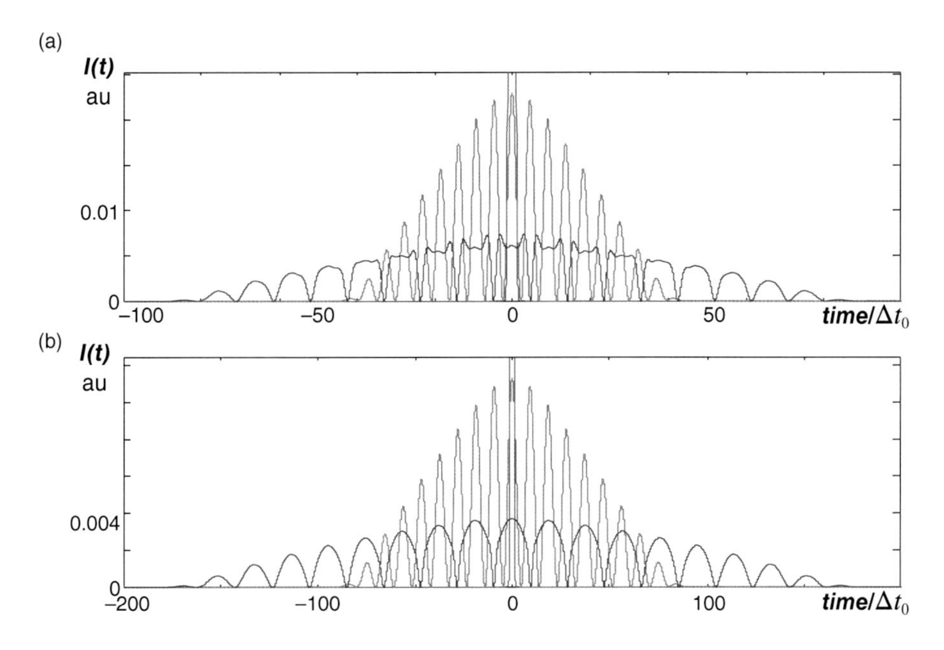

Figure 6.32 Two examples of a train self-interacted in fiber. The intensity profiles of a train shaped by superposition of NL-similariton with its replica (green), and ones for self-interacted train in the fiber (red) are shown (red). Simulation parameters: $T = 2.5$, $\mu = 1$, $R = 50$, $L_f/L_D = 5$ (a), and $L_f/L_D = 10$ (b).

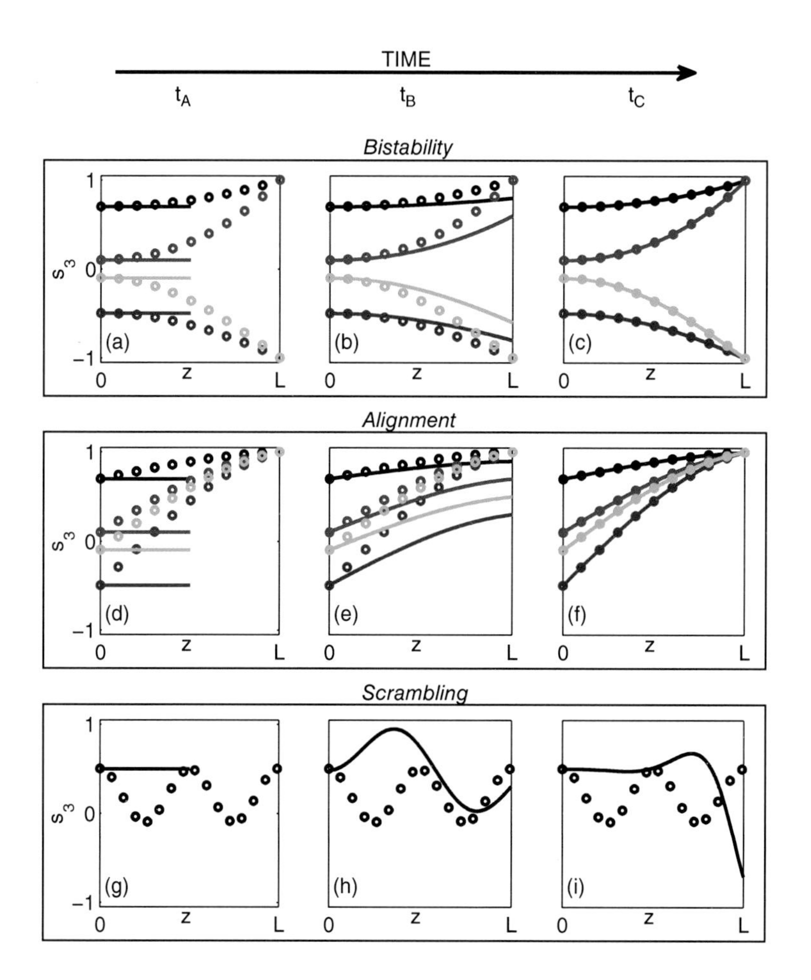

Figure 8.4 Spatial evolution along the fiber length of the normalized Stokes component s_3 (solid lines) in the 3 regimes of the Omnipolarizer. Three consecutive instants are represented: $t_A < L/c$ (left panels a,d,g); $t_B > L/c$ (central panels b,e,h); $t_C \gg L/c$ (right panels c,f,i). The corresponding stationary solutions ($s_{3,stat}$) are represented with circles. Panels a–c: bistable regime. Four different cases are represented that correspond to the input polarizations $s_3(z = 0) = +0.7$ (black lines and circles), $s_3(z = 0) = +0.1$ (blue lines and circles), $s_3(z = 0) = -0.1$ (green lines and circles), $s_3(z = 0) = -0.5$ (red lines and circles). In all cases s_3 asymptotically converges in time towards the corresponding stable stationary solution $s_{3,stat}$. Panels d–f: same as in panels a-c but in the case of the attraction regime. Here the attraction SOP at the fiber output is $s_{3,stat}(z = L) = +1$. However, it is freely tunable by means of R and ρ. Panels g–i: scrambling regime. The stationary solution is unstable and cannot thus play the role of attractor. Therefore, $\mathbf{s}$ does not converge towards $\mathbf{s}_{stat}$ but fluctuates in time without reaching a fixed state. Consequently, the output SOP $\mathbf{s}(z = L, t)$ turns out to be temporally scrambled.

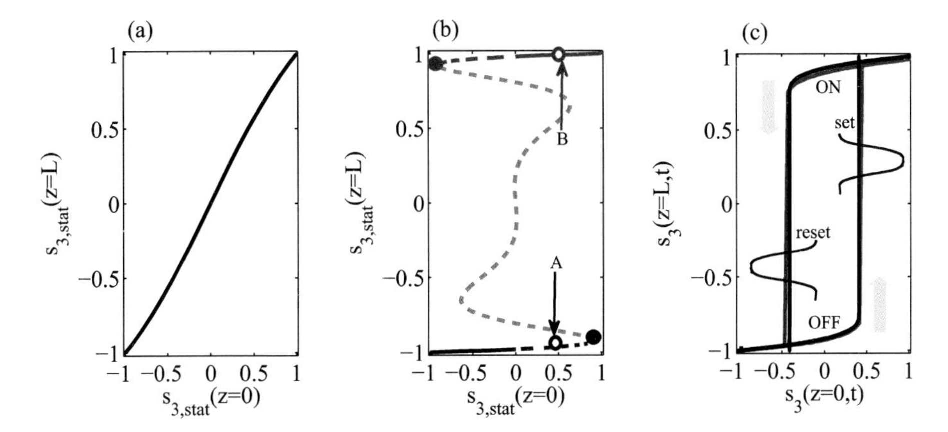

Figure 8.5 (a): $s_{3,stat}(z = L)$ versus $s_{3,stat}(z = 0)$ for the stationary solutions of Eq. (8.1) when $\rho = 1$, $R = diag([1, 1, 1])$, null losses and $N = 0.4$ (weak nonlinear regime). All points of the curve correspond to stable stationary solutions. (b): same as in (a) but when N=4 (high nonlinear regime). Solid black and red lines correspond to the stable states such that $s_{3,stat}(z = L) \approx -1$ and $s_{3,stat}(z = L) \approx +1$, respectively. Dashed black and red lines correspond to the metastable states, whereas the dashed green line corresponds to the set of unstable states. The black and red spots located in $s_3(z = 0) = \pm 0.85$ identify the extremities of the line of unstable states. The empty circle A and B indicate respectively the metastable and the stable state corresponding to $s_{3,stat}(z = 0) = 0.5$. (c): Hysteresis cycle recorded experimentally at the output of the Omnipolarizer (solid black line) when the input wave consists in a 100-GHz polarized incoherent signal with an average power of 570 mW. The reflection coefficient ρ is nearly 0.9. The component s3(z=0,t) varies in time from −1 to +1 and vice versa with a ramping time of 200 ms. The solid red line represents the numerical solution of Eqs. (8.1) when the experimental parameters are introduced in the simulation.

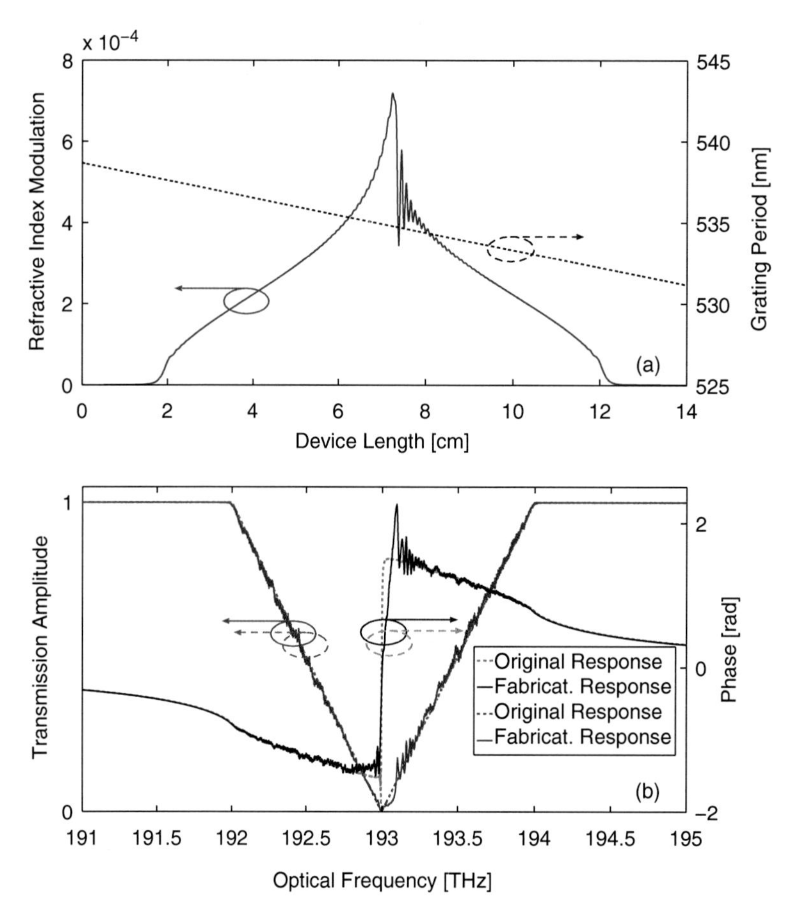

Figure 9.5 First-order optical differentiator based on an FBG in transmission: (a) Apodization (red solid line) and period (black dashed line) profiles considering realistic spatial resolution (limited to 0.3 mm) and linearly chirped phase mask; (b) Specified transmission spectral response, amplitude (dotted blue line) and phase (dotted green line) compared to fabrication-constrained transmission spectral response, amplitude (solid red line) and phase (solid black line). *Source*: María 2013 [89].

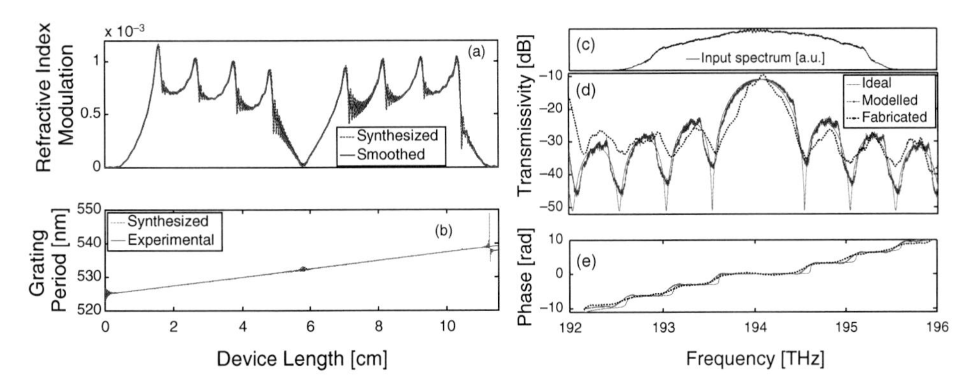

Figure 9.8 Flat-top pulse shaper based on an FBG in transmission: Obtained apodization profile (a) and period (b) from synthesis (blue line) and smooth profiles adapted to fabrication constrains (red line); (c) Input signal spectrum; (d) tranmissivity and (e) transmission phase response: ideal (blue line), simulated considering fabrication constraints (red line) and measured from fabricated device (dotted black line). *Source*: María 2013 [91].

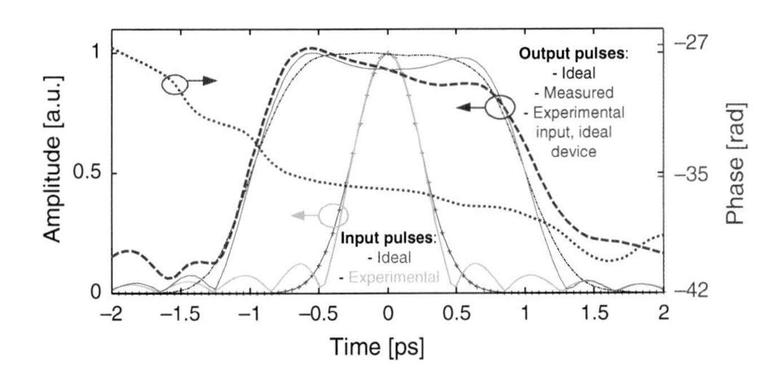

Figure 9.9 Flat-top pulse shaper based on an FBG in transmission: Temporal input response: ideal (blue line) and employed in the experiment (green line); Temporal output: ideal (black line) and experimentally measured (red line). *Source*: María 2013 [91].

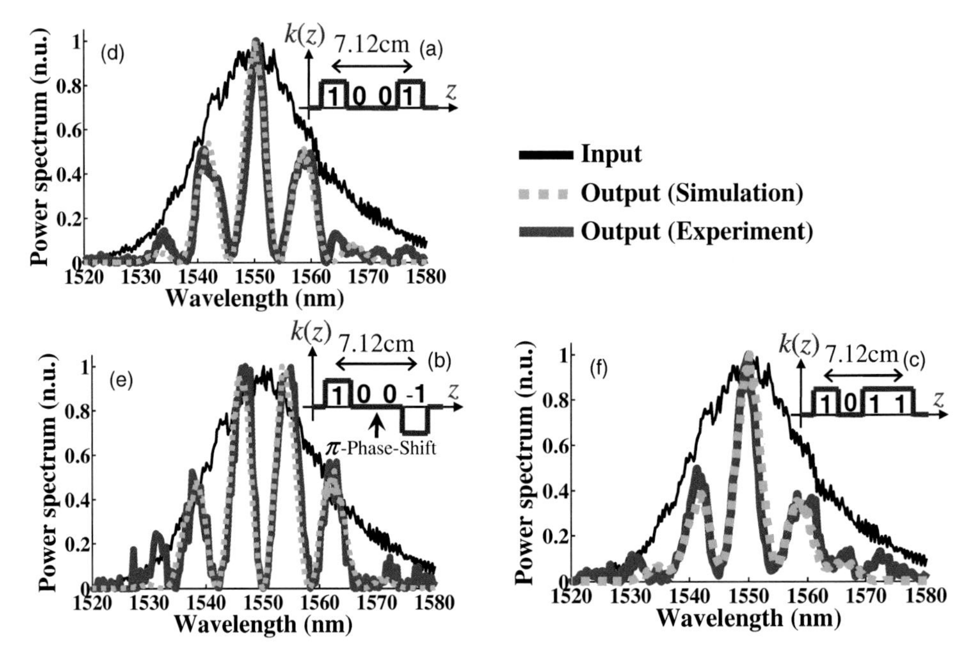

Figure 9.15 (a–c) Fabricated fiber LPG designs for generation of 4-bit data streams, i.e. ″1″0″0″1, ″1″0″0″−1″ and ″1″0″1″1″, respectively, with a target speed of ~3.5-Tbit/s. (d–f) The corresponding experimental spectrum measurements of the femtosecond optical pulse from the OPO laser before (solid black curves) and after (solid blue curves) propagation through the fabricated LPGs, compared with the simulated linear spectral responses of the LPGs (dotted green curves) to the same input OPO laser pulse spectrum. The spectra are represented in normalized units (n.u.). *Source*: Ashrafi 2013 [101].

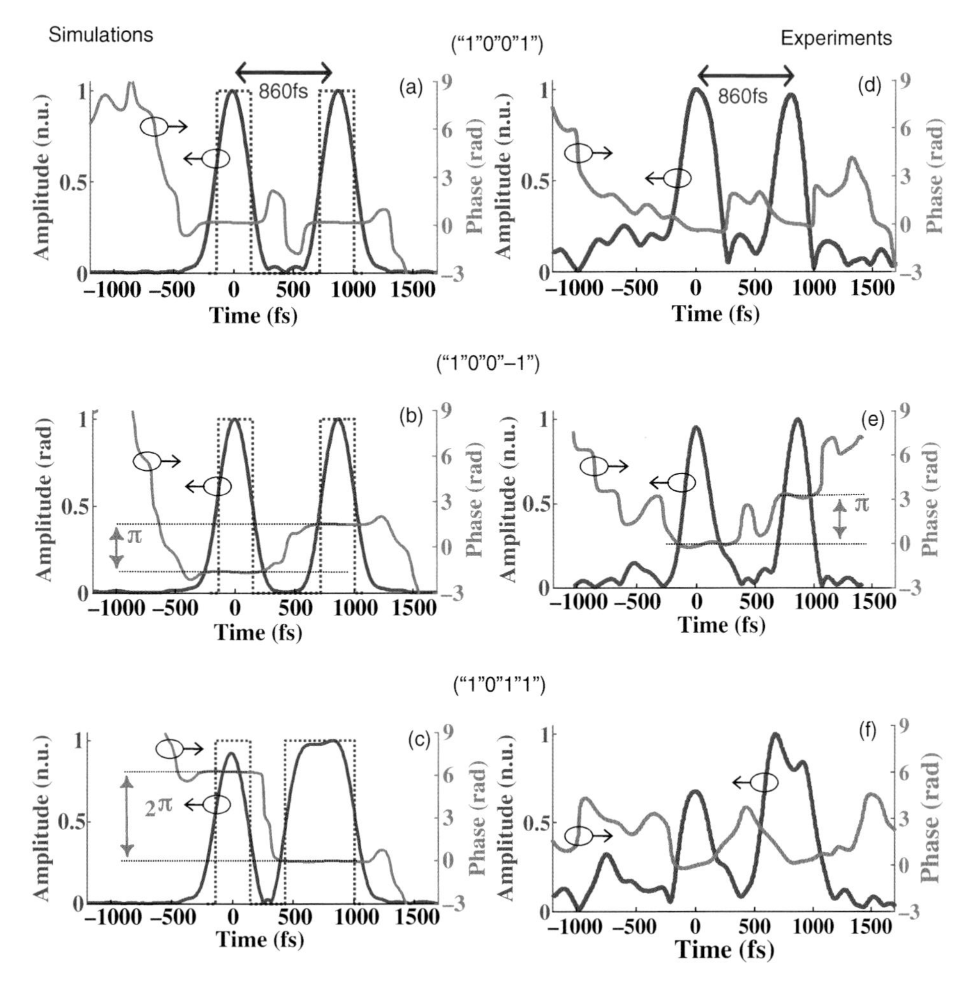

Figure 9.16 (a–c) Simulation results for the temporal amplitude (solid blue curves) and phase (solid green curves) responses of the designed LPGs to the OPO laser pulse data. The dotted, red curves in (a–c) represent the ideal time-domain impulse response amplitudes, obtained by using the predicted superluminal space-to-time scaling of the ideal space-domain profiles. (d–f) Measured output time-domain amplitude (solid blue curves) and phase (solid green curves) responses of the fabricated LPGs. *Source*: Ashrafi 2013 [101].

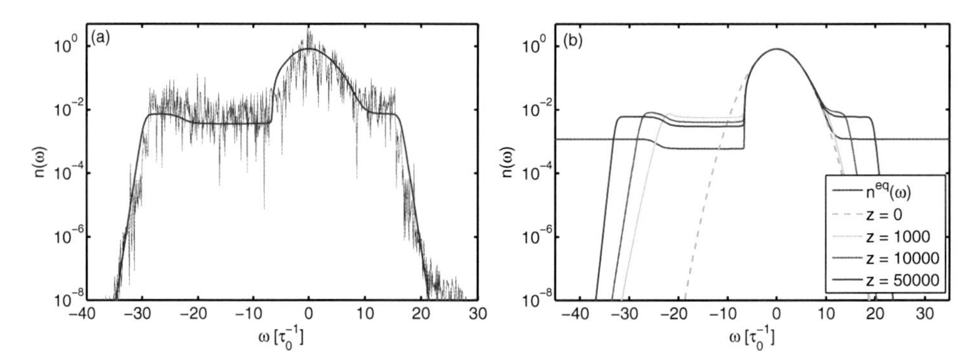

Figure 12.6 Anomalous thermalization of incoherent waves: (a) Spectral evolution obtained by integrating numerically the NLSE with third-order dispersion (blue) and the corresponding WT kinetic equation (red) at $z = 20000$ for $\tilde{\alpha} = 0.05$ (a). (b) Numerical simulations of the WT kinetic equation showing the spectral profile $n(z, \omega)$ at different propagation lengths z: a constant spectral pedestal emerges in the tails of the spectrum ($\tilde{\alpha} = 0.05$). The spectrum slowly relaxes toward the equilibrium state $n^{loc}(\omega)$ given by Eq. (12.10) (blue). *Source*: Michel 2011 [88].

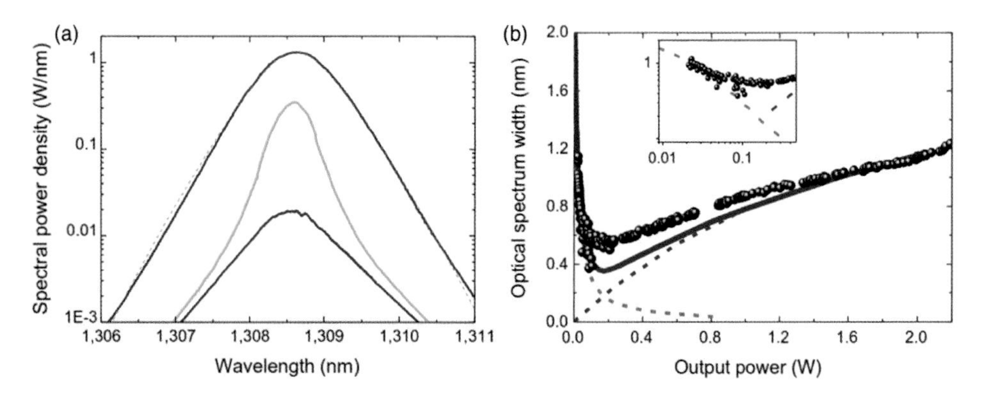

Figure 12.9 Nonlinear kinetic description of the random fiber laser optical spectrum. (a) Optical spectrum measured experimentally: near the generation threshold (blue curve, laser power = 0.025W), slightly above the generation threshold (green curve, 0.2W) and well above the generation threshold (red curve, 1.5W). The optical spectrum predicted by the local wave kinetic equation, for laser power 1.5W is shown by dashed red line. (b) Spectrum width as a function of the laser's output power in theory and experiment. Experimental data are shown by black circles. The prediction for the spectrum broadening from the nonlinear kinetic theory based on the local wave kinetic equation (blue dashed line). The prediction for the spectral narrowing from the modified linear kinetic Schawlow-Townes theory (dashed green line). The red line denotes the sum of nonlinear and linear contributions. The inset shows the spectral narrowing near the threshold in log-scale. For more details see [123]. *Source*: Churkin 2015 [123]. Reproduced with permission of Nature Publishing Group.

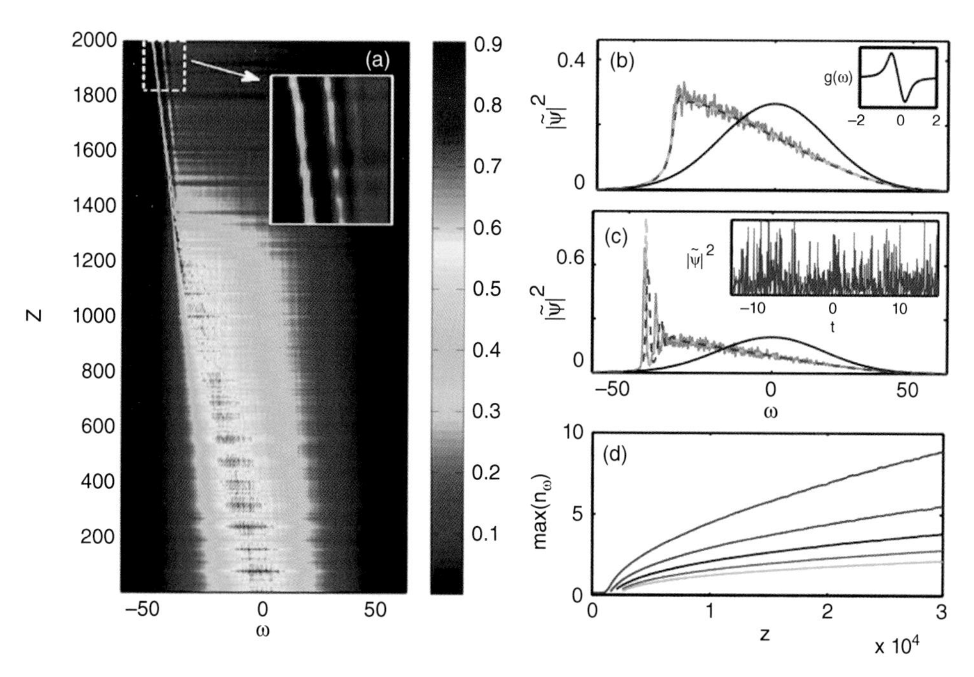

Figure 12.12 Incoherent dispersive shock waves with a Raman-like response function: (a) Numerical simulation of the NLS Eq. (12.16): The stochastic spectrum $|\tilde{\psi}|^2(\omega, z)$ develops an incoherent shock at $z \simeq 1200L_{nl}$ ($\tau_R = 3\tau_0, \eta = 1$). Snapshots at $z = 1040L_{nl}$ (b), $z = 1400L_{nl}$ (c): NLS (12.16) (gray) is compared with WT Langmuir Eq. (12.18) (green), singular kinetic equation (Eq. (12.21)) (dashed line), and initial condition (solid black). (d) First five maxima of n_ω vs z in the long-term post-shock dynamics: the spectral peaks keep evolving, revealing the non-solitonic nature of the incoherent dispersive shock wave. Insets: (b) gain spectrum $g(\omega)$, note that $\Delta\omega_g$ is much smaller than the initial spectral bandwidth of the wave [black line in (b)]. (c) corresponding temporal profile $|\psi(t)|^2$ showing the incoherent wave with stationary statistics. *Source*: Garnier 2013 [49]. Reproduced with permission of American Physical Society.

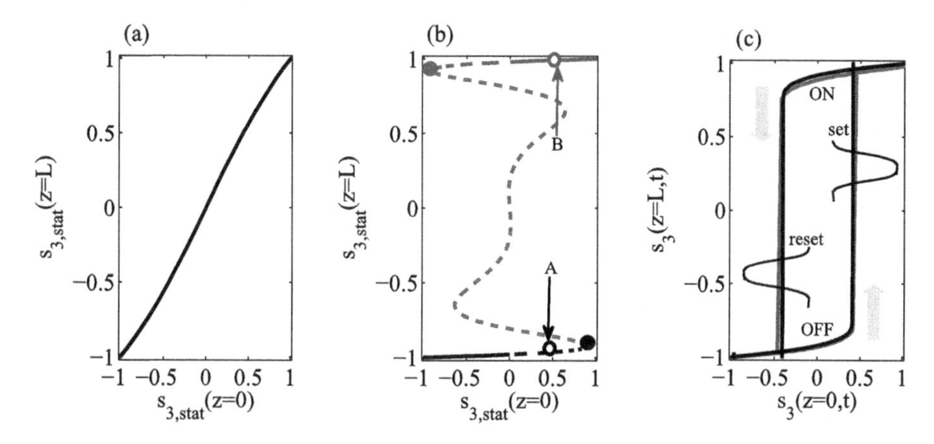

Figure 8.5 (a): $s_{3,stat}(z = L)$ versus $s_{3,stat}(z = 0)$ for the stationary solutions of Eq. (8.1) when $\rho = 1$, $R = diag([1, 1, 1])$, null losses and $N = 0.4$ (weak nonlinear regime). All points of the curve correspond to stable stationary solutions. (b): same as in (a) but when N=4 (high nonlinear regime). Solid black and red lines correspond to the stable states such that $s_{3,stat}(z = L) \approx -1$ and $s_{3,stat}(z = L) \approx +1$, respectively. Dashed black and red lines correspond to the metastable states, whereas the dashed green line corresponds to the set of unstable states. The black and red spots located in $s_3(z = 0) = \pm 0.85$ identify the extremities of the line of unstable states. The empty circle A and B indicate respectively the metastable and the stable state corresponding to $s_{3,stat}(z = 0) = 0.5$. (c): Hysteresis cycle recorded experimentally at the output of the Omnipolarizer (solid black line) when the input wave consists in a 100-GHz polarized incoherent signal with an average power of 570 mW. The reflection coefficient ρ is nearly 0.9. The component s3(z=0,t) varies in time from −1 to +1 and vice versa with a ramping time of 200 ms. The solid red line represents the numerical solution of Eqs. (8.1) when the experimental parameters are introduced in the simulation. For a color version of this figure please see color plate section.

Note in Figure 8.5 (b) that the black and red solid lines correspond to $s_3(z = L) \approx -1$ and $s_3(z = L) \approx +1$, respectively. This confirms what was anticipated in the previous section: in a highly nonlinear regime, here characterized by $N = 4$, the stable stationary states are such that $s_{3,stat}(z = L) \approx sign[s_{3,stat}(z = 0)]$ [54]. This means that if a constant-in-time input polarization $\mathbf{s}(z = 0)$ that is in the north (south) hemisphere of the Poincaré sphere is injected into the fiber, that is to say $s_3(z = 0) > 0$ ($s_3(z = 0) < 0$), then the corresponding output $\mathbf{s}(z = L, t)$ will relax towards the north (south) pole of the sphere, that is $s_3(z = L, t) \approx +1$ ($s_3(z = L, t) \approx -1$) for $t \gg 1$.

Along with the stable and the unstable states previously introduced, another class of stationary solutions exists that are called metastable. These states play an important role in the system dynamics when the input SOP $\mathbf{s}(z = 0, t)$ is time varying, and are indicated by the black and red dashed lines in Figure 8.5 (b). Basically, a metastable state is stable under small perturbations and can be reached starting from a stable state. Let us consider, for example, an input field $\mathbf{s}(z)$ corresponding to a stable stationary state of the system and whose input $s_3(z = 0)$ is such that $-1 \leq s_3(z = 0) \leq 0$, so that the output $s_3(z = L) \approx -1$ (black solid line in Figure 8.5 (b)). If now we slowly vary in time the input SOP $s_3(z = 0, t)$ towards +1, after having covered the black solid line of stable states we enter in the black dashed line of metastable states which starts in correspondence of $s_{3,stat}(z = 0) = 0$. Moving further, we find the first vertical tangent, which is indicated by the black spot in Figure 8.5 (b) and that corresponds to the first unstable state at $s_{3,stat}(z = 0) = 0.85$. For an input $0 \leq s_3(z = 0, t) \leq 0.85$ the output $s_3(z = L, t)$ abruptly jumps toward the opposite branch of stable states, that is the red solid line in Figure 8.5 (b).

The switching value for $s_3(z = 0, t)$ strictly depends on how quickly it varies: the more its transition towards +1 is slow, the larger is the switching value, but in any case the upper limit is represented by the value corresponding to the first unstable state, which is $s_3(z = 0, t) = 0.85$ in the case under analysis. As an example, in Figure 8.5 (b) the switching value is $s_3(z = 0, t) = 0.5$ and gives rise to the jump of the output $s_3(z = L, t)$ from the point A, which is close to −1 and is related to a metastable state, to the point B, which is close to +1 and corresponds to a stable stationary state.

Clearly, a similar behavior occurs if the input $s_3(z = 0, t)$ moves from the red solid line of stable states toward −1. In this case, after having covered the whole red solid line of stable states, we enter in the red dashed line of metastable state and we find the vertical tangent indicated by the red spot in correspondence of $s_{3,stat}(z = 0) = -0.85$. For a switching value $-0.85 \leq s_3(z = 0, t) \leq 0$ the output $s_3(z = L, t)$ suddenly jumps toward the opposite branch of stable states, that is the black solid line in Figure 8.5 (b).

The abrupt transitions between the upper and the lower branches of the curve in Figure 8.5 (b) are a signature of a hysteresis phenomenon. This behavior, here described in the ideal case of zero losses and unitary reflection coefficient, is, however, still observed for typical weak propagation losses of 0.2 dB/km and $0.8 \leq \rho \leq 1$. Figure 8.5(c) displays a complete hysteresis cycle obtained in experiments when employing a dispersion shifted fiber (NZ-DSF) characterized by a length $L = 4$ km, a chromatic dispersion of −1.16 ps/nm/km, a Kerr coefficient $\gamma = 1.7\ \mathrm{W^{-1}\ km^{-1}}$ and a PMD coefficient of 0.05 ps/km$^{1/2}$. The input average power is 570 mW (27.5 dBm) and the corresponding nonlinear length is 1.03 km, so that $N \approx 4$. A polarized 100-GHz incoherent signal is injected into the fiber and undergoes an adiabatic transition so that its input SOP slowly moves from $s_3(z = 0, t) = -1$ toward +1 with a ramping time of 200 ms thanks to an opto-electronic input polarization controller. We clearly observe a switching of the output $s_3(z = L, t)$ from the lower to the upper branch of the hysteresis when the input $s_3(z = 0, t) \approx 0.4$. Similarly, when the input $s_3(z = 0, t)$ moves from +1 toward −1, then we observe a switching of the output $s_3(z = L, t)$ from the upper to the lower branch of the hysteresis when the input $s_3(z = 0, t) \approx -0.4$. Note that the output $s_3(z = L, t)$ always maintains its value close to −1 or +1 and that transitions are remarkably sharp: this is due to the presence of a strong nonlinearity and as consequence the hysteresis cycle turns to be well opened. Note also the excellent agreement with numerical results, obtained by solution of Eqs. (8.1) including exact experimental parameters, which is an important confirmation of the validity of Eqs. (8.1) for the description of the polarization dynamics in the Omnipolarizer.

An interesting application of the hysteresis loop is the implementation of a polarization-based flip-flop memory [56], whose principle of operation can be understood by analyzing Figure 8.5 (c). When a set pulse on the input s_3 is injected into the fiber, its peak value switches the output s_3 to a state, let us say ON state, close to +1. Afterward, the input set pulse vanishes to 0 but the system has stored the ON state because of the hysteresis properties. On the other hand, the output s_3 drops back to the initial state close to −1, let us say OFF state, when a reset pulse is applied on the input s_3.

Figure 8.6 illustrates the experimental proof-of-principle observation of the proposed flip-flop memory, which is loaded and cleared by a train of set and reset pulses on the

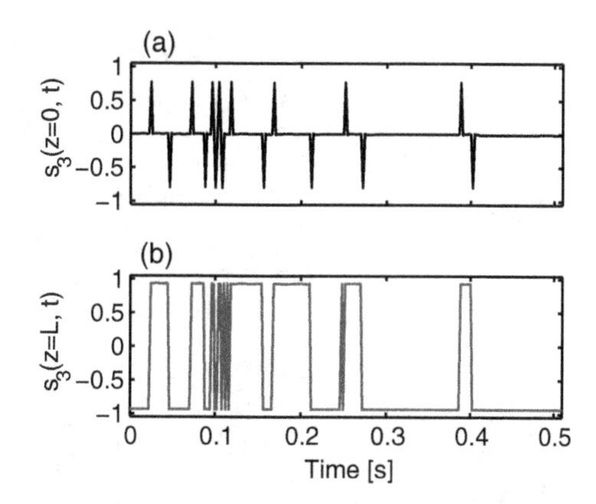

Figure 8.6 Experimental proof-of-principle of a flip-flop memory based on the hysteresis process of the Omnipolarizer. The employed NZ-DSF fiber is 4-km long; the 100-GHz polarized incoherent input signal has an average power of 570 mW; the reflection coefficient is $\rho = 0.9$. (a): Input $s_3(z = 0, t)$ sequence injected into the fiber (width 2 ms, rise time 20 μs). (b): Corresponding output $s_3(z = L, t)$.

input s_3, respectively. The fiber and the average input power employed in the experiment are close to those employed when reproducing the hysteresis cycle of Fig. 8.4(c); consequently the peaks of the set and reset input pulses should be respectively larger than +0.4 and lower than −0.4 in order to lead to the transition of the output s_3 from the OFF to the ON state and vice versa. Therefore, in the experiment the peak of the pulses are ±0.7; furthermore, the pulse width is 2 ms and its raise time is 20 μs. The train of pulses is imprinted on a 100-GHz polarized incoherent wave by means of the input opto-electronic polarization controller. Afterward, this signal is amplified up to 570 mW (average power). Input and output sequences of the pulses on the s_3 component are shown in Figure 8.6: we can clearly observe that the system perpetually stores the OFF/ON state until a new input set/reset control pulse appears, thus demonstrating the storage capacity of this polarization-based flip-flop memory. Sharp edges are observed on the output wave $s_3(z = L, t)$, which represent the temporal counterpart of the sharp transitions in the hysteresis cycle of Figure 8.5 (c) and are thus linked to the strong system nonlinearity: as a rule of thumb, the larger the system nonlinearity, the sharper the edges.

A further exploitation of the hysteresis properties of the system may consist in the implementation of an all-optical polarization switching operation [56]. Similarly to the case of the flip-flop memory, the basic principle lies in the sharp transitions when set and reset events are applied on the input s_3. The experimental proof-of-principle follows that of the flip-flop memory previously discussed, but in this case the partially incoherent signal is replaced by a 10-Gbit/s RZ signal. An arbitrary sequence of 2-ms set/reset s_3 spikes is imprinted on the incident 10-Gbit/s signal by means of the opto-electronic polarization controller. A polarization beam splitter (PBS) is finally inserted at the output of the Omnipolarizer to monitor the intensity profile on both orthogonal axes, which characterizes the switching efficiency: basically, the first axis of the PBS detects the magnitude $|s_3(z = L, t)|$, whereas the second axis detects the conjugate wave $1 - |s_3(z = L, t)|$.

In Figure 8.7 we display the experimental input s_3 set/reset spikes imprinted on the 10-Gbit/s signal and detected on a low bandwidth oscilloscope beyond the PBS. The input signal does not exhibit any polarization segregation and thus the eye-diagram shows a combination of both polarization PBS channels. On the contrary, the combined effects of polarization digitalization and associated hysteresis cycle may switch the whole energy of the output optical data from one axis of the PBS to the other with

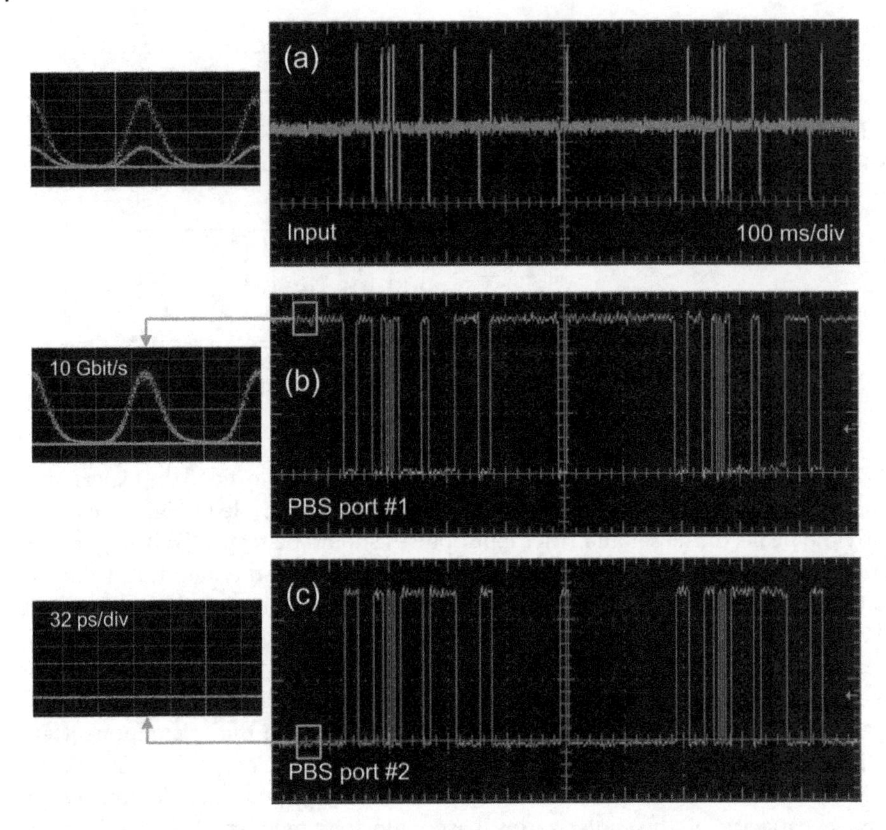

Figure 8.7 Experimental proof-of-principle of data-packet switching operation based on the hysteresis properties of the Omnipolarizer. Parameters of the system are the same as in Figure 8.6, but a 10-Gbit/s RZ input signal is employed instead of the 100-GHz polarized incoherent input wave. Right panels display the temporal evolution of the intensity profile recorded beyond a PBS and monitored on a low-bandwidth oscilloscope of the input $s_3(z = 0, t)$ (a), of the output $|s_3(z = L, t)|$ on port–1 (b) and of the output $1 - |s_3(z = L, t)|$ on port–2 of the PBS (c). Corresponding 10-Gbit/s eye-diagrams are displayed in the left panels.

an excellent discrimination. We thus observe on both axes of the PBS in Figure 8.7 (b), (c) the appearance of conjugate data packets corresponding to the set/reset initial sequence. Furthermore, the corresponding eye-diagrams exhibit a wide opening as well as a high polarization digitalization efficiency with an extinction ratio between each orthogonal axis above 20 dB.

8.6 Alignment Regime

If an amplification is applied to the backward replica in the reflective loop so that $1.2 \leq \rho \leq 2$, then the Omnipolarizer enters into the alignment regime, which is characterized by an attraction of the output SOP toward an unique point over the Poincaré sphere. In this regime, the device acts as an ideal nonlinear polarizer: all input polarization fluctuations are canceled by the device and consequently the output SOP $\mathbf{s}(z = L, t)$

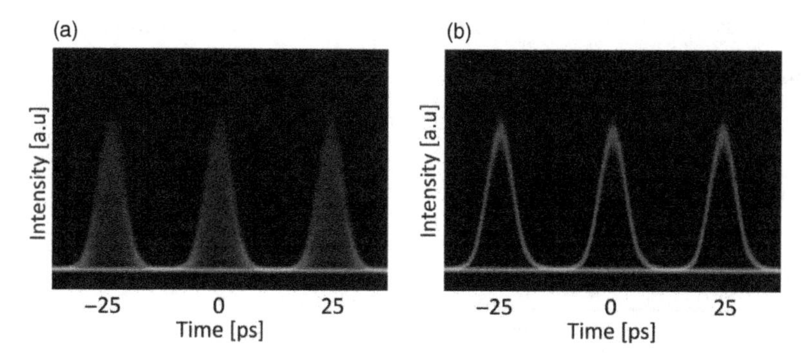

Figure 8.8 Eye-diagram of the 40-Gbit/s signal monitored behind an inline polarizer at the input of the Omnipolarizer (a) and at its output (b).

remains trapped around a single point over the Poincaré sphere (see Figure 8.1). This functionality is of fundamental interest and could find several applications in photonics. For instance, an all-optical repolarization capability could be a powerful means for the implementation of polarization sensitive devices such as silicon-based high-contrast integrated optics or photonic-crystal waveguides for future transparent networks.

As previously anticipated, the position of the attraction SOP over the sphere is determined by the angle of polarization rotation in the reflective loop, and thus can be selected by means of the polarization controller.

The efficiency of the polarization alignment in the temporal domain is clearly visible from experimental results displayed in Figure 8.8. A 40 Gbit/s depolarized signal with average power of 500 mW (27 dBm) is injected into a 6.2-km-long Non-Zero Dispersion-Shifted Fiber (NZDSF) characterized by a chromatic dispersion $D = -1.5$ ps/nm/km at 1550 nm, a Kerr coefficient $\gamma = 1.7\ \mathrm{W^{-1}\ km^{-1}}$, a PMD coefficient of 0.05 ps/km$^{1/2}$. Figure 8.8 (a) displays the eye-diagram of the input signal detected after an inline polarizer. Due to the initial depolarization, all input polarization fluctuations are transferred into the time domain, thus inducing large intensity variations and a complete closure of the eye-diagram.

On the contrary, we observe in Figure 8.8 (b) that if the average power of the backward replica is set to 28 dBm ($\rho \approx 1.2$), then the alignment regime is reached and the resulting eye-diagram of the output signal is outstandingly opened. Indeed, no intensity fluctuations can be observed beyond the polarizer, which confirms the efficiency of the self-induced repolarization process undergone by the signal within the Omnipolarizer.

More importantly, as displayed in Figure 8.9, the corresponding bit-error-rate measurements, performed after a polarizer as a function of the incoming power on the receiver, show that the Omnipolarizer is able to maintain the maximum of performance of data-processing through any polarization-dependent transmission component. Indeed, whereas the input BER is limited to a threshold of about 10^{-2} due to scrambling of the input signal which closes the eye-diagram through the polarizer, the output signal SOP turns out to be nearly constant in time so as to enable a complete recovery of the transmitted data with an error-free measurement.

We point out that the output BER is characterized by a 1 dB sensitivity improvement with respect to back-to-back measurements, which is due to the pulse reshaping properties of the device. Indeed, to be efficient, the alignment regime imposes a high level

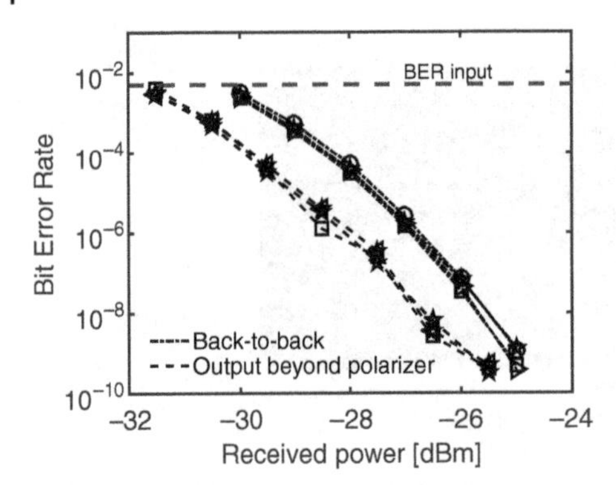

Figure 8.9 Bit-error-rate measurements beyond a polarizer as a function of received power performed after electrical demultiplexing of the 40 Gbit/s signal into 4 tributary 10 Gbit/s channels.

of average power close to 500 mW. Consequently, the optical pulses of the RZ 40-Gbit/s signal propagate in the NZDSF with a peak power around 3 W, and thus undergo a large self-phase modulation (SPM) effect leading to a wide spectral broadening. The broadening is illustrated in Figure 8.10 (a) where input (dashed line) and output (dash-dotted line) spectra are shown. This SPM effect combined with the normal chromatic dispersion of the fiber induces a nonlinear reshaping of the pulses toward a parabolic shape first and then to a square shape [64], as shown in Figure 8.10 (b) with the eye-diagram recorded at the output of the NZDSF. In order to retrieve the initial shape of the pulses and improve the extinction ratio, we thus carry out a 170-GHz Gaussian shape offset filtering (see solid line representing the filtered output spectrum in Figure 8.10 (a)) by using the principle of a Mamyshev regenerator [65–69], which leads to the eye-diagram of Figure 8.8 (b).

It is, however, worth noting that the strong nonlinear regime of propagation undergone in the Omnipolarizer can also be a limitation for practical implementations to high-bit-rate signals. Indeed, careful design of the fiber has to be done, especially for OOK signals, in order to avoid complex soliton dynamics in anomalous dispersive fibers [70] or wave breaking phenomena in the normal dispersion regime [70, 71].

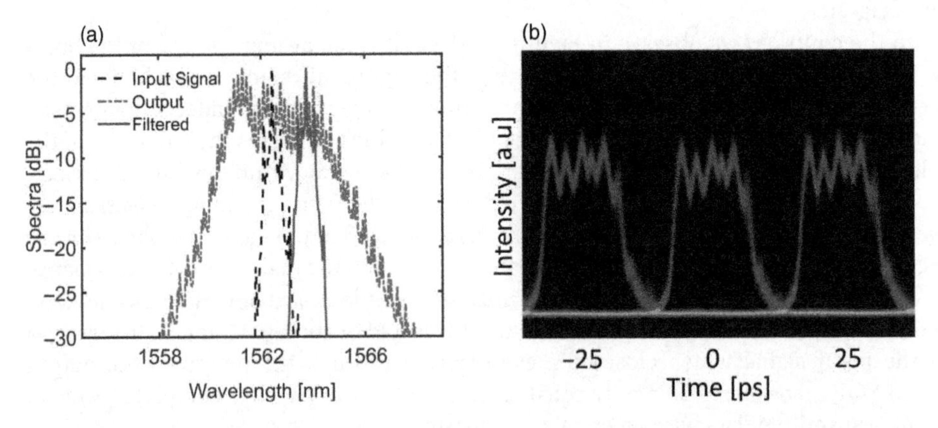

Figure 8.10 (a) Optical spectra of the 40-Gbit/s signal measured at the input of the NZSDF fiber (dashed line), at the output of the NZDSF fiber (dash-dotted line) and after the offset filtering (solid line) (b) Eye-diagram of the 40-Gbit/s signal directly recorded at the output of the NZDSF fiber in the alignment regime for an input power of 500 mW and monitored without filtering operation.

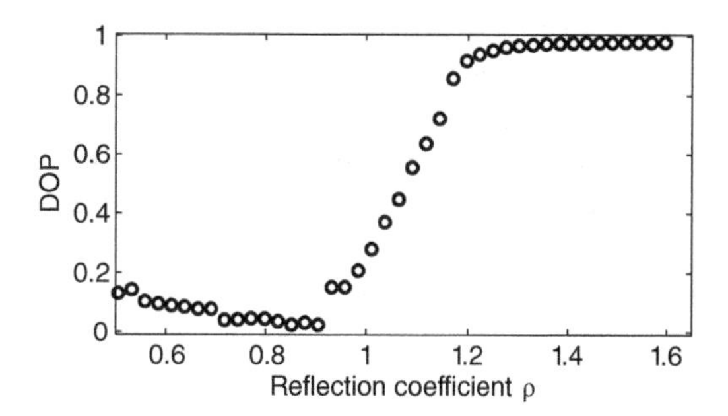

Figure 8.11 Experimental output DOP as a function of ρ for a 40 Gbit/s RZ (average power of 27 dBm) signal transmitted into the Omnipolarizer.

Besides the visual representation of the output eye-diagram, an important quantitative parameter which measures the strength of the alignment process is represented by the degree of polarization (DOP), defined as follows:

$$DOP = \sqrt{(\langle s_1 \rangle^2 + \langle s_2 \rangle^2 + \langle s_3 \rangle^2)} \tag{8.3}$$

where the brackets $\langle\rangle$ denote a temporal averaging. According to Eq. (8.3), a signal which is uniformly scrambled all over the Poincaré sphere, as the input represented in Figure 8.1 (a), has a DOP close to 0. Note, however, that also the bistable output displayed in Figure 8.1 has a DOP close to 0, because its SOP samples are localized around two opposite attraction points. On the other hand, when the attraction SOP is unique, as it is in the case of the output signal in the alignment regime, then the DOP is close to 1.

This behavior is confirmed by results displayed in Figure 8.11, which illustrate the experimental DOP as function of ρ when a 40 Gbit/s depolarized signal with average power of 27 dBm is transmitted in the NZSDF fiber previously introduced. We highlight that the input signal DOP is kept not null by purpose: more precisely, the input DOP is about 0.2, which corresponds to a not fully depolarized input SOP. This identifies the bistable regime, which is characterized by a strong attraction toward two opposite poles and thus by a minimum of DOP close to 0. We observe that the minimum is reached for $\rho \approx 0.9$ and that more generally in all the range $0.8 \leq \rho \leq 1$ values of DOP are low, as an effective bistable dynamics takes place. On the other hand, the DOP rapidly grows up to nearly 1 beyond $\rho = 1.2$, where the alignment regime is reached.

8.7 Chaotic Regime and All-Optical Scrambling for WDM Applications

In this section we discuss the third operational regime of the Omnipolarizer, which implements an all-optical fully chaotic polarization scrambler. Indeed, as underlined in Section 8.2, for a large power imbalance between the forward beam and its backward replica, the system exhibits a genuine chaotic dynamics so that even a well-polarized incident signal becomes scrambled all over the Poincaré sphere at the fiber output [62, 63].

The starting point of our analysis is the stability condition discussed in Section 8.2, namely, the monotonicity of the stable stationary states. We have seen in Section 8.5 that in absence of losses the stationary system Eq. (8.2) admits as a constant of motion the vector $\mathbf{K} = \mathbf{S} - \gamma D\mathbf{J}$; when $\rho \gg 1$ then $|\mathbf{J}| \gg |\mathbf{S}|$ and therefore we may approximate

$|\mathbf{K}| \approx |\gamma D\mathbf{J}| \equiv \gamma(8/9)|\mathbf{J}|$. Taking into account that in absence of losses $|\mathbf{J}(z)| = |\mathbf{J}(z = L)| = \rho|\mathbf{S}(z = L)|$, we may rewrite $|\mathbf{K}| = \gamma(8/9)\rho|\mathbf{S}(z = L)|$. As anticipated in Section 8.5, the stationary solutions **S** and **J** of Eq. (8.2) have spatial period $2\pi/|\mathbf{K}|$. In order to have monotonicity, a quarter of the period should be longer than the fiber length, that is to say $\pi/(2|\mathbf{K}|) > L$. Taking into account the aforementioned approximation for $|\mathbf{K}|$ we easily infer that the stability condition can be written as $\gamma\rho|\mathbf{S}(z = L)|L < (9\pi/16)$, which brings to the following threshold ρ_{st} for the reflection coefficient:

$$\rho_{st} = (9\pi/16)(\gamma P_S L)^{-1}, \tag{8.4}$$

where $P_S = |\mathbf{S}(z = L)|$. We point out that the estimation Eq. (8.4), deduced in the case of null losses, is still valid when small propagation losses are taken into account (typically some fraction of dB-per-km) if the term P_S refers to the forward signal power at fiber output.

When ρ is below the threshold ρ_{st} then the system exhibits stationary states that are stable and thus play the role of attractor. Consequently, any signal which is fully polarized in input (DOP=1) relaxes toward a stationary state and is still polarized in output. On the other hand, differently from the alignment regime ($1.2 \leq \rho \leq 2$), where all input signals are attracted in output towards an unique SOP, when $2 \leq \rho \leq \rho_{st}$ the output SOP is not unique but strictly depends on the input SOP [62].

If ρ is greater than the threshold ρ_{st}, then the stationary states do not play the role of attractor and consequently the output SOP $\mathbf{s}(z = L, t)$ does not reach a constant-in-time value but oscillates indefinitely as illustrated in Figure 8.4 (g)–(i). However, the temporal dynamics of $\mathbf{s}(z = L, t)$ exhibits a different behavior depending on the value of ρ. Massive numerical simulations show indeed the existence of a second threshold characterizing the system and beyond which the output SOP $\mathbf{s}(z = L, t)$ becomes chaotic in time and completely scrambled all over the sphere. We find that this threshold, here named ρ_{ch}, is typically $\rho_{ch} \approx 5\rho_{st}$. For intermediate values $\rho_{st} \leq \rho \leq \rho_{ch}$ the output signal undergoes a complex dynamics that depends on its input SOP as well as the rotation imposed by the reflective loop, and that may result in a periodic trajectory over the Poincaré sphere or a partially scrambled trajectory.

In any case, what is important is that a genuine scrambling of the output SOP is reached whenever $\rho > \rho_{ch}$ and that under this condition the Omnipolarizer behaves as an efficient polarization scrambler: for this reason we refer to this case as scrambling regime. Being $\rho_{ch} \approx 5\rho_{st}$, then the condition $\rho > \rho_{ch}$ can be rewritten as $P_S\rho > (45\pi)/(16L\gamma)$, where $P_S\rho$ corresponds to the total backward power P_J in the fiber. If we consider a typical fiber with $\gamma = 2$ W^{-1} km^{-1} and $L = 4$ km, we find that P_J should be greater than 1100 mW, that is to say, a backward power in the order of 1 W is required to reach the scrambling regime.

In Figure 8.12, experimental results displaying the typical dynamics of the Omnipolarizer when $\rho < \rho_{st}$, $\rho_{st} \leq \rho \leq \rho_{ch}$ and $\rho > \rho_{ch}$ are shown. The forward signal employed in the experiment is a 100-GHz incoherent wave with a fixed and arbitrary polarization state as well as a constant average power $P_S(z = 0) = 15$ dBm. The fiber is 5.3 km-long ; losses and Kerr coefficient are 0.24 dB/km and $\gamma = 1.7$ W^{-1} km^{-1}, respectively. By means of Eq. (8.4) we can thus estimate both the thresholds $\rho_{st} \approx 8$ and $\rho_{ch} \approx 40$.

When the backward power $P_J(z = L) = 20$ dBm (Figure 8.12 (a)) then $\rho \approx 4$ and thus the condition $\rho < \rho_{st}$ applies; consequently the forward signal relaxes towards a stable stationary state and the output SOP is a fixed point over the sphere. The RF spectrum of

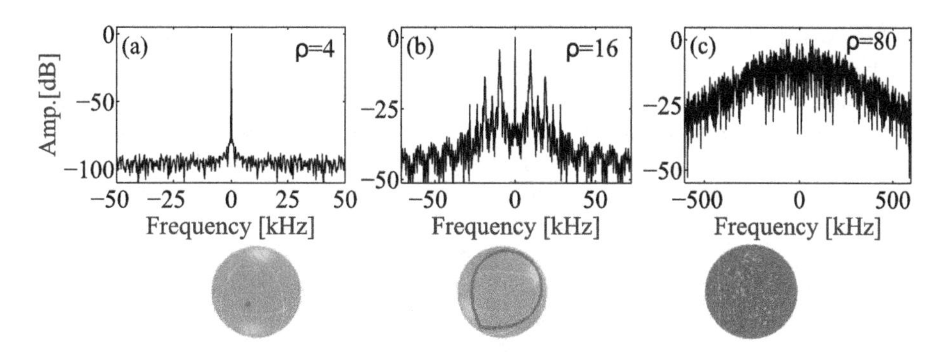

Figure 8.12 Experimental RF spectrum of the temporal intensity profile of the component $S_1(z = L, t)$. The input power injected in the fiber is $P_S(z = 0) = 15$ dBm. Panels (a), (b) and (c) display the spectrum of $S_1(z = L, t)$ and the output SOP distribution $\mathbf{s}(z = L, t)$ on the Poincaré sphere when the backward power $P_J(z = L)$ is 20 dBm ($\rho = 4$), 25 dBm ($\rho = 12$) and 33 dBm ($\rho = 80$), respectively. Similar behavior is found for the components $S_2(z = L, t)$ and $S_3(z = L, t)$, here not represented.

each component of $\mathbf{S}(z = L, t)$ exhibits a single narrow peak centered in f = 0 Hz, which corresponds to a constant-in-time value in the temporal domain and thus a constant output SOP.

When the backward power is increased up to $P_J(z = L) = 26$ dBm then $\rho \approx 16$ and thus $\rho_{st} \leq \rho \leq \rho_{ch}$: in the example of Figure 8.12 (b) the spectrum of each component of $\mathbf{S}(z = L, t)$ exhibits discrete and equi-spaced harmonics corresponding to a closed and periodic trajectory on the Poincaré sphere. On the other hand, as anticipated, the trajectory is not always periodic and may even be semi-chaotic depending on the particular choice of ρ and on the rotation R.

Finally, when the backward power is $P_J(z = L) = 33$ dBm, then $\rho = 80 > \rho_{ch}$ and the scrambling regime is reached: we can clearly see a continuum of frequencies in the spectrum of the components of $\mathbf{S}(z = L, t)$, corresponding to an almost uniform coverage of the Poincaré sphere and thus an efficient polarization scrambling of the output signal. Most important, this behavior is observed whatever ρ and R are.

The efficiency of the scrambling is preserved even when the input incoherent signal is replaced by a modulated telecom signal centered at 1550-nm, which makes the scrambling regime interesting for telecom applications. Figure 8.13 (a) displays the

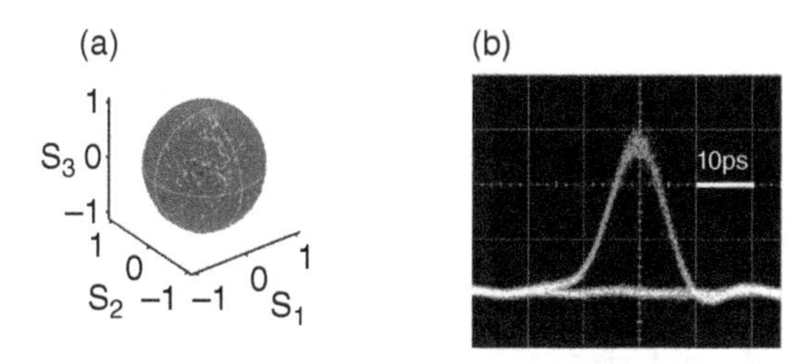

Figure 8.13 (a) Distribution of the output SOP $\mathbf{s}(z = L, t)$ over the Poincaré sphere of the 10-Gbit/s OOK signal injected with average power $P_S(z = 0) = 15$ dBm. The backward power $P_J(z = L) = 30$ dBm ($\rho = 80$), thus the system is in the scrambling regime. (b) Corresponding eye-diagram for the output intensity profile $|\mathbf{S}(z = L, t)|$: despite the output SOP is completely scrambled all over the sphere, however, the shape of $|\mathbf{S}(z = L, t)|$ is well preserved.

output SOP distribution of a 10 Gbit/s OOK signal which is injected into the fiber with fixed polarization and average power of 15 dBm. The backward power is 33 dBm, thus the scrambling regime is reached and the output SOP distribution covers uniformly the Poincaré sphere, indicating an effective scrambling at the fiber output. It is interesting to note that the scrambling dynamics does not degrade the intensity profile of the output pulses: Fig. 8.12 (b) shows that their shape is remarkably maintained with a wide-opened eye, which validates the efficiency of the scrambling regime for RZ modulated signals.

An important parameter for the evaluation of the scrambling efficiency is represented by the scrambling speed $v = Lim_{\partial t \to 0} |\partial \mathbf{s}/\partial t|$, which defines the average angle covered by **s** in 1 second over the Poincaré sphere. Interestingly enough, we find that this speed is directly proportional to the reflection coefficient and can be estimated by the following analytical formula:

$$v = (4/9) c_0 \gamma \rho P_s(z = 0) exp[-(5/3)\alpha L], \tag{8.5}$$

where c_0 is the speed of light in vacuum. Furthermore, the time scale of the output polarization fluctuations is well estimated by 1/v. The possibility to control the scrambling speed–and thus the timescale of the output SOP fluctuations–by tuning the backward power makes the scrambling regime truly appealing for the implementation of an efficient all-optical scrambler in the telecom framework.

In order to test the performances of the Omnipolarizer in this scrambling regime, we have carried out a first series of experiments by considering an initial polarized 100-GHz incoherent wave injected with a power of 15 dBm with a fixed arbitrary polarization state (DOP=1). As previously anticipated, in this case, the thresholds reads $\rho_{st} = 8$ and $\rho_{ch} = 40$.

The experimental output DOP as function of ρ in Figure 8.14 (a) well confirms the theoretical estimation of these thresholds: when $\rho < 8$ the DOP remains nearly unitary because the output SOP is still fully polarized. Conversely, for $\rho > 40$ the system enters into the genuine scrambling regime so that the output DOP drops to low values, typically below 0.3, corresponding to an efficient scrambling of the output SOP over the whole Poincaré sphere.

Figure 8.14 (b) displays the Lyapunov coefficient of the output SOP, evaluated upon the procedure described in [72]: we highlight that this coefficient, when positive, is indicative of a chaos behavior of the system, and that in the experiment under analysis

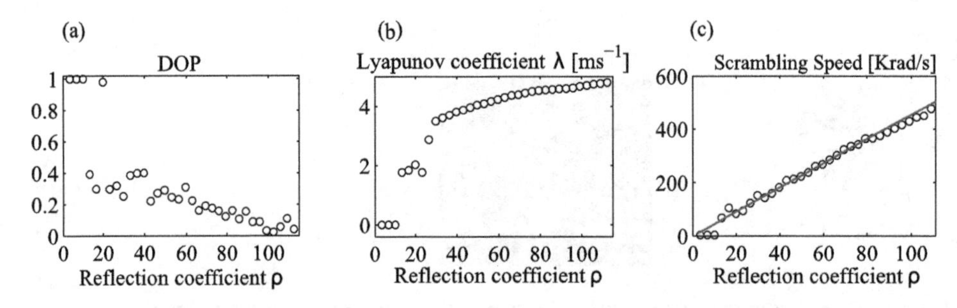

Figure 8.14 Experimental scrambling performances as a function of the reflection coefficent ρ. The input power is set to 15 dBm. (a) DOP of the output signal. (b) Corresponding Lyapunov coefficient. (c) Corresponding scrambling speed. The solid line represents the analytical estimation given in Eq. (8.5).

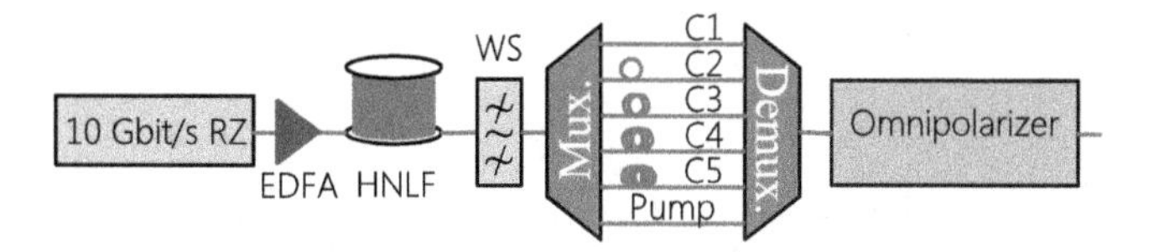

Figure 8.15 Experimental setup for the generation of the 5 WDM channels.

it becomes clearly positive for $\rho > 40$, thus confirming that the scrambling regime is strictly linked to a chaotic dynamics in the Omnipolarizer.

Figure 8.14 (c) displays the scrambling speed: we can see that it is truly almost directly proportional to ρ and that its estimation given by Eq. (8.5) is well accurate.

We highlight that detrimental Rayleigh back-scattering and propagation losses ultimately limit the scrambling performances when a large backward power (typically > 2W) is employed [62]. Consequently, the maximum achievable scrambling speed is around 400 krad/s. Note, however, that this value matches the scale of fast polarization changes encountered in high-speed fiber optic systems: the output scrambled signal from the Omnipolarizer may thus be employed to emulate polarization fluctuations undergone by a signal that is transmitted in a real fiber system, which makes the proposed chaotic regime of practical interest.

To evaluate the potential of this all-optical polarization scrambler in the framework of wavelength division multiplexing (WDM) transmission, we experimentally implemented the setup illustrated in Figure 8.15. A 10-GHz mode-locked fiber laser delivering 2-ps pulses at 1551 nm is encoded at 10 Gbit/s under an OOK RZ modulation format and is next amplified up to 30 dBm. A supercontinuum is then generated within a 500-m long dispersion-flattened highly non-linear fiber (DF-HNLF) characterized by a chromatic dispersion of −1 ps/nm/km, a dispersion slope of 0.006 ps^2/nm/km and $\gamma = 10.5$ W^{-1} km^{-1}. Five 10-Gbit/s, 12-GHz bandwidth, WDM channels centered at 1540.2 nm (C1), 1542 nm (C2), 1543.45 nm (C3), 1545 nm (C4) and 1546.2 nm (C5), as well as an additional pump channel, centered at 1550 nm, are finally sliced into this resulting continuum thanks to a programmable optical filter (Waveshaper WS). All the channels are finally decorrelated both in the time and polarization domains by means of a couple of demultiplexer/multiplexer with different paths and polarization rotations for each channel.

The resulting 10-Gbit/s WDM grid is injected into the Omnipolarizer with a fixed average power of 15 dBm. A 100-GHz optical bandpass filter is inserted into the reflective-loop in order to only counter-propagate the 1550-nm pump channel as a backward signal with average power of 29 dBm. The purpose of this spectral filtering is twofold. First, a single signal counter-propagates and thus imposes a unique and efficient scrambling process for all the transmitted channels. Second, because of the strong level of power used for the backward signal, this frequency offset pump channel limits the deleterious impact of back Rayleigh scattering on the five other transmitted channels.

At the output of the Omnipolarizer, the WDM channels are characterized in polarization and in the time domain by means of an eye-diagram monitoring and bit-error-rate measurements. The output Poincaré spheres for channels C1, C3 and C5 are represented in Figure 8.16 (a)–(c). The two other channels C2 and C4 exhibit similar behaviors. Interestingly enough, in spite of the fact that all the channels are initially decorrelated and

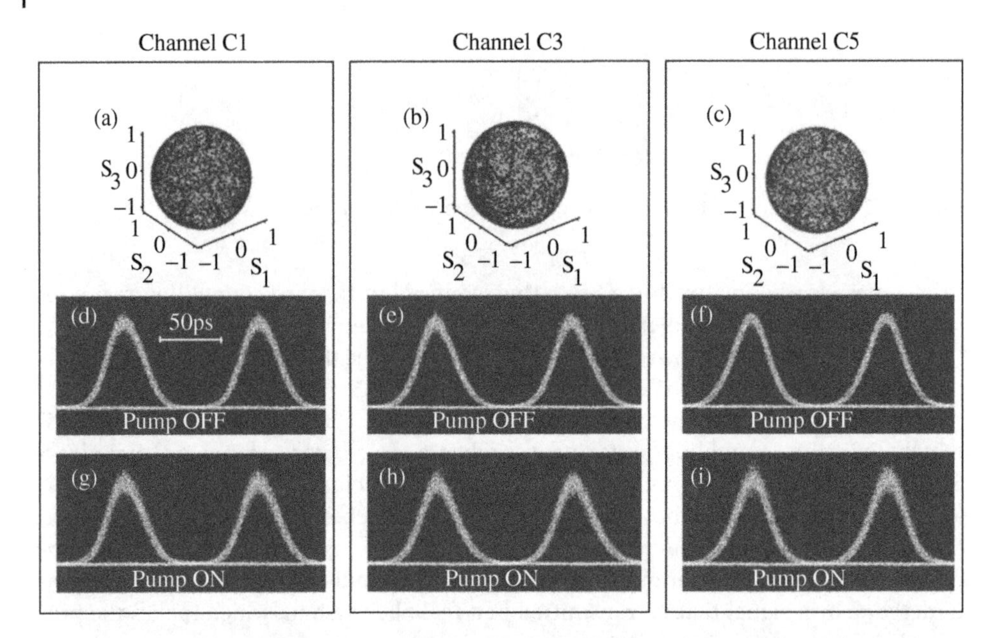

Figure 8.16 Experimental WDM performances of the Omnipolarizer in the scrambling regime. The input power is fixed to 15 dBm and the reflected 1550-nm pump channel is amplified to 29 dBm (a-c) Output SOP distribution for channels C1, C3 and C5, respectively. (d-f) Output eye-diagrams in pump-off modality. (d-f) Output eye-diagrams in pump-on modality.

enter into the system with different fixed SOPs, the device is still able to randomize the polarization of the whole WDM grid. Indeed, the output SOP of each channel is uniformly distributed over the entire Poincaré sphere and is therefore characterized by a low value of DOP (here below 0.2). Moreover, we find that the polarization trajectories undergone by all the individual channels are correlated in time and characterized by roughly the same scrambling speed close to 140 krad/s, thus confirming that the chaotic dynamics is here clearly imposed by the counter-propagating pump wave onto the other channels [62]. Note that such a speed is close to the one measured in the previous single channel configuration for similar involved average powers. Indeed, this all-optical scrambler is mainly sensitive to the average power of the counter-propagative beam.

Similar to the case of the single channel configuration (see Fig. 8.13), the efficient scrambling of each WDM channel does not prevent it remarkably preserving the output temporal intensity profile of each channel, which in principle enables high bit-rate signal processing. Figure 8.16 (d)–(f), show the high quality of the output eye-diagrams for channels C1, C3 and C5 when the backward pump channel is switched off, so that no scrambling occurs on the output polarization. Figure 8.16 (g–i), by contrast, show the corresponding eye-diagrams when the backward pump channel is switched on at an average power of 29 dBm, so that the scrambling process operates efficiently, from which we note that the eye-diagrams are well preserved with a wide opening.

On the other hand, a weak degradation of the temporal profiles can still be observed, which is due to the Rayleigh back-scattering provided by the spectrally broadened backward pump channel as well as to a weak Raman depletion effect caused by the pump

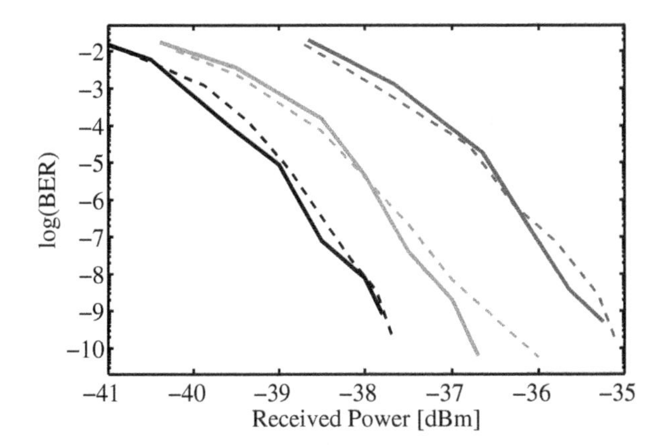

Figure 8.17 Bit-error-rate measurements for 10-Gbit/s WDM channels C1 (black line), C3 (pale gray lines) and C5 (dark gray lines) in pump-off modality (solid lines) and in pump-on modality (dashed-line).

onto the signal. In order to quantify the degradation induced by these impairments we measure the bit-error-rate (BER) in pump off and pump on configurations. These measurements are summarized in Figure 8.17 and show that low penalties are introduced by the scrambling process. Indeed, roughly 0.2 dB of power penalty have been estimated for the whole WDM channels at a BER of 10^{-9} when comparing the pump on/off configuration. Note that different performances among the channels are due to different Optical Signal to Noise Ratio (OSNR) values of each channel at the device input.

8.8 Future Perspectives: Towards an All-Optical Modal Control in Fibers

In Sections 8.5 and 8.6, we depicted the remarkable process of self-induced repolarization undergone by a signal in the bistability and alignment regime. A natural question that arises is if this concept of self-organization may be extended to the spatial modes of a multimode fiber. This paves the way to a more general idea of self-organization, where the involved modes may be either polarization or spatial modes of a fiber. A new scenario emerges, where light could self-organize its modal state, namely, the power distribution among the modes as well as their relative phase.

We highlight that several important applications could benefit from such a phenomenon of modal self-organization. For example, it could lead to the development of an all-optical signal processing technology in the framework of spatial-division-multiplexing schemes, which are emerging as the most promising solution to face the feared capacity crunch of current single-mode fiber systems [73]. Furthermore, it could provide an efficient and all-optical way to counteract parasitic and uncontrolled modal coupling effects in short multimode fibers, which are nowadays widely employed in several contexts, e.g., to carry a large amount of data from large-area telescopes to spectrometers. Finally, modal self-organization could be exploited as a means for phase-synchronization of the cores in multicore fiber lasers, in such a way to focus the energy in the fundamental supermode of the multicore fiber. For this purpose, we point out that

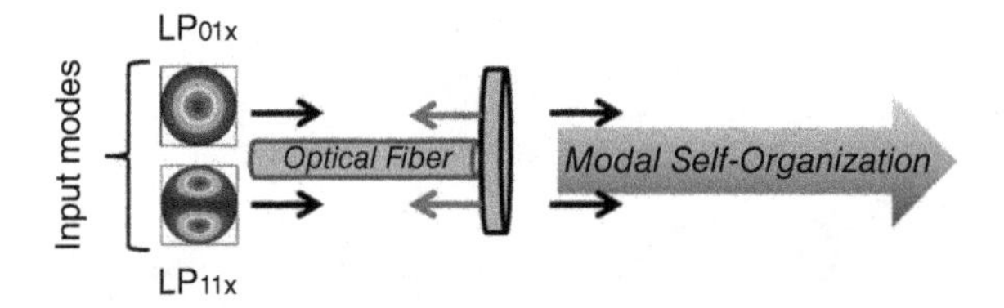

Figure 8.18 Schematic setup for modal self-organization in a bimodal isotropic fiber. A combination of LP_{01} and LP_{11} modes (here both x–polarized) is injected into the fiber (right-facing arrows) and nonlinearly interacts with the backward replica (left-facing arrows) generated at the fiber output by means of a reflecting device.

the idea of exploiting nonlinear phenomena in order to lead to self-phase organization has already been explored [74], however, a clear theoretical understanding is still lacking and could be the key to synchronizing an unprecedented number of cores.

In this section we want to discuss a simple but clear example of modal self-organization which discloses its potential applications. We will not provide a comprehensive picture of the modal self-organization dynamics, which is truly intriguing but quite completely unexplored and will thus deserve several studies in the near future.

With the aim of providing a simple example, here we consider the case of a bimodal isotropic fiber where only two copolarized spatial modes propagate and interact with their backward replica generated by a perfectly reflecting mirror ($\rho = 1$), as it is shown in the schematic setup of Figure 8.18.

An analysis of the modal dynamics in a telecommunication fiber would be more interesting in view of telecom applications. On the other hand, the spatial dynamics in a telecom fiber is complex to analyse even in the simple bimodal case, which is due to the presence of a randomly varying birefringence along with polarization and spatial mode dispersion effects [39]. It will thus be the subject of further research in the future.

We indicate with E_1 the electric field envelope of the forward fundamental LP_{01} mode, whereas with E_2 the envelope of the second mode LP_{11}. Similarly, F_1 and F_2 indicate the envelopes of the backward modes. Assuming that all the modes are polarized along the same direction, e.g., the x-axis, and following a treatment similar to that discussed in [53], we find that the spatiotemporal dynamics of the modal envelopes are ruled by the following CNLSEs:

$$c_n^{-1}\partial_t E_n + \partial_z E_n = 2C_{01}E_m F_m F_n^* + (C_{nn}|E_n|^2 + 2C_{01}|E_m|^2 + 2C_{nn}|F_n|^2 + 2C_{01}|F_m|^2)E_n, \tag{8.6}$$

being c_n the group velocity for the mode-n ($n = \{1, 2\}$, $m = \{1, 2\}$, $n \neq m$).

A similar equation is found for F_n by mutually exchanging E with F and ∂z with $-\partial z$. The boundary condition $E_n(z = L, t) = -F_n(z = L, t)$ accounts for 100% reflection imposed by the perfect mirror at fiber end. The nonlinear coupling coefficients read as $C_{nn} = n_2\omega/(c_0 A_{nn})$ and $C_{01} = n_2\omega/(c_0 A_{01})$, where $n_2 = 3.2 \cdot 10^{-16}$ cm^2/W is the silica nonlinear index, c_0 is the speed of light in vacuum, ω is the central frequency of the signal injected in the fiber, $A_{nn} = (\int_{xy} M_n^2 dxdy)^2/(\int_{xy} M_n^4 dxdy)$ is the intramodal effective area of mode-n and $A_{01} = (\int_{xy} M_0^2 dxdy)(\int_{xy} M_1^2 dxdy)/(\int_{xy} M_0^2 M_1^2 dxdy)$ is the intermodal effective area involving the LP_{01} and LP_{11} modes, whose modal transverse profiles are $M_0(x, y)$ and $M_1(x, y)$, respectively.

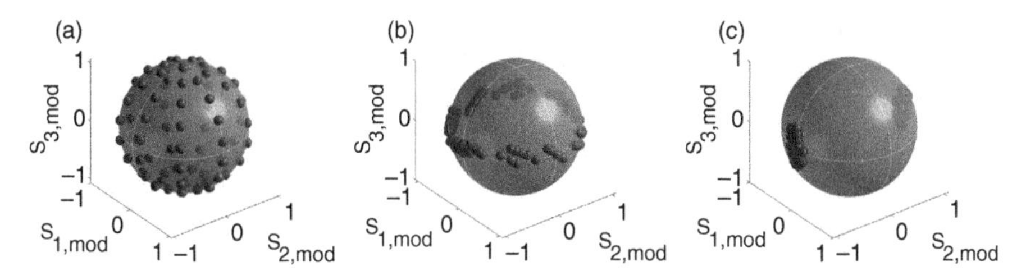

Figure 8.19 (a) Input distribution of the modal Stokes vector $\mathbf{s}_{mod}$ over the modal Poincaré sphere. (b) Output distribution when $N = 4$ and $C_{00} = C_{11} = 10C_{01}$. (c) Output distribution when $N = 4$ and $C_{00} = C_{11} = 1.75C_{01}$.

Note that C_{00} plays the role of the Kerr nonlinear coefficient γ in a single-mode fiber and that the term $2C_{01}E_m F_m B_n^*$ in Eq. (8.6) is responsible for the energy exchange between the two fiber modes. By analogy with polarization phenomena, we define the modal Stokes parameters $\mathbf{S}_{mod} = [|E_1|^2 - |E_2|^2, E_1E_2^* + E_2E_1^*, -iE_1E_2^* + iE_2E_1^*]$ and $\mathbf{J}_{mod} = [|F_1|^2 - |F_2|^2, F_1F_2^* + F_2F_1^*, -iF_1F_2^* + iF_2F_1^*]$, as well as the nonlinear length $L_{nl} = 1/(C_{00}P_S)$, being P_S the total forward power that in this case reads as $P_S = |\mathbf{S}_{mod}| = |E_1|^2 + |E_2|^2$ and where the subscript *mod* is employed to differentiate modal Stokes vectors from polarization ones. Furthermore, we highlight that a modal Poincaré sphere, completely analogue to the polarization Poincaré sphere, can conveniently display the position of the normalized unitary modal Stokes vector $\mathbf{s}_{mod} = \mathbf{S}_{mod}/|\mathbf{S}_{mod}|$.

In order to study the system dynamics in a highly nonlinear regime we solve Eq. (8.6) by setting the number of nonlinear length $N = L/L_{nl} = 4$. Similar to what done when studying polarization alignment and bistability phenomena, here different CW input conditions $\{E_1(z = 0), E_2(z = 0)\}$ are employed that correspond to an uniform coverage of the modal Poincaré sphere by $\mathbf{s}_{mod}(z = 0)$, as depicted in Figure 8.19 (a). The output vectors $\mathbf{s}_{mod}(z = L, t)$ are finally computed at a large time t such that the system has relaxed toward an asymptotic stable stationary state of Eq. (8.6).

A great difference with respect to the case of single-mode fibers is that in this case three distinct nonlinear coefficients are present that can be tailored over a wide range of values by means of an appropriate fiber engineering [75]. This paves the way to new and unexplored self-organization phenomena where the distribution of the output modal Stokes vectors can be controlled by means of the nonlinear coefficients.

As an example, let us consider the case in which $C_{00} = C_{11} = 10C_{01}$: in Figure 8.19 (b) we can see that the output modal Stokes vectors tend to self-organize along the circle $s_{1,mod}^2 + s_{2,mod}^2 = 1$; on the other hand, when we increase the coupling coefficient C_{01} in such a way that $C_{00} = C_{11} = 1.75C_{01}$, we clearly observe an attraction toward the two poles $s_{2,mod} = \pm 1$ (Figure 8.19 (c)) which corresponds to $s_{1,mod} \approx 0$ and thus to an equipartition of the energy among the two modes at the fiber output, namely, $|E_1(z = L, t)|^2 \approx |E_2(z = L, t)|^2$.

We expect that several different self-organization patterns of the modal state may be achieved at the fiber output by tuning the nonlinear coefficients, by adding an amplification stage in the reflective loop similar to what is done in the alignment regime of polarization (see Section 8.3) or by employing different kind of fibers. Such modal patterns may be also found when employing fibers where more than two modes propagate.

The early results outlined in this section seems thus to indicate that a new scenario is going to arise, that is the nonlinear self-organization of the modal state in multimode and multicore optical fibers, and we expect it will provide exciting results in the near future.

8.9 Conclusion

In this chapter, we reviewed our recent work dealing with the all-optical control of the polarization state in a device called the Omnipolarizer. The principle of operation of this device is based on a Kerr nonlinear cross-polarization interaction occurring in a km-long telecom optical fiber between an arbitrary polarized forward signal and its backward replica generated at the fiber end by means of an amplified reflective loop setup. Depending on the reflection coefficient ρ, which defines the power ratio between the backward and forward waves at the fiber exit, three distinct operating regimes have been identified.

When $0.8 \leq \rho \leq 1$, the Omnipolarizer is characterized by a bistable behavior associated with a hysteresis cycle of the output SOP. Our theoretical predictions of the hysteresis are in excellent agreement with our experimental measurements and show a wide opening and sharp transitions in the cycle. This bistable regime has been exploited to provide two proof-of-principles for optical signal processing. First, an all-optical flip-flop memory has been implemented, triggered by a sequence of set/reset polarization spikes imprinted onto the input signal. These results show that a binary information can be perpetually stored or erased thanks to this self-digitalization process of light SOP. Second, a 10-Gbit/s switching operation for data packets has been demonstrated. Experimental monitoring of the eye-diagrams confirms the switching capability of our device along two orthogonal polarization channels with an extinction ratio above 20 dB.

When the reflective coefficient is increased above unity, typically $1.2 \leq \rho \leq 2$, the Omnipolarizer enters into its alignment regime and turns out to act as a polarization funnel for which a self-induced polarization tracking occurs. In this regime, the SOP of an arbitrary polarized input signal is trapped in a unique pool of attraction onto the Poincaré sphere. We have successfully exploited this operating mode in order to self-repolarize a 40-Gbit/s RZ OOK signal. More precisely, we have been able to track the output polarization of a 40-Gbit/s signal in such a way as to align its SOP on the axis of a polarizer with very weak degradation of the eye-diagram quality, thus allowing a polarization-independent error-free reception.

The third operating regime, described as scrambling regime, is based on a strong power imbalance between the two counter-propagating waves ($\rho \gg 1$). In this regime the Omnipolarizer exhibits a genuine chaotic dynamics for which the temporal evolution of the polarization state at the fiber output becomes fully random and thus scrambled all over the Poincaré sphere. This regime has been successfully implemented in order to depolarize a 10-Gbit/s OOK WDM signal, while preserving a well-opened eye-diagram. In particular, we have experimentally shown that in the scrambling regime the Omnipolarizer is able to simultaneously scramble the polarization of several WDM channels with a scrambling speed up to some hundreds of krad/s, whilst providing very low impairments on the signal quality.

From a practical point of view, we highlight that despite the fact that all the present results have been demonstrated around 1550 nm in the C band, however, the nonlinear interaction involved in the Omnipolarizer is independent of wavelength. The only requirement is the implementation of high power amplifiers centered at the signal wavelength. We could thus imagine extending this device to the Ytterbium or Thulium bands around 1 μm or 2 μm, respectively.

On the other hand, the counter-propagative configuration of the Omnipolarizer combined with the cross-polarization interaction imposes a strong nonlinear regime of propagation with levels of powers around 500 mW in a km-long telecom fiber, which is high above the standards of telecommunication systems. Therefore, our device could not be applied to weak input signals and particular attention has to be taken in order to avoid large signal impairments during propagation. For example, nonlinear pulse reshaping is required for high bit-rates return-to-zero signals so as to preserve the signal quality while high levels of nonlinear phase shift could be seen as a limitation for phase shift keying modulation formats and coherent detection.

We also highlight that despite the quasi-instantaneous response of Kerr nonlinearity in silica fibers, the response time of the Omnipolarizer has been found to be much slower, typically around a few μs in km-long standard fibers [44]. This is directly related to the counter-propagating configuration and therefore to the distributed interaction along the fiber to establish a polarization equilibrium. Nevertheless, the response time of the Omnipolarizer could be reduced to sub-microseconds by employing optical fibers with stronger nonlinearities, such as Chalcogenide, Tellurite, Bismuth or lead Silicate fibers.

In the last section of this chapter, we generalized the concept of self-organization of light to the case of the spatial and polarization modes of a multimode fiber. We discussed a simple but effective example of all-optical modal self-organization in a bimodal fiber, which is based on the nonlinear interaction between two forward spatial modes of the fiber with their replicas generated by a perfect reflection at the fiber output. This simple example let us envisage a plethora of new applications based on self-organization phenomena in multimode fibers, which may play a key-role in the next generation of optical communication systems based on mode division multiplexing technology.

Acknowledgments

This research is funded by the European Research Council under Grant Agreement 306633, ERC PETAL (https://www.facebook.com/petal.inside). We thank the financial support of the Conseil Régional de Bourgogne in the framework of the Photcom project, FEDER and the Labex ACTION (ANR-11-LABX-0001-01). We also thank P. Morin, P-Y Bony, M. Gilles, A. Picozzi, D. Sugny, H. R. Jauslin, G. Millot, S. Wabnitz and especially S. Pitois for fruitful discussions.

References

1 Geisler, T. (2006) Low PMD transmission fibres. Paper Mo.3.3.1 in European Conference on Optical Communications, ECOC 2006.

2 Barlow, A.J., Ramskov-Hansen, J.J., and Payne D.N. (1981) Birefringence and polarization mode-dispersion in spun singlemode fibers. *Applied Optics*, **20**, 2962–2968.

3 Li, M.J. and Nolan, D.A. (1998) Fiber spin-profile designs for producing fibers with low polarization mode dispersion. *Optics Letters*, **23**, 1659–1661.

4 Palmieri, L. (2006) Polarization properties of spun single-mode fibers. *IEEE Journal of Lightwave Technology*, **24**, 4075–4088.

5 Boroditsky, M., Brodsky, M., Frigo, N.J., Magill, P., and Rosenfeldt, H. (2005) Polarization dynamics in installed fiber optic systems. *IEEE LEOS Annual Meeting Conference Proceedings (LEOS)*, 413–414.

6 Krummrich P.M. and Kotten, K. (2004) Extremely fast (microsecond scale) polarization changes in high speed long hail WDM transmission systems. In *Proceedings of Optical Fiber Communications Conference*, Los Angeles, USA.

7 Rashleigh, S.C. (1983) Origins and control of polarization effect in single-mode fibers. *IEEE Journal of Lightwave Technology*, **1**, 312–331.

8 Ulrich, R. and Simon, A. (1979) Polarization optics of twisted single-mode fibers. *Applied Optics*, **18**, 2241–2251.

9 Simon, A. and Ulrich, R. (1977) Evolution of polarization along a single-mode fiber. *Applied Physics Letters*, **31**, 517–520.

10 Galtarossa, A., Palmieri, L., Schiano, M., and Tambosso, T. (2001) Measurement of birefringence correlation length in long single-mode fibers. *Optics Letters*, **26**, 962–964.

11 Poole, C.D. and Wagner, R.E. (1986) Phenomenological approach to polarization dispersion in long single-mode fibers. *Electronics Letters*, **22**, 1029–1030.

12 Galtarossa, A. and Menyuk, C.R. (Eds.). (2005) *Polarization Mode Dispersion*. Springer-Verlag, New York.

13 Garnier, J., Fatome, J., and Le Meur, G. (2002) Statistical analysis of pulse propagation driven by polarization-mode dispersion. *Journal of the Optical Society of America B*, **19**, 1968–1977.

14 Wai, P.K.A. and Menyuk, C.R. (1996) Polarization mode dispersion, decorrelation, and diffusion in optical fibers with randomly varying birefringence. *IEEE Journal of Lightwave Technlogy*, **14**, 148–157.

15 Galtarossa, A., Palmieri, L., and Schenato, L. (2006) Simplified phenomenological model for randomly birefringent strongly spun fibers. *Optics Letters*, **31**, 2275–2277.

16 Gisin, N. and Huttner, B. (1997) Combined effects of polarization mode dispersion and polarization dependent losses in optical fibers. *Optics Communications*, **142**, 119–125.

17 Sperti, D., Serena, P., and Bononi, A. (2011) Optical solutions to improve PDM-QPSK resilience against cross-channel nonlinearities: A comparison. *IEEE Photonics Technology Letters*, **23**, 667–669.

18 Mollenauer, L.F. Gordon, J.P., and Heismann, F. (1995) Polarization scattering by soliton–soliton collisions. *Optics Letters*, **20**, 2060–2062.

19 Renaudier, J., Charlet, G., Salsi, M., *et al.* (2008) Linear fiber impairments mitigation of 40-gbit/s polarization-multiplexed QPSK by digital processing in a coherent receiver. *IEEE Journal of Lightwave Technology*, **26**, 36–42.

20 Charlet, G. (2008) Coherent detection associated with digital signal processing for fiber optics communications. *C. R. Physique*, **9**, 1012–1030.

21 Charlet, G., Renaudier, J., Salsi, M., *et al.* (2007) Efficient mitigation of fiber impairments in an ultra-long haul transmission of 40gbit/s polarization-multiplexed

data, by digital processing in a coherent receiver. Paper PDP17 in Optical Fiber Communication Conference and Exposition.

22 Hansryd, J., Andrekson, P.A., Westlund, M., Li, J., and Hedekvist, P. (2002) Fibre-based optical parametric amplifiers and their applications. *IEEE Journal of Selected Topics in Quantum Electronics*, **8**, 506–520.

23 Inoue, K. (1994) Polarization independent wavelength conversion using fiber four-wave mixing with two orthogonal pump lights of different frequencies. *IEEE Journal of Lightwave Technology*, **12**, 1916–1920.

24 Wong, K.K.Y., Marhic, M.E., Uesaka K., and Kazovsky, L.G. (2002) Polarization-independent one-pump fiber-optical parametric amplifier. *IEEE Photonics Technology Letters*, **14**, 1506–1508.

25 Fukuda, H., Yamada, K., Tsuchizawa, T. *et al.* (2008) Silicon photonic circuit with polarization diversity. *Optics Express*, **16**, 4872–4880.

26 Martinelli, M., Martelli, P., and Pietralunga, S.M. (2006) Polarization stabilization in optical communications systems. *IEEE Journal of Lightwave Technology*, **24**, 4172–4183.

27 Koch, B., Noe, R., Sandel, D., and Vitali Mirvoda, V. (2014) Versatile endless optical polarization controller/tracker/demultiplexer. *Optics Express*, **22**, 8259–8276,

28 Koch, B., Noe, R., Mirvoda, V. *et al.* (2010) Record 59-krad/s polarization tracking in 112-Gb/s 640-km PDM-RZ-DQPSK transmission. *IEEE Photonics Technology Letters*, **22**, 1407–1409.

29 Aarts, W.H.J. and Khoe, G. (1989) New endless polarization control method using three fiber squeezers. *IEEE Journal of Lightwave Technology*, **7**, 1033–1043.

30 Koch, B., Noé, R., Mirvoda, V., and Sandel, D. (2011) 100-krad/s endless polarisation tracking with miniaturised module card. *Electronics Letters*, **47**, 813–814.

31 Martinelli, M., Cirigliano, M., Ferrario, M., Marazzi, L., and Martelli, P. (2009) Evidence of Raman-induced polarization pulling. *Optics Express*, **17**, 947–955.

32 Ursini, L., Santagiustina, M., and Palmieri, L. (2011) Raman nonlinear polarization pulling in the pump depleted regime in randomly birefringent fibers. *IEEE Photonics Technology Letters*, **23**, 1041–1135.

33 Galtarossa, A., Palmieri, L., Santagiustina, M., Schenato, L., and Ursini, L. (2008) Polarized brillouin amplification in randomly birefringent and unidirectionally spun fibers. *IEEE Photonics Technology Letters*, **20**, 1420–1422,

34 Muga, N.J., Ferreira, M.F.S., and Pinto, A.N. (2011) Broadband polarization pulling using Raman amplification. *Optics Express*, **19**, 18707–18712.

35 Kozlov, V., Nuno, J., Ania-Castanon, J.D., and Wabnitz, S. (2011) Theoretical study of optical fiber Raman polarizers with counterpropagating beams. *IEEE Journal of Lightwave Technology*, **29**, 341–347.

36 Thevenaz, L., Zadok, A., Eyal, A., and Tur, M. (2008) All-optical polarization control through Brillouin amplification. Paper OML7 presented at Optical Fiber Communication Conference, OFC'08.

37 Shmilovitch, Z., Primerov, N., Zadok, A., *et al.* (2011) Dual-pump push-pull polarization control using stimulated Brillouin scattering. *Optics Express*, **19**, 25873–25880.

38 Stiller, B., Morin, P., Nguyen, D.M., *et al.* (2012) Demonstration of polarization pulling using a fiber-optic parametric amplifier. *Optics Express*, **20**, 27248–27253.

39 Guasoni, M., Kozlov, V., and Wabnitz, S. (2012) Theory of polarization attraction in parametric amplifiers based on telecommunication fibers. *Journal of Optical Society of America B*, **29**, 2710–2720.

40 Pitois, S. and Haelterman, M. (2001) Optical fiber polarization funnel. Paper MC79-1. Nonlinear guided waves and their applications, NLGW'01.

41 Fatome, J., Pitois, S., Morin, P., and Millot, G. (2010) Observation of light-by-light polarization control and stabilization in optical fibre for telecommunication applications. *Optics Express*, **18**, 15311–15317.

42 Kozlov, V.V., Nuno, J., and Wabnitz, S. (2011) Theory of lossless polarization attraction in telecommunication fibers. *Journal of Optical Society of America B*, **28**, 100–108.

43 Turitsyn, K. and Wabnitz, S. (2013) Stability analysis of polarization attraction in optical fibers. *Optics Communications*, **307**, 62–66.

44 Kozlov, V.V., Fatome, J., Morin, P., *et al.* (2011) Nonlinear repolarization dynamics in optical fibers: transient polarization attraction. *Journal of Optical Society of America B*, **28**, 1782–1791.

45 Assémat, E., Lagrange, S., Picozzi, A., Jauslin, H.R., and Sugny, D. (2010) Complete nonlinear polarization control in an optical fiber system. *Optics Letters*, **35**, 2025–2027.

46 Lagrange, S., Sugny, D., Picozzi, A., and Jauslin, H.R. (2010) Singular tori as attractors of four-wave-interaction systems. *Physical Review E*, **81**, 016202.

47 Assémat, E., Picozzi, A., Jauslin, H.R., and Sugny, D. (2012) Hamiltonian tools for the analysis of optical polarization control. *Journal of the Optical Society of America B*, **29**, 559–571.

48 Assémat, E., Dargent, D., Picozzi, A., Jauslin, H.R., and Sugny, D. (2011) Polarization control in spun and telecommunication optical fibers. *Optics Letters*, **36**, 4038–4040.

49 Morin, P., Fatome, J., Finot, C., *et al.* (2011) All-optical nonlinear processing of both polarization state and intensity profile for 40 Gbit/s regeneration applications. *Optics Express*, **19**, 17158–17166.

50 Barozzi, M. and Vannucci, A. (2014) Lossless polarization attraction of telecom signals: application to all-optical OSNR enhancement. *Journal of Optical Society of America B*, **31**, 2712–2720.

51 Costa Ribeiro, V., Luis, R.S., Mendinueta, J. M. D *et al.* (2015) All-optical packet alignment using polarization attraction effect. *IEEE Photonics Technology Letters*, **27**, 541–544.

52 Morin, P., Pitois, S., and Fatome, J. (2012) Simultaneous polarization attraction and Raman amplification of a light beam in optical fibers. *Journal of Optical Society of America B*, **29**, 2046–2052.

53 Pitois, S., Picozzi, A., Millot, G., Jauslin, H.R., and Haelterman, M. (2005) Polarization and modal attractors in conservative counterpropagating four-wave interaction. *Europhysics Letters*, **70**, 88–94.

54 Fatome, J., Pitois, S., Morin, P., *et al.* (2012) A universal optical all-fiber omnipolarizer. *Scientific Reports*, **2**, 938.

55 Bony, P.Y., Guasoni, M., Morin, P. *et al.* (2014) Temporal spying and concealing process in fibre-optic data transmission systems through polarization bypass. *Nature Communications*, **5**, 4678.

56 Bony, P.-Y., Guasoni, M., Assémat, E., *et al.* (2013) Optical flip-flop memory and data packet switching operation based on polarization bistability in a telecomunnication optical fiber. *Journal of Optical Society of America B*, **30**, 2318–2325.

57 Gaeta, A.L., Boyd, R.W., Ackerhalt, J.R., and Milonni, P.W. (1987) Instabilities and chaos in the polarizations of counterpropagating light fields. *Physical Review Letters*, **58**, 2432–2435.

58 Gauthier, D.J., Malcuit, M.S., and Boyd, A.R., (1987) Polarization Instabilities of counterpropagating laser beams in sodium vapor. *Physical Review Letters*, **61**, 1828–1830.

59 Gauthier, D.J., Malcuit, M.S., Gaeta, A.L., and Boyd, R.W. (1990) Polarization bistability of counterpropagating laser beams. *Physical Review Letters*, **64**, 1721–1724.

60 Trillo, S. and Wabnitz, S. (1987) Intermittent spatial chaos in the polarization of counterpropagating beams in a birefringent optical fiber. *Physical Review A*, **36**, 3881–3884.

61 Tratnik, M.V. and Sipe, J.E. (1987) Nonlinear polarization dynamics. II. Counterpropagating-beam equations: New simple solutions and the possibilities for chaos. *Physical Review A*, **35**, 2976–2988.

62 Guasoni, M., Bony, P.Y., Gilles, M., Picozzi, A., and Fatome, J. (2015) Fast and chaotic fiber-based nonlinear polarization scrambler. arXiv:1504.03221.

63 Guasoni, M., Bony, P.-Y., Pitois, S., *et al.* (2014) Fast polarization scrambler based on chaotic dynamics in optical fibers. Paper Tu.1.4 in European Conference on Optical Communications ECOC 2014, 5.

64 Fatome, J., Finot, C., Millot, G., Armaroli, A., and Trillo, S. (2014) Observation of optical undular bores in multiple four-wave mixing fibers. *Physical Review X*, **4**, 021022.

65 Mamyshev, P.V. (1998) All-optical data regeneration based on self-phase modulation effect. Paper at European Conference on Optical Communication, ECOC'98, 475–476.

66 Matsumoto, M. (2012) Fiber-based all-optical signal regeneration. *IEEE Journal on Selected Topics in Quantum Electronics*, **18**, 738–752.

67 Provost, L., Finot, C., Mukasa, K., Petropoulos, P., and Richardson, D.J. (2007) Design scaling rules for 2R-Optical Self-Phase Modulation-based regenerators 2R regeneration. *Optics Express*, **15**, 5100–5113.

68 Matsumoto, M. (2004) Performance analysis and comparison of optical 3r regenerators utilizing self-phase modulation in fibers. *IEEE Journal of Lightwave Technology*, **22**, 1472–1482.

69 Finot, C., Nguyen, T.N., Fatome, J., *et al.* (2008) Numerical study of an optical regenerator exploiting self-phase modulation and spectral offset filtering at 40 Gbit/s. *Optics Communications*, **281**, 252–2264.

70 Agrawal, G.P. (2007) *Nonlinear Fiber Optics*, 4th ed. Academic Press, New York.

71 Finot, C., Kibler, B., Provost, L., and Wabnitz, S. (2008) Beneficial impact of wave-breaking for coherent continuum formation in normally dispersive nonlinear fibers. *Journal of Optical Society of America B*, **25**, 1938–1948.

72 Rosenstein, M.T., Collins, J.J., and De Luca, C.J. (1993) A practical method for calculating largest Lyapunov exponents from small data sets. *Physica D*, **65**, 117–134.

73 Richardson, D.J., Fini, J.M., and Nelson, L.E. (2013) Space-division multiplexing in optical fibres. *Nature Photonics*, **7**, 354–362.

74 Bochove, E.J., Cheo, P.K., and King, G.G. (2003) Self-organization in a multicore fiber laser array. *Optics Letters*, **28**, 1200–1202.

75 Guasoni, M., Kozlov, V.V. and Wabnitz, S. (2013) Theory of modal attraction in bimodal birefringent optical fibers", *Optics Letters*, **38**, 2029–2031.

9

All-Optical Pulse Shaping in the Sub-Picosecond Regime Based on Fiber Grating Devices

María R. Fernández-Ruiz,[1] Alejandro Carballar,[2] Reza Ashrafi,[1,3] Sophie LaRochelle,[4] and José Azaña[1]

[1] *Energy, Materials & Telecommunications Research Centre, INRS, Montreal, Quebec, Canada*
[2] *Department of Electronic Engineering, University of Seville, Seville, Spain*
[3] *Department of Electrical and Computer Engineering, McGill University, Montreal, Quebec, Canada*
[4] *Center for Optics, Photonics and Lasers, Laval University, Quebec, Canada*

9.1 Introduction

Optical pulse shaping involves synthesizing the desired shape of the complex (amplitude and phase) temporal envelope of an optical electromagnetic wave, and it plays a fundamental role in communication and computation systems. Pulse shaping techniques are required to generate advanced pulse waveforms that optimize the overall performance of a communication system, e.g., with the aim of extending the transmission reach of the communication link, achieving optical multiplexing at highest spectral efficiency, or limiting nonlinear distortions. Optical pulse shapers are also important building blocks to implement specific source or channel encoding strategies. Moreover, pulse shaping has been also employed for the generation of optimized control signals in nonlinear-based signal processing blocks, aimed at switching and routing of information in the optical domain.

In order to generate a particular optical pulse form, one needs to be able to reliably define the amplitude and phase profile of an optical field. Typically, the pulsed laser sources generate a train of transform-limited pulses with a Gaussian-like complex envelope. However, for some applications it is necessary to modify the shape of these pulses, i.e., their amplitude and/or phase profiles. Examples of desired optical shapes for optical signal processing applications are the following:

1. *Flat-top (rectangular) shape:* One of its main applications is their use in nonlinear optical switching, most prominently in the context of temporal de-multiplexing of optical time-division multiplexing (OTDM) systems [1, 2]. The approximately constant intensity of flat-top pulses defines a clean switching time window avoiding the problem of pulse breakup, which is a main reason of degradation in the

Shaping Light in Nonlinear Optical Fibers, First Edition. Edited by Sonia Boscolo and Christophe Finot.

performance of optical temporal switches [3]. Another application of rectangular pulses is wavelength conversion by optical time gating, where they are employed as pump pulses. The duration of the gate opening time depends on the pump pulse width and the wavelength tunability depends on their pump pulse power [4]. Additionally, flat-top pulses are required on orthogonal frequency division multiplexing (OFDM), where the system data is encoded onto subcarriers with a rectangular shaped impulse response [5].

2. *Parabolic shape:* Parabolic pulses are of interest to achieve ultra-flat self-phase modulation (SPM)-induced spectral broadening in super continuum generation experiments [5, 6]. They are also attractive for nonlinear signal processing methods based on the use of linearly chirped pump pulses, such as for implementation of nonlinear pulse retiming or time-lens processes [7, 8].
3. *Triangular or saw-tooth pulses:* These are widely employed for several applications, such as the implementation of tunable delay lines, time-domain add-drop multiplexing, wavelength conversion, doubling of optical signals, time-to-frequency mapping of multiplexed signals, etc. [6, 9].
4. *Sinc-shape:* They are of high interest for orthogonal time division multiplexing, where the symbols are transmitted by Nyquist pulses (i.e., Nyquist pulse shaping) [10, 11].
5. *High-order modulation codes:* The generation of fixed several-symbol codes using high-order modulation formats is particularly of interest for optical code-division multiple access (OCDMA) and optical-label-switching communications [12].

9.2 Non-Fiber-Grating-Based Optical Pulse Shaping Techniques

A straightforward method for shaping optical waveforms involves the use of high-speed electronics to directly drive an external electro-optical modulator (EOM). The modulating signal is an electronic waveform with the targeted shape that carves the information in an optical carrier [13, 14]. The speed limitation of the electronics typically limits the generated optical waveforms to frequency contents below a few tens of GHz.

Pulse shaping methods based on Fourier optics are well known and have been widely applied. These methods, also referred to as "spectral shaping," routinely offer temporal resolutions better than 100 fs. In the original technique, the temporal information is converted into a one-dimensional spatial domain waveform through a diffraction grating. The resulting waveform is shaped by a simple linear filtering process based on the concatenation of two Fourier-transform systems in a configuration called 4-f_l system, with f_l being the focal length of the involved thin lens. In the central plane of the system, the so-called Fourier plane, an amplitude or phase mask is placed, which acts as the optical filter implementing modulation of the different spatial frequency components of the input beam, as depicted in Figure 9.1. The resulting shaped wave is converted back to the time-domain via a second diffraction grating. To date, conventional picosecond and femtosecond pulse shaping techniques have been implemented by replacing the amplitude or phase mask by advanced devices, such as liquid crystal spatial light modulators [15–17], acousto-optic modulators [15], or electro-optical phase arrays [17], that impart user-specified spectral amplitude and phase modulations on the pulse in a programmable fashion.

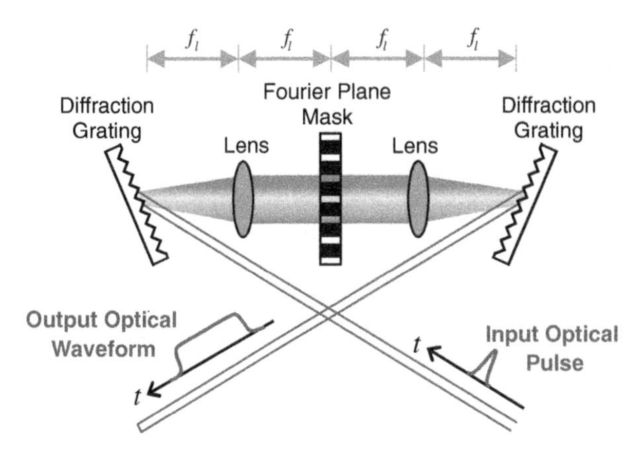

Figure 9.1 Optical arbitrary waveform generation by means of a spatial-domain pulse shaper. The output waveform is determined by the Fourier transform of the pattern transferred from the spatial mask onto the input waveform spectrum.

The main drawback of Fourier-optics-based pulse shaping methods is the need for bulky optical components, which require strict tolerances in their alignment and have limited integration with waveguide devices. Also, the need to couple the shaped waveforms back into a single-mode fiber (SMF) introduces loss and further limits the temporal extent of the generated pulse shapes [18]. To overcome this practical complexity problem, an important body of research has focused on the search for alternative implementations of spectral shaping using optical fiber or integrated-waveguide platforms.

To give a first relevant example, an integrated version of the original Fourier-optics pulse shaping concept has been implemented using arrayed waveguide gratings (AWGs), which have been used to spatially distribute the temporal frequency components of an input waveform. The amplitude and phase profile of the spatially-dispersed pulse is subsequently manipulated with an array of modulators and combined again using a second AWG [19, 20]. AWG-based pulse shapers are compact, integrated devices, but they are typically limited to time resolutions above 10 ps.

A pulse shaping approach similar to the previous one has recently been developed [21]. However, in this case, the employed input signal is an optical frequency comb (OFC). Each line of the OFC is separately modulated using in-phase and quadrature-phase (IQ) modulators and then, the spectral slices are combined to produce a target waveform. This technique is able to generate broad bandwidth signals (scalable to THz frequencies) from narrow spectral slices with bandwidths accessible by current electronics. However, the complexity, cost and power consumption of the required electronic circuits quickly scale up with the bandwidth of the output signal.

With an understanding of the space-time duality (mathematical equivalence between free-space diffraction and narrowband temporal dispersion [22]), the equivalent of pulse-shapers based on spatial-domain optical filtering has been implemented using in-line fiber-optics components. In particular, the spatial 4-f_l system can be substituted by two all-fiber dispersive elements with opposite dispersion and a single EOM in between the dispersive elements. This implementation based on pure fiber optics offers the advantages of smaller size, lower loss, better stability and higher potential for integration [18]. Different schemes have been proposed based on this configuration, using either an electro-optical intensity modulator [23,24] or a phase modulator [25–27], all of them with a resolution reaching the sub-picosecond range.

All the previously mentioned approaches, essentially based on Fourier optics, are programmable, as the filter spectral response is generally generated through an electrical bit pattern generator or arbitrary waveform generator.

Simpler, more compact optical pulse shapers have been also explored using fiber-optics or integrated-waveguide linear optical filters with customized spectral responses. A relevant example is that of optical lattice filters [28–30]. They are a good all-purpose solution due to their ability to generate complicated spectral transfer functions by cascading identical units cells. In a widely used configuration each unit cell employs a combination of a ring resonator and a Mach-Zehnder interferometer, and it contributes a separately controllable pole and zero pair [30]. However, the requirement of a phase-shifter as part of the interferometer in each unit cell makes it difficult to achieve operation bandwidths above a few GHz.

Alternative pulse shaping techniques have been also demonstrated using the concept of temporal coherence synthesization. They are based on the synthesis of the desired output pulse shape by coherently combining a set of input pulse replicas with different time delays [31]; or similarly, using coherent overlapping of different (first and high-order) derivatives of the input optical pulse with specific relative weights [32]. Programmability can be achieved by properly programming the time-delays (first approach) or relative weights using amplitude and/or phase optical modulators (second approach).

Finally, another category of pulse shapers includes nonlinear elements that allow the controlled generation of frequency components outside the frequency spectrum of the input pulse-form. A comprehensive revision of nonlinear pulse shaping approaches can be found in the Chapter 5 of the present book. Nonlinear optical signal-processing devices enable all-optical operations at ultrahigh speeds, i.e., in the femtosecond regime. However, the current level of technological maturity of pulse shaping methods based on nonlinear effects, together with the high degree of control required on the specifications of the input pulse, e.g., power, central frequency, state of polarization, etc., translate into critical drawbacks for their application in practical, real-world systems. In general, optical pulse shaping based on optical nonlinearities is not energy-efficient, high optical powers are commonly required and for now, the pulse shaping operation is limited to a few intensity-only optical shapes [33,34].

9.3 Motivation of Fiber-Grating Based Optical Pulse Shaping

Fiber gratings are periodic perturbations of the refractive index along the core of an optical fiber, typically generated by the exposition of an optical fiber to a spatially varying pattern of ultraviolet light. Since their discovery in 1978, fiber gratings have been extensively studied for their application as linear passive filters, with interesting advantages such a relatively low cost, low losses, polarization insensitivity and full compatibility with fiber-optics systems [35,36]. The main advantage of fiber gratings is their flexibility to implement nearly any desired linear optical filtering functionality, only constrained by practical fabrication limitations. Mathematically, the perturbation of the effective refractive index of the guided mode of interest along the fiber length z is described as

$$n_{FG}(z) = n_{eff} + \Delta n(z) \cdot \cos\left\{ \int_0^z \frac{2\pi}{\Lambda(z')} dz' \right\}, \tag{9.1}$$

where $\Delta n(z)$ is the envelope of the induced refractive index change, also defined as the apodization profile, and $\Lambda(z)$ is the period variation along the grating length.

Several theories have been used to model and describe the functionality of fiber gratings, but the most widely extended is the Coupled Mode theory [37,38]. This theory considers the fiber grating as a device able to couple optical power between two modes (with propagation constants, β_1 and β_2) when the grating period verifies the phase-matching condition [35]

$$\beta_1 - \beta_2 = m\frac{2\pi}{\Lambda}; \quad m = 1, 2, 3, ... \tag{9.2}$$

being m the Bragg-order. Hence, fiber gratings can be broadly classified into two types:

- *Fiber Bragg gratings* (FBGs, also called short-period gratings), which typically have a sub-micron period and act to couple light from the forward-propagating fundamental core-mode to the counter-propagating (backward) core-mode of the optical fiber, i.e, $\beta_2 = -\beta_1$. Single-mode operation is typically assumed. This coupling dominates at a specific wavelength, defined by the Bragg phase-matching condition (Eq. (9.2) considering $m = 1$). FBGs operate as a band-pass filter in reflection and, consequently, as band-stop filter in transmission (Figure 9.2).
- *Long period gratings* (LPGs), which have a period typically in the range of 100s of μm, and induce coupling between the propagating fundamental core-mode and higher-order cladding-modes (Figure 9.3). In LPGs, phase matching occurs at discrete wavelengths determined by the phase matching condition so the wavelength-dependent attenuation resonances occur at a range of different wavelengths, associated with the excitation of different specific higher-order cladding modes [35].

As linear, passive filters, fiber gratings have the properties of linearity and time invariance (LTI). Therefore, their functionality can be described using techniques based on Fourier transform analysis [39]. FBGs can then be characterized in both reflection and

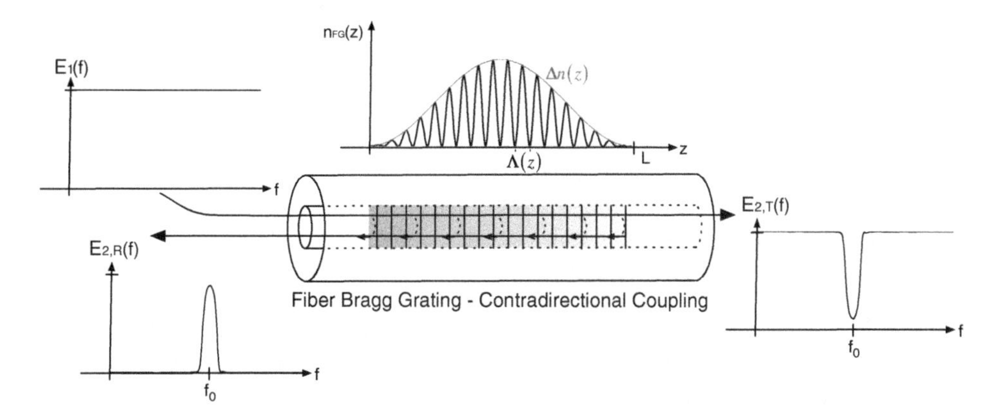

Figure 9.2 Schematic diagram of a fiber Bragg grating (FBG), indicating two operation possibilities: the reflection mode and the transmission mode.

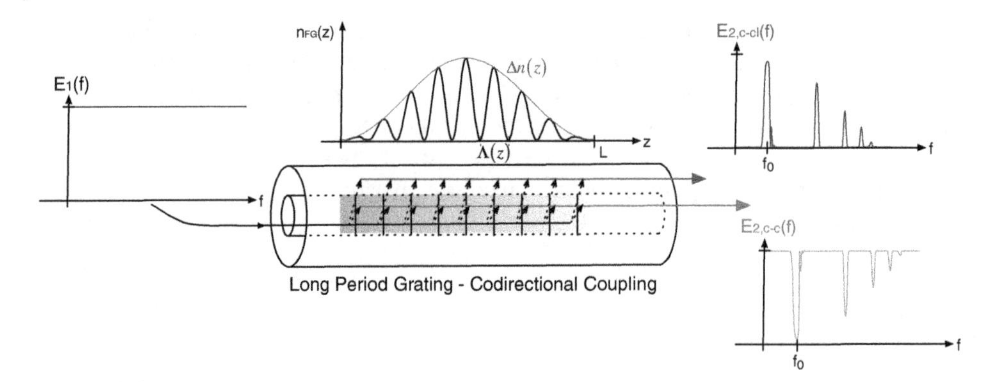

Figure 9.3 Schematic diagram of a long-period fiber grating (LPG), indicating two operation possibilities: the core-to-core mode and the core-to-cladding mode.

transmission operation by their frequency responses, $H_R(f)$ and $H_T(f)$, i.e., the spectral transfer function, or the corresponding temporal impulse responses, $h_R(t)$ and $h_T(t)$):

$$H_R(f) = \frac{E_{2,R}(f)}{E_1(f)}; \quad h_R(t) = \mathcal{F}^{-1}[H_R(f)]; \tag{9.3}$$

$$H_T(f) = \frac{E_{2,T}(f)}{E_1(f)}; \quad h_T(t) = \mathcal{F}^{-1}[H_T(f)], \tag{9.4}$$

where $\mathcal{F}$ denotes Fourier transformation, t is the time variable, and $f = f_{opt} - f_0$, with f_{opt} being the optical frequency variable and f_0 being the carrier frequency.

On the other hand, LPGs work in transmission offering two possibilities: the core-to-cladding coupling mode or core-to-core coupling, where single-mode operation is typically assumed in the fiber as well. For LTI analysis, it is assumed that the grating is operated over a relatively limited bandwidth around one of its resonance wavelengths so that one can assume that the coupling mainly occurs between the core mode and the single cladding mode corresponding to the relevant spectral resonance. Under this assumption, similarly to FBGs, the analogous spectral transfer functions and temporal impulse responses for the two interacting LPG modes are obtained as:

$$H_{c-cl}(f) = \frac{E_{2,c-cl}(f)}{E_1(f)}; \quad h_{c-cl}(t) = \mathcal{F}^{-1}[H_{c-cl}(f)]; \tag{9.5}$$

$$H_{c-c}(f) = \frac{E_{2,c-c}(f)}{E_1(f)}; \quad h_{c-c}(t) = \mathcal{F}^{-1}[H_{c-c}(f)]. \tag{9.6}$$

Let us assume an input optical waveform analytically defined as $e_1(t) = x(t) \cdot \exp(j2\pi f_0 t)$, where $x(t)$ is the input waveform's complex envelope and f_0 is the optical carrier frequency. By launching this optical waveform at the input of a fiber grating-based pulse shaper with a well-defined response around f_0, the output optical waveform can be expressed as $e_2(t) = y(t) \cdot \exp(j2\pi f_0 t)$. In this expression, $y(t)$ is the output waveform's complex envelope and it can be mathematically obtained as:

$$y(t) = h(t) * x(t); \tag{9.7}$$

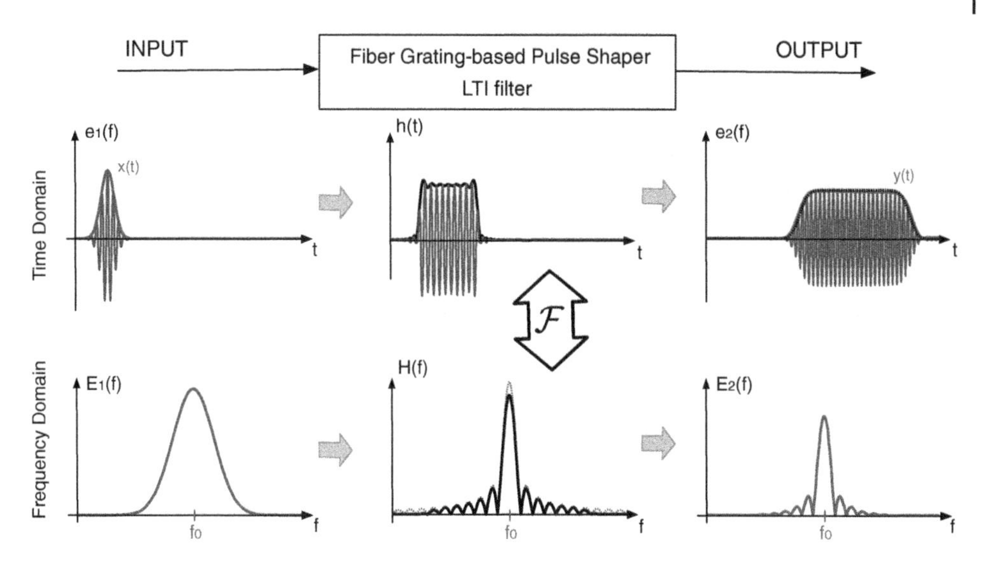

Figure 9.4 Schematic diagram for the operation principle of a fiber grating-based pulse shaper in the time and frequency domain.

where $h(t)$ is the filter's impulse response complex envelope (at f_0) and $*$ denotes the convolution operation. By applying the Fourier transform to Eq. (9.7), we then have:

$$Y(f) = H(f) \cdot X(f) \tag{9.8}$$

where $X(f)$, $Y(f)$ and $H(f)$ are the Fourier transforms of $x(t)$, $y(t)$ and $h(t)$, respectively. The frequency transfer function $H(f)$, or correspondingly the temporal impulse response $h(t)$, of the fiber grating can be designed to generate a target output complex envelop $y(t)$, within the frequency-bandwidth and time-window limitations of the linear filtering device, as depicted in Figure 9.4.

To design the targeted pulse shaper, both the spectral response of the input signal $X(f)$ and the desired output signal $Y(f)$ must be specified. As with any LTI filter, the transfer function of the pulse shaper is obtained as $H(f) = Y(f)/X(f)$. In order to achieve the targeted grating spectral response, nearly all the grating physical parameters can be varied: induced index change $\Delta n(z)$, length L, period chirp $\Lambda(z)$, and whether the grating supports counterpropagating or copropagating coupling at a desired frequency [35]. Intensive research has been carried out to develop design tools that provide a direct relationship between the refractive index perturbation and the resulting grating spectral response. Thus, analysis tools [35–38], i.e., tools that provide the grating spectral response $H(f)$ from the refractive index profile $n_{FG}(z)$; and synthesis tools, i.e., tools that provide the required $n_{FG}(z)$ from a targeted $H(f)$, have been developed and they nowadays possess a high level of technical maturity [40–43].

The performance of the fiber grating-based pulse shapers is typically characterized by two parameters: the time-bandwidth product (TBP) and the energy efficiency. The TBP is obtained from the measurement of the cross-correlation coefficient, C_C between the actual, $y(t)$, and the ideal, $y_{ideal}(t)$, output waveforms. It can be obtained by:

$$C_C[\%] = \frac{\int_{-\infty}^{+\infty} y(t) \cdot y_{ideal}(t) \cdot dt}{\sqrt{\int_{-\infty}^{+\infty} y^2(t)dt \cdot \int_{-\infty}^{+\infty} y_{ideal}^2(t)dt}} \times 100\% \tag{9.9}$$

The TBP is calculated as the ratio between the maximum and the minimum temporal full-width at half maximum (FWHM) of an input Gaussian pulse for which the C_C is higher than a predefined value, e.g., 90% (corresponding to the ratio between the maximum and minimum 3dB-bandwidth of an input Gaussian pulse for which the C_C holds higher than the predefined value). On the other hand, the energy efficiency is defined as the output-to-input pulse energy ratio.

An important limitation of fiber grating solutions in their application as pulse shapers is the lack of reconfigurability. Fortunately, there are a broad range of applications for which a particular fixed temporal shape is required, e.g., rectangular or parabolic pulses. In those cases, a simpler, high efficient and easily reproducible component would be highly recommended over the schemes presented in Section 9.2. Some interesting pulse shaping components have been proposed and demonstrated simply based on passive linear filters, such as FBG or LPG [7,9,44–79]. Moreover, intense study is being done to be able to integrate these periodic structures on photonics integrated chips (PICs) [80–86].

9.3.1 Fiber Bragg Gratings (FBGs)

FBG is a fundamental component for many applications in fiber-optic communications, which has been extensively used and studied for a very wide range of applications based on linear optical filtering [7,9,44–74]. The fundamental advantage of FBGs is their flexibility to achieve almost any arbitrary spectral/temporal responses for optical pulse shaping when operating in reflection [35,36].

9.3.1.1 Fiber Bragg Gratings Operating in Reflection

The more direct approach of use of an FBG as a frequency-filtering stage for pulse shaping applications consists of operating the grating in reflection within the limit of the first-order Born approximation (weak-coupling condition) [7, 9, 44–47]. Accordingly, if the grating is sufficiently weak, then the corresponding reflection impulse response $h_R(t)$ is directly related to the spatial profile of the index-modulation depth $\Delta n(z)$. This simple approach permits the synthesis of nearly any desired optical waveform with resolutions in the picosecond regime. As relevant examples of this design approach, we can mention the implementation of flat-top [45, 46], parabolic [7] and saw-tooth (triangular) [9] pulse shapers based on superstructured FBGs. However, working within the constraints of the Born approximation limits the maximum duration of the temporal waveforms that can be synthesized. In particular, for gratings that are longer than a few centimeters (this corresponds to temporal durations of the order of hundreds of picoseconds), the approach starts to fail even for very weak index modulations. The reason is that the Born approximation is strictly valid when $\kappa L \ll 1$, where κ is the coupling coefficient, related to the index modulation depth as $\kappa = \pi\Delta n/\lambda$ (λ is the wavelength, $\lambda = c_0/f$, with c_0 being the speed of light in vacuum). When the grating is too long, the required Δn may be too weak to ensure light penetrates the full length so that the whole grating contributes identically to the reflected signal, challenging its practical realization.

Another approach for FBG-based pulse shaping based on the Born approximation was proposed to overcome the limitation in the duration of the output pulses of the previous method. The technique is based on a space-to-frequency-to-time mapping [48], and it enables the synthesis of arbitrary waveforms in the picosecond/nanosecond regime, although with a lower energy efficiency. This method consists of using a specially designed apodized linearly chirped FBG, in the regime of weak coupling condition. If the FBG chirp-induced dispersion $\ddot{\Phi}$ (s^2) is sufficiently large so that $\ddot{\Phi} \gg \Delta t_1^2/4$, with Δt_1 being the total temporal duration of the shaper impulse response, the amplitude of this impulse response is proportional to the amplitude of the spectral response of the filter. If besides, the chirped FBG works within the Born approximation, the magnitude of the grating reflection spectral response $H_R(f)$ is approximately proportional to $\Delta n(z)$. Therefore, under these conditions, the shape of the grating apodization profile directly determines the temporal as well as the spectral shape of the output pulse. The desired temporal (or spectral) waveform only needs to be spatially "recorded" in the apodization mask used to write the grating. The application of this approach is limited to amplitude-only pulse shaping, and the time duration of the output pulses is limited by the length of the grating that can be written, typically in the nanosecond regime (corresponding to a physical fiber-grating length shorter than ~10 cm).

More advanced solutions for the use of FBGs as pulse shapers involve the use of general grating-synthesis algorithms, such as inverse-scattering or layer-peeling techniques [41–43], in order to determine the amplitude and phase profiles of the refractive-index modulation that are necessary to obtain a given temporal response. These methods allow one to design high-reflectivity FBGs to achieve the desired pulse-shaping operation, leading to increased energy efficiency.

Employing these gratings design techniques, FBGs have been designed to implement several optical signal-processing devices, which have been in turn employed, directly or indirectly, as optical pulse shapers:

Optical Differentiator An optical differentiator is a signal processing device that provide at its output the derivative of the complex envelope of an input optical signal. The ideal transfer function of an arbitrary order differentiator can be expressed as

$$H_{diff}(f) \propto j \cdot (f_{opt} - f_0)^N, \tag{9.10}$$

where $j = (-1)^{1/2}$ and N is the differentiator order. A number of different approaches for first- and high-order optical differentiators have been proposed based on FBGs operating in reflection, each of them trying to improve their operation bandwidth or spectral resolution within the practical limitations of the technology [49–52]. Optical differentiators have been employed for the generation of a first- and high-order Hermite-Gaussian (HG) optical pulses [53]. HG polynomials form a complete set of orthogonal temporal functions, a property that can readily be exploited in advanced coding for network access applications. Moreover, based on this property, a platform has been proposed for programmable arbitrary optical pulse shaping involving the addition of weighted first- and high-order derivatives of an input Gaussian-like pulse [32].

Optical Integrator Another important pulse processor whose implementation has been widely investigated based on FBG technology is the all-optical integrator. The optical integrator performs the cumulative time integral of the complex temporal envelope of an input arbitrary optical waveform. Its transfer function is

$$H_{int}(f) \propto \frac{1}{j \cdot (f_{opt} - f_0)^N}. \tag{9.11}$$

As with the case of the optical differentiator, a significant effort has been devoted to the development of broadband, high-resolution arbitrary order (N^{th}) optical integrators, and implementations based on passive and active configurations have been proposed and, in some cases, also experimentally demonstrated [54–59]. FBG-based optical integrators can readily be employed for the generation of unit-step time-domain functions and the implementation of flat-top pulse shapers, enabling reconfigurability of the flat-top temporal width [54, 57].

Optical Hilbert Transformer FBG-based Hilbert transformers (HT) have been also proposed and experimentally demonstrated [60–63]. A photonic HT is a pulse processor that provides the Hilbert transform of an incident optical pulse. The transfer function of an arbitrary order photonic HT is defined as

$$H_{HT}(f) \propto \begin{cases} e^{-jP\pi/2} & \text{if } f_{opt} > f_0, \\ 0 & \text{if } f_{opt} = f_0, \\ e^{jP\pi/2} & \text{if } f_{opt} < f_0, \end{cases} \tag{9.12}$$

where P is the fractional order. In the particular case of $P = 1$, the HT is integer. HTs, also known as phase shifters, are important components for a wide range of applications in the fields of computing and communications. As pulse shapers, HTs are particularly useful for generation of phase-shifted pulse doublets, where the order of the HT allows one to control the amplitude ratio between the two resulting pulse lobes.

Optical Coders/Encoders FBGs have been employed as time-domain coder/decoders for OCDMA systems [64–70]. Different approaches have been investigated, e.g., schemes based on arrays of uniform FBGs for generation of bipolar codes, requiring accurate controlled phase between individual gratings [64], or more compact schemes such as those based on superstructured FBGs [65–67] or step-chirped FBGs [68]. A reconfigurable scheme has also been proposed based on a simple uniform FBG with localized heating elements applied through thin resistive wires. These wires are able to apply different phase shifts by inducing impermanent chirp within a small confined segment of the FBG, eventually generating a desired phase code [69].

9.3.1.2 Fiber Bragg Gratings Operating in Transmission

Several implementations of optical pulse processors, and in particular optical pulse shapers, have been proposed recently using an FBG operating in transmission [71]. The use of FBG operating in transmission provides interesting advantages: transmissive FBGs avoid the requirement of an optical circulator or optical coupler to separate the output signal from the input, what reduces the complexity and cost of the processor unit. Also, they are more robust against fabrication errors due to the weak interaction between the signal optical field and the grating when the signal is simply transmitted through the device (instead of reflected), so that imperfections in the grating are not directly "impressed" upon the signal field [87]. However, FBGs in transmission have an important drawback for their application as processing units based on the implementation of a prescribed complex spectral transfer function. The amplitude of the grating's linear spectral response in transmission $H_T(f)$ uniquely determines its phase by means of the Kramers-Kronig relationship, that is, the transmission spectral response is a minimum phase (MP) function [88]. Therefore, there is no freedom in choosing the complex spectral response to be implemented. In principle, only temporal shapes whose spectrum is an MP function can be implemented using this configuration.

Using the transmissive configuration, flat-top pulse shapers have been numerically proposed [72], as their required spectral response is an MP function

$$H_{ft}(f) \propto \text{sinc}\,(f \cdot \tau_{FWHM}), \tag{9.13}$$

where the sinc function is defined as $\sin(\pi f)/(\pi f)$, and τ_{FWHM} is the duration of the desired output flat-top pulse. Also, optical differentiators based on FBGs in transmission were proposed [73]. However, unpractical complex apodization profiles with high peak coupling strengths and a large number of precisely located discrete phase shifts are needed when the processing bandwidth exceeds ~100 GHz.

Alternatively, flat-top, parabolic and rectangular pulse shapers based on phase-modulated FBGs operating in transmission have been implemented [74]. This method relies on engineering the period profile $\Lambda(z)$ (instead of the apodization profile) to achieve the desired pulse shaping functionality in the transmissive spectral response of the FBG. This technique facilitates the grating fabrication process, as the coupling strength remains constant along most of the grating length. However, the fabrication constraints, such as the limitation in the spatial resolution, are transferred to the implementation of a user-defined relatively complex phase mask.

All the previously presented approaches for conventional FBG-based optical signal processing, and in particular optical pulse shaping, present a stringent operation bandwidth limitation. The processing bandwidth of FBGs is limited by the achievable spatial resolution of available fabrication technologies for tailoring the coupling coefficient profile along the grating length [36]. For instance, considering a typically feasible sub-millimeter resolution for the fiber grating apodization profiles, FBG-based pulse shaping implementations have been limited to temporal resolutions of at least several picoseconds, i.e., corresponding to a few hundreds of GHz in terms of frequency bandwidth of the generated output optical pulse shapes [36].

9.3.2 Long Period Gratings (LPGs)

Fiber LPGs are grating-assisted co-directional couplers, which induce coupling between the propagating core-mode and co-propagating cladding-modes of the optical fiber [35, 36]. These devices have recently attracted a great deal of interest for linear optical pulse shaping and processing applications. Thus similar building blocks as those ones introduced in the previous sections for FBGs, e.g., optical differentiators, integrators, Hilbert transformers and ultrafast pulse coders/decoders have been designed and implemented based on LPGs [75–79]. Those building blocks have been employed for a wide variety of signal-processing functionalities, including pulse-shaping applications. The large bandwidth typical for these fiber filters allows scaling the filtering technique to the THz-bandwidth regime.

With regard to pulse processing work using LPGs, some interesting LPG-based optical code generation designs were first reported in [77] but they proved to be limited to the synthesis of temporally symmetric and binary intensity-only (on-off-keying, OOK) optical codes. Also, LPG-based differentiators have been reported and used for (sub-)picosecond optical pulse shaping [78, 79]. However, the pulse shaping based on this original approach is strictly limited to the generation of very specific waveform shapes, more prominently the generation of (sub-)picosecond flat-top and first- and high-order Hermite-Gaussian pulse waveforms from Gaussian-like optical pulses.

9.4 Recent Work on Fiber Gratings-Based Optical Pulse Shapers: Reaching the Sub-Picosecond Regime

Fiber gratings solutions offer interesting advantages that make them adequate technological components in recent and future communication systems. However, as reviewed in Section 9.3, they have also important drawbacks, such as the limited operation bandwidth and the lack of reconfigurability. In the last few years, a great deal of research has been carried out to find solutions that overcome those drawbacks, especially the limitation in the operation bandwidth. In this section, we describe the recent advances in the design and implementation of fiber grating-based pulse shapers that have enabled the achievement of ultrafast operation speed, i.e., well in the THz regime, corresponding to time resolutions down to the sub-picosecond regime.

9.4.1 Recent Findings on FBGs

FBGs have previously been used for many interesting optical time-domain linear signal-processing operations, such as optical pulse shapers, differentiators, integrators or Hilbert transformers. In general, FBG-based optical filters usually work in reflection. As introduced in Section 9.3.1, FBGs working in transmission directly overcome two of the constraints of reflective FBGs, i.e., avoiding the need for additional devices (e.g. circulators) and being more robust against fabrication errors. Recently, it has been demonstrated how transmissive FBGs are able to operate well in the THz regime without increasing the spatial resolution of the apodization profile, as detailed below. However, their main restriction is that the spectral transfer function is necessarily MP [88], imposing a very tight relationship (Hilbert transform) between the real and imaginary parts of the filter spectral response. Fortunately, an important, broad set of optical pulse processors (e.g. time differentiators, integrators) and shapers (e.g., for flat-top, triangular, parabolic pulse generation, among others) can still be realized using MP optical filters.

The design of a linear optical filter to be implemented in an FBG operating in transmission requires the specification of a prescribed minimum-phase spectral transfer function, $H_{ideal}(f)$. Thus, $H_T(f)$ must be proportional to $H_{ideal}(f)$ over the target operation bandwidth. In the notation used, $|H(f)|$ corresponds to the magnitude of the spectral response and $\arg(H(f))$ is the spectral phase.

In an MP filter (e.g., FBG in transmission), the filter's spectral phase response is necessarily determined by the desired spectral amplitude response by means of the

Kramers-Kronig relations. In addition, in an FBG, the transmissivity ($T = |H_T(f)|^2$) and reflectivity ($R = |H_R(f)|^2$) are necessarily related by $T = 1 - R$, due to the principle of conservation of energy. Thus, the specifications of $|H_T(f)|$ uniquely impose the functions $\arg(H_T(f))$ and $|H_R(f)|$. Therefore, from the specifications of the desired transmission amplitude spectral response, $|H_T(f)|$, the design problem reduces to synthesizing an FBG providing the reflection amplitude spectral response, $|H_R(f)|$, with no additional constraints on the FBG reflection spectral phase response, $\arg(H_R(f))$. Hence, the FBG reflection spectral phase can be suitably fixed to achieve the simplest grating design according to the target specifications. Among different possibilities, minimum-phase, maximum-phase, linear phase, quadratic phase, cubic phase profiles (and so on) can be used [71].

It has been demonstrated that the use of a reflective quadratic spectral phase enables the synthesis of arbitrary amplitude spectral responses over bandwidths well in the THz range using feasible and remarkably simple FBG apodization profiles [89]. Recall that a quadratic spectral phase corresponds to a linear group-delay profile. The desired linear group-delay in the FBG reflection response can be induced by introducing a quasi-linear grating-period variation (or grating-period chirp) along the device length. As the reflected frequency components along the grating are related to its period, the bandwidth of the spectral response in reflection, i.e. the corresponding rejected bandwidth in transmission, can then be significantly higher than in the uniform grating-period case.

Using this strategy, the FBG reflection spectral response to be synthesized can be mathematically expressed by

$$H_R(f) = W(f)\sqrt{R_{max}(1 - |H_T(f)|^2)} \cdot \exp\left\{j\left(\frac{1}{2}\ddot{\Phi}(2\pi f)^2 + (2\pi f)\tau_d\right)\right\}, \tag{9.14}$$

where R_{max} is the maximum peak reflectivity; $\ddot{\Phi}$ is the dispersion parameter (s^2) or equivalently, the slope of the group-delay curve as a function of the angular frequency; $W(f)$ represents a windowing function, which is introduced considering that the reflective response of an FBG must be a limited band-pass filtering function; and τ_d is a time delay introduced to make the device causal, generally equal to half the duration of the device impulse response, i.e. $\Delta t_1/2$. The dispersion parameter $\ddot{\Phi}$ will determine the minimum grating length L, according to the relationship:

$$\ddot{\Phi} = \frac{n_{eff}L}{\pi B c_0}, \tag{9.15}$$

being n_{eff}, the effective refractive index of the grating, and B the full-width reflection bandwidth of the device (Hz). From the target specifications defined by Eq. (9.14), the grating perturbation to be implemented, i.e., the apodization profile, can be obtained from a layer-peeling FBG synthesis algorithm, e.g., based on Coupled Mode theory combined with the Transfer Matrix method [41].

The dispersion parameter $\ddot{\Phi}$ is a fundamental design parameter which can be properly selected to ensure that the resulting grating apodization specifications match the fabrication constraints. In particular, a higher dispersion value translates into a more relaxed spatial resolution and a lower refractive index modulation peak, but requiring a longer device. It can easily be tuned by accordingly changing the slope of the linear grating-period variation (for a fixed operation bandwidth).

9.4.1.1 Minimum Phase Functionalities

The introduced approach has been exploited to design two relevant ultra-fast (THz-bandwidth) all-optical signal processing devices, in particular, first- and high-order differentiators [89,90] and a flat-top pulse shaper [89,91].

Optical Differentiators A first-order all-optical differentiator with a processing bandwidth of $B = 2$ THz (full-width at 0.1% of the maximum spectral amplitude) has been recently designed using the approach described above [89]. The dispersion parameter $\ddot{\Phi}$ was set to 80 ps^2, and R_{max} was set to 99.9999%. In this case, no windowing is required ($W(f) = 1$) as the reflective transfer function is inherently a band-pass function.

A synthesis tool [41] was employed to obtain the grating profile of the desired FBG-based optical differentiator from the specifications in Eq. (9.14), which is plotted in Figure 9.5 (a), in the color plate section. The synthesized grating device is readily

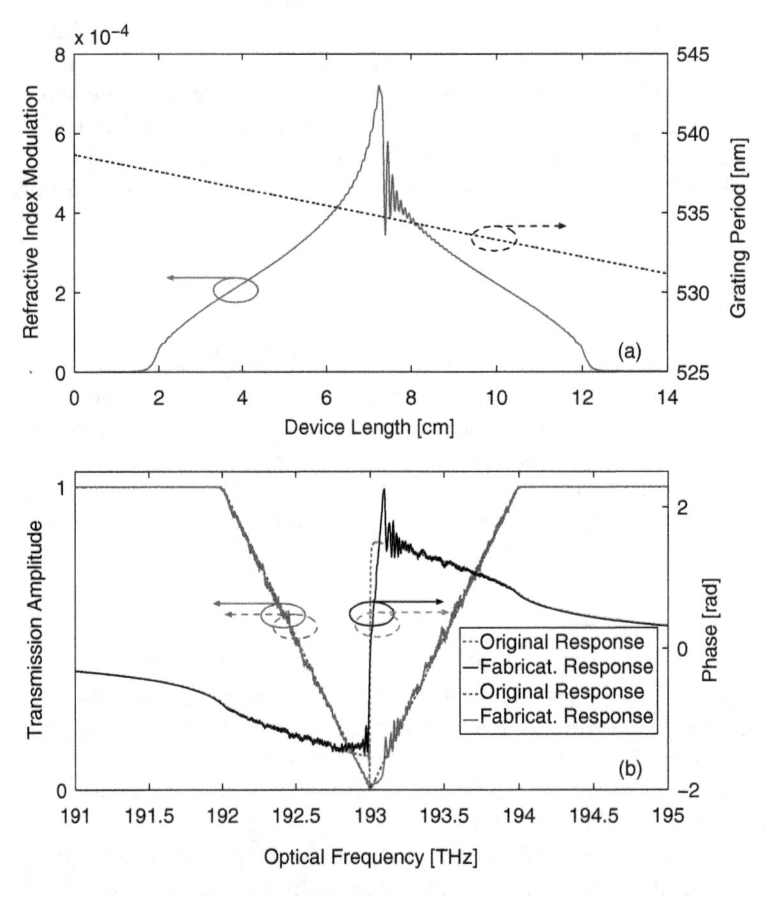

Figure 9.5 First-order optical differentiator based on an FBG in transmission: (a) Apodization (red solid line) and period (black dashed line) profiles considering realistic spatial resolution (limited to 0.3 mm) and linearly chirped phase mask; (b) Specified transmission spectral response, amplitude (dotted blue line) and phase (dotted green line) compared to fabrication-constrained transmission spectral response, amplitude (solid red line) and phase (solid black line). *Source*: María 2013 [89]. For a color version of this figure please see color plate section.

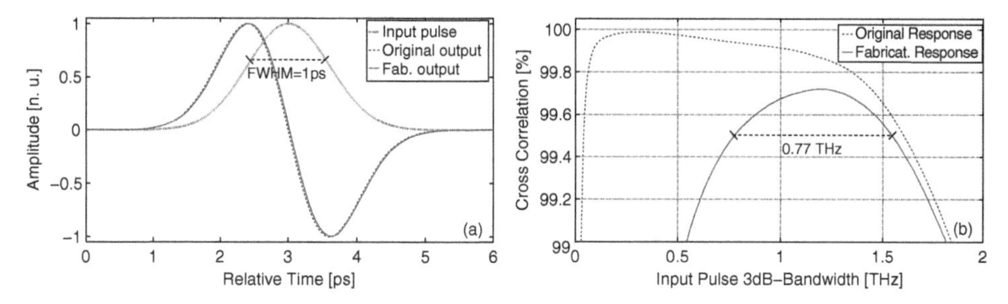

Figure 9.6 First-order optical differentiator based on an FBG in transmission: (a) Input Gaussian-like pulse (dotted black line) and comparison between the original output (dashed blue line) and the numerically obtained output from a differentiator adapted to fabrication constrains (red solid line); (b) Cross-correlation value between the ideal and the numerically obtained temporal output waveform as a function of the 3 dB-bandwidth of an input Gaussian pulse for the original design (blue dashed line) and the perturbed design considering fabrication limitations (red solid line). *Source*: María 2013 [89].

feasible with current fabrication technology, with (i) an effective length of 10.37 cm (as expected from Eq. (9.15)), (ii) a chirp value of −0.52 nm/cm, (iii) an amplitude-only apodization function with a peak refractive index modulation $\Delta n_{max} = 0.72 \times 10^{-3}$, and (iv) an average spatial resolution of the ripples in the apodization profile of ~ 0.3 mm. Therefore, the fabrication parameters were significantly relaxed as compared with previous designs based on uniform-period transmission gratings [73], while the achievable processing speed was increased in one order of magnitude. The robustness of the design was analyzed by evaluating the sensitivity of the grating transmission spectrum to realistic fabrication constraints, i.e., sub-millimeter resolution of the apodization profile and maximum refractive index modulation $< 10^{-3}$, shown in Figure 9.5 (b) in the color plate section.

Figure 9.6 (a) presents the comparison between the original expected output signal and the output from an FBG with typical fabrication constrains, showing an excellent match between the two curves. Besides, as observed in Figure 9.6 (b), the differentiator performance, estimated by its TBP, is mainly degraded for wider temporal input pulses (corresponding to frequency bandwidths narrower than 0.5 THz) while preserving a coefficient higher than 99.5% when the input pulse has a 3 dB-bandwidth ranging between 0.78 and 1.55 THz (TBP ~ 2). This fact is predominantly attributed to the deviations between the original and fabrication-constrained device spectral responses around the transmission resonance notch.

The design of 2 THz higher order (up to $N = 4$) optical differentiators was also proposed based on the same design methodology, shown in Figure 9.7 [90]. All the design parameters were set to similar values to those ones from the previous example. Only the specified transmission deep varies from −60 dB from the first order to −90 dB for the fourth order. Numerical simulations are used to evaluate the output temporal waveforms when a Gaussian-like 850 fs-FWHM input pulse is launched at the input of the FBG. Figure 9.7 (b) shows the first-to-fourth order derivatives of the input pulse, exhibiting an excellent agreement with the expected output waveforms of the ideal differentiators.

Optical Pulse Shapers Picosecond flat-top (rectangular-like) optical pulse shapers, over full-width spectra in the THz regime, have been designed and implemented using

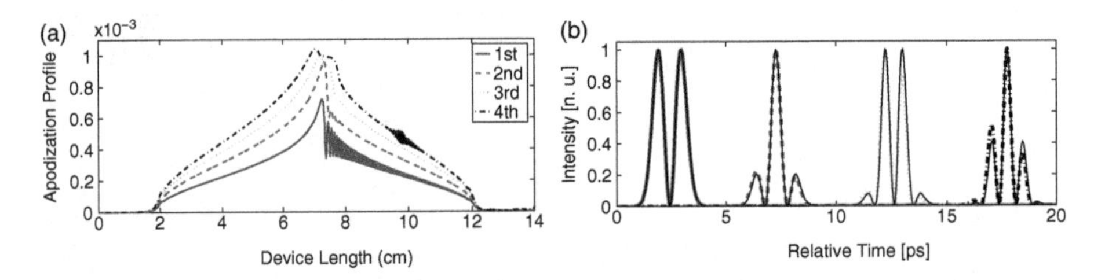

Figure 9.7 N^{th}-order optical differentiators based on FBGs in transmission: (a) Apodization profiles considering realistic spatial resolution (limited to 0.5 mm); (b) Intensity outputs from practical FBG-based differentiators. *Source*: María 2012 [90]. Reproduced with permission from IEEE.

transmission FBGs [89,91]. Assuming a sufficiently short Gaussian input pulse, the output will be a flat-top pulse if the filter provides an spectral transfer function approaching the ideal sinc function (Eq. (9.13)).

For the fabrication of the linearly-chirped-FBG, a 14.5 cm-long phase mask was used, with a period of 1064.05 nm (i.e., grating Bragg wavelength of 1544.87 nm) and a chirp of 2.5 nm/cm. These parameters fixed the dispersion-induced in the grating to $\ddot{\Phi} = 33.77$ ps^2. The target output temporal waveform was set to be a 2-ps flat-top optical pulse, and therefore, $\tau_{FWHM} = 2$ ps. In this case, a raised cosine function was employed as a window function $W(f)$. The full-bandwidth of $W(f)$ (measured at 0.1% of the maximum reflectivity) was chosen to be 5 THz. The output pulse spectrum was then fixed to extend over a full width of ~5 THz, thus including a few sidelobes of the sinc function. R_{max} was set to 99.99% (transmission dip of −40 dB).

Applying the synthesis tool to this prescribed reflection spectral response, the apodization and period profiles shown in blue in Figures 9.8 (a) and (b) in the color plate section were obtained. The synthesized device has an effective length of 11 cm with a peak refractive index of $\Delta n_{max} = 1.2 \times 10^{-3}$. Due to the expected limitation in the fabrication resolution, the apodization profile has been smoothed to have a conservative spatial resolution of 1 mm. The smoothed grating-apodization profile used

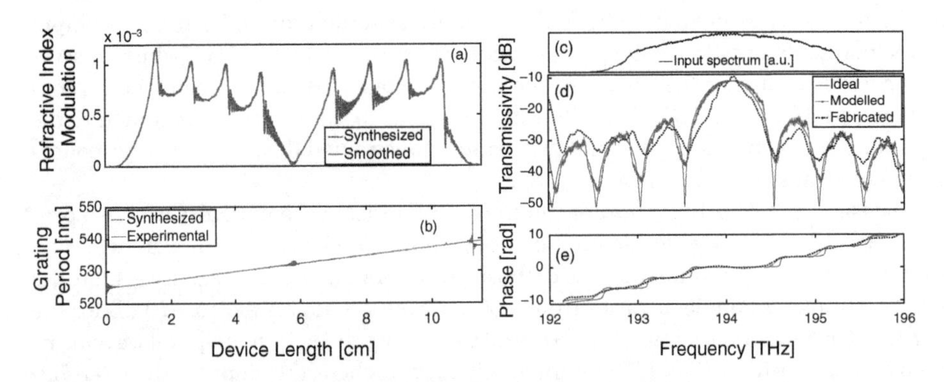

Figure 9.8 Flat-top pulse shaper based on an FBG in transmission: Obtained apodization profile (a) and period (b) from synthesis (blue line) and smooth profiles adapted to fabrication constrains (red line); (c) Input signal spectrum; (d) tranmissivity and (e) transmission phase response: ideal (blue line), simulated considering fabrication constraints (red line) and measured from fabricated device (dotted black line). *Source*: María 2013 [91]. For a color version of this figure please see color plate section.

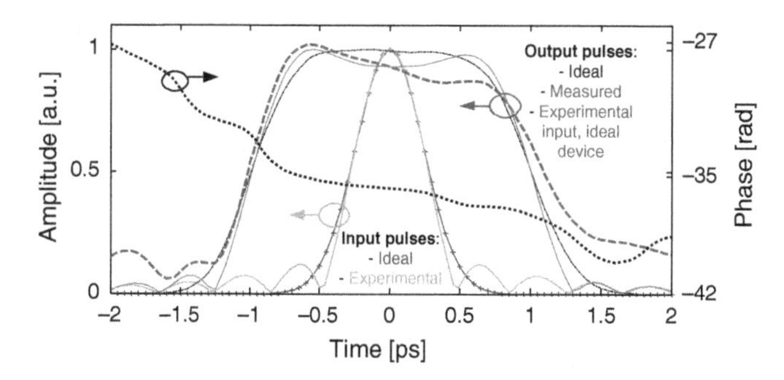

Figure 9.9 Flat-top pulse shaper based on an FBG in transmission: Temporal input response: ideal (blue line) and employed in the experiment (green line); Temporal output: ideal (black dashed-dotted line) and experimentally measured (red dashed line). *Source*: María 2013 [91]. For a color version of this figure please see color plate section.

for the linearly-chirped-FBG fabrication, as well as the linear period profile provided by the employed phase mask, are shown in red in Figures 9.8 (a) and (b). Notice that no additional phase variations are needed on top of the linear grating period chirp.

Based on the calculated refractive index profiles, the linearly-chirped-FBG were fabricated via UV illumination of hydrogen-loaded specialty single-mode fiber (UVS-INT from Coractive) by a frequency-doubled argon-ion laser operating at 244 nm through the described linearly-chirped phase mask. The designed refractive index apodization was implemented by dithering the phase mask during the fiber scan in order to control the visibility of the interference pattern while keeping the average refractive index constant. Figures 9.8 (d) and (e) show the resulting transmissivity and transmission phase compared with the specified ideal ones.

For full time-domain characterization, a Fourier transform spectral interferometry (FTSI) [92] scheme was implemented. The input and output temporal waveforms are plotted in Figure 9.9 in the color plate section. The device is optimized for an input Gaussian-like pulse with a FWHM of 400 fs, depicted in blue in Figure 9.9. A comparison between the measured data (red line) and the target data (black line) shows a good agreement between the ideal and the experimentally obtained pulse, confirming the capability of the fabricated grating for the target ultrafast pulse-shaping application. The FWHM of the measured output pulse's amplitude is 2.1 ps. The 5-THz operation bandwidth provides a rising/decaying time (measured between the 10% and the 90% of the maximum amplitude) of ~0.5 ps.

9.4.1.2 Non-Minimum Phase Functionalities (Arbitrary Optical Pulse Processors)

In spite of the important advantages of using FBGs operating in transmission, they still have a fundamental limitation, which is the fact that their linear spectral response is necessarily MP. An important body of research has been carried out to explore the possibility of implementing non-MP functionalities based on transmissive BGs [71,93]. However, the extremely restrictive MP condition has severely limited the functionalities susceptible to be implemented using this configuration.

Recently, a general approach that enables the implementation of an arbitrary linear optical pulse processing functionality, generally requiring a non-MP filtering response (in which the amplitude and phase spectral profiles need to be tailored independently) has been presented using a MP optical filter [94]. This approach is based on the fact that any causal temporal function with a dominant peak around or close to the origin will be either an MP function or close to one [39,95,96]. Based on this general property, the proposed design procedure involves creating a temporal impulse response that satisfies the MP condition by simply introducing an instantaneous component, e.g., a Dirac delta function $\delta(t)$, before the target non-MP impulse response [95,96]. The resulting impulse response is

$$h_{MP}(t) = K_1\delta(t) + K_2 h_{NMP}(t - \tau_G), \tag{9.16}$$

where τ_G is the time delay between the instantaneous component and the target non-MP impulse response, and K_i, $i = 1, 2$ are the weights given to each of the components of $h_{MP}(t)$, which determine the signal energy distribution at the system output. The spectral transfer function corresponding to this impulse response is MP and given by

$$H_{MP}(f) = K_1 + K_2 H_{NMP}(f)\exp\{-j2\pi f\tau_G\}. \tag{9.17}$$

Notice that the resulting spectrum from Eq. (9.17) looks like an interferometric pattern where the spectral phase information is coded in the phase variations of the resulting cosine-like spectral profile. A formal proof on the MP property of the transfer function defined by $H_{MP}(f)$ can be found in [96].

$H_{MP}(f)$ can be then synthesized by means of an FBG operating in transmission. Hence, $H_T(f)$ must approach $H_{MP}(f)$ over the device's operation bandwidth. To completely define $H_T(f)$, the constants K_1 and K_2 have to be calculated considering the grating physical constraints. On the one hand, since an FBG is a passive device, the maximum transmissivity $T_{max} = 1$, and consequently

$$|H_T(f)| \leqslant \sqrt{T_{max}} \rightarrow K_1 + K_2 \leqslant 1. \tag{9.18}$$

On the other hand, the maximum reflectivity R_{max} achieved by the device imposes a limitation on the minimum of $|H_T(f)|$

$$|H_T(f)| \geqslant \sqrt{1 - R_{max}} \rightarrow K_1 - K_2 \leqslant \sqrt{1 - R_{max}}. \tag{9.19}$$

This inequality becomes strict if $R_{max} = 1$ to avoid singular points in $H_{MP}(f)$. As a 100% reflectivity peak cannot be usually achieved in practice, K_1 and K_2 are obtained by solving the equation system defined by Eqs. (9.18) and (9.19) with the equality signs, thus optimizing the amount of energy that is transferred into the non-MP portion of the output signal. Nearly 50% of the output signal energy could be transferred into the non-MP portion as the maximum reflectivity approaches 100%.

This manipulation of the system response can generally be carried out as long as the target functionality can be restricted over a well-defined, finite time window, e.g., such as arbitrary short pulse processing and re-shaping operations. The signal at the output of the FBG is composed of two terms, one is a scaled version of the input signal, and the second one is proportional to the output of the non-MP processor, i.e., the desired processed waveform. In order to be able to recover the desired processed waveform, these two components must be properly separated in time. Therefore, the time delay τ_G between the delta function and $h_{NMP}(t)$ in Eq. (9.16) must be suitably designed, which essentially imposes a limitation on the input and output signal temporal durations. This design restriction will translate into a constraint in the maximum processing temporal window of the device. The processing time window is directly related to the separation between the two components of the MP impulse response, which is only limited by the spatial resolution of the apodization profile. Depending on the target application, the desired processed waveform may need to be extracted through an additional temporal modulation process.

Optical Hilbert Transformer To demonstrate the potential of the proposed method for encoding an arbitrary phase shift in the device response, the realization of two HTs, an integer, i.e., $P = 1$ and a fractional one, $P = 0.81$ (see Eq. (9.12)) has been experimentally demonstrated [97]. In both cases, a phase mask with a grating chirp of 1.25 nm/cm and a length of 7 cm was employed for fabrication. $W(f)$ was chosen to be a raised cosine function with a full-width bandwidth (measured at 1% of the maximum amplitude) of 3.4 THz. However, different maximum reflectivities R_{max} were imposed for the two devices, in order to evaluate the influence of grating strength or peak reflectivity in the HT device performance, as discussed below. Thus, the integer HT has $K_1 = 0.88$ and $K_2 = 0.11$, corresponding to a peak reflectivity $R_{max} = 0.4$. The fractional HT has $K_1 = 0.97$ and $K_2 = 0.02$, corresponding with $R_{max} = 0.12$. The reflectivity and group delay in reflection of the integer HT are plotted in Figures 9.10 (a) and (b), dashed black curves. The power spectral response follows the anticipated interferogram-like profile with a nearly uniform envelope, corresponding to the constant amplitude spectral response of the all-pass HT filter; the phase shift in the middle of the sinusoidal interferogram profile corresponds to the target discrete shift in the phase spectral response of the HT filter. To facilitate the practical implementation of the resulting profile, the desired spectral response was practically achieved using two superimposed, unapodized, linearly chirped FBGs in a Fabry-Perot configuration [98], i.e., where the gratings are suitably spatially shifted with respect to each other. In particular, the induced Δn_{max} was $\backsim 2.5 \times 10^{-4}$ and the induced shift between gratings was $\backsim$1.7 mm. This value of shift impose $\tau_G \backsim 17$ ps. Besides, one of the gratings that compose the Fabry-Perot structure must have a phase transition of π rad (integer case) and 0.81π rad (fractional case) at the

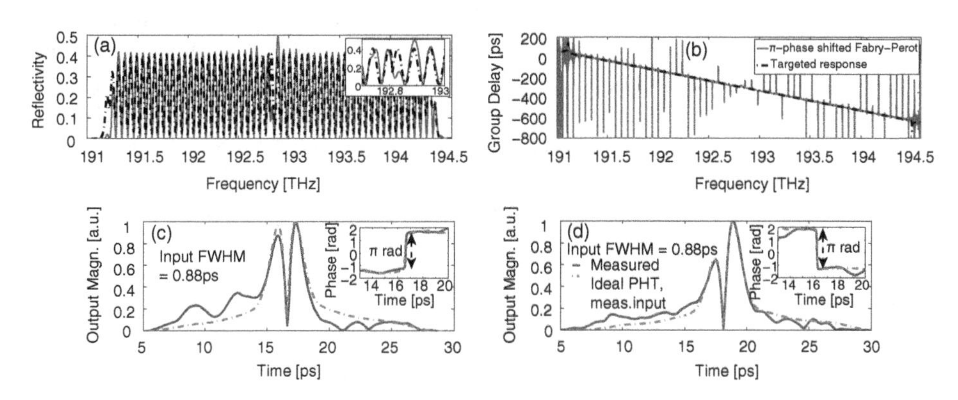

Figure 9.10 (a) Reflectivity and (b) group delay of a HT implemented in an FBG in transmission. The dashed black line represents the specified spectral response and the red line represents the spectral response of an FBG-based π-phase shifted Fabry-Perot structure. Measured output component of the fabricated integer (c) and fractional (d) HTs (solid line) for an input Gaussian-like pulse with a FWHM of 0.88 ps, as compared with the ideally expected output (dashed line). *Source:* María 2015 [97].

center of its length. The comparison between the target spectrum for the integer photonic HT and the reflective spectral response of the FBG-based Fabry-Perot structure is plotted in Figures 9.10 (a) and (b).

The gratings were fabricated following the same procedure as in the previously presented flat-top pulse shaper. In this case, the UV beam first swept a linearly-chirped phase mask with the required phase shift at the middle of its length. Then, a second linearly-chirped phase mask without phase shift was scanned by the UV beam with the same average power and sweep time to achieve the same refractive index modulation as the first grating.

The full time-domain characterization of the fabricated devices was realized by means of an FTSI scheme [92]. The employed input optical source consists of an optical parametric oscillator (OPO) followed by a tunable Gaussian-like band-pass filter. Figures 9.10 (c) and (d) (solid curves) provide the temporal output profiles measured by the FTSI method for input pulse width of 0.88 ps at FWHM for the cases (c) integer HT and (d) 0.81-order HT. The dotted-dashed lines in Figure 9.10 (c) and (d) are for the numerically simulated ideal outputs from the HTs. There is a good agreement between the obtained temporal waveforms and the output of the ideal HTs. The observed leading-edge tails in the integer HT is attributed to the non-uniform, irregular envelope of the interferogram-like pattern of the device transmissive spectral response. On the other hand, the 0.81-order fractional HT offers an improved performance, as observed in its time-domain characterization result, where the leading-edge trails are much less pronounced than for the integer HT. This fact is attributed to the lower reflectivity imposed for this grating.

In this case, the TBP is ~15, which is in line with previously presented photonic HTs based on FBG in reflection [60–63], but providing an operation bandwidth one order of magnitude higher.

9.4.2 Recent Findings on LPGs

The focus of this section is on a recently developed THz-bandwidth *arbitrary* optical pulse shaping theory and technique using fiber LPG devices [99–101]. This solution enables one to synthesize/process optical waveforms with temporal resolutions well into femtosecond range, i.e., faster operation speeds (bandwidths) than conventional FBG-based optical waveform generation and processing schemes. The most striking feature of the introduced novel theory, referred to as *superluminal space-to-time mapping* in LPGs, is that the achievable temporal resolution of the proposed LPG-based optical waveform generation and processing scheme is not limited by the spatial resolution of the grating fabrication technologies.

In all fiber grating devices, the processing or synthesis of faster temporal features necessarily requires the use of smaller spatial resolutions in the grating complex apodization profile [35, 36]. This space-time relationship is more evident in FBGs working in reflection under weak-coupling condition (first-order Born approximation). As discussed in Section 9.3, in this case the output time-domain filter response $h(t)$ (complex envelope) is directly proportional to the complex grating apodization profile $\Delta n(z)$ with a space-to-time scaling factor directly determined by the classical light propagation laws through the medium. However, as expected for any light propagation-based process, in this scheme, the ratio (v) between the space (Δz) and time (Δt) variables is necessarily

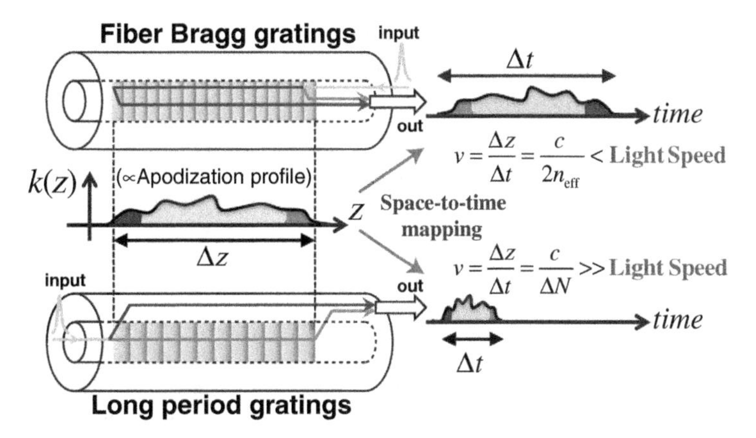

Figure 9.11 Illustration of the space-to-time mapping phenomena in FBG and LPG cases, respectively. *Source*: Ashrafi 2013 [99].

lower than the propagation speed of light in vacuum (c_0) [37], i.e. $v = \Delta z/\Delta t < c_0$, in particular $v = c_0/2n_{eff}$, where n_{eff} is the average effective refractive index of the propagating mode in the FBG (see Figure 9.11). As a result, the achievable temporal resolution must be necessarily larger than the light-wave propagation time through the minimum spatial feature of the grating apodization profile, as determined by the spatial resolution of the used technology. That leads to FBG devices for optical signal processing/synthesis being limited to temporal resolutions of at least several picoseconds, considering a typically feasible sub-millimeter resolution for the fiber grating apodization profiles.

It has recently been found [99] that a similar space-to-time mapping phenomenon can be observed under weak-coupling condition (Born approximation) in the case of LPGs, see illustrations in Figure 9.11. In contrast to FBG devices, it has been demonstrated that the space-to-time mapping speed ($v = \Delta z/\Delta t$) in a LPG device can be significantly higher than the speed of light in vacuum. This new finding is referred to as superluminal space-to-time mapping [99]. An obvious consequence of this phenomenon is that LPGs can be designed for processing/synthesis of optical waveforms with temporal features orders of magnitude faster (shorter) than those achievable using counter-directional coupling devices assuming the same practical spatial resolution limitations in grating fabrication.

For the analysis of the space-to-time mapping approach in LPGs, it is assumed that the LPG-induced coupling occurs predominantly between the fiber core mode and a single cladding mode. A comprehensive analysis of this approach can be found in [99]. The basis of the concept is briefly revisited in what follows. Under the weak-coupling condition, the cross-coupling spectral transfer function $H_{c-cl}(f)$ at the output end of the LPG (i.e., at Δz) is

$$H_{c-cl}(f) \simeq j \int_0^{\Delta z} |k(z)| \mathrm{e}^{-j(2\pi f)z\Delta N/c_0} \mathrm{d}z. \tag{9.20}$$

In Eq. (9.20), $k(z)$ is the complex coupling coefficient defined as $k(z) = |k(z)| \exp\{j\Phi(z)\}$, where $|k(z)| = (\Delta n(z)\pi f_0)/c_0$ accounts for variations in the coupling strength along the waveguide axial (light propagation) direction (z) and $\Phi(z) = 2\pi \int (1/\Lambda(z) - 1/\Lambda_0)\mathrm{d}z$ accounts for local grating period variations along z.

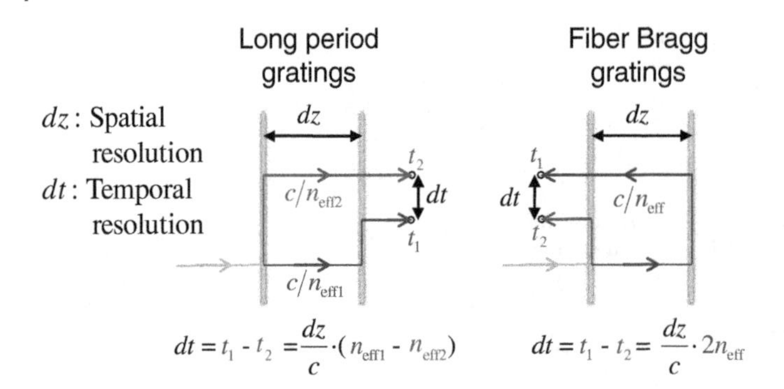

Figure 9.12 Illustration of the temporal resolutions (*dt*) that are associated with a prescribed spatial resolution (*dz*) in the grating apodization profile for the FBG and LPG cases. In each case, the temporal resolution is defined by the difference between the arrival times of the impulses coupled at the input and output ends of the corresponding spatial-resolution element. *Source*: Ashrafi 2013 [99].

Also, $\Delta N = n_{eff1} - n_{eff2}$, where n_{eff1} and n_{eff2} are the average effective refracting indices of the two interacting modes. On the other hand, based on the Fourier transform theory, $H_{c-cl}(f)$ can be also expressed as

$$H_{c-cl}(f) = \int_0^{\Delta t_1} h_{c-cl}(t) e^{-j(2\pi f)t} dt, \tag{9.21}$$

where Δt_1 is the full time-width of the cross-coupling impulse response. By comparing Eqs. (9.20) and (9.21), it can be inferred that

$$h_{c-cl}(t) \propto \{|k(z)| e^{j\Phi(z)}\}_{z=tc_0/\Delta N}. \tag{9.22}$$

Therefore under weak-coupling condition, the cross-coupling temporal impulse response of the fiber LPG is approximately proportional to the variation of the complex coupling coefficient $k(z)$ as a function of the grating propagation distance z after a suitable space-to-time scaling. In particular, the space-to-time mapping speed (v), is obtained as $v = c_0/\Delta N$. Clearly, ΔN can be made much smaller than 1 and consequently, the resulting speed can be made significantly higher than the speed of light in vacuum.

Figure 9.12 illustrates the physical principle of the obtained space-to-time mapping law in FBGs as compared with the case of LPGs. The described phenomenon enables a direct mapping of the grating apodization profile along the device's temporal impulse response. This leads to straightforward, simple grating designs providing the desired linear temporal response. This design strategy should be particularly useful for devices aimed at re-shaping an ultra-short optical pulse (temporal impulse launched at the device input). Most importantly, on the basis of the superluminal space-to-time scaling law, ultrashort temporal features can be achieved using greatly relaxed spatial resolutions, i.e., time features much faster than those intrinsically imposed by the fundamental light propagation laws through the medium, by properly designing the effective refractive index difference between the two interacting modes.

An important practical consideration concerns the fact that the input and output signals must be carried by two different waveguide modes. In an integrated-waveguide approach [102], the device could be practically implemented by using two physically separated single-mode optical waveguides. All-fiber LPGs are typically based on coupling between the core mode (easily extracted from the fiber) and a cladding mode. Traditionally, in order to transfer the cross-coupling signal in the fiber cladding-mode to the fiber core-mode, it was necessary to concatenate a core-mode blocker and a short, strong uniform LPG [75]. In a recent experimental implementation of the LPG-based pulse shaping approach, a novel alternative technique to extract the cross-coupling signal from the LPGs by splicing a suitably misaligned fiber has been employed [101].

9.4.2.1 Triangular and Parabolic Pulse Shapers

The LPG to perform the optical pulse shaping must properly apodized along its length and operates in the cross-coupling mode (e.g., single-mode fiber LPG working in the core-to-cladding operation mode). The complex apodization profile, i.e., $\Delta n(z)$, must be a spatial-domain scaled version of the target output temporal pulse waveform, i.e. $h_{c-cl}(t)$, with a space-to-time scaling law defined by Eq. (9.22). The capability of the introduced optical pulse shaping approach for the generation of two ultra-fast optical arbitrary waveforms has been demonstrated in [99], namely saw-tooth and parabolic pulses, down to the femtosecond regime using feasible LPG designs, i.e., with mm resolutions.

Standard single-mode fiber (Corning SMF-28) has been considered the optical waveguide platform. Also the same LPG design parameters as the well experimentally characterized LPG made on SMF-28 in [103] have been considered. The grating period is $\Lambda = 430\ \mu m$, which corresponds to coupling of the fiber core mode into the LP06 cladding mode at a central wavelength of 1550 nm [103]. The following wavelength dependence has been assumed for the effective refractive indices of the two coupled modes [103]: $n_{0,1}(\lambda) = 1.4884 - 0.031547\lambda + 0.012023\lambda^2$ for the core mode and $n_{0,6}(\lambda) = 1.4806 - 0.025396\lambda + 0.009802\lambda^2$ for the cladding mode, where $1.2 < \lambda < 1.7$ is the wavelength variable in μm. The spectral responses of the LPG designs are numerically simulated using the Coupled Mode theory combined with a Transfer Matrix method [35].

The LPG apodization profiles along the grating structure for implementation of the sub-picosecond optical triangular and parabolic waveform generations are shown in Figure 9.13 (a) and (b) respectively. Three LPGs with the same length and different maximum refractive index modulation have been simulated for each pulse shaper (LPG 1-3 for triangular and LPG 4-6 for parabolic pulse shapers). The corresponding simulation results for power spectral responses of the designed LPGs are presented in Figure 9.13 (c) and (d) respectively. The corresponding temporal responses of the designed LPGs to an ultrashort input Gaussian pulse with 100 fs full-width at 10% of the peak amplitude, are shown in Figures 9.13 (e) and (f) respectively. The simulation results reveal that the designed LPGs implement very nearly the desired pulse re-shaping operations even when the weak-coupling strength condition is clearly not satisfied, see the results corresponding to the design cases with the highest maximum coupling coefficient in Figure 9.13, making a nearly 100% cross-coupling peak.

9.4.2.2 Tsymbol/s Phase Coding

Optical pulse encoding and decoding techniques have been extensively investigated for a wide range of optical processing applications. These techniques are particularly interesting for applications requiring the generation of time-limited data streams (composed

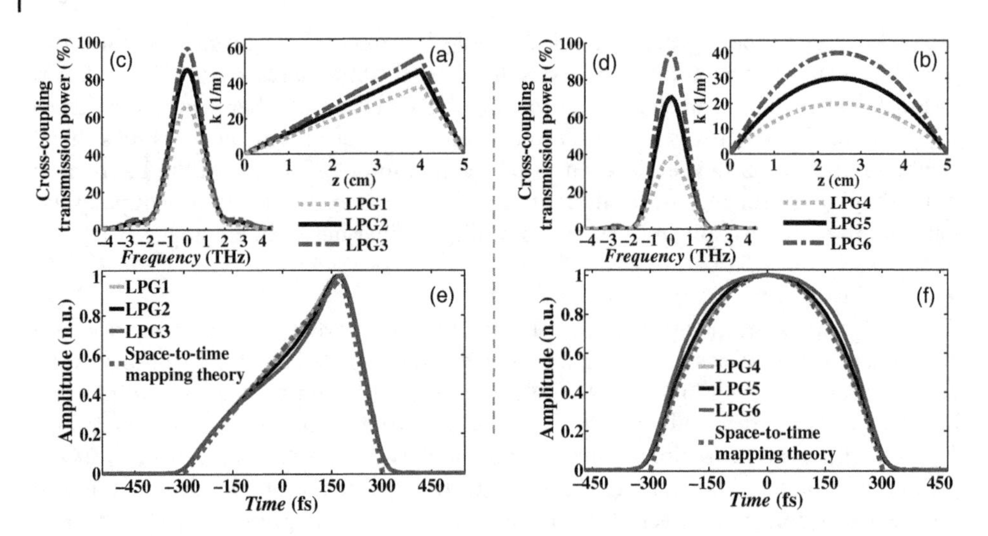

Figure 9.13 (a,b) The apodized-LPG designs for femtosecond optical saw-tooth and parabolic waveform generation respectively. (c,d) The corresponding simulation results for spectral power responses of the designed LPGs. (e,f) The corresponding simulation results for temporal responses (complex envelopes) of the designed LPGs to an ultrashort (~100fs) Gaussian pulse. *Source*: Ashrafi 2013 [99].

of a few consecutive symbols), such as OCDMA and optical label-switching communications [68,70].

The capability of the superluminal space-to-time mapping approach for generation of customized serial optical communication streams under any desired complex coding format well in the Tsymbol/s range (femtosecond resolutions) was shown in [99], where readily feasible LPG designs, e.g., with grating apodization resolutions above the millimeter range were used.

Figure 9.14 shows the numerical simulation of an LPG design for generation of 8-symbol optical QPSK signals with a speed of 4 Tsymbol/s (4 TBaud), from an input ultra-short optical Gaussian pulse with a (full-width at 10% of the peak amplitude) duration of 100 fs. The simulation method and the LPG design parameters is the one used in

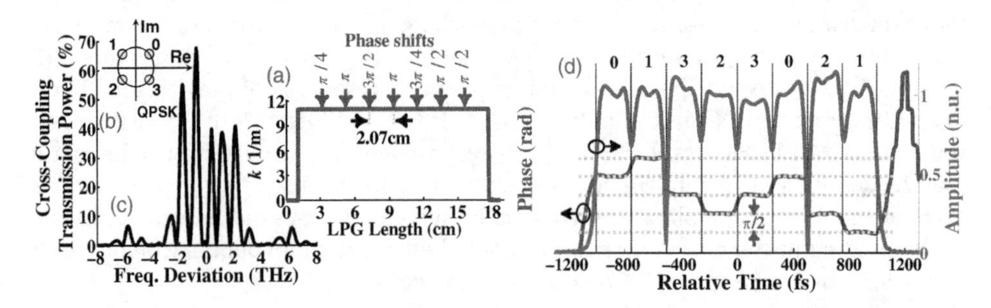

Figure 9.14 Simulation results of the designed LPG (a) to generate 8-symbol optical QPSK (b) data stream patterns, i.e. "0"1"3"2"3"0"2"1", with a speed of 4 Tsymbol/s from an input (full width at 10% of the peak amplitude) 100 fs optical Gaussian pulse. (c) The corresponding spectral power responses of the designed LPGs. (d) The corresponding output temporal amplitude and phase responses. *Source*: Ashrafi 2013. Reproduced with permission of Scientific Research.

previous section. In particular, Figure 9.14 (a) shows the designed amplitude and phase grating-apodization profiles for the target QPSK coding operation. The grating design is relatively straightforward and simple, just being spatial-domain mapped versions of the respective targeted complex time-domain optical data stream. Figure 9.14 (d) shows the amplitude and phase profile of the time-domain waveforms at the output of the simulated LPG design, demonstrating accurate generation of the targeted 4 Tsymbol/s data streams, as per the coding formats defined in Figure 9.14 (b), in excellent agreement with the inscribed grating-apodization profile (Figure 9.14 (a)).

The spectral power response of the designed LPG filter is shown in Figure 9.14 (c). It is worth mentioning that the complexity of this spectral response (also for the phase, not shown here) increases for a larger number of symbols (and for a higher number of amplitude and phase levels in each symbol), making it very challenging for practical implementation using a frequency-domain filter synthesis approach, e.g. conventional programmable linear wave-shapers with limited spectral bandwidth and resolution. In contrast, the proposed superluminal space-to-time mapping approach bypasses the mentioned spectral complexity by directly mapping the simpler filter time-domain impulse response along the grating spatial profile.

Following the previous design methodology, several data pulse sequences have been experimentally demonstrated. As shown in Figure 9.11, the input and output signals must be carried by two different waveguide modes, namely the core-mode and a cladding-mode in a typical fiber-optics configuration. The LPGs are designed for generation of different 4-bits pulse data streams, i.e. ″1″0″0″1, ″1″0″0″−1″ and ″1″0″1″1″, respectively, with a target speed of ~3.5 Tbit/s. In each case, the target temporal pattern is directly mapped along the grating complex coupling-coefficient (apodization) profile of the corresponding LPG, see Figures 9.15 (a)–(c) in the color plate section. The three LPGs revisited here were fabricated via UV illumination of hydrogen-loaded single-mode fiber (Corning SMF28) by a frequency-doubled argon-ion laser operating at 244 nm through an amplitude mask [104].

The reported patterns are specifically designed to demonstrate the anticipated space-to-time mapping process for both amplitude and phase profiles. Whereas the patterns ″1″0″0″1 and ″1″0″1″1″ exhibit a different amplitude variation, the target is that all the pulses in these streams are in phase, requiring in-phase LPGs. Additionally, a pattern ″1″0″0″−1″, where the two pulses at the extremes of the pulse code sequence need to be exactly out of phase, was targeted. The required π-phase-shift has been implemented by use of an amplitude mask having a half-period shift in the middle of its length. In all the reported designs, the grating period is 430 μm which corresponds to coupling of the fundamental core-mode into the LP06 cladding-mode at a central wavelength of 1550 nm [103].

The average effective refractive indices of the fundamental core-mode and LP06 cladding-mode in the SMF-28 fiber around 1550 nm are $n_{eff1} = 1.4684$ and $n_{eff2} = 1.4648$, respectively [103]. The predicted space-to-time mapping speed in the fabricated LPGs is then $v = c_0/\Delta N = 833 \times 10^8$ m/s, which is about ~278 times larger than the speed of light in vacuum. The separation length between the first and last apodization-bits of the fabricated LPGs is ~7.12 cm, see Figure 9.16 (a)–(c), and the total length of the LPG devices is ~9.5 cm, which corresponds to the target bit-rate of 3.5 Tbit/s. According to numerical simulations, LPGs exhibit refractive index modulation amplitudes (half of the peak-to-peak value) of 9.9×10^{-5}, 1.1×10^{-4} and 6.7×10^{-5} for the cases shown in

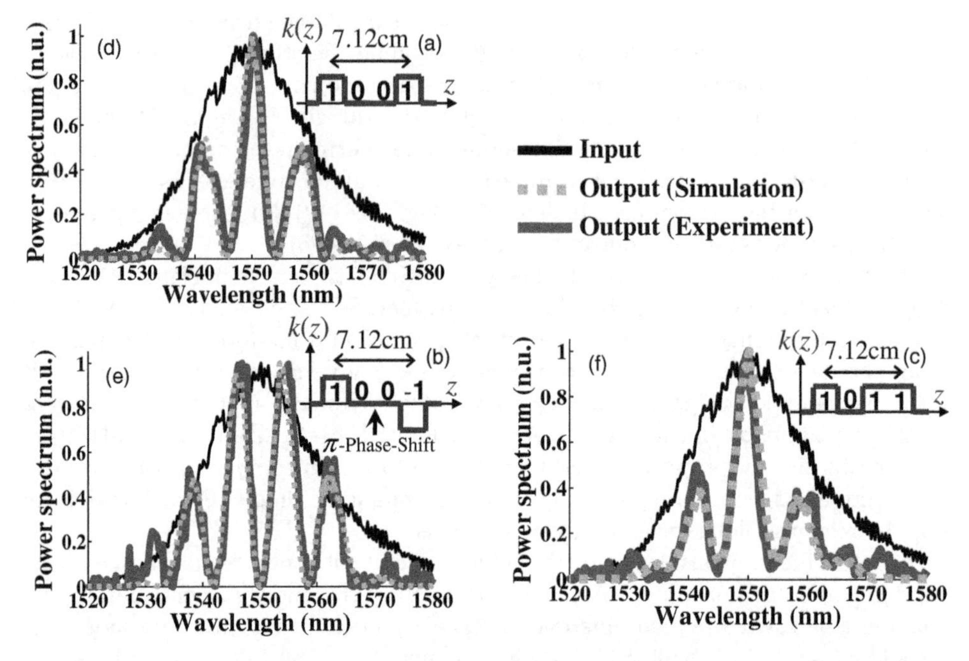

Figure 9.15 (a–c) Fabricated fiber LPG designs for generation of 4-bit data streams, i.e. "1"0"0"1, "1"0"0"–1" and "1"0"1"1", respectively, with a target speed of ~3.5-Tbit/s. (d–f) The corresponding experimental spectrum measurements of the femtosecond optical pulse from the OPO laser before (solid black curves) and after (solid blue curves) propagation through the fabricated LPGs, compared with the simulated linear spectral responses of the LPGs (dotted green curves) to the same input OPO laser pulse spectrum. The spectra are represented in normalized units (n.u.). *Source*: Ashrafi 2013 [101]. For a color version of this figure please see color plate section.

Figures 9.15 (a)–(c), respectively. The peak transmission power of the core-to-cladding spectral responses of the fabricated LPGs is as high as 80% (i.e., corresponding to measured resonance depths of ~7 dB in core-to-core transmissions). Whereas the space-to-time mapping theory is strictly valid under weak-coupling strength condition [99], this is clearly not satisfied in these designs. Simulations have revealed that for the target outputs, LPGs with a strong coupling strength can still implement very nearly the desired pulse shaping operations while improving the devices' energy efficiency.

The experimental setup for time-domain characterization of the fabricated LPGs is based on FTSI, which is used to retrieve the phase and amplitude profiles of the complex temporal waveforms at the LPGs' outputs [105], using the input optical pulse itself as the reference in the measurement setup. The input optical pulse is a nearly transform-limited Gaussian-like optical pulse directly generated from an OPO, which is also used as the reference signal in the interferometric setup. The input pulse from the OPO laser is estimated to have a FWHM time-width of ~200 fs (and a FWHM spectral bandwidth of ~22 nm). The carrier wavelength of the optical pulse is tuned at the resonance wavelength of the LPGs, i.e. 1550 nm. The normalized measured spectra of the optical pulse before and after the fabricated LPGs are shown in Figure 9.15 (d)–(f). The experimentally recovered time-domain amplitude and phase profiles at the output of the LPGs

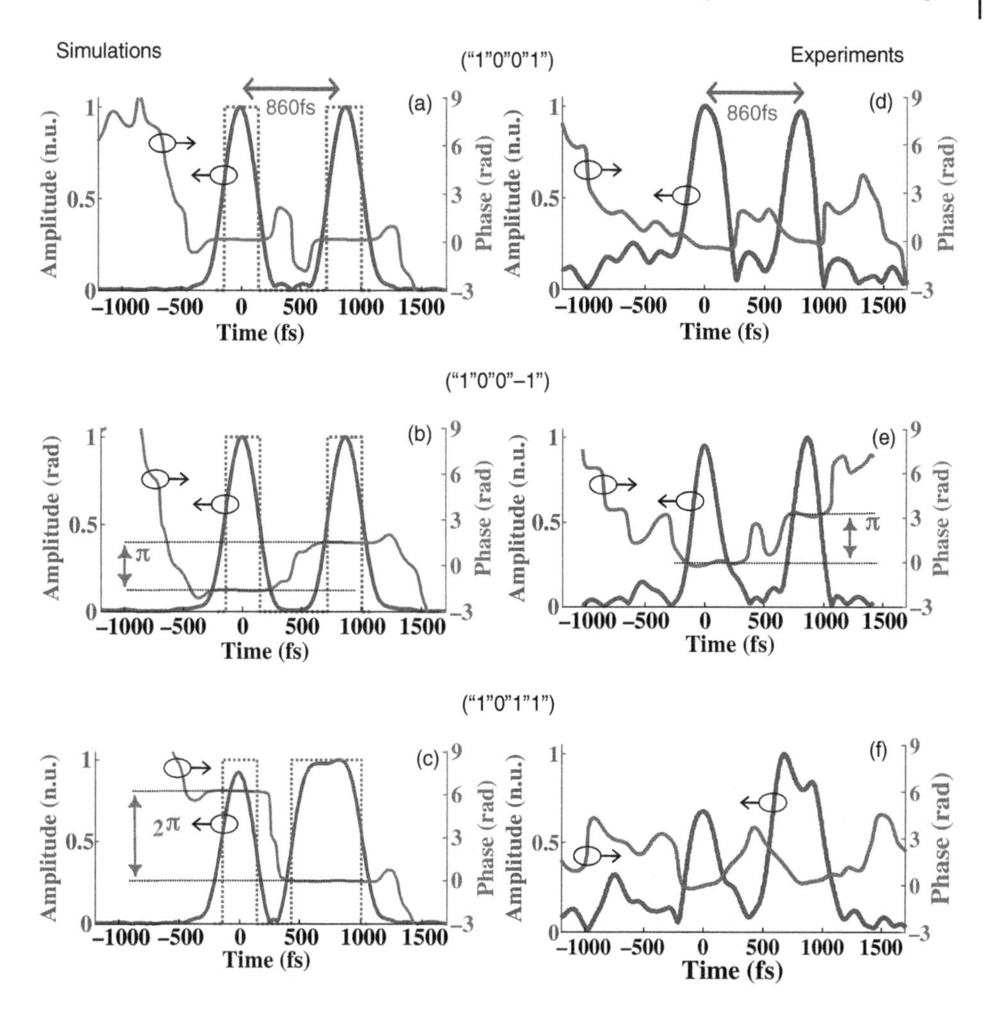

Figure 9.16 (a–c) Simulation results for the temporal amplitude (solid blue curves) and phase (solid green curves) responses of the designed LPGs to the OPO laser pulse data. The dotted, red curves in (a–c) represent the ideal time-domain impulse response amplitudes, obtained by using the predicted superluminal space-to-time scaling of the ideal space-domain profiles. (d–f) Measured output time-domain amplitude (solid blue curves) and phase (solid green curves) responses of the fabricated LPGs. *Source*: Ashrafi 2013 [101]. For a color version of this figure please see color plate section.

are shown in Figure 9.16 (d)–(f) in the color plate section. The responses of the LPGs are also simulated using the same grating design parameters and numerical simulation method used in the previous cases. The simulated spectral and temporal responses to the OPO laser pulse (numerically reconstructed from its spectrum data and assuming a linear spectral phase), are also presented in Figures 9.15 (d)–(f) and Figures 9.16 (a)–(c) respectively. There is a fairly good agreement between the numerically simulated and experimental results. Moreover, in the three reported examples, the obtained temporal responses at the system output closely follow the corresponding ideal impulse response that is predicted by the superluminal space-to-time mapping theory, both in amplitude (ideal impulse responses represented with dotted, red curves) and in phase. The smoothing of the measured temporal responses with respect to the ideal impulse responses are due to the limited bandwidth of the input optical pulse.

Considering the superluminal space-to-time mapping value in the fabricated LPGs ($\backsim 833 \times 10^8$ m/s), the separation time of 860 fs between the first and last bits, shown in Figure 9.16, corresponds to a fairly large spatial length of ~7.12 cm, shown in Figures 9.15 (a)–(c). Thus, as anticipated, time resolutions in the femtosecond regime (e.g., for the inter-bit amplitude transitions and discrete phase jumps) can be achieved based on readily feasible sub-centimeter grating spatial resolutions.

In addition to optical pulse shapers based on LPGs, similar to FBG-based processing devices, recently two LPG-based signal processors, namely THz-bandwidth all-optical temporal differentiators and Hilbert transforms, have been realized and numerically demonstrated based on feasible fiber LPG designs [106, 107].

9.5 Advances towards Reconfigurable Schemes

As discussed in this chapter, one of the main limitations of fiber gratings solutions as pulse shapers is the lack of reconfigurability. However, several approaches can be considered in order to obtain reconfigurable pulse shapers based on this technology.

Fiber gratings can be employed as components in more complex configurations, where the programmability is achieved by controlling parameters unrelated with the gratings. For example, FBG-based optical differentiators have been suggested for their use in multi-arm configuration as optical pulse shapers [32]. In this approach, the desired (arbitrary) optical pulse shape is synthesized by coherently overlapping different successive time derivatives of an input optical pulse (not necessarily a Gaussian-shape pulse), including the input pulse itself, with suitable relative weights. Different output pulse shapes, e.g., rectangular, parabolic or triangular shapes, could be generated from the same platform by properly programming the relative weights among the different pulse derivatives. Strikingly, the effective bandwidth of the output pulse waveform is not necessarily limited by the input pulse bandwidth but rather it depends on the highest derivative order used for the pulse synthesis [32]. Also, FBG-based second order optical integrators have been recently used in a configuration denominated *time-delay to intensity mapping* [58]. A precise point-by-point control of the optical amplitude (intensity) of the waveform generated is achieved at the output of a second order optical integrator by simply tuning the input time-delay vector in an arrayed time-delay system. Using this approach, triangular, rectangular, and bright and dark parabolic pulses have been experimentally generated with temporal resolutions in the sub-picosecond regime. The performance of the pulse shaper is directly related with the number of taps employed in the circuit.

Programmable Bragg gratings have been proposed based on fiber and waveguide-based implementations [69, 108]. In fiber, reconfigurable coder/decoder for OCDMA applications have been implemented by using fine heating elements [69]. Still, the achieved performance needs to be improved to match the required standards for potential commercial purposes. In the case of integrated Bragg gratings, the designed devices use one-dimensional waveguide Bragg grating filled with an electrically controlled dielectric [108]. Arbitrary optical pulse shaping in the subhundred to a few hundred picosecond regime has been theoretically proposed based on this technique, where the output pulses with desired shapes can be obtained by using a genetic algorithm providing the required voltages in each electrode.

Recently a wide range of tunable LPGs based on mechanically [109, 110], thermally [111], or electro-optically [112] controlled modulation of the refractive index along different waveguides/fibers has been reported. These technologies would potentially enable the implementation of the proposed femtosecond optical pulse shaping techniques based on LPGs in a programmable fashion. The development and implementation of such a programmable ultrafast optical pulse-shaping device would represent a significant advance in the field, and could potentially impact a very broad range of applications.

9.6 Conclusion

Fiber gratings are a stable and mature technology for the implementation of all-optical pulse shapers. They act as passive linear optical filters capable of reshaping any input pulse waveforms for optical signal processing, and are competent solutions whenever a fixed pulse shape is required, e.g., flat-top, parabolic, fixed complex modulated code, etc. Apart from their high degree of scalability and capacity of integration, fiber grating solutions offer an outstanding level of flexibility to implement any complex filter spectral response through variation of their physical characteristics, i.e., refractive index modulation or period profile. Recent advances in the design and implementation of fiber gratings have led to the achievement of operations bandwidths well in the THz regime, for both short and long period gratings. In the case of FBGs, a novel design technique for gratings operating in transmission has been introduced, allowing operation over broader frequency bandwidths without increasing the resolution and/or modulation depth in the grating apodization profile. Limitations in the implementation of minimum-phase-only functionalities have been overcome for the case of pulse processors operating over a limited time window. In the case of LPGs, a new design approach has been introduced, the so-called superluminal space-to-time mapping, that allows one to achieve ultrabroad operation bandwidths regardless of the resolution of the apodization profile in a straightforward manner. By employing these methods, a variety of optical pulse processors and shapers, with resolutions down to the sub-picosecond regime, have been numerically and experimentally demonstrated, including all-optical temporal differentiators, Hilbert transformers, flat-top, parabolic, triangular, and arbitrary code pulse shapers, etc. Advances toward reconfigurable schemes are being pursued nowadays through the use of innovative mechanisms to modify the grating physical features in a programmable fashion or by using fixed grating devices in more complex interferometric structures.

References

1 Lee, J.H., Oxenlowe, L.K., Ibsen, M., *et al.* (2003) All-optical TDM data demultiplexing at 80 Gb/s with significant timing jitter tolerance using a fiber Bragg grating based rectangular pulse switching technology. *Journal of Lightwave Technology*, **21** (11), 2518–2523.

2 Oxenlowe, L.K., Slavik, R., Galili, M., *et al.* (2008) 640 Gbit/s timing jitter tolerant data processing using a long-period fiber grating-based flat-top pulse shaper. *IEEE Journal of Selected Topics in Quantum Electronics*, **14**, 566–572.
3 Weiner, A.M., Silberberg, Y., Fouckhardt, H., *et al.* (1989) Use of femtosecond square pulses to avoid pulse breakup in all-optical switching. *IEEE Journal of Quantum Electronics*, **25** (12), 2648–2655.
4 Dutta, A.K., Dutta, N.K., and Fujiwara, M. (2004) *WDM Technologies: Optical Networks*, Elsevier Academic Press, New York.
5 Wabnitz, S. and Eggleton, B.J. (2015) *All-Optical Signal Processing: Data Communications and Storage Applications*, Springer, Berlin.
6 Iakushev, S.O., Shulika, O.V., Sukhoivanov, I.A., *et al.* (2014) Formation of ultrashort triangular pulses in optical fibers. *Optics Express*, **22** (23), 29 119–29 134.
7 Parmigiani, F., Petropoulos, P., Ibsen, M., and Richardson, D.J. (2006) Pulse retiming based on XPM using parabolic pulses formed in a fiber Bragg grating. *IEEE Photonics Technology Letters*, **18** (7), 829–832.
8 Ng, T.T., Parmigiani, F., Ibsen, M., *et al.* (2008) Compensation of linear distortions by using XPM with parabolic pulses as a time lens. *IEEE Photonics Technology Letters*, **20** (13), 1097–1099.
9 Parmigiani, F., Ibsen, M., Ng, T.T., *et al.* (2008) An efficient wavelength converter exploiting a grating-based saw-tooth pulse shaper. *IEEE Photonics Technology Letters*, **20** (17), 2008.
10 Schmogrow, R., Winter, M., Meyer, M., *et al.* (2012) Real-time Nyquist pulse generation beyond 100 Gbit/s and its relation to OFDM. *Optics Express*, **20** (1), 317–337.
11 Nakazawa, M., Hirooka, T., Ruan, P., and Guan, P. (2012) Ultrahigh-speed orthogonal TDM transmission with an optical Nyquist pulse train. *Optics Express*, **20** (2), 1129–1140.
12 Prucnal, P.R. (2006) *Optical Code Division Multiple Access: Fundamentals and Applications*, CRC Press, Boca Raton, FL.
13 Li, G.L. and Yu, P.K.L. (2003) Optical intensity modulators for digital and analog applications. *Journal of Lightwave Technology*, **19**, 2010–2030.
14 Dagli, N. (1999) Wide-bandwidth lasers and modulators for RF photonics. *IEEE Transactions on Microwave Theory and Techniques*, **47**, 1151–1171.
15 Weiner, A.M. (2011) Ultrafast optical pulse shaping: A tutorial review. *Optics Communications*, **284**, 3669–3692.
16 Weiner, A.M. (2000) Femtosecond pulse shaping using spatial light modulators. *Review of Scientific Instruments*, **71** (5), 1929–1960.
17 Cundiff, S.T., and Weiner, A.M. (2010) Optical arbitrary waveform generation. *Nature Photonics*, **4** (11), 760–766.
18 Saperstein, R.E., Panasenko, A.D., Roktiski, R., and Fainman, Y. (2005) Time-domain waveform processing by chromatic dispersion for temporal shaping of optical pulses. *Journal of the Optical Society of America B*, **22** (11), 2427–2436.
19 Kurokawa, T., Tsuda, H., Okamoto, K., *et al.* (1997) Time-space conversion optical signal processing using arrayed-waveguide grating. *Electronics Letters*, **33** (22), 1890–1891.
20 Muralidharan, B., Balakrishnan, V., and Weiner, A.M. (2006) Design of double-passed arrayed-waveguide gratings for the generation of flat-topped femtosecond pulse trains. *Journal of Lightwave Technology*, **24** (1), 586–597.

21 Fontaine, N.K., Geisler, D.J., Scott, R.P., He, T., Heritage, J.P., and Yoo, S.J.B. (2010) Demonstration of high-fidelity dynamic optical arbitrary waveform generation. *Optics Express*, **18**, 22 988–22 995.

22 Kolner, B.H. (1994) Space-time duality and the theory of temporal imaging. *IEEE Journal of Quantum Electronics*, **30** (8), 1951–1963.

23 Li, M., Han, Y., Pan, S., and Yao, J. (2001) Experimental demonstration of symmetrical waveform generation based on amplitude-only modulation in a fiber-based temporal pulse shaping system. *IEEE Photonics Technology Letters*, **23** (11), 715–717.

24 Chi, H. and Yao, J. (2007) Symmetrical waveform generation based on temporal pulse shaping using an amplitude-only modulator. *Electronics Letters*, **43** (7), 415–417.

25 Azana, J., Berger, N.K., Levit, B., and Fischer, B. (2005) Reconfigurable generation of high-repetition-rate optical pulse sequences based on time-domain phase-only filtering. *Optics Letters*, **30** (23), 3228–3230.

26 Wang, X. and Wada, N. (2007) Spectral phase encoding of ultra-short optical pulse in time domain for OCDMA application. *Optics Express*, **15** (12), 7319–7326.

27 Thomas, S., Malacarne, A., Fresi, F., *et al.* (2009) Programmable fiber-based picosecond optical pulse shaper using time-domain binary phase-only linear filtering. *Optics Letters*, **34** (4), 545–547.

28 Norberg, E.J., Guzzon, R.S., Nicholes, S.C., Parker, J.S., and Coldren, L.A. (2010) Programmable photonic lattice filters in InGaAsP-InP. *IEEE Photonics Technology Letters*, **22** (2), 109–111.

29 Dong, P., Feng, N.N., Feng, D., *et al.* (2010) Ghz-bandwidth optical filters based on high-order silicon ring resonators. *Optics Express*, **18**, 23 784–23 789.

30 Guan, B., Djordjevic, S.S., Fontaine, N.K., *et al.* (2014) CMOS compatible reconfigurable silicon photonic lattice filters using cascaded unit cells for RF-photonic processing. *IEEE Journal of Selected Topics in Quantum Electronics*, **20** (4), 8202 110.

31 Park, Y., Ashgari, M.H., Ahn, T.J., and Azana, J. (2007) Transform-limited picosecond pulse shaping based on temporal coherence synthesization. *Optics Express*, **15** (15), 9584–9599.

32 Asghari, M.H. and Azana, J. (2008) Proposal and analysis of a reconfigurable pulse shaping technique based on multi-arm optical differentiator. *Optics Communications*, **281**, 4581–4588.

33 Dudley, J.M., Finot, C., Millot, G., and Richardson, D.J. (2007) Self-similarity in ultrafast nonlinear optics. *Nature Physics*, **3** (9), 597–603.

34 Boscolo, S. and Finot, C. (2012) Nonlinear pulse shaping in fibres for pulse generation and optical processing. *International Journal of Optics*, **2012**, 1–14.

35 Erdogan, T. (1997) Fiber grating spectra. *Journal of Lightwave Technology*, **15** (8), 1277–1294.

36 Kashyap, R. (1999) *Fiber Bragg Gratings*, Academy Press, New York.

37 Miller, S.E. (1954) Coupled wave theory and waveguide applications. *Bell System Technical Journal*, **33**, 661–719.

38 Yariv, A. (1973) Coupled-mode theory for guided-wave optics. *IEEE Journal of Quantum Electronics*, **9**, 919–933.

39 Oppenheim, A.V. and Schafer, R.W. (1989) *Discrete-Time Signal Processing*, Prentice-Hall, New Jersey, 2nd edn.

40 Peral, E., Capmany, J., and Marti, J. (1996) Iterative solution to the Gel'Fand-Levitan-Marchenko coupled equations and application to synthesis of fiber gratings. *IEEE Journal of Quantum Electronics*, **32**, 2078–2084.

41 Feced, R., Zervas, M.N., and Muriel, M.A. (1999) An efficient inverse scattering algorithm for the design of non uniform fiber Bragg gratings. *IEEE Journal of Quantum Electronics*, **35**, 1105–1115.

42 Poladian, L. (2000) Simple grating synthesis algorithm. *Optics Letters*, **25**, 787–789.

43 Skaar, J., Wang, L., and Erdogan, T. (2001) On the synthesis of fiber Bragg gratings by layer peeling. *IEEE Journal of Quantum Electronics*, **37**, 165–173.

44 Kogelnik, H. (1976) Filter response of nonuniform almost-periodic structures. *Bell System Technical Journal*, **55**, 109–126.

45 Petropoulos, P., Ibsen, M., and Richardson, D.J. (2001) Rectangular pulse generation based on pulse reshaping using a superstructured fiber Bragg grating. *Journal of Lightwave Technology*, **19** (5), 746–752.

46 Parmigiani, F., Petropoulos, P., Ibsen, M., and Richardson, D.J. (2006) All-optical pulse reshaping and retiming systems incorporating pulse shaping fiber Bragg gratings. *Journal of Lightwave Technology*, **24** (1), 357–364.

47 Gatti, D., Fernandez, T.T., Longhi, S., and Laporta, P. (2010) Temporal differentiators based on highly-structured fibre Bragg gratings. *Electronics Letters*, **46**, 943–945.

48 Azana, J. and Chen, L.R. (2003) Synthesis of temporal optical waveforms by fiber Bragg gratings: a new approach based on space-to-time-to-frequency mapping. *Journal of the Optical Society of America B*, **19** (1), 2758–2769.

49 Berger, N.K., Levit, B., Fischer, B., *et al.* (2007) Temporal differentiation of optical signals using a phase-shifted fiber Bragg grating. *Optics Express*, **15** (2), 371–381.

50 Kulishov, M. and Azana, J. (2007) Design of high-order all-optical temporal differentiators based on multiple-phase-shifted fiber Bragg gratings. *Optics Express*, **15** (10), 6152–6166.

51 Rivas, L.M., Singh, K., Carballar, A., and Azana, J. (2007) Arbitrary-order ultrabroadband all-optical differentiators based on fiber Bragg gratings. *IEEE Photonics Technology Letters*, **19** (16), 1209–1211.

52 Li, M., Janner, D., Yao, J., and Pruneri, V. (2009) Arbitrary-order all-fiber temporal differentiator based on a fiber Bragg grating: design and experimental demonstration. *Optics Express*, **17** (22), 19 798–19 807.

53 Park, Y., Slavik, R., and Azana, J. (2007) Ultrafast all-optical differentiators for generation of orthogonal (sub-) picosecond Hermite-Gaussian waveforms, in *Optical Fiber Communication (OFC)*, OThI2.

54 Azana, J. (2008) Proposal of a uniform fiber Bragg grating as an ultrafast all-optical integrator. *Optics Letters*, **33** (1), 4–6.

55 Preciado, M.A. and Muriel, M.A. (2008) Ultrafast all-optical integrator based on a fiber Bragg grating: proposal and design. *Optics Letters*, **33** (12), 1348–1350.

56 Asghari, M.H. and Azana, J. (2008) Design of all-optical high-order temporal integrators based on multiple-phase-shifted Bragg gratings. *Optics Express*, **16** (15), 11 459–11 469.

57 Park, Y., Ahn, T.J., Dai, Y., Yao, J., and Azana, J. (2008) All-optical temporal integration of ultrafast pulse waveforms. *Optics Express*, **16** (22), 17 817–17 825.

58 Ashrafi, R., Dizaji, M.R., Romero Cortes, L., *et al.* (2015) Time-delay to intensity mapping based on a second-order optical integrator: application to optical arbitrary waveform generation. *Optics Express*, **23** (12), 16 209–16 223.

59 Slavik, R., Park, Y., Ayotte, N., *et al.* (2008) Photonic temporal integrator for all-optical computing. *Optics Express*, **16** (22), 18 202–18 214.
60 Asghari, M.H. and Azana, J. (2009) All-optical Hilbert transformer based on a single phase-shifted fiber Bragg grating: Design and analysis. *Optics Letters*, **34** (3), 334–336.
61 Cuadrado-Laborde, C. (2010) Proposal and design of a photonic in-fiber fractional Hilbert transformer. *IEEE Photonics Technology Letters*, **22** (1), 33–35.
62 Li, M. and Yao, J. (2010) All-fiber temporal photonic fractional Hilbert transformer based on a directly designed fiber Bragg grating. *Optics Letters*, **35** (2), 223–225.
63 Li, M. and Yao, J. (2010) Experimental demonstration of a wideband photonic temporal Hilbert transformer based on a single fiber Bragg grating. *IEEE Photonics Technology Letters*, **22** (21), 1559–1561.
64 Grunnet-Jepsen, A., Johnson, A.E., Maniloff, E.S., *et al.* (1999) Demonstration of all-fiber sparse lightwave CDMA based on temporal phase encoding. *IEEE Photonics Technology Letters*, **11**, 1283–1285.
65 Teh, P.C., Petropoulos, P., Ibsen, M., and Richardson, D.J. (2001) A comparative study of the performance of seven- and 63-chip otpical code-division multiple-access encoders and decoders based on superstructured fiber Bragg gratings. *Journal of Lightwave Technology*, **19** (9), 1352–1365.
66 Teh, P.C., Ibsen, M., Lee, J.H., Petropoulos, P., and Richardson, D.J. (2002) Demonstration of a four-channel WDM/OCDMA system using 255-chip 320-Gchip/s quarternary phase coding gratings. *IEEE Photonics Technology Letters*, **14**, 227–229.
67 Ayotte, S., Rochette, M., Magne, J., Rusch, L.A., and LaRochelle, S. (2005) Experimental verification and capacity prediction of FE-OCDMA using superimposed FBG. *Journal of Lightwave Technology*, **23** (2), 724–731.
68 Fang, X., Wang, D.N., and Li, S. (2003) Fiber Bragg grating for spectral phase optical code-division multiple-access encoding and decoding. *Journal of the Optical Society of America B*, **20**, 1603–1610.
69 Mokhtar, M.R., Ibsen, M., Teh, P.C., and Richardson, D.J. (2012) Simple dynamically reconfigurable OCDMA encoder/decoder based on uniform fiber Bragg grating, in *Optical Fiber Communications (OFC).*
70 Chen, L.R. (2006) *Optical Code-Division Multiple Access Enabled by Fiber Bragg Grating Technology, in Optical CDMA: Fundamentals and Applications*, CRC Press, Boca Raton, FL.
71 Skaar, J. (2001) Synthesis of fiber Bragg gratings for use in transmission. *Journal of the Optical Society of America A*, **18** (3), 557–564.
72 Preciado, M.A. and Muriel, M.A. (2009) Flat-top pulse generation based on a fiber Bragg grating in transmission. *Optics Letters*, **34** (6), 752–754.
73 Preciado, M.A. and Muriel, M.A. (2008) Design of an ultrafast all-optical differentiator based on fiber Bragg grating in transmission. *Optics Letters*, **33** (21), 2458–2460.
74 Preciado, M.A., Shu, X., and Sugden, K. (2013) Proposal and design of phase-modulated fiber gratings in transmission for pulse shaping. *Optics Letters*, **38** (1), 70–72.
75 Slavik, R., Kulishov, M., Park, Y., and Azana, J. (2009) Long-period fiber-grating-based filter configuration enabling arbitrary linear filtering characteristics. *Optics Letters*, **34**, 1045–1047.

76 Kulishov, M., Krcmarik, D., and Slavik, R. (2007) Design of terahertz-bandwidth arbitrary-order temporal differentiators based on long-period fiber gratings. *Optics Letters*, **32**, 2978–2980.

77 Kim, S.J., Eom, T.J., Lee, B.H., and Park, C.S. (2003) Optical temporal encoding/decoding of short pulses using cascaded long-period fiber gratings. *Optics Express*, **11**, 3034–3040.

78 Park, Y., Kulishov, M., Slavik, R., and Azana, J. (2006) Picosecond and sub-picosecond flat-top pulse generation using uniform long-period fiber gratings. *Optics Express*, **14**, 12 670–12 678.

79 Slavik, R., Park, Y., Kulishov, M., and Azana, J. (2009) Terahertz-bandwidth high-order temporal differentiators based on phase-shifted long-period fiber gratings. *Optics Letters*, **34**, 3116–3118.

80 Strain, M.J. and Sorel, M. (2008) Integrated III-V Bragg gratings for arbitrary control over chirp and coupling coefficient. *IEEE Photonics Technology Letters*, **20** (22), 1863–1865.

81 Rivas, L.M., Strain, M.J., Duchesne, D., *et al.* (2008) Picosecond linear optical pulse shapers based on integrated waveguide Bragg gratings. *Optics Letters*, **33**, 2425–2427.

82 Strain, M.J. and Sorel, M. (2010) Desing and fabriction of integrated chirped Bragg gratings for on-chip dispersion control. *IEEE Journal of Quantum Electronics*, **46** (5), 774–782.

83 Burla, M., Romero Cortes, L., Li, M., *et al.* (2013) Integrated waveguide Bragg gratings for microwave photonics signal processing. *Optics Express*, **21** (21), 25 120–25 147.

84 Zhang, W., Li, W., and Yao, J. (2014) Optical differentiator based on integrated sidewall phase-shifted Bragg grating. *IEEE Photonics Technology Letters*, **26** (23), 2383–2386.

85 Burla, M., Li, M., Romero Cortes, L., *et al.* (2014) Terahertz-bandwidth photonic fractional Hilbert transformer based on a phase-shifted waveguide Bragg grating on silicon. *Optics Letters*, **39** (21), 6241–6244.

86 Simard, A.D., Strain, M.J., Meriggi, L., Sorel, M., and LaRochelle, S. (2015) Bandpass integrated Bragg gratings in silicon-on-insulator with well-controlled amplitude and phase responses. *Optics Letters*, **40** (5), 736–739.

87 Hinton, K. (1998) Dispersion compensation using apodized Bragg fiber gratings in transmission. *Journal of Lightwave Technology*, **16** (12), 2336–2347.

88 Poladian, L. (1997) Group-delay reconstruction for fiber Bragg gratings in reflection and transmission. *Optics Letters*, **22** (20), 1571–1573.

89 Fernandez-Ruiz, M.R., Carballar, A., and Azana, J. (2013) Design of ultrafast all-optical signal-processing devices based on fiber Bragg gratings in transmission. *Journal of Lightwave Technology*, **31** (10), 1593–1600.

90 Fernandez-Ruiz, M.R., Azana, J., and Carballar, A. (2012) Ultrafast all-optical Nth-order differentiators based on transmission fiber Bragg gratings, in *IEEE Photonics Conference (IPC)*, WCC3, pp. 656–657.

91 Fernandez-Ruiz, M.R., Li, M., *et al.* (2013) Picosecond optical signal processing based on transmissive fiber Bragg gratings. *Optics Letters*, **38** (7), 1247–1249.

92 Park, Y., Li, F., and Azana, J. (2006) Characterization and optimization of optical pulse differentiation using spectral interferometry. *IEEE Photonics Technology Letters*, **18** (17), 1798–1800.

93 Eggleton, J., Lenz, G., Slusher, R.E., and Litchinitser, N.M. (1998) Compression of optical pulses spectrally broadened by self-phase modulation using a fiber Bragg grating in transmission. *Applied Optics*, **37** (30), 7055–7061.
94 Fernandez-Ruiz, M.R., Carballar, A., and Azana, J. (2014) Arbitrary time-limited optical pulse processors based on transmission fiber Bragg gratings. *IEEE Photonics Technology Letters*, **26** (17), 1754–1757.
95 Ozcan, A., Digonnet, M.J.F., and Kino, G.S. (2006) Characterization of fiber Bragg gratings using spectral interferometry based on minimum-phase functions. *Journal of Lightwave Technology*, **24** (4), 1739–1757.
96 Carballar, A. and Janer, C. (2012) Complete fiber Bragg grating characterization using an alternative method based on spectral interferometry and minimum-phase reconstruction algorithms. *Journal of Lightwave Technology*, **30** (16), 2574–2582.
97 Fernandez-Ruiz, M.R., Wang, L., Carballar, A., *et al.* (2015) Thz-bandwidth photonic Hilbert transformers based on fiber Bragg gratings in transmission. *Optics Letters*, **40** (1), 41–44.
98 Slavik, R., Doucet, S., and LaRochelle, S. (2003) High-performance all-fiber Fabry-Perot filters with superimposed chirped Bragg gratings. *Journal of Lightwave Technology*, **21** (4), 1059–1065.
99 Ashrafi, R., Li, M., LaRochelle, S., and Azana, J. (2013) Superluminal space-to-time mapping in grating-assisted co-directional couplers. *Optics Express*, **21**, 6249–6256.
100 Ashrafi, R., Li, M., and Azana, J. (2013) Tsymbol/s optical coding based on long period gratings. *IEEE Photonics Technology Letters*, **25**, 910–913.
101 Ashrafi, R., Li, M., Belhadj, N., *et al.* (2013) Experimental demonstration of superluminal space-to-time mapping in long period gratings. *Optics Letters*, **38**, 1419–1421.
102 Jiang, J., Callender, C.L., Noad, J.P., and Ding, J. (2009) Hybrid silica/polymer long period gratings for wavelength filtering and power distribution. *Applied Optics*, **48**, 4866–4873.
103 Smietana, M., Bock, W.J., Mikulic, P., and Chen, J. (2011) Increasing sensitivity of arc-induced long-period gratings-pushing the fabrication technique toward its limits. *Measurement Science and Technology*, **22**, 015 201-1–015 201-6.
104 O'Regan, B.J. and Nikogosyan, D.N. (2011) Femtosecond UV long-period fibre grating fabrication with amplitude mask technique. *Optics Communications*, **284**, 5650–5654.
105 Lepetit, L., Cheriaux, G., and Joffre, M. (1995) Linear technique of phase measurement by femtosecond spectral interferometry for applications in spectroscopy. *Journal of the Optical Society of America B*, **12**, 2467–2474.
106 Ashrafi, R., Li, M., and Azana, J. (2013) Coupling-strength-independent long-period grating designs for Thz-bandwidth optical differentiators. *IEEE Photonics Journal*, **5**, 7100 311.
107 Ashrafi, R. and Azana, J. (2012) Terahertz bandwidth all-optical Hilbert transformers based on long-period gratings. *Optics Letters*, **37**, 2604–2606.
108 Wu, C. and Raymer, M.G. (2006) Efficient picosecond pulse shaping by programmable Bragg gratings. *IEEE Journal of Quantum Electronics*, **42** (9), 873–884.
109 Zhou, X., Shi, S., Zhang, Z., Zou, J., and Liu, Y. (2011) Mechanically-induced pi-shifted long-period fiber gratings. *Optics Express*, **19**, 6253–6259.

110 Chiang, C.C. (2010) Fabrication and characterization of sandwiched optical fibers with periodic gratings. *Applied Optics*, **49**, 4175–4181.

111 Jin, W., Chiang, K.S., and Liu, Q. (2010) Thermally tunable lithium-niobate long-period waveguide grating filter fabricated by reactive ion etching. *Optics Letters*, **35**, 484–486.

112 Balakrishnan, M., Spittel, R., Becker, M., *et al.* (2012) Polymer-filled silica fibers as a step towards electro-optically tunable fiber devices. *Journal of Lightwave Technology*, **30**, 1931–1936.

10

Rogue Breather Structures in Nonlinear Systems with an Emphasis on Optical Fibers as Testbeds

Bertrand Kibler

Laboratoire Interdisciplinaire CARNOT de Bourgogne, UMR 6303 CNRS-Université de Bourgogne - Franche-Comté, Dijon, France

10.1 Introduction

Nowadays, the human and economic toll from severe damage due to rare and extreme (i.e., rogue) events is a major challenge for public policy and scientific research worldwide. One of the main objectives of international research is to provide knowledge and practical tools that can contribute to the reduction of vulnerability. The capacity to predict the course of events clearly depends on the complexity and understanding of systems related to natural and human activities. The most well-known example of rogue events concerns the infamous giant rogue waves that appear and disappear on the surface of the open ocean. They may occasionally encounter a ship or sea platform, thus leading to catastrophic maritime disasters [1]. In oceanography and other domains, the detailed study of such extreme-value phenomena is usually hampered due to their scarce and tricky measurements. Various approaches are found in rogue waves studies, starting from linear wave analysis, which can explain some of the high amplitudes phenomena but the most comprehensive approach is based on nonlinear physics [2].

The recent observation of similarities between the rogue wave phenomena in hydrodynamic and optical systems led to the development of convenient nonlinear fiber optics-based experimental setups to explore both dynamic and stochastic aspects [3–6]. The specific and simplest class of almost conservative physical systems was one of the first main research interests, in particular related to the numerous existing mathematical developments. Analogies between hydrodynamics and optics have been known since the 1960s and 1970s, thanks to two main findings: (i) the derivation of the soliton solution of the nonlinear Schrödinger equation (NLSE) in the form of the secant-hyperbolic shaped pulse, and (ii) the studies of the Benjamin-Feir/Bespalov-Talanov (or modulation) instability of periodic wave trains (or plane waves) [9–15]. The dynamics of waves in weakly nonlinear dispersive media can be described by the NLSE, such as wave propagation in optical Kerr media and on the surface of deep water. However, only recent experimental studies have confirmed that this correspondence applies even in the limit

Shaping Light in Nonlinear Optical Fibers, First Edition. Edited by Sonia Boscolo and Christophe Finot.

of extreme nonlinear wave localization described by the common mathematical model [7, 8]. Indeed, besides the well-known soliton solution, the NLSE also admits breather solutions on finite background, i.e., pulsating envelopes that can mimic the dynamic of rogue waves. Based on the one-dimensional focusing NLSE, one can simply address the issue of rogue waves in terms of NLS breathers whose entire space-time evolution is analytically described [16]. For that reason and because of their specific dynamical properties (i.e., "pulsating" localized waves), these unstable wave structures are considered the simplest nonlinear prototypes of famous hydrodynamic rogue waves [2], in particular, the doubly localized Peregrine breather (or Peregrine soliton) [17]. Note that such solutions describe localized carrier perturbations with a strong amplification. They provide support to the nonlinear stage of the universal modulation instability (MI) phenomenon [18]. MI was originally known as the precursor of highly localized wave structures through amplification of perturbations (or noise) of the plane wave. But surprisingly, NLS breather solutions have remained untested experimentally for almost 30 years, until recent works in nonlinear fiber optics using high-speed telecommunications-grade components.

The recent worldwide attention received by these pioneering studies in fiber optics shows that the NLSE and modulation instability play a fundamental role in the understanding of the formation of rogue waves in almost conservative systems. In confirming their existence and showing that breather dynamics appear even with controlled initial conditions that do not correspond to the mathematical ideal, optical studies have extended in some way breather validity toward nonlinear wave systems driven by noise (or with a partial degree of coherence). In such cases, both competition and interaction between many unstable modes take place, so that rogue events may appear intermittently or randomly in space and time with associated long-tailed statistics, as signatures of extreme-value phenomena [19,20]. This represents a major step forward toward global understanding of rogue wave emergence in a turbulent environment and wave turbulence in integrable systems. Note that there are two regimes of modulation instability that can be distinguished. First, the noise-driven MI that refers to the amplification of initial noise superposed to the plane wave, it leads to spontaneous pattern formation from stochastic fluctuations. In contrast, the second regime is the coherent seeded MI (or coherent driving of MI), and it refers to the preferential amplification of a specific perturbation (i.e., leading to a particular breather solution) relative to any broadband noise. Whatever the regime, the wave dynamics could be interpreted in terms of breathers and competitive interactions, but a complete physical picture is still lacking. But the latter regime can be specifically used to stabilize and manipulate the output wave, thus providing applications and control. As a consequence, this regime was used to generate and measure breathers in optical fibers. Demonstrations of coherent breather generation are mainly based on a fiber-optic parametric amplifier architecture so that one can expect to open the way to breather-based functionalities in ultrafast optics by benefiting from their specific features.

In this chapter, we focus on the simplest breather solutions on a finite background of the NLSE and we review their recent experimental evidence in nonlinear fiber optics. In Section 10.2, we first recall the main properties of first- and second-order NLS breathers based on their theoretical description. In Section 10.3, we describe the specific experimental setups implemented to observe breather propagation in optical fibers. We analyze the different techniques to seed the modulation instability process and their impact on breather generation through numerical simulations based on the NLSE. Moreover, we also discuss potential issues of deviations from theoretical predictions due to linear fiber losses and partial wave coherence. Section 10.4 presents the different experimental measurements of breather dynamics in optical fiber that confirm extreme wave localization described by the nonlinear wave theory. Finally Section 10.5 provides conclusions and an outlook on possible research directions.

10.2 Optical Rogue Waves as Nonlinear Schrödinger Breathers

The phenomenon of modulation instability is usually studied in its simplistic version based on the pioneering work done in the 1960s; the effect was understood as an instability of the plane wave against the long-wave modulation and associated with the growth of spectral sidebands. A linear stability analysis is then performed to identify the instability criterion and to evaluate the initial growth rate of sidebands. This gives the basic information such as the perturbation frequency that experiences the maximum gain and defines the MI period (i.e., the main repetition rate of emerging structures from noise-driven MI). However, the above analysis only provides snapshots of the initial steps of MI and the whole picture of the dynamical evolution is not available. In particular, the dynamics of cascade of MI gain bands that generates the highly localized wave structures cannot be described. MI clearly exhibits much richer dynamics when one goes beyond the simplistic linear stability analysis [21,22]. As an example, the long space evolution may exhibit the Fermi-Pasta-Ulam (FPU) recurrence phenomenon for the coherent seeded MI. Later, researchers also focused on approximate truncated or purely numerical approaches to address this problem in the 1990s [23–25].

However, the NLSE belongs to the remarkable class of integrable systems [9] and can be solved by using the inverse scattering transform method or other integration techniques. Surprisingly, exact breather solutions were derived during the 1970s and the 1980s, but remained untested. This concerns the simplest solutions that are either periodic in space and localized in time or periodic in time and localized in space; they are referred to as Kuznetsov-Ma breathers and Akhmediev breathers, respectively. Taking the period of both solutions to infinity gives rise to a first-order doubly localized breather on finite background: the Peregrine soliton. In the following we first review their properties, to this purpose we refer to the following scaled form of the self-focussing NLSE that can be used in nonlinear wave theory:

$$i\frac{\partial \psi}{\partial \xi} + \frac{1}{2}\frac{\partial^2 \psi}{\partial \tau^2} + |\psi|^2\psi = 0 \tag{10.1}$$

Equation (10.1) that describes wave propagation in space (not in time) is more convenient for direct comparison with laboratory experiments. Here, ψ is a wave group or wave envelope which is a function of ξ (a scaled propagation distance or longitudinal variable) and τ (a co-moving time, or transverse variable, moving with the wave-group velocity), and subscripted variables stand for partial differentiations.

10.2.1 First-Order Breathers

A general one-parameter breather solution on finite background for the NLSE can be written compactly, as previously suggested in several works [26,27], in particular when one aims to study the most well-known first-order solutions such as Kuznetsov-Ma, Akhmediev and Peregrine breathers.

$$\psi(\xi,\tau) = e^{\imath\xi}\left[1 + \frac{2(1-2a)\cosh(b\xi) + \imath b\sinh(b\xi)}{\sqrt{2a}\cos(\omega\tau) - \cosh(b\xi)}\right] \tag{10.2}$$

Here, the governing parameter a determines the physical behavior of the solution through the function arguments $b = [8a(1-2a)]^{1/2}$ and $\omega = 2(1-2a)^{1/2}$ directly linked to space and time evolution, respectively.

10.2.1.1 Kuznetsov-Ma Breathers

The first breather type solution on finite background for the NLSE was discovered in the 1970s by Kuznetsov, but also by Kawata and Inoue, and later by Ma [28–30]. They solved the inital value problem for the NLSE where the initial state was the plane wave solution perturbed by a large localized bump of soliton-type (such breathers approach the plane wave solution at infinite time). These solutions are periodic in space and localized in time; they are now referred to as the Kuznetsov-Ma Breather (KMB) or soliton on finite background. Such solutions can be considered a limiting case of instability of the plane wave with respect to large perturbations. Equation (10.2) describes such solutions when $a > 1/2$. The parameters ω and b become imaginary such that the hyperbolic trigonometric functions become ordinary circular functions and vice versa. As a result, the spatial period of the KMB is given by $2\pi/|b|$, and the transverse localization is determined by $2\pi/|\omega|$. Figure 10.1 illustrates the KMB solutions for two distinct values of the parameter a. The maximal amplitude occurs at $\xi = 0$. We clearly note that spatial period and temporal localization are inversely proportional to a, whereas the maximal peak amplitude is proportional to a. When $a \to \infty$ the KMB solution tends to the standard soliton solution, a localized sech pulse with stable and uniform propagation. Typical KMB dynamics related to the periodic beating of the soliton structure with the background wave is revealed through the plot of the trajectory of the centered point of breathers (i.e., along the line $\tau = 0$) on the complex plane. The trajectory continuously follows a circle centered on the specific point with coordinates $(-1, 0)$ related to the background amplitude. Each spatial period reproduces a complete rotation. The

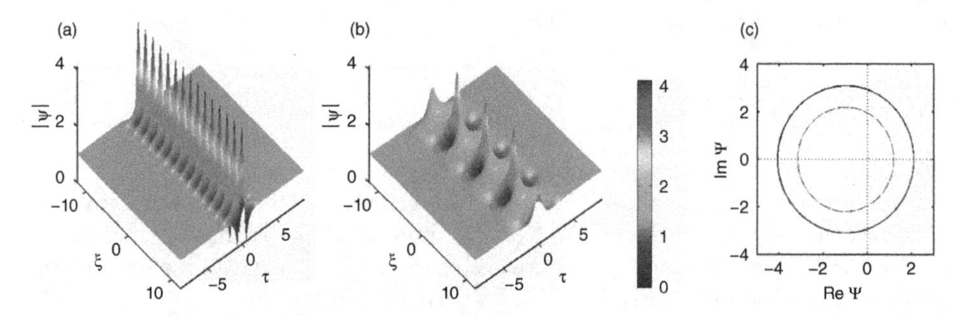

Figure 10.1 Space-periodic KM breathers on a background plane wave $|\psi(\tau \to \pm\infty)| = 1$ for the following values of the governing parameter: (a) $a = 1.2$ and (b) $a = 0.6$. (c) Trajectory of the center point of both KM breathers (i.e., along the line $\tau = 0$) on the complex plane. The trivial phase factor $exp(\imath\xi)$ is omitted. Solid black (dashed) curve corresponds to $a = 1.2$ ($a = 0.6$).

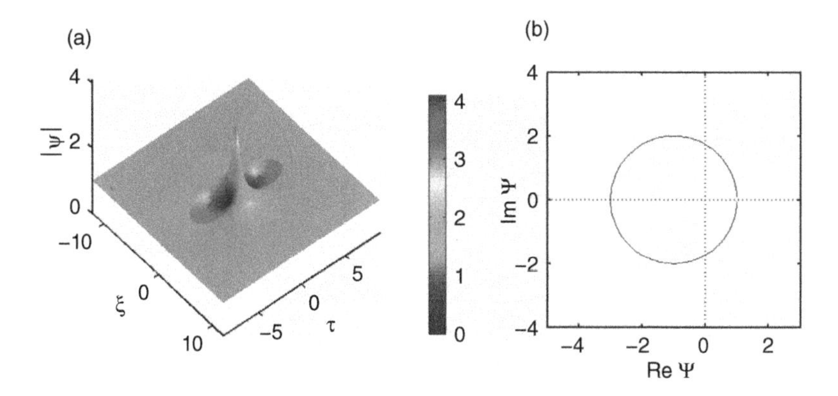

Figure 10.2 (a) Peregrine breather solution. (b) Corresponding trajectory of the center point of PB (i.e., along the line $\tau = 0$) on the complex plane. The trivial phase factor $exp(\imath\xi)$ is omitted.

circle diameter on real axis exactly corresponds to the sum of both maximum and minimum absolute values of breather amplitude. The ratio of the maximum amplitude to the background is always higher than 3.

10.2.1.2 Peregrine Breather

In 1983, D.H. Peregrine found a non-trivial solution in the limit of zero amplitude perturbation of KM breathers. It also corresponds to the infinite-period limiting case of previous solutions (i.e., $a = 0.5$). Apart from a simple exponential factor it is a rational function (see Eq. (10.3)), and it describes an isolated "amplitude peak" in space-time arising out of the plane wave solution [17], as shown in Figure 10.2. The Peregrine breather (PB) exhibits a ratio of the maximum amplitude to the background equal to 3. The corresponding trajectory on the complex plane is one circle due to the single growth-return cycle.

$$\psi(\xi, \tau) = e^{\imath\xi} \left[1 - \frac{4(1 + 2\imath\xi)}{1 + 4\tau^2 + 4\xi^2}\right] \tag{10.3}$$

10.2.1.3 Akhmediev Breathers

Later in the 1980s, Akhmediev *et al.* found a one parameter family of time-periodic solutions with the property that they approach the plane wave solution at infinite propagation; they breathe only once in space [18, 31]. They exhibit FPU-like growth-return evolution and are now widely referred to as the Akhmediev Breather (AB). These solutions are valid for initial periodic and small modulations of the plane wave (i.e., a single sideband perturbation into the MI gain band). Such solutions are given by Eq. (10.2) for $0 < a < 0.5$, two examples are presented in Fig. 10.3. Indeed AB solutions are valid over the range of modulation frequencies that experience MI gain: $2 > \omega > 0$. Note that the parameter $b > 0$ also governs the MI growth, it corresponds to the MI gain calculated from the linear stability analysis. The maximum value $b = 1$ is obtained for $a = 0.25$ ($\omega = \sqrt{2}$). These time-periodic solutions provide, as a first step, a powerful framework with which to describe the full MI dynamics. The temporal period of the AB is given by $2\pi/|\omega|$, and the spatial localization is determined by $2\pi/|b|$. Contrary to the KMB,

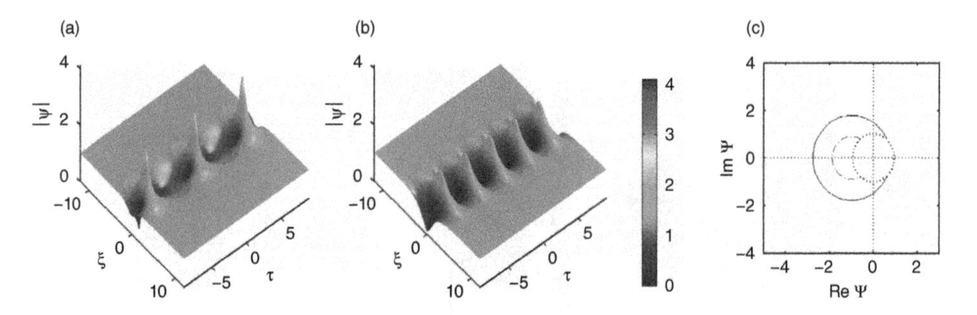

Figure 10.3 Time-periodic Akhmediev breathers on a background plane wave $|\psi(\xi \to \pm\infty)| = 1$ for the following values of the governing parameter: (a) $a = 0.4$ and (b) $a = 0.1$. (c) Trajectory of the center point of AB (i.e., along the line $\tau = 0$) on the complex plane. The trivial phase factor $exp(\imath\xi)$ is omitted. Solid black (dashed) curve corresponds to $a = 0.4$ ($a = 0.1$).

temporal period and spatial localization are here proportional to a, and the maximal peak amplitude is still proportional to a, but lower than 3. When $a \to 0$ the AB solution tends to the plane wave solution, while for $a \to 0.5$ (i.e., infinite-period) the AB solution tends to the Peregrine breather solution. Here, the trajectory of the center point of AB on the complex plane is only part of a circle since initial ($\xi \to -\infty$) and final ($\xi \to +\infty$) states exhibit a phase shift difference (i.e., opposite imaginary coordinates in Figure 10.3 (c)). This nonlinear shift on the phase of the plane wave is accumulated during the AB growth-return cycle, and it is defined by $\Delta\phi = 2\arccos(1 - 4a)$ [32]. In the limit of the Peregrine breather, the phase shift becomes 2π (see Figure 10.2 (b)). Note that a unit circle with its center at the origin can be drawn to define all the initial and final complex amplitudes of the plane wave for $0 < a < 0.5$.

10.2.1.4 Localization Properties

One of the most important property of unstable rogue waves is the maximum amplification of the perturbation with respect to the background wave. Figure 10.4 (a) reports the ratio of the maximum amplitude to the background for first order breathers described

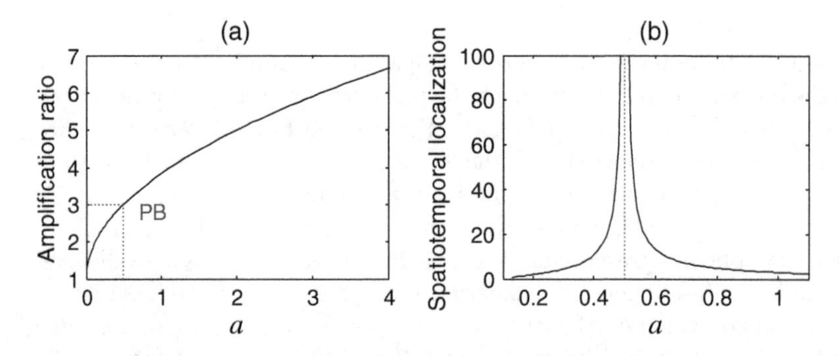

Figure 10.4 (a) Maximum amplification of the perturbation with respect to the background wave for first-order breathers described by Eq. (10.2) as a function of their governing parameter a. (b) Spatio-temporal localization associated to dynamical evolution of first-order breathers as a function of their governing parameter a.

by Eq. (10.2). It clearly appears that the Peregrine breather does not exhibit the highest peak amplitude when including KM solutions. A wide range of a values gives a maximal amplification ratio beyond the factor of two. This limit may be used to partially define the rogue breather waves by extrapolating the usual condition applied on statistical measurements: a rogue wave height is two times the significant wave height of the area [6]. Besides this amplification factor, another key feature of rogue structures that cannot be avoided is related to their localization in space and time. The increased spatio-temporal localization associated with the limiting case of PB is then highlighted in Figure 10.4 (b), thus confirming extreme characteristics of PB emergence from the plane wave. Here we determined the degree of localization in terms of ratios of the temporal and spatial periods ($\tau_{per} = 2\pi/|\omega|$ and $\xi_{per} = 2\pi/|b|$) relative to the individual temporal and spatial peak widths (τ_0 and ξ_0). These can be readily calculated analytically from Eq. (10.2) as a function of parameter a. The profile temporal width τ_0 is found with time coordinates for which the amplitude is zero-valued adjacent to the peak (zeros only appear in the profile for $a > 1/8$). The spatial width ξ_0 is found with space coordinates for which the amplitude is half of the peak amplitude (only valid for profiles with $a < 1.1$). We then defined the spatio-temporal localization as the following product $(\xi_{per}/\xi_0)(\tau_{per}/\tau_0)$, similarly to [7]. A similar localization behavior can be retrieved by using the simple product $(\tau_{per})(\xi_{per})$. In any case, the calculation of spatial period for AB solutions or temporal period for KM breathers appears here to be meaningless, however, it becomes completely relevant when one goes beyond these limiting cases of periodic breathers.

10.2.1.5 Generalized Behavior

A more general description of first-order breather solutions with varying group velocity in the plane (ξ, τ) can be found in recent works [19,33–35]. Some of them are also called quasi-Akhmediev breathers. Figure 10.5 depicts a few examples of these analytical solutions compared to the previous simpler solution. Such breathers are neither periodic in time nor in space, they exhibit double quasi-periodicity that corroborates our above calculation of localization features. Note that the line of maxima is tilted with respect to the line $\xi = 0$, so that one can define the velocity as the tangent of the angle between this line and the temporal axis [33]. However, the amplification of each peak almost remains aligned with the propagation direction. These solutions have a chance to collide (when

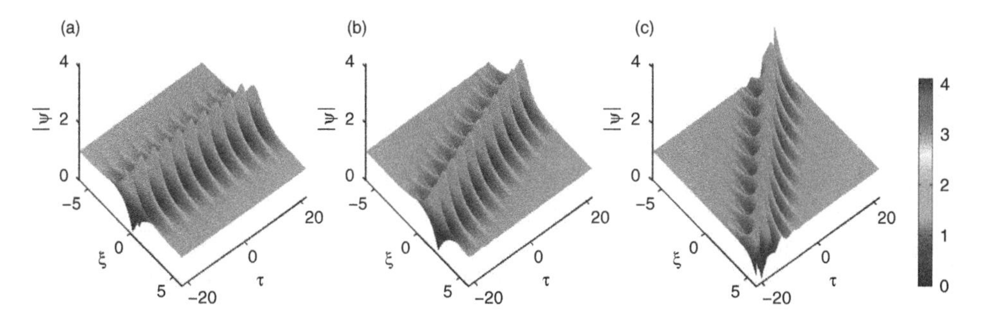

Figure 10.5 First-order breather solutions with varying group velocity in the plane (ξ, τ). (a) AB with zero velocity ($a = 0.25$). (b–c) Similar ABs with increasing non-zero velocity.

choosing appropriate distinct velocities) and generate a giant amplitude wave when synchronization of peaks is satisfied.

Moreover, it is worth mentioning that there exists another large class of first-order breathers that are periodic in both ξ and τ. They are called elliptic or multiphase breathers; some of analytic formula are based on Jacobi elliptic functions [18], or expressed in terms of Riemann theta functions (i.e., multidimensional Fourier series) [36, 37].

10.2.1.6 Spectral Description

In the case of Akhmediev breathers, one can easily find exact analytic expressions for the evolution of spectral components. During the initial stage of propagation, the sidebands experience exponential growth at the expense of the pump, as expected from the linear stability analysis of the modulation instability process. However, the subsequent dynamics of energy exchange between multiple spectral modes are more complex. The AB solution describes growth and decay of a harmonically perturbed plane wave. In addition to the temporal description, expansion in a Fourier series and integration yield exact solutions for the pump and spectral harmonic amplitudes as a function of propagation distance [18, 38]:

$$A_0(\xi) = -1 + \frac{\imath b \sinh b\xi + \omega^2 \cosh b\xi}{\sqrt{\cosh^2 b\xi - 2a}} \tag{10.4}$$

$$A_n(\xi) = \frac{\imath b \sinh b\xi + \omega^2 \cosh b\xi}{\sqrt{\cosh^2 b\xi - 2a}} \times \left[\frac{\cosh b\xi - \sqrt{\cosh^2 b\xi - 2a}}{\sqrt{2a}} \right]^{|n|}$$

where A_0 and A_n are the amplitudes of the pump and the nth sideband ($n = \pm 1, \pm 2$, etc.), respectively, and ignoring factors of constant amplitude and phase. The sum $|A_0|^2 + \Sigma |A_n|^2$ is equal to 1, which is equivalent to the conservation of energy. When $\xi \to \pm\infty$, all the energy is concentrated in the pump, whereas for $\xi \to 0$ the energy of sidebands increases and the pump is progressively depleted. At the maximum spectral broadening (i.e., the maximum compression of the localized structure on finite background), the spectral amplitudes of the pump and the nth sideband are $A_0 = 1 - \omega$ and $A_n = \omega[(1 - (1 - 2a)^{1/2})/(2a)^{1/2}]^{|n|}$, respectively. Equation (10.4) exactly predicts the dynamics of an arbitrary number of sidebands without any assumption of an undepleted pump. Figure 10.6 illustrates the growth–decay cycle of sideband generation related to the spectral evolution of the AB solution for $a = 0.25$. The spectrum is associated with an exponentially decaying energy transfer from the pump frequency, it also reveals a characteristic universal triangular spectral form, when analyzed logarithmically [39]. Moreover, the reciprocal energy exchange between the pump mode and an infinite number of side modes can be related to the FPU recurrence [40, 41].

As the triangular feature of the envelope spectrum on a log scale appears at an early stage of their evolution, this raises the possibility of early detection of breather emergence in chaotic optical fields [42], in particular by considering real-time measurements of optical spectra [43, 44]. The triangular spectral decay of the wings can also be easily calculated for the Peregrine breather [42].

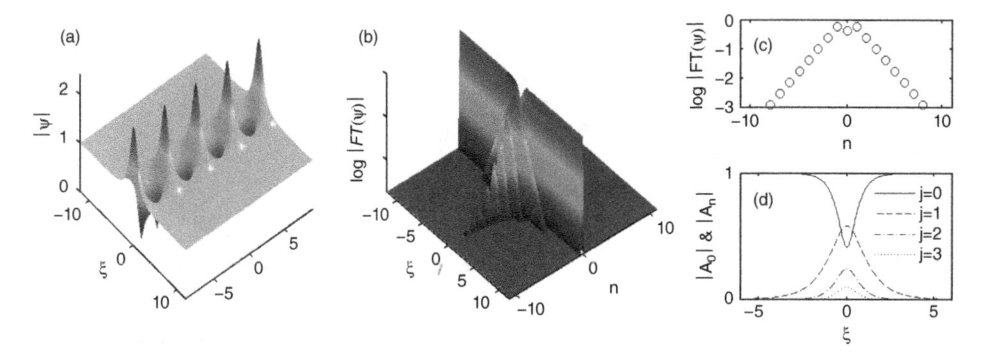

Figure 10.6 Typical evolution of AB spectrum (here $a = 0.25$). (a,b) 3D illustration of both temporal and spectral descriptions from Eqs. (10.2) and (10.4). (c) AB spectrum at maximum sideband generation (i.e., at $\xi = 0$). (d) Evolution of amplitudes of the pump and the most significant sidebands along propagation distance.

10.2.2 Second-Order Breathers

The NLSE also admits higher-order breather solutions on finite background. These higher-order solutions can be interpreted as a nonlinear superposition of two or more breather solutions of first order. The generation of higher-order breathers is of great interest since they offer the possibility of higher energy concentration in space and time, and the generation of waves with giant amplitudes.

Here we consider the nonlinear interaction of two first-order breathers, i.e., a second-order breather solution [31]. By employing the recursive Darboux method [45], one can find an explicit analytic form for the two-breather solution of the NLSE with two independent governing parameters a_j (with $j = 1, 2$) related to each breather, as reported in [46]:

$$\psi_{12}(\xi,\tau) = e^{\imath\xi}\left[1 + \frac{G + \imath H}{D}\right] \tag{10.5}$$

with

$$G = -\left(\omega_1^2 - \omega_2^2\right)\left[\frac{\omega_1^2 b_2}{\omega_2}\cosh(b_1\xi_{s1})\cos(\omega_2\tau_{s2}) - \frac{\omega_2^2 b_1}{\omega_1}\cosh(b_2\xi_{s2})\cos(\omega_1\tau_{s1})\right.$$
$$\left. - \left(\omega_1^2 - \omega_2^2\right)\cosh(b_1\xi_{s1})\cosh(b_2\xi_{s2})\right]$$

$$H = -2\left(\omega_1^2 - \omega_2^2\right)\left[\frac{b_1 b_2}{\omega_2}\sinh(b_1\xi_{s1})\cos(\omega_2\tau_{s2}) - \frac{b_1 b_2}{\omega_1}\sinh(b_2\xi_{s2})\cos(\omega_1\tau_{s1})\right.$$
$$\left. - b_1\sinh(b_1\xi_{s1})\cosh(b_2\xi_{s2}) + b_2\sinh(b_2\xi_{s2})\cosh(b_1\xi_{s1})\right]$$

$$D = 2\left(\omega_1^2 + \omega_2^2\right)\frac{b_1 b_2}{\omega_1\omega_2}\cos(\omega_1\tau_{s1})\cos(\omega_2\tau_{s2}) + 4b_1 b_2[\sin(\omega_1\tau_{s1})\sin(\omega_2\tau_{s2})$$
$$+ \sinh(b_1\xi_{s1})\sinh(b_2\xi_{s2})] - \left(2\omega_1^2 - \omega_1^2\omega_2^2 + 2\omega_2^2\right)\cosh(b_1\xi_{s1})\cosh(b_2\xi_{s2})$$
$$- 2\left(\omega_1^2 - \omega_2^2\right)\left[\frac{b_1}{\omega_1}\cos(\omega_1\tau_{s1})\cosh(b_2\xi_{s2}) - \frac{b_2}{\omega_2}\cos(\omega_2\tau_{s2})\cosh(b_1\xi_{s1})\right]$$

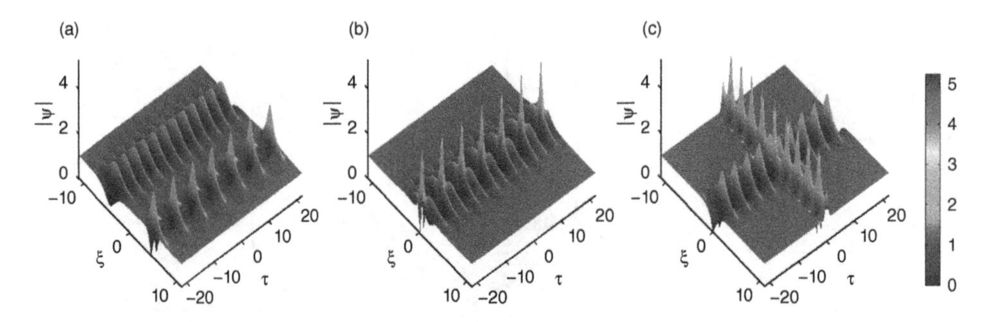

Figure 10.7 Various forms of the second-order breather solution on finite background. (a) Nonlinear superposition of two ABs with $a_1 = 0.1$, $a_2 = 0.4$, $\tau_{1,2} = 0$, $\xi_2 = -\xi_1 = 4$. (b) Same parameters than in case (a) with synchronization at the origin $\xi_{1,2} = 0$. (c) Nonlinear superposition of one AB with one KMB: $a_1 = 0.25$, $a_2 = 1$ (synchronized collision at the origin: $\tau_{1,2} = 0$, $\xi_{1,2} = 0$).

Equation (10.5) describes various nonlinear combinations of Akhmediev breathers and Kuznetsov-Ma breathers with distinct governing parameters $0 < a_{1,2} < 0.5$ or $a_{1,2} > 0.5$. Each first-order breather j in the higher-order solution is described by the governing parameter a_j, the function arguments $\omega_j = 2(1 - 2a_j)^{1/2}$ and $b_j = [8a_j(1 - 2a_j)]^{1/2}$, and a shifted point from the origin (ξ_j, τ_j). The above solution describes the full wave evolution of a second-order breather during its nonlinear propagation, where $\xi_{sj} = \xi - \xi_j$ and $\tau_{sj} = \tau - \tau_j$ are shifted variables. When $\xi_1 = \xi_2 = 0$ and $\tau_1 = \tau_2 = 0$, we consider a synchronized nonlinear superposition at the origin.

Figure 10.7 illustrates a few examples of second-order breathers, in particular by highlighting the importance of the synchronization of the interaction between breathers. The nonlinear interaction strongly depends on the centering of the two elementary solutions in the plane (ξ, τ) [31], i.e., when their peak coincide. If the centers of the elementary solutions are separated by a large distance (time) shift along the $\xi(\tau)$ axis, the superposition tends to be linear, which prevents the emergence of a giant amplitude wave. Another degenerate solution can be found in the limit of equal governing parameters. In such a case, these degenerate solutions consist of two near-parallel lines almost periodic in structure, with only one point of intersection [46]. In the limit of infinite periods, this fully establishes the hierarchy of synchronized second-order breather solutions, ranging from the general case to the higher-order rational breather via degenerate breathers.

General multi-breather solutions with more than two elementary breathers can still be studied, but their analytic expressions become very complex, in particular when introducing breathers with varying group-velocity. For instance, a general N-breather solution on finite background of the focusing NLSE was recently found by using the dressing method [34]. Such theoretical solutions are of considerable importance to fully describe the nonlinear stage of the modulation instability for arbitrary perturbation of the plane wave. However, for a good qualitative description of the higher-order wave structures generated, one may restrict the study to the doubly localized breathers in space and time, also called rogue-wave solutions. They exhibit a unique hierarchical structure with relatively convenient analytical forms. The lowest-order solution is known as the Peregrine breather [17], the second-order solution was first found in 1985 [31] and recently introduced in the context of rogue waves [47]. We then refer the reader to [48] and the references therein for a complete classification of higher-order rogue-wave solutions.

10.3 Linear-Nonlinear Wave Shaping as Rogue Wave Generator

Several theoretical or numerical papers claimed that previous breather solutions can be considered as prototypes to model extreme energy localization in various nonlinear dispersive systems [2,19,49]. However, to confirm such an approach, as a first step, this requires their experimental evidence in real physical systems, even with ideal initial perturbation on the plane wave. Of course, rogue waves are not just an offshoot of such solutions, other mechanisms depending on the physical system must be taken into account, including the statistical approach when noise is present [6]. But it is still of fundamental importance to consider the coherent and deterministic approach to the understanding of rogue wave phenomena. As an example in hydrodynamics, a recent work experimentally investigated the interaction between waves and ships during extreme ocean conditions using such breather solutions [50]. Besides their first evidence in optics, their existence was later confirmed in other fields of physics driven by the NLSE, namely in hydrodynamics and plasma physics for the Peregrine breather [8, 51]. In water wave experiments, the initial wave profiles are generated with a paddle located at one end of a tank. An electric signal, derived from the exact mathematical expression describing the water surface elevation, drives the paddle to directly modulate the surface height. Specific initial modulations such as a ratio of polynomials have been applied to the wave maker to excite rogue wave solutions [52, 53]. However, such ideal perturbations in optics are nontrivial to synthesize in the temporal domain. Therefore, the first experimental studies in optics used non-ideal initial periodic perturbations based on widely accessible techniques in practice, such as the beating of two narrow-linewidth lasers to create an initial low-frequency-modulated wave or by means of usual electro-optic (intensity) modulators at gigahertz frequencies [7,54]. Breather waves are then observed when such small- or large-amplitude perturbations on a high-power continuous wave (cw) become strongly focused due to the nonlinear wave reshaping occurring into an optical fiber. It was also observed that non-ideal initial perturbations sometimes lead to the generation of complex behaviors that may differ from the expected breather [55]. The sensitivity to initial perturbations depends on the complexity or order of the solution (i.e., the order of energy localization). There is always a tradeoff between the simplicity of the initial modulated wave (inherent to experiments) and the degree of accuracy with which we reach the mathematical ideal. Consequently, it was recently proposed introducing the advantages of ultrafast optics technology and programmable optical pulse shaping to study higher-order breathers, since this allows the generation of nearly arbitrarily shaped optical wave forms. The optical processing is then based on spectral line-by-line shaping of a frequency comb source (i.e., a Fourier-transform optical pulse shaping) in order to provide the ideal excitation of breather solutions in terms of phase and amplitude [56, 57].

In the following, we describe in detail the different experimental setups implemented to generate optical breathers on finite background, and we give some physical insights into the corresponding nonlinear dynamics drawn from numerical simulations of the NLSE.

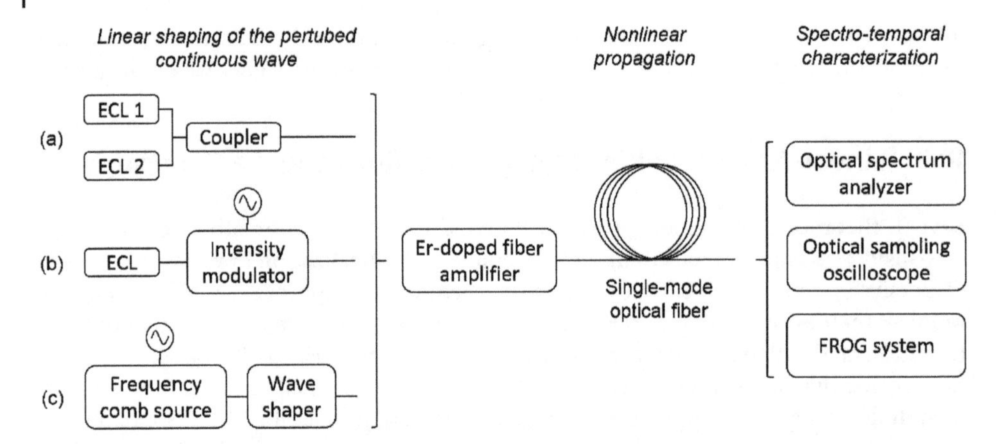

Figure 10.8 Experimental configurations for observing breather dynamics in nonlinear fiber optics and based on three distinct linear shaping of the initial periodically-perturbed continuous wave: (a) the beating of two narrow-linewidth lasers whose perturbation frequency depends on their frequency spacing, (b) the electro-optic (intensity) modulation of a narrow-linewidth laser at a frequency given by a RF clock, (c) the linear shaping of an optical frequency comb source driven by a RF clock. ECL: external-cavity laser. FROG: frequency-resolved optical gating. Note that a phase-modulation stage to mitigate the detrimental effect of Brillouin scattering [58] is also required in the different linear-shaping configurations.

10.3.1 Experimental Configurations

Figure 10.8 depicts three different experimental configurations used to observe breather dynamics in nonlinear fiber optics. All the test-beds are based on commercially-available high-speed telecommunications-grade components.

These experimental configurations only differ in the linear shaping of the initial perturbation imposed to the cw. The beating of two lasers and the electro-optic modulation (see Figures 10.8 (a), (b)) allow us to induce a simple sinusoidal perturbation in the temporal domain whose both amplitude and frequency can be controlled. However, in the spectral domain, the beating configuration corresponds to an asymmetric initial condition with a strong cw (i.e., ECL1 plays the role of the pump) perturbed by a weak single sideband (i.e., ECL2 plays the role of the seed) shifted by Ω from the pump. The amplitude and frequency detuning of the second laser have to be finely controlled to choose the required perturbation. While in the case of the electro-optic modulation (typically a $LiNbO_3$ intensity modulator), the initial condition is a symmetric spectrum with a strong pump perturbed by a pair of sidebands shifted by $\pm\Omega$ from the pump. Here the amplitude and frequency detuning of the perturbation are changed through an external RF signal generator. In any case, these two configurations cannot exactly generate ideal perturbation of the cw derived from previous theoretical expressions. We are limited to periodic modulations and particularly sinusoidal perturbations, and there is no control of the phase difference between the spectral components [59]. The third configuration overcomes this important issue by spectrally shaping both relative phase and intensity of each line of a frequency comb source [60,61]. The programmable pulse shaper allows the generation of nearly arbitrarily shaped optical wave forms, in particular through a time-periodic pattern whose frequency is equal to the spectral separation of the comb lines. High resolution (about 1 GHz) systems are available to select and control individual spectral peaks [62]. The initial frequency comb can be generated by the implementation of variable-repetition-rate pulse source based on the nonlinear compression of an initial sinusoidal signal in a cavity-less optical-fiber-based device (i.e., similar to the second configuration or see [63]). The spectrum of such a pulse source can be approximated as a series of Dirac δ functions separated by the repetition rate. The width of the comb envelope depends on the nonlinear compression of the initial modulated cw and it determines the number of sidebands and their decreasing amplitude. A phase modulator is also introduced in the two first configurations or in the above frequency comb source to prevent the detrimental effect of stimulated Brillouin backscattering. Note that all the configurations restrict the experiments to study periodic perturbations of the cw, or to fit the limited time-window over which we can inscribe localized perturbation (the latter implies that no interaction occurs between neighboring elements of the periodic pattern on the studied distance).

Whatever the linear shaping of the perturbed cw, an erbium doped-fiber amplifier (EDFA) is then used to amplify the average power of the synthesized wave before coupling into a single-mode optical fiber to undergo nonlinear propagation. The average power of the input wave is fixed so that it satisfies the value of the governing parameter a of breathers according to the fiber properties (see below for rescaling in dimensional units). At the fiber output, the optical wave profile is typically characterized using a high-dynamic-range optical spectrum analyser and an ultrafast optical sampling oscilloscope with subpicosecond resolution, or an adapted frequency-resolved optical gating (FROG) technique to retrieve the intensity and phase of periodic pulse trains on finite background fields [64].

To reconstruct the growth-decay intensity dynamics of breathers along propagation, distinct methods were already used. The direct (and destructive) method is based on cutback experiments, the low cost standard single mode fiber SMF-28 is well-suited for such measurements. More recently and inspired by hydrodynamic experiments, an original approach of short propagation sequences for reconstructing the full wave evolution was reported. It benefits from the programmable wave shaper to shape repetitively different initial shaped conditions for a fixed and short nonlinear propagation length [57]. First, we begin with initial conditions fixed from theory at an arbitrary position from the maximal breather amplitude and we record the wave profile at the fiber output. This provides in the next step the initial condition. Repetition of this recording process several times permits a long propagation distance to be reached without detrimental fiber losses.

The correspondence between theory and experiment can be retrieved by recalling that dimensional distance z (m) and time t (s) are related to the previous normalized parameters by $z = \xi L_{NL}$ and $t = \tau t_0$, where the characteristic (nonlinear) length and time scales are $L_{NL} = (\gamma P_0)^{-1}$ and $t_0 = (|\beta_2| L_{NL})^{1/2}$, respectively. The dimensional field envelope $\varepsilon(z,t)$ ($W^{1/2}$) is $\varepsilon = P_0^{1/2}\psi$, P_0 being the average power of the input wave. The modulation frequency ω of a single breather is related to the general governing parameter a by $2a = [1 - (\omega/\omega_c)^2]$, where the critical frequency value of the modulation instability gain is given by $\omega_c^2 = 4\gamma P_0/|\beta_2|$ [65]. Note that $\beta_2 (< 0)$ and γ refer to the group-velocity dispersion and the nonlinear coefficient of the fiber used, respectively [58]. The corresponding dimensional form of the NLSE is written as follows:

$$\imath\frac{\partial \varepsilon}{\partial z} - \frac{\beta_2}{2}\frac{\partial^2 \varepsilon}{\partial t^2} + \gamma|\varepsilon|^2\varepsilon = 0 \tag{10.6}$$

As an example, the observation of the AB with $a = 0.45$ in the standard telecommunication single-mode fiber SMF-28 at 1550 nm ($\beta_2 = -21$ ps^2 km^{-1} and $\gamma = 1.2$ W^{-1} km^{-1}) requires an initially 20-GHz modulated continuous wave with average power equal to $P_0 = 0.7$ W.

10.3.2 Impact of Initial Conditions

We investigate the impact of non-ideal initial conditions as previously described on the generation of breather waves, in the particular case of temporal periodic perturbations. We performed numerical simulations based on the NLSE with three distinct input conditions for the modulated continuous wave in order to generate the AB with $a = 0.25$, namely an ideal perturbation corresponding to the theoretical AB at $\xi = -3$, and two simplified (non-ideal) perturbations with similar amplitudes induced by an intensity modulator or by a weak single sideband. Figure 10.9 (a)–(c) report the evolution dynamics of the corresponding temporal intensity profiles. Even in the presence of very similar

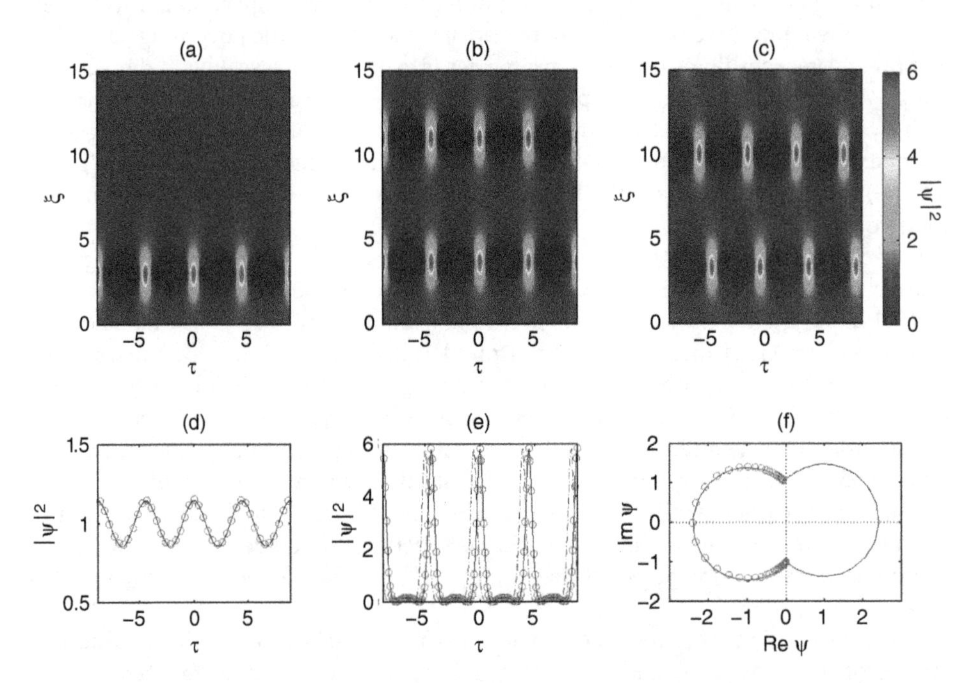

Figure 10.9 Impact of initial conditions on AB dynamics during propagation (here in the case of $a = 0.25$) investigated through NLSE simulations. (a) Intensity evolution for exact initial condition from Eq. (10.2). (b) Intensity evolution for initial 'cos' perturbation induced by an intensity modulator: $\psi = \sqrt{1 + 0.145\cos(\omega\tau)}$. (c) Intensity evolution for initial 'exp' perturbation that corresponds to a weak single sideband: $\psi = 1 + 0.07\exp(-\iota\omega\tau)$. Note that amplitudes of perturbation were chosen to provide an initial contrast of modulation similar to the ideal case as confirmed in subplot (d). Red circles correspond to the ideal initial condition. Solid black (dashed) curve corresponds to cosine (exponential) perturbation. (e) Comparison between the different intensity profiles at maximum compression of the breather. (f) Trajectory of the center point of AB (i.e., along the line $\tau = 0$) on the complex plane (circles). The trivial phase factor $exp(\iota\xi)$ is omitted. Black curve corresponds to the trajectory when cosine perturbation is used. Note that an arbitrary initial phase factor was added to rotate the trajectory and superpose the first growth-decay cycle to the ideal AB trajectory.

intensity modulation as shown in Figure 10.9 (d), we observe that non-ideal initial conditions used in practice yield periodic evolution as a function of propagation in contrast to the exact AB theory. However, each growth-return cycle remains well-described by the analytic AB solution. The profiles of the maximally compressed breathers are in excellent agreement (see Figure 10.9 (e)). Figure 10.9 (f) confirm that the features of AB trajectory in the complex plane are well generated even with non-ideal perturbation of the cw. This agreement was confirmed over a range of modulation amplitudes and for various modulation frequencies across the MI gain curve [65]. Such a periodic evolution of the nonlinear stage of modulation instability as a function of propagation is reminiscent of the fundamental FPU recurrence phenomenon, it may be described by bi-periodic (elliptic) breather solutions.

One can also clearly distinguish that the breather propagates with a certain angle to the line $\tau = 0$ for the single sideband perturbation. Indeed, the inclined trajectory acts in the first steps of the initial perturbation growth whose spectrum asymmetry induces a distinct mean group velocity of AB compared to another symmetric perturbation. Similar dynamics are also observed in the final decay of the breather cycle. It is important to mention that the relative frequency position of the single perturbation to the continuous wave enables the control of the mean group velocity of the AB. The mean group velocity of the AB can be slower or faster than the group velocity of the cw. It is therefore possible to control the group velocity difference between two ABs (with distinct values of a) through a suitable choice of perturbations (with opposite sign of modulation frequency) in order to favour the collision. A simple continuous wave that contains a bi-modulation by means of two spectral distinct sidebands may provide the required conditions for breather collision [33,56], and the generation of a second-order breather. Moreover, the inclined trajectory observed in Figure 10.9 (c) can be associated with the excitation of breather dynamics close to breather solutions with varying group velocity (see Figure 10.5).

The impact of initial conditions to generate the KMB with $a = 1$ is studied in Figure 10.10 through NLSE simulations with three different configurations, namely the ideal perturbation at $\xi = -\pi/|b|$ (i.e., half the period), a simplified (non-ideal) periodic perturbation induced by an intensity modulator, and a secant-hyperbolic (sech) pulse on a finite background with similar amplitude and temporal width as shown in Figure 10.10 (d). The full width at half maximum ΔT_{FWHM} of the time-varying intensity (with maximal intensity $I_{max} = (2\sqrt{2a} - 1)^2$) above background of the ideal perturbation was used to determine numerically both the sech pulse width and the optimal frequency for the cosine modulation, as follows: $\Delta\tau_{sech} = 1.8\Delta T_{FWHM}$ and $f_{cos} = 1/(2.4\Delta T_{FWHM})$. Figures 10.10 (a)–(c) report the corresponding evolution dynamics of the intensity profiles. We observe that the cosine modulation yields periodic evolution in good agreement with the exact KMB behavior, only slight discrepancies appear on the longitudinal period and the maximal peak power as revealed by Figures 10.10 (e)–(f). The strongly modulated wave well approximates the KMB over each modulation cycle, but a better initial approach could be potentially provided by the sech perturbation. In this case, two growth-decay cycles are in excellent agreement with KMB theory, but some instabilities then appear with further propagation, thus inducing significant deviations from the expected breather. Nevertheless, the above analysis confirms that the periodic longitudinal dynamics observed with a large initial modulation on a finite background can be described and interpreted in terms of Kuznetsov-Ma

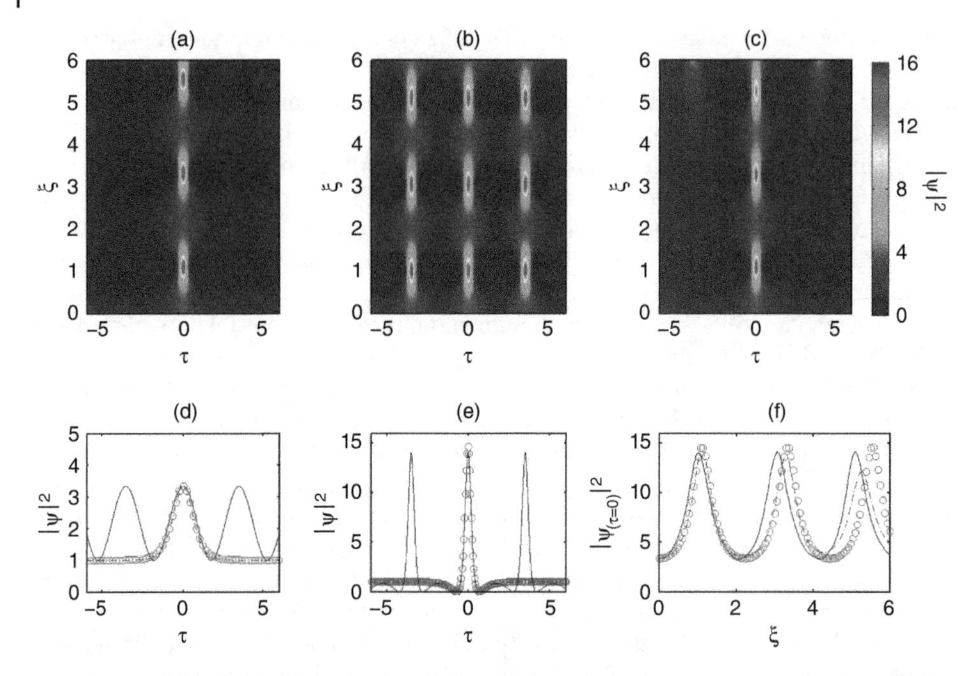

Figure 10.10 Impact of initial conditions on KMB dynamics during propagation (here in the case of $a = 1$) investigated through NLSE simulations. (a) Intensity evolution for exact initial condition from Eq. (10.2). (b) Intensity evolution for initial cosine perturbation induced by an intensity modulator: $\psi = \sqrt{(I_{max}+1)/2 + (I_{max}-1)/2\cos(2\pi f_{cos}\tau)}$. (c) Intensity evolution for the sech pulse on a finite background: $\psi = \sqrt{1 + (I_{max}-1)\,\text{sech}(\tau/\Delta\tau_{sech})^2}$. (d) Comparison of initial conditions. Circles correspond to the ideal case. Solid black (dashed) curve corresponds to cosine (sech pulse) perturbation. (e) Comparison between the different intensity profiles at first maximum compression of the breather. (f) Comparison of the intensity evolution as a function of propagation distance at $\tau = 0$ to reveal the longitudinal periodicity.

breathers [54]. It shows that KMB dynamics can be obtained with very different conditions in NLSE propagation. Such periodic evolution in the NLSE is also an example of FPU recurrence, one may now interpret the KMB as an analytic description of this process.

10.3.3 Higher-Order Modulation Instability

In some cases, non-ideal initial conditions lead to the observation of higher-order modulation instability [55,66]. This higher-order instability arises from the nonlinear superposition of elementary instabilities, associated with initial single Akhmediev breather evolution followed by a regime of complex pulse splitting. The excitation of higher-order MI can observed be readily in experiments using only a single initial frequency modulation on a plane wave, provided that the modulation frequency is below a critical low frequency limit such that multiple instability harmonics fall under the elementary gain curve [67] (i.e., when $a > 0.375$ or $w < 1$). The harmonics actively participate in the evolutionary process since they are forced by the fundamental unstable mode, and then they grow independently at an exponential rate to dominate the global dynamics.

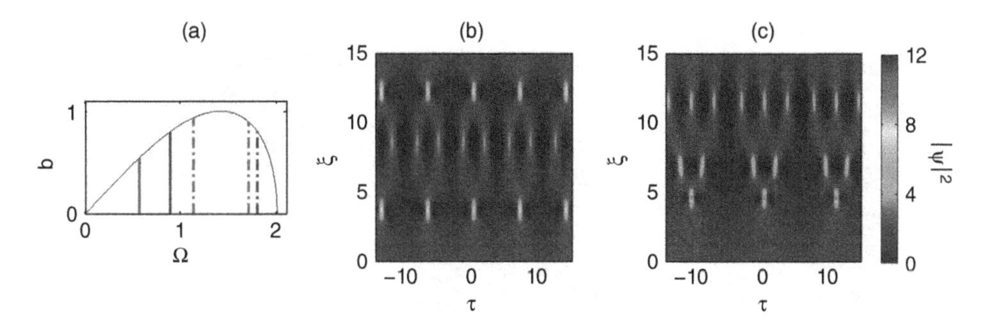

Figure 10.11 Higher-order modulation instability induced by non-ideal excitation of ABs (NLSE simulations). (a) MI gain b curve predicted by the linear stability. Solid lines indicate the perturbation frequencies used in subplots (b,c) to excite ABs corresponding to $a = 0.4$ (left line) and $a = 0.46$ (right line), respectively. Dashed lines indicate their corresponding harmonics that lie within the unstable region. (b,c) Intensity evolution for initial periodic modulation of the cw with a frequency calculated from $a = 0.4$ and 0.46, respectively. In both cases the initial condition is $\psi = \sqrt{1 + 0.15\cos(\omega\tau)}$.

Consequently, the evolution dynamics appear to be a complex composition of elementary breathers. Moreover, in such cases, one cannot extract a well-defined single value of recurrence distance as previously shown in Figure 10.9, a more complex recurrence is observed. Figure 10.11 gives some numerical examples of higher-order MI when using non-ideal excitation of a single AB delivered by an intensity modulator. For $a = 0.4$, only one harmonic of the initial excitation is located within the MI gain band (i.e., unstable mode), which leads to the emergence of two elementary breathers. For $a = 0.46$, two harmonics are now unstable, and then we observe a third-order MI evolution corresponding to the nonlinear superposition of three breathers. Note that the superposition pattern (i.e., the spatial arrangement) is also sensitive to the initial modulation amplitude. Furthermore, it is obvious that the evolution dynamics are also driven by this higher-order instability when multiple sidebands are already present in the initial perturbation, which can be interpreted in terms of higher-order breathers.

10.3.4 Impact of Linear Fiber Losses

Another important issue about the experimental excitation of breathers on finite background is the detrimental effect of linear losses occurring during propagation in optical fibers. We give an overview of the impact of fiber losses on the excitation of ideal ABs in Figure 10.12, based on the dimensional NLSE simulations taking into account a simple dissipative term in the form of $+\alpha\varepsilon/2$ (on the left-hand side of Eq. (10.6)) with $\alpha > 0$ [58]. We compare the evolution dynamics modified by the fiber losses with the free-loss propagation. In particular, we used fiber parameters described above for the standard single-mode fiber SMF-28 and the typical loss is $\alpha = 4.6 \times 10^{-5}$ m^{-1} (i.e., $\alpha_{dB} = 0.2$ dB/km). The corresponding intensity evolutions reported in Figures 10.12 (a) and (b) reveal that the linear loss decreases the maximal peak power of the breather, but also it introduces a longitudinal recurrence with a particular π phase shift. This recurrence exhibits different features from the one introduced by non-ideal excitation shown in Figure 10.9 (see trajectory of the complex plane in Figure 10.12 (f)). However, the first growth-decay cycle remains in good agreement with theory for low fiber losses. Figure 10.12 (c) reports the

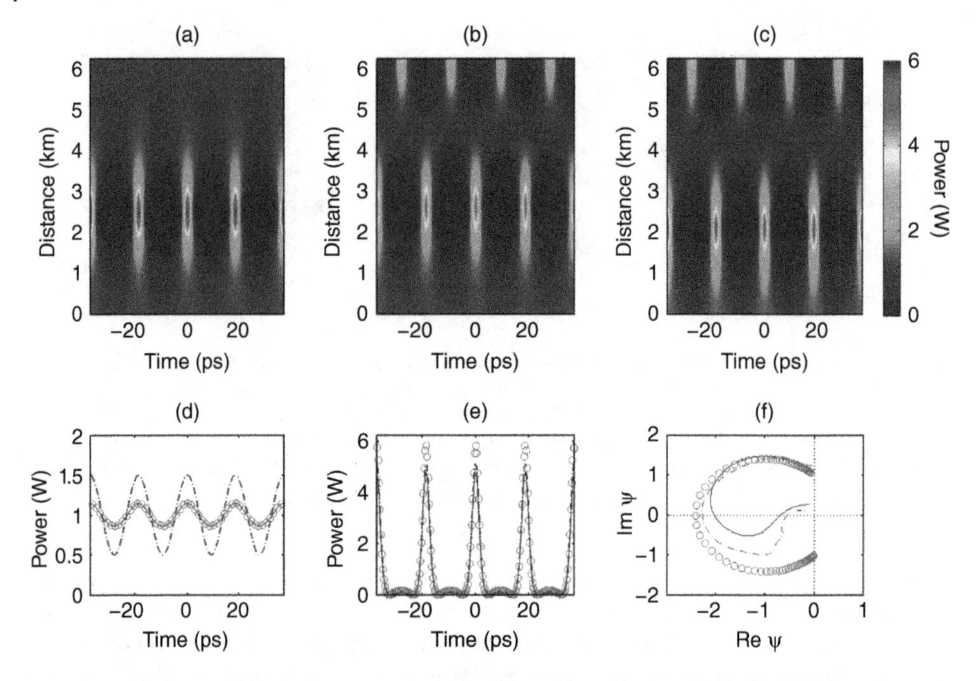

Figure 10.12 Impact of fiber losses on AB dynamics (here $a = 0.25$) investigated through dimensional NLSE simulations. (a) Intensity evolution for an ideal excitation and free-loss propagation. (b) Intensity evolution for an ideal excitation and lossy propagation. (c) Intensity evolution for a significant cosine excitation and lossy propagation. (d) Comparison of initially perturbed continuous waves. Circles correspond to the ideal case. Solid black (dashed) curve corresponds to ideal (cosine) perturbation with lossy propagation. (e) Comparison of maximally compressed breather profiles. (f) Trajectory of the center point of generated breathers (i.e., along the line $\tau = 0$) on the complex plane.

intensity evolution when using non-ideal cosine perturbation with the same lossy propagation. Again we retrieve the π phase shift recurrence, but the main difference is the short distance required to obtain the maximally compressed breather. This minimizes the impact of loss on the maximal peak power as shown in Figure 10.12 (e) at the cost of an initial strong modulation amplitude that significantly differs from the ideal excitation (see Figure 10.12 (d)). The present analysis suggests that the impact of fiber loss can nearly be counteracted by using specific non-ideal conditions; AB dynamics appear more universal than for the specific theoretical conditions. In summary, any kind of disturbance of the NLSE propagation induces a deviation from the expected theoretical trajectory on the complex plane (specific to each breather on finite background), but most of features related to their pulsating dynamics remain well observable. This clearly corroborates the experimental demonstrations described later in this chapter. An essential aspect is related to the sensitivity of the maximal peak amplitude to small deviations of the parameters involved in the propagation. Besides fiber losses, higher order effects linked to pulse propagation such as third-order dispersion, self-steepening, and the Raman effect can also be considered, even theoretically by extending the NLSE [68]. Rogue wave solutions were found in such equations that are integrable in special cases, such as the Sasa–Satsuma or the Hirota equations [69].

10.3.5 Noise and Turbulence

Breathers on finite background may be considered "robust solutions" (but unstable solutions from the mathematical point of view) [68], in the sense that they can be excited or propagated even with non-ideal conditions, and the main features of a localized high amplitude event (i.e., rogue wave) still occur. They provide a coherent and deterministic approach to the understanding of rogue wave phenomena. As a matter of fact, rogue waves are known to spontaneously emerge from an incoherent turbulent state of the system. It is thus of a fundamental importance to study whether first- and higher-order breathers can emerge from a turbulent environment. In this respect, it is worth mentioning that they were identified in the midst of a modulationally unstable chaotic field [19,70] (i.e., the noise-seeded MI regime). However, besides these preliminary numerical observations (also confirmed in [20,71], the problem of the existence of breather waves in a truly turbulent wave system was again considered only numerically in an extended version of the NLSE [72]. The impact of the amount of incoherence in the system on the possible emergence of breather structures was analyzed. In general, the formation of a large-scale coherent structure is only possible if the amount of incoherence in the system is not too large [73–75], a feature that was also observed in the highly incoherent regime of supercontinuum generation in optical fibers [54]. But it was found that complementary temporal and spectral analyses of the turbulent field detected some intermittent or sporadic rogue waves with properties reminiscent of the hierarchy of the breather solutions. Even so, to date, the major challenge still remains the experimental evidence of such numerical predictions.

10.4 Experimental Demonstrations

The first quantitative experimental evidence of breathers on finite background dates from 2010 with both temporal and spectral analyses [7]. It is worth mentioning that former works qualitatively found that the MI-induced formation of arrays of spatial solitons in a planar waveguide, or the onset phase of cw supercontinuum generation in optical fibers, may be described by the usual breather solutions when analyzing the experiments [65, 77]. In addition to this, the ability to generate high-repetition-rate pulse trains through coherent-driven MI in optical fibers was studied earlier, in the 1980s [78,79]. However, until recently, most of these experiments focused on specific initial conditions leading to soliton trains without background, the latter are more suitable for telecommunications applications (this corresponds to $0.1 < a < 0.25$, but strongly modulated initial cw was typically used, i.e., a bichromatic pumping). The design of experimental setups was then based on numerical simulations or empirical laws, all this may explain why the general breather solution was not used. Note that one of the main signature of breather dynamics was reported in 2001 without comparison to theoretical solutions, it concerns the growth-decay cycle of the seeded-MI regime that can be linked to the FPU recurrence phenomenon [41]. In the following, we review the quantitative experimental demonstrations of the different classes of breather solutions. We also point out the impact of non-ideal initial excitation on breather dynamics. Experimental results are compared with corresponding theoretical predictions from breather solutions. Numerical simulations based on the NLSE with experimental initial conditions and fiber losses are not shown, since they are usually indistinguishable from experimental results.

10.4.1 Peregrine Breather

In 2010, Kibler *et al.* used the analytic description of NLSE breather propagation to implement experiments in optical fiber generating femtosecond pulses with strong temporal and spatial localization, and near-ideal temporal Peregrine breather (PB) characteristics. These experiments represent the first amplitude and phase measurements of a nonlinear breather structure in any continuous NLSE soliton-supporting system. The experimental setup was based on the beating of two lasers (see Figure 10.8 (a)) to induce a simple sinusoidal perturbation in the temporal domain whose both amplitude and frequency can be controlled. This technique simply excites the large family of AB solutions by adapting the perturbation frequency and the input power as a function of the fiber characteristics to control the governing parameter a. The evolution toward the Peregrine breather as $a \rightarrow 1/2$ corresponds to the limit where $\omega \rightarrow 0$ is then accessible by decreasing the frequency space between lasers. They reached the divergent regime where near-PB characteristics are observed with the values of $a > 0.4$. A highly nonlinear fiber was used to reduce both fiber length and input power required to observe the Peregrine breather. Detailed temporal measurements using FROG were carried out at the distance where PB features are expected at maximum compression. The retrieved intensity and phase are shown in Figure 10.13 (a). The FROG measurements confirm the expected temporally localized peak surrounded by a non-zero background, and the different signs of the peak and background amplitudes through the measured relative π phase difference in the vicinity of the intensity null. The measured spectral intensity was compared to the analytic spectrum for the ideal PB, the decay of the measured sideband intensities is well reproduced (see Figure 10.13 (c)), as follows: $\exp[-|\Omega|(|\beta_2|/\gamma P_0)^{1/2}]$.

In this work, spatial localization dynamics were studied indirectly by changing the pump-signal detuning to vary a while studying dynamical evolution by varying the input power (recall $\xi = z\gamma P_0$). Later, in 2011, Hammani *et al.* further explored the generation of the PB characteristics in the standard fiber SMF-28, using a much simplified setup

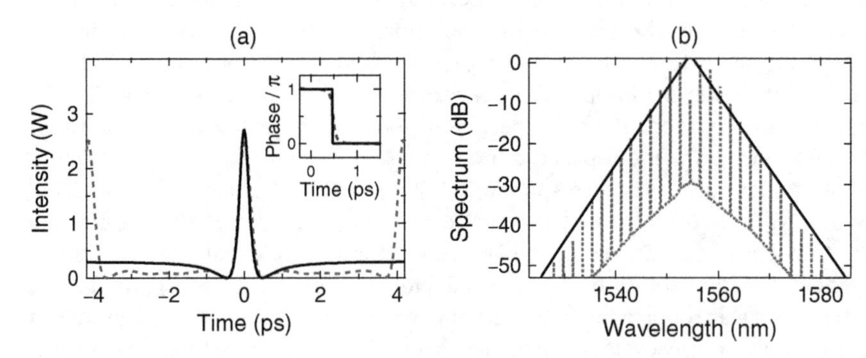

Figure 10.13 Experimental results showing the measured temporal characteristics of the maximally compressed breather, and comparison with the predicted Peregrine breather solution. (a) Intensity and phase from experiment (dots) and for the ideal PB (lines). The maximum peak power from theory is $9P_0 = 2.7W$. (b) Corresponding spectral characteristics from experiment (blue dots) and PB theory (black lines). Note that the theoretical spectrum is calculated for the time-varying envelope component so that the delta-function component at the pump is not shown. *Source*: Adapted from Kibler 2010 [7].

[55]. This setup is based exclusively on commercially available telecommunication-ready components and standard silica SMF-28 fiber. The initially modulated cw is created through direct intensity modulation of a laser diode and the temporal characterization is easily obtained by means of an ultrafast optical sampling oscilloscope. Higher values of a were reached and with cutback measurements, the first direct observation of PB longitudinal evolution dynamics was reported. By showing that Peregrine breather characteristics appear with initial conditions that do not correspond to the mathematical ideal, such results widely impacted on studies of rogue events induced by modulation instability.

10.4.2 Periodic First-Order Breathers

When studying the family of Akhmediev breathers with $0 < a < 0.5$, increased temporal localization is also associated with increasing spatial localization. The modulation-instability recurrence period increases asymptotically as $a \to 1/2$. Two-dimensional localization dynamics were experimentally investigated in [7] in order to to find the regime where the AB approches the Peregrine limit, even with non-ideal excitation. Based on extensive autocorrelation measurements, the degree of spatio-temporal localization was retrieved in good agreement with NLSE simulations for a large range of a values.

As previously described, the exact theory describing the frequency domain evolution of ABs can be compared with experiments. Such experiments measuring pump and multiple sideband generation over a growth–return cycle of MI were performed in 2011 to quantitatively test the theory for an arbitrary value of a (i.e., for arbitrary gain) [38]. The setup is similar to the simplified configuration used to observe the PB dynamics. The excellent ratio allowed to compare experiment and theory out to more than 10 spectral sidebands and over a 30 dB dynamic range, as shown in Fig. 10.14. This was the most

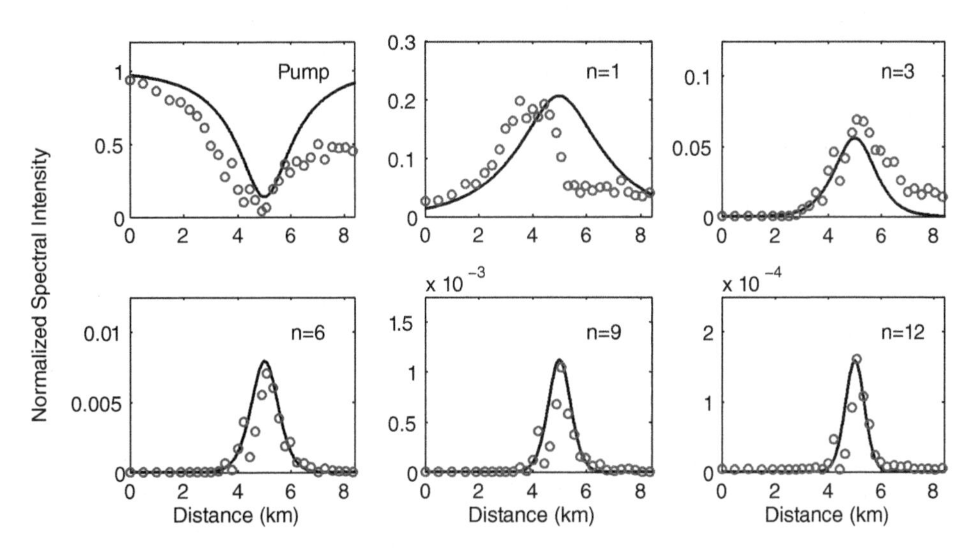

Figure 10.14 Spectral dynamics as a function of propagation distance for the pump and sideband orders as shown. Experiment (circles) is compared with the exact AB theory from Eq. (10.4) (curve). *Source*: Hammani 2011 [38]. Reproduced with permission of Nature Publishing Group.

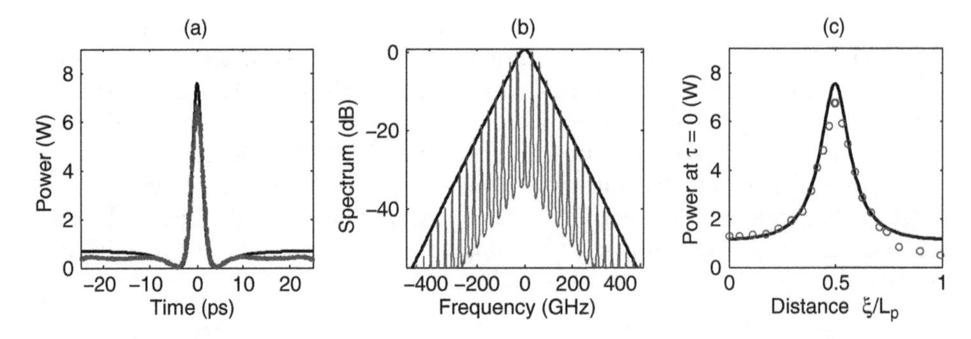

Figure 10.15 (a,b) Comparison of experimental (blue curve) and theoretical (black curve) KM soliton profiles for maximum temporal compression in time and frequency domains, respectively. Note that the theoretical spectrum is calculated for the time-varying envelope component so that the delta-function component at the pump is not shown. (c) Evolution of the maximal power (i.e., at $\tau = 0$) as a function of normalised distance for both experiment (circles) and theory (line). *Source*: Adapted from Kibler 2012 [54].

complete and highest signal of noise characterization of MI dynamics in any NLSE-governed system. We clearly observe the good agreement about the depletion dynamics of the injected modulated continuous wave pump and the near-recovery towards the initial state. This detailed analysis reveals that the temporal compression is associated with an increased spatial localization of the energy transfer to higher sideband orders. These results also confirm the validity of AB theory in describing the dynamical evolution from growth to decay over an MI cycle. Note that imperfect initial conditions of AB excitation and fiber losses here deviate the dynamics from the perfect recurrence.

Next the counterpart Kuznetsov-Ma breathers that are periodic in space were studied experimentally in 2012 [54]. This work reported the first experimental confirmation of the pioneering theoretical studies of the KMB solution of the NLSE. The experimental configuration is similar to AB studies with the direct modulation of a single cw laser. This simple setup only generates non-ideal excitation of KMB, however, it was confirmed that seeding the modulation instability process with strong modulation amplitude allows the observation of KM dynamics, as previously shown in Figure 10.10. One cycle of the initial periodic modulation approximates the ideal KMB solution at a point of minimal intensity in its evolution. The evolving temporal profile was measured with propagation distance by means of fiber cutback experiments. Figure 10.15 shows the direct comparison between the generated temporal profile (at half the KMB period, $L_p/2$) and the longitudinal evolution of the power (at the center of the modulation cycle), and the KMB theory. The agreement between experiment and theory confirms that KM solutions describe the evolution of individual modulation cycles. KMB dynamics then appear more universal than for the specific conditions considered in the original theory. The experimental analysis of the longitudinal periodic behavior was limited to one growth-decay cycle due to detrimental fiber losses.

Note that AB and KMB dynamics were recently studied with exact initial conditions in water waves. The experiments conducted in a water wave flume showed results that are in good agreement with theoretical predictions, thus confirming that such breather solutions can explain the generation of rogue waves in diverse nonlinear dispersive media [80].

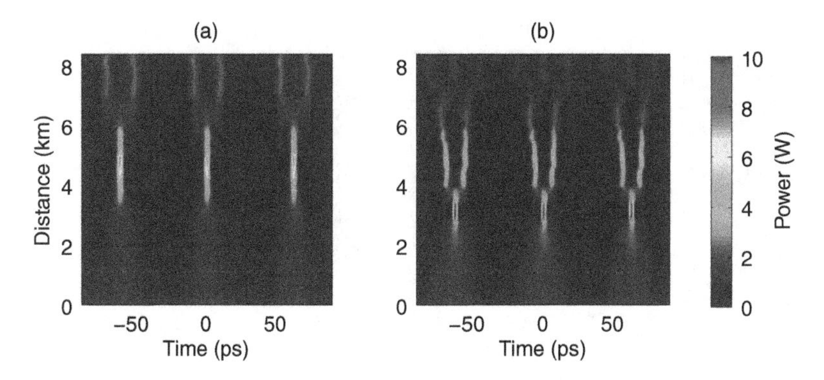

Figure 10.16 Experimental measurements of spatio-temporal evolution of modulated cw undergoing higher-order MI for $a = 0.42$ (a) and $a = 0.464$ (b). *Source*: Adapted from [55] and [66].

10.4.3 Higher-Order Breathers

The experimental observation of higher-order breather dynamics can be classified into two categories that depend on breather excitation. The first relies on a single-frequency non-ideal perturbation [55,66], whereas the second category includes the simultaneous seeding of modulation instability by multiple frequencies [56,57].

Figure 10.16 reports the nonlinear superposition of two or three first-order breathers, observed when a single breather (for $a > 0.375$) is excited with non-ideal perturbation. The experimental setup was based on the direct cosine modulation of a cw laser to excite an AB with parameter $a = 0.42$ or 0.464 close to the Peregrine regime. The initial evolution is only driven by the excited unstable mode, but next impacted by the second and third unstable modes with frequencies corresponding to the harmonics of initial excitation. The generated pattern in the plane (t, z) looks like a complex pulse splitting, it was explained in terms of higher-order MI since the multiple instability harmonics fall under the MI gain curve (see similar pattern in Figure 10.11). The expected return to the initial state for an ideal AB is not observed. Corresponding theoretical predictions can easily be retrieved by using second or third-order breather solutions with a suitable choice of shifted variables ξ_{sj} and τ_{sj} (see Figure 10.7).

To describe giant-amplitude rogue waves, we have to consider the synchronized interaction of first-order breathers that correspond to their collision, i.e., higher-order breather solutions with shifted variables equal to zero. This requires a multiple-sideband perturbation of the continuous wave. The collision of two ABs was experimentally studied in 2013 with non-ideal excitation of the two breathers [56]. An efficient collision occurs during propagation with proper initial phase and velocity differences between breathers. The easiest way to generate two ABs was the seeding of MI process at two distinct frequencies with only two spectral sidebands. The initial cw then contains a bimodulation in the temporal domain. By controlling the initial asymmetry of the sideband amplitudes, their relative phase difference, and their relative frequency spacing from the pump (with opposite sign to favor the collision), we can find specific conditions where the two ABs collide efficiently at a specific distance. Corresponding numerical simulations (without fiber losses) are shown in Figures 10.17 (a)–(c) to illustrate the overall dynamics. These spatio-temporal dynamics are very well described

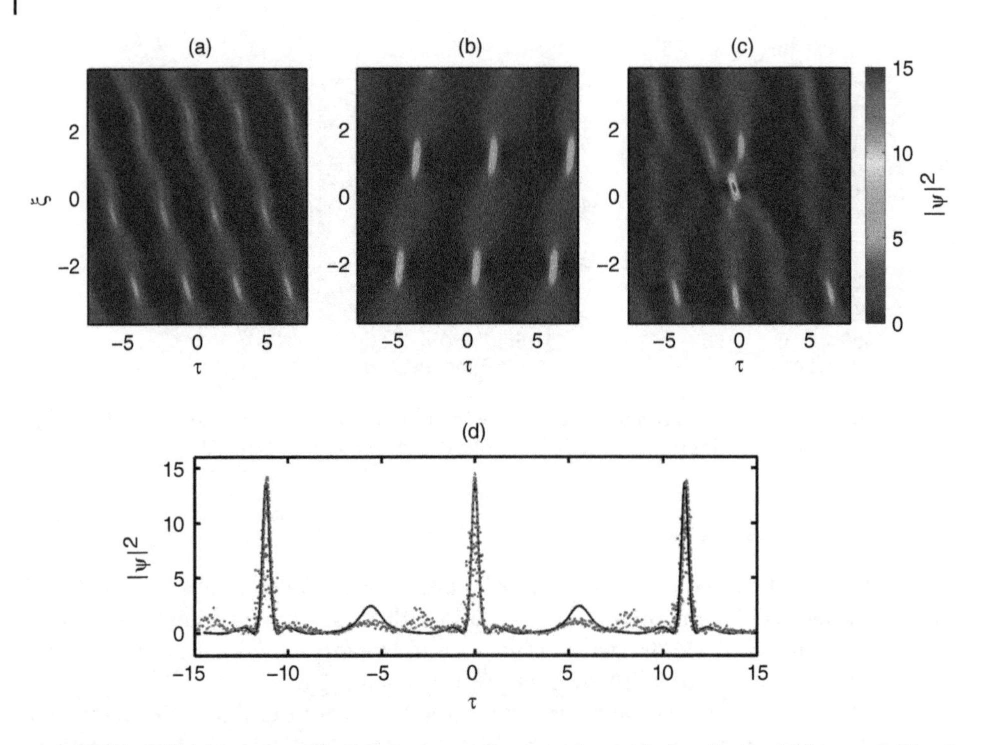

Figure 10.17 NLSE simulations: Single breather evolution with nonideal excitation (a) for $a_1 = 0.14$ and (b) for $a_2 = 0.34$. (c) Collision of the two ABs when excited simultaneously by using the superposition of previous initial conditions. (d) Experimental collision (dots) profile compared to ideal theory (curve) and NLSE simulations with non-ideal excitation(dashed line) by using normalized variables.
Source: Adapted from [56]. http://journals.aps.org/prx/abstract/10.1103/PhysRevX.3.041032 Used under CC-BY-SA 3.0 https://creativecommons.org/licenses/by/3.0/.

by higher-order breather theory. In this work, such a control of spectral sidebands was created through the spectral shaping of three optical comb lines, i.e., the pump and two sidebands (see Figure 10.8). The experimental observation of the giant wave profile generated at the predicted collision distance is reported in Figure 10.17 (d). Obviously, when one of the initial sidebands was switched off, the evolution of a single breather was recovered with a lower peak power. Besides the presence of secondary side lobes attributed to non-ideal excitation of ABs, the main discrepancy with the ideal second-order breather solution is the maximal peak power that can be reached. A better agreement would be conceivable if one could excite the collision on a shorter propagation distance (here 3.8 km) to minimize the impact of fiber losses.

Soon thereafter the experimental demonstration of exact higher-order breather generation was reported by combining this programmable pulse shaping technique (to shape the exact initial excitation) with short propagation sequences to reconstruct the full wave evolution (to overcome the fiber loss issue) [57]. This two-stage linear-nonlinear shaping of an optical frequency comb can be now considered an optical rogue wave-solution generator. As an example, the explicit analytical form of a synchronized two-breather solution of the NLSE was applied as a linear spectral filter to shape the ideal modulation of a continuous wave. Relative amplitude and phase differences of 25 comb lines were managed. The additional nonlinear propagation of the tailored wave provided

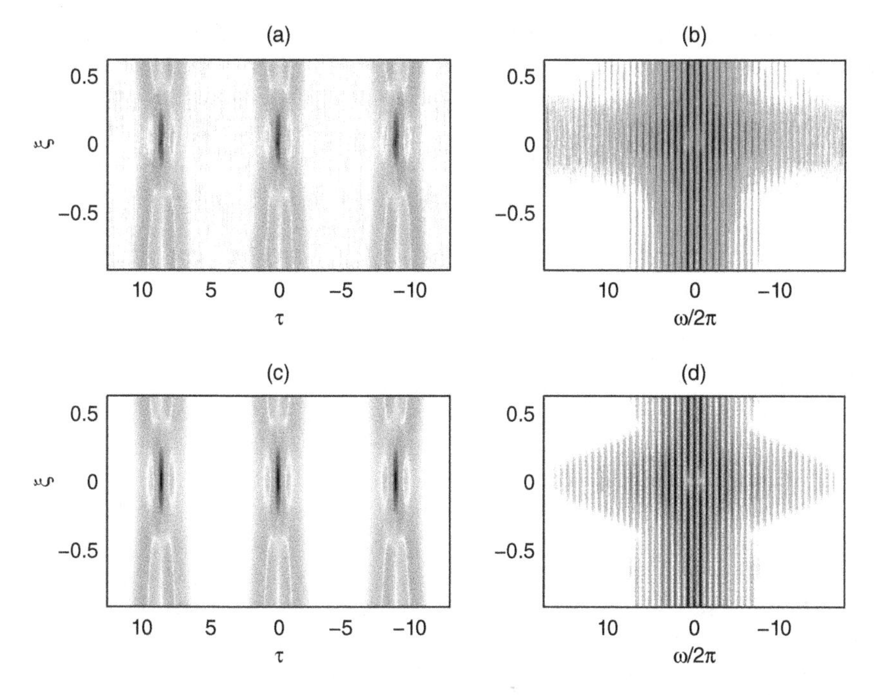

Figure 10.18 Evolution of a second-order periodic breather in both temporal and spectral domains, respectively, as a function of propagation distance. (a,b) Experiments. (c,d) Analytic solution given by Eq. (10.5) $a_1 = 0.2294$ and $a_2 = 0.4323$. *Source*: Adapted from [57]. http://journals.aps.org/pra/abstract/10.1103/PhysRevA.89.023821 Used under CC-BY-SA 3.0 https://creativecommons.org/licenses/by/3.0/.

the first complete experimental evidence of both the growth and decay of this kind of fundamental breather solution in excellent agreement with theory, as reported in Figure 10.18. A misfit parameter between the experimental shape and theory was calculated and found to be below 4%, only due to the residual fiber losses.

10.5 Conclusion

This chapter has reviewed the features of the simplest breather solutions on the finite background of the NLSE, and particularly the main conditions required for their observation in optical fibers and related to the well-known modulation instability process. The recent pioneering experimental studies reported the first complete proofs of existence and control of this untested class of nonlinear waves. Such breather waves are of fundamental importance since they contribute to fully describe the growth-decay cycle of extreme localized waves emerging from modulation instability. These studies were extremely successful and attracted much worldwide attention. However, they have been restricted so far to an essentially reduced class of breathers and in the absence of complex dynamics. To go beyond the frontier in terms of testing and manipulation of the rich dynamics of breathers on finite background, current experimental issues mainly rely on the development of ultrafast optical arbitrary wave generation and characterization, and the design of more complex optical systems without loss (i.e., multi-variable and nonhomogeneous propagation, as well as wave turbulence and perturbed NLS-based systems).

In this context, a recent work investigated the dynamics of ABs in an optical fiber with a longitudinally tailored dispersion that nearly freezes the breather evolution near their point of maximal compression [81]. From a general point of view, the direct analogy drawn within the narrowband approximation of the NLS model between ocean waves and light wave propagation in optical fibers, or even with other physical domains such as plasma physics, makes this research area very challenging for both fundamental and applied aspects.

The results presented here are confined to an essentially ideal class of rogue waves in the absence of dissipation and more complex dynamics such as internal feedback. Another complementary and very active research area aims to address the extended problems of rogue waves in dissipative optical systems. Typically, optical workbenches based on nonlinear cavity systems can be now accurately developed with controllable complexity. But in such complex systems, there is no analytical description of the full spatio-temporal wave dynamics. Very few cases within the very large landscape potential dynamical regimes were recently studied in the context of optical rogue waves (see some of the pioneering works [82–88]). Researchers have revisited and explored the dynamics of such nonlinear systems studied in the 1980s and the 1990s [89], such as optical turbulence and chaos, in order to deal with instability issues in depth and provide a complete understanding of the appearance of "noise-like" intensity fluctuations or "turbulent-like" spatial patterns. A strong motivation for studying dissipative optical rogue waves lies in the wider potential interactions with other scientific domains, such as hydrodynamics, chemical systems, biology and climate where effects such as dissipation and feedback are central features of the dynamics.

During the last year some major experimental developments have been achieved in the observation of rogue breather solutions. First, the evidence of vector rogue wave solutions arising from polarization modulation instability is expected to be a crucial progress in explaining extreme waves in multicomponent systems [90, 91]. Secondly, real-time single-shot measurements of spontaneous breather structures were finally reported. They can emerge from the nonlinear propagation of initially coherent or incoherent waves, and be associated with long-tailed statistics [92, 93]. To conclude, the widest class of creation and annihilation dynamics of modulation instability, also called superregular breathers, was recently and simultaneously observed in optics and hydrodynamics [94].

Acknowledgments

I gratefully acknowledge past and present collaborators in the above-described experimental works: B. Frisquet, K. Hammani, J. Fatome, C. Finot, A. Chabchoub, G. Millot and J. M. Dudley, as well as funding from the French National Research Agency and the Conseil Régional de Bourgogne.

References

1 Osborne, A.R. (2013) *The Sinking of the Prestige: Were Rogue Waves Responsible?* Elsevier, Oxford.

2 Osborne, A.R. (2010) *Nonlinear Ocean Waves and the Inverse Scattering Transform.* Academic Press, San Diego, CA.

3 Solli, D.R., Ropers, C., Koonath, P., and Jalali, B. (2007) Optical rogue waves. *Nature* **450**, 1054–1057.
4 Akhmediev, N. and Pelinovsky, E. (2010) Rogue waves-towards a unifying concept. *European Physics Journal of Special Topics*, **185** (Special issue).
5 Akhmediev, N., Dudley, J.M., Solli, D.R., and Turitsyn, S.K. (2013) Recent progress in investigating optical rogue waves. *Journal of Optics*, **15**, 060201.
6 Onorato, M., Residori, S., Bortolozzo, U., Montina, A., and Arecchi, F.T. (2013) Rogue waves and their generating mechanisms in different physical contexts. *Physics Reports*, **528**, 47–89.
7 Kibler, B., Fatome, J., Finot, C. *et al.* (2010) The Peregrine soliton in nonlinear fibre optics, *Nature Physics*, **6**, 790–795.
8 Chabchoub, A., Hoffmann, N.P., and Akhmediev, N. (2011) Rogue wave observation in a water wave tank. *Physical Review Letters*, **106**, 204502.
9 Zakharov, V.E. and Shabat, A.B. (1971) Exact theory of two-dimensional self-focusing and one-dimensional self-modulation of waves in nonlinear media. *Zh. Eksp. Teor. Fiz.* **61**, 118, [*Soviet Physics JETP*, **34**, 62, 1972].
10 Lighthill, M.J. (1965) Contribution to the theory of waves in non-linear dispersive systems. *Journal of the Institute of Mathematics and its Applications*, **1**, 269–306.
11 Whitham, G.B. (1965) A general approach to linear and nonlinear dispersive waves using a Lagrangian. *Journal of Fluid Mechanics*, **22**, 273–283.
12 Bespalov, V.I. and Talanov, V.J. (1966) Filamentary structure of light beams in nonlinear liquids. *JETP Letters*, **3**, 307–310.
13 Benjami, T.B. and Feir, J.E. (1967) The disintegration of wave trains on deep water. Part 1: Theory. *Journal of Fluid Mechanics*, **27**, 417–430.
14 Benjamin, T.B. (1967) Instability of periodic wavetrains in nonlinear dispersive systems. *Proceedings of Royal Society A*, **299**, 59–75.
15 Zakharov, V.E. (1968) Stability of periodic waves of finite amplitude on a surface of deep fluid. *Journal of Applied Mechanics and Technical Physics*, **9**, 190–194.
16 Akhmediev, N. and Ankiewicz, A. (1997) *Solitons, Nonlinear Pulses and Beams.* Chapman and Hall, London.
17 Peregrine, D.H. (1983) Water waves, nonlinear Schrödinger equations and their solutions. *Journal of Australian Mathematical Society, Series B*, **25**, 16–43.
18 Akhmediev, N. and Korneev, V.I. (1986) Modulation instability and periodic solutions of the nonlinear Schrödinger equation. *Theoretical and Mathematical Physics*, **69**, 1089–1093.
19 Akhmediev, N. Soto-Crespo, J.M., and Ankiewicz, A. (2009) Extreme waves that appear from nowhere: On the nature of rogue waves. *Physics Letters A*, **373**, 2137–2145.
20 Dudley, J.M., Dias, F., Erkintalo, M., and Genty, G. (2014) Instabilities, breathers and rogue waves in optics. *Nature Photonics*, **373**, 755–764.
21 Lake, B.M., Yuen, H.C., Rungaldier, H., and Ferguson, W.E. (1977) Nonlinear deep-water waves: theory and experiment. Part 2. Evolution of a continuous wave train. *Journal of Fluid Mechanics*, **83**, 49–74.
22 Yuen. H.C. and Lake, B.M. (1982) Nonlinear dynamics of deep-water gravity waves. *Advances in Applied Mechanics*, **22**, 67–129.
23 Ablowitz, M.J. and Herbst, B.M. (1990) On homoclinic structure and numerically induced chaos for the nonlinear Schrödinger equation. *SIAM Journal of Applied Mathematics*, **50**, 339–351.

24 Cappellini, G. and Trillo, S. (1991) Third-order three-wave mixing in single-mode fibers: exact solutions and spatial instability effects. *Journal of Optical Society of America B*, **8**, 824–838.

25 Trillo, S. and Wabnitz, S. (1991) Dynamics of the nonlinear modulational instability in optical fibers. *Optics Letters*, **16**, 986–988.

26 Akhmediev, N., Eleonskii, V.M., and Kulagin, N.E. (1987) Exact first-order solutions of the nonlinear Schrödinger equation. *Theoretical and Mathematical Physics*, **72**, 809–818.

27 Dysthe, K.B. and Trulsen, K. (1999) Note on breather-type solutions of the NLS as models for freak-waves. *Physica Scripta*, **82**, 48–52.

28 Kuznetsov, E.A. (1977) Solitons in a parametrically unstable plasma, *Soviet Physics Dokl.* **22**, 507–508.

29 Kawata, T. and Inoue, H. (1978) Inverse scattering method for the nonlinear evolution equations under nonvanishing conditions. *Journal of Physical Society of Japan*, **44**, 1722–1729.

30 Ma, Y.C. (1979) The perturbed plane-wave solutions of the cubic Schrödinger equation. *Studies in Applied Mathematics*, **60**, 43–58.

31 Akhmediev, N., Eleonskii, V.M., and Kulagin, N.E. (1985) Generation of periodic trains of picosecond pulses in an optical fiber: exact solutions. *Soviet Physics JETP*, **62**, 894-899.

32 Devine, N., Ankiewicz, A., Genty, G., Dudley, J.M. and Akhmediev, N. (2011) Recurrence phase shift in Fermi–Pasta–Ulam nonlinear dynamics. *Physics Letters A*, **375**, 4158–4161.

33 Akhmediev, N., Soto-Crespo, J. M., and Ankiewicz, A. (2009) How to excite a rogue wave. *Physical Review A*, **80**, 043818.

34 Zakharov, V.E. and Gelash, A.A. (2013) Nonlinear stage of modulation instability. *Physical Review Letters*, **111**, 054101.

35 Gelash, A.A. and Zakharov, V.E. (2014) Superregular solitonic solutions: a novel scenario for the nonlinear stage of modulation instability. *Nonlinearity*, **27**, 1–39.

36 Smirnov, A.O. (2012) Solution of a nonlinear Schrödinger equation in the form of two-phase freak waves. *Theoretical and Mathematical Physics*, **173**, 1403–1416.

37 Smirnov, A.O. (2013) Periodic two-phase rogue waves. *Mathematical Notes*, **94**, 897–907.

38 Hammani, K., Wetzel, B., Kibler, B., *et al.* (2011) Spectral dynamics of modulation instability described using Akhmediev breather theory. *Optics Letters*, **36**, 2140–2142.

39 Akhmediev, N., Ankiewicz, A., Soto-Crespo, J.M. and Dudley, J.M. (2011) Universal triangular spectra in parametrically-driven systems. *Physics Letters A*, **375**, 775–779.

40 Akhmediev, N. (2001) Nonlinear physics. Déjà vu in optics. *Nature*, **413**, 267–268.

41 Van Simaeys, G., Emplit, P., and Haelterman, M. (2001) Experimental demonstration of the Fermi-Pasta-Ulam recurrence in a modulationally unstable optical wave. *Physical Review Letters*, **87**, 033902.

42 Akhmediev, N., Ankiewicz, A., Soto-Crespo, J.M., and Dudley, J.M. (2011) Rogue wave early warning through spectral measurements? *Physical Letters A*, **375**, 541–544.

43 Solli, D.R., Herink, G., Jalali, B., and Ropers, C. (2012) Fluctuations and correlations in modulation instability. *Nature Photonics*, **6**, 463–468.

44 Wetzel, B., Stefani, A., Larger, L., *et al.* (2012) Real-time full bandwidth measurement of spectral noise in supercontinuum generation. *Scientific Reports.* **2**, 882.
45 Matveev, V.B. and Salle, M. (1991) *Darboux Transformations and Solitons.* Springer-Verlag, Berlin.
46 Kedziora, D.J., Ankiewicz, A., and Akhmediev, N. (2012) Second-order nonlinear Schrödinger equation breather solutions in the degenerate and rogue wave limits. *Physical Review E,* **85**, 066601.
47 Akhmediev, N., Ankiewicz, A., and Taki, M. (2009) Waves that appear from nowhere and disappear without a trace. *Physics Letters A,* **373**, 675–678.
48 Kedziora, D.J., Ankiewicz, A., and Akhmediev, N. (2013) Classifying the hierarchy of nonlinear-Schrödinger-equation rogue-wave solutions. *Physical Review E,* **88**, 013207.
49 Shrira, V.I. and Geogjaev, V.V. (2010) What makes the Peregrine soliton so special as a prototype of freak waves? *Journal of Engineering Mathematics,* **67**, 11–22.
50 Onorato, M., Proment, D., Clauss, G., and Klein, M. (2013) Rogue waves: from nonlinear Schrödinger breather solutions to sea-keeping test. *PLoS One,* **8**, e54629.
51 Bailung, H., Sharma, S.K., and Nakamura, Y. (2011) Observation of Peregrine solitons in a multicomponent plasma with negative ions. *Physical Review Letters,* **107**, 255005.
52 Chabchoub, A., Hoffmann, N.P., Onorato, M., and Akhmediev, N. (2012) Super rogue waves: observation of a higher-order breather in water waves. *Physics Review X,* **2**, 011015.
53 Chabchoub, A., Hoffmann, N., Onorato, M., *et al.* (2012) Observation of a hierarchy of up to fifth-order rogue waves in a water tank. *Physical Review E,* **86**, 056601.
54 Kibler, B., Fatome, J., Finot, C., *et al.* (2012) Observation of Kuznetsov-Ma soliton dynamics in optical fibre. *Scientific Reports,* **2**, 463.
55 Hammani, K., Kibler, B., Finot, C., *et al.* (2011) Peregrine soliton generation and breakup in standard telecommunications fiber. *Optics Letters,* **36**, 112–114.
56 Frisquet, B., Kibler, B., and Millot, G. (2013) Collision of Akhmediev breathers in nonlinear fiber optics. *Physics Review X,* **3**, 041032.
57 Frisquet, B., Chabchoub, A., Fatome, J., *et al.* (2014) Two-stage linear-nonlinear shaping of an optical frequency comb as rogue nonlinear-Schrödinger-equation-solution generator. *Physics Review A,* **89**, 023821.
58 Agrawal, G.P. (2010) *Nonlinear Fiber Optics,* 5th ed. Academic Press, Oxford.
59 Erkintalo, M., Genty, G., Wetzel, B., and Dudley, J.M. (2011) Akhmediev breather evolution in optical fiber for realistic initial conditions. *Physics Letters A,* **375**, 2029–2034.
60 Cundiff, S.T. and Weiner, A.M. (2010) Optical arbitrary waveform generation. *Nature Photonics,* **4**, 760–766.
61 Weiner, A.M. (2011) Ultrafast optical pulse shaping: A tutorial review. *Optics Communications,* **284**, 3669–3692.
62 Clarke, A.M., Williams, D.G., Roelens, M.A.F. and Eggleton, B.J. (2010) Reconfigurable optical pulse generator employing a Fourier-domain programmable optical processor, *Journal of Lightwave Technology,* **28**, 97–103.
63 El Mansouri, I., Fatome, J., Finot, C., Lintz, M., and Pitois, S. (2011) All-fibered high-quality stable 20- and 40-GHz picosecond pulse generators for 160-Gb/s OTDM applications. *IEEE Photonics Technology Letters,* **23**, 1487–1489.

64 Dudley, J.M., Thomson, M.D., Gutty, F., *et al.* (1999) Complete intensity and phase characterisation of optical pulse trains at terahertz repetition rates. *Electronics Letters*, **35**, 2042–2044.

65 Dudley, J.M., Genty, G., Dias, F., Kibler, B., and Akhmediev, N. (2009) Modulation instability, Akhmediev breathers and continuous wave supercontinuum generation. *Optics Express*, **17**, 21497–21508.

66 Erkintalo, M., Hammani, K., Kibler, B., *et al.* (2011) Higher order modulation instability in nonlinear fiber optics. *Physical Review Letters*, **107**, 253901.

67 Yuen, H.C. and Ferguson, W.E. (1978) Relationship between Benjamin–Feir instability and recurrence in the nonlinear Schrödinger equation. *Physics of Fluids*, **21**, 1275.

68 Ankiewicz, A., Devine, N., and Akhmediev, N. (2009) Are rogue waves robust against perturbations? *Physics Letters A*, **373**, 3997–4000.

69 Ankiewicz, A., Soto-Crespo, J.M., Chowdhury, M.A., and Akhmediev, N. (2013) Rogue waves in optical fibers in presence of third-order dispersion, self-steepening, and self-frequency shift. *Journal of Optical Society of America B*, **30**, 87–94.

70 Akhmediev, N., Ankiewicz, A., and Soto-Crespo, J.M. (2009) Rogue waves and rational solutions of the nonlinear Schrödinger equation. *Physical Review E*, **80**, 026601.

71 Toenger, S., Godin, T., Billet, C., *et al.* (2015) Emergent rogue wave structures and statistics in spontaneous modulation instability. *Scientific Reports*, **5**, 10380.

72 Kibler, B., Hammani, K., Michel, C., Finot, C., and Picozzi, A. (2011) Rogue waves, rational solitons and wave turbulence theory. *Physics Letters A*, **375**, 3149–3155.

73 B. Rumpf and Newell, A.C. (2001) Coherent structures and entropy in constrained, modulationally unstable, nonintegrable systems. *Physical Review Letters*, **87**, 054102.

74 Rumpf, B. and Newell, A.C. (2003) Localization and coherence in nonintegrable systems. *Physics D*, **184**, 162–191.

75 Hammani, K., Kibler, B., Finot, C., and Picozzi, A. (2010) Emergence of rogue waves from optical turbulence. *Physical Letters A*, **374**, 3585–3589.

76 Kibler, B., Barviau, B., Michel, C., Millot, G., and Picozzi, A. (2012) Thermodynamic approach of supercontinuum generation. *Optical Fiber Technology*, **18**, 257–267.

77 Cambournac, C., Maillotte, H., Lantz, E., Dudley, J.M., and Chauvet, M. (2002) Spatiotemporal behavior of periodic arrays of spatial solitons in a planar waveguide with relaxing Kerr nonlinearity. *Journal of Optical Society of America B*, **19**, 574–585.

78 Hasegawa, A. (1984) Generation of a train of soliton pulses by induced modulational instability in optical fibers. *Optics Letters*, **9**, 288–290.

79 Tai, K., Tomita, A., Jewell, J.L., and Hasegawa, A. (1986) Generation of subpicosecond solitonlike optical pulses at 0.3 THz repetition rate by induced modulational instability. *Applied Physics Letters*, **49**, 236–238.

80 Chabchoub, A., Kibler, B., Dudley, J.M., and Akhmediev, N. (2014) Hydrodynamics of periodic breathers. *Philosophical Transactions of the Royal Society A*, **372**, 20140005

81 Bendahmane, A., Mussot, A., Szriftgiser, P., *et al.* (2014) Experimental dynamics of Akhmediev breathers in a dispersion varying optical fiber. *Optics Letters*, **39**, 4490–4493.

82 Montina, A., Bortolozzo, U., Residori, S., and Arecchi, F.T. (2009) Non-Gaussian statistics and extreme waves in a nonlinear optical cavity. *Physical Review Letters*, **103**, 173901.

83 Bonatto, C., Feyereisen, M., Barland, S., *et al.* (2011) Deterministic optical rogue waves. *Physical Review Letters*, **107**, 053901.

84 Pisarchik, A.N., Jaimes-Reátegui, R., Sevilla-Escoboza, R., Huerta-Cuellar, G., and Taki, M. (2011) Rogue waves in a multistable system. *Physical Review Letters*, **107**, 274101.
85 Soto-Crespo, J.M., Grelu, P., and Akhmediev, N. (2011) Dissipative rogue waves: Extreme pulses generated by passively mode-locked lasers. *Physical Review E*, **84**, 016604.
86 Lecaplain, C., Grelu, P., Soto-Crespo, J.M., and Akhmediev, N., (2012) Dissipative rogue waves generated by chaotic pulse bunching in a mode-locked laser. *Physical Review Letters*, **108**, 233901.
87 Zamora-Munt, J., Garbin, B., Barland, S., *et al.* (2013) Rogue waves in optically injected lasers: Origin, predictability, and suppression. *Physical Review A*, **87**, 035802.
88 Cavalcante, H.L.D. de S., Oriá, M., Sornette, D., Ott, E., and Gauthier, D.J. (2013) Predictability and suppression of extreme events in a chaotic system. *Physical Review Letters*, **111**, 198701.
89 Gauthier, D.J. (1994) The dynamics of optical systems: A renaissance of the 1990s. *Nonlinear Science Today*, **4**, 2–11.
90 Frisquet, B., Kibler, B., Fatome, J., Morin, P., Baronio, F., Conforti, M., Millot, G., and Wabnitz, S. (2015) Polarization modulation instability in a Manakov fiber system. *Phys. Rev. A*, **92**, 053854.
91 Frisquet, B., Kibler, B., Morin, P., Baronio, F., Conforti, M., Millot, G., and Wabnitz, S. (2016) Optical dark rogue wave. *Sci. Rep.*, **6**, 20785.
92 Suret, P., El Koussaifi, R., Tikan, A., Evain, C., Randoux, S., Szwaj, C., and Bielawski, S. (2016) Single-shot observation of optical rogue waves in integrable turbulence using time microscopy. *Nat. Commun.*, **7**, 13136.
93 Närhi, M., Wetzel, B., Billet, C., Toenger, S., Sylvestre, T., Merolla, J.-M., Morandotti, R., Dias, F., Genty, G., and Dudley, J. M. (2016) Real-time measurements of spontaneous breathers and rogue wave events in optical fibre modulation instability. *Nat. Commun.*, **7**, 13675.
94 Kibler, B., Chabchoub, A., Gelash, A., Akhmediev, N., and Zakharov, V. (2015) Superregular breathers in optics and hydrodynamics: omnipresent modulation instability beyond simple periodicity. *Phys. Rev. X*, **5**, 041026.

11

Wave-Breaking and Dispersive Shock Wave Phenomena in Optical Fibers

Stefano Trillo[1] *and Matteo Conforti*[2]

[1] *Department of Engineering, University of Ferrara, Ferrara, Italy*
[2] *Univ. Lille, CNRS, UMR 8523–PhLAM–Physique des Lasers Atomes et Molécules, Lille, France*

11.1 Introduction

Wave-breaking in optical fibers was first investigated and observed in the 1980s [1–6]. Along with the concept of optical fiber solitons, where the Kerr-effect induced self-phase modulation (SPM) balances the anomalous group-velocity dispersion (GVD), it was realized that, in the normal GVD regime, the SPM enforces the dispersive broadening leading to the steepening of the pulse fronts. It was experimentally confirmed that such steepening occurs symmetrically on both the pulse fronts [1], at variance with regimes where it is only one front that steepens due to mechanism of intensity dependent velocity [7, 8]. The steepening in fibers also features the onset of fast oscillations [3, 5]. At that time, however, it went completely unnoticed that this is the manifestation in optics of a general principle, namely the formation of dispersive shock waves (DSWs), i.e., oscillating wavetrains which are formed owing to the dispersive regularization of a gradient catastrophe [9], the study of which was pioneered in the context of plasmas and fluids. In the 1960s, in fact, Sagdeev and coworkers pointed out the oscillatory nature of shock waves occurring in extremely rarefied (collisionless) plasmas [10], while Zabuski and Kruskal [11] investigated the dispersive breaking of a sinusoidal wave in the framework of the Korteweg-de Vries (KdV) equation. Undulatory structures were studied also in water waves under the denomination of undular bores [12]. Then, in the 1970s, dispersive breaking was observed in ion acoustic waves [13] and water waves [14], while the first construction of a DSW solution of a nonlinear dispersive model (again the KdV) was proposed by Gurevich and Pitaevskii [15], by applying Whitham modulation theory [16] to the asymptotic evolution of a step-like initial condition (shock). Soon thereafter this type of approach was extended to the nonlinear Schrödinger (NLS) equation [17–19]. In the 1990s, fiber optics further stimulated the studies of shock waves [20–26]. However,

Shaping Light in Nonlinear Optical Fibers, First Edition. Edited by Sonia Boscolo and Christophe Finot.

it is only recently that a more general understanding of the dispersive wave-breaking phenomenon in optics has emerged, thanks to several contributions in the area of spatial beam propagation (for which we refer the reader to Chapter 13 in this volume by Gentilini and Conti) and pulse propagation in fibers. Advances in the latter area encompass the prediction and observation of DSWs in multiple four-wave mixing, the control of shock dynamics via simple wave excitation, the competition of wave breaking and MI, the complete theory of the radiation emitted by DSWs, and the role of DSWs in passive fiber cavities. In this chapter we review such studies.

It is worth pointing out from the beginning that, often, shock waves in fibers are associated with the steepening term that arises as a higher-order correction to SPM [2, 27]. This shock-driving term, however, becomes effective only for ultrashort (sub-ps) pulses. Conversely Kerr-induced SPM is effective for longer time scales (ps or longer pulses or tens of GHz modulations). Indeed, SPM, which taken alone does not affect the temporal waveform of pulses, becomes responsible for strong steepening when acting in conjunction with weak group-velocity dispersion (GVD). The effect of initially weak GVD then becomes quite pronounced close to the steepened fronts, inducing dispersive wave-breaking characterized by smooth evolutions toward strongly oscillating envelopes. This mechanism is the leading-order effect that is at the basis of the phenomena discussed in this short survey.

This chapter is organized as follows. In Section 11.2, we review briefly basic concept of classical shocks and their regularization. In Section 11.3, we discuss the details of wave breaking phenomena in fibers, with emphasis on recent advances. In Section 11.4, we briefly show that, in certain regimes wavebreaking can compete with modulational instability. In Section 11.5, we discuss the regimes in which DSWs start to radiate. Section 11.6 briefly considers the case of a passive fiber resonator.

11.2 Gradient Catastrophe and Classical Shock Waves

Let us start by recalling the fundamental concepts related to shock waves, by considering the simplest nonlinear extension of the linear unidirectional linear wave equation $u_z + cu_t = 0$, i.e., the Hopf or inviscid Burgers equation

$$u_z + uu_t = 0 \qquad \Leftrightarrow \qquad u_z + f_t(u) = 0;\ f(u) = \frac{u^2}{2}, \tag{11.1}$$

which implies a velocity c proportional to the local wave elevation, i.e., $c = c(u) = u$, and takes the form of a scalar conservation law characterized by a particular form of the flux $f(u)$. Equation (11.1) can be solved with the method of characteristics [9, 28], i.e., given a generic initial waveform $u_0(t) = u(t, z = 0)$, the generic value $u_0 = u_0(t_0)$ is transported along the characteristics $t(z) = t_0 + c(u_0(t_0))z$. Along a negative slope front these are oblique convergent lines that determine a gradient catastrophe at the finite breaking distance z_b where they first crosses ($z_b = 1/(-m)$, where m is the maximal negative slope of $u_0(t)$ [9]). Beyond this distance the field becomes multi-valued, as shown, as an example, in Figure 11.1 (a) for a Gaussian input.

A jump is introduced to overcome the problem of the multi-valued u, as shown by the dashed vertical line in Figure 11.1 (a). A classical shock wave is a piecewise smooth

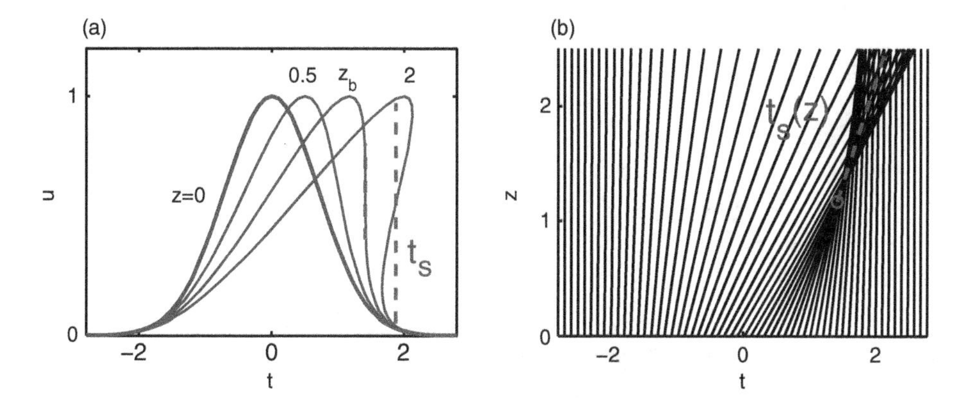

Figure 11.1 (a) Gaussian input developing a gradient catastrophe at distance $z = z_b$ according to the Hopf equation. The dashed line is the classical shock wave which has (temporal) location $t_s = t_s(z)$; (b) Corresponding shock dynamics in the plane (t, z). The first point where characteristic lines intersect stands for the gradient catastrophe point occurring at the breaking distance $z = z_b$. The dashed curve $t_s(z)$ emanating from this point corresponds to the classical shock wave path.

solution of the general conservation law $u_z + f_t = 0$ which contains such a jump. The jump moves along a path $t_s(z)$ according to the so-called Rankine-Hugoniot (RH) condition, which gives the shock velocity

$$V_s = \frac{dt_s(z)}{dz} = \frac{[f]}{[u]}, \tag{11.2}$$

where [...] is the contracted notation for the difference of the quantity inside parenthesis across the jump. Equation (11.2) is derived by considering that across the jump the differential form of the conservation law is undefined, whereas the more general integral form

$$\frac{d}{dz}\int_{t_1}^{t_2} u(t,z)dt = f(t_1,z) - f(t_2,z), \tag{11.3}$$

continues to hold for any t_1, t_2 and reduces to Eq. (11.2), assuming $t_1 \le t_s \le t_2$ [9].

On this basis, the initial jump from $u(t \le 0) = u_L$ to $u(t > 0) = u_R$ is a classical shock wave that moves, according to Eq. (11.2), with velocity $V_s = (u_L^2/2 - u_R^2/2)/(u_L - u_R) = \frac{u_L + u_R}{2}$ (= 1/2 for a unit amplitude jump to zero). We briefly recall that the additional condition $u_L > u_R$ (entropy condition) must hold for a shock solution to be valid. In the opposite case $u_L < u_R$, the solution that turns out to be compatible with the conservation law is a rarefaction wave (see Figure 11.2 (b) for an visual example). In general, the RH condition is equivalent to selecting one curve in the cone of intersecting characteristics, as shown in Figure 11.1 (b).

11.2.1 Regularization Mechanisms

Classical shock waves constitute an important tool in many problems of fluid dynamics. However, in several applications the impact of dissipative or dispersive phenomena

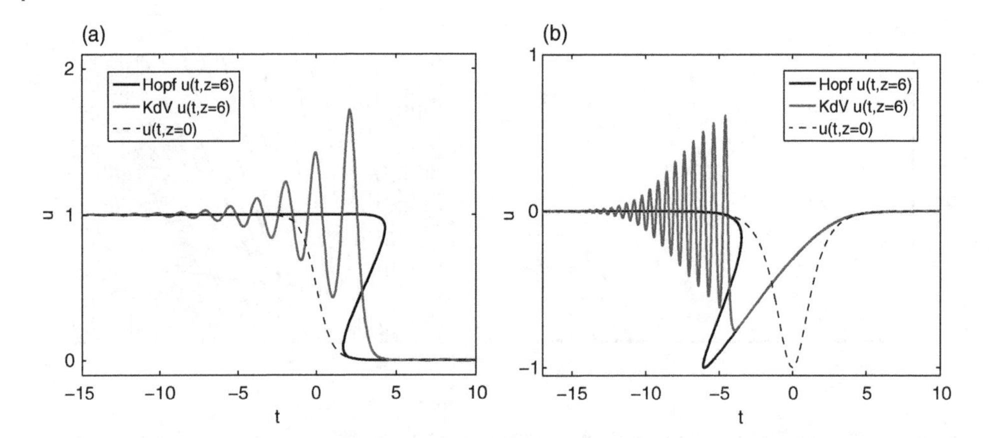

Figure 11.2 Snapshots comparing the post-breaking evolutions ruled by the dispersionless (Hopf) and dispersive (KdV) models: (a) DSW formed by input $u(t,0) = [1 - \tanh(t)]/2$ (dashed), $\varepsilon^2 = 0.05$. (b) input $u(t,0) = -\text{sech}(t)$ (dashed), $\varepsilon^2 = 0.005$ giving rise to a DSW and a rarefaction wave on the negative and positive slope front, respectively.

cannot be ignored. One can include dissipation in Eq. (11.1) by adding the lowest order even derivative, which leads to the famous Burgers equation

$$u_z + uu_t = \alpha u_{tt}, \tag{11.4}$$

which possesses, assuming boundary conditions $u(t = -\infty) = 1$ and $u(t = \infty) = 0$, the shock solution $u = \frac{1}{2}\{1 - \tanh[\frac{1}{4\alpha}(t - \frac{1}{2}z)]\}$. This represents a smooth shock moving with velocity equal to $1/2$. Therefore the viscous effect introduces a finite width 4α of the shock, while the velocity remains equal to the velocity obtained in the inviscid case. Indeed, in the limit $\alpha \to 0$, such solution reduces to the unit jump (classical shock wave) solution of Eq. (11.1) with the same velocity $v_s = 1/2$ predicted by Eq. (11.2). Therefore the classical shock wave is the zero dissipation limit of the shock wave propagating in the presence of losses.

Dispersive regularization, however, is more interesting. By considering the lowest order dispersive correction to Eq. (11.1), namely the odd derivative u_{ttt} weighted by the small coefficient ε^2 (the first odd derivative u_t is trivially removable by introducing a shifted coordinate), the well-known KdV equation is obtained

$$u_z + uu_t + \varepsilon^2 u_{ttt} = 0. \tag{11.5}$$

Even if $\varepsilon^2 \ll 1$, the dispersion drastically alters the dynamics. In Figure 11.2, we compare the evolution ruled by the dispersionless [Eq. (11.1)] and the weakly dispersive [Eq. (11.5)] models, respectively. As shown in Figure 11.2 (a), for a negative slope smooth front, the three-valued field ruled by Eq. (11.5) is regularized by dispersion through the onset of spontaneous oscillations, i.e., a DSW that links the lower and upper states. In Figure 11.2 (b) we show the case of $u(t,0) = -\text{sech}(t)$, which for the KdV stands for an initial value which contains no solitons. In this case, the DSW is accompanied by the formation of a rarefaction wave on the positive slope front. Importantly, (i) the period of the oscillations scales with dispersion being $O(\varepsilon)$, as evident by comparing Figures 11.2 (a) and (b); (ii) the oscillations expand in a region bounded by the characteristic velocities of the leading and trailing edges of the DSW, which replace the single velocity of the classical shock. These velocities can be obtained in the framework of Whitham modulation theory [15,16,29], which describes the DSW as a *cn*–oidal wave with slowly varying parameters. Only by averaging over the oscillations one recover the dispersionless limit.

11.3 Shock Formation in Optical Fibers

In general, a sufficiently accurate model to describe the propagation along an optical fiber of a pulse that modulates the carrier frequency ω_0 is the following envelope equation (a more general model is employed to properly describe the fiber supercontinuum, but the following is sufficient for our goals) [27]

$$iE_Z + ik'E_T - \frac{k''}{2}E_{TT} + \gamma\left[|E|^2E + \frac{i}{\omega_0}(|E|^2E)_T - T_R E(|E|^2)_T\right] = 0. \tag{11.6}$$

The nonlinear terms correspond, in order, to the Kerr effect (SPM), the so-called self-steepening or shock term due to the dispersion of nonlinearity (also the notation $T_{shock} = 1/\omega_0$ is used for its coefficient, which we find misleading since shock waves originate from SPM as well; note that the effect of this term can be especially important in photonic crystal waveguides [30]), and the intrapulse Raman scattering with coefficient T_R (the Raman response time). For the time being we restrict ourselves to the effect of second-order dispersion k'' (higher-order effects will be discussed in Section 11.5). When the GVD is weak, the correct way to derive a hydrodynamic limit of Eq. (11.6), is to make use of the WKB (or Madelung, or geometric optics) transformation applied to the equation cast in the following semiclassical form

$$i\varepsilon\psi_z - \beta_2\frac{\varepsilon^2}{2}\psi_{tt} + |\psi|^2\psi + i\varepsilon\gamma_S(|\psi|^2\psi)_t - \gamma_R\psi(|\psi|^2)_t = 0, \tag{11.7}$$

where we have introduced the normalized variables $t = (T - k'Z)/T_0$ and $z = Z/\sqrt{L_d L_{nl}}$, $\psi = E/\sqrt{P}$, $\beta_2 = \text{sign}(k'') = k''/|k''|$, as well as the smallness dispersion parameter and the normalized nonlinear coefficients

$$\varepsilon = \sqrt{\frac{L_{nl}}{L_d}} = \frac{1}{\sqrt{N}}, \qquad \gamma_S = \frac{1}{\omega_0}\sqrt{\frac{\gamma P_0}{|k''|}}, \qquad \gamma_R = \frac{T_R}{T_0}, \tag{11.8}$$

where $L_{nl} = (\gamma P)^{-1}$ and $L_d = T_0^2/|k''|$ are the nonlinear and dispersion length, respectively, $N = L_d/L_{nl}$ is the soliton order, $P = max(|E(Z=0,T)|^2)$ and T_0 are the peak power and the duration of the input envelope $E(Z=0,T)$. By inserting in Eq. (11.7) the WKB ansatz $\psi(t,z) = \sqrt{\rho(t,z)}\exp[iS(t,z)/\varepsilon]$, and introducing the chirp $u = -S_t$, we obtain

$$\rho_z + \left[\beta_2\rho u + \frac{3}{2}\gamma_S\rho^2\right]_t = 0, \tag{11.9}$$

$$u_z + \beta_2 u u_t + \left[\rho + \gamma_S(\rho u) - \gamma_R\rho_t\right]_t = \varepsilon^2\frac{\beta_2}{4\sqrt{\rho}}\left[\frac{\rho_t}{\sqrt{\rho}}\right]_t. \tag{11.10}$$

Equations (11.9 and 11.10) without approximations are fully equivalent to the NLS Eq. (11.7). The "dispersionless limit" of the NLS (analogous of the Hopf equation for the

KdV) is obtained by neglecting the RHS of Eq. (11.10), also known as a quantum pressure term, which is of higher-order [$O(\varepsilon^2)$] with respect to the LHS of the equations that correspond to order $O(\epsilon^1)$ and $O(\epsilon^0)$, respectively. With this approximation Eqs. (11.9 and 11.10) have the form of the evolution equations in a Eulerian fluid, with ρ and u playing the role of the fluid density (or elevation) and velocity, respectively. Potentially, all the nonlinear terms can contribute to forming shock singularities. However, one must realize that the coefficients of steepening and Raman terms are such that $\gamma_S, \gamma_R \ll 1$, i.e., they are much smaller than the Kerr coefficient (scaled to 1), unless the regime of ultrashort pulses (sub-ps down to few fs) is considered. In particular, both the steepening term and the Raman response indeed affects the wave-breaking dynamics of ultrashort pulses [2,27,31,32], a regime which, however, will not be discussed further in this short survey.

Nonetheless, the general form of Eqs. (11.9) and (11.10) is extremely useful to understand the impact of GVD over the formation of shock waves. In particular, if we formally consider the limit of vanishing GVD ($\beta_2 = 0$), they reduces to the following system

$$\rho_z + 3\gamma_S \rho \rho_t = 0; \qquad u_z + \left[\rho + \gamma_S(\rho u) - \gamma_R \rho_t\right]_t = 0, \tag{11.11}$$

which shows that the power ρ is decoupled from the chirp (phase) dynamics, and obeys a shock-bearing Hopf equation. In the strict limit of zero GVD, therefore, one can conclude that the shock formation in the intensity profile is solely driven by the steepening term, thus occurring along the negative slope front of the pulse, as in the generic example of Figure 11.1 (a) [2, 27]. This is consistent with the fact that, in this limit, the Kerr SPM [i.e., the term ρ_t present in the second of Eqs. (11.11)] only induces a chirp (phase modulation), thereby not affecting the temporal profile of the intensity. However, such conclusion becomes incorrect in the presence of an arbitrarily small dispersion [i.e., a GVD of order ε^2 compared with the Kerr effect of order $O(1)$, as in Eq. (11.7)]. In fact, even neglecting the RHS of Eq. (11.10), i.e., in the dispersionless limit of the NLS (11.7), the GVD still appears in Eqs. (11.9) and (11.10) through terms which retain the same order of the nonlinear terms. These terms are extremely important because they couple the SPM-induced chirp to the power ρ. Therefore, when GVD is arbitrarily small but non-vanishing, the steepening of the pulse fronts can occur through the Kerr term, which becomes the dominant one whenever $\gamma_S, \gamma_R \ll 1$. In this regime, the formation of the shock can be described in the framework of the reduced system [17,33]

$$\rho_z + \beta_2(\rho u)_t = 0; \qquad u_z + \beta_2 u u_t + \rho_t = 0. \tag{11.12}$$

Shock formation occurs when Eqs. (11.12) are hyperbolic, i.e., in the regime of normal GVD, where $\beta_2 = 1$. In this case, the tendency to overtake is caused by the formation of a non-monotonic chirp which drives also the steepening of the power profile [21], until the gradients becomes so large that the dispersive effects associated with the term denoted as quantum pressure [RHS in Eq. (11.10)] set in. Below, we will discuss in more detail the mechanisms of wave-breaking in this regime.

11.3.1 Mechanisms of Wave-Breaking in the Normal GVD Regime

In the normal GVD regime ($\beta_2 = 1$), Eqs. (11.12) are identical to the Saint Venant or shallow water equations or the so-called p-system for an isentropic gas with pressure

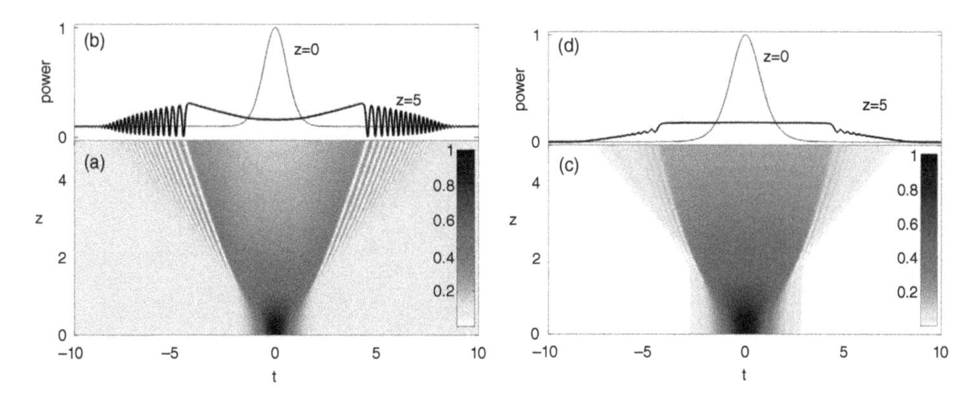

Figure 11.3 False color level plot (a,c) and snapshots of power $\rho = |\psi|^2$ (b,d) of DSWs ruled by NLS Eqs. (11.14), $\varepsilon = 0.05$: (a-b) input $\psi_0 = 0.01 + 0.99\exp(-t^2)$, Gaussian with 10 % power pedestal; (c-d) input $\psi_0 = \text{sech}(t)$, with no pedestal.

law $p = \rho^2/2$. They can be cast in the diagonal form in terms of Riemann invariants $r^{\pm} = u \pm 2\sqrt{\rho}$

$$r_z^{\pm} + V^{\pm}(r^{\pm})r_t^{\pm} = 0, \tag{11.13}$$

with real eigenvelocities $V^{\pm} = (3r^{\pm} + r^{\mp})/4 = u \pm \sqrt{\rho}$. Equations (11.13) entail two different breaking mechanisms depending on the input pulse shape. For bright bell-shaped pulses on a finite background, steepening occurs symmetrically (unlike the steepening ruled by the Hopf equation) until two gradient catastrophies occur. The strong gradients are regularized by the quantum pressure term, which is equivalent to reconsidering the NLS equation

$$i\varepsilon\psi_z - \frac{\varepsilon^2}{2}\psi_{tt} + |\psi|^2\psi = 0. \tag{11.14}$$

The dynamics ruled by Eq. (11.14) is smooth at any distance and exhibits the spontaneous formation of oscillations, starting slightly before the breaking distance (predicted in the dispersionless limit). The expanding oscillations form two symmetric DSWs, as shown in Figure 11.3 (a,b) for a Gaussian input. A mathematical description of a DSW can be obtained by starting from a *cn*-oidal wave, allowing its parameters to vary slowly across the oscillating part of the pulse. The evolution of the parameters is found to obey the Whitham modulation equations. A rarefaction simple wave solution of such equations describes the DSW [17, 18, 29]. On this basis, it turns out that the contrast of the DSW oscillations decreases with the pulse background. This is clear from Figure 11.3 (c,d) which shows the evolution relative to the input $\psi_0 = \text{sech}(t)$ (for a more rapidly decaying pulse such as the Gaussian considered in Figure 11.3 (b), the oscillations would be barely visible in the limit of vanishing background). This is the reason why early measurements in fibers [5] reported a much lower contrast compared with later spatial experiments performed with non-zero background [34].

When a dark input of the type $\psi_0 = \tanh(t)$ is considered, the breaking occurs in $t = 0$. In contrast with the previous breaking mechanism (only one Riemann invariant breaks at each catastrophe point), the latter is a non-generic mechanism which implies

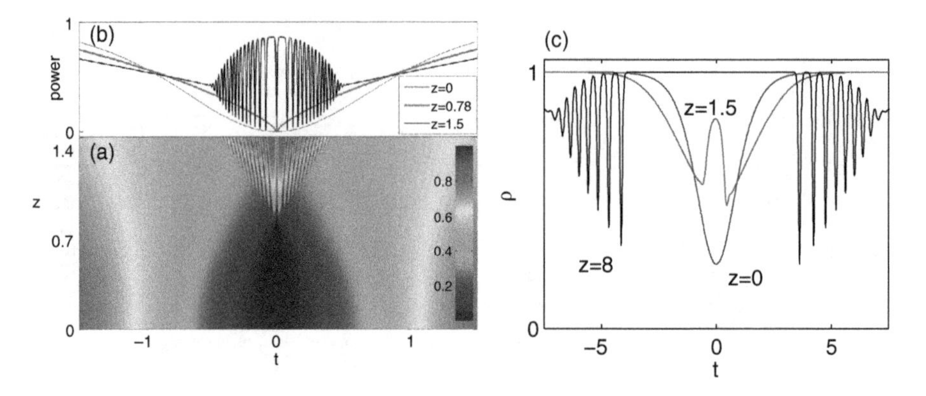

Figure 11.4 (a,b) Wave breaking occurring from $\psi_0 = \tanh(t)$, $\epsilon = 0.005$: (a) False color level plot; (b) snapshots at $z = 0$ (input), $z = 0.78$ (breaking), $z = 1.5$ (DSW). (c) Snapshots of DSW formation from a gray input $\psi_0 = w\tanh(wt) + iv$, $w^2 + v^2 = 1$. Here $v = 0.5$, $\epsilon = 0.05$.

breaking of both Riemann invariants in the null intensity point $t = 0$ [35]. The emerging DSW exhibits a single fan with a narrower central black (zero-velocity) soliton and symmetric gray pairs around it [see Figures 11.4 (a) and (b)]. The intermediate stage shown in Figures 11.4 (a) and (b) shows the typical features of a DSW. However, in this case $\psi_0 = \tanh(t)$ turns out to be a reflectionless potential for $1/\epsilon$ integer, which contains $2/\epsilon - 1$ solitons (i.e., a central black solitons and symmetric gray pairs) which asymptotically separate ($z \sim 1/\epsilon$) on the same background [36]. In this case the DSW is a multisoliton solutions where the solitons start to emerge after the catastrophe in the dispersionless limit [37,38]. Similarly a gray input still generates $2/\epsilon - 1$ (narrower) gray solitons. However, in this case, the field breaks in two distinct points [see snapshot at $z = 1.5$ in Figure 11.4 (c)], from which two asymmetric DSWs emerge. These wave-trains (solitonic DSWs) are asymptotically composed of $1/\epsilon - 1$ and $1/\epsilon$ gray solitons, respectively [39].

As discussed above, the process of breaking depends on the shape of the input pulse, as shown in Figures 11.3 and 11.4. In particular, one can also find pulse shapes that do not lead to breaking. For instance, parabolic pulses are wave-breaking free [40], and led to the concept of similaritons, pulses which can be amplified while retaining their parabolic shape [41]. On the other hand, other situations where wave-breaking phenomena leads to interesting developments involve a continuous wave with chirp modulation [42,43]. For instance, this leads to the generation of flaticon pulses, recently demonstrated in [44].

Finally it is worth emphasizing that, in the spectral (Fourier) domain, the wave-breaking process is characterized by a strong spectral broadening. Indeed the steepening process corresponds to the generation of high frequencies in the spectrum. Beyond the breaking distance the spectrum does not substantially reshape while the DSW spread (see Figure 11.10 (b), for an example of this behavior). In this regime, roughly speaking, the spectrum extends up to the highest frequencies of the (generally non-monochromatic) oscillations. Such as a dramatic spectral broadening has been shown to be highly beneficial for several applications such as supercontinuum generation and comb spectroscopy [45–47].

11.3.2 Shock in Multiple Four-Wave Mixing

While in Section 11.3.1 we considered wave-breaking generated from bright or dark pulses, it was recently shown that DSWs can be generated also from periodic waves, i.e., amplitude modulated waves which gives rise, via the Kerr effect, to the phenomenon of multiple four-wave mixing (mFWM). This indicates the generation of multiple side-band orders at $\omega_0 \pm n\Omega/2$, n odd integer, which is produced via the Kerr term from a dual-frequency input $\omega_0 \pm \Omega/2$ [48, 49]. Similarly an amplitude modulated carrier frequency ω_0, whose spectrum contains frequencies $\omega_0, \omega_0 \pm \Omega$ generates mFWM products $\omega_0 \pm n\Omega$, n integer. When such processes occur in the strong nonlinear regime which involves the generation of several mFWM orders, the field undergoes breaking. The wave-breaking mechanism turns out to be similar to the one illustrated in Figure 11.4 dark solitons. Let us consider, for instance, the NLS (11.14) subject to the initial condition $\psi_0 = \sqrt{2}\cos(\pi t/2)$ (i.e., a Zabusky and Kruskal type of experiment though performed for the NLS instead of the KdV [11]) where, without loss of generality, the normalized frequency detuning is fixed to $\Omega T_0 = \pi$ by choosing $T_0 = 1/2\Delta f = \pi/\Omega$. The dynamics is ruled, in this case, by the single parameter $\varepsilon = \Delta f\sqrt{4k''/(\gamma P)}$ [50].

When $\varepsilon \ll 1$ such input exhibits multiple points of breaking at the nulls of the power profile $|\psi_0|^2 = 2\cos^2(\pi t/2)$, i.e., at $t = (2k+1)$, $k = 0, 1, 2, \ldots$, as shown in Figures 11.5 (a) and (b). The field at the wave-breaking points, displayed in Figure 11.5 (b), is strongly reminiscent of the one generated in the hyperbolic tangent case shown in Figure 11.4 (a). A remarkable difference is that, in the periodic case, the DSW emerging from each breaking point collides with the adjacent ones forming multi-phase structures [see Figures 11.5 (a) and (c)]. The elastic nature of the collision as well as the relationship between the darkness and the velocity of the single filaments that compose the DSW suggests that they behave as dark solitons. While, solitons cannot exists in the strict sense due to the periodic nature of the problem, numerical results show that the

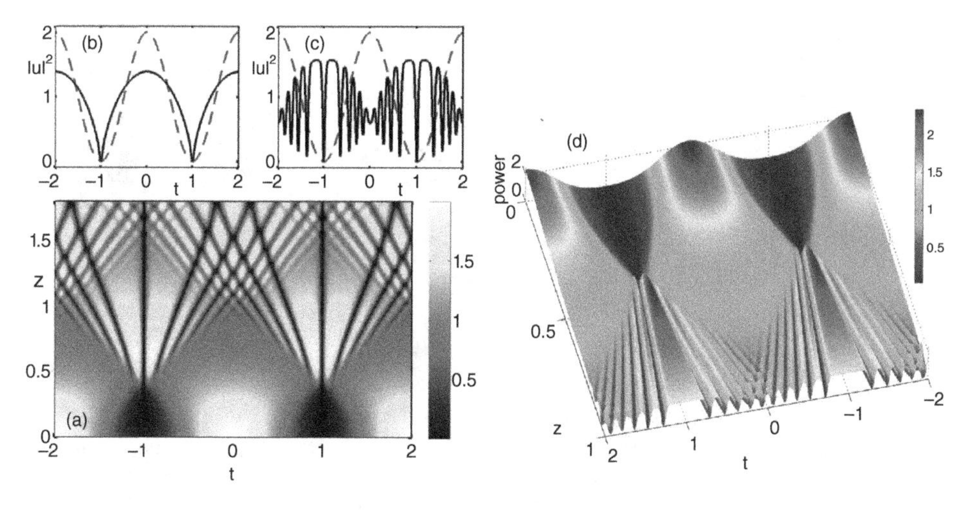

Figure 11.5 (a,b,c) Breaking of a input cosine according to NLS Eq. (11.14): (a) level colorplot of power $|\psi|^2$; (b-c) snapshots of power $\rho = |\psi|^2$ at breaking distance $z = 0.34$ (b) and $z = 0.95$ (c) compared with the input (dashed). (d) Breaking for an input modulated field $\psi_0 = \sqrt{\eta} + \sqrt{2(1-\eta)}\cos(\pi t)$, $\eta = 0.8$. Here $\varepsilon = 0.04$.

finite-band solutions that arise in the scattering problem shrinks in the limit of small ε, resembling true solitons of the infinite line problem in the limit $\varepsilon \to 0$.

Noteworthy, when the modulation has no zeros (i.e., for an imbalanced dual-frequency or triple-frequency input), the temporal locations of the breaking becomes non-degenerate, and breaking occurs at two distinct instants around all the minima of the input modulation, as shown in Figure 11.5 (d). In this case, two symmetric DSWs emerge from each double breaking point, still giving rise to multiphase regions.

The phenomenon of DSW in mFWM was observed in a recent fiber experiment, by employing the Picasso platform at Laboratorie Interdisciplinaire Carnot de Bourgogne in Dijon [51]. The setup achieves the necessary power of the input modulation at 28 GHz, without resorting to pulses, which would hamper the visibility of the DSW of the periodic case. Furthermore, since one cannot easily measure the field along fibers which are several km long, in the experiment the evolution of the DSW is reconstructed at finite physical propagation length $L = 6$ km, by increasing the input power of the modulated field. Changing the power P at fixed length L (and fiber parameters k'' and γ, as well as fixed modulation frequency) amounts indeed to changing the normalized length $z = L/\sqrt{L_d L_{nl}} = L\sqrt{k''\gamma P}/T_0$.

The results are shown in Figure 11.6. In particular the top row figures show the measured temporal traces versus the input power for three different configurations: corresponding to: (i) sinusoidal input (suppressed carrier, balanced dual frequency input); (ii) dual-frequency (suppressed carrier) input with imbalanced power fractions; (iii) modulated carrier corresponding to a triple frequency input. In particular, Figure 11.6 (a) confirms the scenario illustrated in Figures 11.5 (a)–(c). In the other two cases two breaking points arise around the minima of the modulated input, with preserved symmetry (in time) in the triple-frequency case [see Fig. 11.5 (c)] and broken symmetry for the imbalanced dual-frequency (carrier suppressed) case [see Fig. 11.5 (b)]. The simulations based on the dimensional NLS equation shows

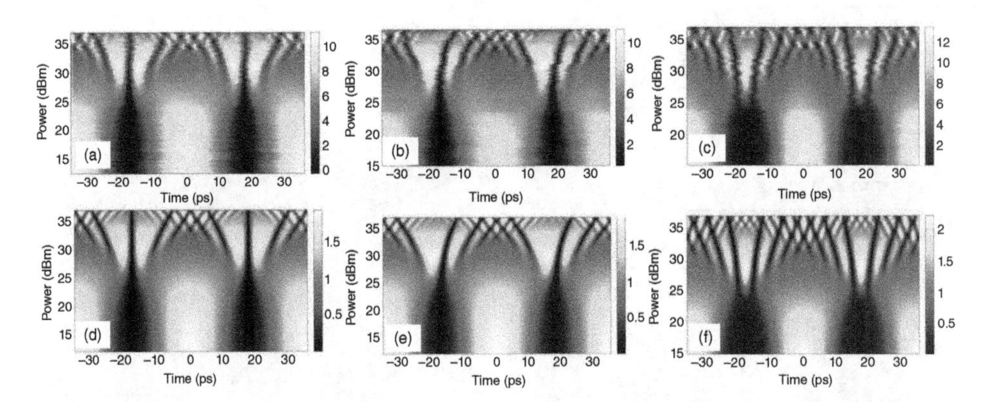

Figure 11.6 DSWs arising from mFWM, showing the color level plot of the temporal profile of the modulated field against the input power P: experiment (a,b,c) vs. simulations (d,e,f) based on NLS Eq. (11.6), neglecting steepening and Raman terms. (a-d) sinusoidal (suppressed carrier balanced dual-frequency) input; (b-e) imbalanced dual frequency (suppressed carrier) input; (c,f) modulated carrier (triple frequency) input. The power is given in log units, $P(dBm) = 10\log_{10} P(mW)$.
Source: From Ref. [51]. http://journals.aps.org/prx/abstract/10.1103/PhysRevX.4.021022 Used under CC-BY-SA 3.0 https://creativecommons.org/licenses/by/3.0/.

a remarkable agreement in all cases [see Figures 11.6 (d)–(f)], without using any fitting parameter. Noteworthy, the NLS does not need any higher-order corrections (steepening, Raman, higher-order dispersion) to correctly describe the experiment.

11.3.3 The Focusing Singularity

We have extensively discussed the normal GVD regime case. However, it is worth mentioning that also the anomalous GVD regime, which yields a *focusing* NLS equation, can exhibit singularity formation. In this case the WKB reduction (11.12) with $\beta_2 = -1$, takes the following form, obtained with the transformation $u \to -u$, [33]

$$\rho_z + (\rho u)_t = 0; \qquad u_z + uu_t - \rho_t = 0, \tag{11.15}$$

which shows the presence of the equivalent negative pressure term $-\rho_t$. As a consequence, this dispersionless limit turns out to be elliptic, which reflects the fact that the full NLS equation exhibits modulational instability (MI). In this case the catastrophe implies breaking around a focus point and is termed elliptic umbilic [52]. An example, which corresponds to the implicit solution of Eqs. (11.15) for the initial datum $\psi_0 = \text{sech}(t)$ discussed in [19, 33], is illustrated in Figure 11.7. In this case, the SPM induces a chirp which is initially linear around the origin and then increases its slope until eventually determines an abrupt change of sign around $t = 0$ [see Figure 11.7 (a); similar to the $\psi_0 = \tanh(t)$ initial datum in the defocusing NLS [38])]. This, in turn, represents a compressional wave which induces a cusp formation at a finite critical distance [see Figure 11.7 (b)]. We point out that this dynamics is caused by the strongly dominant nonlinearity in the NLS equation, thus being of different nature from the collapse phenomenon in the critical NLS (dimensions 1+2D), which occurs right above the threshold (critical power) where nonlinearity and diffraction mutually balance (soliton or Townes

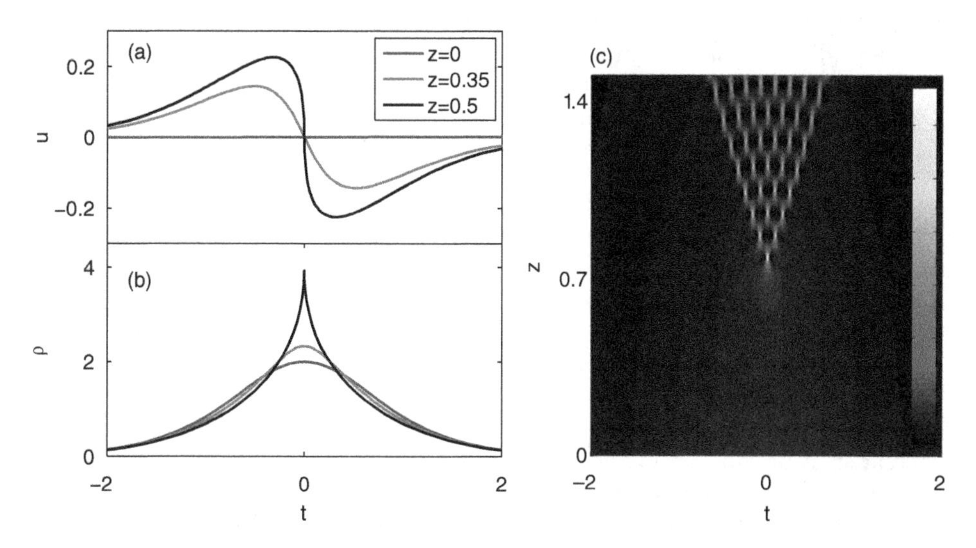

Figure 11.7 Focusing catastrophe: (a,b) snapshots of chirp (hydrodynamical velocity) u (a) and power ρ (b) obtained from the dispersionless equations [Eq. (11.15)] with initial data $\rho_0 = \text{sech}^2(t)$, $u_0 = 0$, corresponding to NLS initial data $\psi_0 = \text{sech}(t)$; (c) corresponding NLS dynamics, $\varepsilon = 0.05$.

profile). In the 1+1D NLS, the post-catastrophe dynamics is usually very complicated (see Figure 11.7 (c) for an example) and exhibits zones of multiphase oscillations separated by caustics ruled by a large ensemble of bright solitons [53, 54]. Nevertheless, under particular conditions, the observed oscillating structure represents the focusing analog of the DSWs featured by the defocusing case [55, 56].

In the presence of noise, due to both high gain and large bandwidth which characterize the MI in the semiclassical (i.e., strongly nonlinear or weakly dispersive) limit, the impact of MI-amplified fluctuations could dramatically affect the breaking process [57]. In this regime the breaking dynamics would strongly benefit from MI-suppressing mechanisms (such as the nonlocality of the nonlinearity). The latter, however, is not effective in fibers, at least for relatively long pulses. Overall, wave breaking in the anomalous dispersion regime remains an experimentally open problem.

11.3.4 Control of DSW and Hopf Dynamics

We finish this section by returning to the Hopf equation [Eq. (11.1)] and discuss its validity in the context of (moderately long) pulse propagation ruled by the NLS equation (11.14). If we consider pulses where the chirp (equivalent hydrodynamical velocity u) is not arbitrary, but rather linked to the power $\rho = |\psi|^2$ as $u = \pm\sqrt{2\rho}$, one of the two Riemann invariants in Eqs. (11.13) identically vanishes. Under this constraint, the solution of Eqs. (11.13) is called a *simple wave* solution (more generally a simple wave solution is such that one of the Riemann invariants is constant in the region of interest). In particular, in this case, Eqs. (11.13) reduce to the following single equation [58]

$$\rho_z \pm 3\sqrt{\rho}\rho_t = 0 \qquad \Leftrightarrow \qquad a_z \pm a a_t = 0, \quad a = 3\sqrt{\rho}, \tag{11.16}$$

i.e., a canonical Hopf equation for the normalized envelope amplitude $a = a(t, z)$. It is worth noting that, in this case, the Hopf equation holds for the pulse envelope (modulating the carrier ω_0), and not for the wave amplitude itself as in the dispersionless limit of the KdV discussed in Section 11.1. Moreover, at variance with Eq. (11.11), where the steepening dominates the dynamics, in Eq. (11.11) it is the Kerr effect that dominates and the validity extends to the regimes of relatively long pulses (i.e., ps to to nsec).

According to Eq. (11.16), the dynamics of a pre-chirped pulse with phase initially locked to $\phi(t, z = 0) = \pm 2\epsilon^{-1} \int_{-\infty}^{t} \sqrt{\rho(t', z = 0)}dt'$ is ruled by the implicit solutions $\rho = \rho_0(t \mp 3\sqrt{\rho}z)$, $u = \pm 2\sqrt{\rho_0(t \mp 3\sqrt{\rho}z)}$ up to the point where the catastrophe occurs and dispersive effects appear. In this case the symmetry in time implicit in the cases shown in Figures 11.3 and 11.4 (a) is broken, since steepening occurs over one of the front controlled by the choice of the upper or lower sign in the above solutions. For instance, Figure 11.8 (a) and (b) show the case of the breaking of an input power profile $\rho_0 = \tanh^2(t)$ with chirp $u_0 = +\sqrt{\rho_0}$, which breaks on the trailing (positive slope) front. Such a phenomenon is not restricted to any specific form of the pulse. Moreover, by combining different signs of chirp on the negative and positive temporal semiaxis, one can also enhance the rapidity of breaking compared with the unchirped case (case not shown, see [58]) or suppress the shock formation [see Figure 11.8 (c)]. Therefore, not only in this regime the SPM induces a pure Hopf dynamics, but also it gives extra degrees of freedom to control the occurrence of breaking. Furthermore, in the presence of a tapering of the GVD (or equivalently third-order dispersion and frequency shifting), extreme

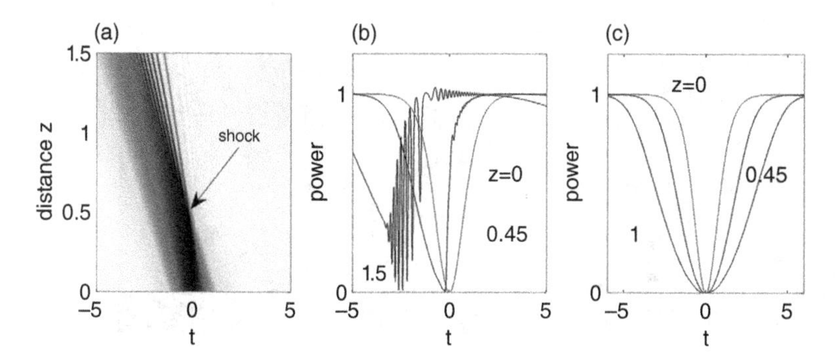

Figure 11.8 (a,b) Evolution and snapshot of a chirped input with amplitude profile $|\psi_0| = \tanh(t)$, undergoing a gradient catastrophe on the trailing edge at $z \sim 0.45$ (a rarefaction wave develops on the leading edge) and the subsequent development into a DSW; (c) shock suppression by a suitable choice of input chirp.

high intensity compressed pulses can develop, which have been termed optical tsunamis by analogy to a shoaling tsunami [43].

11.4 Competing Wave-Breaking Mechanisms

Dispersive nonlinear wave propagation gives rise to different universal mechanisms of breaking. In addition to the dispersive shock formation discussed so far, probably MI, namely, breaking of a carrier wave due to growth of low frequency modulations (see [59] for a review and Chapter 1 by Hansson *et al.* in this book), is the most universal mechanism. For the scalar NLS equation, these two mechanisms are mutually exclusive. In fact, the gradient catastrophe occurs in the defocusing regime characterized indeed by a hyperbolic dispersionless limit. Conversely, MI takes place in the focusing regime where the gradient catastrophe is precluded, reflecting the elliptic dispersionless limit of Eqs. (11.15), discussed in Section 11.3.3. However, it has been recently shown that in more complex setup the two mechanisms can compete. For example, they can coexist in the presence of higher order dispersion [60] or when nonlinearly coupled modes are considered as, for example, in the case of second harmonic generation [61].

Another model that permits to observe the interplay between MI and DSW is the defocusing vector NLS (VNLS), that can be written in weakly dispersing form as

$$i\varepsilon\psi_{jz} - \frac{\varepsilon^2}{2}\psi_{jtt} + \left(|\psi_j|^2 + X|\psi_{3-j}|^2\right)\psi_j = 0, \tag{11.17}$$

where $j = 1, 2$. In contrast with the scalar case ($X = 0$) which is modulationally stable, the plane-wave solutions $u_j = \sqrt{P_j}\exp[i(P_j + XP_{3-j})z/\varepsilon]$ of Eqs. (11.17) are modulationally unstable [62], provided $X > 1$. The MI gain reads as $g = \varepsilon^{-1}\sqrt{K^2[\sqrt{(P_1 - P_2)^2 + 4X^2P_1P_2} - (P_1 + P_2) - K^2]}$, with $K^2 = (\varepsilon q)^2/2$. On the other hand Eqs. (11.17) admit a hyperbolic dispersionless limit, as it is immediately clear in the symmetric case $\psi_1 = \psi_2$ for which Eqs. (11.17) reduce to a scalar NLS equation with nonlinear coefficient $(1 + X)$ [23]. Therefore a competition between DSW and MI

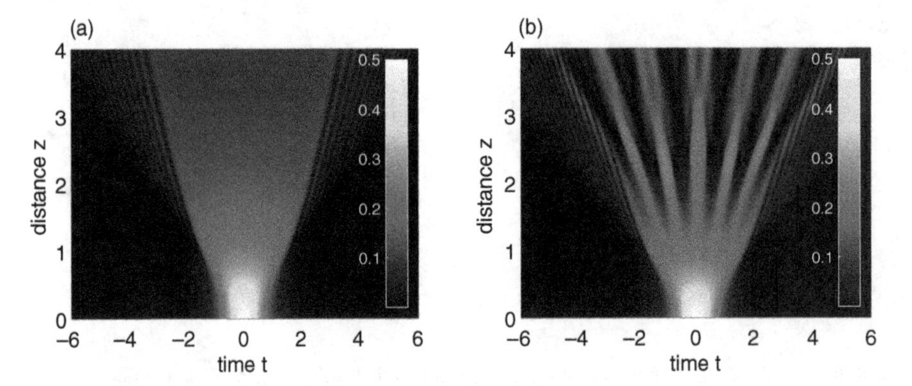

Figure 11.9 Color level plot of $|\psi_1|^2$ ($|\psi_2|^2$ is similar) evolving according to Eqs. (11.17), comparing stable DSW formation for $X = 1$ (a), with the MI unstable case for $X = 2$ (b). Here $\varepsilon = 0.04$ and $\psi_1(0,t) = \psi_2(0,t) = [0.1 + 0.9\exp(-t^2)]/\sqrt{2}$.

should be expected in this case [20, 23]. Restricting for definiteness to the symmetric excitation of Gaussians on pedestal, Fig. 11.9 shows that this is indeed the case. Here, a crossover behavior is induced by changing the cross-phase modulation coefficient X at fixed $\varepsilon = 0.04$. While for $X = 1$ (so-called Manakov case) the system exhibits the formation of stable DSW (Figure 11.9 (a)) in both modes, when $X > 1$ the onset of MI [Fig. 11.9 (b)] leads to additional oscillations appearing on the flat top of the beams which dramatically affect the coherence of the DSW at large propagation distances.

11.5 Resonant Radiation Emitted by Dispersive Shocks

Bright solitons propagating in standard or photonic crystal fibers close to the zero-dispersion wavelength (ZDW) are known to emit resonant radiation (RR) in the region of normal GVD. The underlying mechanism is the resonant coupling with linear dispersive waves (DW) induced by higher-order dispersion [63]. The emission of RR is usually thought to be a prerogative of solitons, but quite recently experimental observations [64] and theoretical investgations [60,65–67], proved that indeed this is not the case. In particular, the DSWs, which develop in the regime of weak dispersion, resonantly amplify DW at frequencies given by a specific phase-matching selection rule. In fact, the strong spectral broadening that accompanies wave-breaking seeds linear waves, which may be resonantly amplified thanks to the well-defined velocity of the shock front.

Consider, for example, a standard telecom fiber (Corning MetroCor) with nonlinear and dispersion parameters as follows: $\gamma = 2.5$ W^{-1} km^{-1}, $k_2 = 6.4$ ps^2/Km, $k_3 = 0.134$ ps^3/Km, and $k_4 = -9 \times 10^{-4}$ ps^4/Km (higher-order terms are negligible), which gives a ZDW $\lambda_{ZDW} = 1625$ nm [64]. Figure 11.10 shows the temporal and spectral propagation of a hyperbolic secant pulse in the normal dispersion regime of the fiber, obtained from the numerical solution of generalized NLS (gNLS) equation. The pulse undergoes a steepening of the leading and trailing edge, which leads to wave breaking at a propagation distance of 20 meters. After the breaking, two DSWs develop with broken symmetry (in time) due to the presence of third order dispersion. The spectrum is the

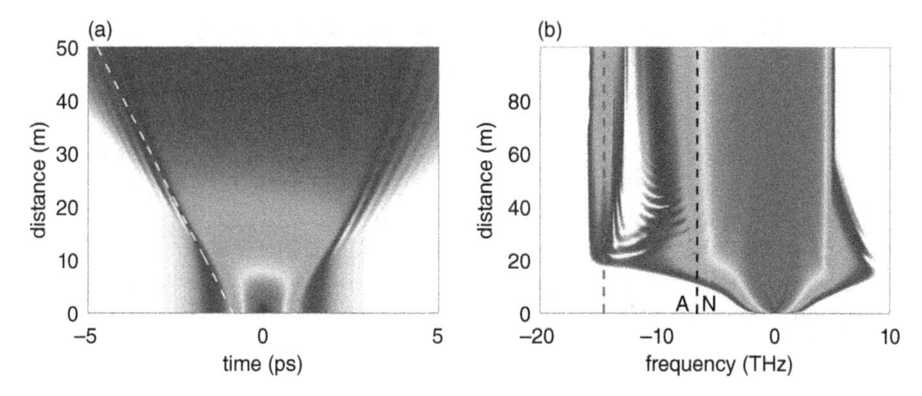

Figure 11.10 Temporal (a) and spectral (b) evolution of an input sech pulse $P_0 = 600$ W, $T_0 = 850$ fs, at $\lambda_p = 1568.5$ nm (normal GVD). A/N labels anomalous/normal GVD regions, and the dashed lines stand for the DW detuning predicted by Eq. (11.21) with velocity given by oblique dashed line in (a).

broadest at the breaking point, and clearly shows a narrowband peak in the anomalous dispersion region. This peak can be interpreted as a resonant radiation emitted by the leading edge of the pulse. In Section 11.5.1, we will show how to predict the spectral position of this resonant radiation.

11.5.1 Phase Matching Condition

We develop our analysis starting from the NLS equation suitably extended to account for the effects of higher order dispersion (HOD). In particular we extend the semiclassical form of NLS in Eq. (11.7) to include HOD terms, while we safely neglect Raman and self-steepening due to the pulse duration and power range that we consider. By defining the dispersion coefficients as $\beta_n = \partial_\omega^n k/\sqrt{(L_{nl})^{n-2}(\partial_\omega^2 k)^n}$, we recover the defocusing NLS in the weakly dispersing form (with $\beta_2 = 1$)

$$i\varepsilon\partial_z\psi + d(i\varepsilon\partial_t)\psi + |\psi|^2\psi = 0, \tag{11.18}$$

$$d(i\varepsilon\partial_t) = \sum_{n\geq 2}\frac{\beta_n}{n!}(i\varepsilon\partial_t)^n = -\frac{\varepsilon^2}{2}\partial_t^2 - i\frac{\beta_3\varepsilon^3}{6}\partial_t^3 + \frac{\beta_4\varepsilon^4}{24}\partial_t^4 + \cdots$$

where we considered normal GVD. Note that the normalized dispersive operator $d(i\varepsilon\partial_t)$ has progressively smaller terms, weighted by powers of the parameter $\varepsilon \ll 1$ and coefficients β_n.

The process of wave-breaking ruled by Eq. (11.18) can be described by applying the Madelung transformation $\psi = \sqrt{\rho}\exp(iS/\varepsilon)$. At leading-order in ε, we obtain a quasi-linear hydrodynamic reduction, with $\rho = |\psi|^2$ and $u = -S_t$ equivalent density and velocity of the flow, which can be further cast in the form [60]

$$\rho_z + \left[\beta_2\,\rho u + \frac{\beta_3}{2}\,\rho u^2 + \frac{\beta_4}{6}\,\rho u^3 + \cdots\right]_t = 0, \tag{11.19}$$

$$(\rho u)_z + \left[\beta_2\,\rho u^2 + \frac{\beta_3}{2}\,\rho u^3 + \frac{\beta_4}{6}\,\rho u^4 + \cdots + \frac{1}{2}\rho^2\right]_t = 0, \tag{11.20}$$

of a conservation law $\mathbf{q}_z + [\mathbf{f}(\mathbf{q})]_t = 0$ for mass and momentum, with $\mathbf{q} = (\rho, \rho u)$. Since Eqs. (11.19) and (11.20) turn out, for small $\beta_{3,4}$, to be hyperbolic, they admit weak solutions in the form of classical shock waves, i.e., traveling jumps from left (ρ_l, u_l) to right (ρ_r, u_r) values, whose velocity V_c can be found from the generalized RH condition (i.e., the natural extension of Eq. (11.2) discussed in the scalar case in Section 11.2): $V_c(\mathbf{q}_l - \mathbf{q}_r) = [\mathbf{f}(\mathbf{q}_l) - \mathbf{f}(\mathbf{q}_r)]$ [28]. However, the jump is regularized by GVD in the form of a DSW. In this regime, the shock velocity can be identified with the velocity V_s of the steep front near the deepest oscillation (DSW leading edge), which differs from V_c and can be determined numerically. For very specific initial contitions, it can be determined analytically [60]. The strong spectral broadening that accompanies steep front formation can act as an efficient seed for DW which are phase-matched to the shock in its moving frame at velocity V_s.

In order to calculate the frequency of the DW, we assume an input pump $\psi_0 = \psi(t, z = 0)$ with central frequency $\omega_p = 0$ [i.e., in real-world units ω_p coincides with ω_0, around which $d(i\varepsilon\partial_t)$ in Eq. (11.18) is expanded]. Let us denote as $V_s = dt/dz$ the "velocity" of the SW near a wave-breaking point and as $\tilde{d}(\varepsilon\omega) = \sum_n \frac{\beta_n}{n!}(\varepsilon\omega)^n$ the Fourier transform of $d(i\varepsilon\partial_t)$. Linear waves $\exp(ik(\omega)z - i\omega t)$ are resonantly amplified when their wavenumber in the shock moving frame, which reads as $k(\omega) = \frac{1}{\varepsilon}[\tilde{d}(\omega) - V_s(\varepsilon\omega)]$ equals the pump wavenumber $k_p = k(\omega_p = 0) = 0$. Denoting also as k_{nl} the difference between the nonlinear contributions to the pump and RR wavenumber (the nonlinear contribution to the wawenumber of the resonant radiation is induced by cross-phase modulation with a non-zero background, on top of which RR propagates), respectively, the radiation is resonantly amplified at frequency detuning $\omega = \omega_{RR}$ that solves the explicit equation [60]

$$\sum_n \frac{\beta_n}{n!}(\varepsilon\omega)^n - V_s(\varepsilon\omega) = \varepsilon k_{nl}. \tag{11.21}$$

We show below that Eq. (11.21) correctly describes the RR emitted by a DSW. At variance with solitons of the focusing NLS where $V_s(\omega_p = 0) = 0$ [63], DSWs possess non-zero velocity V_s, which must be carefully evaluated, having a great impact on the determination of ω_{RR}.

11.5.2 Step-Like Pulses

We consider first a step initial value that allows us to calculate analytically the velocity. Without loss of generality, we take $\beta_3 < 0$. Specifically, we consider the evolution of an initial jump from the "left" state $\rho_l, u_l = 0$ for $t < 0$ to the "right" state $\rho_r (< \rho_l), u_r = 2(\sqrt{\rho_r} - \sqrt{\rho_l})$ for $t > 0$[29, 60]. The leading edge of the resulting DSW can be approximated by a gray soliton, whose velocity can be calculated as $V_l = \sqrt{\rho_l} + u_r = 2\sqrt{\rho_r} - \sqrt{\rho_l}$ [29, 60].

If we account for $k_{nl} = k_{nl}^{sol} - k_{nl}^{RR} = -\frac{1}{\varepsilon}\rho_l$ arising from the soliton $k_{nl}^{sol} = \rho_l/\varepsilon$ and the cross-induced contribution $k_{nl}^{RR} = 2\rho_l/\varepsilon$ to the RR, Eq. (11.21) explicitly reads as

$$\frac{\beta_3}{6}(\varepsilon\omega)^3 + \frac{\beta_2}{2}(\varepsilon\omega)^2 - V_s(\varepsilon\omega) + \rho_l = 0. \tag{11.22}$$

Real solutions $\omega = \omega_{RR}$ of Eq. (11.22) correctly predict the RR as long as $|\beta_3| \lesssim 0.5$, as shown by the NLS simulation in Figure 11.11. The DSW displayed in Figure 11.11 (a)

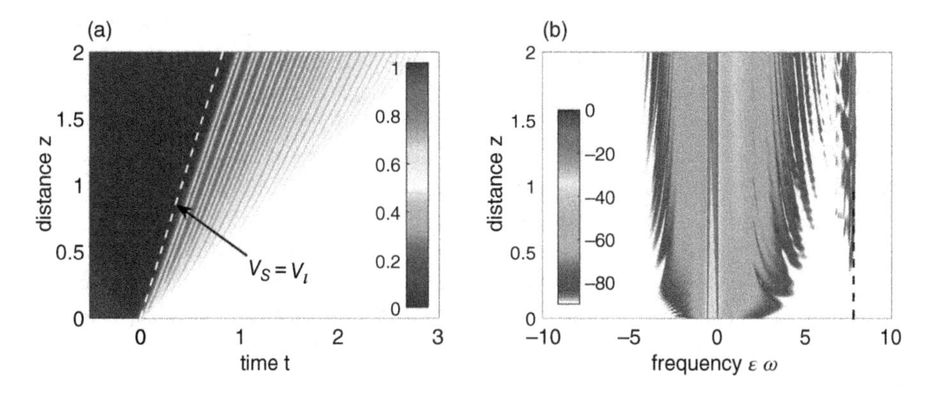

Figure 11.11 Radiating DSW ruled by NLS Eq. (11.18) with $\varepsilon = 0.03$, input step $\rho_l, \rho_r = 1, 0.5$, and 3-HOD $\beta_3 = -0.35$: (a) Color level plot of density $\rho(t, z)$ (the dashed line gives the DSW leading edge velocity V_l); (b) corresponding spectral evolution.

clearly exhibits a spectral RR peak besides spectral shoulders due to the oscillating front, as shown by the spectral evolution in Figure 11.11 (b). Perfect agreement is found between the RR peak obtained in the numerics and the prediction [dashed vertical line in Figure 11.11 (b)] from Eq. (11.22) with velocity $V_s = V_l$. We also point out that k_{nl} represents a small correction, so ω_{RR} can be safely approximated by dropping the last term in Eq. (11.22) to yield $\varepsilon\omega_{RR} = \frac{3}{2\beta_3}(-\beta_2 \pm \sqrt{\beta_2^2 + 8V_s\beta_3/3})$, that can be reduced to the simple formula $\varepsilon\omega_{RR} = -\frac{3\beta_2}{\beta_3}$ [64] only when $\beta_3 V_s \to 0$.

11.5.3 Bright Pulses

The behaviors of step initial data are basically recovered for pulse waveforms that are more manageable in experiments. As shown in Figure 11.12, RR occurs also in the limit of vanishing background, allowing us to conclude that a bright pulse does not need to be a soliton (as in the focusing NLS, $\beta_2 = -1$) to radiate. In fact, resonant

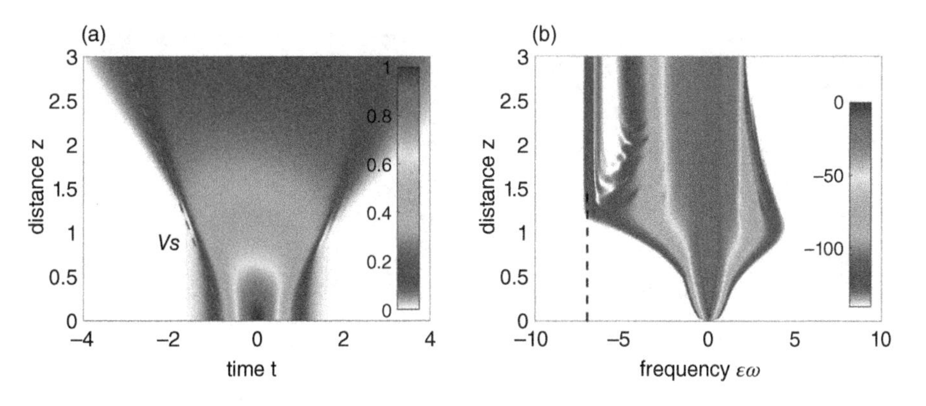

Figure 11.12 (a,b) Temporal and spectral evolution of a Gaussian pulse without background emitting RR, for $\beta_3 = 0.35$, and $\varepsilon = 0.03$.

amplification of linear waves occurs via SWs also in the opposite regime where the nonlinearity strongly enforces the effect of leading-order dispersion, the only key ingredients being a well-defined velocity of the front and the spectral broadening that seeds the RR at phase-matching. Experimental evidence for such RR scenario was reported quite recently [64], corresponding to numerical simulations reported in Figure 11.10. The physical parameters used to obtain Figure 11.10, give normalized parameters $\varepsilon \simeq 0.07$ and $\beta_3 \simeq 0.37$, typical of the wave-breaking regime ($\varepsilon \ll 1$) with perturbative 3-HOD. Since $\beta_3 > 0$, the radiating shock turns out to be the one on the leading edge ($t < 0$), and its velocity $V_s = -0.75$, inserted in Eq. (11.21), gives a negative frequency detuning $\Delta f_{RR} = \omega_{RR} T_0^{-1}/2\pi \simeq 13$ THz, in excellent agreement with the value reported in Ref. [64]. A detailed numerical study of this particular case, including Raman effects, is reported in [65].

11.5.4 Periodic Input

We are interested in the evolution ruled by Eq. (11.18) subject to the dual-frequency initial condition

$$\psi_0 = \sqrt{\eta}\exp(i\omega_p t/2) + \sqrt{1-\eta}\exp(-i\omega_p t/2), \tag{11.23}$$

where we fix the normalized frequency $\omega_p \equiv \Omega T_0 = \pi$ consistently with the notation introduced in Section 11.3.2. Here η accounts for the possible imbalance of the input spectral lines. In this case a frequency comb is generated thanks to mFWM. Formation of DSWs, occurring in the regime of weak normal dispersion ($\beta_2 = 1$) as discussed in Section 11.3.2, enhances the broadening of the comb towards high orders of mFWM. When such DSWs are excited sufficiently close to a ZDW, they are expected to generate RR, owing to phase-matching with linear waves induced by higher-order dispersion [66]. An example of the RR ruled by TOD (we set $\beta_3 = 0.3$) is shown in Figure 11.13 for an imbalanced input [$\eta = 0.3$ in Eq. (11.23)]. The color level evolution in Figure 11.13 (a) clearly shows that the initial waveform undergoes wave breaking around $z \sim 0.4$. The mechanism of breaking has been analyzed in detail in [50, 51] and involves two gradient catastrophes occurring across each minimum of the injected modulation envelope. The

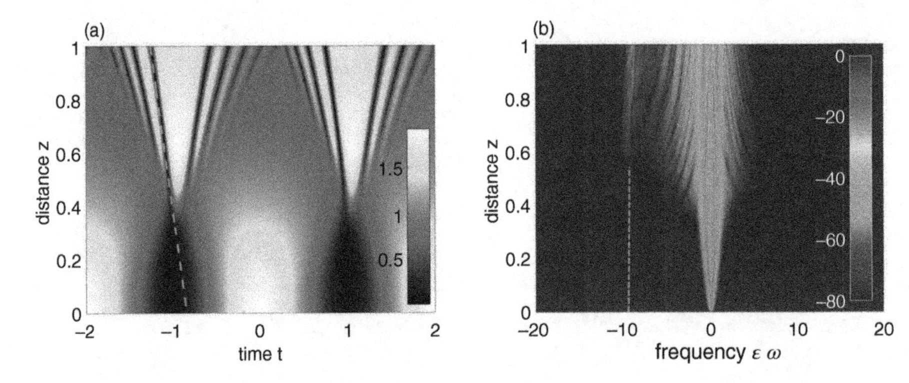

Figure 11.13 RR emitted by shock with asymmetric pumping $\eta = 0.3$, and $\varepsilon = 0.04$: (a) temporal and (b) spectral evolution for $\beta_3 = 0.3$; Dashed line in (a) highlights the DSW edge velocity $V = -0.4$. Vertical dashed line in (b) indicates ω_{RR} from Eq. (11.21) with $k_{nl} = -|\psi_0|^2/\varepsilon$.

GVD regularizes the catastrophes leading to the formation of two DSWs, where the individual oscillations in the trains exhibit dark soliton features, moving with nearly constant darkness and velocity inversely proportional to it. Importantly, the breaking scenario is weakly affected by TOD; However, one can notice that the darkest soliton-like oscillation emits RR. This radiation has much higher frequency than the comb spacing and turns out to be generated over the CW plateau of the leading edge labeled ψ_0. This is clear from Figure 11.13 (b), which shows the enhancement of such frequency at the distance of breaking where the strong spectral broadening associated with the shock acts as a seed for the phase-matched (resonant) frequency.

11.6 Shock Waves in Passive Cavities

Externally driven passive nonlinear cavities, implemented both in fiber rings [68] and monolithic microresonators [69], are exploited for fundamental studies as well as impactful applications such as the formation of wide-span frequency combs. Such resonators exhibit extremely rich dynamics characterized by a host of phenomena such as bistability, modulational instability and soliton formation. It has been recently shown that passive cavities admits a novel dynamical behavior featuring the formation of dispersive-dissipative shock waves [70]. These phenomena are well captured by a mean field approach which yields a driven damped NLS [71], often referred to as the Lugiato-Lefever equation (LLE) [72]. By accounting for HOD, a generalized LLE can expressed in dimensionless units as

$$i\varepsilon\psi_z + d(i\varepsilon\partial_t)\psi + |\psi|^2\psi = [\delta - i\alpha]\,\psi + iS, \tag{11.24}$$

where we adopt the normalization introduced in [67,70]. We just recall that the parameter $\varepsilon = \sqrt{L/L_d} \ll 1$ (L and $L_d = T_0^2/k''$ are the fiber (cavity) length and the dispersion length associated with time scale T_0 and GVD) quantifies the smallness of the GVD and the HOD introduced through the operator $d(i\varepsilon\partial_t) = \sum_{n\geq 2}\beta_n(i\varepsilon\partial_t)^n/n! = -\beta_2\varepsilon^2\partial_t^2/2 - i\beta_3\varepsilon^3\partial_t^3/6 + \cdots$, where the coefficients $\beta_n = \partial_\omega^n k/\sqrt{(L)^{n-2}(\partial_\omega^2 k)^n}$ [note that $\beta_2 = \text{sign}(\partial_\omega^2 k)$] are related to real-world HOD $\partial_\omega^n k$.

Let us first ignore the HOD terms ($\beta_n = 0, n > 2$). Equation 11.24 can exhibit a bistable response with two coexisting stable branches of CW solutions. A DSW can be seen as a fast oscillating modulated wavetrain that connects two sufficiently different quasi-stationary states. Starting from a cavity biased on the lower state, one can easily reach a different state on the upper branch by using an addressing external pulse with moderate power. In this regime the intracavity pulse edges undergo initial steepening, which is mainly driven by the Kerr effect, tending to form shock waves. The strong gradient associated with the steepened fronts enhances the impact of GVD, which ends up inducing the formation of wavetrains that connect the two states of the front. An example is reported in Figure 11.14, by using the injected field $S(t) = \sqrt{P} + \sqrt{P_p}\text{sech}(t)$ with $P = 0.0041$ and $P_p = 0.16P$.

As shown for the conservative case in the previous section, the presence of HOD may lead the shock to radiate [67]. We concentrate on the first relevant dispersive perturbation, i.e., third order dispersion (TOD $\beta_3 \neq 0$), but the scenario is qualitatively

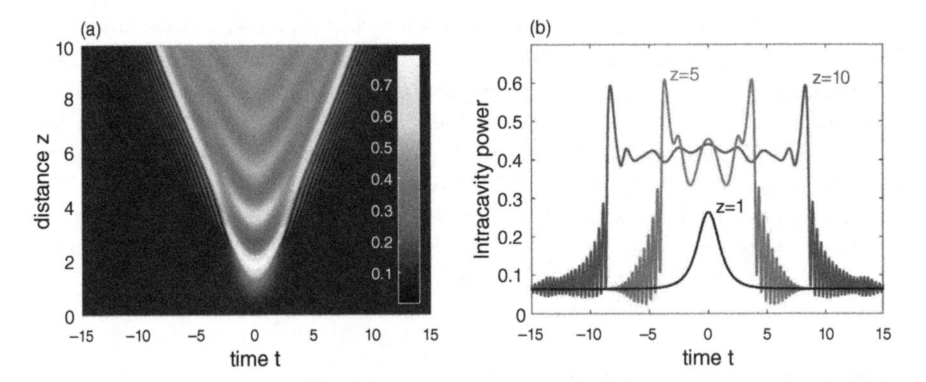

Figure 11.14 DSW generation ruled by Eq. (11.24). (a) Color level plot of intracavity power $|\psi(z,t)|^2$. (b) Snapshots of intracavity power at difference distances. Here $\varepsilon = 0.1$, $\delta = \pi/10$, $\alpha = 0.03$

similar for others order of HOD. An example is shown in Figure 11.15, for the same injected field used for the example reported Figure 11.14. In the temporal domain the effect of TOD is to induce an asymmetry between the leading and trailing fronts. The most striking feature is visible in the spectral propagation shown in Figure 11.15 (b), where an additional frequency component, well detached from the shock spectrum, is generated starting at a distance $z \approx 2$. The frequency of this radiation can be found by means of a perturbation approach [67], similar to those developed for the conservative case. If in the limit of small losses α, we find that the frequency of the RR must satisfy the following equation:

$$\left[\beta_3 \frac{(\varepsilon\omega)^3}{6} + \beta_2 \frac{(\varepsilon\omega)^2}{2} - \frac{(\varepsilon\omega)}{V_s} - \delta\right] + 2P_{uH} = 0, \tag{11.25}$$

where P_{uH} is the power of the higher state of the front, propagating with velocity V_s where RR is shed. This equation is very similar to Eq. 11.22, but it contains the cavity detuning δ as an additional parameter.

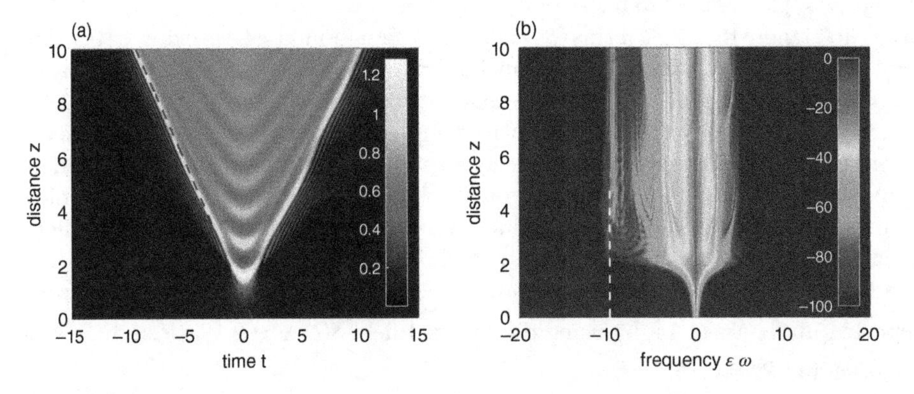

Figure 11.15 Temporal (a) and spectral (b) evolution of a shock. Parameters: $\beta_2 = 1$, $\varepsilon = 0.1$, $\delta = \pi/6$, $\beta_3 = 0.25$ and $\alpha = 0.03$. Dashed line in (a) represents the front velocity $V_s = -1$; dashed line in (b) stands for the value of $\varepsilon\omega_{RR} = -9.6$ calculated from Eq. (11.25).

11.7 Conclusion

In summary, we have shown that the area of fiber optics represents an ideal ground for investigating wave-breaking phenomena. Recent theoretical studies have permitted substantial progress in the full understanding of the phenomenon. Moreover, challenging experiments have been carried out that have demonstrated the accessibility of wave-breaking phenomena in different new contexts, ranging from multiple four-wave mixing to pulse shaping experiment, and observation of radiative effects. Yet, a lot of new experiments can be envisaged that could aim at unveiling new fundamental scenarios discussed in this chapter as well as to exploit wave-breaking in fiber optics applications. Furthermore optical experiments are important also for the understanding of similar hydrodynamical phenomena (e.g. tidal bores in river estuaries), for which the experimental implementation in the lab is more challenging.

Acknowledgments

The authors gratefully acknowledge collaborations with A. Armaroli, F. Baronio, G. Bellanca, S. Malaguti, J. Fatome, C. Finot, G. Millot, and funding from Labex CEMPI (ANR-11-LABX-0007-01), Equipex FLUX (ANR-11-EQPX-0017), by the projects NoAWE (ANR-14-ACHN-0014), and PRIN 2012BFNWZ2.

References

1 Nakatsuka, H., Grischkowsky, D., and Balant, A.C. (1981) Nonlinear picosecond-pulse propagation through optical fibers with positive group velocity dispersion. *Physical Review Letters*, **47**, 910–913.

2 Anderson, D. and Lisak, M. (1983) Nonlinear asymmetric self-phase modulation and self-steepening of pulses in long optical waveguides. *Physical Review A*, **27**, 1393–1398.

3 Tomlinson, W.J., Stolen, R.H., and Johnson, A.M. (1985) Optical wave breaking of pulses in nonlinear optical fibers. *Optics Letters*, **10**, 467–469.

4 Hamaide, J.P. and Emplit, P. (1988) Direct observation of optical wave breaking of picosecond pulses in nonlinear single-mode optical fibres. *Electronics Letters*, **24** (13), 818–819.

5 Rothenberg, J.E. and Grischkowsky, D. (1989) Observation of the formation of an optical intensity shock and wave breaking in the nonlinear propagation of pulses in optical fibers. *Physical Review Letters*, **62** (5), 531.

6 Rothenberg, J.E. (1989) Femtosecond optical shocks and wave breaking in fiber propagation. *Journal of the Optical Society of America B*, **6** (12), 2392–2401.

7 DeMartini, F., Townes, C.H., Gustafson, T.K., and Kelley, P.L. (1967) Self-steepening of light pulses. *Physical Review*, **164**, 312–323.

8 Grischkowsky, D., Courtens, E., and Armstrong, J.A. (1973) Observation of self-steepening of optical pulses with possible shock formation. *Physical Review Letters*, **31**, 422–425.

9 Whitham, G.B. (1974) *Linear and Nonlinear Waves*, vol. 42, John Wiley & Sons, New York.

10 Moiseev, S. and Sagdeev, R. (1963) Collisionless shock waves in a plasma in a weak magnetic field. *Journal of Nuclear Energy. Part C, Plasma Physics, Accelerators, Thermonuclear Research*, **5** (1), 43.

11 Zabusky, N.J. and Kruskal, M.D. (1965) Interaction of "solitons" in a collisionless plasma and the recurrence of initial states. *Physical Review Letters*, **15** (6), 240.

12 Peregrine, D. (1966) Calculations of the development of an undular bore. *Journal of Fluid Mechanics*, **25** (02), 321–330.

13 Taylor, R., Baker, D., and Ikezi, H. (1970) Observation of collisionless electrostatic shocks. *Physical Review Letters*, **24** (5), 206.

14 Hammack, J.L. and Segur, H. (1974) The korteweg-de vries equation and water waves. part 2. comparison with experiments. *Journal of Fluid Mechanics*, **65** (02), 289–314.

15 Gurevich, A. and Pitaevskii, L. (1974) Nonstationary structure of a collisionless shock wave. *Soviet Physics JETP*, **38**, 291.

16 Whitham, G. (1965) Non-linear dispersive waves. *Proceedings of the Royal Society of London A*, **283** (1393), 238–261.

17 Gurevich, A. and Krylov, A.L. (1987) Dissipationless shock waves in media with positive dispersion. *Soviet Physics JETP*, **65**, 944–953.

18 El, G.A., Geogjaev, V.V., Gurevich, A.V., and Krylov, A.L. (1995) Decay of an initial discontinuity in the defocusing nls hydrodynamics. *Physica D*, **87**, 186–192.

19 Kamchatnov, A. (2000) *Nonlinear Periodic Waves and Their Modulations: An Introductory Course*, World Scientific, Singapore.

20 Rothenberg, J.E. (1991) Observation of the buildup of modulational instability from wave breaking. *Optics Letters*, **16** (1), 18–20.

21 Anderson, D., Desaix, M., Lisak, M., and Quiroga-Teixeiro, M.L. (1992) Wave breaking in nonlinear-optical fibers. *Journal of the Optical Society of America B*, **9** (8), 1358–1361.

22 Kodama, Y. and Wabnitz, S. (1995) Analytical theory of guiding-center nonreturn-to-zero and return-to-zero signal transmission in normally dispersive nonlinear optical fibers. *Optics Letters*, **20** (22), 2291–2293.

23 Kodama, Y., Maruta, A., and Wabnitz, S. (1996) Minimum channel spacing in wavelength-division-multiplexed nonreturn-to-zero optical fiber transmissions. *Optics Letters*, **21** (22), 1815–1817.

24 Forest, M. and McLaughlin, K.R. (1998) Onset of oscillations in nonsoliton pulses in nonlinear dispersive fibers. *Journal of Nonlinear Science*, **8** (1), 43–62.

25 Forest, M.G., Kutz, J.N., and McLaughlin, K.R.T. (1999) Nonsoliton pulse evolution in normally dispersive fibers. *Journal of the Optical Society of America B*, **16** (11), 1856–1862.

26 Kodama, Y. (1999) The Whitham equations for optical communications: Mathematical theory of nrz. *SIAM Journal on Applied Mathematics*, **59** (6), 2162–2192.

27 Agrawal, G.P. (2013) *Nonlinear Five Optics*, vol. 5th ed, Academic Press, New York.

28 LeVeque, R.J. (2004) *Finite-Volume methods for Hyperbolic Problems*, Cambridge University Press, Cambridge.

29 Hoefer, M., Ablowitz, M., Coddington, I., *et al.* (2006) Dispersive and classical shock waves in Bose-Einstein condensates and gas dynamics. *Physical Review A*, **74** (2), 023 623.

30 Husko, C. and Colman, P. (2015) Giant anomalous self-steepening in photonic crystal waveguides. *Physical Review A*, **92**, 013 816.
31 Quiroga-Teixeiro, M. (1995) Raman-induced asymmetry of wave breaking in optical fibers. *Physica Scripta*, **51** (3), 373.
32 Conti, C., Stark, S., Russell, P.S.J., and Biancalana, F. (2010) Multiple hydrodynamical shocks induced by the Raman effect in photonic crystal fibers. *Physical Review A*, **82**, 013 838.
33 Gurevich, A. and Shvartsburg, A. (1970) Exact solutions of the equations of nonlinear geometric optics. *Soviet Physics JETP*, **31** (6), 1084–1089.
34 Wan, W., Jia, S., and Fleischer, J.W. (2007) Dispersive superfluid-like shock waves in nonlinear optics. *Nature Physics*, **3** (1), 46–51.
35 Moro, A. and Trillo, S. (2014) Mechanism of wave breaking from a vacuum point in the defocusing nonlinear Schrödinger equation. *Physical Review E*, **89** (2), 023 202.
36 Fratalocchi, A., Conti, C., Ruocco, G., and Trillo, S. (2008) Free-energy transition in a gas of noninteracting nonlinear wave particles. *Physical Review Letters*, **101**, 044 101.
37 Dutton, Z., Budde, M., Slowe, C., and Hau, L.V. (2001) Observation of quantum shock waves created with ultra-compressed slow light pulses in a Bose-Einstein condensate. *Science*, **293** (5530), 663–668.
38 Conti, C., Fratalocchi, A., Peccianti, M., Ruocco, G., and Trillo, S. (2009) Observation of a gradient catastrophe generating solitons. *Physical Review Letters*, **102** (8), 083 902.
39 Fratalocchi, A., Armaroli, A., and Trillo, S. (2011) Time-reversal focusing of an expanding soliton gas in disordered replicas. *Physical Review A*, **83**, 053 846.
40 Anderson, D., Desaix, M., Karlsson, M., Lisak, M., and Quiroga-Teixeiro, M.L. (1993) Wave-breaking-free pulses in nonlinear-optical fibers. *Journal of the Optical Society of America B*, **10** (7), 1185–1190.
41 Fermann, M.E., Kruglov, V.I., Thomsen, B.C., Dudley, J.M., and Harvey, J.D. (2000) Self-similar propagation and amplification of parabolic pulses in optical fibers. *Physical Review Letters*, **84**, 6010–6013.
42 Biondini, G. and Kodama, Y. (2006) On the Whitham equations for the defocusing nonlinear Schrödinger equation with step initial data. *Journal of Nonlinear Science*, **16** (5), 435–481.
43 Wabnitz, S., Finot, C., Fatome, J., and Millot, G. (2013) Shallow water rogue wavetrains in nonlinear optical fibers. *Physical Letters A*, **377** (12), 932–939.
44 Varlot, B., Wabnitz, S., Fatome, J., Millot, G., and Finot, C. (2013) Experimental generation of optical flaticon pulses. *Optics Letters*, **38** (19), 3899–3902.
45 Finot, C., Kibler, B., Provost, L., and Wabnitz, S. (2008) Beneficial impact of wave-breaking for coherent continuum formation in normally dispersive nonlinear fibers. *Journal of the Optical Society of America B*, **25** (11), 1938–1948.
46 Liu, Y., Tu, H., and Boppart, S.A. (2012) Wave-breaking-extended fiber supercontinuum generation for high compression ratio transform-limited pulse compression. *Optics Letters*, **37** (12), 2172–2174.
47 Millot, G., Pitois, S., Yan, M., *et al.* (2016) Frequency-agile dual-comb spectroscopy. *Nature Photonics*, **10**, 27–30.
48 Thompson, J.R. and Roy, R. (1991) Nonlinear dynamics of multiple four-wave mixing processes in a single-mode fiber. *Physical Review A*, **43**, 4987–4996.

49 Trillo, S., Wabnitz, S., and Kennedy, T.A.B. (1994) Nonlinear dynamics of dual-frequency-pumped multiwave mixing in optical fibers. *Physical Review A*, **50**, 1732–1747.

50 Trillo, S. and Valiani, A. (2010) Hydrodynamic instability of multiple four-wave mixing. *Optics Letters*, **35** (23), 3967–3969.

51 Fatome, J., Finot, C., Millot, G., Armaroli, A., and Trillo, S. (2014) Observation of optical undular bores in multiple four-wave mixing. *Physical Review X*, **4** (2), 021 022.

52 Dubrovin, B., Grava, T., and Klein, C. (2009) On universality of critical behavior in the focusing nonlinear Schrödinger equation, elliptic umbilic catastrophe and the tritronquée solution to the painlevé-i equation. *Journal of Nonlinear Science*, **19** (1), 57–94.

53 Miller, P.D. and Kamvissis, S. (1998) On the semiclassical limit of the focusing nonlinear Schrödinger equation. *Physical Letters A*, **247** (1-2), 75–86.

54 Kamvissis, S., McLaughlin, K.D.T.R., and Miller, P. (2003) *Semiclassical Soliton Ensembles for the Focusing Nonlinear Schrödinger Equation*, Princeton University Press, Princeton, NJ.

55 El, G.A. and Gurevich, A. (1993) Modulational instability and formation of a nonlinear oscillatory structure in a "focusing" medium. *Physical Letters A*, **177** (4-5), 357–361.

56 El, G.A., Khamis, E.G., and Tovbis, A. (2015) Dam break problem for the focusing nonlinear Schrödinger equation and the generation of rogue waves. *Nonlinearity*, **29** (9).

57 Ghofraniha, N., Conti, C., Ruocco, G., and Trillo, S. (2007) Shocks in nonlocal media. *Physical Review Letters*, **99** (4), 043 903.

58 Malaguti, S., Corli, A., and Trillo, S. (2010) Control of gradient catastrophes developing from dark beams. *Optics Letters*, **35** (24), 4217–4219.

59 Zakharov, V.E. and Ostrovsky, L.A. (2009) Modulation instability: The beginning. *Physica D*, **238** (5), 540–548.

60 Conforti, M., Baronio, F., and Trillo, S. (2014) Resonant radiation shed by dispersive shock waves. *Physical Review A*, **89**, 013 807.

61 Conforti, M., Baronio, F., and Trillo, S. (2013) Competing wave-breaking mechanisms in quadratic media. *Optics Letters*, **38** (10), 1648–1650.

62 Berkhoer, A.L. and Zakharov, V.E. (1970) Self excitation of waves with different polarizations in nonlinear media. *Soviet Journal of Experimental and Theoretical Physics*, **31** (3), 486–490.

63 Akhmediev, N. and Karlsson, M. (1995) Cherenkov radiation emitted by solitons in optical fibers. *Physical Review A*, **51**, 2602–2607.

64 Webb, K.E., Xu, Y.Q., Erkintalo, M., and Murdoch, S.G. (2013) Generalized dispersive wave emission in nonlinear fiber optics. *Optics Letters*, **38** (2), 151–153.

65 Conforti, M. and Trillo, S. (2013) Dispersive wave emission from wave breaking. *Optics Letters*, **38** (19), 3815–3818.

66 Conforti, M. and Trillo, S. (2014) Radiative effects driven by shock waves in cavity-less four-wave mixing combs. *Optics Letters*, **39** (19), 5760–5763.

67 Malaguti, S., Conforti, M., and Trillo, S. (2014) Dispersive radiation induced by shock waves in passive resonators. *Optics Letters*, **39** (19), 5626–5629.

68 Leo, F., Coen, S., Kockaert, P., Gorza, S.P., Emplit, P., and Haelterman, M. (2010) Temporal cavity solitons in one-dimensional Kerr media as bits in an all-optical buffer. *Nature Photonics*, **4** (7), 471–476.

69 Del'Haye, P., Schliesser, A., Arcizet, O., Wilken, T., Holzwarth, R., and Kippenberg, T.J. Optical frequency comb generation from a monolithic microresonator. *Nature*, (7173), 1214–1217.

70 Malaguti, S., Bellanca, G., and Trillo, S. (2014) Dispersive wave-breaking in coherently driven passive cavities. *Optics Letters*, **39** (8), 2475–2478.

71 Haelterman, M., Trillo, S., and Wabnitz, S. (1992) Additive-modulation-instability ring laser in the normal dispersion regime of a fiber. *Optics Letters*, **17** (10), 745–747.

72 Lugiato, L.A. and Lefever, R. (1987) Spatial dissipative structures in passive optical systems. *Physical Review Letters*, **58**, 2209–2211.

12

Optical Wave Turbulence in Fibers

Antonio Picozzi,[1] Josselin Garnier,[2] Gang Xu,[1] and Guy Millot[1]

[1] Laboratoire Interdisciplinaire CARNOT de Bourgogne, UMR 6303 CNRS-Université de Bourgogne - Franche-Comté, Dijon, France
[2] Centre de Mathématiques Appliquées, Ecole Polytechnique, Palaiseau, France

12.1 Introduction

The coherence properties of partially incoherent optical waves propagating in nonlinear media have been studied since the advent of nonlinear optics in the 1960s, because of the natural poor degree of coherence of laser sources available at that time. However, it is only recently that the dynamics of incoherent nonlinear optical waves received a renewed interest. The main motive for this renewal of interest is essentially due to the first experimental demonstration of incoherent solitons in photorefractive crystals [1,2]. The formation of an incoherent soliton results from the *spatial self-trapping* of incoherent light that propagates in a highly noninstantaneous response nonlinear medium [3,4]. This effect is possible because of the noninstantaneous photorefractive nonlinearity that averages the field fluctuations provided that its response time, τ_R, is much longer than the correlation time t_c that characterizes the incoherent beam fluctuations, i.e., $t_c \ll \tau_R$. The remarkable simplicity of experiments realized in photorefractive crystals has led to a fruitful investigation of the dynamics of incoherent nonlinear waves. Different theoretical approaches have been also developed to describe these experiments [5–8], which have been subsequently shown to be formally equivalent to each others [9, 10].

In this way, the field of incoherent optical solitons has become a blooming area of research, as illustrated by several important achievements, e.g., the existence of incoherent dark solitons [11, 12], the modulational instability of incoherent waves [13, 14], incoherent solitons in periodic lattices [15, 16], in resonant interactions [17, 18], in liquid crystals [19], in nonlocal nonlinear media [20–22], or spectral incoherent solitons in optical fibers [23, 24]. Nowadays, statistical nonlinear optics constitutes a growing field of research covering various topics of modern optics, e.g., supercontinuum generation [25], filamentation [26], random lasers [27], or extreme rogue wave events emerging from optical turbulence [28–32].

From a broader perspective, statistical nonlinear optics is fundamentally related to fully developed turbulence [33, 34], a subject which still constitutes one of the most challenging problems of theoretical physics [35, 36]. In its broad sense, the kinetic wave

Shaping Light in Nonlinear Optical Fibers, First Edition. Edited by Sonia Boscolo and Christophe Finot.

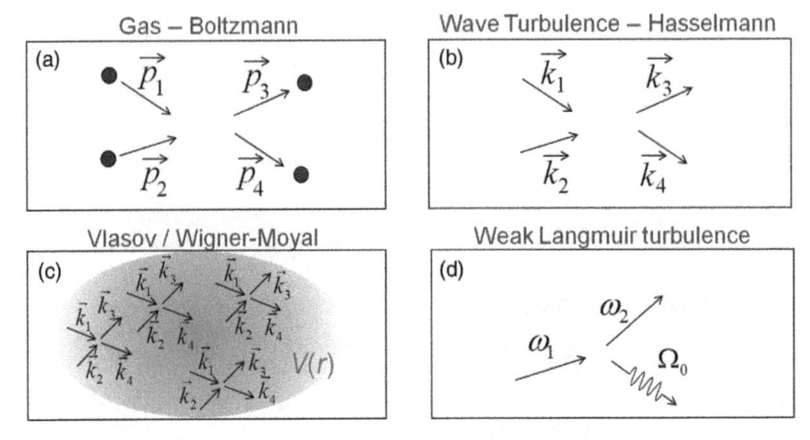

Figure 12.1 Analogy between a system of classical particles and the propagation of an incoherent optical wave in a Kerr medium. (a) As described by the kinetic gas theory, collisions between particles are responsible for an irreversible evolution of the gas towards thermodynamic equilibrium. (b) In complete analogy, the WT kinetic equation and the underlying four-wave mixing describe an irreversible evolution of the incoherent optical wave toward the thermodynamic Rayleigh-Jeans equilibrium state. (c) When the incoherent optical wave exhibits an inhomogeneous statistics, the quasi-particles feel the presence of an effective self-consistent potential, $V(r)$, which prevents them from relaxing to thermal equilibrium: The dynamics is described by a Vlasov-like kinetic equation. (d) In the presence of a noninstantaneous Raman-like nonlinearity, the causality condition inherent to the response function changes the physical picture, and the dynamics of the incoherent wave can be described by the weak Langmuir turbulence formalism.

theory provides a nonequilibrium thermodynamic description of developed turbulence. We schematically report in Figure 12.1 (a) qualitative and intuitive physical insight into the analogy which underlies the kinetic wave approach and the kinetic theory relevant to gas systems. We may note that the wave turbulence theory occupies a rather special place on the road-map of modern science, at the interface between applied mathematics, fluid dynamics, statistical physics and engineering. It has potential applications and implications in a diverse range of subjects including oceanography, plasma physics and condensed matter physics. The review article [37] was aimed at showing that the kinetic wave theory appears as the appropriate theoretical framework to formulate statistical nonlinear optics. In this chapter we illustrate the applicability of the kinetic wave theory to the specific example of optical fiber systems by considering three fundamentally different formalisms.

1. ***Wave turbulence formalism:*** Consider the nonlinear propagation of a partially coherent optical wave characterized by fluctuations that are statistically homogeneous in time (or in space). By complete analogy with a system of classical particles, the incoherent optical field evolves, owing to nonlinearity, towards a thermodynamic equilibrium state, as schematically illustrated in Figure 12.1 (a) and (b). A detailed theoretical description of the process of dynamical thermalization constitutes a difficult problem. However, a considerable simplification occurs when the dynamics is essentially dominated by linear dispersive effects, so that a weakly nonlinear description of the field becomes possible [33, 35, 36]. The weak- (or wave-)turbulence (WT) theory has been the subject of a lot of investigations in the context of plasma physics,

in which it is often referred to the so-called "random phase-approximation" approach [33, 38, 39]. This approach may be considered a convenient way of interpreting the results of the more rigorous technique based on a multi-scale expansion of the cumulants of the nonlinear field, as originally formulated in [40, 41], and recently reviewed in [42]. The so-called "random phase approximation" may be considered justified when phase information becomes irrelevant to the wave interaction due to the strong tendency of the waves to decohere. The random phases can thus be averaged out to obtain a weak turbulence description of the incoherent wave interaction, which is formally based on irreversible kinetic equations [33]. The result is that, in spite of the formal reversibility of the equation governing wave propagation, the kinetic equation describes an irreversible evolution of the field to thermodynamic equilibrium. This equilibrium state refers to the fundamental Rayleigh-Jeans spectrum, whose tails are characterized by an equipartition of energy among the Fourier modes. The mathematical statement of such irreversible process relies on the H-theorem of entropy growth, whose origin is analogous to the Boltzmann's H-theorem relevant for gas kinetics.

Note that, besides this *nonequilibrium* kinetic approach, the equilibrium properties of a random nonlinear wave may be studied on the basis of *equilibrium* statistical mechanics by computing appropriate partition function [43–45]. In this way, the statistical properties of incoherent fields in random lasers have been analyzed by applying methods inherited from spun-glass theory [46]. In Section 12.2.1 we will review different processes of optical wave thermalization on the basis of the WT theory, while in Section 12.2.2 some mechanisms responsible for its inhibition will be discussed. In particular, we will see how the phenomenon of supercontinuum generation can be interpreted, under certain conditions, as a consequence of the natural thermalization of the optical field toward the thermodynamic equilibrium state. On the other hand, when a wave system is driven away from equilibrium by an external source, it no longer relaxes toward the Rayleigh-Jeans equilibrium distribution. A typical physical example of forced system can be the excitation of hydrodynamic surface waves by the wind. We will review in Section 12.2.3 recent important efforts aimed at providing a description of the turbulent dynamics of active and passive fiber cavities by means of a nonequilibrium kinetic formulation based on the wave turbulence equation.

2. ***Weak Langmuir turbulence formulation:*** When the incoherent wave propagates in a noninstantaneous Raman-like nonlinear medium, the dynamics turn out to be strongly affected by the causality property inherent to the nonlinear response function (see Figure 12.1). The kinetic wave theory reveals in this case that the appropriate description is provided by a formalism analogous to that used to describe weak Langmuir turbulence in plasmas [23, 47]. A major prediction of the theory is the existence of spectral incoherent solitons [23, 24, 48]. This incoherent soliton is of a fundamental different nature from the incoherent solitons discussed here above. In particular, it does not exhibit a confinement in the spatiotemporal domain, but exclusively in the frequency domain. In Section 12.3, we will review the properties of these stable nonequilibrium incoherent states, as well as those associated to the formation of a novel form of spectral incoherent shock singularities [49].
3. ***Vlasov formalism:*** When the nonlinear material is characterized by a highly-noninstantaneous response (temporal nonlocality), the dynamics of the incoherent

wave turn out to be essentially governed by an effective nonlinear potential $V(r)$. This potential is self-consistent in the sense that it depends itself on the averaged intensity distribution of the random field, as schematically illustrated in Figure 12.1 (c). Actually, the mechanism underlying the formation of an incoherent localized soliton state finds its origin in the existence of such self-consistent potential, which is responsible for the self-trapping of the incoherent optical wave. In Section 12.4, we will review recent works in which the phenomena of incoherent modulational instability and incoherent localized temporal structures have been described in the framework of a Vlasov formalism. We note that such a Vlasov formalism differs from the traditional Vlasov equation considered for the study of incoherent modulational instability and incoherent solitons in plasmas [50, 51], hydrodynamics [52] and optics [8, 53–55]. The structure of this Vlasov equation is in fact analogous to that recently used to describe systems of particles with long-range interactions [56]. For this reason we will term this equation "long-range Vlasov" equation. It is important to underline that the long-range nature of a highly nonlocal nonlinear response prevents the wave system from reaching thermal equilibrium [22]. This fact can be interpreted intuitively in analogy with gravitational long-range systems and the Vlasov-like description of the dynamics of galaxies in the universe [56].

12.2 Wave Turbulence Kinetic Equation

In this section we illustrate the WT formalism by considering the process of optical wave thermalization through SC generation, as well as different mechanisms responsible for the inhibition of thermalization. We also review recent important efforts aimed at developing an appropriate kinetic description of the turbulent dynamics in optical cavities.

12.2.1 Supercontinuum Generation

The phenomenon of SC generation is characterized by a dramatic spectral broadening of the optical field during its propagation. This process has been extensively studied and different regimes have been identified, which essentially depend on whether the highly nonlinear photonic crystal fiber (PCF) is pumped in the normal or anomalous dispersion regimes, or with short (subpicosecond) or long (picosecond, nanosecond, and quasi-CW) pump pulses [25, 57].

As a rather general rule, the process of spectral broadening inherent to SC generation is interpreted through the analysis of the following main nonlinear effects: the four-wave mixing effect, the soliton fission, the Raman self-frequency shift and the generation of dispersive waves [57]. Due to such a multitude of nonlinear effects involved in the process, a complete and satisfactory theoretical description of SC generation is still lacking. However, there is a growing interest in developing new theoretical tools aimed at describing SC generation in more details. Besides the theories describing the interaction between individual soliton pulses and dispersive waves [58], we can quote the effective three-wave mixing theory and the underlying first-Born approximation successfully applied to describe femtosecond SC generation in different configurations [59, 60]. We also mention recent works aimed at providing a complete characterization of the

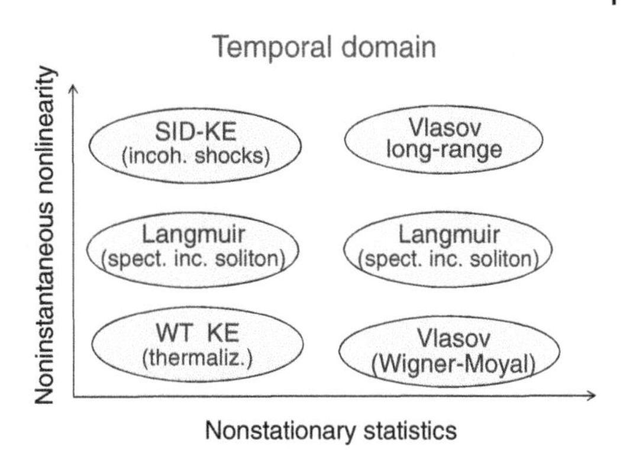

Figure 12.2 Schematic illustration of the validity of the fundamental kinetic equations. When the incoherent wave is characterized by fluctuations that are statistically stationary in time, the relevant kinetic description is provided by the WT kinetic equation. If the incoherent wave exhibits a nonstationary statistics, the relevant description is provided by the Vlasov formalism. If the response time of the nonlinearity can no longer be ignored, then the dynamics are ruled by the weak Langmuir turbulence formalism. If the random wave exhibits a non-stationary statistics in the presence of a highly noninstantaneous response, then the dynamics are governed by a long-range Vlasov formalism.

coherence properties of SC light by using second-order coherence theory of nonstationary light [61–63].

Incoherent Turbulent Regime of SC Generation The general physical picture of SC generation in PCFs can be summarized as follows. When the PCF is pumped with long pulses in the anomalous dispersion regime, MI is known to lead to the generation of a train of soliton-like pulses, which in turn lead to the emission of Cherenkov radiation in the form of spectrally shifted dispersive waves. These optical solitons are known to exhibit a self-frequency shift toward longer wavelengths as a result of the Raman effect. One encounters the same picture if the PCF is characterized by two zero dispersion wavelengths. In this case the Raman frequency shift of the solitons is eventually arrested in the vicinity of the second zero dispersion wavelengths. The SC spectrum then is essentially bounded by the corresponding dispersive waves [25, 58, 64–66]. The important aspect to underline here is that in all these regimes the existence of coherent soliton structures plays a fundamental role in the process of SC generation.

This physical picture of SC generation changes in a significant way when one considers the regime in which long and intense pump pulses are injected into the PCF. Indeed, in this highly nonlinear regime, the spectral broadening process is essentially dominated by the combined effects of the Kerr nonlinearity and higher-order dispersion, i.e., by four-wave mixing processes [67]. In this regime the optical field exhibits rapid and random temporal fluctuations, which prevent the formation of robust and persistent coherent soliton structures. It turns out that the optical field exhibits an incoherent turbulent dynamics, in which coherent soliton structures do not play any significant role. In the following we shall term this regime the “incoherent regime of SC generation” [68].

Wave Turbulence Approach to SC Generation In these last few years a nonequilibrium thermodynamic interpretation of this incoherent regime of SC generation has been formulated [24, 48, 68–70] on the basis of the WT theory. This WT description can be introduced through the analysis of the numerical simulation in Figure 12.3 (a). It reports a typical evolution of the spectrum of the optical field in the incoherent regime of SC generation. It is obtained by integrating numerically the generalized nonlinear Schrödinger (NLS) equation (see Eq. (12.1)), with the dispersion curve reported in Figure 12.3 (b).

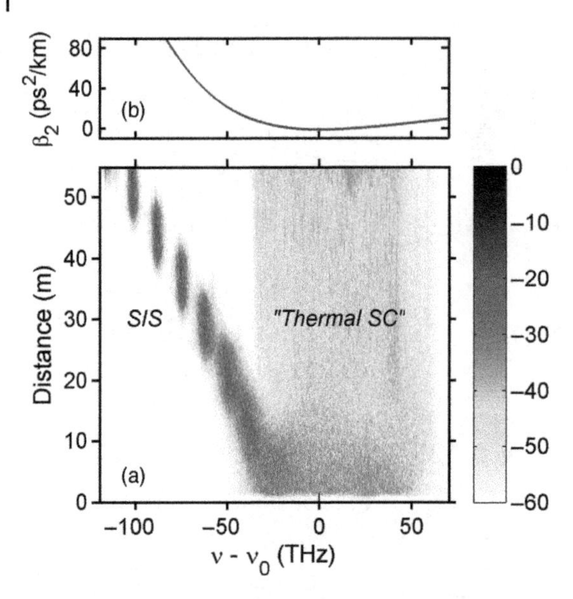

Figure 12.3 Incoherent turbulent regime of SC generation: (a) Numerical simulations of the generalized NLS Eq. (12.1) reporting the spectral evolution as a function of propagation distance in a 50m long PCF, for an input CW power equal to 200W ($\gamma = 0.05\ W^{-1}m^{-1}$). The corresponding dispersion curve of the PCF used in the simulations is illustrated in (b). The optical spectrum is characterized by two main features: (i) A broad central part governed by the four-wave mixing process that exhibits a process reminiscent of thermalization. (ii) A narrower low-frequency branch governed by the Raman effect that self-organizes into a continuous, and subsequently a discrete, spectral incoherent soliton (see Section 12.3).

The initial condition is a high-power (200 W) continuous wave whose carrier frequency $\nu_0 = 283$ THz ($\lambda_0 = 1060$ nm) lies in the anomalous dispersion regime and thus leads to the development of the modulational instability process.

We note in Figure 12.3 (a) that the spectrum of the field essentially splits into two components during the propagation:

(i) On the one hand, one notices a broad central part whose evolution is essentially governed by the dispersion effects and the Kerr nonlinearity. These effects are inherently conservative effects and lead to a process of wave thermalization through SC generation, a feature that has been discussed in [68–70] using the WT theory. Accordingly, the saturation of SC spectral broadening can be ascribed to the natural tendency of the optical field to reach an equilibrium state. Note, however, that, as will be discussed below, the phenomenon of wave thermalization through SC generation is not achieved in a complete fashion, in the sense that the tails of the numerical spectra exhibit some discrepancy with the corresponding expected tails of the Rayleigh-Jeans distribution. While this discrepancy can be simply ascribed to a limited propagation length in the PCF, another possible physical origin of such discrepancy will be discussed in Section 12.2.2.1. This WT approach also reveals the existence of an unexpected phase-matching process that can be interpreted on the basis of thermodynamic properties.

(ii) On the other hand, one notices in Figure 12.3 (a) that a low-frequency spectral branch moves away from the central part of the spectrum. This low-frequency branch is essentially governed by the dissipative Raman effect, whose noninstantaneous nonlinear nature is responsible for the generation of spectral incoherent solitons, a feature that will be discussed in Section 12.3.

12.2.1.1 Generalized NLS Equation

The generalized NLS equation is known to provide an accurate description of the propagation of an optical field in a PCF [57,71],

$$i\frac{\partial\psi(z,t)}{\partial z} + \sum_{j\geq 2}^{m} \frac{i^j\beta_j}{j!}\frac{\partial^j\psi(z,t)}{\partial t^j} + \gamma\left(1 + i\tau_s\frac{\partial}{\partial t}\right)\psi(z,t)$$
$$\times \int_{-\infty}^{+\infty} R(t')\,|\psi(z,t-t')|^2\,dt' = 0, \qquad (12.1)$$

where we note that γ refers to the nonlinear coefficient and $R(t) = (1-f_R)\delta(t) + f_R h_R(t)$ to the usual nonlinear response function of silica fibers, which accounts for both the instantaneous Kerr effect and the non-instantaneous Raman response function $h_R(t)$ $[\tilde{\psi}(\omega,z) = (2\pi)^{-1/2}\int \psi(t,z)\exp(i\omega t)\,dt]$ [71]. The linear dispersion relation of Eq. (12.1) reads $k(\omega) = \sum_{j\geq 2}^{m} \frac{\beta_j\omega^j}{j!}$.

As discussed above through Figure 12.3 (a), wave thermalization is driven by the combined effects of dispersion and Kerr nonlinearity, which are inherently conservative effects. On the other hand, the Raman effect [$f_R \neq 0$ in Eq. (12.1)] is a dissipative effect and prevents the establishment of a thermodynamic equilibrium state (see Section 12.3). Note that this is consistent with the fact that the Raman effect breaks the Hamiltonian structure of Eq. (12.1). In this section we will thus ignore the dissipative Raman effect.

We report in Figure 12.4 (a) exactly the same numerical simulation as that reported in Figure 12.3 (a), except that we removed the Raman effect, $f_R = 0$ in Eq. (12.1). We also removed in this simulation the influence of the shock term ($\tau_s = 0$), whose influence has

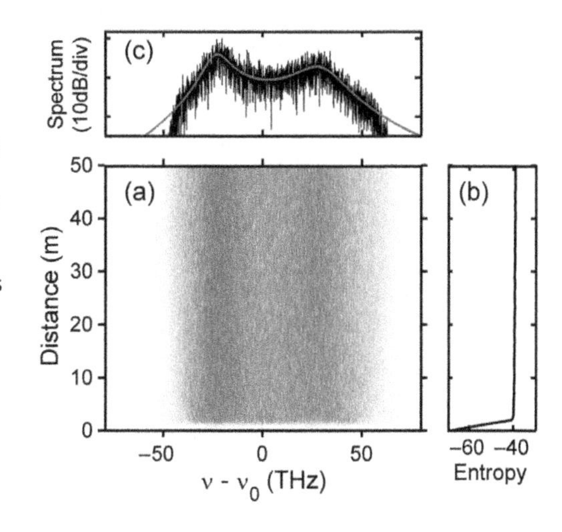

Figure 12.4 Optical wave thermalization through SC generation. (a) Same as in Figure 12.3 (a), except that the Raman effect and the shock term have been neglected, $f_R = \tau_s = 0$: this simulations thus refers to the numerical integration of the instantaneous NLS Eq. (12.2). (b) Optical wave thermalization is characterized by a process of entropy production, which saturates to a constant level once the equilibrium state is reached, as described by the *H*–theorem of entropy growth. (c) Comparison of the the thermodynamic Rayleigh-Jeans equilibrium spectrum $n^{eq}(\omega)$ (Eq. (12.5)) (continuous line), and the numerical spectrum corresponding to an averaging over the last 20m of propagation. A good agreement is obtained without adjustable parameters – note, however, a discrepancy in the tails of the spectrum (see the text for discussion).

been considered in [70]. Comparison of Figures 12.3 (a) and 12.4 (a) clearly shows that the essential role of the Raman effect is to lead to the generation of a spectral incoherent soliton in the low-frequency branch in the SC spectrum. Besides spectral incoherent solitons, a peculiar feature revealed by Figure 12.4 (a) is that the spectral broadening inherent to SC generation saturates during the propagation, a feature related to the thermalization of the optical field.

12.2.1.2 Wave Thermalization through Supercontinuum Generation

Neglecting the Raman effect and the shock term, the generalized NLS Eq. (12.1) reduces to

$$i\frac{\partial\psi}{\partial z} + \sum_{j\geqslant 2}^{m} \frac{i^j\beta_j}{j!}\frac{\partial^j\psi}{\partial t^j} + \gamma|\psi|^2\psi = 0. \tag{12.2}$$

We recall that, if only the second-order dispersion effect is retained ($m = 2$), Eq. (12.2) recovers the completely integrable 1D-NLS equation. The corresponding infinite number of conserved quantities prevents the thermalization of the wave toward thermodynamic equilibrium, though the system still exhibits a relaxation toward an equilibrium state of a different nature [72, 73]. This aspect will be discussed in Section 12.2.2.

If one includes the influence of third-order dispersion ($m = 3$), the system exhibits a process of anomalous thermalization [74,75], which is characterized by an irreversible evolution toward an equilibrium state of a fundamental different nature than the thermodynamic equilibrium state. This previous work [75] can be important to study the evolution of an incoherent wave in a PCF characterized by a single zero dispersion wavelength, and will be discussed in Section 12.2.2.

If one includes dispersion effects up to the fourth-order ($m = 4$), the simulations reveal the existence of a phenomenon of "truncated thermalization": The incoherent wave exhibits an irreversible evolution toward the Rayleigh-Jeans thermodynamic equilibrium state characterized by a compactly supported spectral shape. This aspect will be discussed in Section 12.2.2.

Thermodynamic Equilibrium Spectrum In the following we consider realistic dispersion curves of PCFs characterized by two zero dispersion wavelengths, whose accurate description requires a high-order Taylor expansion of the dispersion relation ($m > 4$ and even). Starting from the high-order dispersion NLS Eq. (12.2), one can derive the irreversible WT kinetic equation governing the evolution of the averaged spectrum of the field $n(z,\omega)$ $[\langle\tilde{\psi}(z,\omega_1)\tilde{\psi}^*(z,\omega_2)\rangle = n(z,\omega_1)\,\delta(\omega_1-\omega_2)]$:

$$\partial_z n(z,\omega_1) = Coll[n], \tag{12.3}$$

with the collision term

$$\begin{aligned} Coll[n] = \iiint d\omega_2\, d\omega_3\, d\omega_4\; n(\omega_1)n(\omega_2)n(\omega_3)n(\omega_4) \\ \times W\,[n^{-1}(\omega_1) + n^{-1}(\omega_2) - n^{-1}(\omega_3) - n^{-1}(\omega_4)] \end{aligned} \tag{12.4}$$

where "$n(\omega)$" stands for "$n(z,\omega)$" in Eq. (12.4). As usual in the WT kinetic equation, the phase-matching conditions of energy and momentum conservation are expressed by the presence of Dirac δ–functions in $W = \frac{\gamma^2}{\pi}\,\delta(\omega_1+\omega_2-\omega_3-\omega_4)\,\delta[k(\omega_1)+k(\omega_2)-k(\omega_3)-k(\omega_4)]$, where $k(\omega)$ refers to the linear dispersion relation. Equation (12.3) conserves the power density $N/T_0 = \int n(z,\omega)\,d\omega$, the density of kinetic energy $E/T_0 = \int k(\omega)\,n(z,\omega)\,d\omega$ and the density of momentum $P/T_0 = \int \omega\, n(z,\omega)\,d\omega$, where T_0 refers to the considered numerical time window. It also exhibits a H-theorem of entropy growth, $\partial_z S \geqslant 0$, where the nonequilibrium entropy reads $S(z) = \int \log[n(z,\omega)]\,d\omega$. The Rayleigh-Jeans equilibrium distribution is obtained by maximizing the entropy under the constraints imposed by the conservation of the energy, momentum and power, which gives

$$n^{eq}(\omega) = \frac{T}{k(\omega)+\lambda\omega-\mu}, \qquad (12.5)$$

where T and μ are by analogy with thermodynamics the temperature and the chemical potential of the incoherent wave at equilibrium. The three parameters (T,μ,λ) are calculated from the conserved quantities (E,N,P) by substituting the equilibrium spectrum Eq. (12.5) into the definitions of E, N and P. One thus obtains an algebraic system of three equations for three unknown parameters, which can be solved numerically. A unique triplet solution (T,μ,λ) for a given set (E,N,P) is usually obtained, a feature which is consistent with the fact that a "closed" (conservative and Hamiltonian) system should exhibit a unique thermodynamic equilibrium state.

The meaning of the parameter λ becomes apparent through the analysis of the group-velocity v_g of the optical field $[k'(\omega) \equiv \partial k/\partial\omega = 1/v_g(\omega)]$. Indeed, recalling the definition of an average, $\langle A\rangle_{eq} = \int A\, n^{eq}(\omega)\,d\omega / \int n^{eq}(\omega)\,d\omega$ and making use of the equilibrium spectrum Eq. (12.5), one readily obtains

$$< k'(\omega) >_{eq} = -\lambda. \qquad (12.6)$$

According to relation (12.6), the parameter λ denotes the average of the inverse of the group-velocity of the optical field at equilibrium. We report in Figure 12.4 (c) the comparison of the theoretical prediction (12.5) with the results of the numerical simulations of the high-order NLS Eq. (12.2). A quantitative agreement is obtained between the simulations and the theory Eq. (12.5), without using adjustable parameters. The Rayleigh-Jeans spectrum is characterized by a double-peaked structure, which results from the presence of two zero dispersion wavelengths in the dispersion curve of the PCF. The relaxation toward thermal equilibrium is also corroborated by the saturation of the process of entropy production illustrated in Figure 12.4 (b). Note, however, that a notable discrepancy is visible in the tails of the spectrum in Figure 12.4 (c), as if the thermalization process were not achieved in a complete fashion. Actually, the simulations reveal that the tails of the spectrum exhibit a very slow process of spectral broadening, which apparently tends to evolve toward the expected Rayleigh-Jeans tails – though the required propagation length is extremely large. This aspect will be discussed in more detail in Section 12.2.2 in the particular case where the dispersion relation is truncated to the fourth-order ($m = 4$ in the dispersion relation). The good agreement between the theory and the simulations has been obtained in a variety of configurations, e.g., under cw or incoherent pumping, as discussed in detail in [69, 70].

Thermodynamic Phase-Matching The thermodynamic equilibrium spectrum given in Eq. (12.5) is characterized by a double peak structure, which originates from the two zero dispersion wavelengths that characterize the PCF dispersion curve. It is important to underline, however, that the frequencies (ω_1, ω_2) of the two peaks of $n^{eq}(\omega)$ do not simply correspond to the minima of the dispersion relation, i.e. $k'(\omega_{1,2}) \neq 0$. To further analyze this aspect, let us write the thermodynamic equilibrium spectrum in the form $n^{eq}(\omega) = T/\mathcal{F}(\omega)$, with $\mathcal{F}(\omega) = k(\omega) + \lambda\omega - \mu$. The two frequencies (ω_1, ω_2) which maximize the equilibrium spectrum (12.5) satisfy $\mathcal{F}'(\omega_1) = \mathcal{F}'(\omega_2) = 0$, i.e., $k'(\omega_1) = k'(\omega_2) = -\lambda$. This observation reveals that the two frequencies (ω_1, ω_2) of the double peaked equilibrium spectrum (12.5) are selected in such a way that the corresponding group-velocities coincide with the average group-velocity of the optical wave,

$$v_g(\omega_1) = v_g(\omega_2) = 1/\langle k'(\omega)\rangle_{eq} = -1/\lambda. \tag{12.7}$$

It can be shown that there exists, in principle, a unique pair of frequencies (ω_1, ω_2) satisfying the conditions given by Eq. (12.7). In other terms, for a given thermodynamic equilibrium spectrum (12.5), there exists a *unique* pair of frequencies (ω_1, ω_2) that leads to a matched group-velocity of the double peaked spectrum [70]. In this sense, Eq. (12.7) can be regarded as a thermodynamic phase-matching condition.

The thermodynamic phase-matching given by Eq. (12.7) then imposes a matching of the group-velocities of the two spectral peaks of the SC spectrum. The fact that different wave-packets naturally tend to propagate with the same group-velocity was discussed in [76]. This can be interpreted by analogy with basic equilibrium thermodynamic properties, namely that an isolated system can only exhibit a uniform motion of translation (and rotation) as a whole, while any macroscopic internal motion is not possible at thermodynamic equilibrium [77]. In this way, it was shown that a velocity locking is required, in the sense that it prevents "a macroscopic internal motion in the wave system." We refer the interested reader to [37, 70] for more details on this aspect.

12.2.2 Breakdown of Thermalization

A fundamental assumption of statistical mechanics is that a closed system with many degrees of freedom ergodically samples all equal energy points in phase space. In order to analyze the limits of this assumption, Fermi, Pasta and Ulam (FPU) considered in the 1950s a one-dimensional chains of particles with anharmonic forces between them [78]. They argued that, owing to the nonlinear coupling, an initial state in which the energy is in the first few lowest modes would eventually relax to a state of thermal equilibrium where the energy is equidistributed among all modes on the average. However, they observed that, instead of leading to the thermalization of the system, the energy transfer process involves only a few modes and exhibits a reversible behavior, in the sense that after a sufficiently (long) time the system nearly goes back to its initial state. Fundamental mathematical and physical discoveries, like the Kolmogorov-Arnold-Moser theorem and the formulation of the soliton concept, have led to a better understanding of the Fermi-Pasta-Ulam problem, although it is still the subject of intense research activity, see e.g., [78–80].

We should note that, in spite of the large number of theoretical studies, experimental demonstrations of FPU recurrences have been reported in very few systems. In particular, the FPU recurrences associated with modulational instability of the NLS equation have been experimentally studied in deep water waves [81], and, more recently, in magnetic feedback rings [82] and optical wave systems [79,83–85]. By considering the one-dimensional NLS equation, we present in this section three different mechanisms that inhibit the process of optical wave thermalization toward the Rayleigh-Jeans distribution. Depending on whether the dispersion relation is truncated up to the second, third, or fourth-order, the wave system exhibits different types of relaxation processes. Provided that the interaction occurs in the weakly nonlinear regime, the WT theory provides an accurate description of such mechanisms of breakdown of thermalization.

12.2.2.1 Fourth-Order Dispersion: Truncated Thermalization

We consider here the 1D NLS equation in which the dispersion relation is truncated to the fourth-order. In this case, the WT theory reveals the existence of an irreversible evolution toward a Rayleigh-Jeans equilibrium state characterized by a compactly supported spectral shape [86]. This phenomenon of truncated thermalization may explain the physical origin of the abrupt SC spectral edges discussed above in Section 12.2.1.2. More generally, it can shed new light on the mechanisms underlying the formation of bounded spectra in SC generation [64–66, 87]. Besides its relevance in the context of SC generation, this phenomenon is also important from a fundamental point of view. Indeed, it unveils the existence of a genuine frequency cut-off that arises in a system of classical waves described by the generalized NLS equation, a feature of importance considering the well-known ultraviolet catastrophe of ensemble of classical waves [37].

The starting point is the NLS Eq. (12.2) accounting for third- and fourth-orders dispersion effects, as well as the corresponding WT kinetic Eq. (12.3). The kinetic theory reported in [86] reveals that the process of thermalization to the Rayleigh-Jeans spectrum (12.5) is not achieved in a complete way, but turns out to be truncated within a specific frequency interval defined by the bounds, $\omega \in [\omega_-, \omega_+]$, with

$$\omega_\pm = -\frac{\tilde{\alpha}}{4\tilde{\beta}\tau_0} \pm \frac{\sqrt{21}}{12\tilde{\beta}\tau_0}\sqrt{3\tilde{\alpha}^2 + 8\tilde{\beta}}, \tag{12.8}$$

where $\tilde{\alpha}$ and $\tilde{\beta}$ refer to the normalized third- and fourth-orders dispersion parameters, namely $\tilde{\alpha} = L_{nl}\beta_3/(6\tau_0^3)$, and $\tilde{\beta} = L_{nl}\beta_4/(24\tau_0^4)$, where $\tau_0 = \sqrt{\beta_2 L_{nl}/2}$ is the corresponding healing time, i.e., the characteristic time for which linear and nonlinear effects are of the same order of magnitude [37].

The confirmation of this process of truncated thermalization by the numerical simulations has not been a trivial task. This is due to the fact that in the usual configurations of SC generation discussed above, the cascade of MI side-bands generated by the cw pump in the early stage of propagation spreads beyond the frequency interval predicted by the theory. As already discussed, the MI process is inherently a coherent nonlinear phase-matching effect which is not described by the WT kinetic equation (Eqs. (12.3) and (12.4)). This explains why the numerical simulations reported above (or in [69,70]) did not evidence a precise signature of this phenomenon of truncated thermalization.

In order to analyze the theoretical predictions in more detail, one needs to decrease the injected pump power so as to maintain the (cascaded) MI side-bands within the frequency interval (12.8). Intensive numerical simulations of the NLS equation in this regime of reduced pump power have been performed in [86]. This study reveals that the nonlinear dynamics slows down in a dramatic way, so that the expected process

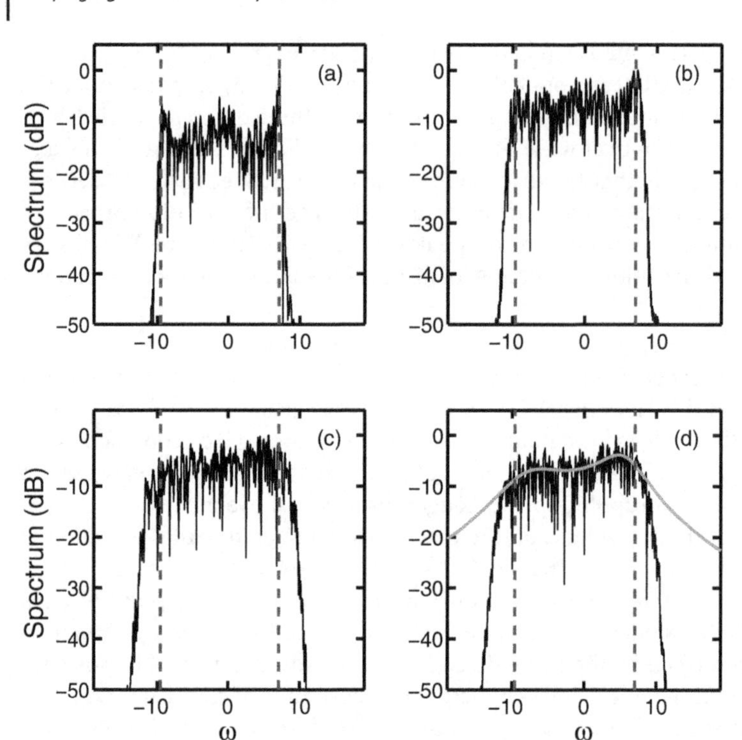

Figure 12.5 Truncated thermalization of incoherent waves: Spectra $|\tilde{\psi}|^2(\omega, z)$ obtained by solving the NLS Eq. (12.2) with solely third and fourth-order dispersion effects ($\tilde{\alpha} = 0.1$, $\tilde{\beta} = 0.02$): (a) $z = 200$, (b) $z = 10^4$, (c) $z = 5 \times 10^5$, (d) $z = 10^6$. After a long transient, the wave relaxes toward a truncated Rayleigh-Jeans distribution (Eq. (12.5), continuous line) (d). The dashed lines denote the frequencies $\omega_\pm$ in Eq. (12.8) – ω is here in units of τ_0^{-1}. *Source*: Barviau 2013 [86]. Reproduced with permission of American Physical Society.

of thermalization requires huge nonlinear propagation lengths. This results from the fact that the normalized parameters $\tilde{\alpha}$ and $\tilde{\beta}$ decrease as the pump power decreases, so that the NLSE approaches the integrable limit, which does not exhibit thermalization [73] (see Section 12.2.2.3). We report in Figure 12.5 the wave spectra at different propagation lengths obtained by solving the NLS equation with $\tilde{\alpha} = 0.1$ and $\tilde{\beta} = 0.02$. In the early stage of propagation, $z \sim 200$, the spectrum remains confined within the frequency interval $[\omega_-, \omega_+]$ predicted by the theory (Eq. (12.8)), although the spectrum exhibits a completely different spectral profile than the expected Rayleigh-Jeans distribution. As a matter of fact, the process of thermalization requires enormous propagation lengths, as illustrated in Figure 12.5 (d), which shows that the wave spectrum eventually relaxes toward a truncated Rayleigh-Jeans distribution. For more details on these numerical simulations, we refer the reader to [37].

12.2.2.2 Third-Order Dispersion: Anomalous Thermalization

Here we discuss another mechanism that inhibits the natural process of thermalization. We consider the 1D NLS equation by truncating the dispersion relation up to the third order. We will see that the incoherent wave exhibits an irreversible evolution toward an equilibrium state of a different nature than the conventional

Rayleigh-Jeans equilibrium state. The WT kinetic equation reveals that this effect of anomalous thermalization is due to the existence of a local invariant in frequency space J_ω, which originates in degenerate resonances of the system [74,75]. In contrast to conventional integral invariants that lead to a generalized Rayleigh-Jeans distribution, here, it is the local nature of the invariant J_ω that makes the new equilibrium states different from the usual Rayleigh-Jeans equilibrium states. We remark that local invariants and the associated process of anomalous thermalization have also been identified in the 1D vector NLS equation, a configuration in which optical fiber experiments have been also performed, see [74].

The starting point is the NLS Eq. (12.2) accounting for third-order dispersion effects, as well as the corresponding WT kinetic Eq. (12.3). A refined analysis of the WT kinetic equation reveals a remarkable property, namely the existence of a local invariant in frequency space:

$$J(\omega) = n(\omega, z) - n(q - \omega, z), \tag{12.9}$$

where $q = 2s\omega_*$, ω_* being the zero-dispersion angular frequency, and $s = \text{sign}(\beta_2)$ [74,75]. This invariant is "local" in the sense that it is verified for each frequency ω individually, $\partial_z J(\omega) = 0$. It means that the subtraction of the spectrum by the reverse of itself translated by q, remains invariant during the whole evolution of the wave. The invariant (12.9) finds its origin in the following degenerate resonance of the phase-matching conditions: a pair of frequencies $(\omega, q - \omega)$ may resonate with any pair of frequencies $(\omega', q - \omega')$, because $k(\omega) + k(q - \omega) = sq^2/3$ does not depend on ω. Because of the existence of this local invariant, the incoherent wave relaxes toward an equilibrium state of fundamental different nature from the expected thermodynamic Rayleigh-Jeans spectrum:

$$n^{loc}(\omega) = \frac{J_\omega}{2} + \frac{1}{\lambda}\left(1 + \sqrt{1 + \left(\frac{\lambda J_\omega}{2}\right)^2}\right). \tag{12.10}$$

Here, the parameter λ is determined from the initial condition through the conservation of the power. We remark that the equilibrium distribution Eq. (12.10) vanishes exactly the collision term of the kinetic equation, i.e., it is a stationary solution. The equilibrium distribution is characterized by a remarkable property: it exhibits a constant spectral pedestal, $n^{loc}(\omega) \to 2/\lambda$ for $|\omega| \gg |\omega_*|$. We remark in this respect that in the tails of the spectrum ($|\omega| \gg |\omega_*|$), the invariant J_ω vanishes, so that a constant spectrum ($n_\omega = const$) turns out to be a stationary solution of the WT kinetic equation. The existence of the process of anomalous thermalization has been confirmed by the numerical simulations of both the NLS equation and the WT kinetic equation, as illustrated in Figure 12.6 in the color plate section. For more details on theoretical and numerical simulations of anomalous thermalization, we refer the reader to [37,75,88].

12.2.2.3 Second-Order Dispersion: Integrable Limit

In this section we consider the case where the dispersion relation of the NLS equation is truncated to the second-order, so that the equation recovers the completely integrable NLS equation which is known to admit genuine soliton solutions. During the past fifty years, the question of the interaction among solitons has been extensively studied

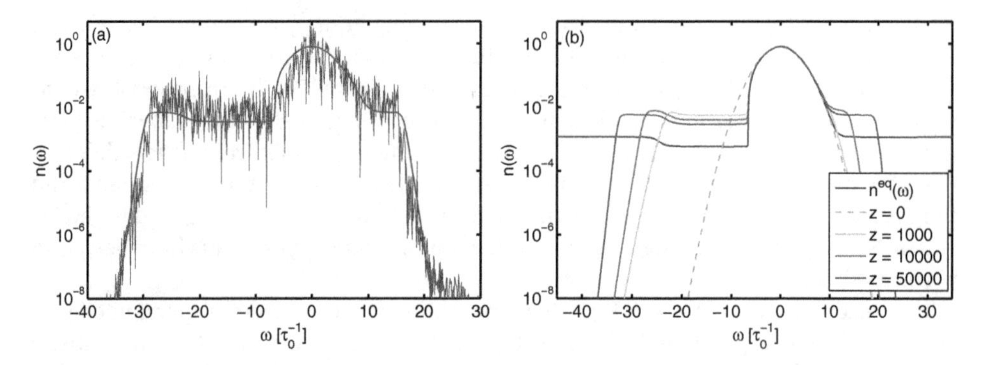

Figure 12.6 Anomalous thermalization of incoherent waves: (a) Spectral evolution obtained by integrating numerically the NLSE with third-order dispersion (blue) and the corresponding WT kinetic equation (red) at $z = 20000$ for $\tilde{\alpha} = 0.05$ (a). (b) Numerical simulations of the WT kinetic equation showing the spectral profile $n(z, \omega)$ at different propagation lengths z: a constant spectral pedestal emerges in the tails of the spectrum ($\tilde{\alpha} = 0.05$). The spectrum slowly relaxes toward the equilibrium state $n^{loc}(\omega)$ given by Eq. (12.10) (blue). *Source*: Michel 2011 [88]. For a color version of this figure please see color plate section.

by using the method of the inverse scattering transform (see, e.g., [89]). From a different perspective, the formation and the dynamics of shock-waves in the defocusing regime have been studied in different experimental circumstances (see, e.g., [90–92]). The evolution of a dense gas of uncorrelated NLS solitons has been also examined in [93], in which a general method to derive kinetic equations describing the evolution of the spectral distribution function of solitons has been proposed.

The non-integrability of the model equation is usually considered a prerequisite for the applicability of WT theory: Applying the conventional WT procedure to the integrable NLS equation, one finds that all collision terms in the kinetic equation vanish identically at any order [94]. Accordingly, the traditional WT procedure predicts that the spectrum of a weakly nonlinear wave does not evolve during the propagation. We note in this respect that accurate experiments have been performed in optical fibers since 2006 [95], which revealed that a significant evolution of the spectrum of the wave occurs beyond the weakly nonlinear regime of propagation. This issue was subsequently addressed in [72,73], in which a generalized WT kinetic equation was proposed by considering that the fourth-order moment of the field is not necessarily a stationary quantity. It is important to note that similar generalizations of the WT kinetic equation were originally developed in the context of hydrodynamic waves (see [96] for a review), and are still important when one considers the early stage of the evolution of the turbulent system, see, e.g., [97]. From a broader perspective, the study of wave turbulence in the framework of integrable, e.g., NLS, equations is an important field of research which is attracting a growing interest in relation, e.g., with the phenomenon of intermittency [98], or the formation of extreme rogue wave events [32,99].

The starting point is the NLS Eq. (12.2) accounting solely for second-order dispersion. Since exact resonant interactions lead to a vanishing collision term in the conventional kinetic equation, the idea here is to derive a generalized kinetic equation by considering quasi-resonant four-wave interactions, i.e., interactions that do not verify the phase-matching condition, $\Delta k = k(\omega_1) + k(\omega_2) - k(\omega_3) - k(\omega_4) = 0$, where

$k(\omega) = \beta_2\omega^2/2$. In this way, one can derive a generalized form of the kinetic equation:

$$\partial_z n_{\omega_1}(z) = \frac{\gamma^2}{\pi^2}\int_0^z dz' \int\int\int Q[n(z')]\cos[\Delta k(z'-z)] \times \delta(\omega_1+\omega_2-\omega_3-\omega_4)\, d\omega_2 d\omega_3 d\omega_4. \tag{12.11}$$

where $Q[n] = n_{\omega_1}(z)n_{\omega_3}(z)n_{\omega_4}(z) + n_{\omega_2}(z)n_{\omega_3}(z)n_{\omega_4}(z) - n_{\omega_1}(z)n_{\omega_2}(z)n_{\omega_3}(z) - n_{\omega_1}(z) n_{\omega_2}(z)n_{\omega_4}(z)$. Because of the presence of the cosine function, the collision term of this kinetic equation no longer vanishes, despite the degenerate phase-matching interaction. Actually, the evolution of the spectrum described by Eq. (12.11) is characterized by the rapid growth of spectral tails that exhibit damped oscillations, until the whole spectrum ultimately reaches a statistically stationary state. The kinetic equation provides an analytical expression of the damped oscillations, which is found in agreement with the numerical simulations of both the NLS and kinetic equations [73]. An interesting experiment aimed at observing this phenomenon of irreversible relaxation in the limit of the integrable NLS equation has been reported in [73], see Figure 12.7.

We conclude this paragraph by reminding that the applicability of the generalized WT kinetic equation to the description of the dynamics of the integrable NLS equation is constrained by the usual assumption of weakly nonlinear interaction. A rigorous mathematical treatment of the evolution of the incoherent wave beyond this weakly nonlinear regime would require the application of the inverse scattering machinery (see, e.g., [100,101]), a feature which is also of interest considering the recent Hanbury Brown and Twiss experiment [102,103].

12.2.3 Turbulence in Optical Cavities

The phenomenon of wave thermalization can be characterized by a self-organization process, in the sense that it is thermodynamically advantageous for the system to generate a large-scale coherent structure in order to reach the most disordered equilibrium state. A remarkable example of this counterintuitive phenomenon is provided by wave

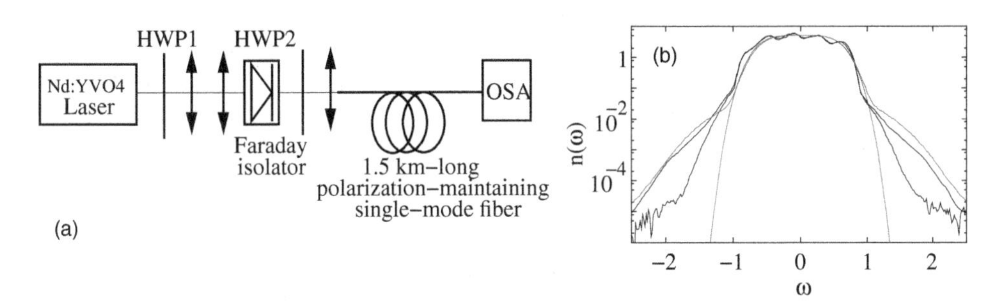

Figure 12.7 Irreversible relaxation process of the integrable NLS equation. (a) Shematic representation of the experimental setup. HWP: half-wave plate. OSA: optical spectrum analyzer. (b) Spectra recorded in experiments (black lines) and obtained from numerical simulations (red lines) of the integrable NLS equation. Note that numerical simulations of a reduced form of the kinetic Eq. (12.11) provide identical results (not represented here, see [73]). The narrow spectrum plotted in the black line is the spectrum of the Nd:YVO4 laser launched inside the polarization maintaining fiber. The wide spectrum plotted in the black line is the spectrum recorded at the output of the polarization maintaining fiber. For more details on the experiments and the simulations, see [73]. *Source*: Suret 2011 [73].

condensation [36, 104–108], whose thermodynamic equilibrium properties are analogous to those of quantum Bose-Einstein condensation [105]. Classical wave condensation can be interpreted as a redistribution of energy among different modes, in which the (kinetic) energy is transferred to small scales fluctuations, while an inverse process increases the power (i.e., number of "particles") into the lowest allowed mode, thus leading to the emergence of a large-scale coherent structure. The phenomenon of condensation has been recently interpreted within a broader perspective in different active and passive optical cavity configurations [109–116]. This raises important questions, such as e.g., the relation between laser operation and the phenomenon of Bose-Einstein condensation. As a matter of fact, these questions are still the subject of vivid debate – we refer the reader to [117–119] for some recent discussions on this important problem.

An important analogy with condensation has been also discussed in the dynamics of active mode-locked laser systems in the presence of additive noise source [111, 118, 120]. On the basis of their previous works [121, 122], the authors showed that the formation of coherent pulses in actively mode-locked lasers exhibits in certain conditions a transition of the laser mode system to a light pulse state that is similar to the Bose-Einstein condensation, in the sense that it is characterized by a macroscopic occupation of the fundamental mode as the laser power is increased. The analysis is based on statistical light-mode dynamics with a mapping between the distribution of the laser eigenmodes to the equilibrium statistical physics of noninteracting bosons in an external potential. Another analogy with condensation has been pointed out to interpret the radiation emitted by a random laser system in [109]. In this work, the analogy with condensation is supported by the fact that the random laser linewidth is ruled by a nonlinear differential equation, which is the equivalent of the Schwalow-Townes law in standard lasers, and is formally identical to the NLS (Gross-Pitaevskii) equation with a trapping potential.

12.2.3.1 Wave Turbulence in Raman Fiber Lasers

The dynamics of Raman fiber lasers have been also shown to exhibit some interesting analogies with condensation-like phenomena [110, 115, 116]. Here we discuss in more detail these systems in light of the WT theory that has been developed to describe their turbulent dynamics. For more details, we refer the interested reader to [114] for an overview on the WT description of Raman fiber lasers (also see the more recent work [123]).

In [124], the Raman fiber laser is modeled as a turbulent system whose optical power spectrum results from a weakly nonlinear interaction among the multiple modes of the cavity. Performing a mean field approach in which the Raman Stokes field does not evolve significantly over one cavity round trip, the authors of [124] first establish a differential equation for the evolution of the complex amplitude E_n of the nth longitudinal mode

$$\tau_{rt}\frac{dE_n}{dt} - \frac{1}{2}(g - \delta_n)E_n(t) = -\frac{i}{2}\gamma L \sum_{l\neq 0} E_{n-l}(t) \times \sum_{m\neq 0} E_{n-m}(t)E^*_{n-m-l}(t)\exp(2i\beta\, ml\Delta^2\, ct). \tag{12.12}$$

In their approach, the time evolution of E_n is determined by the Raman gain g, the dispersion of the fiber, the losses δ_n of the fiber and of the cavity mirrors, and the four-wave mixing process. γ is the Kerr coupling coefficient and β represents the second-order dispersion coefficient of the cavity fiber. $\Delta = 1/\tau_{rt} = c/2L$ is the free spectral range of the Fabry-Perot cavity that has a length L. Gain, losses and dispersive effects occurring inside the whole laser cavity are supposed to influence the formation of the optical power spectrum through their dependence in frequency-space. In particular, fiber Bragg grating mirrors are considered as spectral filters introducing parabolic losses in frequency space ($\delta_n = \delta_0 + \delta_2(n\Delta)^2$). Dispersive effects occurring inside the laser cavity are supposed to be dominantly governed by the second-order dispersion β of the cavity fiber. It must be emphasized that Eq. (12.12) refers to the discretized version of the one-dimensional NLS equation, in which gain and losses terms have been added [125]. In other words, the approach developed by the authors of [124] amounts to apply a WT treatment to a one-dimensional NLS equation, whose integrability is broken by the presence of gain and loss terms.

Assuming an exponential decay for the correlation function among the modes, $\langle E_n(t)E_n^*(t')\rangle = I_n \exp(-|t-t'|/\tau)$, the following WT kinetic equation that governs the temporal evolution of the intracavity spectrum was derived [124]

$$\tau_{rt}\frac{dI(\Omega)}{dt} = (g - \delta(\Omega))I(\Omega) + S_{\mathrm{FWM}}(\Omega), \tag{12.13}$$

where $I(\Omega) = \langle E_n E_n^*\rangle/\Delta$. The mathematical expression of the collision term $S_{\mathrm{FWM}}(\Omega)$ can be separated into two parts

$$S_{\mathrm{FWM}}(\Omega) = -\delta_{\mathrm{NL}} I(\Omega) + (\gamma L)^2 \int \frac{\mathcal{F}[I]\, d\Omega_1\, d\Omega_2}{(3\tau_{rt}/\tau)[1 + (4\tau L\beta/3\tau_{rt})^2\Omega_1^2\Omega_2^2]}, \tag{12.14}$$

where the functional reads $\mathcal{F}[I] = I(\Omega-\Omega_1)I(\Omega-\Omega_2)I(\Omega-\Omega_1-\Omega_2)$, while the nonlinear term responsible for four-wave-mixing-induced losses δ_{NL} reads

$$\delta_{\mathrm{NL}} = (\gamma L)^2 \int \frac{\mathcal{G}[I]\, d\Omega_1\, d\Omega_2}{(3\tau_{rt}/\tau)[1 + (4\tau L\beta/3\tau_{rt})^2\Omega_1^2\Omega_2^2]}, \tag{12.15}$$

where $\mathcal{G}[I] = [I(\Omega-\Omega_1) + I(\Omega-\Omega_2)]I(\Omega-\Omega_1-\Omega_2) - I(\Omega-\Omega_1)I(\Omega-\Omega_1)$. A stationary solution of the WT kinetic Eq. (12.13) has been obtained by Babin et al. in [124], which exhibits the following hyperbolic-secant structure, $I(\Omega) = 2I/\big(\pi\Gamma\cosh(2\Omega/\Gamma)\big)$, where Γ is the width of the intracavity laser power spectrum. This analytical solution is in very good agreement with spectra recorded in experiments in which the fiber laser operates well above threshold, in various different configurations, even in regimes in which the mean field approximation should no longer hold [126]. Although the WT approach developed in [124] has undoubtedly provided a new insight into the physics of Raman fiber lasers, some other numerical and experimental works have raised some interesting questions concerning the applicability of the WT approach to the description of the spectral broadening phenomenon. In particular, numerical simulations of the mean field equations introduced in [124] revealed that the shape of the laser optical power spectrum strongly depends on the sign of the second-order dispersion coefficient [110]. This cannot be captured by the WT theory, which is inherently insensitive to the sign of the second-order dispersion parameter. As pointed out in [125,127], the formation of the Stokes spectrum is also deeply influenced both by dispersive effects and by the spectral shape of the fiber Bragg grating mirrors used to close the laser cavity.

Laminar-Turbulent Transition in Raman Fiber Lasers Fast recording techniques have recently been exploited for the experimental characterization of a laminar-turbulent transition in Raman fiber lasers [116]. The fiber laser used in these experiments has been specifically designed. It is made with dispersion-free ultra-wideband super-Gaussian fiber grating mirrors. Slightly changing the pump power, an abrupt transition with a sharp increase in the width of laser spectrum has been observed, together with an abrupt change of the statistical properties of the Stokes radiation. The laminar state observed before the transition is associated with a multimode Stokes emission with a relatively narrow linewidth and relatively weak fluctuations of the Stokes power. On the other hand, the turbulent state corresponds to a high multimode operation with a wider spectrum and stronger fluctuations of the Stokes power. The laminar-turbulent transition has been also studied by means of intensive numerical simulations (see Figure 12.8 and [110,115,116]). The simulations reveal that, by increasing the pump power, the mechanism underlying the laminar-turbulent transition relies on the generation of an increasing number of dark (or gray) solitons. This experimental work opens new fields of investigations, in particular as regard the impact of phase-defects on the turbulent dynamics of purely 1D wave systems.

Wave Kinetics of Random Fiber Lasers Random lasers are a rapidly growing field of research, with implications in soft-matter physics, light localization, and photonic devices [27,128,129]. Considering a different perspective, the authors of [123] described

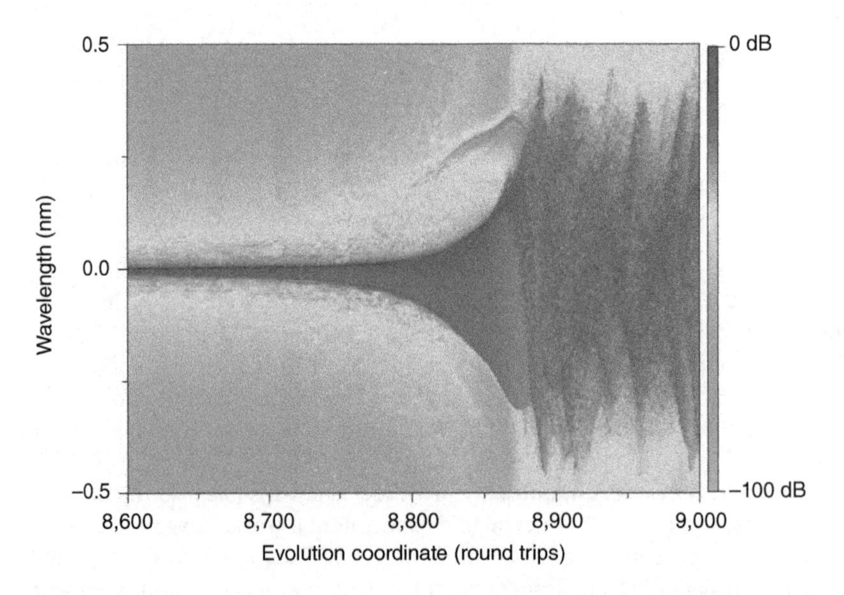

Figure 12.8 Numerical simulations evidencing the laminar-turbulent transition in a Raman fiber laser. The evolution of the laser optical power spectrum is plotted as a function of number of round trips inside the laser cavity. *Source*: Turitsyna 2013 [116]. Reproduced with permission of Nature Publishing Group.

the cyclic wave dynamics inherent to laser systems by considering weakly dissipative modifications of the integrable NLS equation. In this way, a "local kinetic equation" describing the turbulent dynamics of a random fiber laser system is derived [123]. The key property of this kinetic equation is that the δ−function reflecting energy conservation at each elementary four-wave interaction is replaced by an effective Lorentzian function that involves a frequency dependent gain. As a remarkable result, the collision term of the local kinetic equation does not vanish in spite of the trivial resonant conditions inherent to the 1D four-wave interaction with a purely quadratic dispersion relation (see Section 12.2.2.3). From this point of view, the local kinetic equation exhibits properties reminiscent of those considered in [38, 130], although the equations are different, e.g., as regard the renormalization of the dispersion relation by the nonlinearity and the additional nonlinear damping. Then at variance with the purely conservative (Hamiltonian) system, in active cyclic laser systems, the interactions are mediated by a non-homogenous gain, which leads to an effective interaction over the finite interval of the evolution coordinate. We also note that the local kinetic equation is derived under a double separation of scales, i.e., the turbulent regime is dominated by dispersive effects as compared to gain effects, and the gain itself is much larger than gain variation over the typical spectral width of the radiation. Furthermore, the authors confirm their theoretical work by means of direct experimental measurements in random fibre lasers: In the high-power regime, the equilibrium spectrum of the random laser measured experimentally is found in good agreement with the nonequilibrium stationary solution of the local kinetic equation, see Figure 12.9 in the color plate section. Finally, the theory is also completed by means of a generalization of the linear kinetic Schawlow-Townes theory. For more details, we refer the reader to [123].

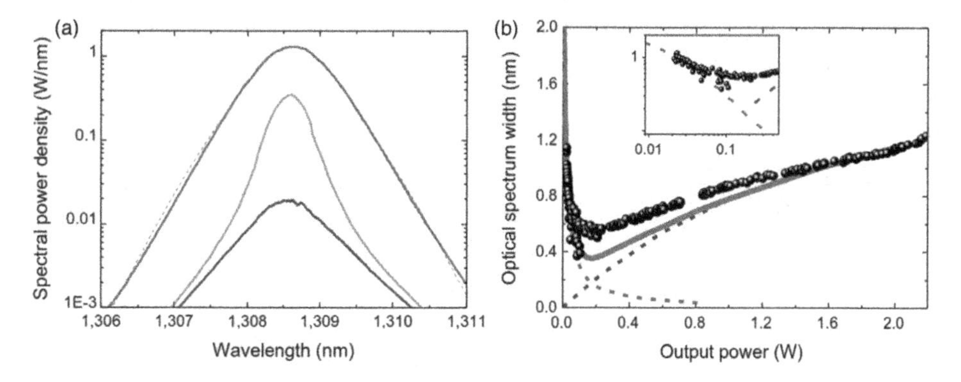

Figure 12.9 Nonlinear kinetic description of the random fiber laser optical spectrum. (a) Optical spectrum measured experimentally: near the generation threshold (blue curve, laser power = 0.025W), slightly above the generation threshold (green curve, 0.2W) and well above the generation threshold (red curve, 1.5W). The optical spectrum predicted by the local wave kinetic equation, for laser power 1.5W is shown by dashed red line. (b) Spectrum width as a function of the laser's output power in theory and experiment. Experimental data are shown by black circles. The prediction for the spectrum broadening from the nonlinear kinetic theory based on the local wave kinetic equation (blue dashed line). The prediction for the spectral narrowing from the modified linear kinetic Schawlow-Townes theory (dashed green line). The red line denotes the sum of nonlinear and linear contributions. The inset shows the spectral narrowing near the threshold in log-scale. For more details see [123].
Source: Churkin 2015 [123]. Reproduced with permission of Nature Publishing Group. For a color version of this figure please see color plate section.

12.2.3.2 Turbulent Dynamics in Passive Optical Cavities

As commented above, a classical wave can exhibit a genuine process of wave condensation as it propagates in a 2D conservative Kerr material, [36, 105, 107]. Actually, a phenomenon completely analogous to such conservative condensation process can occur in an incoherently pumped passive optical cavity, despite the fact that the system is inherently dissipative [113]. For this purpose, let us consider a *passive optical cavity* pumped by an incoherent optical wave, whose time correlation, t_c, is much smaller than the round trip time, $t_c \ll \tau_{rt}$. In this way, the optical field from different cycles are mutually incoherent with one another, which makes the optical cavity *non-resonant.* Because of this property, the cavity does not exhibit the widely studied dynamics of pattern formation [131, 132]. Instead, the dynamics of the cavity exhibit a turbulent behavior that can be characterized by an irreversible process of thermalization toward energy equipartition. A mean-field WT equation was derived in [113], which accounts for the incoherent pumping, the nonlinear interaction and both the cavity losses and propagation losses. In spite of the dissipative nature of the cavity dynamics, the intracavity field undergoes a condensation process below a critical value of the incoherence (kinetic energy) of the pump. This phenomenon is illustrated in Figure 12.10 (a), which shows the temporal evolution of the condensate fraction in the intracavity field: After a transient, the fraction of power condensed in the fundamental transverse mode of the cavity saturates to a constant value, which is found in agreement with the theory. Figure 12.10 (b) reports the condensation curve, i.e., the fraction of condensed power at equilibrium vs the kinetic energy of the injected pump wave. This latter quantity reflects the degree of coherence of the pump wave and plays the role of the control parameter of the transition to wave condensation in the cavity configuration. We remark in Figure 12.10 (b) that the condensate fraction in this dissipative optical cavity is found in agreement with the theory inherited from the conservative Hamiltonian NLS equation, without using adjustable

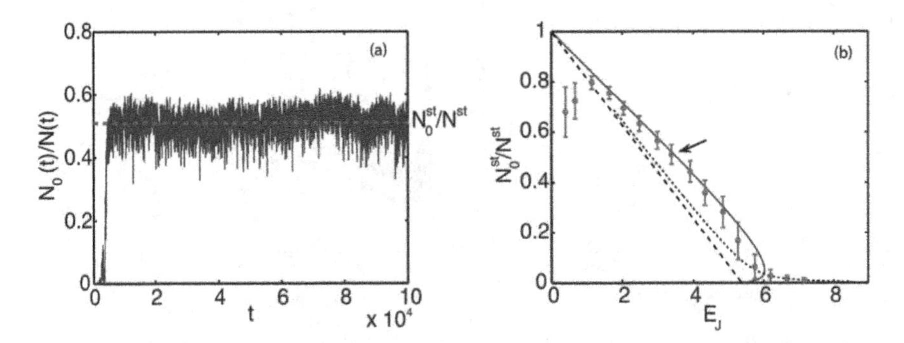

Figure 12.10 Wave condensation in an incoherently pumped passive optical cavity. (a) Evolution of the fraction of condensed power $N_0(t)/N(t)$ vs time t: The condensate growth saturates to a constant value N_0^{st}/N^{st}, which is in agreement with the theory [113]. (b) Condensation curve: fraction of condensed power in the stationary equilibrium state N_0^{st}/N^{st} vs the kinetic energy of the pump E_J. The condensation curve is computed for a fixed value of the pump intensity J_0, while E_J is varied by modifying the degree of coherence of the pump (i.e., its spectral width). The solid line refers to the (Bogoliubov) strong condensation regime. The dotted line refers to the weak condensation regime beyond the thermodynamic limit ($\mu \neq 0$), while the dashed line refers to the thermodynamic limit ($\mu \to 0$). The points correspond to the NLS numerical simulations with the cavity boundary conditions. For more details, see [113]. *Source*: Michel 2011 [113]. Reproduced with permission of American Physical Society.

parameters. For more detail on the simulations and the theory, we refer the reader to [113].

Let us note an important difference that distinguishes the thermalization and condensation processes discussed here with those reported in the quantum photon context in [112, 133]. In these works the thermalization process is achieved thanks to the presence of dye molecules, which thus play the role of an external thermostat. Conversely, in the passive cavity configuration considered here, the process of thermalization solely results from the four-wave interaction mediated by the intracavity Kerr medium, while the "temperature" is controlled by varying the kinetic energy (degree of coherence) of the injected pump.

In a recent experimental work [134], the incoherently pumped passive cavity has been implemented in a fully integrated optical fiber system, close to the zero-dispersion wavelength of the fiber. The dynamics of the cavity exhibit a quasi-soliton turbulent behavior which is reminiscent of the turbulent dynamics of the purely Hamiltonian wave system considered in [135,136]. The analysis reveals that, as the coherence of the injected pump wave is degraded, the cavity undergoes a transition from the coherent quasi-soliton regime toward the highly incoherent (weakly nonlinear) turbulent regime characterized by short-lived and extreme rogue wave events. This transition can then be interpreted by analogy with a phenomenon of quasi-soliton condensation. The experiments realized in the incoherently pumped passive optical cavity have been characterized by means of complementary spectral and temporal PDF measurements [134].

An unexpected result of [134] is that quasi-soliton condensation can take place efficiently, even in the presence of a low cavity finesse, in contrast to wave-condensation in 2D defocusing media discussed here above, which requires a high finesse [113]. This can be interpreted as a consequence of the fact that the process of thermalization of an optical wave constitutes a prerequisite for the phenomenon of wave-condensation in a defocusing medium, while wave thermalization is known to require a high cavity finesse. There is another important difference which distinguishes wave-condensation and (quasi-)soliton condensation. Wave-condensation is known to exhibit a property of long-range order and coherence, in the sense that the correlation function of the field amplitude does not decay at infinity, $\lim_{|\mathbf{r}-\mathbf{r}'|\to\infty}\langle A(\mathbf{r})A^*(\mathbf{r}')\rangle \neq 0$, a property consistent with the idea that the coherence length of a plane-wave diverges to infinity [105]. This is in contrast with the spatial localized character of a (quasi-)soliton, which naturally limits the range of coherence to the characteristic spatial width of the (quasi-)soliton structure. Wave-condensation then appears to be more sensitive to the "boundary conditions" of the system, and thus is less robust than (quasi-) soliton condensation when considered in an optical cavity system. The understanding of this aspect requires further investigation.

12.3 Weak Langmuir Turbulence Formalism

In this section we study the temporal evolution of a partially coherent wave that propagates in a nonlinear medium characterized by a noninstantaneous response. As discussed in Section 12.1 through Figure 12.2, a delayed nonlinearity leads to a kinetic description which is formally analogous to the weak Langmuir turbulence kinetic equation, irrespective of the nature of the fluctuations that may be either stationary or non-stationary. In the presence of a temporal long-range response and a stationary statistics of the incoherent wave, the weak Langmuir turbulence formalism reduces to a family of singular integro-differential kinetic equations (e.g., Benjamin-Ono equation) that describe incoherent dispersive shock waves and incoherent collapse singularities in the spectral evolution of the random wave.

12.3.1 NLS Model

A typical example of noninstantaneous nonlinear response in one dimensional systems is provided by the Raman effect in optical fibers [71]. We consider the standard 1D NLS equation to account for a noninstantaneous nonlinear response function

$$i\partial_z\psi + \beta\partial_{tt}\psi + \gamma\psi\int_{-\infty}^{+\infty} R(t-t')\,|\psi|^2(z,t')\,dt' = 0, \tag{12.16}$$

where the response function $R(t)$ is constrained by the causality condition. In the following we use the convention that $t > 0$ corresponds to the leading edge of the pulse, so that the causal response will be on the trailing edge of a pulse, i.e., $R(t) = 0$ for $t > 0$. We will write the response function in the form $R(t) = H(-t)\bar{R}(-t)$, where $\bar{R}(t)$ is a smooth function from $[0, \infty)$ to $(-\infty, \infty)$, while the Heaviside function $H(-t)$ guarantees the causality property. As we will see, this convention will allow us to easily compare the dynamics of temporal and spatial incoherent solitons. Because of the causality property, the real and imaginary parts of the Fourier transform of the response function

$$\tilde{R}(\omega) = \tilde{U}(\omega) + i g(\omega), \tag{12.17}$$

are related by the Kramers-Krönig relations. We recall that $\tilde{U}(\omega)$ is even, while the gain spectrum $g(\omega)$ is odd. The causality condition breaks the Hamiltonian structure of the NLS equation, so that Eq. (12.16) only conserves the total power ("number of particles") of the wave $\mathcal{N} = \int |\psi|^2(t,z)\,dt$. The typical temporal range of the response function $R(t)$ denotes the response time, τ_R. Note that $\beta = -\frac{1}{2}\partial^2_\omega k(\omega)$ in Eq. (12.16), so that $\beta > 0$ ($\beta < 0$) denotes the regime of anomalous (normal) dispersion.

12.3.2 Short-Range Interaction: Spectral Incoherent Solitons

The dynamics are ruled by the comparison of the response time, τ_R, with the 'healing time', $\tau_0 = \sqrt{|\beta| L_{nl}}$. We note that the weakly nonlinear regime of interaction refers to the regime in which linear dispersive effects dominate nonlinear effects, i.e., $L_d/L_{nl} \ll 1$ where $L_d = t_c^2/|\beta|$ and $L_{nl} = 1/(|\gamma|\rho)$ refer to the dispersive and nonlinear characteristic lengths respectively, t_c being the correlation time of the partially coherent wave. We consider here the case of a noninstantaneous nonlinearity characterized by a short-range response time, i.e., $\tau_R \sim \tau_0$. In this regime, it can be shown that the kinetic equation governing the evolution of the incoherent wave takes a form analogous to the WT Langmuir kinetic equation [23, 37]:

$$\partial_z n_\omega(z) = \frac{\gamma}{\pi}\, n_\omega(z)\int_{-\infty}^{+\infty} g(\omega-\omega')\, n_{\omega'}(z)\,d\omega', \tag{12.18}$$

where we have implicitly assumed that the incoherent wave exhibits fluctuation that are statistically stationary (homogeneous) in time—a generalized WT Langmuir equation can be obtained for a non-stationary statistics [37]. We first note that this equation does not account for dispersion effects (it does not involve the parameter β), although the role of dispersion in its derivation is essential in order to verify the criterion of weakly nonlinear interaction, $L_d/L_{nl} \ll 1$. The fact that the dynamics ruled by the WT Langmuir equation do not depend on the sign of the dispersion coefficient has been verified by direct numerical simulations of the NLS Eq. (12.16) [48]. The kinetic Eq. (12.18) conserves the power of the field $N = \frac{1}{2\pi}\int n_\omega(z)\,d\omega$. Moreover, as discussed above for the Vlasov equation, the WT Langmuir Eq. (12.18) is a formally reversible equation [it is invariant under the transformation $(z,\omega) \rightarrow (-z,-\omega)$], a feature which is consistent with the fact that it also conserves the non-equilibrium entropy $S = \frac{1}{2\pi}\int \log[n_\omega(z)]\,d\omega$.

The WT Langmuir equation admits solitary wave solutions [23, 47, 48, 54]. This may be anticipated by remarking that, as a result of the convolution product in (12.18), the odd spectral gain curve $g(\omega)$ amplifies the low-frequency components of the wave at the expense of the high-frequency components, thus leading to a global red-shift of the spectrum. We note that these incoherent solitons are termed "spectral" because they can only be identified in the spectral domain, since in the temporal domain the field exhibits stochastic fluctuations at any time, t.

12.3.2.1 Numerical Simulations

Typical spectral incoherent soliton behaviors are reported in Figure 12.11. The initial condition is an incoherent wave characterized by a Gaussian spectrum with δ−correlated random spectral phases, so that the initial wave exhibits stationary fluctuations. The Gaussian spectrum is superposed on a background of small noise of averaged intensity $n_0 = 10^{-5}$. This is important in order to sustain a steady soliton propagation, otherwise the soliton undergoes a slow adiabatic reshaping so as to adapt its shape to the local value of the noise background. The relative intensity of the background noise with respect to the average power of the wave plays an important role in the dynamics of discrete spectral incoherent solitons. Indeed, the *continuous* spectral incoherent soliton is known to become narrower (i.e., of higher amplitude) as the intensity of the background noise decreases. Accordingly, a transition from a continuous to a discrete spectral incoherent soliton behavior occurs as the relative intensity of the background noise is decreased: as the spectral soliton becomes narrower than ω_R, the leading edge of the tail of the spectrum will be preferentially amplified, thus leading to the formation of a discrete spectral incoherent soliton. In order to test the validity of the WT Langmuir theory, we reported in Figure 12.11 a direct comparison with NLS simulations. We emphasize that an excellent agreement has been obtained between the simulations of the NLS equation and the WT Langmuir equation, without using any adjustable parameter [48].

Note that if the background noise level increases in a significant way and becomes of the same order as the amplitude of the spectral soliton, the incoherent wave enters a novel regime [137]. This regime is characterized by an oscillatory dynamics of the incoherent spectrum which develops within a spectral cone during the propagation. Such spectral dynamics exhibit a significant spectral blueshift, which is in contrast with the expected Raman-like spectral redshift.

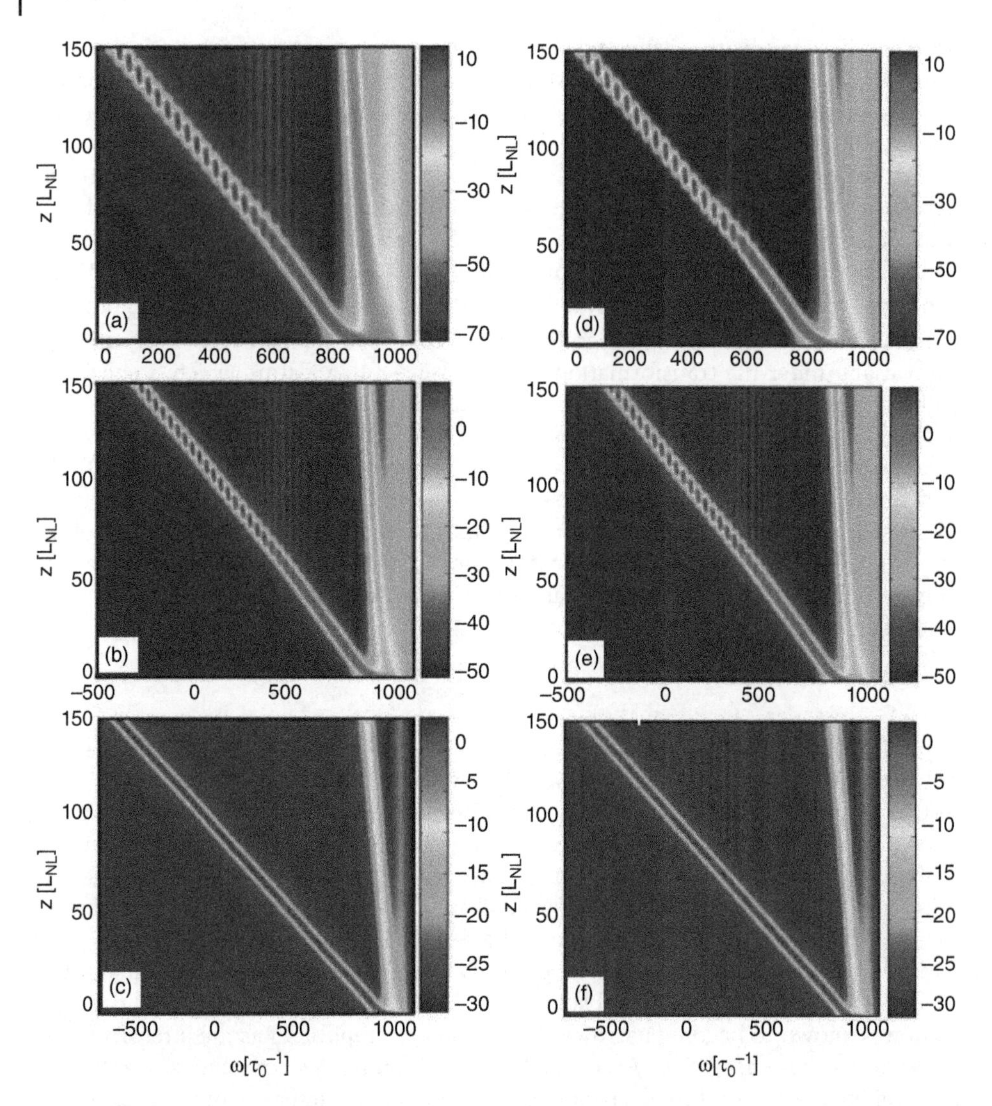

Figure 12.11 Spectral incoherent solitons: Transition from discrete to continuous solitons. Left column (a–c): Evolution of the non-averaged spectrum of the optical field, $|\tilde{\psi}|^2(z,\omega)$ (in dB-scale), obtained by integrating numerically the NLS Eq. (12.16) for three different values of the noise background, $n_0 = 10^{-7}$ (a), $n_0 = 10^{-5}$ (b), $n_0 = 10^{-3}$ (c). Right column (d–f): Corresponding evolution of the averaged spectrum, $n(z,\omega)$ (in dB-scale), obtained by solving the Langmuir WT Eq. (12.18). The comparison reveals a quantitative agreement, without using adjustable parameters. We considered the typical Raman-like gain spectrum, $g(\omega)$ ($\beta\gamma < 0$). *Source*: Michel 2011 [48]. Reproduced with permission of American Physical Society.

12.3.2.2 Analytical Soliton Solution

The WT Langmuir kinetic equation (12.18) admits analytical soliton solutions [37, 54, 138]. More precisely, it is possible to compute the width and velocity of the soliton given its peak amplitude n_m in the regime $n_m \gg n_0$, where n_0 denotes the spectral amplitude of the background noise. We introduce the antiderivative of the spectral gain $G(\omega) = -\int_\omega^\infty g(\omega')d\omega'$. The gain spectrum $g(\omega)$ is characterized by its typical gain amplitude g_i and its typical spectral width ω_i. Regardless of the details of the gain curve $g(\omega)$, g_i and ω_i can be assessed by two characteristic quantities, namely the gain slope at the origin $\partial_\omega g(0)$ and the total amount of gain $G(0) = -\int_0^\infty g(\omega)d\omega$. A dimensional analysis allows to express g_i and ω_i in terms of these two quantities, $g_i = \frac{1}{\sqrt{2}}(-\partial_\omega g(0))^{1/2}$ $\left[-\int_0^\infty g(\omega)d\omega\right]^{1/2}$, $\omega_i = \sqrt{2}\left[-\int_0^\infty g(\omega)d\omega\right]^{1/2}/\left[-\partial_\omega g(0)\right]^{1/2}$. With these definitions, the function $G(\omega)$ can be written in the following normalized form $G(\omega) = g_i\omega_i h(\omega/\omega_i)$, where the dimensionless function $h(x)$ verifies $h(0) = 1$, $h'(0) = 0$, and $h''(0) = -2$. Proceeding as in [138], the profile of the soliton in the regime $n_m \gg n_0$ is of the form [54], $\log\left(\frac{n_\omega(z)}{n_0}\right) = \log\left(\frac{n_m}{n_0}\right)h\left(\frac{\omega - Vz}{\omega_i}\right)$, or equivalently:

$$n_\omega(z) - n_0 = (n_m - n_0)\exp\left[-\log\left(\frac{n_m}{n_0}\right)\frac{(\omega - Vz)^2}{\omega_i^2}\right], \tag{12.19}$$

where the velocity of the soliton is

$$V = -\frac{n_m - n_0}{\log^{3/2}\left(\frac{n_m}{n_0}\right)}\frac{\gamma g_i \omega_i^2}{\sqrt{\pi}}, \tag{12.20}$$

and its full width at half maximum is $\omega_{sol} = 2\omega_i \log^{1/2}(2)/\log^{1/2}(n_m/n_0)$.

Spectral incoherent solitons have recently been generalized in the framework of the generalized NLS equation accounting for the self-steepening term and a frequency dependence of the nonlinear Kerr coefficient [139]. Such nonlinear dispersive effects are shown to strongly affect the dynamics of the incoherent wave. A generalized WT Langmuir kinetic equation is derived and its predictions have been found in quantitative agreement with the numerical simulations of the NLS equation, without adjustable parameters [139].

The structure of discrete spectral incoherent solitons can also be interpreted with an analytical soliton solution of the *discretized* WT Langmuir equation derived in [140]. In this way, discrete frequency bands of the soliton are modelled as coupled Dirac δ-functions in frequency space (δ−peak model). However, the simulations show that, when injected as initial condition into the WT Langmuir equation with a Raman-like gain spectrum, the analytical soliton solution rapidly relaxes during the propagation toward a discrete spectral incoherent solution [48]. This property reveals the incoherent nature of discrete spectral incoherent solitons.

We finally note that the emergence of continuous and discrete spectral incoherent solitons has been identified experimentally owing to the Raman effect in photonic crystal fibers in the context of supercontinuum generation, a feature discussed in detail in [24].

12.3.3 Long-Range Interaction: Incoherent Dispersive Shock Waves

In this section we present the procedure which allows one to derive appropriate reduced kinetic equations from the WT Langmuir equation in the long-range limit, i.e., the limit of a highly noninstantaneous nonlinear response, $\tau_R \gg \tau_0$. As discussed here above, the causality condition leads to a gain spectrum $g(\omega)$ that decays algebraically at infinity, a property which introduces singularities into the convolution operator of the WT Langmuir Eq. (12.18). The mathematical procedure consists in accurately addressing these singularities, see [49]. It reveals that, as a general rule, a singular integro-differential operator arises systematically in the derivation of the reduced kinetic equation [49,141]. The resulting singular integro-differential kinetic equation then originates in the causality property of the nonlinear response function.

These singular integro-differential kinetic equations find a direct application in the description of dispersive shock waves, i.e., shock waves whose singularity is regularized by dispersion effects instead of dissipative (viscous) effects [142] – see Chapter 11 of this book. Dispersive shock waves have been constructed mathematically [143] and observed in ion acoustic waves [144] long ago, though it is only recently that they emerged as a general signature of singular fluid-type behavior in areas as different as Bose-Einstein condensed atoms [145,146], nonlinear optics [90–92,147–149], quantum liquids [150], and electrons [151]. We remark that dispersive shock waves have been also recently studied in the presence of structural disorder of the nonlinear medium [92,152,153].

These previous studies on dispersive shock waves have been discussed for coherent, i.e., deterministic, amplitudes of the waves. Through the analysis of the WT Langmuir equation, we will see that incoherent waves can exhibit dispersive shock waves of a different nature that their coherent counterpart. They manifest themselves as a wave breaking process ("gradient catastrophe") in the spectral dynamics of the incoherent field [49]. Contrary to conventional shocks which are known to require a strong nonlinear regime, these incoherent shocks develop into the weakly nonlinear regime. This WT kinetic approach also reveals unexpected links with the 3D vorticity equation in incompressible fluids [154], or the integrable Benjamin-Ono equation [155], which was originally derived in hydrodynamics, and recently considered in the field of quantum liquids [150].

12.3.3.1 Damped Harmonic Oscillator Response

The derivation of singular integro-differential kinetic equations has been developed for a general form of the response function (see the Supplement of [49]). Here we illustrate the theory by considering two physically relevant examples of response functions, which, respectively, induce and inhibit the formation of incoherent shock waves.

Let us first consider the example of the damped harmonic oscillator response, $\bar{R}(t) = \frac{1+\eta^2}{\eta\tau_R}\sin(\eta t/\tau_R)\exp(-t/\tau_R)$. Figure 12.12 in the color plate section reports a typical evolution of the spectrum of the incoherent wave obtained by numerical simulations of the NLS Eq. (12.16). Here we considered the highly incoherent limit, $\Delta\omega \gg \Delta\omega_g$ ($t_c \ll \tau_R$). We see that the low frequency part of the spectrum exhibits a self-steepening process, whose wave-breaking is ultimately regularized by the development of large amplitude and rapid spectral oscillations typical of a dispersive shock wave. This behavior has been described by deriving a singular integro-differential kinetic equation from the WT Langmuir equation in the long-range regime ($\tau_R \gg \tau_0$):

$$\tau_R^2 \partial_z n_\omega = \gamma(1+\eta^2)\left(n_\omega \partial_\omega n_\omega - \frac{1}{\tau_R} n_\omega \mathcal{H}\partial_\omega^2 n_\omega\right), \tag{12.21}$$

where the singular operator $\mathcal{H}$ refers to the Hilbert transform, $\mathcal{H}f(\omega) = \frac{1}{\pi}\mathcal{P}\int_{-\infty}^{+\infty} \frac{f(\omega-u)}{u} du$, where we recall that $\mathcal{P}$ denotes the Cauchy principal value. This kinetic

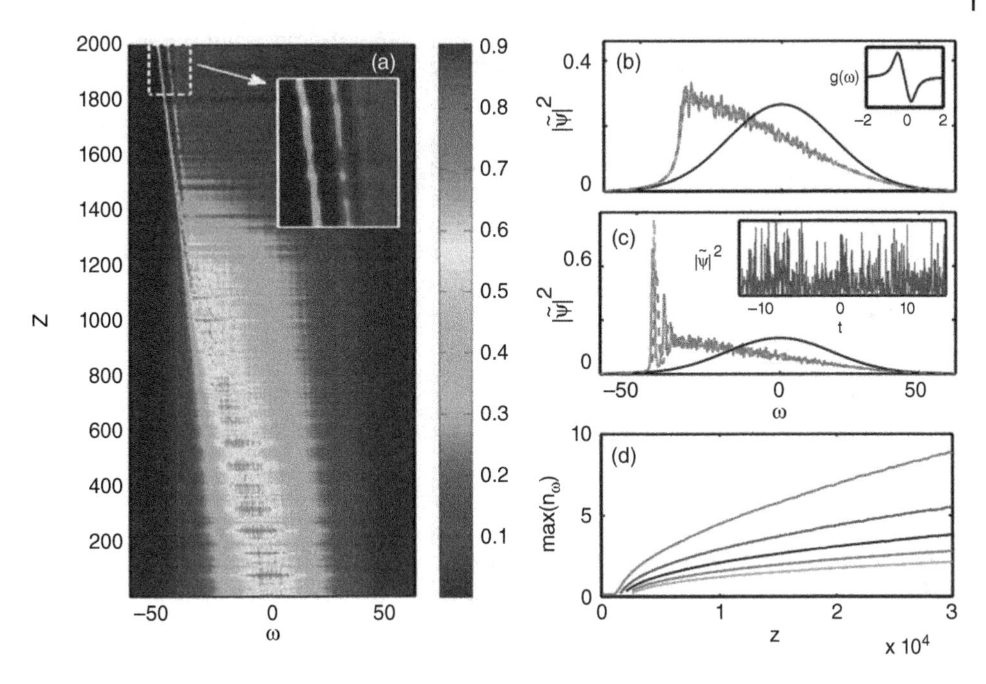

Figure 12.12 Incoherent dispersive shock waves with a Raman-like response function: (a) Numerical simulation of the NLS Eq. (12.16): The stochastic spectrum $|\tilde{\psi}|^2(\omega, z)$ develops an incoherent shock at $z \simeq 1200L_{nl}$ ($\tau_R = 3\tau_0, \eta = 1$). Snapshots at $z = 1040L_{nl}$ (b), $z = 1400L_{nl}$ (c): NLS (12.16) (gray) is compared with WT Langmuir Eq. (12.18) (green), singular kinetic equation (Eq. (12.21)) (dashed line), and initial condition (solid black). (d) First five maxima of n_ω vs z in the long-term post-shock dynamics: the spectral peaks keep evolving, revealing the non-solitonic nature of the incoherent dispersive shock wave. Insets: (b) gain spectrum $g(\omega)$, note that $\Delta\omega_g$ is much smaller than the initial spectral bandwidth of the wave [black line in (b)]. (c) corresponding temporal profile $|\psi(t)|^2$ showing the incoherent wave with stationary statistics. *Source*: Garnier 2013 [49]. Reproduced with permission of American Physical Society. For a color version of this figure please see color plate section.

equation describes the essence of incoherent dispersive shock waves: The leading-order Burgers term describes the formation of the shock, which is subsequently regularized by the nonlinear dispersive term involving the Hilbert operator. We note in Figure 12.12 that a quantitative agreement is obtained between the simulations of Eq. (12.21) and those of the NLS and WT Langmuir equations, without adjustable parameters. Also note that in the presence of a strong spectral background noise, the derived singular equation coincides with the Benjamin-Ono equation, which is a completely integrable equation [49].

12.3.3.2 Exponential Response: Spectral Collapse Singularity

As described by the general theory reported in [49], the previous scenario of incoherent dispersive shock waves changes in a dramatic way when the response function is not continuous at the origin, as it occurs for a purely exponential response function, $\bar{R}(t) = \exp(-t/\tau_R)/\tau_R$. In this case, considering the limit $\tau_R/\tau_0 \gg 1$, the singular kinetic equation takes the form:

$$\tau_R \partial_z n_\omega = -\gamma n_\omega \mathcal{H} n_\omega - \frac{\gamma}{\tau_R} n_\omega \partial_\omega n_\omega + \frac{\gamma}{2\tau_R^2} n_\omega \mathcal{H} \partial_\omega^2 n_\omega. \quad (12.22)$$

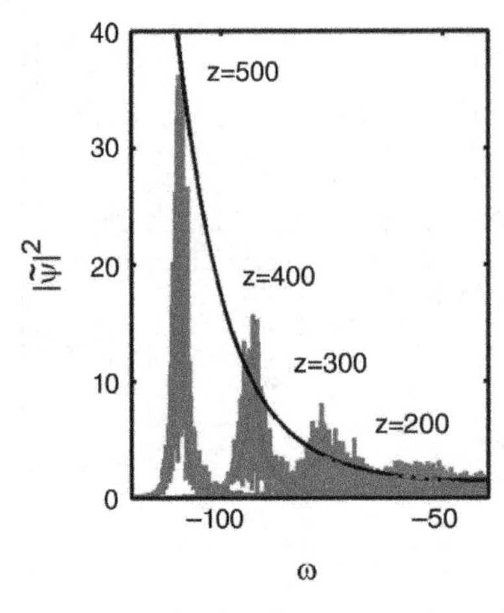

Figure 12.13 Inhibition of incoherent shocks with an exponential response function. Without background spectral noise the spectrum exhibits a collapse-like behavior: NLS (12.16), gray; singular kinetic Eq. (12.22), dashed ($\tau_R = 5\tau_0$). The dark continuous line denotes the theoretical behavior $\sim 1/[z^2 n^0(\omega = \tilde{c}z)]$, with $\tilde{c} = -\gamma N/\tau_R$, predicted from the first term of Eq. (12.22) and the corresponding analytical solution Eq. (12.23).
Source: Garnier 2013 [49]. Reproduced with permission of American Physical Society.

Interestingly, the first term of Eq. (12.22) was considered as a one-dimensional model of the vorticity formulation of the 3D Euler equation of incompressible fluid flows [154]. In this work, the authors found an explicit analytical solution to the equation $\tau_R \partial_z n_\omega = -\gamma n_\omega \mathcal{H} n_\omega$. For a given initial condition $n_\omega(z=0) = n^0_\omega$ the solution has the form

$$n_\omega(z) = \frac{4n^0_\omega}{\left(2 + (\gamma z/\tau_R)\mathcal{H}n^0_\omega\right)^2 + (\gamma z/\tau_R)^2 (n^0_\omega)^2}. \tag{12.23}$$

There is blow up if and only if there exists ω such that $n^0_\omega = 0$ and $\mathcal{H}n^0_\omega < 0$. Then the blow up distance z_c is given by $z_c = -2\tau_R/[\gamma \mathcal{H}n^0_{\omega=\omega_0}]$, where ω_0 is such that $n^0_{\omega_0} = 0$. It can be shown [49] that, if the initial condition decays faster than a Lorentzian, the spectrum exhibits a collapse-like dynamics, which is ultimately arrested by a small background noise. In this process, the spectrum moves at velocity $\tilde{c}$, while its peak amplitude increases according to $\sim 4\tau_R^2/[\gamma^2 z^2 n^0(\omega = \tilde{c}z)]$. This property is confirmed by the simulations of the NLS equation, as illustrated in Figure 12.13.

Impact of Self-Steepening The previous results of incoherent spectral singularities have been recently extended by considering the generalized NLS equation accounting for the self-steepening term [141]. The analysis reveals that self-steepening significantly affects the spectral singularities: (i) It leads to a delay in the development of incoherent dispersive shocks, and (ii) it arrests the incoherent collapse singularity. Furthermore, the spectral collapse-like behavior can be exploited to achieve a significant enhancement (by two orders of magnitudes) of the degree of coherence of the optical wave as it propagates in the fiber.

12.4 Vlasov Formalism

In Section 12.3.3, we considered the role of long-range temporal responses for an incoherent wave characterized by a stationary (homogeneous) statistics. In this section we consider statistically non-stationary random waves in the presence of a highly noninstantaneous nonlinear response, $\tau_R \gg \tau_0$. As discussed in the introductory section (see

Figure 12.2), in this regime, the dynamics of the incoherent wave is ruled by a completely different Vlasov-like formalism. This formalism models the incoherent wave as an ensemble of particles that evolve in a reduced phase-space (i.e., spectrogram), (t, ω). The corresponding particle density is obtained from a Wigner-like transform of the auto-correlation function of the field, $n_\omega(t, z) = \int B(t, \tau, z) \exp(i\omega\tau)\, d\tau$, where $B(t, \tau, z) = \langle \psi(t - \tau/2, z)\psi^*(t + \tau/2, z)\rangle$. Starting from the NLS Eq. (12.16) and considering a highly noninstantaneous nonlinear response, the temporal long-range Vlasov kinetic equation takes the form [14, 37, 156]:

$$\partial_z n_\omega(t, z) + \partial_\omega \tilde{k}_\omega(t, z)\, \partial_t n_\omega(t, z) - \partial_t \tilde{k}_\omega(t, z)\, \partial_\omega n_\omega(t, z) = 0, \tag{12.24}$$

where the generalized dispersion relation reads

$$\tilde{k}_\omega(t, z) = k(\omega) + V(t, z), \tag{12.25}$$

with $k(\omega) = \beta\omega^2$ and the effective potential

$$V(t, z) = -\gamma \int R(t - t')N(t', z)\, dt'. \tag{12.26}$$

The intensity profile of the incoherent wave is $N(t, z) = B(t, \tau = 0, z) = (2\pi)^{-1} \int n_\omega(t, z)\, d\omega$. Equation (12.24) conserves $\mathcal{N} = (2\pi)^{-1} \iint n_\omega(t, z) d\omega dt$, and more generally $\mathcal{M} = \iint f[n] d\omega dt$ where $f[n]$ is an arbitrary functional of n. Because of the causality property of $R(t)$, Eq. (12.24) is no longer Hamiltonian [37]. As a consequence, the incoherent wave-packet exhibits a spectral shift, as revealed by the analysis of the total momentum, $\mathcal{P}(z) = (2\pi)^{-1} \iint \omega\, n_\omega(t, z)\, dt\, d\omega$, which is related to the barycenter of the spectrum, $\langle\omega\rangle = \mathcal{P}/\mathcal{N}$. An equation for the momentum can easily be obtained from the Vlasov Eq. (12.24)

$$\partial_z \mathcal{P}(z) = \int V_G(t, z)\, \partial_t N(t, z)\, dt, \tag{12.27}$$

where $V_G(t, z) = -\gamma \int G(t - t')N(t', z)\, dt'$ and $G(t) = \frac{i}{2\pi} \int g(\omega) \exp(-i\omega t)\, d\omega$ [37]. Remarking furthermore that $G(t)$ can expressed in terms of the response function, $G(t) = \frac{1}{2}\left[R(t) - R(-t)\right]$, it becomes easy to see that a focusing (defocusing) nonlinearity leads to a spectral red-shift, $\partial_z \mathcal{P} < 0$ (blue-shift, $\partial_z \mathcal{P} > 0$). We can remark that this dependence of the spectral shift on the sign of the nonlinearity is also apparent in the WT Langmuir (see Eq. (12.18)). Note that a spectral blue-shift induced by a defocusing delayed nonlinearity is known to occur, e.g., in plasma [26, 157], in which, however, the total power of the wave is no longer conserved (see Chapter 3 in this book).

It is also interesting to analyze the position of the wave-packet in the time domain, $\mathcal{T}(z) = (2\pi)^{-1} \iint t\, n_\omega(t, z)\, dt\, d\omega$, which is related to the barycenter by $\langle t\rangle = \mathcal{T}/\mathcal{N}$. The evolution of $\mathcal{T}(z)$ can easily be obtained from the Vlasov Eq. (12.24)

$$\partial_z \mathcal{T}(z) = 2\beta \mathcal{P}(z), \tag{12.28}$$

so that $\mathcal{T}(z) = \mathcal{T}(0) + 2\beta \int_0^z \mathcal{P}(z')\, dz'$. Accordingly, propagation in the normal (anomalous) dispersion regime in the presence of a focusing (defocusing) nonlinearity leads to an acceleration of the wave-packet toward $t > 0$. Equation (12.28) also reveals that there is a close relation between the spectral shift and the temporal shift of a wave-packet. This can easily be interpreted by remarking that a spectral shift combined with group-velocity dispersion leads to an acceleration of the wave-packet.

12.4.1 Incoherent Modulational Instability

The Vlasov formalism is known to predict the existence of an incoherent MI, a feature which is in some sense an unexpected result. Indeed, as discussed above in this section in the framework of the WT Langmuir formalism, one would expect that a statistically stationary incoherent wave would exhibit a Raman-like spectral red-shift during its propagation. However, a *highly* noninstantaneous response leads to a genuine process of incoherent MI of the wave, which is characterized by the growth of two symmetric MI bands within the spectrum of the incoherent wave [14].

The details of the incoherent MI analysis through a linearization of the long-range Vlasov equation can be found in [14, 37]. In substance, one assumes that the incident field exhibits a stationary statistics, except for small perturbations that depend on t and z. Note that any homogeneous stationary distribution, n^0_ω, is a solution of the Vlasov equation, that is, $\partial_z n^0_\omega = 0$. Perturbing this stationary solution according to $n_\omega(t,z) = n^0_\omega + \delta n_\omega(t,z)$, linearizing the Vlasov equation and then solving it by means of a Fourier-Laplace transform, $\tilde{\delta n}_\omega(\Omega,\lambda) = \int_0^\infty dz \int_{-\infty}^{+\infty} dt \exp(-\lambda z - i\Omega t)\, \delta n_\omega(t,z)$, gives the following MI growth-rate:

$$\lambda(\Omega) = -2\Delta\omega|\beta\Omega| + |\Omega|\sqrt{2\beta\gamma N_0 \tilde{R}(\Omega)}, \tag{12.29}$$

where we assumed an initial Lorentzian-shaped spectrum, $n^0_\omega = 2N_0\Delta\omega/(\omega^2 + (\Delta\omega)^2)$ [i.e., $N_0 = (2\pi)^{-1}\int n^0_\omega d\omega$]. The corresponding incoherent MI gain reads $g_{\mathrm{MI}}(\Omega) = 2\Re[\lambda(\Omega)]$. This expression of MI gain is formally analogous to the expression considered in the spatial case, see Eq. (12.28), in [37]. However, because of the causality property of the response function $R(t)$, its Fourier transform is complex, $\tilde{R}(\omega) = \tilde{U}(\omega) + ig(\omega)$. Recalling that $\tilde{U}(\omega)$ is even and $g(\omega)$ odd, the MI gain $g_{\mathrm{MI}}(\Omega)$ is always even, which means that incoherent MI is characterized by the growth of two symmetric side-bands. Another consequence of the fact that $\tilde{R}(\omega)$ is complex is that incoherent MI can also occur in the normal dispersion regime, i.e., for $\gamma\beta < 0$ [37].

Difference with Incoherent MI in Instantaneous Response Nonlinear Media We underline that incoherent MI in noninstantaneous nonlinear media is of fundamental different nature with respect to incoherent MI in *instantaneous* media [158]. In the limit $\tau_R \to 0$, incoherent MI can only take place if the spectral width of the incoherent wave is smaller than the MI frequency, $\Delta\omega \ll \omega_{\mathrm{MI}}$ [158]. This means that temporal modulations associated to MI are more rapid than the time correlation, $t_c \gg \tau_0$, i.e., MI modulations take place within each individual fluctuation of the incoherent wave. This is in contrast with the incoherent MI discussed here, in which the optimal MI frequency gets much smaller than the spectral bandwidth ($\omega_{\mathrm{MI}} \ll \Delta\omega$) as the nonlinearity becomes noninstantaneous, i.e., as τ_R increases. This means that incoherent MI manifests itself by a slow modulation of the whole random wave profile, i.e., the modulation frequency is smaller than the spectral bandwidth, $\omega_{\mathrm{MI}} \ll \Delta\omega$. This feature has been confirmed by the numerical simulations of the NLS Eq. (12.16) and the corresponding long-range Vlasov Eq. (12.24) in [14].

12.4.2 Incoherent Solitons in Normal Dispersion

The study of the existence of incoherent soliton states in the temporal domain revealed an unexpected remarkable result [156]. In contrast to a usual soliton, which is known to require a focusing nonlinearity with anomalous dispersion, a highly non-instantaneous nonlinear response leads to incoherent soliton structures which require the inverted situation. In the focusing regime (and anomalous dispersion) the incoherent wave-packet experiences an unlimited spreading, whereas in the defocusing regime (still with anomalous dispersion) the incoherent wave-packet exhibits a self-trapping [156]. This remarkable result has been demonstrated by means of numerical simulations of both NLS and Vlasov equations, and a quantitative agreement has been obtained between them without adjustable parameters. We refer the reader to [156] for details concerning the numerical simulations.

The unexpected existence of localized soliton states in the defocusing regime is explained in detail by the long-range Vlasov Eqs. (12.24)–(12.26). The Vlasov simulations reveal indeed that, after a transient, the wave-packet adopts an invariant profile characterized by a linear spectral shift, which in turn induces a constant IS acceleration (parabolic trajectory) in the temporal domain. More specifically, let us denote by α_0 the soliton velocity in frequency space. As discussed above through Eq. (12.27), the momentum evolves linearly as $\mathcal{P}(z) = \mathcal{N}\alpha_0 z$. Then Eq. (12.28) explicitly shows that, when combined with group-velocity dispersion, this linear spectral shift induces an acceleration of the incoherent soliton in the temporal domain given by Eq. (12.28)

$$\mathcal{T}(z) = \mathcal{T}(0) + \beta\mathcal{N}\alpha_0 z^2. \tag{12.30}$$

Both phenomena of linear spectral shift and constant acceleration of the incoherent soliton are clearly visible in the numerical simulations, as illustrated in Figure 12.14.

- ***Temporal vs Spatial Incoherent Solitons*** To discuss the mechanism underlying the formation of the incoherent soliton, it is instructive to comment first on an analogy with a *nonlocal spatial response.* Contrary to temporal effects, nonlocal spatial effects are not constrained by the causality condition, so that the spatial response function $U(x)$ is even, as well as the self-consistent potential $V(x) = -\gamma U * N$. Then in the focusing regime ($\gamma > 0$), the optical beam induces an attractive potential $V(x) < 0$, so that the beam is guided by its own induced potential. Conversely, in the defocusing regime ($\gamma < 0$) the repelling potential leads to the expected beam spreading, see Figures 12.15 (a) and (b) [22].
- ***Vlasov Approach: Noninertial Reference Frame*** As a result of the causality property of $R(t)$, the self-consistent potential $V(t)$ is shifted toward $t < 0$ in the temporal domain (see Figures 12.15 (c) and (d). Moreover, as commented above through Figure 12.14, the spectral-shift of the wave-packet, with spectral velocity α, leads to a constant acceleration of the IS. It thus proves convenient to study the dynamics of the wave-packet in its own accelerating reference frame, $\xi = z$, $\tau = t - \alpha\beta z^2$, $\Omega = \omega - \alpha z$. In this non-inertial reference frame the Vlasov Eq. (12.24) reads $\partial_\xi n_\Omega(\tau,\xi) + 2\beta\Omega\partial_\tau n_\Omega(\tau,\xi) - \partial_\tau V_{\text{eff}}(\tau,\xi)\,\partial_\Omega n_\Omega(\tau,\xi) = 0$. This equation remarkably reveals the existence of an effective self-consistent potential

$$V_{\text{eff}}(\tau,\xi) = V(\tau,\xi) + \alpha\tau, \tag{12.31}$$

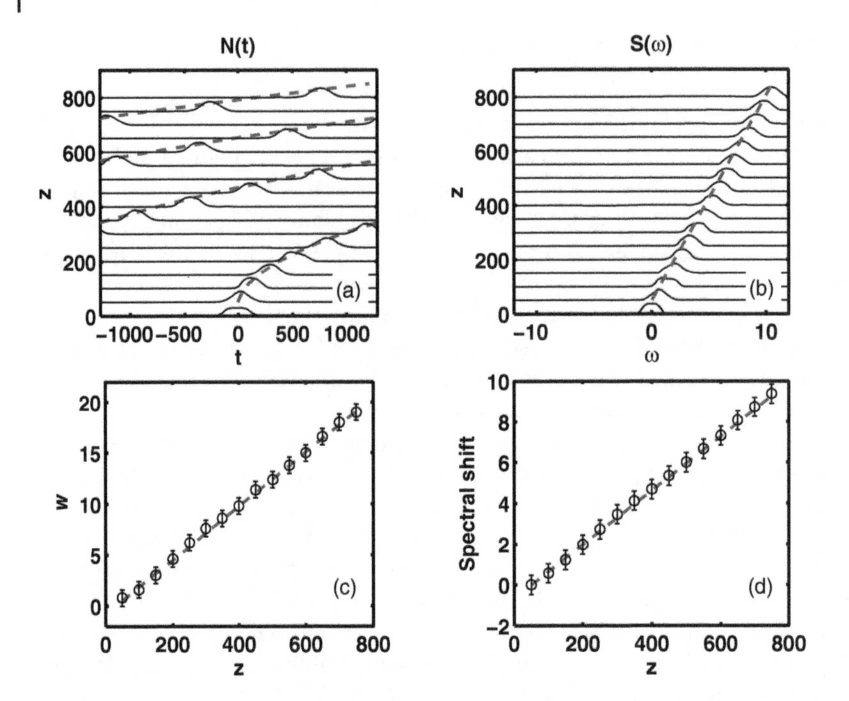

Figure 12.14 Spatial and temporal behaviors of temporal incoherent solitons. The soliton is characterized by a constant acceleration: Parabolic trajectory of the intensity profile $N(t, z) = (2\pi)^{-1} \int n_\omega(t, z) d\omega$ (a), and evolution of the spectral profile $S(\omega, z) = \int n_\omega(t, z) dt$ (b). The linear increase of the incoherent soliton velocity w (constant acceleration) (c), results from the linear spectral shift of the incoherent soliton (d). *Source*: Michel 2012 [156]. Reproduced with permission of American Physical Society.

where $V(\tau, \xi) = -\gamma \int_{-\infty}^{+\infty} R(\tau - \tau') N(\tau', \xi) \, d\tau'$. The linear part of the potential in Eq. (12.31) finds its origin in the fictitious force which results from the non-inertial nature of the reference frame. It is this fictitious force which prevents the IS structure from dispersing toward the direction of increasing τ. Note that this force is analogous to the effective gravity mimicked by an elevator, an analogy that was commented on in [159].

This fictitious force due to the accelerating reference frame explains both phenomena of self-trapping with a defocusing nonlinearity, as well as the inhibition of self-trapping with a focusing nonlinearity. Let us first discuss the defocusing regime. Recalling that the potential is induced by the wave-packet itself, an IS can only form provided that *the self-induced potential* $V_{\rm eff}(\tau)$ has a local minimum at the pulse center, i.e., at $\tau = 0$ in the accelerating reference frame of the IS. Contrary to the spatial case (Figure 12.15 (a)), it seems that this condition cannot be satisfied in the temporal case, since the causality condition shifts the potentials toward $\tau < 0$. However, in the defocusing regime, a local minimum can be restored at $\tau = 0$ thanks to the fictitious force due to the non-inertial reference frame, as illustrated in Figure 12.15 (d). More precisely, one can Taylor expand the effective potential $V_{\rm eff}(\tau) = a + (b + \alpha)\tau + c\tau^2 + \mathcal{O}(\tau^3)$ at $\tau = 0$, where $b < 0$ in the defocusing regime and $c > 0$ if the nonlinear response is slow

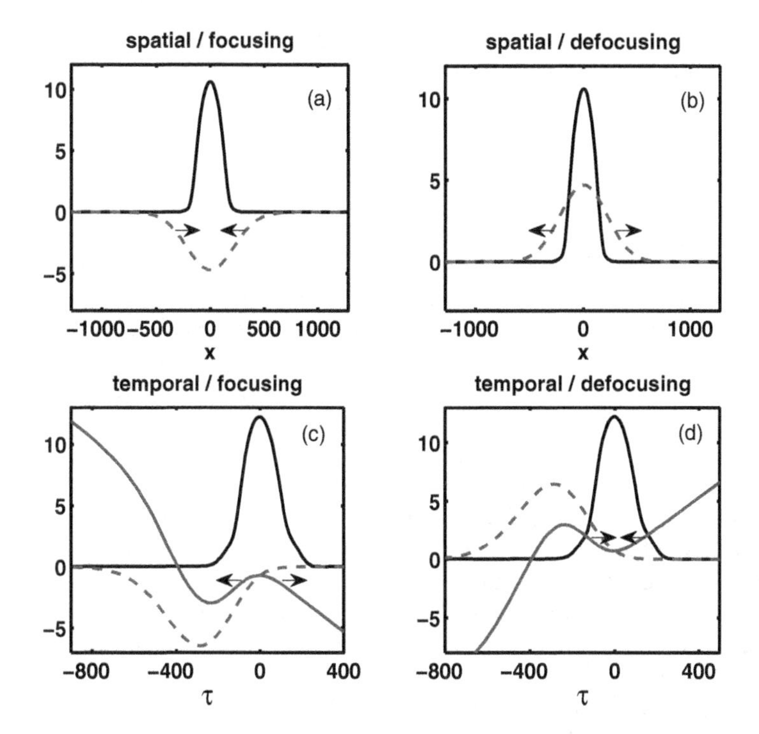

Figure 12.15 Mechanism underlying the formation of temporal incoherent solitons: Intensity profile $N(x)$ (continuous dark line) and corresponding self-consistent potential $V(x) = -\gamma U * N$ (dashed line), in the case of a spatial nonlocal nonlinearity in the focusing (a), and defocusing (b), regimes. Intensity profile $N(\tau)$ obtained by integrating numerically the Vlasov equation (continuous dark line), corresponding self-consistent potential $V(\tau) = -\gamma R * N$ (dashed line), and effective potential $V_{\text{eff}}(\tau)$ [Eq. (12.31)] (continuous line) in the accelerating reference frame, in the case of a temporal noninstantaneous nonlinearity in the focusing (c), and defocusing (d), regimes. The arrows indicate the "particle motions" in the effective self-consistent potentials $V_{\text{eff}}(\tau)$: The non-inertial fictitious force inhibits (c) (induces (d)) the self-trapping in the defocusing (focusing) regime. *Source*: Michel 2012 [156]. Reproduced with permission of American Physical Society.

enough (see Figure 12.15 (d)). In these conditions the particular choice $\alpha_0 = -b$ guarantees that $V_{\text{eff}}(\tau)$ has a local minimum at $\tau = 0$ (see Figure 12.15 (d)). In other words, the system spontaneously selects the amount of spectral shift, $\alpha_0 = -\partial_\tau V|_{\tau=0}$, and hence the amount of soliton acceleration, $2\beta\alpha_0$, in such a way that the effective self-consistent potential $V_{\text{eff}}(\tau)$ admits a local minimum at $\tau = 0$. This is confirmed by the numerical simulations of the Vlasov equation reported in Figure 12.15 (d) (for more details see [156]).

Let us now discuss the focusing regime, which is characterized by a redshift of the wave-packet, $\alpha < 0$. Following the same reasoning as above and remarking that we now have $b > 0$ and $c < 0$, the choice $\alpha_0 = -b$ still leads to an extremum of $V_{\text{eff}}(\tau)$ at $\tau = 0$. However, contrary to the defocusing regime, this extremum refers to a local maximum, as illustrated in Figure 12.15 (c). Actually, in the focusing regime, there is no value of α such that $V_{\text{eff}}(\tau)$ has a local minimum at $\tau = 0$. The local maximum around $\tau = 0$ then plays the role of a repelling potential, which explains the temporal broadening of the incoherent pulse: the "unstable particles" located near by $\tau = 0$ are either attracted toward the local minimum at $\tau < 0$, or either pushed toward $\tau > 0$ by the non-inertial force (see Figure 12.15 (c)). For more details on the dynamics and the interaction of temporal incoherent solitons, we refer the reader to [156, 160].

12.5 Conclusion

In this section we briefly comment on some open interesting issues related to optical wave turbulence in fibers. An interesting problem concerns a proper description of the emergence of extreme rogue waves from a turbulent environment. As discussed in Chapter 10 in this book, a rather commonly accepted opinion is that RWs can be conveniently interpreted in the light of exact analytical solutions of integrable nonlinear wave equations, the so-called Akhmediev breathers, or more specifically their limiting cases of infinite spatial and temporal periods, the rational soliton solutions, such as Peregrine and higher-order solutions of the integrable 1D NLSE – see the recent reviews [30, 31]. Rational soliton solutions can be regarded as a coherent and deterministic approach to the understanding of RW phenomena. On the other hand, RWs are known to spontaneously emerge from an incoherent turbulent state [32, 98, 161–164]. This raises a difficult problem, since the description of the turbulent system requires a statistical WT approach, whereas rational soliton solutions are inherently coherent deterministic structures. This problem was addressed in the optical fiber context in [135, 136] by considering a specific NLSE model that exhibits a quasi-soliton turbulence scenario, a feature that can be interpreted by analogy with wave condensation, see Section 12.2.3.2. It was shown that the deterministic description of rogue wave events in terms of rational soliton solutions is not inconsistent with the corresponding statistical WT description of the turbulent system [136]. It is important to stress that the emergence of RW events was shown to solely occur close to the transition to (quasi-)soliton condensation. From a different perspective, the fluctuations of the condensate fraction in 2D wave condensation have recently been computed theoretically, revealing that large fluctuations solely occur near by the transition to condensation, while they are significantly quenched in the strongly condensed Bogoliubov regime (small "temperature"), and almost completely suppressed in the weakly nonlinear turbulent regime (high "temperature"). This result is consistent with the general idea that nearby second-order phase-transitions, physical systems are inherently sensitive to perturbations and thus exhibit large fluctuations. One can then address a possible alternative point of view on the question of the spontaneous emergence of rogue waves from a conservative turbulent environment: Is it possible to interpret the sporadic emergence of RW events as the natural large fluctuations inherent to the phase transition to soliton condensation? This issue may pave the way for a statistical mechanics approach based on the idea of scaling and universal theory of critical phenomena to the description of RWs.

An other interesting problem concerns a proper statistical description of nonlinear partially polarized optical waves in the framework of the WT theory. In particular, a remarkable phenomenon of nonlinear repolarization without loss of energy is discussed in this book in Chapter 8. This effect of "polarization attraction" has been shown to exhibit different properties depending on the experimental configuration, and on the type of the optical fiber (isotropic, highly birefringent, as well as randomly birefringent fibers), see Chapter 8. Although significant progress has been achieved in order to interpret this phenomenon on the basis of Hamiltonian singularities (see, e.g., [165]), so far, no statistical description has been developed to describe polarization attraction with partially polarized waves. This problem raises important difficulties, since the phenomenon of repolarization is inherently associated to the emergence of a mutual correlation between orthogonal polarization components, while the basic WT theory does not account for such phase-correlation effects. It would be interesting to construct a generalized WT formulation of random nonlinear waves that would account for a possible mutual correlation between distinct incoherent wave components.

Finally, in this chapter we pointed out several remarkable properties of turbulent wave systems that exhibit long-range interactions, by complete analogy with collective behaviors in self-gravitational systems (e.g., formation of galaxies) or 2D geophysical fluids (e.g., Jupiter red hot spot) [56]. Recently, experiments in the spatial domain have been performed in solutions of graphene nano-flakes in which the range of nonlocal interaction can be tuned over more than one order of magnitude. A different experimental setting for the study of long-range wave turbulence in the temporal domain could be hollow-core fibers (see Chapter 3 in this book). Indeed, one may exploit the easily tailorable non-instantaneous response via the well-known Raman effect, as well as other recently investigated mechanisms involving liquid-cores or photo-ionizable noble gases and surface plasmon polaritons [166–168].

Acknowledgments

The authors are grateful to many collaborators for fruitful discussions: P. Aschieri, S.A. Babin, J. Barré, B. Barviau, P. Béjot, F. Biancalana, S. Coen, C. Conti, D. Faccio, C. Finot, M. Guasoni, R. Kaiser, B. Kibler, M. Haelterman, K. Hammani, H.R. Jauslin, M. Lisak, C. Michel, C. Montes, S. Nazarenko, M. Onorato, S. Pitois, E.V. Podivilov, M. Segev, S. Trillo, S. Wabnitz, V.E. Zakharov. The authors also acknowledge P. Suret, S. Randoux, D. Churkin and S. Turitsyn for their deep insights and valuable contributions in the understanding of optical turbulence, in particular in dissipative laser systems or near by the limit of integrability. A.P. is especially grateful to S. Rica for several illuminating discussions on the wave turbulence theory, and to T. Dauxois and S. Ruffo for their introduction to the physics of long-range interactions.

A.P. and G.M. acknowledge support from the French National Research Agency (ANR-12-BS04-0011 OPTIROC), as well as from the Labex ACTION (ANR-11-LABX-01-01) program. A.P. also acknowledges support from the European Research Council under the European Community's Seventh Framework Programme (FP7/20072013 Grant Agreement No. 306633, PETAL project).

References

1 Mitchell, M., Chen, Z., Shih, M., and Segev, M. (1996) Self-trapping of partially spatially incoherent light. *Physical Review Letters*, 77, 490.

2 Mitchell, M. and Segev, M. (1997) Self-trapping of incoherent white light. *Nature (London)*, **387**, 880.

3 Segev, M. and Christodoulides, D. (2001) Incoherent solitons. In Eds. S. Trillo and W. Torruellas, Spatial Solitons, Springer; Berlin.
4 Kivshar, Y. and Agrawal, G. (2003) *Optical Solitons: From Fibers to Photonic Crystals*, Academic Press, New York.
5 Pasmanik, G.A. (1974) Self-interaction of incoherent light beams. *Soviet Physics JETP*, **39**, 234.
6 Mitchell, M., Segev, Coskun, M.T., and Christodoulides, D. (1997) Theory of self-trapped spatially incoherent light beams. *Physical Review Letters*, **79**, 4990.
7 Christodoulides, D., Coskun, T., Mitchell, M., and Segev, M. (1997) Theory of incoherent self-focusing in biased photorefractive media. *Physical Review Letters*, **78**, 646.
8 Hall, B., Lisak, M., Anderson, D., Fedele, R., and Semenov, V.E. (2002) Statistical theory for incoherent light propagation in nonlinear media. *Physical Review E*, **65**, 035 602.
9 Christodoulides, D.N., Eugenieva, E.D., Coskun, T.H., Segev, M., and Mitchell, M. (2001) Equivalence of three approaches describing partially incoherent wave propagation in inertial nonlinear media. *Physical Review E*, **63**, 035 601.
10 Lisak, M., Helczynski, L., and Anderson, D. (2003) Relation between different formalisms describing partially incoherent wave propagation in nonlinear optical media. *Optics Communications*, **220**, 321.
11 Christodoulides, D.N., Coskun, T.H., Mitchell, M., Chen, Z., and Segev, M. (1998) Theory of incoherent dark solitons. *Physical Review Letters*, **80**, 5113.
12 Chen, Z., Mitchell, M., Segev, M., Coskun, T.H., and Christodoulides, D.N. (1998) Self-trapping of dark incoherent light beams. *Science*, **280**, 889.
13 Soljacic, M., Segev, M., Coskun, T., Christodoulides, D., and Vishwanath, A. (2000) Modulation instability of incoherent beams in noninstantaneous nonlinear media. *Physical Review Letters*, **84**, 467.
14 Kibler, B., Michel, C., Garnier, J., and Picozzi, A. (2012) Temporal dynamics of incoherent waves in noninstantaneous response nonlinear Kerr media. *Optics Letters*, **37**, 2472.
15 Buljan, H., Cohen, O., Fleischer, J.W., *et al.* (2004) Random-phase solitons in nonlinear periodic lattices. *Physical Review Letters*, **92**, 223 901.
16 Cohen, O., Bartal, G., Buljan, H., *et al.* (2005) Observation of random phase lattice solitons. *Nature (London)*, **433**, 500.
17 Picozzi, A. and Haelterman, M. (2001) Parametric three-wave soliton generated from incoherent light. *Physical Review Letters*, **86**, 2010–2013.
18 Picozzi, A., Haelterman, M., Pitois, S., and Millot, G. (2004) Incoherent solitons in instantaneous response nonlinear media. *Physical Review Letters*, **92**, 143 906.
19 Peccianti, M. and Assanto, G. (2001) Incoherent spatial solitary waves in nematic liquid crystals. *Optics Letters*, **26** (22), 1791–1793.
20 Cohen, O., Buljan, H., Schwartz, T., Fleischer, J.W., and Segev, M. (2006) Incoherent solitons in instantaneous nonlocal nonlinear media. *Physical Review E*, **73**, 015 601.
21 Rotschild, C., Schwartz, T., Cohen, O., and Segev, M. (2008) Incoherent spatial solitons in effectively-instantaneous nonlocal nonlinear media. *Nature Photonics*, **2**, 371.
22 Picozzi, A. and Garnier, J. (2011) Incoherent soliton turbulence in nonlocal nonlinear media. *Physical Review Letters*, **107** (23), 233 901.

23 Picozzi, A., Pitois, S., and Millot, G. (2008) Spectral incoherent solitons: A localized soliton behavior in frequency space. *Physical Review Letters*, **101**, 093 901.

24 Kibler, B., Michel, C., Kudlinski, A., *et al.* (2011) Emergence of spectral incoherent solitons through supercontinuum generation in a photonic crystal fiber. *Physical Review E*, **84**, 066 605.

25 Dudley, J.M. and Taylor, J.R. (2010) *Supercontinuum Generation in Optical Fibers*, Cambridge University Press, Cambridge.

26 Couairon, A. and Mysyrowicz, A. (2007) Femtosecond filamentation in transparent media. *Physics Reports*, **441** (2), 47–189.

27 Wiersma, D.S. (2008) The physics and applications of random lasers. *Nature Physics*, **4** (5), 359–367.

28 Akhmediev, N. and Pelinovsky, E. (2010) Editorial–introductory remarks on discussion & debate: Rogue waves–towards a unifying concept. *The European Physical Journal Special Topics*, **185** (1), 1–4.

29 Ruban, V., Kodama, Y., Ruderman, M., *et al.* (2010) Rogue waves–towards a unifying concept: Discussions and debates. *The European Physical Journal Special Topics*, **185** (1), 5–15.

30 Onorato, M., Residori, S., Bortolozzo, U., Montina, A., and Arecchi, F. (2013) Rogue waves and their generating mechanisms in different physical contexts. *Physics Reports*.

31 Dudley, J.M., Dias, F., Erkintalo, M., and Genty, G. (2014) Instabilities, breathers and rogue waves in optics. *Nature Photonics*, **8** (10), 755–764.

32 Walczak, P., Randoux, S., and Suret, P. (2015) Optical rogue waves in integrable turbulence. *Physical Review Letters*, **114**, 143 903.

33 Zakharov, V.E., L'vov, V.S., and Falkovich, G. (1992) *Kolmogorov Spectra of Turbulence I*, Springer; Berlin.

34 Frisch, U. (1995) *Turbulence. The Legacy of A. N. Kolmogorov*, Cambridge: Cambridge University Press.

35 Newell, A.C. and Rumpf, R. (2011) Wave turbulence. *Annual Review of Fluid Mechanics*, **43**, 59–78.

36 Nazarenko, S. (2011) *Wave Trubulence*, Lecture Notes in Physics 825, Springer, Berlin.

37 Picozzi, A., Garnier, J., Hansson, T., *et al.* (2014) Optical wave turbulence: Towards a unified nonequilibrium thermodynamic formulation of statistical nonlinear optics. *Physics Reports*, **542** (1), 1–132.

38 Dyachenko, S., Newell, A.C., Pushkarev, A., and Zakharov, V.E. (1992) Optical turbulence: weak turbulence, condensates and collapsing filaments in the nonlinear Schrödinger equation. *Physica D*, **57**, 96.

39 Zakharov, V., Dias, F., and Pushkarev, A. (2004) One-dimensional wave turbulence. *Physics Reports*, **398**, 1.

40 Benney, D.J. and Saffman, P.G. (1966) Nonlinear interactions of random waves in dispersive medium. *Proceedings of the Royal Society of London Ser. A*, **289**, 301.

41 Benney, D.J. and Newell, A. (1969) Random wave closure. *Studies in Applied Mathematics*, **48**, 29.

42 Newell, A.C., Nazarenko, S., and Biven, L. (2001) Wave turbulence and intermittency. *Physica D*, **152-153**, 520–550.

43 Rumpf, R. and Newell, A.C. (2001) Coherent structures and entropy in constrained, modulationally unstable, nonintegrable systems. *Physical Review Letters*, **87**, 054 102.

44 Rumpf, R. and Newell, A.C. (2003) Localization and coherence in nonintegrable systems. *Physica D*, **184**, 162.
45 Rumpf, B. (2004) Simple statistical explanation for the localization of energy in nonlinear lattices with two conserved quantities. *Physical Review E*, **69**, 016 618.
46 Leuzzi, L., Conti, C., Folli, V., Angelani, L., and Ruocco, G. (2009) Phase diagram and complexity of mode-locked lasers: From order to disorder. *Physical Review Letters*, **102**, 083 901.
47 Musher, S., Rubenchik, A., and Zakharov, V. (1995) Weak Langmuir turbulence. *Physics Reports*, **252**, 177.
48 Michel, C., Kibler, B., and Picozzi, A. (2011) Discrete spectral incoherent solitons in nonlinear media with noninstantaneous response. *Physical Review A*, **83**, 023 806.
49 Garnier, J., Xu, G., Trillo, S., and Picozzi, A. (2013) Incoherent dispersive shocks in the spectral evolution of random waves. *Physical Review Letters*, **111** (11), 113 902.
50 Zakharov, V.E., Musher, S.L., and Rubenchik, A.M. (1985) Hamiltonian approach to the description of non-linear plasma phenomena. *Physics Reports*, **129**, 285.
51 Hasegawa, A. (1975) Dynamics of an ensemble of plane waves in nonlinear dispersive media. *Physics of Fluids*, **18**, 77.
52 Onorato, M., Osborne, A., Fedele, R., and Serio, M. (2003) Landau damping and coherent structures in narrow-banded 1 + 1 deep water gravity waves. *Physical Review E*, **67**, 046 305.
53 Garnier, J., Ayanides, J.P., and Morice, O. (2003) Propagation of partially coherent light with the Maxwell-Debye equation. *Journal of the Optical Society of America B*, **20**, 1409.
54 Garnier, J. and Picozzi, A. (2010) Unified kinetic formulation of incoherent waves propagating in nonlinear media with noninstantaneous response. *Physical Review A*, **81**, 033 831.
55 Dylov, D. and Fleischer, J. (2008) Observation of all-optical bump-on-tail instability. *Physical Review Letters*, **100**, 103 903.
56 Campa, A., Dauxois, T., Fanelli, D., and Ruffo, S. (2014) *Physics of Long-Range Interaction Systems*, Oxford University Press, University.
57 Dudley, J.M., Genty, G., and Coen, S. (2006) Supercontinuum generation in photonic crystal fiber. *Review of Modern Physics*, **78**, 1135.
58 Skryabin, D. and Gorbach, A. (2010) Colloquium: Looking at a soliton through the prism of optical supercontinuum. *Review of Modern Physics*, **82**, 1287.
59 Kolesik, M., Wright, E.M., and Moloney, J.V. (2004) Dynamic nonlinear X-waves for femtosecond pulse propagation in water. *Physical Review Letters*, **92**, 253 901.
60 Kolesik, M., Tartara, L., and Moloney, J.V. (2010) Effective three-wave-mixing picture and first born approximation for femtosecond supercontinua from microstructured fibers. *Physical Review A*, **82**, 045 802.
61 Genty, G., Surakka, M., Turunen, J., and Friberg, A. (2011) Complete characterization of supercontinuum coherence. *Journal of the Optical Society of America B*, **28**, 2301.
62 Erkintalo, M., Surakka, M., Turunen, J., Friberg, A.T., and Genty, G. (2012) Coherent-mode representation of supercontinuum. *Optics Letters*, **37** (2), 169–171.

63 Korhonen, M., Friberg, A.T., Turunen, J., and Genty, G. (2013) Elementary field representation of supercontinuum. *Journal of the Optical Society of America B*, **30** (1), 21–26.

64 Mussot, A., Beaugeois, M., Bouazaoui, M., and Sylvestre, T. (2007) Tailoring cw supercontinuum generation in microstructured fibers with two-zero dispersion wavelengths. *Optics Express*, **15** (18), 11 553–11 563.

65 Martin-Lopez, S., Abrardi, L., Corredera, P., Herraez, M.G., and Mussot, A. (2008) Spectrally-bounded continuous-wave supercontinuum generation in a fiber with two zero-dispersion wavelengths. *Optics Express*, **16** (9), 6745–6755.

66 Cumberland, B.A., Travers, J.C., Popov, S.V., and Taylor, J.R. (2008) 29 w high power cw supercontinuum source. *Optics Express*, **16** (8), 5954–5962.

67 Wadsworth, W., Joly, N., Knight, J., *et al.* (2004) Supercontinuum and four-wave mixing with q-switched pulses in endlessly single-mode photonic crystal fibres. *Optics Express*, **12** (2), 299–309.

68 Barviau, B., Kibler, B., Coen, S., and Picozzi, A. (2008) Towards a thermodynamic description of supercontinuum generation. *Optics Letters*, **33**, 2833.

69 Barviau, B., Kibler, B., Kudlinski, A., *et al.* (2009) Experimental signature of optical wave thermalization through supercontinuum generation in photonic crystal fiber. *Optics Express*, **17**, 7392.

70 Barviau, B., Kibler, B., and Picozzi, A. (2009) Wave turbulence description of supercontinuum generation: influence of self-steepening and higher-order dispersion. *Physical Review A*, **79**, 063 840.

71 Agrawal, G. (2012) *Nonlinear Fiber Optics*, 5th Ed., Academic Press, New York.

72 Soh, D.B.S., Koplow, J.P., Moore, S.W., Schroder, K.L., and Hsu, W.L. (2010) The effect of dispersion on spectral broadening of incoherent continuous-wave light in optical fibers. *Optics Express*, **18** (21), 22 393–22 405.

73 Suret, P., Picozzi, A., and Randoux, S. (2011) Wave turbulence in integrable systems: Nonlinear propagation of incoherent optical waves in single-mode fibers. *Optics Express*, **19**, 17 852.

74 Suret, P., Randoux, S., Jauslin, H., and Picozzi, A. (2010) Anomalous thermalization of nonlinear wave systems. *Physical Review Letters*, **104**, 054 101.

75 Michel, C., Suret, P., Randoux, S., Jauslin, H., and Picozzi, A. (2010) Influence of third-order dispersion on the propagation of incoherent light in optical fibers. *Optics Letters*, **35**, 2367.

76 Pitois, S., Lagrange, S., Jauslin, H.R., and Picozzi, A. (2006) Velocity locking of incoherent nonlinear wave packets. *Physical Review Letters*, **97**, 033 902.

77 Landau, L. and Lifchitz, E. (1980) *Statistical Physics*, Pergamon Press, New York, Part 1.

78 Gallavotti, G.E. (2007) *The Fermi-Pasta-Ulam Problem: A Status Report*, Lecture Notes in Physics, Springer.

79 Mussot, A., Kudlinski, A., Droques, M., Szriftgiser, P., and Akhmediev, N. (2014) Fermi-Pasta-Ulam recurrence in nonlinear fiber optics: The role of reversible and irreversible losses. *Physical Review X*, **4**, 011 054.

80 Onorato, M., Vozella, L., Proment, D., and Lvov, Y.V. (2015) Route to thermalization in the α-fermi–pasta–ulam system. *Proceedings of the National Academy of Sciences*, **112** (14), 4208–4213.

81 Lake, B., Yuen, H., Rundgaldier, H., and Ferguson, W. (1977) Nonlinear deep-water waves: theory and experiment. part 2. evolution of a continuous wave train. *Journal of Fluid Mechanics*, **83**, 49.

82 Wu, M. and Patton, C. (2007) Experimental observation of Fermi-Pasta-Ulam recurrence in a nonlinear feedback ring system. *Physical Review Letters*, **98**, 047 202.

83 Simaeys, G., Emplit, P., and Haelterman, M. (2001) Experimental demonstration of the Fermi-Pasta-Ulam recurrence in a modulationally unstable optical wave. *Physical Review Letters*, **87**, 033 902.

84 Cambournac, C., Maillotte, H., Lantz, E., Dudley, J.M., and Chauvet, M. (2002) Spatiotemporal behavior of periodic arrays of spatial solitons in a planar waveguide with relaxing Kerr nonlinearity. *Journal of the Optical Society of America B*, **19** (3), 574–585.

85 Hammani, K., Wetzel, B., Kibler, B., *et al.* (2011) Spectral dynamics of modulation instability described using Akhmediev breather theory. *Optics Letters*, **36** (11), 2140–2142.

86 Barviau, B., Garnier, J., Xu, G., *et al.* (2013) Truncated thermalization of incoherent optical waves through supercontinuum generation in photonic crystal fibers. *Physical Review A*, **87** (3), 035 803.

87 Travers, J.C., Rulkov, A.B., Cumberland, B.A., Popov, S.V., and Taylor, J.R. (2008) Visible supercontinuum generation in photonic crystal fibers with a 400w continuous wave fiber laser. *Optics Express*, **16** (19), 14 435–14 447.

88 Michel, C., Garnier, J., Suret, P., Randoux, S., and Picozzi, A. (2011) Kinetic description of random optical waves and anomalous thermalization of a nearly integrable wave system. *Letters In Mathematics and Physics*, **96** (1), 415–447.

89 Ablowitz, M. and Segur, H. (1981) *Solitons and the Inverse Scattering Transform*, SIAM, Society for Industrial and Applied Mathematics, 1981.

90 Ghofraniha, N., Conti, C., Ruocco, G., and Trillo, S. (2007) Shocks in nonlocal media. *Physical Review Letters*, **99**, 043 903.

91 Conti, C., Fratalocchi, A., Peccianti, M., Ruocco, G., and Trillo, S. (2009) Observation of a gradient catastrophe generating solitons. *Physical Review Letters*, **102**, 083 902.

92 Ghofraniha, N., Gentilini, S., Folli, V., DelRe, E., and Conti, C. (2012) Shock waves in disordered media. *Physical Review Letters*, **109**, 243 902.

93 El, G. and Kamchatnov, A. (2005) Kinetic equation for a dense soliton gas. *Physical Review Letters*, **95**, 204 101.

94 Zakharov, V. (2009) Turbulence in integrable systems. *Studies in Applied Mathematics*, **122**, 219.

95 Barviau, B., Randoux, S., and Suret, P. (2006) Spectral broadening of a multimode continuous-wave optical field propagating in the normal dispersion regime of a fiber. *Optics Letters*, **31** (11), 1696–1698.

96 Janssen, P. (2003) Nonlinear four-wave interactions and freak waves. *Journal of Physical Oceanography*, **33**, 864.

97 Annenkov, S.Y. and Shrira, V.I. (2006) Role of non-resonant interactions in the evolution of nonlinear random water wave fields. *Journal of Fluid Mechanics*, **561**, 181–207.

98 Randoux, S., Walczak, P., Onorato, M., and Suret, P. (2014) Intermittency in integrable turbulence. *Physical Review Letters*, **113**, 113 902.

99 Costa, A., Osborne, A.R., Resio, D.T., Alessio, S., Chrivì, E., Saggese, E., Bellomo, K., and Long, C.E. (2014) Soliton turbulence in shallow water ocean surface waves. *Physical Review Letters*, **113**, 108 501.

100 Derevyanko, S.A. and Prilepsky, J.E. (2008) Random input problem for the nonlinear Schrödinger equation. *Physical Review E*, **78**, 046 610.

101 Derevyanko, S.A. (2011) Appearance of bound states in random potentials with applications to soliton theory. *Physical Review E*, **84**, 016 601.

102 Bromberg, Y., Lahini, Y., Small, E., and Silberberg, Y. (2010) Hanbury Brown and Twiss interferometry with interacting photons. *Nature Photonics*, **4**, 721.

103 Derevyanko, S. and Small, E. (2012) Nonlinear propagation of an optical speckle field. *Physical Review A*, **85**, 053 816.

104 Davis, M., Morgan, S., and Burnett, K. (2001) Simulations of Bose fields at finite temperature. *Physical Review Letters*, **87**, 160 402.

105 Connaughton, C., Josserand, C., Picozzi, A., Pomeau, Y., and Rica, S. (2005) Condensation of classical nonlinear waves. *Physical Review Letters*, **95**, 263 901.

106 During, G., Picozzi, A., and Rica, S. (2009) Breakdown of weak-turbulence and nonlinear wave condensation. *Physica D*, **238**, 1524.

107 Aschieri, P., Garnier, J., Michel, C., Doya, V., and Picozzi, A. (2011) Condensation and thermalization of classsical optical waves in a waveguide. *Physical Review A*, **83**, 033 838.

108 Picozzi, A. and Rica, S. (2012) Condensation of classical optical waves beyond the cubic nonlinear Schrödinger equation. *Optics Communications*, **285**, 5440.

109 Conti, C., Leonetti, M., Fratalocchi, A., Angelani, L., and Ruocco, G. (2008) Condensation in disordered lasers: Theory, 3d+1 simulations, and experiments. *Physical Review Letters*, **101**, 143 901.

110 Turitsyna, E., Falkovich, G., Mezentsev, V., and Turitsyn, S. (2009) Optical turbulence and spectral condensate in long-fiber lasers. *Physical Review A*, **80**, 031 804.

111 Weill, R., Fischer, B., and Gat, O. (2010) Light-mode condensation in actively-mode-locked lasers. *Physical Review Letters*, **104** (17), 173 901.

112 Klaers, J., Schmitt, J., Vewinger, F., and Weitz, M. (2010) Bose-Einstein condensation of photons in an optical microcavity. *Nature*, **468**, 545.

113 Michel, C., Haelterman, M., Suret, P., *et al.* (2011) Thermalization and condensation in an incoherently pumped passive optical cavity. *Physical Review A*, **84**, 033 848.

114 Turitsyn, S.K., Babin, S.A., Turitsyna, E.G., *et al.* (2013) *Optical Wave Turbulence*, Eds. V. Shira and S. Nazarenko, *Wave Turbulence*, World Scientific Series on Nonlinear Science Series A: Volume 83.

115 Turitsyna, E., Falkovich, G., El-Taher, A., *et al.* (2012) Optical turbulence and spectral condensate in long fibre lasers. *Proceedings of Royal Society A*, **468**, 2145.

116 Turitsyna, E., Smirnov, S., Sugavanam, S., *et al.* (2013) The laminar-turbulent transition in a fibre laser. *Nature Photonics*, **7** (10), 783–786.

117 Kirton, P. and Keeling, J. (2013) Nonequilibrium model of photon condensation. *Physical Review Letters*, **111**, 100 404.

118 Fischer, B. and Weill, R. (2012) When does single-mode lasing become a condensation phenomenon? *Optics Express*, **20** (24), 26 704–26 713.

119 Carusotto, I. and Ciuti, C. (2013) Quantum fluids of light. *Review of Modern Physics*, **85**, 299–366.

120 Weill, R., Levit, B., Bekker, A., Gat, O., and Fischer, B. (2010) Laser light condensate: experimental demonstration of light-mode condensation in actively mode locked laser. *Optics Express*, **18** (16), 16 520–16 525.

121 Gordon, A. and Fischer, B. (2002) Phase transition theory of many-mode ordering and pulse formation in lasers. *Physical Review Letters*, **89**, 103 901.

122 Weill, R., Rosen, A., Gordon, A., Gat, O., and Fischer, B. (2005) Critical behavior of light in mode-locked lasers. *Physical Review Letters*, **95**, 013 903.

123 Churkin, D.V., Kolokolov, I.V., Podivilov, E.V., *et al.* (2015) Wave kinetics of random fibre lasers. *Nature Communications*, **2**.

124 Babin, S.A., Churkin, D.V., Ismagulov, A.E., Kablukov, S.I., and Podivilov, E.V. (2007) Four-wave-mixing-induced turbulent spectral broadening in a long Raman fiber laser. *Journal of the Optical Society of America B*, **24**, 1729.

125 Dalloz, N., Randoux, S., and Suret, P. (2010) Influence of dispersion of fiber Bragg grating mirrors on formation of optical power spectrum in Raman fiber lasers. *Optics Letters*, **35**, 2505–2507.

126 Randoux, S., Dalloz, N., and Suret, P. (2011) Intracavity changes in the field statistics of Raman fiber lasers. *Optics Letters*, **36** (6), 790–792.

127 Turitsyna, E.G., Turitsyn, S.K., and Mezentsev, V.K. (2010) Numerical investigation of the impact of reflectors on spectral performance of Raman fibre laser. *Optics Express*, **18**, 4469.

128 Turitsyn, S.K., Babin, S.A., El-Taher, A.E., *et al.* (2010) Random distributed feedback fibre laser. *Nature Photonics*, **4**, 231.

129 Turitsyn, S.K., Babin, S.A., Churkin, D.V., *et al.* (2014) Random distributed feedback fibre lasers. *Physics Reports*, **542** (2), 133–193. Random Distributed Feedback Fibre Lasers.

130 L'vov, V.S., L'vov, Y., Newell, A.C., and Zakharov, V. (1997) Statistical description of acoustic turbulence. *Physical Review E*, **56**, 390–405.

131 Leo, F., Coen, S., Kockaert, P., *et al.* (2010) Temporal cavity solitons in one-dimensional Kerr media as bits in an all-optical buffer. *Nature Photonics*, **4**, 471.

132 Arecchi, T., Boccalettic, S., and Ramazza, P. (1999) Pattern formation and competition in nonlinear optics. *Physics Reports*, **318**, 1.

133 Klaers, J., Vewinger, F., and Weitz, M. (2010) Thermalization of a two-dimensional photonic gas in a white wall photon box. *Nature Physics*, **6** (7), 512–515.

134 Conforti, M., Mussot, A., Fatome, J., *et al.* (2015) Turbulent dynamics of an incoherently pumped passive optical fiber cavity: Quasisolitons, dispersive waves, and extreme events. *Physical Review A*, **91**, 023 823.

135 Hammani, K., Kibler, B., Finot, C., and Picozzi, A. (2010) Emergence of rogue waves from optical turbulence. *Physics Letters A*, **374**, 3585.

136 Kibler, B., Hammani, K., Finot, C., and Picozzi, A. (2011) Rogue waves, rational solitons and wave turbulence theory. *Physics Letters A*, **375**, 3149.

137 Xu, G., Garnier, J., Trillo, S., and Picozzi, A. (2013) Spectral dynamics of incoherent waves with a noninstantaneous nonlinear response. *Optics Letters*, **38** (16), 2972–2975.

138 Montes, C. (1979) Photon soliton and fine structure due to nonlinear compton scattering. *Physical Review A*, **20**, 1081.

139 Xu, G., Garnier, J., Conforti, M., and Picozzi, A. (2014) Generalized description of spectral incoherent solitons. *Optics Letters*, **39** (14), 4192–4195.

140 Musher, S., Rubenchik, A., and Zakharov, V. (1976) Weak Langmuir turbulence of an isothermal plasma. *JETP*, **42**, 80.

141 Xu, G., Garnier, J., Trillo, S., and Picozzi, A. (2014) Impact of self-steepening on incoherent dispersive spectral shocks and collapselike spectral singularities. *Physical Review A*, **90**, 013 828.

142 Whitham, G. (1974) *Linear and Nonlinear Waves*, Wiley, New York.

143 Gurevich, A. and Pitaevskii, L. (1974) Nonstationary structure of a collisionless shock wave. *Soviet Physics JETP*, **38**, 291.

144 Taylor, R., Baker, D., and Ikezi, H. (1970) Observation of collisionless electrostatic shocks. *Physical Review Letters*, **24**, 206.

145 Hoefer, M.A., Ablowitz, M.J., Coddington, I., *et al.* (2006) Dispersive and classical shock waves in Bose-Einstein condensates and gas dynamics. *Physical Review A*, **74**, 023 623.

146 Chang, J.J., Engels, P., and Hoefer, M.A. (2008) Formation of dispersive shock waves by merging and splitting Bose-Einstein condensates. *Physical Review Letters*, **101**, 170 404.

147 Wan, W., Jia, S., and Fleischer, J. (2007) Dispersive, superfluid-like shock waves in nonlinear optics. *Nature Physics*, **3**, 46.

148 Ghofraniha, N., Amato, L.S., Folli, V., *et al.* (2012) Measurement of scaling laws for shock waves in thermal nonlocal media. *Optics Letters*, **37** (12), 2325–2327.

149 Conforti, M., Baronio, F., and Trillo, S. (2013) Competing wave-breaking mechanisms in quadratic media. *Optics Letters*, **38** (10), 1648–1650.

150 Wiegmann, P. (2012) Nonlinear hydrodynamics and fractionally quantized solitons at the fractional quantum hall edge. *Physical Review Letters*, **108**, 206 810.

151 Mo, Y.C., Kishek, R.A., Feldman, D., *et al.* (2013) Experimental observations of soliton wave trains in electron beams. *Physical Review Letters*, **110**, 084 802.

152 Fratalocchi, A., Armaroli, A., and Trillo, S. (2011) Time-reversal focusing of an expanding soliton gas in disordered replicas. *Physical Review A*, **83**, 053 846.

153 Gentilini, S., Ghofraniha, N., DelRe, E., and Conti, C. (2013) Shock waves in thermal lensing. *Physical Review A*, **87**, 053 811.

154 Constantin, P., Lax, P., and Majda, A. (1985) A simple one-dimensional model for the three-dimensional vorticity equation. *Communications on Pure and Applied Mathematics*, **38**, 715–724.

155 Fokas, A. and Ablowitz, M. (1983) The inverse scattering transform for the Benjamin-ono equation – a pivot to multidimensional problems. *Studies in Applied Mathematics*, **68**, 1.

156 Michel, C., Kibler, B., Garnier, J., and Picozzi, A. (2012) Temporal incoherent solitons supported by a defocusing nonlinearity with anomalous dispersion. *Physical Review A*, **86** (4), 041 801.

157 Saleh, M.F., Chang, W., Travers, J.C., Russell, P.S.J., and Biancalana, F. (2012) Plasma-induced asymmetric self-phase modulation and modulational instability in gas-filled hollow-core photonic crystal fibers. *Physical Review Letters*, **109**, 113 902.

158 Sauter, A., Pitois, S., Millot, G., and Picozzi, A. (2005) Incoherent modulation instability in instantaneous nonlinear Kerr media. *Optics Letters*, **30**, 2143–2145.

159 Skryabin, D., Luan, F., Knight, J., and Russell, P. (2003) Soliton self-frequency shift cancellation in photonic crystal fibers. *Science*, **301**, 1705.

160 Xu, G., Garnier, J., and Picozzi, A. (2014) Spectral long-range interaction of temporal incoherent solitons. *Optics Letters*, **39** (3), 590–593.

161 Onorato, M., Osborne, A., Serio, M., *et al.* (2006) Extreme waves, modulational instability and second order theory: wave flume experiments on irregular waves. *European Journal of Mechanics-B/Fluids*, **25** (5), 586–601.

162 Akhmediev, N., Soto-Crespo, J., and Ankiewicz, A. (2009) Extreme waves that appear from nowhere: On the nature of rogue waves. *Physics Letters A*, **373** (25), 2137–2145.

163 Ankiewicz, A., Soto-Crespo, J., and Akhmediev, N. (2010) Rogue waves and rational solutions of the Hirota equation. *Physical Review E*, **81** (4), 046 602.

164 Viotti, C., Dutykh, D., Dudley, J.M., and Dias, F. (2013) Emergence of coherent wave groups in deep-water random sea. *Physical Review E*, **87**, 063 001.

165 Sugny, D., Picozzi, A., Lagrange, S., and Jauslin, H. (2009) Role of singular tori in the dynamics of spatiotemporal nonlinear wave systems. *Physical Review Letters*, **103**, 034 102.

166 Conti, C., Schmidt, M.A., Russell, P.S.J., and Biancalana, F. (2010) Highly noninstantaneous solitons in liquid-core photonic crystal fibers. *Physical Review Letters*, **105**, 263 902.

167 Saleh, M.F., Chang, W., Hölzer, P., *et al.* (2011) Theory of photoionization-induced blueshift of ultrashort solitons in gas-filled hollow-core photonic crystal fibers. *Physical Review Letters*, **107** (20), 203 902.

168 Marini, A., Conforti, M., Della Valle, *et al.* (2013) Ultrafast nonlinear dynamics of surface plasmon polaritons in gold nanowires due to the intrinsic nonlinearity of metals. *New Journal of Physics*, **15** (1), 013 033.

13

Nonlocal Disordered Media and Experiments in Disordered Fibers

Silvia Gentilini and Claudio Conti

Institute for Complex Systems (ISC-CNR), Rome, Italy

13.1 Introduction

Disorder radically alters wave propagation in the linear and the nonlinear regime. Even if we know very well the effects of randomness in the linear propagation, particular phenomena remain largely debated in the literature. This is specifically the case of the three-dimensional Anderson localization of light, which, even if claimed by various authors, is still a controversial subject.

The situation is even more intricate in the nonlinear case, where the interplay of various localization phenomena, as solitons and rogue waves, with disorder induced localization is only marginally understood.

On the other hand, defocusing nonlinearity sustains shock waves, and related processes as supercontinuum generation, which transforms regular wave patterns into incoherent-like energy distribution. Understanding the way these extreme events are affected by structural disorder represents a theoretical challenge, and explains the interest in developing novel experiments.

This chapter is aimed at reviewing some theoretical and experimental investigations concerning solitons and shock waves in the presence of randomness. The goal of the developed theoretical analysis is to introduce the disordered internal structure of the material in the nonlinear Schrödinger equation. Our work is dedicated to experimental observables like the localization length and the shock point, and the way they are affected in a measurable way. Correspondingly, we also report the theoretical analysis that is found in agreement with observations. The considered cases show the richness of the interplay of disorder and nonlinearity in optics, and provide examples of the way one can use nonlinear optics to test well-known theories concerning disordered waves and also to unveil novel physical effects. In particular, we report on the transport of Anderson localization by the nonlinear nonlocal interaction, a paradigmatic example of this new physics.

The analysis of nonlinear random waves is an important research direction for several applications, as, for example, ultrafast spectroscopy of random media, and imaging in

Shaping Light in Nonlinear Optical Fibers, First Edition. Edited by Sonia Boscolo and Christophe Finot.

turbid systems. Understanding novel and fundamental physical laws is very relevant to developing these applications.

The chapter is organized as follows: in Section 13.2, we describe the theoretical model, based on highly nonlocal approximation (HNA), that quantitatively explains the effect of an optical nonlinearity of thermal origin on the behavior of localized modes arising in a fiber with a transverse disorder structure. In Section 13.3, we report on the experimental investigation on the nonlinear propagation of light in a fiber with a binary distribution of the refractive index, demonstrating the predictions outlined by the theoretical model. In Sections 13.4 and 13.5, we consider theoretically and experimentally the effect of disorder on another nonlinear phenomenon, i.e., the spatial dispersive shock waves (DSWs), by retrieving a quantitative agreement between the theoretical predictions and the experimental results. Conclusions are drawn in Section 13.6.

13.2 Nonlinear Behavior of Light in Transversely Disordered Fiber

Optical nonlinearity is the change of material optical properties with the intensity, I, of a radiation field. At a high-intensity light, the nonlinear response of a material gives rise to a plethora of phenomena ranging from frequency mixing processes to multi-photons absorption, however, here we limit our discussion to considering specific processes supported by the so-called Kerr effect. This effect is characterized by an intensity dependence of the refractive index of the material given by $n = n_0 + n_2 I$, where n_0 is the linear refractive index and n_2 the nonlinear coefficient.

When considering the nonlinear behavior of materials with a disordered internal structure, the interaction between the nonlinearity and the diffusive nature of the propagating light can be very complex and counter-intuitive. In fact, on one hand, the Kerr nonlinearity, depending on whether the nonlinear coefficient n_2 is negative or positive, can sustain a defocusing beam or a self-phase modulation, resulting in a converging wavefront, that can eventually overcome the diffraction for a sufficiently large intensity. On the other hand, disorder supports a trapping mechanism resulting in the formation of localized modes within a spatially exponentially decaying area, i.e., to the so-called Anderson localization (AL), direct observation of which in three dimensional media is still debated [1–5]. However, in the media in which the refractive index is randomly modulated orthogonally to the direction of propagation, disorder always sustains *transverse* localization [6,7], i.e., non-diffracting beams similar to solitons.

Moreover we emphasize that here we take into account disordered materials presenting a Kerr nonlinearity of thermal origin, where the role of spatial nonlocality, i.e., a nonlinear perturbation that extends far beyond the region of interaction, becomes substantial in the light propagation process. Indeed, if disorder induces exponential localizations and reduces the interactions of distant modes, nonlocality is expected to create some action at a distance dependent on power.

A key quantity to characterize the localization of light is the localization length l, i.e., the length over which the localized states exponentially decay. Since it competes with other characteristic lengths of the light propagation process, i.e., the photon transport mean free path and the wavelength, its determination is fundamental to establish the occurrence of the localization and to study the interaction among different localized states. Here we are interested in analyzing this latter aspect of the localization phenomenon in the presence of a nonlinear and nonlocal response of the disordered medium.

Theoretically the variation of l with the nonlinear nonlocal response can be described by the highly nonlocal approximation (HNA) [8].

Following recently reported work [9–11], we consider a system composed by two materials, polystirene (refractive index n_{PS}) and PMMA (refractive index n_{PMMA}), that are randomly alternated in a two-dimensional (2D) geometry and form a random fiber that is invariant in the propagation direction z. Further details are given in the section dedicated to the experiments.

We model the system by resorting to the paraxial wave propagation

$$2ik\frac{\partial A}{\partial z} + \nabla_{x,y}A + 2k^2\frac{\Delta n}{n_0}A = 0, \tag{13.1}$$

where A is the optical field normalized such that $|A|^2$ is the intensity, $k = 2\pi n_0/\lambda$ is the wave-number with $n_0 = (n_{PS} + n_{PMMA})/2$ the average refractive index. $\Delta n = n_{PS} - n_{PMMA} = \Delta n_L + \Delta n_{NL}$ is the index perturbation including two contributions: (i) a linear perturbation $\Delta n_L(x, y)$ due to the disorder; and (ii) a nonlinear and nonlocal term, dependent of $|A|^2$, which is written as (see, e.g., [12, 13])

$$\Delta n_{NL} = \int K(x - x', y - y')|A|^2(x', y')dx'dy'. \tag{13.2}$$

The nonlinear nonlocal nonlinearity that we consider, is the photothermal action, i.e., the temperature induced refractive index change. In Eq. (13.2), $K(x, y)$ is the Kernel function which, for a thermal nonlinearity, is given by the Green function of the Fourier heat equation, which, in general cannot be written explicitly for a given geometry.

As the temperature varies on a spatial-scale comparable with the fiber size much larger than the wave localization length, a specific localization $|A|^2$ is much more localized than $K(x, y)$, so that the latter can be taken out of the integral in Eq. (13.2), letting $r^2 \equiv x^2 + y^2$ we have

$$\Delta n_{NL} \cong K(x, y) \int |A|^2 d\mathbf{r} \cong P\left(\Delta n_1 + \frac{r^2}{2}\Delta n_2\right). \tag{13.3}$$

In Eq. (13.3), P is the power of the single localized state, and we expanded $K(x, y)$ in a Taylor series centered on the AL location $r = 0$. Here we are interested in the effect of the nonlocal nonlinearity on the localization length, hence we limiting ourselves to consider localized propagation invariant solutions. Since these are disorder induced localizations that decay exponentially, we can further ignore in Eq. (13.3) the term weighted Δn_2, and treat the Kernel function as a constant: $K(x, y) \cong K(0, 0) \equiv \Delta n_1$, being Δn_1 the nonlinear index perturbation at a unitary input power P. Eq. (13.3) shows that the leading effect of the nonlocal nonlinearity Δn_{NL} is modifying the refracting index perturbation Δn_R and hence shifting the eigenvalues of the disorder induced localized states. In the HNA, these are given by the solutions $A(x, y, z) = a(x, y)\exp(i\beta z)$ with

$$-\frac{1}{2k}\nabla_\perp a - \frac{k}{n_0}\Delta n_R(x, y)a = \left(\beta + k\frac{\Delta n_1 P}{n_0}\right)a. \tag{13.4}$$

We can see from Eq. (13.4) that in the absence of nonlinearity, $\Delta n_1 = 0$, the ground state that is exponentially localized with negative eigenvalue $\beta = \beta_L < 0$. The effect of nonlinearity is to shift the eigenvalue in a power dependent way, so that we have $\beta(P) + k\Delta n_1 P/n_0 = \beta_L$.

In the section devoted to the experiments, we will show the quantitative agreement with the above theoretically retrieved behavior. However, before we consider the experiments, we theoretically take into account the effect of nonlocality, an intrinsic feature in a photothermal nonlinearity, on the localization process. Since the nonlocality mediates an action at a distance between the disorder induced states, it is expected to affect the collective interaction of several mode. This action is absent in the absence of nonlinearity, or at low power.

The main result of the theoretical model above is that the disorder-driven trapping mechanism can be enhanced by the nonlinearity in a such a way to resemble the soliton formation. Even though the two phenomena are originated by two different mechanisms, i.e., the disorder for the former and nonlinearity for the latter, on closer inspection, they look similar for various reasons: they are exponentially localized, they correspond to well-defined negative eigenvalues and they may be located in any position in space. All these resemblances enable us to resort to an approach originally developed for solitons [14].

Specifically, we denote the mean position of any of the exponential localizations by a two dimensional vector $\mathbf{r}_p = (x_p, y_p)$, with $p = 1, 2, \ldots, N$ and N the number of localizations. By writing the optical field A as a superposition of ALs, and by using the Ehrenfest theorem of standard quantum mechanics on Eq. (13.1), following the arguments in [14], it turns out that in the presence of a nonlinear perturbation Δn_{NL}, the following equations holds true ($\mathbf{r} = (x, y)$):

$$P_p \frac{d^2\mathbf{r}_p}{dz^2} = \int\int I_p(\mathbf{r} - \mathbf{r}_p)\nabla_{x,y}\frac{\Delta n_{NL}}{n}d\mathbf{r}, \tag{13.5}$$

which describe the motion in the z-direction of the states under the action of the nonlinearity. In Eq. (13.5), $I_p(\mathbf{r} - \mathbf{r}_p)$ is the intensity profile of the AL with index p and power $P_p = \int I_p(\mathbf{r} - \mathbf{r}_p)d\mathbf{r}$. As the various localizations do have different wavelengths, they are incoherent, and Δn_{NL} can be written as the sum of their respective index perturbations $\Delta n_{NL,q}(\mathbf{r} - \mathbf{r}_q)$ with:

$$\Delta n_{NL} = \sum_{q=1}^{N} \Delta n_{NL,q} \cong \sum_{q=1}^{N} \frac{P_q \Delta n_2}{2}(\mathbf{r} - \mathbf{r}_q)^2. \tag{13.6}$$

In Eq. (13.6) after the HNA as in Eq. (13.3), we expand $\Delta n_{NL,q}(\mathbf{r} - \mathbf{r}_q)$ as a Taylor series with respect to the spatial coordinate centered in the localization positions $\mathbf{r_q}$, and omit the constant term Δn_1 as it does not affect the change in the AL location because of the operator $\nabla_{x,y}$ in Eq. (13.5). As we did with Eq. (13.3), we retained the higher order term weighted by $\Delta n_2 < 0$, which induces the action at a distance between localizations. In other words, Δn_2 is negligible when considering the localization length of the single state, but sustain the interactions of distant AL. By substituting Eq. (13.6) in Eq. (13.5) and treating I_p as a Dirac δ with area P_p, because of the strong localization of the AL with respect to the refractive index profile, we have

$$P_p \frac{d^2\mathbf{r}_p}{dz^2} = -\nabla_{x_p,y_p} \sum_{q=1}^{N} \frac{|\Delta n_2| P_q P_p}{2n_0} |\mathbf{r}_p - \mathbf{r}_q|^2. \tag{13.7}$$

Equation (13.7) predicts that the ensemble of disorder induced localizations behaves as a system of interacting particles with pairwise attractive potential [14], and the conservative force between two states is proportional to the product of their two powers, as in a gravitating system the attractive force is proportional to the product of masses according to Newton's law. As a result of this analogy, the more powerful localizations will attract the others, and the whole system will tend to collapse in a specific point. After the collapse and the interaction, as the system conservative, the kinetic energy will convert into potential energy, and the localization will spread again (see Fig. 13.4 (a)).

Now consider the simplest case of only two localizations, with powers P_{probe} and $P_{pump} \gg P_{probe}$, the probe localization, has a displacement $D(z=0)$ at the fiber input from the pump beam. Accordingly with Eq. (13.7), after a propagation distance z, at the lowest order in $|\Delta n_2|$, the displacement is reduced and given by

$$D(z) = D(0)\left(1 - \frac{|\Delta n_2| z^2}{2n_0} P_{pump}\right). \tag{13.8}$$

The above equation shows that the probe position scales linearly with the pump beam power P_{pump}.

The theoretical predictions provided here will be supported by experimental demonstrations reported in the dedicated paragraph below.

13.3 Experiments on the Localization Length in Disordered Fibers

In order to provide an experimental foundation to the theoretical models, based on the highly nonlocal approximation (HNA), presented in the previous paragraph, we review the experimental investigation of light localization in a 2D disordered fiber with a nonlinearity of thermal origin, which is known to be highly nonlocal [12, 15–20].

We show that many exponentially localized states can be simultaneously excited by a broadband laser beam. We measure their localization length, l, in terms of the optical power, and give evidence of the action at a distance between localized states. We also find that, because of nonlocality, the AL migrate from their position and move in a collective way.

The sample used in the experiments is a fiber with a disordered binary distribution of the refractive index, realized by means of Polymethyl-methacrylate (PMMA, refractive index $n_{PMMA} = 1.49$), and Polystyrene (PS, $n_{PS} = 1.59$) resulting in an index contrast $\Delta n = 0.1$.

Figure 13.1 shows the experimental setup that has enabled the observation of the Anderson modes (AM) and, hence, the measurements of the localization length. A

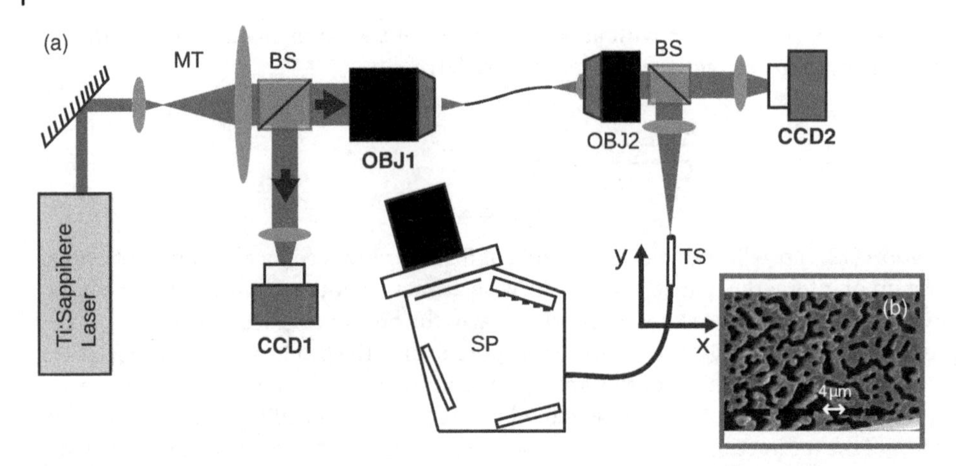

Figure 13.1 (a) Sketch of the experimental setup; (b) scanning electron microscope image of the fiber facet.

broadband laser (Ti:Shappire oscillator model Coherent Mantis; 500 mW power, 80 nm bandwidth, 820 nm center wavelength, 15fs pulse duration) is injected in the fiber through a long working-distance microscopy objective (OBJ1), which produces a spot size between 0.8 and 10 μm at the entrance of the fiber. The output face of the fiber is collected by a second objective (OBJ2), which produces through a beam splitter (BS), two separate images: the first is imaged on a CCD camera (CCD2) and the second is collected by a fiber coupled with a spectrograph, which retrieves the output emission from an area with sub-micrometer spatial extension. Two computer-controlled motors vary the fiber position in the virtual image plane with a spatial resolution below one micron, and enable measurement of the spatial distribution of Anderson localizations at different wavelengths.

We collect the spot at the fiber output by the second CCD camera (CCD2): for an input power of the order of 1 mW, the average measured localization length is 45 ± 5 μm; when increasing the input power to 35 mW, the localization length drops to 30 μm. In order to demonstrate the thermal origin of the retrieved nonlinear response, we analyze the temporal behavior of the intensity at the fiber output. By means of a mechanical shutter (with opening time < ms) to control the input light and the CCD camera (with an exposure of 100 μs) to capture the time evolution of the images maximum intensity, we found a response time of the order of few seconds, i.e., a value consistent with a thermal response. In fact, the PPMA light absorption and thermal conductance are respectively 2500 $\frac{dB}{km}$ and $K_{PMMA} = 0.25$ W/mC°, while PS has 600 $\frac{dB}{km}$ and $K_{PS} = 0.033$ W/mC°), which is temporally slow and inherently nonlocal [12,17]. Indeed, even if PMMA and PS show a nearly identical dependence of the refractive index with temperature, PS absorption is about four times smaller than PMMA, and PMMA thermal conductivity is one order of magnitude larger than PS. This results in the presence of temperature hotspots located in the PMMA, which induce a local decrease of the refractive index and thus an increase of the refractive index mismatch.

Because of the different thermal conductivities of PS and PMMA, the effect of the increased temperature is to enhance the refractive index modulation and hence the

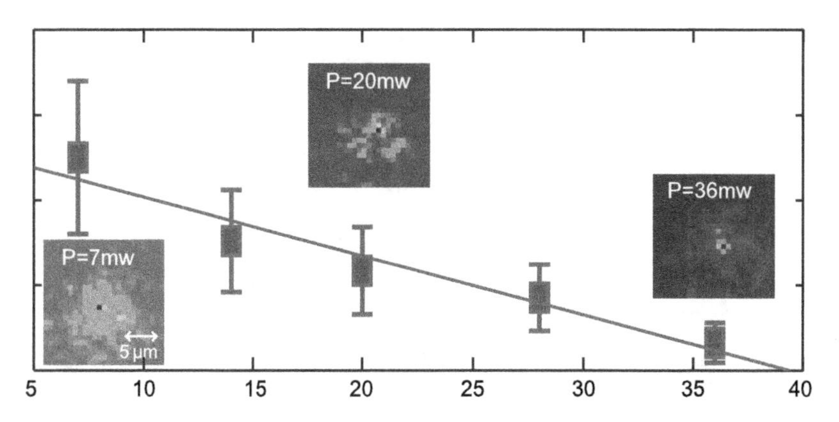

Figure 13.2 Average localization length l estimated by Gaussian fit (error bars are obtained from the statistics over 35 modes). Fit results: $l(0) = 14$ μm, $P_C = 12.5$ mW; fiber length $z = 8$ cm. The insets show spatial distribution for the most intense mode for different input power P.

strength of disorder, so that $\Delta n_1 > 0$; this implies that the negative eigenvalue $\beta(P) = \beta_L - k\Delta n_1 P/n_0$ decreases with P. Since the localization length l scales as [21]

$$1/\sqrt{2k|\beta(P)|},$$

this induces the self-focusing action observed in the experiments and produces the behavior

$$l(P) = l(0)/\sqrt{1 + P/P_C}$$

with $P_C = n_0|\beta_L|/2k\Delta n_1$.

We also numerically calculate the solution of the heat transfer equation by using the above absorption and conductance and we found that an input power of 10 mW produces temperature gradients that increase the refractive index mismatch of an amount of the order of $\Delta n_{NL} = 10^{-5}$. Below we compare this value with that experimentally obtained.

Insets of Figure 13.2 show the mode distribution at the output of the fiber for three different input power. In the big panel we report the average mode size, estimated by a Gaussian fit of the intensity profile, for a set of 35 different modes versus the input laser power. The fit is performed by using a linear approximation of Eq. (13.5):

$$l(P) = l(0)\left(1 - \frac{P}{2P_C}\right),$$

and the retrieved fit parameters are $l(0) \cong 1/\sqrt{2k|\beta_L|} \cong 14$ μm and $P_C \cong 12.5$ mW, which give $\Delta n_1 \cong 10^{-3}$ W^{-1}, so that at power $P = 10$ mW we estimate $\Delta n_{NL} = \Delta n_1 P = 10^{-5}$, in agreement with the estimation from numerical solution of the heat transfer equation.

We now show the effect of nonlocality on the ALs position. To do this we define a new quantity, i.e., the mode-density $\rho(x, y)$ (number of modes per pixel). In order to clarify how we calculate such a function we resort to the inset of Figure 13.2, where it

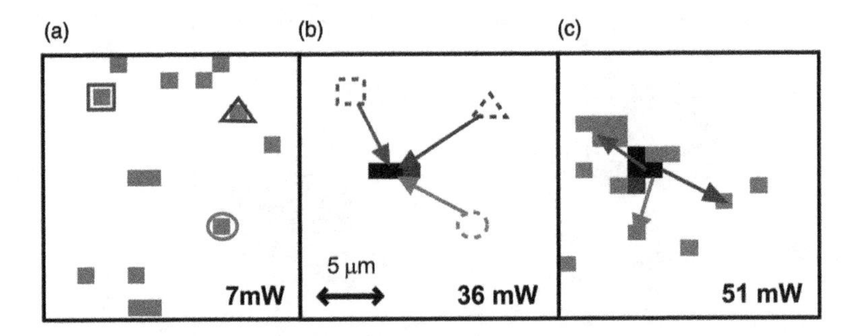

Figure 13.3 Position of several Anderson states at different frequencies (scale bar is 5 μm) at different powers showing the nonlinearity driven motion. Some modes are highlighted by geometrical shapes corresponding to output spectral peaks. Arrows indicate their motion upon increasing power.

is shown that at the maximum degree of localization, most of the intensity of a single mode is collapsed in a single pixel, which can be considered the AL location. Hence the modes position of the 35 most intense peaks of the spectrum is approximated with their intensity maximum, then the mode-density is retrieved counting the number of modes present per pixel: when a single pixel, which gives our spatial resolution, at position (x,y) hosts many modes, $\rho(x, y)$ is high; on the contrary, an empty pixel at (x,y) corresponds to a vanishing ρ. In Figure 13.3 the density ρ is reported: darker pixels represent positions in which many modes at different frequencies are simultaneously present. At low power, ALs are sparsely distributed in space; when increasing the power, they "migrate" toward the peak of optical intensity. Modes translate into an amount greater than 12 μm, i.e., a distance one order of magnitude larger than their spatial extension, when power is increased from 7 mW to 36 mW.

To demonstrate that this collective motion is driven by a nonlocal action at a distance between localized states, we designed an experiment by exciting only two modes. We inject into the fiber two beams: (i) a "probe," low power P_{probe} green laser, at wavelength 532 nm (continuous wave Nd:YAG Ventus Laser Quantum, 500 mW maximum power), and (ii) a "pump," broadband high-power P_{pump} infrared (wavelength 800 nm) laser. Light reflected from the entrance of the fiber is shown in Figure 13.4 (a), and measures the distance between the two input spots, which is about 9 μm. Both the beams excite ALs. Figures 13.4 (a)–(d) show the probe light (a laser line filter in front of the camera allows to collect only light at 532 nm) at the exit of the fiber for various powers of the pump beam (at 800 nm). We observe two effects: (i) the controlled steering of the Anderson localization at 532 nm toward the position of the control beam (the "migration"), (ii) a contraction of its spatial extension, that is an all-optically controlled localization length. These features are quantified in Figure 13.4 (e), which shows the displacement D (open circles) of the probe position and in Figure 13.4 (f), which reports the localization length l (full squares) as a function of the power of the pump beam.

Both the behavior of the displacement D and of the localization length l are well-fitted by Eq. (13.8) and the linear approximation of Eq. (13.5) respectively and hence provide a robust experimental demonstration of the theoretical model presented in the previous section.

The origin of the observed behavior in the fiber with transverse disorder is due to two competing phenomena: (i) a focusing effect due to the interaction between nonlinearity and randomness, and driven by the increasing refractive index mismatch that enhances

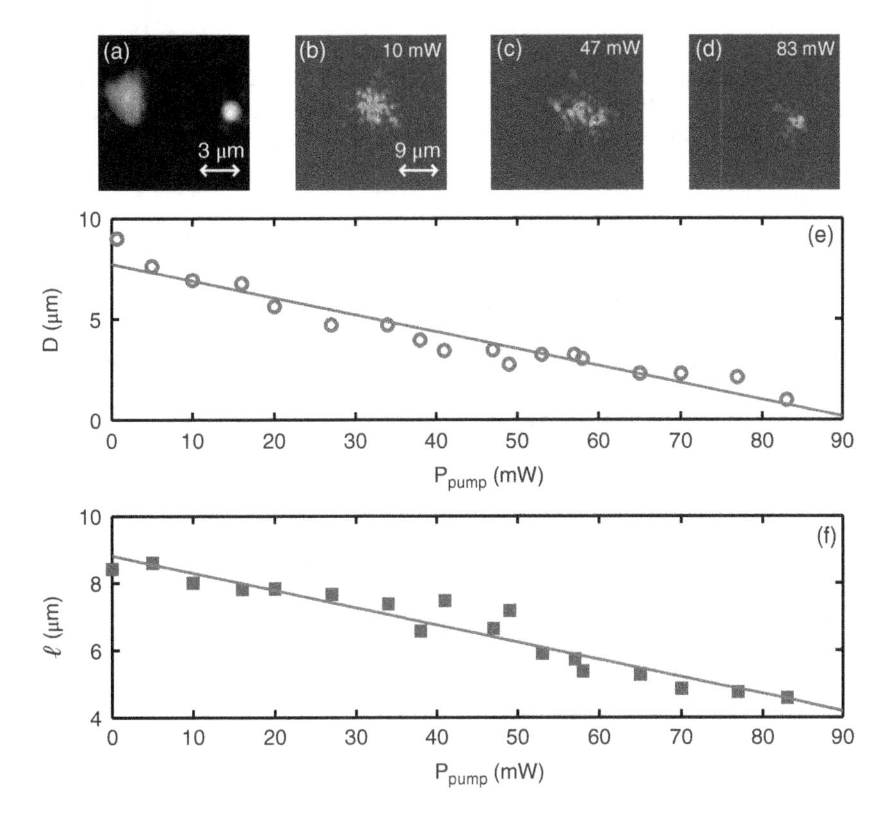

Figure 13.4 (a) image of the entrance of the fiber (scale is 3 μm); (b,c,d) fiber exit corresponding, respectively, to 10, 47 and 83 mW input power, the pump beam has been eliminated by a laser line filter (scale is 9 μm); (e) displacement D versus P_{Pump}, fitted with Eq. 13.7; (f) localization length ℓ as a function of P_{pump} and fitted with Eq. 13.5 as in Figure 13.3.

the strength of the disorder, and (ii) a thermal de-focusing nonlinearity for the bulk PS and PMMA. Hence we have evidence that the interplay between the binary structure and the Anderson localization turns a de-focusing nonlinear response into a focusing one, producing a spatial compression of the localization down to the diffraction limit when input power reaches a certain threshold power (35 mW).

In what follows we will still consider a medium presenting a defocusing Kerr-like nonlinearity of thermal origin and hence intrinsically nonlocal. However, we will take into account another phenomenon that can be sustained by such a nonlinear response, i.e., the dispersive shock waves (DSWs), and we will show how such a phenomenon is very differently affected by the disorder. This is due to the fact that being the DSWs, a coherent phenomenon, the introduction of an even small amount of disorder competes with the nonlinearity and hampers the shock formation.

13.4 Shock Waves in Disordered Systems

The competition between the coherence of the nonlinear waves and scattering has long been an object of interest for many theoretical and experimental studies [22–24].

In this framework, the case of DSWs is particularly relevant [25, 26], as their emergence is characterized by the formation of strongly nonlinear and coherent oscillation, the so-called *undular bore* [27–29], and they are expected to be strongly affected (and eventually inhibited) by disorder, at variance with solitons, which can survive (see, e.g., [30,31]) and, as demonstrated in Section 13.3, can even be enhanced by a certain amount of randomness.

Historically the first evidence of a shock wave can be dated back to 1887 when Ernst Mach published the first visualization of a shock front, generated by the motion of a body traveling at supersonic speed. A shock front originates from the occurrence of a singularity in some quantity involved in the wave propagation, as e.g. the velocity profile. It is the way in which such singularity has overcome by the system that distinguishes two types of shock waves: the viscous and the dispersive shock waves (DSWs). The former displays a steep front associated with a jump in energy or pressure; conversely the dispersive shock waves, occurring in optics, regularize the singularity through the action of dispersion or diffraction, by the formation of the characteristic undular bore. Understood as the general mechanism that originates the wave-breaking phenomena, it is clear that they are excitable in all the wave phenomena where the conditions enable the occurrence of some singularity. Such ubiquity allows, nowadays, the observation of the DSWs wave in many systems, as Bose-Einstein condensate or optical nonlinear defocusing media.

Here we consider the formation of DSWs in a liquid dyed solution displaying a Kerr-like nonlinearity of thermal origin, where is introduced a certain amount of disorder. Neglecting, in a first approximation, the spatial nonlocality [32], the refractive index perturbation in the presence of nonlinearity and disorder to the bulk index n_0 is written as:

$$\Delta n = n_2 I + \Delta n_R(X, Y, Z) \tag{13.9}$$

where $n_2 < 0$ takes into account the considered defocusing Kerr effect, I is the optical intensity and Δn_R represents the random perturbation due to the presence of disorder.

The propagation of light inside the nonlinear medium can be modeled by the nonlinear Schrödinger equation (NLS), that for a TEM_{00} Gaussian beam can be written as the paraxial wave equation for the complex envelope, A, of a monochromatic electric field $E = (\frac{2}{c\epsilon_0 n_0})^{1/2} A \exp(ikZ - i\omega T)$:

$$2ik\frac{\partial A}{\partial Z} + \nabla^2_{X,Y} A + 2k^2 \frac{\Delta n}{n_0} A = 0. \tag{13.10}$$

where $k = 2\pi n_0/\lambda$ is the wave vector, c the velocity of light and $\epsilon_0$is the electric permittivity of free space and A is normalized such that $I = |A|^2$. Indicating with I_0 the input peak intensity, w_0 the input beam waist, we can define two characteristic length scales, the nonlinear length scale $L_{nl} = n_0/(k_0|n_2|I_0)$, and the diffraction length $L_d = kw_0^2$, which rule the regime of propagation of light within the nonlinear medium. By exploiting such lengths we introduce scaled coordinates $x, y, z = X/w_0, Y/w_0, Z/L$, and the normalized field $\psi = A/\sqrt{I_0}$, in order to obtain the following dimensionless equation:

$$i\epsilon\frac{\partial \psi}{\partial z} + \frac{\epsilon^2}{2}\nabla^2_{x,y}\psi - |\psi|^2\psi + U_R\psi = 0, \tag{13.11}$$

where $\epsilon \equiv L_{nl}/L = \sqrt{L_{nl}/L_d}$ and $U_R = \Delta n_R/(n_2 I_0)$. The quantity ϵ measures the strength of the nonlinearity with respect to the diffraction: a small value for ϵ implies negligible diffraction and a pronounced nonlinear response. U_R is the ratio between the perturbation of index due to the disorder and the nonlinearity.

Considering an optical field with a factorized form $\psi = \sqrt{\rho(r,z)}\exp[i\phi(r,z)/\epsilon]$ and retaining only the leading order in ϵ, Eq. (13.11) becomes an equation for the phase ϕ:

$$\phi_z + \frac{1}{2}\left(\phi_x^2 + \phi_y^2\right) + \rho - U_R = 0 \tag{13.12}$$

Limited to consider only one dimension ($\partial_y = 0$), performing the transverse derivative, and defining a *velocity field* equal to the phase chirp, $u \equiv \phi_x$, Eq. (13.12) reduces to:

$$u_z + uu_x + \partial_x(\rho - U_R) = 0. \tag{13.13}$$

In the homogeneous case (ρ =const.) and for an ordered medium ($U_R = 0$), Eq. (13.13) takes the form of the Hopf equation [33], that is a paradigmatic equation in hydrodynamics since it allows solutions able to develop discontinuities in the velocity profile, $u_x \to \infty$, and hence gives rise to the wave-breaking phenomenon.

The initial condition that allows us to retain only the leading term in ϵ is called hydrodynamical approximation and holds true only when $\epsilon \to 0$, i.e., $L_{nl} \ll L_d$, this condition implies a threshold in the nonlinear degree inasmuch the shock can occur. Another threshold arises from the term $U_R = \Delta n_R/(n_2 I_0)$, corresponding to the existence of a critical value for the amount of randomness, above which it is expected that no shock occurs: when the random index perturbation Δn_R becomes comparable with the nonlinearity $n_2 I_0$, the material refractive index fluctuations are so pronounced that the nonlinear effect is totally masked.

The aim of the present review is to provide a theoretical and experimental demonstration of the existence of such a threshold and its determination in terms of light power and disorder strength. In order to do this, we resort to an observable position that could be accessed by both numerics and experiments, such an observable position is the position of the shock formation, Z_s, defined as the position where is originated the singularity in the velocity profile, i.e., $d\phi/dX \to \infty$ [26, 34].

In the following we show two different ways to identify the shock point Z_s in the numerical simulations and in the experiments.

The method of characteristic lines. In order to address the comparison between the theoretical predictions and the experiments, we present numerical simulations based on a particle-like model. The analogy between photons and particles is made possible by applying the characteristic lines method to Eq. (13.13), that is transformed into the motion equation of a particle experiencing a good potential and, in the approximation of weak disorder, it behaves like a white noise that scrambles the potential profile.

Because of the huge number of particles within the intense optical beam and in the hypothesis of low index contrast, $\Delta n_R(x, y, z)$ can be taken as a random Gaussian distribution. In the hydrodynamic limit $\epsilon \to 0$ ($L_{nl} \ll L_d$), this results in the equation of motion of a unitary mass particle (we limit ourselves to the one-dimensional case for the sake of simplicity):

$$\frac{d^2x}{dz^2} = -\frac{dU}{dx} - \frac{dU_R}{dx} = -\frac{dU}{dx} + \eta_R. \tag{13.14}$$

In Eq. (13.14) $U = exp(-x^2/2)$ is the deterministic potential from the nonlinear part due to the Gaussian beam profile and U_R is the disordered term due to the refractive randomness. This latter is treated as a noise over the deterministic potential, hence $\eta_R = -dU_R/dx$ is taken as a Langevin force that we assume with Gaussian distribution, such that $\langle \eta_R(z)\eta_R(z')\rangle = \eta^2\delta(z - z')$, with strength of disorder measured by

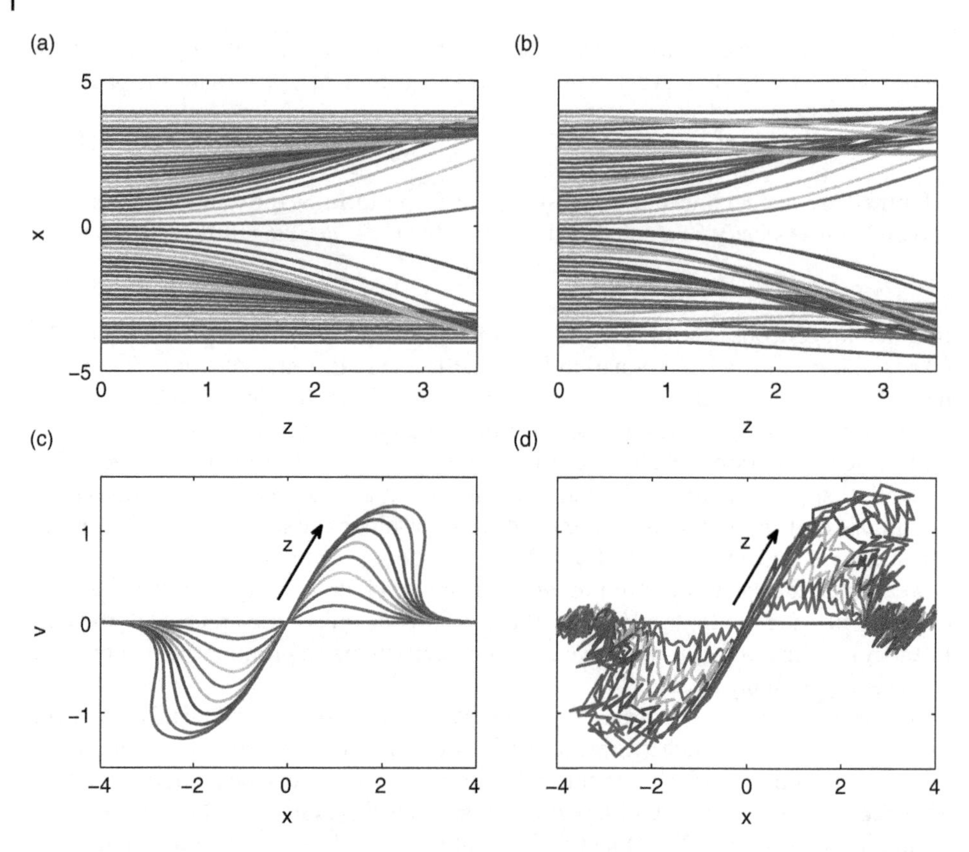

Figure 13.5 Trajectories of colliding particles forming shock versus z: (a) without disorder ($\eta_R = 0$) and, (b), for $\eta_R = 0.1$; (c) shock profile in the phase space for $\eta_R = 0$ and, (d), for $\eta_R = 0.1$ (z varies in the range [0,3]).

$\eta = \langle (dU_R/dr)^2 \rangle^{1/2} \cong < \Delta n_R^2 >^{1/2} / |n_2| I_0$, and the brackets denoting statistical average. We notice that Eq. (13.14) is solved for several values of x taking for each of them an independent realization of the noise $\eta_R(z)$, this ignores the dependence of η_R on x.

In Figure 13.5, we show the result of the numerical solution of Eq. (13.14) as obtained by a stochastic Runge-Kutta algorithm [35]. In panels Figures 13.5 (a) and (b), we report several of these trajectories resulting from the initial uniformly distributed position in the x axis and zero initial velocity $v = dx/dz$ in the ordered ($\eta_R = 0$) and disordered case ($\eta_R = 0.1$) respectively, Figures 13.5 (d) and (e) display the correspondent velocity profiles in phase space. In this simulation the occurrence of shock is signaled by the collision of the trajectories and the point where such collisions occur is defined as the shock point Z_s; in the phase space of v and x (Figure 13.5 (c)), these correspond to the folding of the velocity profile into a multivalued function when increasing z, which induces the wave-breaking phenomenon.

We notice that in the presence of disorder, the particles tend to diffuse, as evident from their trajectories (Figure 13.5 (b)) and in the phase space (Figure 13.5 (d)); correspondingly, the propagation distance before their collisions is greater for their random walk and the shock is delayed in the z-direction.

13.5 Experiments on Shock Waves in Disordered Media

13.5.1 Experimental Setup

As we have seen in the theoretical section, in the hydrodynamic limit, DSWs are expected to occur when the nonlinearity is dominant compared to diffraction; nevertheless the diffraction starts to play a major role in the proximity of the wave-breaking point by regularizing the singularity by means of the characteristic oscillations (see Figure 13.7).

Besides these regularizing oscillations, the singularity in the field phase and amplitude also results in a diffraction enhancement, evident in the funnel shape along the propagation direction (see Fig. 13.8) appearing with the increase of the input power. This shows that the shock involves the spatial spectrum of the beam as detected in far field measurements.

The main signature of the shock occurrence is the appearance of the undular bores, the enhancement of the funnel shape of the propagating beam and, hence, an enlargement of the spectral content. In order to catch all these features we realized an experimental setup that enables access to both near and far field measurements.

Figure 13.6 (a) shows the near-field configuration of the setup, that allows a direct visualization of the propagating beam profile, i.e., the intensity as function of the transverse coordinate, X, and of the propagation direction Z. These measurements enable the identification of the shock point Z_s as the propagation distance at which the maximum chirp occurs. A continuous wave (CW) laser at wavelength $\lambda = 532$ nm is focused inside the sample. The beam waist in the focus is $w_0 = 10$ μm. A 1 cm × 1 cm × 3 cm glass cell is used as sample holder and the laser beam propagates along the 1 cm side. Top images of the fluorescence emission are collected by a *MZ*16 Leica microscope placed perpendicularly to the propagation direction, Z, and recorded by a 1024 × 1392 pixels CCD camera.

Figure 13.6 (b) shows far field setup configuration. The CW laser beam is focused inside the sample ($w_0 = 50$ μm), placed in a 1 mm × 1 cm × 3 cm glass cell. The laser beam propagates along the 1 mm side and the cell is placed in a vertical direction in order to moderate the effect of heat convection. As shown in Figure 13.6 (c) the intensity distribution of the Fourier transform of the transmitted beam is collected by a CCD camera placed at the focal length from the collecting lens. We calibrate the CCD detector by fitting with the Airy function the experimentally obtained Fourier transform of a 500 μm diameter pinhole, placed on the exit face of the cell. The angular spreading θ is related to the transverse wave-vector as $k_{X,Y} = (2\pi/\lambda)\sin(\theta)$.

Figure 13.6 (d) shows the mutual positions of the focus plane, the shock plane and the output plane.

13.5.2 Samples

The considered samples were used in several our experimental works [26, 34, 36] and consist of an absorbing dye doped liquid media presenting a defocusing nonlinearity of thermal origin. The dyed liquid is an aqueous solution of 0.1 mM Rhodamine B.

The disorder is introduced by using monodisperse 1 μm diameter silica (SiO_2) spheres. To vary the degree of disorder, several silica concentrations c are prepared,

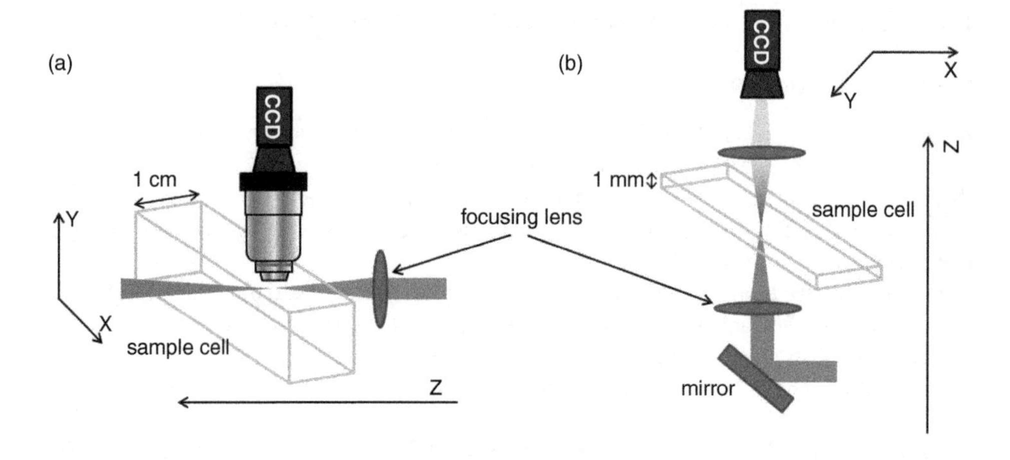

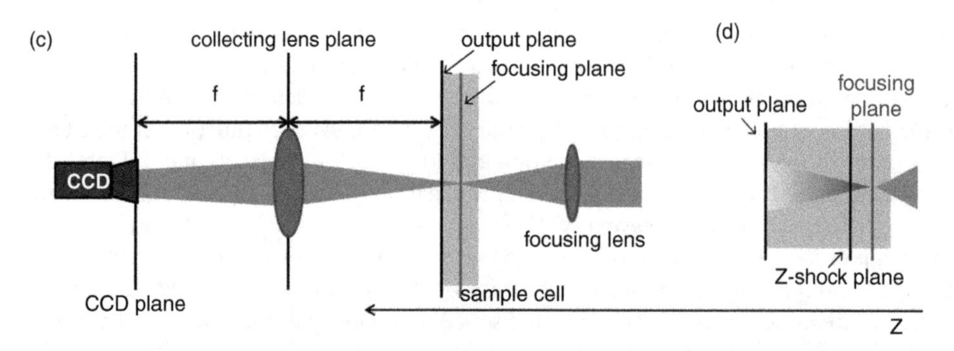

Figure 13.6 (Experimental setup: (a) detection of the top fluorescence emission of the beam; (b) configuration for the collection of the far field intensity; (c) details of the optical setup of panel (b); (d) sketch showing the position of the shock plane inside of the sample.

ranging from 0.005 w/w to 0.03 w/w, in units of weight of silica particles over suspension weight. In terms of refractive index perturbation, the amount od disorder can be estimated by the following relation:

$$\left\langle \Delta n_R^2 \right\rangle^{1/2} = c_{SiO_2} \rho_{H_2O} (n_{SiO_2} - n_{H_2O}) / \rho_{SiO_2}, \tag{13.15}$$

being n_{SiO_2} (n_{H_2O}) and ρ_{SiO_2} (ρ_{H_2O}) the refractive index and the density of the SiO_2 (H_2O), respectively. The angular brackets in Eq. (13.15) denotes volume average. Being the silica (water) density $\rho_{SiO_2} = 2$ g/cm^3 ($\rho_{H_2O} \approx 1$ g/cm^3 at 25°C), for the considered range of c_{SiO_2} concentration, $\langle \Delta n_R^2 \rangle^{1/2}$ varies between $4 \cdot 10^{-4}$ and $32 \cdot 10^{-4}$. Therefore, since from the theory a threshold in the disorder amount is predicted when $\langle \Delta n_R^2 \rangle^{1/2}$ becomes comparable with the nonlinear perturbation $|n_2|I_0 \cong 10^{-3}$, such a threshold is expected for the silica concentration $c_{SiO_2} = 0.030$ w/w as was confirmed by our experiments (see below).

The loss mechanisms to take into account in our samples are absorption and scattering. We estimate the loss length L (absorption + scattering) by fitting with an exponential decay the beam intensity versus propagation distance Z; the obtained values for L are

$\simeq 1.6$ mm for pure dye solution and $\simeq 1.2$ mm for the sample with the highest concentration c (and hence highest losses). These values are obtained. Since we verified that L is always greater than the position of the shock point Z_s as obtained by the near-field measurement, we have confirmation that, at first approximation, we can neglect losses in our theory. In addition, we also find that the scattering mean free path is of the order of millimeters for all the considered samples.

13.5.3 Measurements

In Figure 13.7 we show images of the transmitted beam (on the X-Y transverse plane) obtained with the far-field configuration setup for different input laser powers P and various concentrations c. The profiles display post-shock rings with outer rings being more intense than the inner ones, as typical for DSWs from Gaussian beams. The number and the visibility of the oscillations increase with P and decrease with c, evidencing that nonlinearity and disorder are competing: the undular bores are enhanced with the increase of power and inhibited by disorder.

The effect of disorder on the DSWs along beam propagation is reported in Figure 13.8, where images, collected by the far-field configuration of the setup, of the transverse distribution of the beam intensity versus Z for different input power P and silica concentrations c are shown. Here the hampering effect of disorder is evidenced by the reduction of the beam aperture and the disappearance of the undular bores.

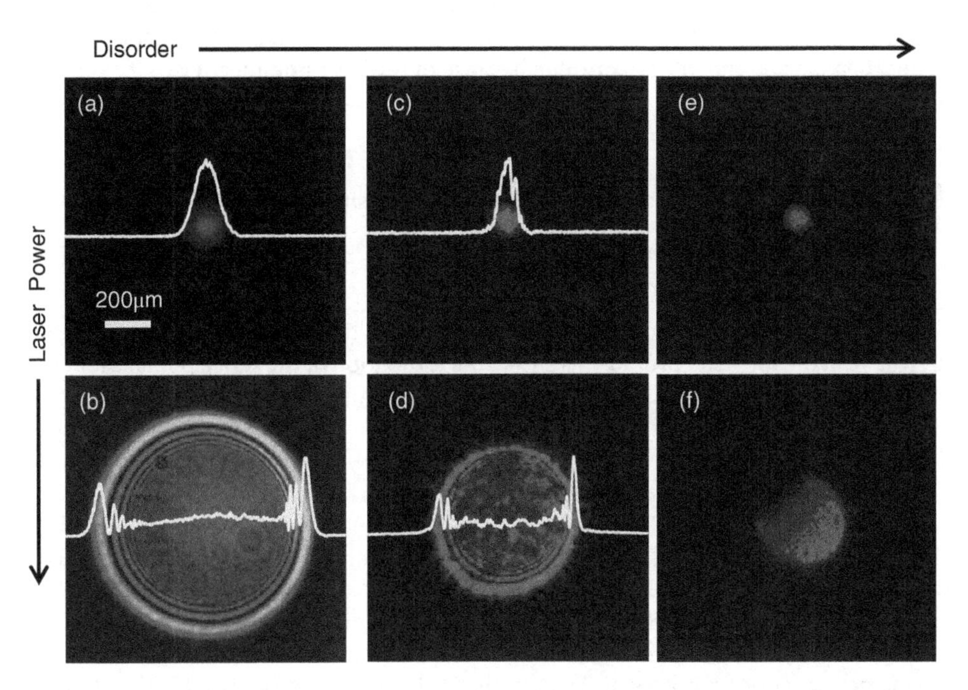

Figure 13.7 Images of transmitted intensity for different input power P and particle concentration c: a) $P = 5$ mW, $c = 0$, b) $P = 400$ mW, $c = 0$, c) $P = 5$ mW, $c = 0.017$ w/w, d) $P = 400$ mW, $c = 0.017$ w/w, e) $P = 5$ mW, $c = 0.030$ w/w, f) $P = 400$ mW, $c = 0.030$ w/w. Superimposed curves show the measured section of the intensity profiles.

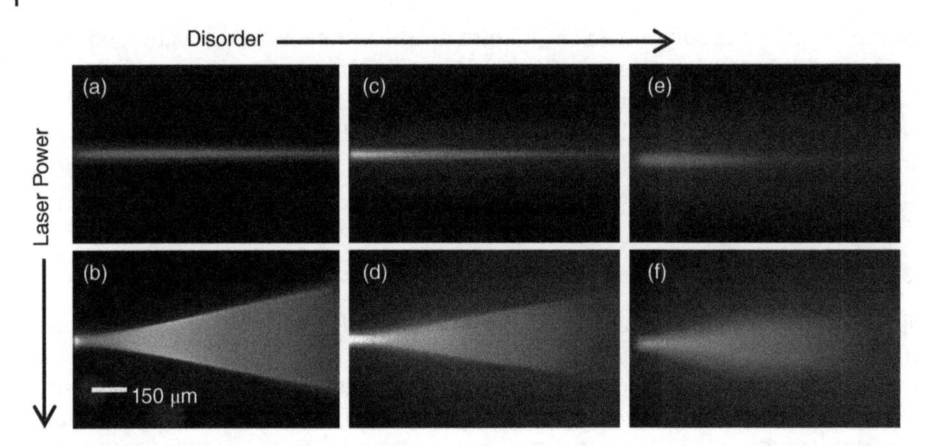

Figure 13.8 Beam propagation as observed from top fluorescence emission for different input power P and particle concentration c: a) $P = 8$ mW, $c = 0$, b) $P = 450$ mW, $c = 0$, c) $P = 8$ mW, $c = 0.017$ w/w, d) $P = 450$ mW, $c = 0.017$ w/w e) $P = 8$ mW, $c = 0.030$ w/w, f) $P = 450$ mW, $c = 0.030$ w/w.

From the near-field measurements reported in Figure 13.8 we experimentally determine the shock point Z_s by using the following procedure, recalling that the shock is originated from a singularity in the phase chirp $|d\phi/dX| \rightarrow \infty$ [12, 32, 34].

In the hydrodynamical approximation ($L_{nl} \ll L_d$), in a regime of negligible losses and diffraction, the laser beam is mainly affected by the defocusing nonlinearity. Hence at first approximation the phase is proportional to the refractive index perturbation, which in turn depends on the intensity profile because of the Kerr nonlinearity:

$$\phi(X, Y, Z) = \frac{k_0 Z}{n_0} \Delta n[I(X, Y, Z)]. \tag{13.16}$$

From Eq. (13.16) we can estimate the occurrence of the singularity in the phase from the intensity profiles, in fact:

$$\nabla_{X,Y}\phi(X, Y, Z) \propto \nabla_{X,Y} I(X, Y, Z). \tag{13.17}$$

Eq. (13.17) shows that the point of maximum phase chirp is given by the maximum derivative in the intensity profile, this allows the estimation of the shock point as follows: we calculate the transverse derivative of the intensity normalized to the peak value, I_N, and we define the steepness $S(Z)$ as the maximum with respect to the transverse coordinates of such a derivative:

$$S(Z) = max_{X,Y}[\nabla_{X,Y} I_N(X, Y, Z)]. \tag{13.18}$$

The shock point, Z_s, is finally defined as the position of the maximum steepness versus Z.

In Figure 13.9 are shown the steepness curves $S(Z)$ at three different laser power P. The point of shock occurs at smaller and smaller propagation distances with the increase of the incident laser beam power, a signature that, for a fixed level of disorder, the increase of the nonlinearity enhances the shock formation.

We report in Figure 13.10 (a) Z_S versus P for different concentrations c. Two effects are evident: 1) when increasing power P, Z_S decreases, corresponding to the speed-up

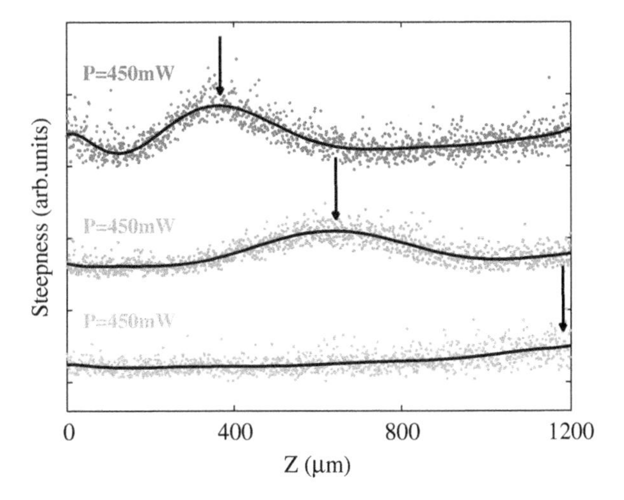

Figure 13.9 Scattered dots are the calculated steepness along the Z direction for three different powers. The solid lines are polynomial fit of the steepness curves to identify their maximum value, indicated by the arrows.

of the shock formation caused by the augmented nonlinearity; 2) when increasing concentration c, Z_S increases, as disorder delays shock formation up to its total cancellation observed for $c = 0.030$ w/w (star symbols). The plateau at low P indicates that shock is not occurring, as the steepness of the profiles increases with Z but does not have any maximum in the observation window $L_o \simeq 1$ mm.

We define the value of P at which Z_S starts to decrease as the threshold power between shock and non-shock regimes and map the phase diagram in Figure 13.10 (b).

Figures 13.11 (a) and (b) show the numerically obtained histograms of the particle positions at various propagation distances. If compared with the ordered case in Figure 13.11 (a), disorder induces a spreading of the particle distribution. We extract the position of the shock $z_s = z_s(\eta_R)$ as that approximately corresponding to the maximum of the histogram (precisely, as the mean value among the positions for which the

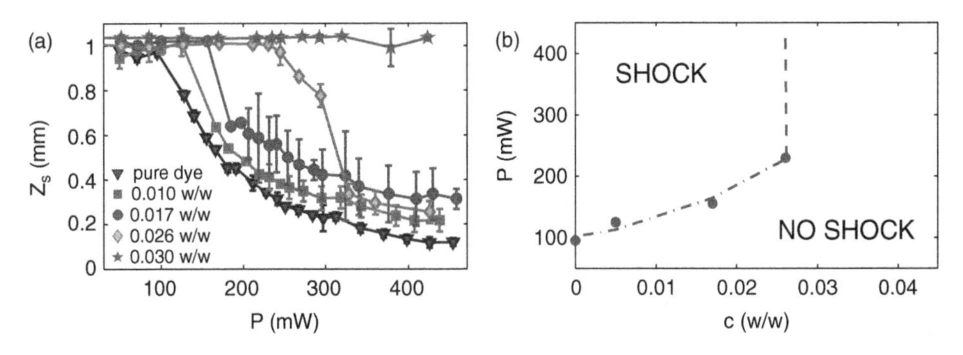

Figure 13.10 (a) Measured shock point Z_S Vs P for various c; (b) disorder-power phase diagram with shock and non-shock regimes obtained from the data in panel (a): dots correspond to the threshold powers, dashed line and the dot-dashed line are the boundaries as estimated by the theory.

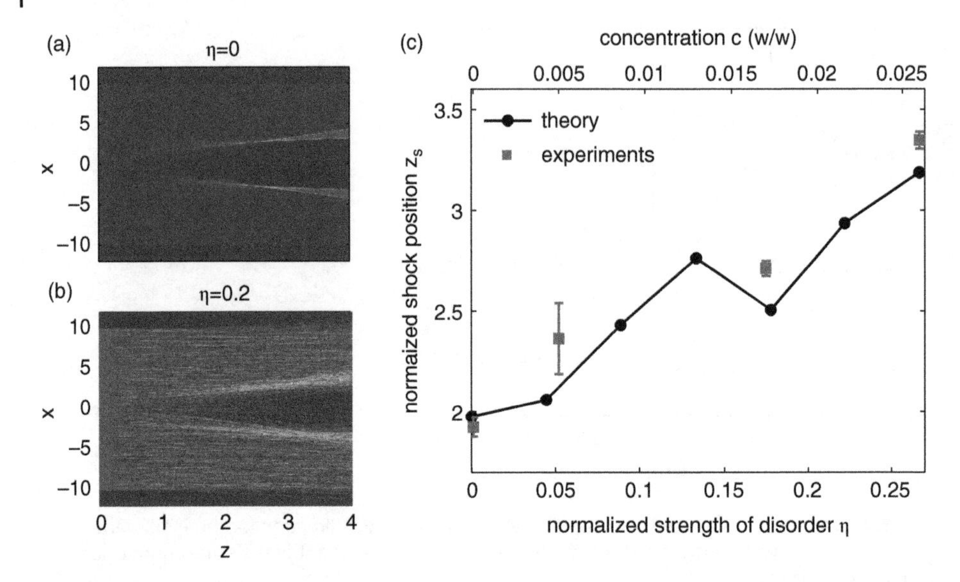

Figure 13.11 Theoretical histograms of particle positions for $\eta_R = 0$ (ordered case, panel a) and for $\eta_R = 0.2$ (panel b); (c) theoretical normalized shock position z_s versus amount of disorder η_R (black continuous line) and comparison with the measured z_s Vs concentration c (red squares).

histogram is above the 90% of its maximum, to limit fluctuations). This allows us to determine z_s for various amounts of disorder η_R (in the ordered case $z_s(0) \cong 2.5$). Figure 13.11 (c) shows $z_s(\eta_R)$ vs. disorder degree for 10^3 particles and demonstrates that the shock process is delayed by disorder.

13.6 Conclusion

In this chapter we have reviewed our theoretical and experimental recent investigations on the interplay between disorder and optical nonlinearity of thermal origin in two particular case: (i) two-dimensional Anderson localization in an optical fibers with a binary disordered structure and (ii) the dispersive shock waves in an aqueous dye solution with a dispersion of silica particles.

The specific cases considered are relevant because they present the main features than can be described theoretically and easily supported by experiments.

The reported results show that nonlinearity radically alters observables like the localization length and the shock point. In particular, the results show that the interplay of the two effects is substantially opposite: in the former case we found that the trapping mechanism sustained by the 2D disorder structure is enhanced by the defocusing optical nonlinearity; in the latter case, conversely, we demonstrate that the nonlinearity is strongly hampered by disorder that leads to a total inhibition of the DSWs' formation.

Moreover in the case of optical fiber we also consider the nonlocality intrinsic in a nonlinearity of thermal origin and that, we have demonstrated, mediates the transport and a distance action among localized states.

The nonlinear effect is still sustained by a defocusing nonlinearity of thermal origin, but here the effect of disorder is very different from that depicted in the case of the optical fiber.

The reported results may have implications in image transmission through multimode fibers, and may also be applied to the temporal case. The use of the extreme localization due to disorder may be employed to secure connections based on random systems, as well as novel superresolved spectroscopy systems.

Acknowledgments

We acknowledge support from the European Research Council projects VANGUARD (grant number 664782) and Complexlight (grant number 201766), and the Templeton Foundation (grant number 58277).

References

1 Anderson, P. (1958) Absence of diffusion in certain random lattices. *Physical Review*, **109**, 1492–1505.
2 John, S. (1987) Strong localization of photons in certain disordered dielectric superlattices. *Physical Review Letters*, **58**, 2486–2489.
3 Wiersma, D.S., Bartolini, P., Lagendijk, A., and Righini, R. (1997) Localization of light in a disordered medium. *Nature*, **390**, 671–673.
4 Sperling, T., Buhrer, W., Aegerter, C., and Maret, G. (2013) Direct determination of the transition to localization of light in three dimensions. *Nature Photonics*, **7** (1), 48–52.
5 Gentilini, S., Fratalocchi, A., Angelani, L., Ruocco, G., and Conti, C. (2009) Ultrashort pulse propagation and the Anderson localization. *Optics Letters*, **34**, 130.
6 Schwartz, T., Bartal, G., Fishman, S., and Segev, M. (2007) Transport and Anderson localization in disordered two-dimensional photonic lattices. *Nature*, **446**, 52–55.
7 Szameit, A., Kartashov, Y., Zeil, P., *et al.* (2010) Wave localization at the boundary of disordered photonic lattices. *Optics Letters*, **35** (8), 1172–1174.
8 Snyder, A.W. and Mitchell, D.J. (1997) Accessible solitons. *Science*, **276**, 1538–1541.
9 Leonetti, M., Karbasi, S., Mafi, A., and Conti, C. (2014) Observation of migrating transverse Anderson localizations of light in nonlocal media. *Physical Review Letters*, **112**, 193902.
10 Leonetti, M., Karbasi, S., Mafi, A., and Conti, C. (2014) Experimental observation of disorder induced self-focusing in optical fibers. *Applied Physics Letters*, **105**, 171102.
11 Leonetti, M., Karbasi, S., Mafi, A., and Conti, C. (2014) Light focusing in the Anderson regime. *Nature Communications*, **5**, 5534.
12 Turitsyn, S.K. (1985) Spatial dispersion of nonlinearity and stability of multidimensional solitons. *Teor. Mat. Fiz.*, **64**, 226.
13 Krolikowski, W., Bang, O., Nikolov, N.I., *et al.* (2004) Modulational instability, solitons and beam propagation in spatially nonlocal nonlinear media. *Journal of Optics B: Quantum and Semiclassical Optics*, **6** (5), S288.
14 Conti, C. (2005) Complex light: Dynamic phase transitions of a light beam in a nonlinear nonlocal disordered medium. *Physical Review E*, **72** (6), 066 620.

15 Bang, O., Krolikowski, W., Wyller, J., and Rasmussen, J.J. (2002) Collapse arrest and soliton stabilization in nonlocal nonlinear media. *Physical Review E*, **66**, 046 619.

16 Conti, C., Peccianti, M., and Assanto, G. (2004) Observation of optical spatial solitons in a highly nonlocal medium. *Physical Review Letters*, **92** (11), 113 902.

17 Rotschild, C., Cohen, O., Manela, O., Segev, M., and Carmon, T. (2005) Solitons in nonlinear media with an infinite range of nonlocality: first observation of coherent elliptic solitons and of vortex-ring solitons. *Physical Review Letters*, **95** (21), 213 904.

18 Conti, C. (2012) Solitonization of the Anderson localization. *Physical Review A*, **86**, 061 801.

19 Petrović, M.S., Aleksić, N.B., Strinić, A.I., and Belić, M.R. (2013) Destruction of shape-invariant solitons in nematic liquid crystals by noise. *Physical Review A*, **87**, 043 825.

20 Izdebskaya, Y.V., Desyatnikov, A.S., and Kivshar, Y.S. (2013) Self-induced mode transformation in nonlocal nonlinear media. *Physical Review Letters*, **111**, 123 902.

21 I.M. Lifshits, S.A. Gredeskul, and L.A. Pastur, *Introduction to the Theory of Disordered Systems* (John Wiley & Sons, New York, 1988).

22 Conti, C., Angelani, L., and Ruocco, G. (2007) Light diffusion and localization in three-dimensional nonlinear disordered media. *Physical Review A*, **75** (3), 033812.

23 Shadrivov, I.V., Bliokh, K.Y., Bliokh, Y.P., Freilikher, V., and Kivshar, Y.S. (2010) Bistability of Anderson localized states in nonlinear random media. *Physical Review Letters*, **104** (12), 123 902.

24 Folli, V. and Conti, C. (2012) Two-level laser by the interaction of self-induced transparency pulses and surface Anderson localizations of light. *Journal of the Optical Society of America B*, **29** (8), 2080–2089.

25 Barsi, C., Wan, W., Sun, C., and Fleischer, J.W. (2007) Dispersive shock waves with nonlocal nonlinearity. *Optics Letters*, **32** (20), 2930–2932.

26 Ghofraniha, N., Gentilini, S., Folli, V., DelRe, E., and Conti, C. (2012) Shock waves in disordered media. *Physical Review Letters*, **109**, 243 902.

27 Deykoon, A.M. and Swartzlander, G.A. (2001) *Journal of the Optical Society of America B*, **18**, 804.

28 Wyller, J., Krolikowski, W., Bang, O., and Rasmussen, J.J. (2002) Generic features of modulational instability in nonlocal Kerr media. *Physical Review E*, **66**, 066 615.

29 Wurtz, G.A., Pollard, R., Hendren, W., *et al.* (2011) Designed ultrafast optical nonlinearity in a plasmonic nanorod metamaterial enhanced by nonlocality. *Nature nano.*, **6**, 107–111.

30 Kartashov, Y.V., Vysloukh, V.A., and Torner, L. (2008) Brownian soliton motion. *Physical Review A*, **77** (5), 051 802.

31 Folli, V. and Conti, C. (2010) Frustrated Brownian motion of nonlocal solitary waves. *Physical Review Letters*, **104** (19), 193 901.

32 Ghofraniha, N., Conti, C., Ruocco, G., and Trillo, S. (2007) Shocks in nonlocal media. *Physical Review Letters*, **99** (4), 043903.

33 G.B. Whitham, *Linear and Nonlinear Waves* (Wiley Interscience, New York, 1974).

34 Ghofraniha, N., Santamaria Amato, L., Folli, V., *et al.* (2012) Measurement of scaling laws for shock waves in thermal nonlocal media. *Optics Letters*, **21**, 612.

35 Honeycutt R. L. (1992) Stochastic Runge-Kutta algorithms. I. White noise. *Phys. Rev. A*, **45**, 600.

36 Gentilini, S., Ghofraniha, N., DelRe, E., and Conti, C. (2012) Shock wave far-field in ordered and disordered nonlocal media. *Optics Express*, **20** (24), 27 369–27 375.

14

Wide Variability of Generation Regimes in Mode-Locked Fiber Lasers

Sergey V. Smirnov,[1] *Sergey M. Kobtsev,*[1] *and Sergei K. Turitsyn*[2]

[1] *Novosibirsk State University, Novosibirsk, Russia*
[2] *Aston Institute of Photonic Technologies, School of Engineering and Applied Science, Aston University, Birmingham, UK*

14.1 Introduction

The evolution of modern laser systems necessarily involves an improvement in the output radiation parameters required by applications, with a parallel simplification of handling and operation. More and more laser products are adopting the black box approach, delivering an adjustment- and maintenance-free user experience with the ability to control the laser parameters electronically. Until very recently, these features were achievable only in lasers with a quite basic optical layout, but today, many hands-off maintenance-free lasers incorporate complex optical configurations, generate ultra-short radiation pulses, and boast very long lifetimes. These features, which were until only recently no more than wishful thinking, very largely became a reality due to fiber-optical technologies, the added benefits of which include relatively high efficiency and a small footprint. This progress is due to both advances in materials science and optical engineering and a better understanding of the complex physics underlying the operation of laser systems. Mode-locked fibre lasers (MLFL) have now become an indispensable tool in fundamental research and an integral part of many technological processes. They are widely used in an extensive application field, including material processing, spectroscopy, two-photon microscopy, ophthalmology, and may other promising applications.

Mode-locked lasers emerged at the very dawn of the laser epoch. The first report of mode locking a He-Ne laser with the help of an acousto-optical modulator was published more than half a century ago [1]. Passive mode locking of a continuous-wave laser via a saturable absorber was not achieved until almost 10 years later [2]. In a dye laser using Rhodamine 6G, spectrally limited 1.5-ps pulses were generated and spectrally tuned within the range of 590–610 nm. A diethyloxadicarbocyanine solution in methanol was employed as a saturable absorber in [2].

One decade later, research was under way in mode-locked fibre lasers. One of the first publications on the subject [3] reported 1-ns pulses with energy of 17 pJ generated in a

Shaping Light in Nonlinear Optical Fibers, First Edition. Edited by Sonia Boscolo and Christophe Finot.

Nd-doped fibre laser. Within a few years, the ensuing attempts to shorten the duration of the generated pulses had led to orders-of-magnitude reduction, first to 90 ps [4], soon followed by a 4-ps result [5].

The past 50 years have witnessed profound progress in mode locking approaches. A large number of passive mode locking techniques were proposed and successfully implemented in lasers of different types: gas, solid-state, dye, diode, fiber lasers. Temporal compression of pulses generated in passively mode-locked solid-state lasers culminated in the achievement of attosecond pulse duration range (1 as = 10^{-18} sec). For fiber lasers, typical pulse duration values lie in the range of several hundred femtoseconds, although the best achieved results reach 10–20 fs. Generation of ultra-short pulses in fiber lasers is anything but simple because of high nonlinearity and large chromatic dispersion of group velocities in fiber resonators. Nevertheless, these lasers feature a number of undeniable advantages in comparison with other laser types, being robust, compact, and efficient coherent light sources. All this draws a great deal of attention to ultra-fast fiber lasers in relation to multiple applications such as supercontinuum generation and optical metrology, spectroscopy, microscopy, processing and characterization of nano- and meta-materials. Fiber beam delivery provides the most convenient way of transporting laser radiation to the point of application, a critically important circumstance in ophthalmology and several other biomedical applications of ultra-short pulses. This stresses once more the importance of studying the science of complex laser systems, treating lasers not only as an engineering device, but also as a nonlinear physical system.

Apart from important practical applications, mode-locked fiber lasers exhibit quite a fascinating physics of formation and propagation of ultra-short pulses in optical fibers. This chapter explores the diversity of generation regimes observed in passively mode-locked fiber lasers, which may take place both in different lasers relying on different configurations for the generation of ultra-short pulses and in a single continuously operating laser device while the parameters of its cavity are adjusted.

Before proceeding to the main subject of the properties and diversity of generation regimes, let us briefly summarize the basic principles and approaches used for passive mode locking in order to generate ultra-short pulses in fiber laser cavities.

By creating optical feedback, a laser resonant cavity selects from electromagnetic radiation so-called modes: intrinsic, or natural, oscillations of the internal electromagnetic field. The frequencies of longitudinal modes are governed by the requirement that the cavity length be a multiple of an integer or half-integer number of wavelengths, and the frequency difference of adjacent longitudinal modes depends on the cavity length. For a linear 10–20-m-long cavity (the typical length of a fiber laser resonator), the frequency difference of neighboring longitudinal modes amounts to 15–7.5 MHz. Provided that all the optical elements of the fiber laser allow generation over the entire spectral gain range (it may be as wide as 100 nm), the number of laser modes present in the laser's generation spectrum may reach enormous values of $\sim 10^6$. If the phases of individual laser radiation modes are independent of each other or of an external common factor, the output of such a laser is continuous and, in relation to statistical properties, approaches that of band-pass-filtered incandescent and other thermal light sources. A great number of continuous fiber lasers operate in this regime.

In order to achieve pulsed generation, it is necessary to either directly create temporally localized light waves inside the resonator (Q-switching, gain switching, or synchronous pumping technologies) or establish a correlation between individual modes of the laser cavity by applying active or passive mode locking. Methods of active mode locking are based on RF-driven acousto-optical modulation of either optical losses or phase incursion of intra-cavity radiation. This approach allows the generation of short (usually in the nano- and pico-second range of duration) laser pulses. In this chapter, however, we will mostly focus on passively mode-locked fiber lasers capable of producing even faster pico- and femto-second pulses. Passive mode locking relies on the introduction into the laser cavity of a saturable absorber, an optical element (or a group of elements), in which the optical losses depend on the intensity of the intra-cavity radiation in such a way that at higher intensities they become lower (saturate). This facilitates formation in the cavity of pulses with relatively high power and ensures the suppression of noise and continuous radiation components. Note that in terms of nonlinear science, the pulse is formed from linear dispersive waves due to the interplay between nonlinearity and dispersion (with the possible involvement of other effects) leading to light localization.

To date, passive mode locking in fiber lasers has been achieved through saturable absorbers based on a great number of different materials including carbon nano-tubes [6–12], graphene [13–15], molybdenum disulphide (MoS_2) [16,17], tungsten disulphide (WS_2) [18,19], topological isolators [20,21], and semiconductor saturable absorber mirrors (SESAM) [22–26].

In order to generate laser pulses with relatively high energy and peak power, it is advantageous to employ artificial saturable absorbers such as nonlinear optical loop mirrors (NOLM) [27]or nonlinear amplifying loop mirrors (NALM) [28] and nonlinear polarization evolution effect (NPE) [29]. In comparison to material-based saturable absorbers, these methods ensure a much longer lifetime and a much higher thermal damage threshold, thus allowing relatively high output pulse energies in passively mode-locked lasers without resorting to additional optical amplification [30–35]. Systems relying on artificial saturable absorbers also offer a unique possibility (discussed in more detail further on) of absorber parameter adjustment. The parameters of the output radiation are controlled through adjusting the power level of the optical pump, as well as the settings of the intra-cavity polarization controllers and/or phase plates. Though often more complex for understanding and handling, these methods potentially provide access to a very broad range of generated pulse parameters.

14.2 Variability of Generation Regimes

In our study of the variety of fiber laser generation regimes, we will rely on the numerical modeling of a ring fiber laser passively mode-locked due to the effect of nonlinear polarization evolution. This particular type of laser configuration was selected as a test bed because of its two polarization controllers defining the action of the artificial saturable absorber and offering a great number of degrees of freedom affecting the laser generation regime. The optical layout of the modeled laser is similar to those used in our previous studies [36–38], see Figure 14.1. The fiber laser has a ring cavity with a wave-division multiplexer (WDM) used to couple in the pump radiation. Either 2-m-long Er- or 8-m-long Yb-doped optical fiber is used in different experiments as the active medium, both of which produce qualitatively similar results at net-normal and all-normal cavity dispersion respectively. Either SMF-28 or normal-dispersion fiber (NDF) is used to elongate the cavity and thus increase pulse energy. Output laser radiation is

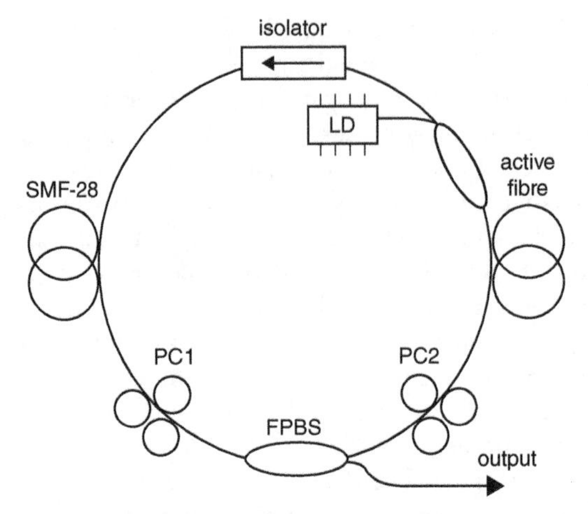

Figure 14.1 Laser layout used in experiments.
Key: LD laser diode, FPBS fibre polarization beam splitter; PC1, 2 fibre polarization controllers.

extracted through a fiber polarization beam splitter (FPBS) or through an additional coupler inserted into the cavity. Mode-locked operation was achieved by adjusting fiber polarization controllers PC1, 2. Operation regimes were studied with the help of a fast oscilloscope, an optical spectrum analyser (OSA), and an optical pulse auto-correlator.

In order to investigate the variety of possible single-pulse lasing regimes (the generation of only one pulse/train during laser cavity round trip) and their properties, we used a well-established numerical model based on a set of modified nonlinear Schrödinger equations for the orthogonal polarization components of the field envelope [39]:

$$\frac{\partial A_x}{\partial z} = i\gamma\left\{|A_x|^2A_x + \frac{2}{3}|A_y|^2A_x + \frac{1}{3}A_y^2A_x^*\right\} + \frac{g_0/2}{1+E/(P_{sat}\cdot\tau)}A_x - \frac{i}{2}\beta_2\cdot\frac{\partial^2A_x}{\partial t^2} \tag{14.1}$$

$$\frac{\partial A_y}{\partial z} = i\gamma\left\{|A_y|^2A_y + \frac{2}{3}|A_x|^2A_y + \frac{1}{3}A_x^2A_y^*\right\} + \frac{g_0/2}{1+E/(P_{sat}\cdot\tau)}A_y - \frac{i}{2}\beta_2\cdot\frac{\partial^2A_y}{\partial t^2} \tag{14.2}$$

where A_x, and A_y are the polarization components of the field envelope, z is the longitudinal coordinate along the fiber, t – time in the retarded frame of reference, γ and β_2 are nonlinear and dispersion coefficients correspondingly, g_0 and P_{sat} stand for unsaturated gain coefficient and saturation power for the active fiber.

Equations (14.1 and 14.2) describe the propagation of radiation along an active fiber. Taking $g_0 = 0$, we can use the same equations to model laser pulse propagation inside a passive resonator fiber. The fiber-optical polarization beam splitter is represented in the model by the following matrix:

$$\hat{T}_{PBS} = \begin{pmatrix} 1 & 0 \\ 0 & 0 \end{pmatrix}. \tag{14.3}$$

Unitary 2×2 matrices must be used to describe polarization controllers. In particular, a polarization controller based on the principle of fiber compression in the direction at an angle φ can be expressed as a matrix introducing phase delay α rotated by angle φ by multiplying it by the corresponding rotation matrix:

$$\hat{T}_{PC1}(\alpha,\varphi) = \begin{pmatrix} \cos\varphi & -\sin\varphi \\ \sin\varphi & \cos\varphi \end{pmatrix}\begin{pmatrix} e^{i\alpha/2} & 0 \\ 0 & e^{-i\alpha/2} \end{pmatrix}\begin{pmatrix} \cos\varphi & \sin\varphi \\ -\sin\varphi & \cos\varphi \end{pmatrix} \tag{14.4}$$

The parameter α in Eq. (14.4) stands for the phase delay introduced by the polarization controller and depends on the transverse fiber deformation. Similarly, a polarization controller using fiber torsion can be expressed through its eigenvector projections: $\hat{T}_{PC2} = e^{i\alpha/2}\hat{P}_{+} + e^{-i\alpha/2}\hat{P}_{-}$, where the circular polarization state projections can be written as

$$\hat{P}_{\pm} = \frac{1}{2}\begin{pmatrix} 1 & \mp i \\ \pm i & 1 \end{pmatrix}. \tag{14.5}$$

In order to model the propagation of laser pulses around the fiber cavity, Eqs. (14.1 and 14.2) can be integrated numerically by the step-split Fourier method [39]. At the required points along the optical path, the polarization transformations (3–5) are applied and losses corresponding to the intra-cavity elements are taken into account.

This modeling cycle is carried out repeatedly until a stationary state is reached. As a rule, it takes between several hundred and several thousand modeled cavity round trips to reach the asymptotic state, depending on the system parameters and the initial conditions, taken here as white noise or seed pulses. In certain cases, a laser may exhibit bi-stability: the limit cycle of the propagation equations may depend on the initial conditions. It is relevant to mention here that a similar phenomenon is observed in experiment as hysteresis in switching between generation regimes [40]. When the limit cycle of the propagation equations corresponds to the "conventional" pulse generation regime, the generation parameters (power, pulse duration, spectrum width, etc.) of the limit cycle are highly stable (of the 10^{-3} order and better). Conversely, when the laser generates double-scale pulses (regular trains of pico- or nanosecond wave packets stochastically filled with femtosecond sub-pulses), the pulse parameters in the generated train fluctuate around their average values by a few percent or even more between two successive cavity round trips. This circumstance can be used in modeling as a basic criterion to tell apart generation regimes. We need to point out as well that the average generation parameters must be insensitive to variation within reasonable limits of the "non-physical" modeling parameters, such as the mesh node count, mesh width, and the step of numerical integration of Eqs. (14.1 and 14.2). This has to be controlled during modeling.

The laser layout depicted in Figure 14.1 includes two fiber polarization controllers, which provide several degrees of freedom, including the ability to trigger the mode-locked operation and then switch between different generation regimes or adjust laser parameters. Since investigation of laser properties in multi-dimensional space is quite complicated, we confine ourselves to a statistical study of different generation regimes. The approach consists of choosing the polarization controller settings (tilt/slew angles) in a random way and integrating Eqs. (14.1 and 14.2) with white noise as the initial conditions. After a certain number (typically $10^2 \ldots 10^4$) of round trips, we may obtain one of the following solutions: (i) quasi-CW laser operation (no mode-lock reached; thus Eqs. (14.1 and 14.2) and their numerical solution are not valid); (ii) multi-pulse mode-locked operation (that is, two or more pulses co-exist in the cavity); and (iii) single-pulse, mode-locked operation. If a single-pulse, mode-locked operation is reached after the fixed amount of cavity round trips, the program saves the results (tilt angles of PC1, 2, pulse duration, energy, optical spectrum, temporal intensity distribution, etc.). Otherwise, no information is kept about this program run. In either case, the program selects a new combination of random PC settings and proceeds with the next run, thus accumulating information about single-pulse operation regimes.

The lasing regimes found in simulation demonstrate both quantitative and qualitative differences. Qualitatively, one can distinguish two main types of single-pulse generation regimes: fully coherent "regular" pulses and partially coherent noise-like (double-scale) pulses with a complex inner structure [37]. The temporal distribution of the radiation intensity for fully coherent pulses features a smooth envelope and can be described by a single parameter, that is, the envelope width. In contrast, the temporal intensity distribution of the radiation of partially coherent pulses is stochastic: inside wave-packets with overall duration of several picoseconds to several nanoseconds; there are fast stochastic variations of radiation intensity with typical time scale of a hundred to several hundred femtoseconds.

Correspondingly, the temporal distribution of radiation intensity for noise-like pulses is defined by two temporal parameters: the pulse-train envelope width and the typical intensity fluctuation time inside the train. Overall parameters of the entire wave packet, such as bandwidth, energy, and duration, fluctuate around their average values within a wide range from one wave packet to another. Intensity fluctuations inside a single wave packet may also vary from relatively small values up to peak intensity of the pulse. Lasing regimes with strong intensity fluctuations are usually classified as noise-like generation.

Noise-like pulses have not yet received a universal designation. Different terms are found across the available literature: noise-like pulses [41, 42], double-scale lumps [37], femtosecond clusters [43], etc. These types of pulses can easily be identified by a singular auto-correlation function shape featuring a narrow (100–200 fs) peak on a broader picosecond pedestal. Significant attention to these pulses is predominantly due to the presence of femtosecond components with high peak power, while the cavities of fiber lasers generating them may have relatively large dispersion.

There is also a series of intermediate possibilities between "regular" laser pulses and noise-like generation which exhibits a relatively small and variable fraction of intensity noise and phase fluctuations on the background of single-scale laser pulses [36]. Remarkably, even inside a single type of lasing (i.e., "regular" or noise-like pulses), the studied laser supports a large variety of sub-regimes that correspond to different settings of PC1, 2 and thus differ vis-à-vis energy, duration, bandwidth, etc. Generated pulse parameters may vary by an order of magnitude or even more, depending on PC settings. Probability density functions (PDF) for rms-duration (see Figures 14.2 (a) and (b)) and rms- bandwidth (see Figure 14.2 (c), (d)) obtained by randomly changing simulation PC settings are shown in Figure 14.2. These PDFs reveal the extent of parameter variability in different realizations of two main single-pulse lasing regimes, namely "regular" lasing (Figures 14.2 (a) and (c)) and noise-like pulse generation (Figures 14.2 (b) and (d)). For instance, rms-bandwidth varied in random simulation runs from 0.2 up to 3.9 nm for "regular" pulses and from 0.3 up to 7.4 nm for noise-like pulses as a function of PC settings. (Note that the given values correspond to the pulse rms-bandwidth, which is usually several times as narrow as the spectrum's full width at half-maximum, FWHM. As an example, for a Π-shaped spectrum, the ratio between spectral FWHM and rms-bandwidth is 3.5. For differently shaped spectra, this ratio may vary.)

In addition, Figures 14.2 (e) and (f) show PDF for output pulse energy in 'regular' and noise-like pulse generation regimes respectively. Similarly to rms pulse duration, pulse energy may also vary by an order of magnitude when intra-cavity PC angles are changed, as can be seen in Figure 14.2.

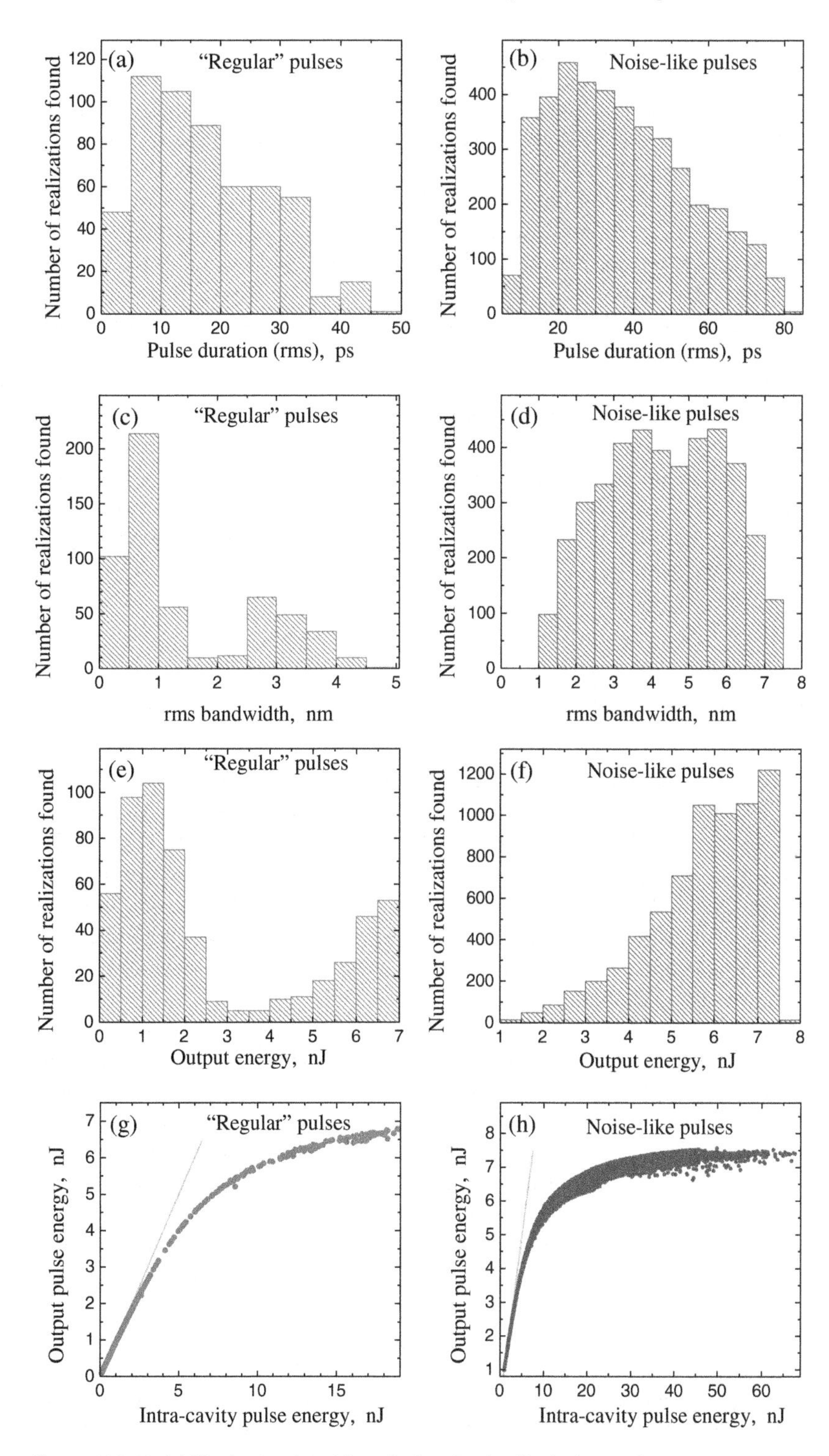

Figure 14.2 Variability in simulated "regular" and noise-like lasing regimes.

Highly different and widely ranging generation parameters in the studied fiber lasers passively mode-locked due to nonlinear polarization evolution (NPE) arise from the ability of the experimentalist to vary at will the properties of the artificial saturable absorber of an operating laser system by adjusting the intra-cavity polarization controllers. As an illustration, dots in Figures 14.2 (g) and (h) correspond to various "regular" (Figure 14.2 (g)) and noise-like (Figure 14.2 (h)) generation regimes identified in modeling and shown in coordinates intra-cavity pulse energy *vs.* output pulse energy. The gray solid line in these figures corresponds to $E_{out} = E_{in}$, which is achieved at 100% out-coupling of the intra-cavity radiation through the PBS. The observed spread of the angular coordinate of dots in these diagrams indicates that it is possible to control the amount of radiation coupled out of the cavity through PBS by adjustment of the intra-cavity polarization controllers.

Another important difference between lasing regimes attainable at different angular settings of intra-cavity PCs is related to the efficiency of nonlinear laser pulse conversion. Recently it was found that the efficiency of second harmonic generation (SHG) may vary by an order of magnitude or even more depending, along with other laser pulse parameters, on PC settings (see [38] and Figures 14.3 (a) and (b)). In order to generalize our conclusions, we study dimension-relative SHG efficiency ζ, which is defined [38] as

$$\zeta = \left(\sum_n \left| \sum_{j+k=n+1} A_{1j} A_{1k} \right|^2 \right) \cdot \left(\sum_j |A_{1j}|^2 \right)^{-2}.$$

In other words, ζ is the SHG relative efficiency which is equal to the ratio of two SH powers, of which the first is obtained when the nonlinear crystal is pumped by a double-scale laser pulse with given mode amplitudes A_{1j} and the second is generated with the use of single-mode monochromatic pumping of the same power $P_1 = \sum |A_{1j}|^2$. The dimensionless SHG relative efficiency ζ does not depend on the power and thickness of the thin nonlinear crystal but is sensitive to mode correlations and fluctuations, thus allowing us to easily compare different lasing regimes from the viewpoint of the efficiency of the nonlinear frequency.

As our simulation shows (see Figure 14.3 (c)), double-scale pulses have comparable or higher SHG-relative efficiency compared with that of single-scale laser pulses of the same duration. Figure 14.3 (d) shows the correlation between the SHG relative efficiency and spectral bandwidth of "regular" and noise-like laser pulses. It should be stressed that here we consider the case of a very thin crystal, so that the SHG efficiency tends to grow with the number of interacting spectral modes (and, as a result, with the spectral bandwidth of laser pulses).

We also conducted a comparative study of Raman scattering spectra generated by single-scale and double-scale laser pulses in a long stretch of extra-cavity fiber. As is illustrated below, notwithstanding the similar duration of single-scale and double-scale pulse envelopes, Raman spectra resulting from these pulses are substantially different.

Shown in Figures 14.4 and 14.5 are the radiation spectra and auto-correlation functions (ACF) of single-scale and double-scale pulses studied in our experiment. Pulses of both types were generated in the laser cavity of Figure 14.1 at different polarization

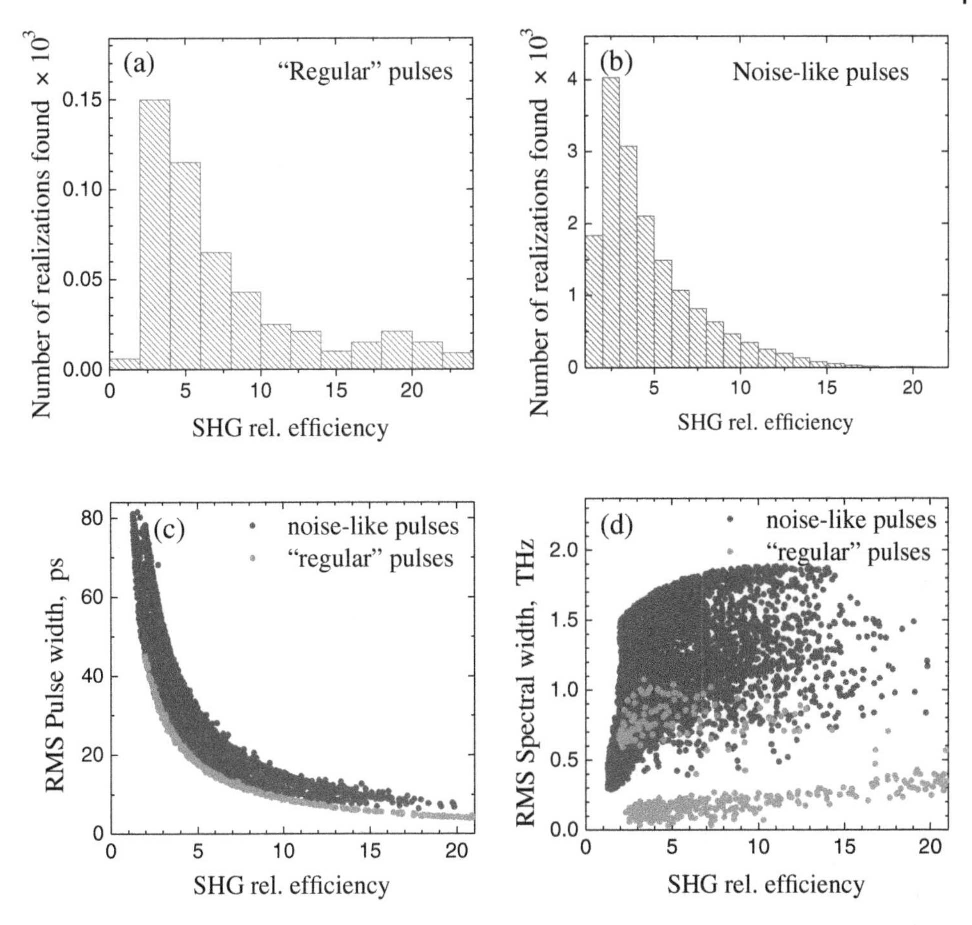

Figure 14.3 Relative SHG efficiency in "regular" (a) and noise-like (b) generation regimes and its correlation with rms pulse duration (c) and spectral width (d).

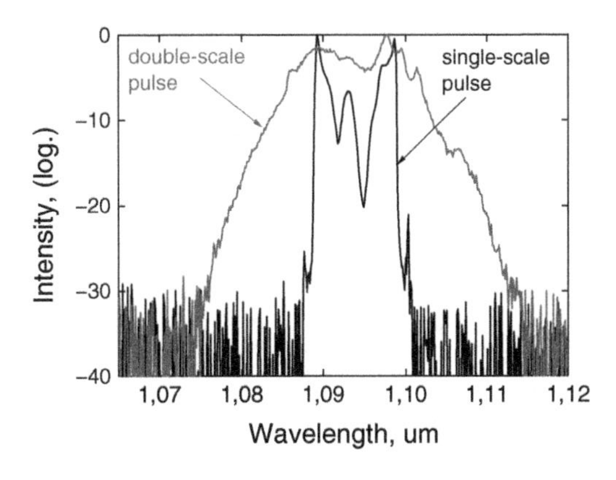

Figure 14.4 Radiation spectra of single-scale and double-scale pulses used in the experiment.

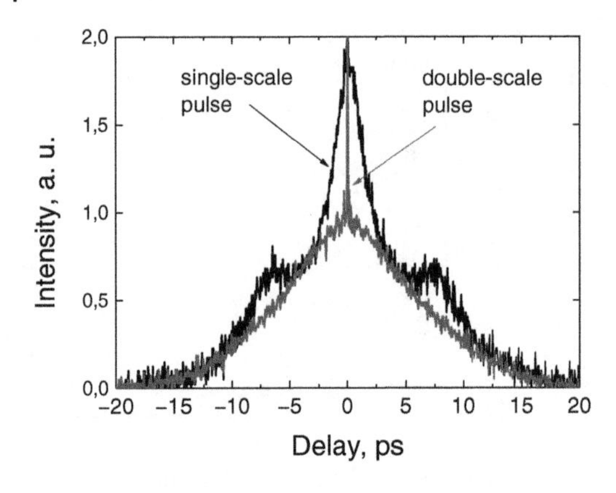

Figure 14.5 Auto-correlation functions of single-scale and double-scale pulses used in the experiment.

controller positions. ACFs were registered using commercial femto- and picosecond scanning auto-correlator "Tekhnoscan FS-PS-Auto." Auto-correlation traces appear to have a certain amount of noise, an effect caused by the optical power being insufficient for reliable detection. Optical spectra were measured with optical spectrum analyser Advantest Q8381A. Laser regimes were self-started and could be triggered by appropriate adjustment of PC and pump power (in the range of 1.3 and 1.5 W). Experimental results are in good qualitative agreement with the numerical simulation. Single-scale pulses have Π-shaped optical spectrum and quasi-bell-shaped ACF, whereas the spectrum of double-scale pulses is much smoother, bell-shaped, and its ACF contains a narrow (180-fs-wide) peak on a 12-ps-wide pedestal.

Figure 14.6 shows the Raman scattering spectra of single-scale (gray curves) and double-scale (black curves) pulses propagating along a 1.2-km long phosphosilicate fiber. In order to produce sufficient spectral conversion, pulses from the master fiber oscillator could be passed through a fiber optic amplifier to reach the average output power of 840 mW. This figure presents broader Raman scattering spectra of double-scale pulses in comparison to analogous spectra for single-scale pulses. Most likely, this is a result of femtosecond components present in the structure of double-scale pulses, which have higher nonlinear conversion efficiency.

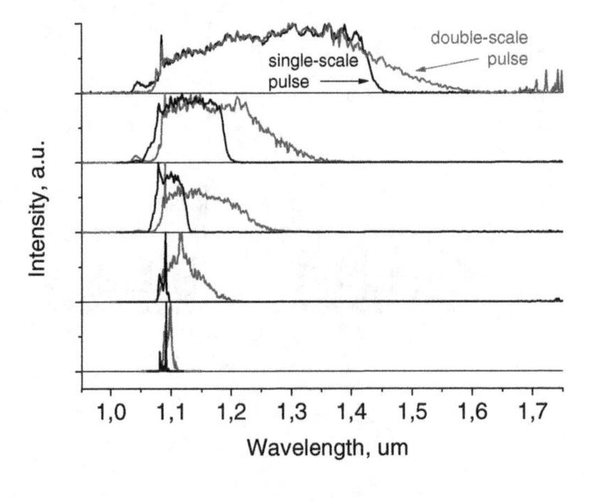

Figure 14.6 Raman radiation spectra from 1.2-km-long phosphosilicate fiber pumped with single-scale and double-scale pulses.

14.3 Phenomenological Model of Double-Scale Pulses

As was mentioned above, lasing regimes generating noise-like pulses may be used in a number of applications due to their particular properties, such as high peak power and low coherence time. The list of promising applications of noise-like pulses includes nonlinear frequency conversion, such as harmonic generation [38], Raman conversion [44, 45] and super-continuum generation [46–51], as well as applications in imaging and sensing systems with high temporal and/or spatial resolution. Excessive noise level intrinsic to noise-like pulses may constitute an obstacle for pulse compression [36, 52]. However, noise is known to play an important role in both supercontinuum generation [53–55] and Raman conversion, including soliton self-frequency shift effect [56–58] where noise may not only deteriorate SC coherence but also boost spectral broadening. It was furthermore demonstrated that the generation of noise-like pulses in long lasers represents a remarkably multiform phenomenon encompassing many nonlinear optical mechanisms, whose interaction may result in the emergence of diverse spatio-temporal coherent structures in laser radiation [59].

In order to benefit from noise-like pulses in various applications of laser physics, extensive numerical simulations are required. Here we propose a fairly simple phenomenological model of noise-like pulses, which is much easier to implement than the full-vector NLSE-based model used earlier. Since this model has a number of free parameters, it allows one to describe a large variety of lasing regimes achievable in experiments. Furthermore, this model helps one to obtain deeper insight into the nature of noise-like pulses by revealing their stochastic and deterministic properties.

The proposed simplified model of "random pulses" is constructed in two steps as follows. (i) Let us consider stochastic continuous radiation resulting from the superposition of a large number of uncorrelated modes within a given bandwidth. (ii) Let us shape the pulse by multiplying this stochastic quasi-CW radiation by a pulse envelope, for example, Gaussian pulse $\exp(-(t/T)^2 \ln 4)$, where T is the pulse full-width at half maximum (FWHM) and ln denotes natural logarithm. As a result, the random pulse envelope can be represented as

$$A(t) \sim \sqrt{P(t)} \cdot \sum_j A_j \exp(i\omega_j t). \tag{14.6}$$

Here, ω_j is the frequency of the j-th mode, t = time, A_j = complex amplitude of the j-th mode, $P(t)$ = temporal profile of the 'random' pulse, and $P(t) = \exp(-(t/T)^2\, 4 \ln 2)$. The phases of complex amplitudes $\arg\{A_j\}$ are taken as independent random variates uniformly distributed from 0 to 2π. Physically "random pulses" Eq. (14.6) may appear, for example, as a result of temporal shaping of spectrally filtered radiation of a light bulb. Let us compare the properties of such "random pulses" with those of noise-like pulses obtained in an NLSE-based model of a fiber laser.

First of all, let us consider the temporal intensity distributions shown in Figures 14.7 (a) and (b) for some random realization of Eq. (14.6) and for some random realization of noise-like pulses obtained in NLSE-based modeling respectively. Both temporal intensity distributions shown in Figure 14.7 (a) and Figure 14.7 (b) consist of numerous subpulses filling a bell-shaped envelope of a wave-packet. Note that in this particular case, the wave-packet envelope is Gaussian, however, any particular pulse shape $P(t)$ known

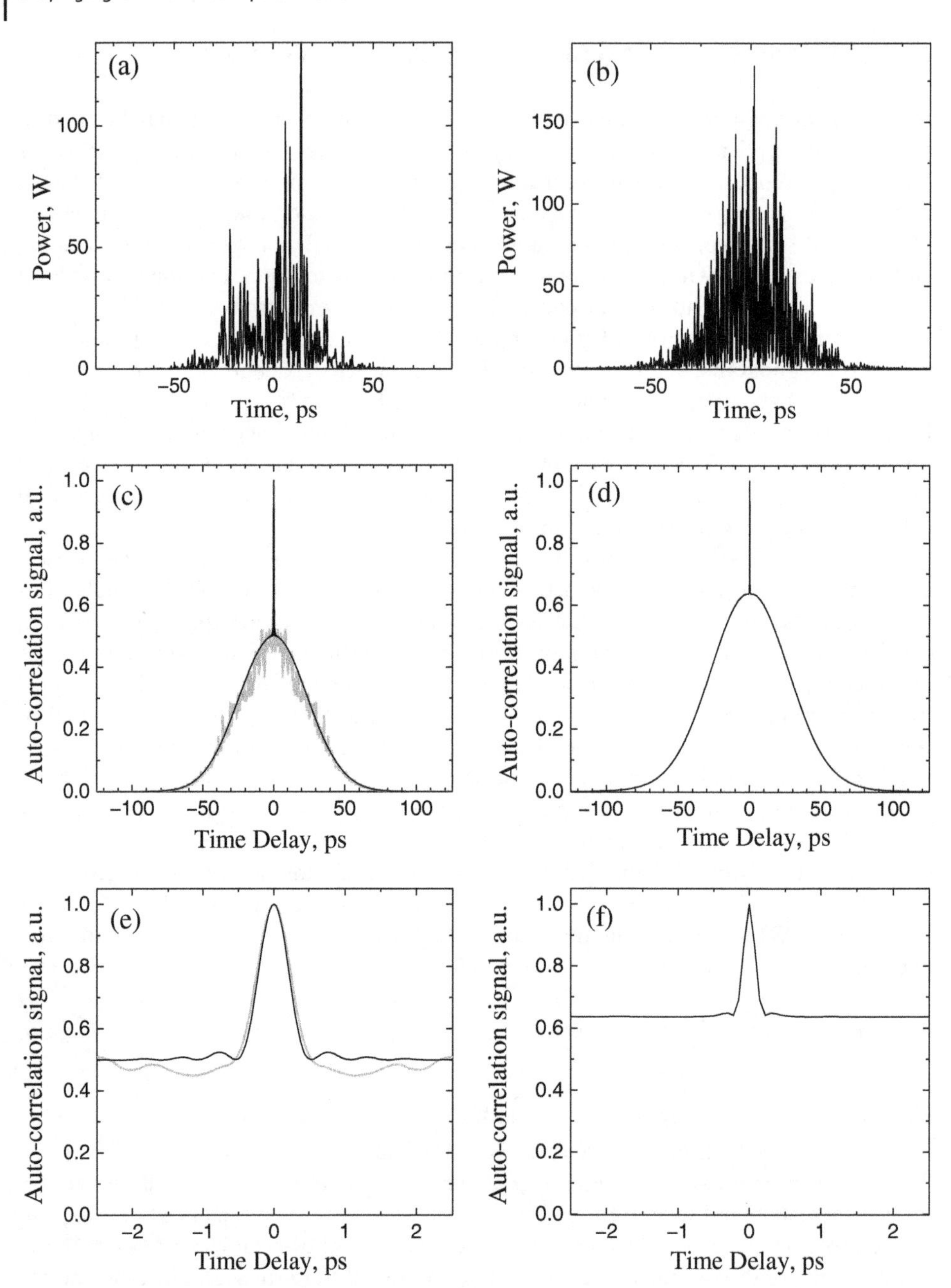

Figure 14.7 Temporal intensity distributions (a, d) and intensity auto-correlation functions (b, c, e, f) of "random" pulses (upper row, a, b, c) and noise-like pulses (lower row, d, e, f). Panels (c) and (f) illustrate the central ACF peak magnified from graphs (b) and (e) respectively.

from experiment may be used within the considered model in Eq. (14.6). Sub-pulses inside a wave-packet are located chaotically having random durations and amplitudes. Note that the duration of these sub-pulses is in the femtosecond range so that they cannot be directly resolved by oscilloscope in experiment. In most experimental configurations, the presence of such sub-pulses inside a wave-packet manifests itself as a narrow (femtosecond) peak on ACF. Figures 14.7 (c, d) show ACFs of a "random" pulse Eq. (14.6) and a noise-like pulse obtained in NLSE-based modeling respectively. The gray line in Figure 14.7 (c) corresponds to a single realization of ACF whereas the black line appears as a result of the ensemble averaging over a large number of random realizations of A_j in Eq. (14.6). It should be noted that both ACFs shown in Figure 14.7 (c) and Figure 14.7 (d) have a bell-shaped pedestal with a narrow peak on it. Central peaks of both ACFs are shown magnified in Figure 14.7 (e) and Figure 14.7 (f). It can readily be seen from Figure 14.7 (e) and Figure 14.7 (f) that there is a relatively minor but qualitative difference between "random" pulses Eq. (14.6) and noise-like pulses obtained within NLSE-based model Eqs. (14.1 and 14.2) which consists of the height of the ACF pedestal. In the proposed model of "random" pulses Eq. (14.6), the height of the ACF pedestal is equal to 0.5 (see Figure 14.7 (e)), indicating the completely random nature of radiation. However, for noise-like pulses obtained within an NLSE-based model (Eqs. 14.1, 14.2), the ACF pedestal is a bit higher (about 0.64 in height), suggesting the presence of some additional mode correlations inherent to noise-like pulses that are not taken into account by the proposed model Eq. (14.6).

Spectra of "random" pulses Eq. (14.6) and noise-like pulses obtained within NLSE-based model (Eqs. 14.1, 14.2) are shown in Figure 14.8 (a) and Figure 14.8 (c) respectively. In both cases, gray lines indicate single random realizations of the pulse spectrum, whereas black lines correspond to the result of ensemble averaging. Similar to the temporal intensity distribution, the optical spectrum of a single pulse is peaky. Experimentally measured spectra, as a rule, are smooth due to averaging over multiple successive pulses in a pulse train generated by a mode-locked laser in experiment. One can easily see completely different forms of ensemble-averaged spectra shown in Figure 14.8. Thus the spectrum of noise-like pulses obtained within the NLSE-based model (Eqs. 14.1, 14.2) and shown in Figure 14.8 (c) has a bell-shaped form whereas the spectrum of "random" pulses Eq. (14.3) shown in Figure 14.8 (a) is rectangular (Π-shaped). It should be stressed that this difference is not a drawback of the proposed model. Indeed, the model Eq. (14.6) can use any spectral shape $|A_j(\omega)|^2$ known from experiment or direct simulation of laser generation. We used a rectangular Π-shaped spectrum in Eq. (14.6) when we plotted Figure 14.8 (a) only for the sake of simplicity.

Finally, let us consider mode correlations that appear in NLSE-based simulations and in the model of "random" pulses, see Eq. (14.6). In what follows, we will use the coefficient of mode correlations γ calculated for a pair of modes in simulations, one of them is the central mode of the spectrum, another one has frequency detuning ν from the central spectral mode:

$$\gamma(\nu) = \gamma(A_\nu, A_0) = \frac{\left|\sum A_\nu A_0^*\right|}{\sqrt{\left(\sum |A_\nu|^2\right) \cdot \left(\sum |A_0|^2\right)}} \tag{14.7}$$

where A_ν and A_0 = complex amplitudes of spectral modes, one (A_ν) with frequency detuning ν from the center of the spectrum, the other (A_0) is the amplitude of the

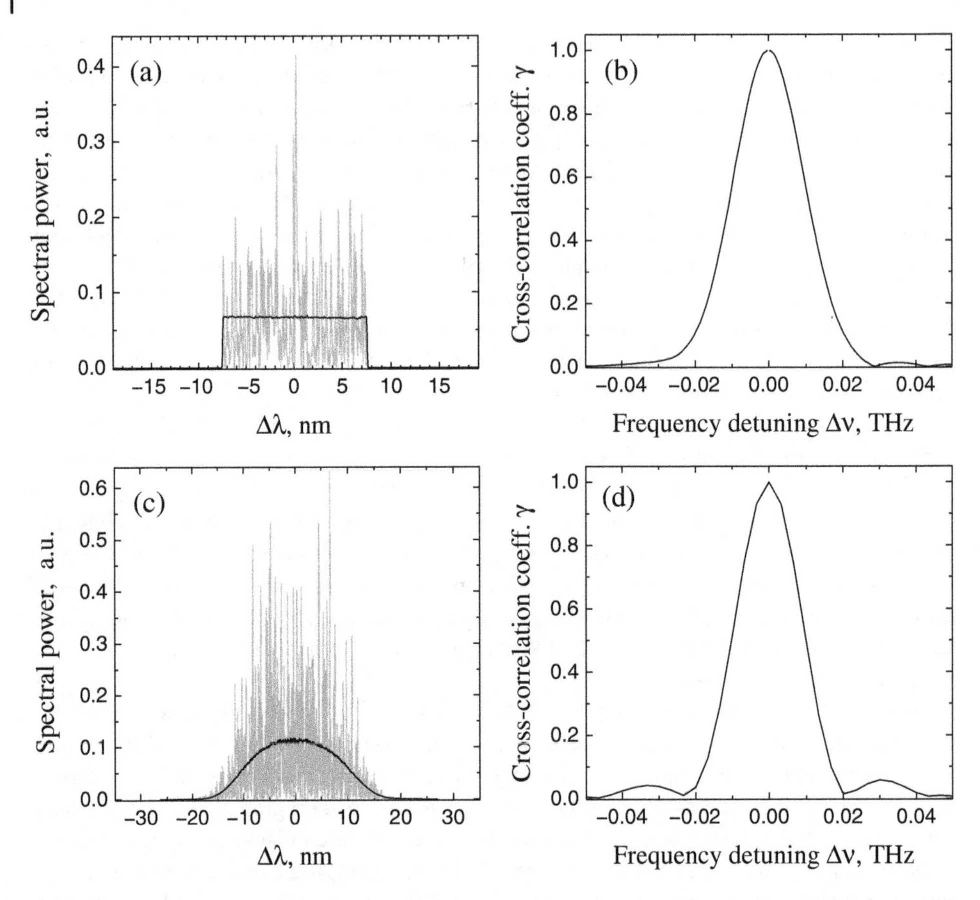

Figure 14.8 Spectra (a, c) and coefficient of mode correlations $\gamma(\Delta\nu,0)$ (b, d) of "random" pulses (upper row a, b) and noise-like pulses (lower row c, d).

central mode in the generated spectrum. The graphs $\gamma\ (\nu)$ for "random" pulses Eq. (14.6) and noise-like pulses obtained within NLSE-based model (Eqs. 14.1, 14.2) are shown in Figures 14.8 (b) and 14.8 (d) respectively. The correlation coefficient of any variate with itself is equal to unity, $\gamma\ (\nu = 0) = 1$. Let us note that both graphs look similar. Adjacent spectral modes of both "random" and noise-like pulses are strongly correlated but the correlation coefficient drops rapidly down to 0 in a spectral area much narrower than the pulse bandwidth. The width of the spectral area of correlated modes is equal to the spectral width of pulse envelope both for "random" and noise-like pulses. It should be noted that different modes A_j in Eq. (14.6) were independent before being multiplied by $P(t)^{1/2}$, which was carefully checked in simulations. Thus, mode correlations of "random" pulses Eq. (14.6) appear as a result of pulse shaping.

14.4 Conclusion

Passively mode-locked fiber lasers based on artificial saturable absorption, such as nonlinear polarization evolution effect, NOLM/NALM, etc., possess a significant number of degrees of freedom (e.g., orientation angles of intra-cavity polarization elements and the pump source power). Owing to these degrees of freedom, these lasers demonstrate a great variety of generation regimes differing from each other both qualitatively and in quantitative parameters (energy and duration of pulses, spectral width, etc.), which latter may vary by an order of magnitude and even more. For instance, along with "regular" pulses, passively mode-locked fiber lasers may also generate noise-like pulses (or double-scale lumps [37]), the terms used to designate wave-packets stochastically filled with short sub-pulses. The duration of these sub-pulses is usually within the femtosecond or sub-picosecond range, while their peak power may be a few times or even by an order of magnitude higher than the average peak power of the wave packet. Modeling of such lasers is more challenging and sensitive to the initial conditions [60]. The peculiar structure and unique properties of noise-like pulses make them very attractive in a number of applications, including nonlinear frequency conversion, imaging, and sensing systems. The parameters of such pulses, including duration, energy, and spectral width, may also vary within very broad limits. Moreover, due to highly nonlinear intra-cavity dynamics, the order of elements in the cavity may be important for the optimization of the laser system performance [61]. The proposed phenomenological model, assuming noise-like pulses to be the result of the superposition of uncorrelated modes with a given (freely chosen) spectral shape and temporal envelope, may be advantageous in modeling practical applications of noise-like pulses. Note that the double-scale clusters of pulses may be advantageous for the ablation applications similar to burst-mode operating lasers [62].

Acknowledgments

This work was supported by grants of the Ministry of Education and Science of the Russian Federation (agreement No. 14.B25.31.0003, order No. 3K 3.889.2017/PCH) the Russian Foundation for Basic Research (agreement No. 16-02-00104).

References

1 Hargrove, L.E., Fork, R.L., and Pollack, M.A. (1964) Locking of He–Ne laser modes induced by synchronous intracavity modulation. *Applied Physics Letters*, **5**, 4–5.
2 Ippen, E.P. Shank, C.V. and Dienes, A. (1972) Passive mode locking of the CW dye laser. *Applied Physics Letters* **21**, 348–350.
3 Alcock, I.P., Ferguson, A.I., Hanna, D.C., and Tropper, A.C. (1986) Mode-locking of a neodymium-doped monomode fibre laser. *Electronics Letters* **22** (5), 268–269.
4 Geister, G. and Ulrich, R. (1988) Neodymium-fibre laser with integrated-optic mode locker. *Optics Communications*, **68** (3), 187–189.
5 Kafka, J.D., Baer, T., and Hall, D.W. (1989) Mode-locked erbium-doped fiber laser with soliton pulse shaping. *Optics Letters*, **14** (22), 1269–1271.
6 Set, Y.S., Yaguchi, H., Tanaka, Y., and Jablonski, M. (2004) Laser mode locking using a saturable absorber incorporating carbon nanotubes. *Journal of Lightwave Technology*, **22**(1), 51–56.
7 Kieu, K. and Mansuripur, M. (2007) Femtosecond laser pulse generation with a fiber taper embedded in carbon nanotube/polymer composite. *Optics Letters*, **32** (15), 2242–2244.

8 Yamashita, S., Inoue, Y., Maruyama, S., *et al.* (2004) Saturable absorbers incorporating carbon nanotubes directly synthesized onto substrates and fibers and their application to mode-locked fiber lasers. *Optics Letters*, **29** (14), 1581–1583.

9 Wang, F., Rozhin, A.G., Scardaci, V., *et al.* (2008) Wideband-tuneable, nanotube mode-locked, fibre laser. *Nature Nanotechnology*, **3** (12), 738–742.

10 Schibli, T.R., Minoshima, K., Kataura, H., *et al.* (2005) Ultrashort pulse-generation by saturable absorber mirrors based on polymer-embedded carbon nanotubes. *Optics Express*, **13** (20), 8025–8031.

11 Kieu, K. and Wise, F.W. (2008) All-fiber normal-dispersion femtosecond laser. *Optics Express*, **16** (15), 11453–11458.

12 Song, Y.W., Yamashita, S., and Maruyama, S. (2008) Single-walled carbon nanotubes for high-energy optical pulse formation. *Applied Physics Letters*, **92** (2), 021115.

13 Bao, Q., Zhang, H., Wang, Y., *et al.* (2009) Atomic-layer graphene as a saturable absorber for ultrafast pulsed lasers. *Advanced Functional Materials*, **19** (19), 3077–3083.

14 Sun, Z., Hasan, T., Torrisi, F., *et al.* (2010) Graphene mode-locked ultrafast laser. *ACS Nano*, **4** (2), 803–810.

15 Zhang, H., Tang, D.Y., Zhao, L.M., Bao, Q.L., and Loh, K.P. (2009) Large energy mode locking of an erbium-doped fiber laser with atomic layer graphene. *Optics Express*, **17** (20), 17630–17635.

16 Zhang, H., Lu, S.B., Zheng, J., *et al.* (2014) Molybdenum disulfide (MoS2) as a broadband saturable absorber for ultra-fast photonics. *Optics Express*, **22** (6), 7249–7260.

17 Du, J., Wang, Q., Jiang, G., *et al.* (2014) Ytterbium-doped fiber laser passively mode locked by few-layer Molybdenum Disulfide (MoS2) saturable absorber functioned with evanescent field interaction. *Scientific Reports*, **4**, 6346.

18 Woodward, R.I. and Kelleher, E.J. (2015) 2D saturable absorbers for fibre lasers. *Applied Science*, **5** (4), 1440–1456.

19 Mao, D., Zhang, S., Wang, Y., *et al.* (2015) WS_2 saturable absorber for dissipative soliton mode locking at 1.06 and 1.55 μm. *Optics Express*, **23** (21), 27509–27519.

20 Zhao, C., Zhang, H., Qi, X., *et al.* (2012) Ultra-short pulse generation by a topological insulator based saturable absorber. *Applied Physics Letters*, **101** (21), 211106.

21 Yu, H., Zhang, H., Wang, Y., *et al.* (2013) Topological insulator as an optical modulator for pulsed solid-state lasers. *Laser Photonics Review*, **7** (6), L77–L83.

22 Keller, U., Miller, D.A.B., Boyd, G.D., *et al.* (1992) Solid-state low-loss intracavity saturable absorber for Nd:YLF lasers: an antiresonant semiconductor Fabry-Perot saturable absorber. *Optics Letters*, **17** (7), 505–507.

23 Keller, U., Weingarten, K.J., Kärtner, F.X., *et al.* (1996) Semiconductor saturable absorber mirrors (SESAM's) for femtosecond to nanosecond pulse generation in solid-state lasers. *IEEE Journal of Selected Topics in Quantum Electronics*, **2** (3), 435–453.

24 Okhotnikov, O.G., Jouhti, T., Konttinen, J., Karirinne, S., and Pessa. M. (2003) 1.5-μm monolithic GaInNAs semiconductor saturable-absorber mode locking of an erbium fiber laser. *Optics Letters*, **28** (5), 364–366.

25 Okhotnikov, O.G., Gomes, L., Xiang, N., Jouhti, T., and Grudinin, A.B. (2003) Mode-locked ytterbium fiber laser tunable in the 980–1070-nm spectral range. *Optics Letters*, **28** (17), 1522–1524.

26 Okhotnikov, O., Grudinin, A., and Pessa, M. (2004) Ultra-fast fibre laser systems based on SESAM technology: new horizons and applications. *New Journal of Physics*, **6** (1), 177–198.

27 Doran, N.J. and Wood, D. (1988) Nonlinear-optical loop mirror. *Optics Letters*, **13** (1), 56–58.

28 Fermann, M.E., Haberl, F., Hofer, M., and Hochreiter, H. (1990) Nonlinear amplifying loop mirror. *Optics Letters*, **15** (13), 752–754.

29 Matsas, V.J., Newson, T.P., Richardson, D.J., and Payne, D.N. (1992) Self-starting, passively mode-locked fibre ring soliton laser exploiting non-linear polarisation rotation. *Electronics Letters*, **28** (15), 1391–1393.

30 Kobtsev, S., Kukarin, S., and Fedotov, Y. (2008) Ultra-low repetition rate mode-locked fiber laser with high-energy pulses. *Optics Express*, **16** (26), 21936–21941.

31 Nyushkov, B.N., Denisov, V.I., Kobtsev, S.M., *et al.* (2010) Generation of 1.7-μJ pulses at 1.55 μm by a self-modelocked all-fiber laser with a kilometers-long linear-ring cavity. *Laser Physics Letters*, **7** (9), 661–665.

32 Kobtsev, S.M., Kukarin, S.V., Smirnov, S.V., and Fedotov, Y.S. (2010) High-energy mode-locked all-fiber laser with ultralong resonator. *Laser Physics*, **20** (2), 351–356.

33 Ivanenko, A., Turitsyn, S., Kobsev, S., and Dubov, M. (2010) Mode-locking in 25-km fiber laser. *European Conference on Optical Communication (ECOC)*, 1–3.

34 Fedotov, Y.S., Ivanenko, A.V., Kobtsev, S.M., and Smirnov, S.V. (2014) High average power mode-locked figure-eight Yb fibre master oscillator. *Optics Express*, **22** (25), 31379–31386.

35 Smirnov, S.V., Kobtsev, S.M., Kukarin, S.V. and Turitsyn, S.K. (2011) *Mode-Locked Fibre Lasers with High-Energy Pulses*. InTech, Moscow. Chapter 3.

36 Smirnov, S., Kobtsev, S., Kukarin, S., and Ivanenko, A. (2012) Three key regimes of single pulse generation per round trip of all-normal-dispersion fiber lasers mode-locked with nonlinear polarization rotation. *Optics Express*, **20** (24), 27447–27453.

37 Kobtsev, S., Kukarin, S., Smirnov, S., Turitsyn, S., and Latkin, A. (2009) Generation of double-scale femto/pico-second optical lumps in mode-locked fiber lasers. *Optics Express*, **17** (23), 20707–20713.

38 Smirnov, S., Kobtsev, S., and Kukarin, S. (2014) Efficiency of non-linear frequency conversion of double-scale pico-femtosecond pulses of passively mode-locked fiber laser. *Optics Express*, **22** (1), 1058–1064.

39 G. P. Agrawal (2001) *Nonlinear Fiber Optics*, 3rd ed., Academic Press, San Diego, CA.

40 Komarov, A., Leblond, H., and Sanchez, F. (2005) Multistability and hysteresis phenomena in passively mode-locked fiber lasers. *Physical Review A*, **71**, 053809.

41 Horowitz, M., Barad, Y., and Silberberg, Y. (1997) Noise-like pulses with a broadband spectrum generated from an erbium-doped fiber laser. *Optics Letters*, **22** (11), 799–801.

42 Pottiez, O., Grajales-Coutiño, R., Ibarra-Escamilla, B., Kuzin, E.A., and Hernández-García, J.C. (2011) Adjustable noise-like pulses from a figure-eight fiber laser. *Applied Optics*, **50** (25), E24–E31.

43 Nie, B., Parker, G., Lozovoy, V.V., and Dantus, M. (2014) Energy scaling of Yb fiber oscillator producing clusters of femtosecond pulses. *Optical Engineering*, **53** (5), 051505.

44 Kobtsev, S., Kukarin, S., Smirnov, S., and Ankudinov, I. (2014) Cascaded SRS of single- and double-scale fiber laser pulses in long extra-cavity fiber. *Optics Express.* **22** (17), 20770–20775.

45 Kobtsev, S., Kukarin, S., and Smirnov, S. (2015) Supercontinuum from single- and double-scale fiber laser pulses in long extra-cavity P2O5-doped silica fiber. *Proceedings of the SPIE*, **9347**, 93471X-1.

46 Kobtsev, S., Kukarin, S., and Smirnov, S. (2010) All-fiber high-energy supercontinuum pulse generator. *Laser Physics*, **20** (2), 375–378.

47 Hernandez-Garcia, J.C., Pottiez, O., and Estudillo-Ayala, J.M. (2012) Supercontinuum generation in a standard fiber pumped by noise-like pulses from a figure-eight fiber laser. *Laser Physics*, **22** (1), 221–226.

48 Zaytsev, A., Lin, C., You, Y., *et al.* (2013) Supercontinuum generation by noise-like pulses transmitted through normally dispersive standard single-mode fibers. *Optics Express*, **21** (13), 16056–16062.

49 Lin, S., Hwang, S., and Liu, J. (2014) Supercontinuum generation in highly nonlinear fibers using amplified noise-like optical pulses. *Optics Express*, **22** (4), 4152–4160.

50 Kobtsev, S., Kukarin, S., Smirnov, S., and Fedotov, Y. (2010) Ultra-wide-tunable fibre source of femto- and picosecond pulses based on intracavity Raman conversion. *Proceedings of the SPIE*, **7580**, 758023.

51 Smirnov, S.V., Ania-Castanon, J.D., Ellingham, T.J., *et al.* (2006) Optical spectral broadening and supercontinuum generation in telecom applications. *Optical Fiber Technology*, **10** (2), 122–147.

52 Smirnov, S.V., Kobtsev, S.M., and Kukarin, S.V. (2015) Linear compression of chirped pulses in optical fibre with large step-index mode area. *Optics Express*, **23** (4), 3914–3919.

53 Kobtsev, S.M. and Smirnov, S.V. (2005) Modelling of high-power supercontinuum generation in highly nonlinear, dispersion shifted fibers at CW pump. *Optics Express*, **13** (18), 6912–6918.

54 Kobtsev, S.M. and Smirnov, S.V. (2008) Temporal structure of a supercontinuum generated under pulsed and CW pumping. *Laser Physics*, **18** (11), 1260–1263.

55 Kobtsev, S.M. and Smirnov, S.V. (2008) Influence of noise amplification on generation of regular short pulse trains in optical fibre pumped by intensity-modulated CW radiation. *Optics Express*, **16** (10), 7428–7434.

56 Kobtsev, S.M., and Smirnov, S.V. (2006) Coherent properties of super-continuum containing clearly defined solitons. *Optics Express*, **14** (9), 3968–3980.

57 Kobtsev, S.M., Kukarin, S.V., Fateev, N.V. and Smirnov, S.V. (2005) Coherent, polarization and temporal properties of self-frequency shifted solitons generated in polarization-maintaining microstructured fibre. *Applied Physics B*, **81** (2), 265–269.

58 Kobtsev, S.M., Kukarin, S.V., Fateev, N.V., and Smirnov, S.V. (2004) Generation of self-frequency-shifted solitons in tapered fibers in the presence of femtosecond pumping. *Laser Physics*, **14** (5), 748–751.

59 Churkin, D.V., Sugavanam, S., Tarasov, N., *et al.* (2015) Stochasticity, periodicity and localized light structures in partially mode-locked fibre lasers. *Nature Communications*, **6**, 7004.

60 Yarutkina, I.A., Shtyrina, O.V., Fedoruk, M.P., and Turitsyn, S.K. (2013) Numerical modeling of fiber lasers with long and ultra-long ring cavity. *Optics Express*, **21**, 12942–12950.

61 Shtyrina, O.V., Yarutkina, I.A., Skidin, A., Fedoruk, M.P., and Turitsyn, S.K. (2015) Impact of the order of cavity elements in all-normal dispersion ring fiber lasers. *IEEE Photonics Journal*, 7(2) 1501207.

62 Kalaycıoğlu, H., Akçaalan, Ö., Yavaş, S., Eldeniz, Y.B., and Ilday, F.Ö. (2015) Burst-mode Yb-doped fiber amplifier system optimized for low-repetition-rate operation. *Journal of the Optical Society of America B*, **32**, 900–906.

15

Ultralong Raman Fiber Lasers and Their Applications

Juan Diego Ania-Castañón[1] and Paul Harper[2]

[1] *Instituto de Óptica, CSIC, C/Serrano 121, Madrid, Spain*
[2] *Aston Institute of Photonic Technologies, School of Engineering and Applied Science, Aston University, Birmingham, UK*

15.1 Introduction

In 1923, Austrian theoretical physicist Adolf Smekal predicted the inelastic scattering of photons [1]. This effect was later experimentally proven by C.V. Raman and K.S. Krishnan [2, 3], initially in liquids, and independently by G. Landsberg and L. Mandelstam in crystals [4]. The phenomenon would become known as the Raman effect, and C.V. Raman was awarded the 1930 Nobel Prize in Physics for its discovery. Raman scattering, like Brillouin scattering, is a photon scattering process mediated by phonons, but whereas in the case of Brillouin scattering these phonons are low-frequency ones associated with acoustic waves traveling through the medium, in the case of Raman scattering we are talking of high-frequency phonons (usually in the range of 10^{12} to 10^{14} Hz) associated with vibrational energy states of the medium molecules. Since the energy of such states can be used to unequivocally identify a given molecule, Raman scattering has traditionally been used in spectroscopy. Following the usual absorption and emission naming convention, scattered photons that lose energy are referred to as Stokes photons, whereas those which gain energy are called Anti-Stokes photons. Since Anti-Stokes emission requires that the molecule be already in an excited state when the incident photon arrives, the emission of Stokes photons is much more probable. In practical terms in optical fibers, Raman scattering results in the frequency shift of the radiation propagating through the waveguide, with the probability of shifting to different frequencies depending on the vibrational energy state distribution of the molecules of the medium. If high enough numbers of photons are present at a given Stokes frequency, then the probability of scattering at such frequencies is enhanced, and then process becomes stimulated. Spontaneous Raman scattering shows no direction preference and is polarization-preserving. Stimulated Raman scattering is, similarly, highly polarization-dependent, and produces a polarization-dependent gain. In the case of standard Silica optical fibers with a Ge-doped core, the maximum probability of scattering corresponds to a Stokes frequency shift of about 13.2×10^{12} Hz.

Shaping Light in Nonlinear Optical Fibers, First Edition. Edited by Sonia Boscolo and Christophe Finot.

Stimulated Raman Scattering (SRS) has already been discussed in greater detail elsewhere in this book, so in this chapter we will restrict ourselves to some basic considerations upon which to build the fundamentals of ultralong Raman fiber lasers. We will start from the simplest possible configuration for the study of SRS in an optical fiber, in which radiation from a single-frequency depolarized Raman laser pump experiences Raman scattering in an optical fiber, stimulated by a signal co-propagating with the pump at its optimal Stokes frequency. Ignoring thermal noise terms and Rayleigh backscattering, the interaction between the optical powers of the pump (*Pp*) and signal (*Ps*) is given by the following set of ODEs:

$$\frac{dP_s}{dz} = g_R P_P P_S - \alpha_S P_S \tag{15.1}$$

$$\frac{dP_P}{dz} = -\frac{\nu_P}{\nu_S} g_R P_P P_S - \alpha_S P_P \tag{15.2}$$

where the subindexes P and S denote respectively pump and Stokes, α represents the fiber attenuation coefficient, ν is the frequency and g_R is the Raman gain coefficient from ν_P to ν_S, which is related to the imaginary part of the third-order nonlinear susceptibility of the medium and inversely proportional to the effective area of the fiber.

Under the additional assumption of small signal (i.e. low pump depletion, with $P_P \gg P_S$), this set of ODEs can be solved analytically to obtain the Stokes power at the end of a fiber of length L:

$$P_S(L) = P_S(0)e^{(g_R P_0 L_{eff} - \alpha_S L)} \tag{15.3}$$

where P_0 is the input pump power at $z = 0$, L_{eff} being the effective interaction length of the fiber at the pump frequency, defined as

$$L_{eff} = \int_0^L \frac{P_P(z)}{P_0} = \frac{1 - e^{(-\alpha_P L)}}{\alpha_P} \tag{15.4}$$

15.2 Raman Amplification

From Eq. (15.3), it becomes obvious that SRS can be used to induce amplification in an optical fiber, transferring energy from a pump to a signal located at a frequency close to the Stokes scattering resonance. It is entirely beyond the scope of this short introduction to review the main characteristics, advantages and performance limitations of Raman amplifiers, but several excellent review textbooks have been published over the past few years covering this topic, from which the reader can obtain a much more complete and thorough view [5, 6]. Raman amplification can be obtained from either co-propagating pumps, counter-propagating pumps or both, and can be achieved through a single Stokes shift (first-order) or several (higher-order). It can be used to provide periodical, lumped amplification, or as a tool to transform the whole transmission fiber into a distributed amplifier. The latter offers the advantage of an improved noise figure when compared to the former, so distributed Raman amplification is the most common choice. Indeed, Raman distributed amplifiers offer two clear advantages over lumped amplification based on doped fiber transitions (such as Erbium-doped amplifiers), namely:

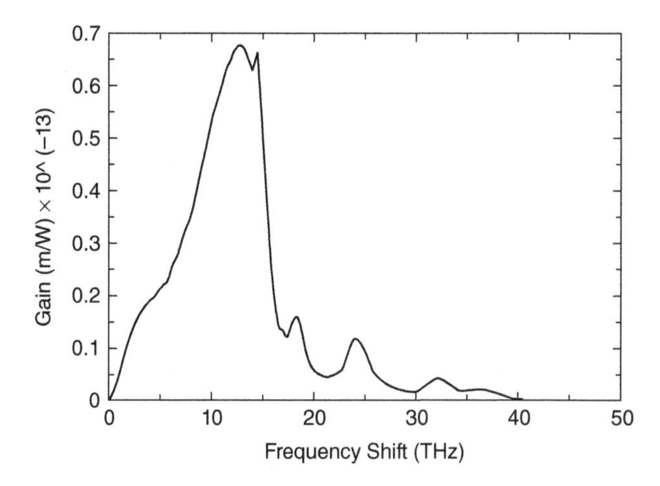

Figure 15.1 Raman gain profile for standard silica fiber, referred to a pump at 1455 nm.

- *Improved ASE noise performance.* As mentioned above, distributing the total gain across the transmission span reduces the build-up of noise. The noise figure (*NF*) of an amplifier is defined as the ratio between the input and output signal-to-noise ratios (SNR) of the signal. In the case of a distributed amplifier, it is typical to define the equivalent noise figure (NF_{eq}) as the noise figure that would correspond to a discrete amplifier placed at the end of the span, with identical gain and generated noise to that of the distributed amplifier. When derived from either the amplified field fluctuations model or from amplified spontaneous emission theory [7], the equivalent ASE noise figure takes the form:

$$NF_{eq} = \frac{1}{G_R} + \frac{2P_{ASE}}{G_R h\nu_S B_0} \tag{15.5}$$

where h is Planck's constant, B_0 is the optical bandwidth, P_{ASE} is the power of the ASE noise generated in the amplifier:

$$P_{ASE}(L) = (G(L) - 1)h\nu_s B_0 + \alpha_s G(L) h\nu_s B_0 \int_0^L [1/G(z)]\, dz \tag{15.6}$$

where $G(z)$ represents the Raman net gain, and G_R is the Raman on-off gain, which are related by:

$$G(L) = \exp\left(-\alpha_S L + g_R \int_0^L P_P(z)dz\right) = G_R \exp(-\alpha_S L) \tag{15.7}$$

where $P_P(z)$ represents pump power.
In the small-signal approximation, $G_R = \exp(g_R P_0 L_{eff})$, and

$$NF_{eq} = \exp\left(-g_R P_0 L_{eff}\right)\left(1 + \frac{2P_{ASE}}{h\nu_S B_0}\right) \tag{15.8}$$

which can even take values lower than 1 (or negative in a *dB* scale), showcasing the notable advantage of distributed Raman amplification in terms of noise performance in comparison to that of an lumped amplifier providing an equivalent on-off gain.

- *Improved gain bandwidth and flatness.* The fact that the Raman effect can shift all frequencies propagating down the fiber, instead of being limited to a given frequency range associated with a certain rare earth transition means that it is possible to extend the amplification bandwidth over the whole fiber transmission window, provided adequate pumps are found. A flat profile spectral profile can be achieved by using multiple simultaneous pumps at different wavelengths [8].

Unfortunately, the use of distributed Raman amplification also has some particular problems associated with its very nature, which are not present or play a less important role in traditional lumped amplifiers. These are:

- *RIN noise transfer from the pumps to the signal.* Raman gain is a very fast process which takes places at subpicosecond time scales, hence, fluctuations in the pump power (usually referred as relative-intensity-noise, or RIN) translate directly into gain oscillations that end up imprinted in the amplified signal [9, 10]. The RIN transfer function vs. oscillation frequency of a given amplifier will depend on the actual amplifier configuration and fiber base. Transfer from counter-propagating pumps is reduced at high frequencies as oscillations are averaged over the length of the fiber. In the case of co-propagating pumps, averaging takes place only due to the different propagation velocities of pump and signal (chromatic dispersion), and hence reduces RIN transfer only for oscillations at considerably higher frequencies. High-power fiber lasers, such as the ones used for pumping in long-distance Raman amplification schemes, are inherently very noisy, and thus RIN transfer can be a very limiting impairment.
- *Double Rayleigh backscattering noise.* Rayleigh scattering is the elastic scattering of electromagnetic radiation by particles much smaller than its wavelength. In optical fibers, density fluctuations during the fabrication process distribute scattering centers randomly through the fiber, that are responsible for the main contribution to its nearly-constant attenuation coefficient. Rayleigh-scattered photons can be deflected in any direction, so a small fraction of them will be backscattered, and an even smaller fraction will fulfill the critical angle condition to be confined into the fiber in a counter-propagating direction to that of the transmitted signal . Some of these will be back-reflected again at a random scattering point down the fiber and end up propagating with the signal [11]. Since the phenomenon is elastic, these noisy photons share the frequency of the signal and cannot be spectrally filtered out.

  In a system without distributed amplification, the small capture fraction of backscattered photons is low enough so that double Rayleigh backscattering (DRB) noise is not an issue. But with distributed amplification, backscattered radiation is amplified both in the co- and counter-propagating direction, and DRB noise can be a very important source of noise.
- *Gain and noise spectral tilt.* As a consequence of the non-null Raman gain coefficient even for small frequency shifts, signals or pumps at short wavelengths will amplify those at longer wavelengths, even if these are not at the optimal Stokes shift for maximum gain. In addition, the amount of ASE generated will depend on the frequency difference between the pump and the signal. Both effects combine to create a tilt on the gain and noise spectra, in which longer wavelengths experience a higher gain and noise build-up. Although the effect can be small over short distances, it can become very relevant after transmission through a large number of periodic cells.

15.3 Ultralong Raman Fiber Lasers Basics

15.3.1 Theory of Ultralong Raman Lasers

15.3.1.1 Concept of URFL and Basic Theory

Cascading in high-order Raman amplification helps to push signal gain further into the span, which translates into a more efficient distribution of gain, and a lower overall variation of the effective gain-loss coefficient experienced by the signal over the whole transmission span. A perfectly even distribution of the gain along the propagation distance would lead to ideally lossless transmission. Lossless transmission is not necessarily the optimal solution for every situation, since it can lead to strong nonlinear impairments from relatively low signal powers, but it has nevertheless been a long-term goal of optical communications, as it would bring with it a minimization of the amplified spontaneous emission (ASE) noise build-up. Indeed, in situations in which ASE can be considered the only meaningful contribution to noise, the fundamental limit for the ratio between optical signal-to-noise ratio (OSNR) and path-averaged signal power is approached as signal power variations along the transmission line are minimized. A quasi-lossless transmission system would be ideal both for systems in which ASE is the limiting source of noise and for systems in which high nonlinearities play a positive role, such as those based on conventional solitons.

With this in mind, the idea of an ultra-long Raman fiber laser was first proposed in 2004 [12] as a simple scheme to achieve high-order pumping from a single wavelength. In the proposed design, schematically depicted in Figure 15.2, two equal-power primary pumps at 1366 nm are launched from both extremes of a periodic transmission cell comprised of standard single-mode fiber (SMF). This bi-directional pumping structure is combined with two fiber Bragg grating-reflectors positioned at both ends of the cell. The central wavelength of the gratings is chosen to be 1455 nm, in the vicinity of the primary pumps' Stokes peak, so that the pair of gratings and the fiber create a cavity for radiation at this wavelength. If the primary pumps power is above the threshold necessary to overcome the attenuation of the first Stokes, a stable secondary pump at 1455 nm is generated in the cavity from the ASE at this wavelength. This secondary pump, that is used to amplify the signal centered at 1550 nm, presents a nearly constant combined forward- and backward-propagating power, and can therefore provide a nearly constant gain that can be adjusted to closely match the signal attenuation at every step of the propagation. The choice of wavelengths in the proposed example has been dictated by the adoption of the typical transmission wavelength of about 1550 nm for our

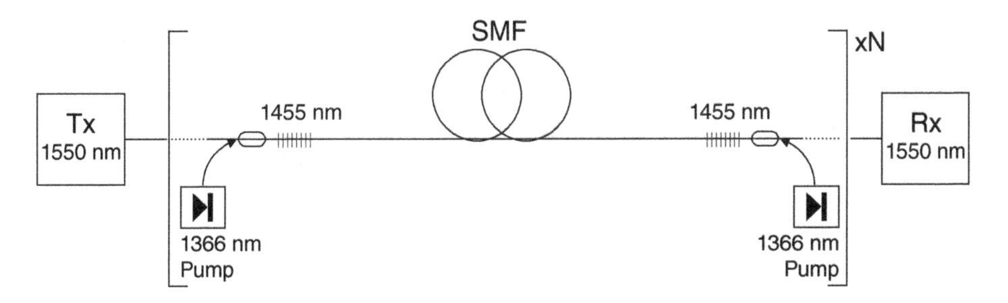

Figure 15.2 Schematic depiction of a typical URLF amplifier. *Source*: Ania-Castañón 2004 [12].

signal. Although a wavelength slightly higher than 1366 nm for the primary pumps would have been desirable in order to maximize the Raman conversion to the 1455 nm component, the proximity of the water peak in the fiber attenuation profile (in the vicinities of 1370 nm) informed our choice.

The structure of such an amplification scheme is essentially that of a standard cavity Raman fiber laser with a single cascade, but with a cavity extending over the whole transmission length, and design focused not on optimizing the output radiation or conversion efficiency from the primary pump to the generated Stokes, but on providing optimal amplification to a signal propagating itself at a frequency close to the Raman Stokes resonance of the characteristic frequency of the laser.

The system can be described in terms of average power evolution by the corresponding steady-state laser equations, which take into account pump depletion, ASE and double Rayleigh scattering (DRS) noise:

$$\frac{dP_{P1}^{\pm}}{dz} = \mp\alpha_1 P_{P1}^{\pm} \mp g_1 \frac{\nu_1}{\nu_2} P_{P1}^{\pm}\left(P_{P2}^{+} + P_{P2}^{-} + 4h\nu_2\Delta\nu_2\left(1 + \frac{1}{e^{h(\nu_1-\nu_2)/K_BT} - 1}\right)\right) \pm \epsilon_1 P_{P1}^{\mp} \tag{15.9}$$

$$\begin{aligned}\frac{dP_{P2}^{\pm}}{dz} &= \mp\alpha_2 P_{P2}^{\pm} \mp g_2 \frac{\nu_2}{\nu_S} P_{P2}^{\pm}\left(N_S^{+} + N_S^{-} + P_S + 4h\nu_S\Delta\nu_S\left(1 + \frac{1}{e^{h(\nu_2-\nu_S)/K_BT} - 1}\right)\right)\\ &\quad \pm g_1\left(P_{P2}^{\pm} + 2h\nu_2\Delta\nu_2\left(1 + \frac{1}{e^{h(\nu_1-\nu_2)/K_BT} - 1}\right)\right) \pm \epsilon_2 P_{P2}^{\mp}\end{aligned} \tag{15.10}$$

$$\frac{dP_S}{dz} = -\alpha_S P_S + g_2 P_S\left(P_{P2}^{+} + P_{P2}^{-}\right) \tag{15.11}$$

$$\frac{dN_S^{+}}{dz} = -\alpha_S N_S^{+} + g_2\left(N_S^{+} + 2h\nu_S\Delta\nu_S\left(1 + \frac{1}{e^{h(\nu_2-\nu_S)/K_BT} - 1}\right)\right)\left(P_{P2}^{+} + P_{P2}^{-}\right) + \epsilon_S N_S^{-} \tag{15.12}$$

$$\frac{dN_S^{-}}{dz} = \alpha_S N_S^{-} - g_2\left(N_S^{-} + 2h\nu_S\Delta\nu_S\left(1 + \frac{1}{e^{h(\nu_2-\nu_S)/K_BT} - 1}\right)\right)\left(P_{P2}^{+} + P_{P2}^{-}\right) - \epsilon_S(P_S + N_S^{+}) \tag{15.13}$$

Here, the (+) and (−) superscripts represent forward and backward propagation respectively, the 1, 2, and S subscripts identify the first-order pump, second-order pump and signal respectively, P_{Pi} are the pump powers, P_S are the signal powers and N_S are the noise powers at the frequency of the signal, the i are the corresponding frequencies of the pumps and signal, the i represent the effective bandwidths of the secondary pump and the signal (limited, in the case of the secondary pump, by the bandwidth of the fiber Bragg gratings), the gi are the corresponding Raman gain coefficients (divided by the effective area) for each of the Raman transitions, the i are the fiber attenuations at each respective frequency, h is Planck's constant, K_B is Boltzmann's constant, T is the absolute temperature of the fiber and the ϵ_i are the Rayleigh backscattering coefficients of the fiber at each particular frequency. Please note that all noise contributions have been grouped into the N_S equations, but ASE and Rayleigh backscatter terms present different

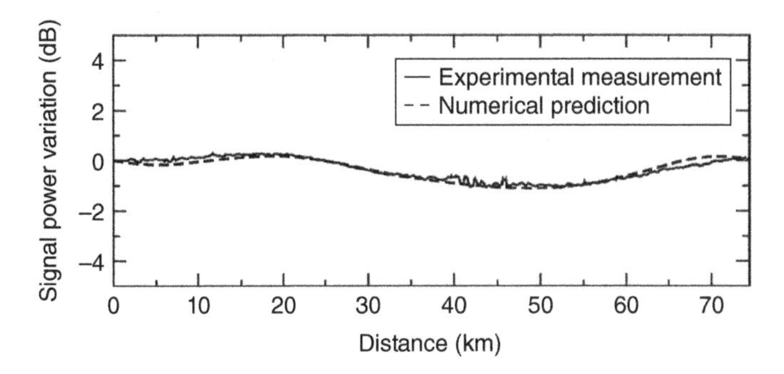

Figure 15.3 Predicted and experimentally measured signal power evolution in a 75 km URFL. *Source*: Ania-Castañón 2006 [13]. Reproduced with permission of American Physical Society.

statistics, and hence it might be of use to treat them separately depending on the actual problem being studied.

The boundary conditions for the problem defined in Eqs. (15.9)–(15.13) are:

$$P_{P1}^{+}(0) = P_{P1}^{-}(L) = P_0; P_{P2}^{+}(0) = R_1 P_{P2}^{-}(0); P_{P2}^{-}(L) = R_2 P_{P2}^{+}(L);$$
$$N_S^{+}(0) = N_0; N_S^{-}(L) = 0; P_S(0) = P_{IN} \quad (15.14)$$

where R_1 and R_2 are the reflectivities of the fiber Bragg gratings at the beginning and end of the periodic cell respectively (typically close to 99 %), and L is the length of the transmission span. The model safely assumes that the effect of frequencies far from the first and second Stokes of the primary pump can be considered negligible.

The evolution of the signal, pumps and noise powers in the fiber can be precisely obtained by solving numerically Eqs. (15.9)–(15.13) with boundary conditions Eq. (15.14), at room temperature (25 °C) and typical parameter values for standard-single mode fiber. The first experimental demonstration of ultralong Raman fiber lasers for amplification [13] showcased the excellent agreement between theory and experiment, as can be seen in Figure 15.3, which compares the model prediction and the experimentally measured (via OTDR) power evolution for a 1 mW 1550 nm signal propagating through a 75 km standard single-mode fiber.

Note that this simplified model does not account for the Raman gain from the primary pump at 1366 nm to the signal, since this is considered negligible when compared to the direct gain from the 1455 nm Stokes. As we will see in subsequent sections, this term plays an important role in extending the available amplification bandwidth in multichannel transmission, and opens up the possibility of achieving simultaneous spatial-spectral transparency in medium-length URFL cavities. Similarly, the set of equations describes an URFL with a single cascade, but the model can easily be extended to include higher-order cascading (with pumps at wavelengths of 1280 nm or shorter), which would provide even more efficient gain distribution at the cost of reduced conversion efficiency [14].

15.3.1.2 Mode Structure and Dephasing Mechanisms

Taking advantage of the reduced attenuation offered by standard optical fiber in the telecommunication spectral window, it is possible to reach very long cavity lengths while still retaining a resolvable cavity mode structure that can be observed in the laser RF spectrum through the fixed spectral separation between intermode beating peaks, confirming the formation of an ultralong standing electromagnetic wave.

Of course, such an enormous resonant cavity presents a similarly extraordinary number of longitudinal cavity modes [15]. Indeed, the observed spectral separation between modes clearly follows the classical formula $\Delta\nu = c/2nL$, where n is the refractive index of the fiber core, c the speed of light and L the cavity length, which for a typical grating bandwidth of 100 or 200 GHz, brings the number of modes to the hundreds of millions. These modes are broadened and eventually washed out as they interact with each other through intensity-dependent, turbulent-like, four-wave mixing processes in the fiber, meaning that in order for the mode-structure to be resolvable at such extended cavity lengths, Stokes wave intensity must be kept low. Note that other manifestations of the Kerr nonlinearity such as self-phase or cross-phase modulation (SPM, XPM) do not lead to the broadening as they change the mode phases synchronously.

Weak wave turbulence is a fundamental nonlinear phenomenon that occurs in a variety of nonlinear dispersive physical systems, and has been shown to describe efficiently the weak four-wave-mixing interaction between a large number of neighboring longitudinal modes in Raman fiber lasers [15], which leads, as intra-cavity power increases, to the broadening of the laser spectra and the washing-out of the modes themselves, if the spectral separation between modes becomes equal to or smaller than the individual mode bandwidth.

As predicted by the weak-turbulence model presented in [17], the broadening caused by the FWM interaction between the hundreds of millions of modes is mostly dependent on intra-cavity power at the lasing wavelength, and only weakly dependent cavity length itself. This means that the mode structure of URFLs can be resolved in the RF spectrum only for a given range of powers, ranging from the laser threshold to the point at which intra-cavity Stokes power reaches a certain critical value. Assuming a stable intracavity power of around 10 mW, it can be estimated that modal structure in an SMF-based URFL would be impossible to observe for lengths above 1000 km.

Figure 15.4 shows the experimental values for spectral width of the RF peaks as a function of the total intracavity power at 1455 nm for cavity lengths of 6.6, 22, 44, and 84 km and the corresponding mode spacing values marked by dots 22 km, dash dots 44 km, and dashes 84 km; 15.5 kHz spacing at 6.6 km lies beyond the graph. b is maximum power with resolved mode structure as a function of the cavity length; solid lines are a linear fit and bA/x fit.

Two physical effects contribute to the washing out of these modes and the eventual loss of coherence: their interaction through the aforementioned intensity-dependent, turbulent-like, four-wave mixing processes in the fiber, and the presence of Rayleigh backscattering [16]. The random back-reflection of Stokes photons by Silica molecules along the optical fiber forms a family of overlapping cavities of randomly varying length. Solving Eqs. (15.9)–(15.13) one can find that for a system based on standard single-mode fiber, the amount of radiation reflected in these random scatterings and the amount of radiation reflected at the ultra-long laser cavity gratings themselves become comparable when the length of fiber is of the order of 270 km, in agreement with the observed experimental limit for mode resolution. Random distributed feedback Raman fiber lasers (RDFLs) display different transients and spectral characteristics from cavity URFLs, which make them more flexible than cavity lasers for applications such as multiwavelength generation [18], generally at the cost of presenting a much lower

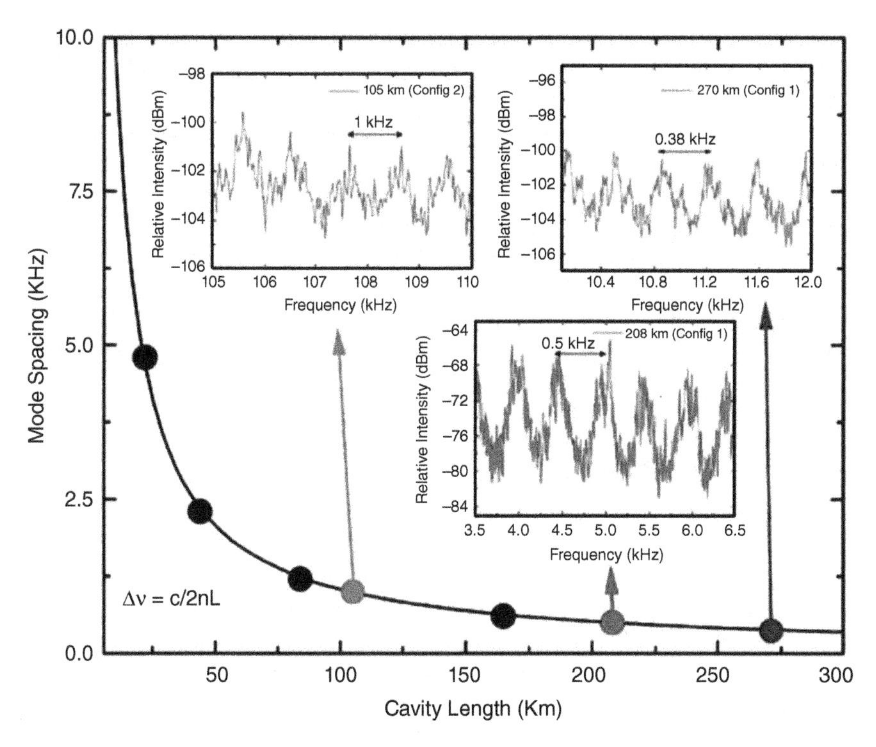

Figure 15.4 Mode spacing vs. cavity length in URFLs. Inset: intensity beating spectra for 105, 208 and 270 km length URFLs. *Source*: Turitsyn 2009 [16]. Reproduced with permission of American Physical Society.

conversion efficiency. Shorter lengths RDFLs are easily achievable by disposing of the cavity altogether, by removing one or both FBG reflectors.

15.3.1.3 Kinds of Ultralong Lasers

Considering all of the above, and from the viewpoint of their coherence preservation and gain distribution characteristics, we can distinguish three more or less clear regimes in single-cascade URFLs:

- Mid-length URFLs, with maximum distances of the order of a few 10s of km, depending on the fiber base, in which lasing leads to very clearly defined modes, and gain can be distributed evenly enough to achieve virtual transparency over a broad range of frequencies. These lasers allow for minimum ASE noise growth. They also provide a nearly ideal laboratory testbed for the experimentation with virtually non-dissipative, integrable nonlinear optical systems and can be of use, for example, in the implementation of classical soliton systems. Short lengths allow the use of lower-powered pump lasers, which can help avoid RIN transfer issues and reduce DRB noise in amplification applications. They can also be used for the efficient generation of supercontinuum.
- Long URFLs, with cavity lengths for SMF in the ranges of 40 to 270 km. In these lasers, mode structure is still preserved at low pump powers. They are typically used as amplifiers in telecommunications and distributed sensing schemes, which benefit from the increased contrast provided by efficiently distributed amplification. Although they

still offer a clear advantage over other amplification systems in terms of ASE, their reliance on high-power pumps means they are affected by issues such as RIN, DRB noise and power-dependent reflectivity.

- RDFLs [19], with either open-ended cavities or cavity lengths too long for longitudinal modes to be observed due to the decoherence caused by the Rayleigh effect. These schemes have lower conversion efficiencies, but can still find interesting applications, for example, in the development of multi-wavelength fiber laser sources, or by providing low-RIN, long-distance amplification through specific designs.

In Section 15.3 we will review a number of applications for URFLs on the different aforementioned categories. But for now, we will focus on the basics of URFL-based amplification.

15.3.2 Amplification Using URFLs

15.3.2.1 URFL Optimization for Quasi-Lossless Transmission

As we have seen in Section 15.3.1.1, URFLs can distribute Raman gain across the transmission fiber as efficiently as a second-order Raman amplification system, while relying exclusively on single-wavelength pumps. Under quasi-lossless transmission, relatively low signal powers may lead to strong nonlinear impairments, but a quasi-lossless system would also offer the best trade-off between accumulated nonlinearities and optical signal-to-noise ratio (OSNR) in any system in which amplified spontaneous emission (ASE) can be considered as the only significant source of noise, i.e., as long as additional sources of noise, such as RIN transfer from the pumps and double Rayleigh scattering are not significant [20].

When considering the design of an URFL for amplification, conventional fiber laser lore would suggest that the use of high reflectivity gratings is a logical step in a situation where pump conversion efficiency is potentially an issue. However, the best pump efficiency does not necessarily translate into the lowest possible signal power excursion. In the case of URFLs, where the laser cavity is also the transmission medium, and the transmitted signal introduces a directional element of depletion and asymmetry, it is possible that reducing the reflectivity of the FBGs can improve the signal power excursion without having a dramatic impact on the power efficiency of the scheme under investigation. This can be understood if we consider that varying the gratings' reflectivity is equivalent to finely adjusting the effective "input" pump power of the secondary pump, as in a more conventional second-order amplification scheme relying on four bi-directional pumps.

In order to verify the above statement, a series of simulations and experiments were performed. Using the theoretical model of the URFL from Eqs. (15.9)–(15.13), it is possible to obtain a full chart detailing the behavior of the proposed scheme for any given signal power and transmission fiber in the plane defined by grating reflectivity (normalized so a value of 1 indicates 100% reflectivity) and span length. Figure 15.5 depicts the evolution of the signal power excursion in such a plane when using an average power of 0 dBm and SSMF as the main transmission medium. Experimental signal variations were obtained by using the OTDR technique described in [21] for different values of the span length and grating reflectivities in order to confirm the theoretical predictions. These values appear as dots on the contour plot, showing excellent agreement with the

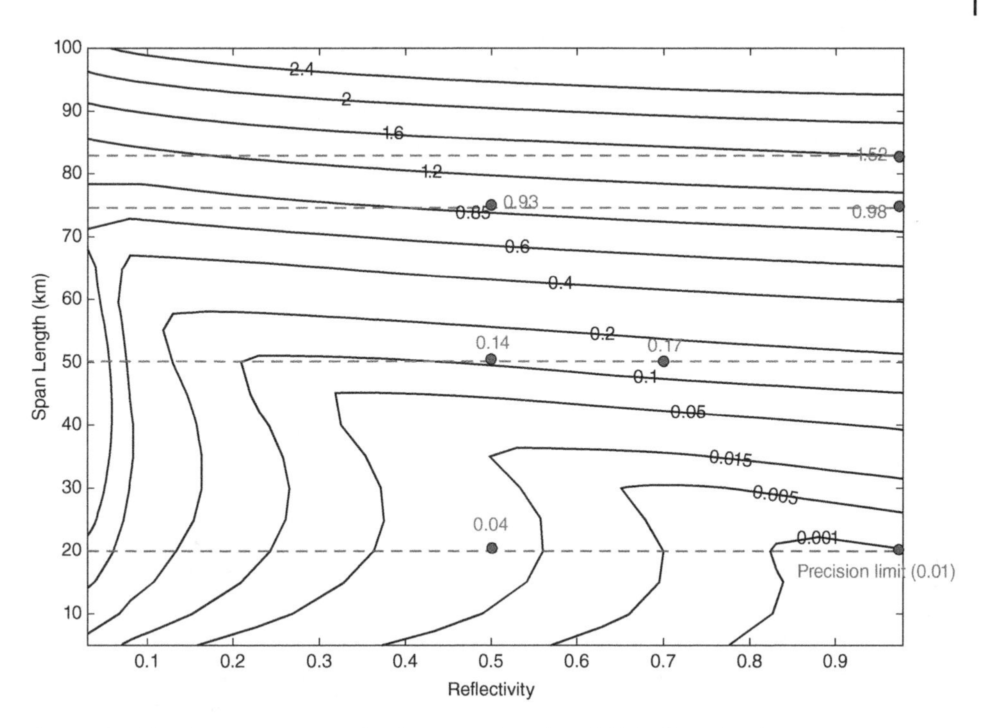

Figure 15.5 Signal power excursion (in dB) vs. span length and FBG reflectivity. The dots represent experimental data. *Source*: Karalekas 2007 [22]. Reproduced with permission of Elsevier.

numerical results. Although the power excursions on Figures 15.2 and 15.3 have been numerically calculated and experimentally measured for a signal at 1550 nm, performances of a similar order can be expected over a certain bandwidth centered on this wavelength (as illustrated by the results in [21]). By examining Figure 15.6, two very different regimes become apparent. The first corresponds to the short span case and the second to the long span case. For short lengths, a reduction of the FBG reflectivity has a clear negative impact on signal power excursion, which can vary, in the case of the 20-km span, from negligible at high reflectivities to 0.05 dB for reflectivities of about 35%, and all the way up to 0.4 dBs for reflectivities closer to 5%. For long spans,on the other hand, reduced power excursion is obtained by lowering the FBG reflectivity, although the relative impact of this change is much less significant than in the case of short spans. This allows, for example, a power excursion reduction from 0.98 to 0.93 dB by lowering reflectivity from 99% to 50%. It is worth noting that in the case of an 83-km cavity length with the highest reflectivity (98%), the variation is approximately equal to 1.51 dB. This figure can be compared with the 16.6 dB variation the signal would experience under a conventional lumped amplification scheme.

To continue with our analysis, the dependence of pump power on system parameters was investigated both numerically and experimentally, in order to determine the impact of grating reflectivity variations on the power efficiency of our system. The results are summarized in Figure 15.6, where the evolution of the total required pump power (that is, the sum of the pump powers at both ends of the span) for full signal recovery at the end of the span is plotted. As with the Figure 15.5, the plane of interest is defined by the

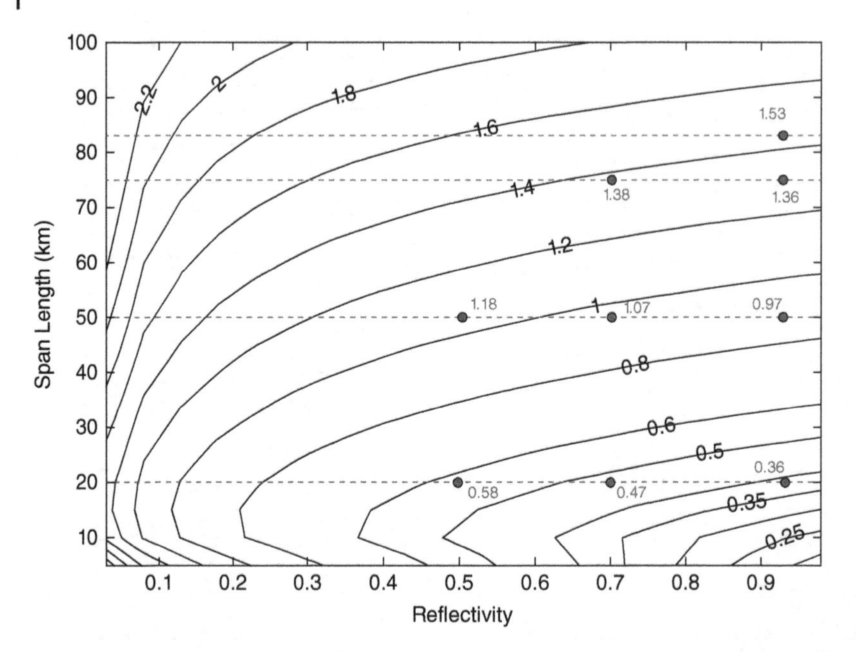

Figure 15.6 Required pump power (in *W*) for full amplification vs. span length and FBG reflectivity. The measured pump powers are represented with dots. *Source:* Karalekas 2007 [22]. Reproduced with permission of Elsevier.

normalized FBG reflectivity and the span length. Experimental results are included in the form of dots for comparison, as in the case of Figure 15.5.

As can be expected, the pump power requirement decreases with decreasing span length and increasing grating reflectivity. Furthermore, and in particular at long span lengths, the power efficiency is not greatly compromised by moderate changes on the FBG reflectivity, which means that the signal power excursion can be improved without significant efficiency penalties. Even more importantly, the required pump powers over the whole range of reflectivities considered (from 3% to 98%) and span lengths (from 5 to 100 km) are well within the capabilities of modern pump lasers.

15.3.2.2 ASE Noise in URFLs

From Eqs. (15.5)–(15.8), it is easy to see that OSNR is subjected to minimal possible growth in the limit of perfect gain distribution provided by a virtually-lossless amplified system, as long as its main source is ASE. Experimental confirmation of the advantages of quasi-lossless transmission in terms of reduced ASE growth can be found in Figure 15.7, in which the measured OSNR vs. transmission distance is shown for a four channel WDM system [22] for URFL amplification and EDFA-only amplification. The channels occupy a bandwidth of 3 nm, ranging from 1550.6 to 1553.6 nm with 1 nm spacing between each channel. The total span length was around 75 km of SMF, introducing a loss of 15 dB at 1550 nm. The OSNR improvement for URFL amplification measured at 4000 km was 4.3 dB over conventional lumped EDFA amplification. The total power at the input of the span in both of the configurations was 6 dBm.

15.3.2.3 Other Sources of Noise in URFLs

As in any Raman amplified system, two additional sources of optical noise, in addition to ASE, potentially limit the performance of URFLs as optical amplifiers: relative intensity

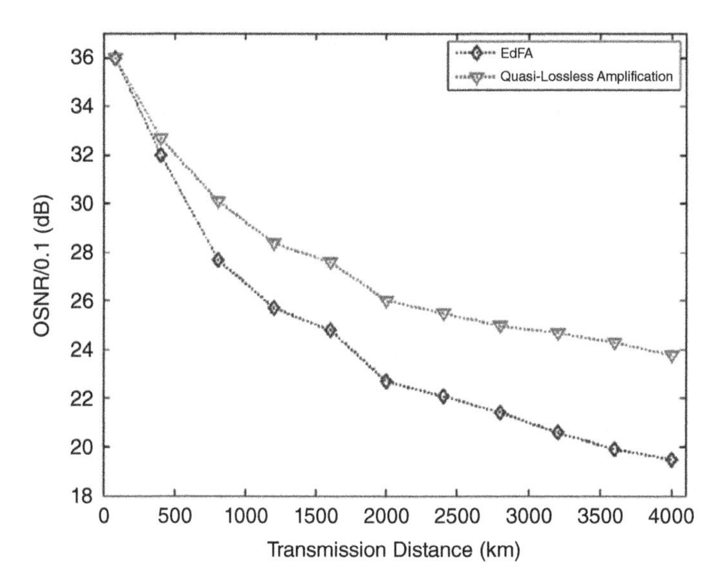

Figure 15.7 OSNR evolution vs. transmission distance in the example EDFA-based and URFL-based transmission schemes. *Source*: Karalekas 2007 [22]. Reproduced with permission of Elsevier.

noise (RIN) and double Rayleigh backscattering (DRB) noise. The contribution of latter is typically noticeably lower than that of ASE in common URFL setups [21], thanks to the efficient distribution of gain, that avoids sudden increases in signal power along the line, which lead to the growth of DRB noise in discrete Raman amplifiers. The relative contribution of ASE and DRB can be derived from the numerical solution of Eqs. (15.9)–(15.13), and experimentally evaluated using approaches such as the optical time-domain extinction technique [11]. Although DRB noise is rarely a limiting factor in the performance of URFL amplifiers, Rayleigh backscattering takes on a critical role as a dephasing mechanism in very long URFL cavities, and provides the distributed feedback mechanism in random distributed feedback URFL configurations, as we have seen in section 15.3.1.2. RIN transfer from the pumping sources to the amplified signal, on the other hand, can be a limiting factor on the performance of URFL-based amplified systems. Due to the high power requirements, pump sources used in URFLs are typically fiber-based and inherently noisy. In addition, as shown in [23], the RIN transfer function of cavity URFLs can be considerably higher than that of an equivalent multi-pump second-order amplification setup, particularly for mid-length URFLs. Hence, the use of low-noise (below −120 db/Hz) fiber-based or semiconductor Raman pumps is highly recommended.

15.3.2.4 Simultaneous Spatio-Spectral Transparency

As mentioned above, it is possible for mid-length URFLs to reproduce the conditions for virtual transparency (close to perfect match between gain and attenuation across the whole link) in standard optical fiber. Moreover, these schemes can induce such transparency simultaneously in several frequencies over a rather broad (more than 20 nm) range, thanks to the overlapping of the first and second-order Raman gain spectra (see Figure 15.7).

By properly adjusting pump powers, as was shown in [24], a system with close-to-zero effective attenuation across the frequency-distance plane is achievable with standard fiber URFL links of moderate lengths (of up to 25 km). Such fiber spans act, in terms of intensity variations, as perfectly transparent media for a broad range of transmitted intensities and could provide, for example, the basis for a new class of multiwavelength optical processing devices in which interchannel nonlinear interactions can be exploited within a nondissipative, integrable framework. Moreover, such a transparent medium can be advantageous both for linear and nonlinear transmission and processing regimes, thanks to the reduction in ASE noise caused by the mutual cancelation of the dissipative and gain terms along the spatiospectral plane.

Following the general steady-state equations for a bidirectionally pumped URFL link presented in Section 15.2.1, ignoring noise terms, assuming negligible spectral broadening of the first Stokes and taking into account the contribution of the primary pump to signal gain (usually ignored when cascading between Stokes waves, but not here as our signal is multiwavelength), it is easy to arrive at an expression for the spatial derivative of the signal power:

$$\frac{dP_S(\nu,z)}{dz} = \left[-\alpha(\nu) + g_R(\mu_2,\nu)\left(P_2^+(z)+P_2^-(z)\right) + g_R(\mu_1,\nu)\left(P_1^+(z)+P_1^-(z)\right)\right] P_S(z) \tag{15.15}$$

where + and - represent, respectively, copropagating and counterpropagating directions, ν represents the frequency of the particular signal component, $g_R(\mu_i,\nu)$ represents the fiber Raman gain coefficient from frequency μ_i to frequency ν, and P_1 and P_2 represent, respectively, the primary pump power and the power of the trapped first Stokes component, each assumed monochromatic. From 15.15 it follows that for cross-domain transparency to occur for our signal in the transmission fiber, a continuous interval in the frequency domain must exist such that for all $(\nu,z) \in [f_1,f_2] \times [0,L]$:

$$\alpha(\nu) \approx +g_R(\mu_2,\nu)\left(P_2^+(z)+P_2^-(z)\right) + g_R(\mu_1,\nu)\left(P_1^+(z)+P_1^-(z)\right) \tag{15.16}$$

where L is the length of the URFL transmission link. In general, P_1 and P_2 are both dependent on each other and on P_S, so 15.15 cannot be rearranged in a more explicit analytical form in terms of the initial values of P_S, but it is possible to do so by assuming that neither P_1 nor P_2 suffer pump depletion from the signal (very low signal approximation), and thus are not dependent on P_S. Under this consideration, it is straightforward to rewrite

$$P_s(\nu,Z) = P_s(\nu,0)e^{-\alpha_{eff}(\nu,Z)Z} \tag{15.17}$$

where

$$\alpha_{eff}(\nu,Z) = \frac{1}{Z}\int_0^Z \left[\alpha(\nu) - g_R(\mu_2,\nu)\left(P_2^+(z)+P_2^-(z)\right) - g_R(\mu_1,\nu)\left(P_1^+(z)+P_1^-(z)\right)\right] dz \tag{15.18}$$

represents the effective gain-loss coefficient, which must be zero in order to achieve transparency. From 15.18, it can be seen that this condition is equivalent to achieving $\alpha_{eff}(\nu,Z) = 0$ over the specified spatio-spectral domain. Please note that the assumed undepleted approximation can be expected to give accurate results only for $P_S \ll P_i$, where i can be either 1 or 2.

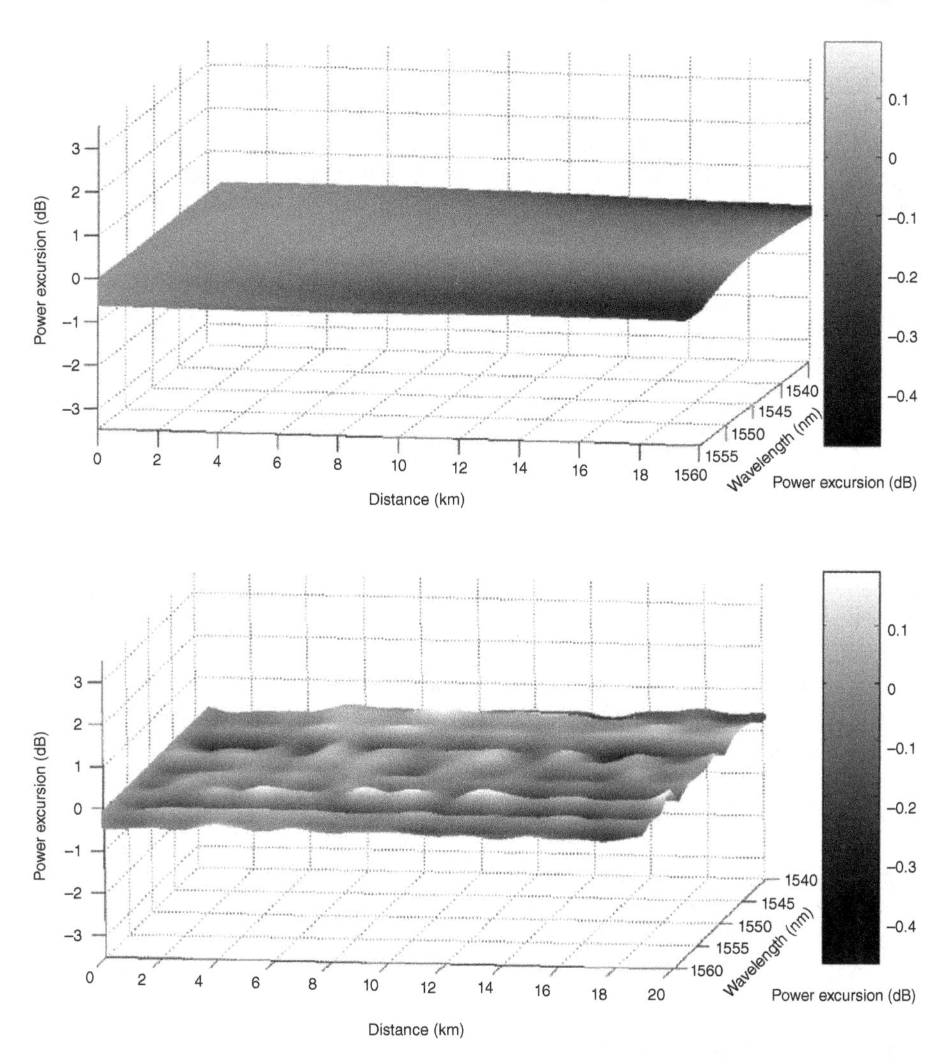

Figure 15.8 Top - Numerical signal power excursion in the spectral-spatial plane for a 20 km link. Bottom - Experimental signal power excursion in the spectral-spatial plane for a 20 km link. The wavelength range is 20 nm. Right - OSNR as a function of the transmission distance for the two transmission schemes. *Source*: Ania-Castañón 2008 [24]. Reproduced with permission of American Physical Society.

If pump depletion is to be considered, as would be required for highly nonlinear transmission, then full simulations can be carried out to evaluate the powers of the spectral components at different positions. This can be achieved by resorting to a revised set of Eqs. (15.9)–(15.13), with additional gain and depletion terms to account for the direct Raman transfer from P_1 to P_S. Comparison between experiments in standard single-mode fiber and such numerical simulations shows excellent agreement, as shown in Figure 15.8.

The depicted experimental results were obtained with a basic URFL design such as the one depicted in Figure 15.2. The two Raman pump lasers acting as primary pump sources have central wavelengths of 1365.4 and 1364.2 nm, respectively, for the right and left pump. Two fiber Bragg gratings (FBGs) with reflectivities of 95% at a central wavelength of 1455 nm were utilized to trap the radiation of the Stokes wave, generating a cavity in the standard single-mode optical fiber link. FBG reflectivity was chosen to be high in order to reduce losses in the system and increase power efficiency in our experiment. The full-width at half-maximum bandwidth of the FBGs was 125 GHz.

The measurement setup relies on a tunable laser source to provide a variable wavelength probe signal whose power evolution along the transmission line can then be monitored using a standard optical time-domain reflectometer. Average signal power in the depicted figure was 0.126 mW. Signal powers are kept much lower than that of the generated 1455 nm Stokes wave, so that operation can take place with no noticeable depletion.

From Figure 15.8 it is obvious that the 20 km URFL constitutes a fully transparent transmission link, in which the combined action of the two primary pumps and the 1455 nm laser wave can provide constant amplification to a number of transmitted signals over a bandwidth as broad as 20 nm. For a low-powered signal such as the one used to obtain our OTDR traces, total required 1365 nm pump power is 370 mW, in full agreement with the predictions of the numerical model, and the combined + and − power of the 1455 nm Stokes wave at either end of the cavity is 130 mW. It is important to point out that the apparent noise in the experimental trace is inherent to the OTDR measurements and of the same order as the device precision. The total absolute signal power variation measured along the transmission link is <0.4 dB, very close to the theoretical prediction of 0.35 dB (7.8% intensity) variation over the 1540–1560 nm range. Variation over the central 10 nm range between 1545 and 1555 nm is even smaller, of the order of 0.14 dB (3.1%) according to the theory, and below the precision of our OTDR setup.

To provide a meaningful comparison between the transparency performance of URFLs with different cavity lengths, we introduce the dimensionless power excursion parameter δP, defined as the quotient between the maximal value of signal power excursion ΔP in the whole spatial-spectral window and the total attenuation in the span, both in dB.

$$\delta P = max\left[\Delta P_S(\nu, z)/(\alpha_S L)\right] \tag{15.19}$$

where α_S is considered to be 0.2 dB/km over the whole bandwidth of our spectral window.

δP is a figure of merit indicating how closely the local attenuation of the fiber can be compensated by the Raman gain over the length of the fiber span and the wavelength range measured, relative to the total loss of the span. A value of 0 corresponds to no variation in signal power along the fiber, whereas a value of 1 would correspond to lumped amplification at the end of the span, or in general to a complete lack of distributed gain. Figure 15.9 shows the predicted evolution of the δP parameter vs. cavity length for a 0 dBm signal power, over both a 10 nm and a 20 nm bandwidth.

These results show that URFL-based transparent links are resilient to high transmitted powers, which makes them attractive building blocks for multiwavelength nonlinear processing devices which could be described as nonlinear integrable systems without non-Hamiltonian dissipative terms.

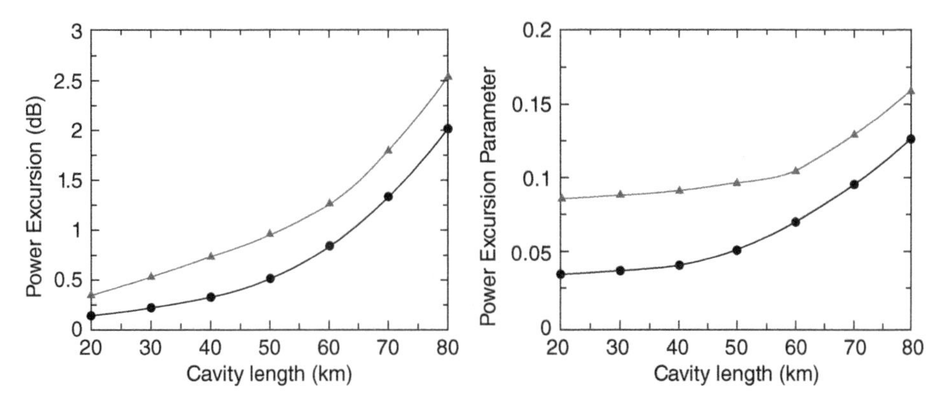

Figure 15.9 Evolution of the power excursion (left panel) and Δ P parameter (right panel) vs. cavity length for 0 dBm total signal power. Triangles, 20 nm bandwidth; circles, 10 nm bandwidth. *Source:* Ania-Castañón 2008 [24]. Reproduced with permission of American Physical Society.

15.3.2.5 Long-Distance Soliton Transmission

In most practical optical transmission systems, the intrinsic fiber loss makes unperturbed soliton transmission over long distances impossible even for low-loss media, as the integrity of a soliton pulse relies on the balance between the power-dependent self-phase modulation in the transmission fiber and the constant anomalous chromatic dispersion of the fiber. Prior to the use of URFLs, classical soliton transmission had been demonstrated in recirculating-loop distributedly amplified spans. Still, until recently, such demonstrations had been restricted to the case where fiber spans were shorter than the corresponding soliton period, until the almost exact cancelation of effective gain-loss in the fiber link provided by URFLs was first used in 2009 [25] to achieve transmission of fundamental solitons over multiple soliton periods in conventional optical fiber with negligible amplitude or phase distortion. Monitoring second harmonic generation frequency resolved optical gating spectrograms at various points along 22 km and 72 km spans (see Figure 15.10) indicated that the pulse was propagating without changes to the temporal width or spectrum.

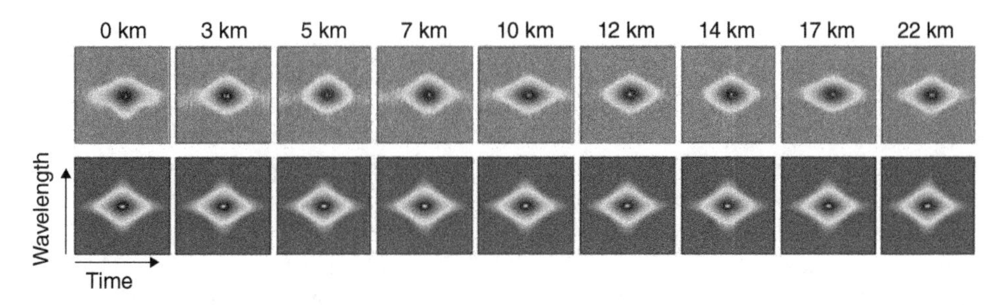

Figure 15.10 Set of experimental (top row) and numerical simulation (bottom row) frequency-resolved optical gating spectrograms at different locations along the fiber span. *Source:* Alcon-Camas 2009 [25].

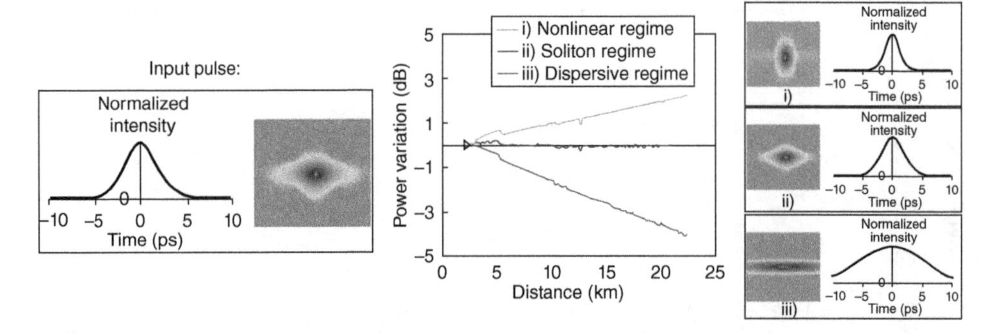

Figure 15.11 Left - Spectrogram and temporal intensity profile of the launched pulse. Center - Power variation profile along a 22 km fiber span for the (i) nonlinear regime (SPM dominates), 0.83 W Raman power; (ii) soliton regime (virtually nondissipative), 0.56 W Raman power; and (iii) dispersive regime (dispersion dominates), 0.29 W Raman power. Right - Spectrogram and temporal intensity profile of the output pulse for the (i) nonlinear regime, (ii) soliton regime, and (iii) dispersive regime. *Source*: Alcon-Camas 2009 [25].

URFLs provide an excellent platform in which to design and test schemes that operate in different regimes, including ideally lossless nonlinear regimes describable mathematically through integrable systems. This is exemplified in the experimental Figure (15.11), displaying the output shape and spectrogram of a sech-shape input soliton pulse being transmitted through a linearly amplified link (nonlinear regime), a virtually lossless link (soliton regime) and a lossy link (dispersive regime).

15.4 Applications of Ultralong Raman Fiber Lasers

Since the first experimental demonstration of URFL-based amplification [13], the number of applications for URFLs has grown steadily year after year. Although initially conceived as a simple, energy-efficient amplification scheme for long-haul transmission in optical fiber communication systems, the unique characteristics of URFLs as a transmission medium with a controllable effective gain-loss coefficient has spawned multiple applications in other areas, from flat supercontinuum generation in conventional fiber to improved distributed fiber sensing. Several of these applications will be swiftly reviewed in Section 15.4.1.

As can be expected, each different application sets its own design constraints and specific requirements for URFL design, as well as its own particular set of parametres to be optimized. For example, URFLs used in unrepeatered transmission with extremely long spans will not aim for quasi-losslessness, or even for net zero gain, but will instead be optimized for the best balance between noise and nonlinear impairments for the specific modulation format being transmitted, as would any other amplification solution. On the other hand, for example, in the case of distributed temperature sensing based on Brillouin optical time domain analysis (BOTDA) assisted by URFL amplification, the aim will not be to recover a particular transmitted signal at the end of the sensing span, but to provide the best possible contrast improvement in the region in which the sensor sensitivity drops the most, or to increase the overall sensing length according to the required specifications of the sensing solution, and so on.

15.4.1 Applications in Telecommunications

15.4.1.1 Amplification in Fiber Optic Communication Systems

As we have seen in Section 15.2, distributed Raman amplification schemes based on ultra-long Raman fiber lasers have proved to be a very useful tool for the reduction of ASE noise impairments in optical communications. Moreover, the enhanced spatial gain distribution offered by these solutions in quasi-lossless configurations allows for a better management of the balance between noise and nonlinear impairments, and the extended bandwidth achievable from a single pump, which has been shown to cover up to 57 nm with the use of two optimally tuned sets of FBG reflectors [26] (see Figure 15.12) makes them particularly attractive for dense multichannel transmission. Even

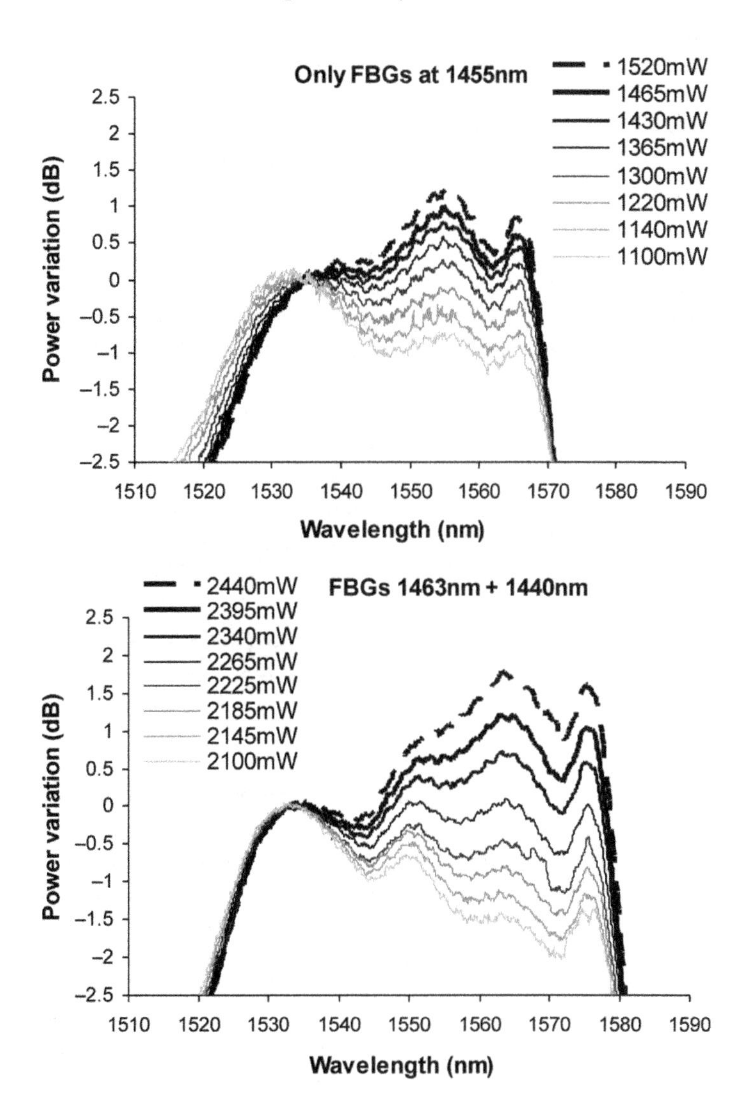

Figure 15.12 Spectral profile variation for increasing pump power for Top - single pair of FBGs at 1455 nm and Bottom - a combination of the FBGs at 1463 nm and 1440 nm. *Source*: Ref. [26].

with a single grating [27] URFLs were shown to provide efficient amplification over the whole C-band without the need for gain-flattening filters, improving upon the performance of gain-flattened C-band Erbium-doped fiber amplifiers. Moreover, the improved noise performance leads to optimal signal launch powers that are typically 6 dB lower than required for equivalent EDFA-amplified systems.

In 2009, the potential for combining optical-phase conjugation (OPC) techniques with quasi-lossless URFL transmission to compensate simultaneously for dispersion and deterministic nonlinear impairments was demonstrated [28], paving the way for the transmission at capacities exceeding the nonlinear-Shannon limit in 2014 [29] using a similar idea. The pioneering experiment carried out at Aston University used all-fiber polarization-insensitive OPC and URFL amplification to transmit 8x114 Gbit/s dual polarization quadrature phase shift-keyed signal to 10400 km, with an increase on the nonlinear threshold of around 3 dB, corresponding to a 70 % compensation of nonlinearity. In this experiment the URFL-based amplification technique was used to provide a near symmetric power profile along the transmission which is a requirement for effective compensation of the nonlinear transmission impairments via OPC.

In all these solutions, RIN transfer from the pumps to the signal seems to be a limiting factor to the potential of URFL performance, stressing the need for low-RIN pumps. Recently, a novel scheme based on the use of "open-cavity" random distributed feedback URFLs has demonstrated the possibility of efficient RIN transfer mitigation without a large impact on system efficiency [30]. This, in combination with recent studies focusing on optimizing symmetry in ultralong laser spans for improved OPC, [31,32] could hold the key to further capacity improvements.

URFLs have also demonstrated their potential in unrepeatered transmission systems. In contrast to long-haul submarine transmission systems, in which the goal is to achieve maximum capacity at extremely long distances, resorting to periodic amplification, unrepeatered transmission aims to offer a cost-efficient solution for short links, without the need for in-line active elements. In this context, the extra noise margins provided by distributed amplification, and in particular by URFLs, can be of great use. Unrepeatered transmission through URFLs was first proposed in 2008 [33] and has since been shown to be a highly effective solution with a number of transmission formats, including, in recent times, Nyquist PDM-16QAM superchannel transmission [34,35].

15.4.1.2 Secure Key Exchange Using URFLs

The main weakness of shared-key encripcion protocols is the secure distribution of the secret keys used in the protocol. In 2006, Scheuer and Yariv [36] proposed an intriguing new concept for secure key distribution based on establishing laser oscillations between the sender and receiver which, compared to quantum mechanics based systems, offered significantly higher key-stablishing rates at long distances. The concept relied on the random switching of the mirrors delimiting a very long (a "giant laser") cavity in which lasing was achieved through Erbium-doped fiber amplification. The authors called this technique "giant fiber laser secure key distribution." Given the parallelism between the "giant" and "ultralong" laser concept, in further work this kind of setup has come to be referred to as ultra-long fiber laser secure key distribution. In recent years, a number of different variant schemes have been proposed, including some relying on Raman amplification and using the whole fiber link as the lasing medium. Using such URFL-based scheme, recent works [37] have managed to extend secure key distribution to 500 km.

Other interesting variants recently demonstrated but not based on Raman amplification, include using variations in the free spectral range of the ultralong laser for secret key distribution [38].

15.4.2 Applications in Sensing

In recent times, optical fiber sensors have become widespread thanks to their unique properties, which include their small size and weight, immunity to electromagnetic noise, environmental ruggedness, and a chemically inert nature. This kind of sensor is particulary interesting for the monitoring of civil infrastructures, and even more so when such infrastructures are located in electrically noisy environments or when the distance to be monitored is long. Distributed sensors constitute a particular set of fiber sensors in which the fiber used to carry the optical information acts itself as the optical sensor, allowing the monitoring of physical parameters at different points along the fiber simultaneously, instead of relying on a number of lumped sensing points located at specified positions along the fiber.

Whether the sensing scheme is distributed, quasi-distributed or based on point sensors, in those sensors in which the optical fiber link is long, the retrieved signal can suffer from important attenuation. In this context, Raman amplification can provide a simple solution to improve sensitivity and overall system performance.

URFL amplification, in particular, has found a number of applications in the context of distributed sensing. In 2011 [39], URFL amplification was shown to extend the measurement range of temperature-sensing Brillouin optical-domain analyzers (BOTDAs) from 50 km to 100 km with a 2 m resolution, without the need for additional enhancement techniques, such as pulse coding, setting a new reference in the field. Constant progress has continued to be made in the area of distributed sensing, with recent works demonstrating extended measurement range of up to 125 km in vibration sensing via URLF-assisted phase-sensitive optical time domain reflectometry [40]. Random distributed feedback ultralong fiber lasers have also been used in BOTDA sensing in combination with low-noise laser-diode based 1st-order pumping to reduce RIN transfer, achieving a sensing distance of 154 km with a 5 m spatial resolution [41].

Similarly, point and quasi-distributed sensors can benefit from URFL and random distributed feedback Raman ultralong laser amplification, with several works reporting successful implementation of such schemes with arrays of FBGs or fiber loop mirrors for temperature, vibration or strain monitoring [42, 43].

Raman amplification based on URFL has been also been proposed as a solution to improve the length, and hence the performance of fiber Sagnac interferometres for the detection of seismic rotational events, showing the potential for a five-fold increase in sensitivity [44].

15.4.3 Supercontinuum Generation

Although similar structurally to those used in amplification, the URFLs used in supercontinuum generation [45] have to address different needs than those designed for optical communications. Gain flatness is not an issue here, but conversion efficiency is, and fiber characteristics such as fiber dispersion play a critical role. Shorter cavities, of no more than 15 km, are favored in these designs, and given the typically high powers used, special attention has to be paid to broadening of the first Stokes, the impact of nonlinearities on the effective reflectivity of the fiber Bragg gratings and other issues that do

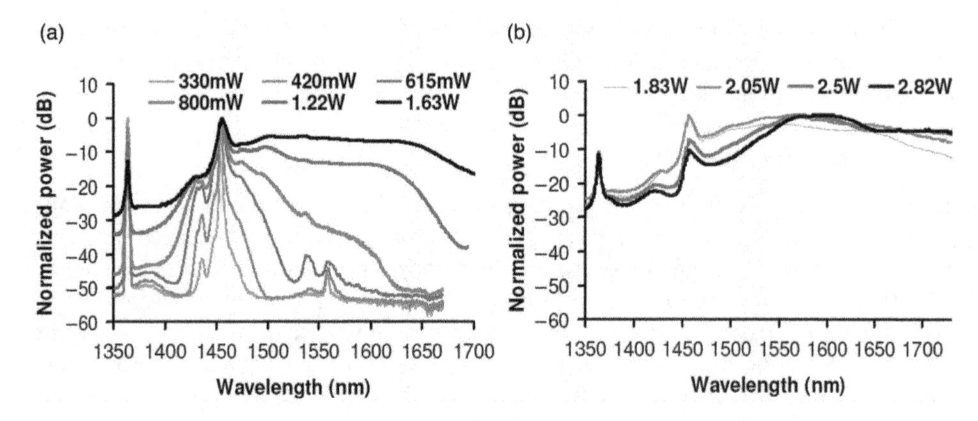

Figure 15.13 Supercontinuum spectra for an ultra-long Raman fiber cavity combining TW fiber with HNLF as a function of pump power (a) up to power 1.63 W (b) from 1.83 to 2.82 W. *Source*: El-Taher 2009 [45].

not typically have the same relevance in periodic cell amplification. Using an optimized cavity combining different conventional (not based on photonic crystal) fibers, SC generation with a spectral flatness of less than 1 dB over 180 nm was achieved in 2009 from a single 1.63 W 1366 nm CW pump (see Figure 15.13). The increased flatness, extended bandwidth, and improved generation efficiency demonstrated using URFLs, combined with the practicality and possibility of using only low cost conventional optical fibers, make this an attractive technique for supercontinuum generation.

15.5 Conclusion

Despite extraordinary advances in laser science, only recently the fundamental limits of laser cavity length have become an area of exploration. Since they were first proposed little more than a decade ago, ultralong Raman fiber lasers have, thanks to their unique properties, shown their potential in a number of applications, including efficient amplification and nonlinear compensation in high-speed optical communications, secure key exchange, improving the performance of fiber-based optical sensing schemes or generating highly efficient supercontinua in conventional fibers. Moreover, the study of the mechanisms limiting coherence in cavity ultralong fiber lasers has paved the way to the development of another breed of ultralong Raman fiber lasers, namely the random distributed feedback Raman fiber lasers [19,46], which display a combination of the characteristics of both traditional fiber lasers and random lasers. An ultralong fiber laser cavity represents a unique transmission medium offering possibilities that we are still beginning to explore, and it is quite possible that the number of applications and technologies derivative from URFL will continue to expand in the near future.

References

1 Smekal, A. (1923) Zur Quantentheorie der Dispersion. *Naturwissenschaften*, **11** (43), 873–875, DOI:10.1007/BF01576902. URL http://dx.doi.org/10.1007/BF01576902.

2 Raman, C.V. (1928) A new radiation. *Indian Journal of Physics*, **2**, 387–398.
3 Raman, C.V. and Krishnan, K.S. (1928) A new type of secondary radiation. *Nature*, **121**, 501–502.
4 Landsberg, G. and Mandelstam, L. (1928) Eine neue Erscheinung bei der Lichtzerstreuung in Krystallen. *Naturwissenschaften*, **16** (28), 557–558, DOI:10.1007/BF01506807. URL http://dx.doi.org/10.1007/BF01506807.
5 Islam, M. (2004) *Raman Amplifiers for Telecommunications*, Springer, New York.
6 Headley, C. (2005) *Raman Amplification in Fiber Optical Communication Systems*, Elsevier, San Diego.
7 Bristiel, B., Gallion, P., Jaouen, Y., and Pincemin, E. (2004) Intrinsic noise figure derivation for fiber Raman amplifiers from equivalent noise figure measurement, in *Proceedings of the Lightwave Technologies in Instrumentation and Measurement Conference, 2004.* pp. 135–140, DOI:10.1109/LTIMC.2004.1371010.
8 Emori, Y. and Namiki, S. (1999) 100 nm bandwidth flat gain Raman amplifiers pumped and gain-equalized by 12-wavelength-channel WDM high power laser diodes, in *Optical Fiber Communication Conference, 1999, and the International Conference on Integrated Optics and Optical Fiber Communication. OFC/IOOC '99. Technical Digest*, vol. Supplement, PD19/1–PD19/3 Suppl., DOI:10.1109/OFC.1999.766052.
9 Fludger, C.R.S., Handerek, V., and Mears, R.J. (2001) Pump to signal RIN transfer in Raman fiber amplifiers. *Journal of Lightwave Technology*, **19** (8), 1140–1148, DOI:10.1109/50.939794.
10 Bristiel, B., Jiang, S., Gallion, P., and Pincemin, E. (2006) New model of noise figure and RIN transfer in fiber Raman amplifiers. *IEEE Photonics Technology Letters*, **18** (8), 980–982, DOI:10.1109/LPT.2006.873551.
11 Lewis, S.A.E., Chernikov, S.V., and Taylor, J.R. (2000) Characterization of double Rayleigh scatter noise in Raman amplifiers. *IEEE Photonics Technology Letters*, **12** (5), 528–530, DOI:10.1109/68.841275.
12 Ania-Castanón, J.D. (2004) Quasi-lossless transmission using second-order Raman amplification and fiber Bragg gratings. *Optics Express*, **12** (19), 4372–4377, DOI:10.1364/OPEX.12.004372. URL http://www.opticsexpress.org/abstract.cfm?URI=oe-12-19-4372.
13 Ania-Castanón, J.D., Ellingham, T.J., Ibbotson, R., *et al.* (2006) Ultralong Raman fiber lasers as virtually lossless optical media. *Physical Review Letters*, **96**, 023 902, DOI:10.1103/PhysRevLett.96.023902. URL http://link.aps.org/doi/10.1103/PhysRevLett.96.023902.
14 Jia, X.H., Rao, Y.J., Wang, Z.N., *et al.* (2012) Detailed theoretical investigation on improved quasi-lossless transmission using third-order Raman amplification based on ultralong fiber lasers. *Journal of the Optical Society of America B*, **29** (4), 847–854, DOI:10.1364/JOSAB.29.000847. URL http://josab.osa.org/abstract.cfm?URI=josab-29-4-847.
15 Babin, S.A., Karalekas, V., Podivilov, E.V., *et al.* (2008) Turbulent broadening of optical spectra in ultralong Raman fiber lasers. *Physical Review A*, **77**, 033 803, DOI:10.1103/PhysRevA.77.033803. URL http://link.aps.org/doi/10.1103/PhysRevA.77.033803.
16 Turitsyn, S.K., Ania-Castanón, J.D., Babin, S.A., *et al.* (2009) 270-km ultralong Raman fiber laser. *Physical Review Letters*, **103**, 133 901, DOI:10.1103/PhysRevLett.103.133901. URL http://link.aps.org/doi/10.1103/PhysRevLett.103.133901.

17 Babin, S.A., Churkin, D.V., Ismagulov, A.E., Kablukov, S.I., and Podivilov, E.V. (2007) Four-wave-mixing-induced turbulent spectral broadening in a long Raman fiber laser. *Journal of the Optical Society of America B*, **24** (8), 1729–1738, DOI:10.1364/JOSAB.24.001729. URL http://josab.osa.org/abstract.cfm?URI=josab-24-8-1729.

18 El-Taher, A.E., Harper, P., Babin, S.A., *et al.* (2011) Effect of Rayleigh-scattering distributed feedback on multiwavelength Raman fiber laser generation. *Optics Letters*, **36** (2), 130–132, DOI:10.1364/OL.36.000130. URL http://ol.osa.org/abstract.cfm?URI=ol-36-2-130.

19 Turitsyn, S.K., Babin, S.A., El-Taher, A.E., *et al.* (2010) Random distributed feedback fiber laser. *Nature Photonics*, **4** (4), 231–235, DOI:10.1038/nphoton.2010.4. URL http://dx.doi.org/10.1038/nphoton.2010.4.

20 Bouteiller, J.C., Brar, K., and Headley, C. (2002) Quasi-constant signal power transmission, in *Optical Communication, 2002. ECOC 2002. 28th European Conference on*, vol. 3, vol. 3, pp. 1–2.

21 Ellingham, T.J., Ania-Castanon, J.D., Ibbotson, R., *et al.* (2005) Quasi-lossless spans for broadband transmission and data processing, in *Optical Communication, 2005. ECOC 2005. 31st European Conference on*, vol. 1, vol. 1, pp. 65–66 vol. 1, DOI:10.1049/cp:20050380.

22 Karalekas, V., Ania-Castanón, J., Pérez-González, J., *et al.* (2007) Performance optimization of ultra-long raman laser cavities for quasi-lossless transmission links. *Optics Communications*, **277** (1), 214–218, DOI:http://dx.doi.org/10.1016/j.optcom.2007.05.009. URL http://www.sciencedirect.com/science/article/pii/S0030401807004877.

23 Alcón-Camas, M. and nón, J.D.A.C. (2010) RIN transfer in 2nd-order distributed amplification with ultralong fiber lasers. *Opt. Express*, **18** (23), 23 569–23 575, DOI:10.1364/OE.18.023569. URL http://www.opticsexpress.org/abstract.cfm?URI=oe-18-23-23569.

24 Ania-Castanón, J.D., Karalekas, V., Harper, P., and Turitsyn, S.K. (2008) Simultaneous spatial and spectral transparency in ultralong fiber lasers. *Physical Review Letters*, **101**, 123 903, DOI:10.1103/PhysRevLett.101.123903. URL http://link.aps.org/doi/10.1103/PhysRevLett.101.123903.

25 Alcon-Camas, M., El-Taher, A.E., Wang, H., Harper, P., Karalekas, V., Harrison, J.A., and nón, J.D.A.C. (2009) Long-distance soliton transmission through ultralong fiber lasers. *Optics Letters*, **34** (20), 3104–3106, DOI:10.1364/OL.34.003104. URL http://ol.osa.org/abstract.cfm?URI=ol-34-20-3104.

26 null (2010) Gain bandwidth optimisation and enhancement in ultra-long Raman fiber laser based amplifiers, in *Optical Communication (ECOC), 2010 36th European Conference and Exhibition on*, pp. 1–3, DOI:10.1109/ECOC.2010.5621094.

27 null (2010) Extended bandwidth for long haul DWDM transmission using ultra-long Raman fiber lasers, in *Optical Communication (ECOC), 2010 36th European Conference and Exhibition on*, pp. 1–3, DOI:10.1109/ECOC.2010.5621427.

28 Minzioni, P., Harper, P., Pusino, V., *et al.* (2009) Optical phase conjugation for dispersion and nonlinearity compensation in a 1600 km, 42 gb/s quasi-lossless system, in *Advances in Optical Sciences Congress*, Optical Society of America, p. NThA5, DOI:10.1364/NLO.2009.NThA5. URL http://www.osapublishing.org/abstract.cfm?URI=NLO-2009-NThA5.

29 Phillips, I., Tan, M., Stephens, M.F., *et al.* (2014) Exceeding the nonlinear-Shannon limit using Raman laser based amplification and optical phase conjugation, in *Optical Fiber Communication Conference*, Optical Society of America, p. M3C.1, DOI:10.1364/OFC.2014.M3C.1. URL http://www.osapublishing.org/abstract.cfm?URI=OFC-2014-M3C.1.

30 Tan, M., Rosa, P., Iqbal, M.A., *et al.* (2015) RIN mitigation in second-order pumped Raman fiber laser based amplification, in *Asia Communications and Photonics Conference 2015*, Optical Society of America, p. AM2E.6, DOI:10.1364/ACPC.2015.AM2E.6. URL http://www.osapublishing.org/abstract.cfm?URI=ACPC-2015-AM2E.6.

31 Rosa, P., Le, S.T., Rizzelli, G., Tan, M., and Ania-Casta nón, J.D. (2015) Signal power asymmetry optimisation for optical phase conjugation using Raman amplification. *Optics Express*, **23** (25), 31 772–31 778, DOI:10.1364/OE.23.031772. URL http://www.opticsexpress.org/abstract.cfm?URI=oe-23-25-31772.

32 Rosa, P., Rizzelli, G., Tan, M., Harper, P., and Ania-Castanon, J.D. (2015) Characterisation of random DFB Raman laser amplifier for WDM transmission. *Optics Express*, **23** (22), 28 634–28 639, DOI:10.1364/OE.23.028634. URL http://www.opticsexpress.org/abstract.cfm?URI=oe-23-22-28634.

33 Ania-Castanón, J. and Turitsyn, S. (2008) Unrepeatered transmission through ultra-long fiber laser cavities. *Optics Communications*, **281** (23), 5760–5763, DOI:http://dx.doi.org/10.1016/j.optcom.2008.08.021. URL http://www.sciencedirect.com/science/article/pii/S0030401808008110.

34 Rosa, P., Ania-Casta nón, J.D., and Harper, P. (2014) Unrepeatered DPSK transmission over 360 km SMF-28 fiber using URFL based amplification. *Optics Express*, **22** (8), 9687–9692, DOI:10.1364/OE.22.009687. URL http://www.opticsexpress.org/abstract.cfm?URI=oe-22-8-9687.

35 Galdino, L., Tan, M., Lavery, D., *et al.* (2015) Unrepeatered Nyquist PDM-16QAM transmission over 364 km using raman amplification and multi-channel digital back-propagation. *Optics Letters*, **40** (13), 3025–3028, DOI:10.1364/OL.40.003025. URL http://ol.osa.org/abstract.cfm?URI=ol-40-13-3025.

36 Scheuer, J. and Yariv, A. (2006) Giant fiber lasers: A new paradigm for secure key distribution. *Physical Review Letters*, **97**, 140 502, DOI:10.1103/PhysRevLett.97.140502. URL http://link.aps.org/doi/10.1103/PhysRevLett.97.140502.

37 El-Taher, A., Kotlicki, O., Harper, P., Turitsyn, S., and Scheuer, J. (2014) Secure key distribution over a 500 km long link using a Raman ultra-long fiber laser. *Laser & Photonics Reviews*, **8**(3), 436–442, DOI:10.1002/lpor.201300177. URL http://dx.doi.org/10.1002/lpor.201300177.

38 Tonello, A., Barthelemy, A., Krupa, K., Kermene, V., Desfarges-Berthelemot, A., Shalaby, B.M., Boscolo, S., Turitsyn, S.K., and Ania-Castanon, J.D. (2015) Secret key exchange in ultralong lasers by radiofrequency spectrum coding. *Light Sci Appl*, **4**, e276–. URL http://dx.doi.org/10.1038/lsa.2015.49.

39 Martin-Lopez, S., Alcon-Camas, M., Rodriguez, F., Corredera, P., non, J.D.A.C., Thévenaz, L., and Gonzalez-Herraez, M. (2010) Brillouin optical time-domain analysis assisted by second-order raman amplification. *Optics Express*, **18** (18), 18 769–18 778, DOI:10.1364/OE.18.018769. URL http://www.opticsexpress.org/abstract.cfm?URI=oe-18-18-18769.

40 Martins, H.F., Martin-Lopez, S., Corredera, P., Ania-Castanon, J.D., Frazao, O., and Gonzalez-Herraez, M. (2015) Distributed vibration sensing over 125 km with enhanced SNR using PHI-OTDR over a urfl cavity. *Journal of Lightwave Technology*, **33** (12), 2628–2632, DOI:10.1109/JLT.2015.2396359.

41 Jia, X.H., Rao, Y.J., Yuan, C.X., *et al.* (2013) Hybrid distributed raman amplification combining random fiber laser based 2nd-order and low-noise ld based 1st-order pumping. *Optics Express*, **21** (21), 24 611–24 619, DOI:10.1364/OE.21.024611. URL http://www.opticsexpress.org/abstract.cfm?URI=oe-21-21-24611.

42 Martins, H., Marques, M.B., and ao, O.F. (2011) 300 km-ultralong Raman fiber lasers using a distributed mirror for sensing applications. *Optics Express*, **19** (19), 18 149–18 154, DOI:10.1364/OE.19.018149. URL http://www.opticsexpress.org/abstract.cfm?URI=oe-19-19-18149.

43 Fernandez-Vallejo, M., Bravo, M., and Lopez-Amo, M. (2013) Ultra-long laser systems for remote fiber Bragg gratings arrays interrogation. *IEEE Photonics Technology Letters*, **25** (14), 1362–1364, DOI:10.1109/LPT.2013.2265916.

44 Nuño, J. and Ania-Casta nón, J. (2014) Fiber sagnac interferometers with ultralong and random distributed feedback raman laser amplification. *Optics and Lasers in Engineering*, **54**, 21–26, DOI:http://dx.doi.org/10.1016/j.optlaseng.2013.09.007. URL http://www.sciencedirect.com/science/article/pii/S0143816613002832.

45 El-Taher, A.E., nón, J.D.A.C., Karalekas, V., and Harper, P. (2009) High efficiency supercontinuum generation using ultra-long raman fiber cavities. *Optics Express*, **17** (20), 17 909–17 915, DOI:10.1364/OE.17.017909. URL http://www.opticsexpress.org/abstract.cfm?URI=oe-17-20-17909.

46 Turitsyn, S.K., Babin, S.A., Churkin, D.V., *et al.* (2014) Random distributed feedback fiber lasers. *Physics Reports*, **542** (2), 133–193, DOI:http://dx.doi.org/10.1016/j.physrep.2014.02.011. URL http://www.sciencedirect.com/science/article/pii/S0370157314001215, random Distributed Feedback Fiber Lasers.

16

Shaping Brillouin Light in Specialty Optical Fibers

Jean-Charles Beugnot and Thibaut Sylvestre

Department of Optics, FEMTO-ST Institute, CNRS-Université de Bourgogne - Franche-Comté, Besançon, France

16.1 Introduction

Brillouin scattering in optical fibers is a remarkable physical phenomenon that results from the coherent interaction between light and acoustic waves. Due to their strong light confinement capabilities, optical fibers were soon recognized as ideal waveguides to investigate and exploit the Brillouin scattering [1,2]. In optical fibers, light excites and interacts with bulk shear and longitudinal acoustic waves, giving rise to forward and backward scattering, respectively known as guided acoustic wave Brillouin scattering (GAWBS) and stimulated Brillouin scattering (SBS) [3, 4]. Both these effects have often been regarded as detrimental to fiber-based optical communication systems, since SBS is ultimately responsible for restricting the light power that can be transmitted in optical fibers and GAWBS is considered as a phase noise source [3, 5]. However, they also offer new and exciting potential applications, such as photonic devices in which light can be controlled by hypersound. An increasingly important number of applications in various domains such as telecommunications and optical sensing rely on SBS, which can become rather significant, widely tailorable, and also sensitive to temperature and strain [6–9].

Brillouin scattering thus plays a central role in nonlinear optical technologies, and has recently been the subject of a renewed and widespread interest in tiny optical waveguides such as small-core photonic crystal fibers (PCFs), tapered optical fibers (TOFs), and integrated chip optical waveguides [10–14]. Unlike standard optical fibers, light propagating in such ultra-thin optical waveguides can generate hybrid and surface acoustic waves due to strong coupling with the waveguide boundaries. This has been recently shown in small-core PCF, where multiple downshifted acoustic modes have been demonstrated, giving rise to multimode and broadband Brillouin spectra and higher threshold than those theoretically predicted [10, 12]. Brillouin scattering from surface acoustic waves has also recently been demonstrated in subwavelength-diameter silica microwires [15]. Simultaneously, GAWBS

Shaping Light in Nonlinear Optical Fibers, First Edition. Edited by Sonia Boscolo and Christophe Finot.

can be strongly enhanced and stimulated in PCFs and integrated photonic circuits, opening the way for new applications using forward stimulated Brillouin scattering (FSBS) [16].

This chapter is intended to provide an overview of Brillouin light scattering in specialty optical fibers in a way that is simultaneously accessible and technically comprehensive. The objective of this chapter is not to provide another review or a book chapter on the stimulated Brillouin scattering, as was done in [1, 2], but rather to focus on the new physics of Brillouin scattering in both tapered optical fibers and photonic crystal fibers. First, we will depict a short history and describe some basic concepts of forward and backward Brillouin scattering, so as to understand the origin of one of the most important nonlinear effects in optical fibers. We will also discuss the most important applications arising from fiber-based Brillouin scattering. Then we will present a recent theory of Brillouin scattering based on the elastodynamics equation. This theory allows the acoustic waves and displacements to be accurately determined in arbitrarily small waveguides, specifically when the plane wave approximation is no longer valid. This model provides in particular a unified theoretical modeling of both forward and backward Brillouin scattering. We will further apply it to the case of optical microwires and then to solid-core microstructured optical fibers, revealing all new types of acoustic waves including shear, surface and hybrid waves. Experimental measurements of both the forward and backward Brillouin spectra will also be presented and checked against numerical simulations. Finally, we will draw some conclusions and outlooks for the future prospects of light-sound interactions at the micro and nanoscale.

16.2 Historical Background

Historically, inelastic light scattering by acoustic waves was first predicted in 1922 by the French physicist, Léon Nicolas Brillouin [17]. Specifically, Brillouin theoretically showed that the scattered optical wave is made up of the sum of three components: one at the frequency of the incident wave (ω_1), the two others at frequencies located relative to it ($\omega_1 \pm \Omega$) (Brillouin doublet), their separation is dependent on the scattering angle and on the elastic properties of the material. Leonid Mandelstam is also believed to have recognized the possibility of such a theory of light scattering as early as 1918, but he published it only in 1926 [18]. The process of stimulated Brillouin scattering (SBS) using intense light beam was later observed by Chiao *et al.* in 1964 [4], after the invention of the laser in 1960. In optical fibers, SBS was first observed in 1972 by Ippen and Stolen [19]. Since its first observation, SBS has been extensively investigated in optical fibers for both fundamental and applied interests, ranging from slow and fast light to microwave photonics [20, 21]. Nowadays, the research on Brillouin scattering has turned to other photonic platforms including optical microwires, photonic crystal fibers, wisphering-gallery-mode (WGM) resonators and silicon photonics [22–28]. In Section 16.3 we will derive the theory based on the elastodynamics equation which governs the light-sound interaction in such tiny optical waveguides while pointing out the impact of waveguide's boundaries on the elastic wave spectrum.

16.3 Theory

16.3.1 Elastodynamics Equation

Brillouin scattering process is a three-wave parametric interaction whereby an optical pump wave with angular frequency ω_1 and wavenumber k_1 generates a weak Stokes wave with angular frequency ω_2 and wavenumber k_2 when interacting with an acoustic wave of frequency Ω and wavenumber K. In such a parametric interaction, energy and momentum are both conserved such as $\omega_1 = \omega_2 + \Omega$ and $k_1 = k_2 + K$, respectively. The total incident light's electric field can be written as the sum of the pump and Stokes fields

$$E(r,z) = E_1(r,z)e^{i(\omega_1 t - k_1 z)} + E_2(r,z)e^{i(\omega_2 t - k_2 z)}, \tag{16.1}$$

where $k_1 = k(\omega_1)$ and $k_2 = k(\omega_2)$ satisfy the dispersion relations of the optical waveguide where the waves are propagating. Figure 16.1 shows the two main types of the three-wave Brillouin interaction. When the two optical waves counter-propagate, this refers to backward stimulated Brillouin scattering (SBS) [4], where the acoustic wavevector is almost twice the optical wavevector, $K = 2k_1$. Whereas if the two optical waves propagate in the same direction, this refers to forward guided acoustic wave Brillouin scattering (GAWBS), with the acoustic wavevector $K \approx 0$ [3]. In both cases, the elastic wave can be written as the form of a plane wave

$$u_i(r,z) = u_i(r)e^{i(\Omega t - Kz)}, \tag{16.2}$$

with $u_i(r)$ the mechanical displacements associated with the elastic wave, which are actually the unknowns of the problem to be determined. Assuming that the hypersound acoustic frequency (a few GHz) is small and negligible compared to optical frequencies (several hundreds of THz), we can write the elastodynamics equation that governs the opto-acoustic interaction in the following form [29]

$$\rho \frac{\partial^2 u_i}{\partial t^2} - \frac{\partial}{\partial x_j}\left(c_{ijkl}\frac{\partial u_k}{\partial x_l}\right) = \frac{\partial}{\partial x_j}\epsilon_0 \chi_{klij}\left[E_k E_l^*\right], \tag{16.3}$$

with $\chi_{klij} = \epsilon_{im}\epsilon_{jn}p_{klmn}$, ϵ_{ij} is the dielectric tensor, ϵ_0 is the permittivity of the vacuum, c_{klij} is the elastic tensor, and p_{klmn} is the photoelastic tensor. ρ is the material density ($\rho = 2.203$ g.cm^{-3} for silica). We used contracted tensorial notations to keep summation over repeated indices [30]. The last term of Eq. (16.3) accounts for the electrostrictive stress tensor, which includes the pump and Stokes electric fields. The electrostriction stress tensor is defined by the optical modal distributions, which is calculated beforehand using a finite element analysis (COMSOL Multiphysics). The variational problem is then solved for all displacements u_i, $(i = x, y, z)$ by setting K to a given value and

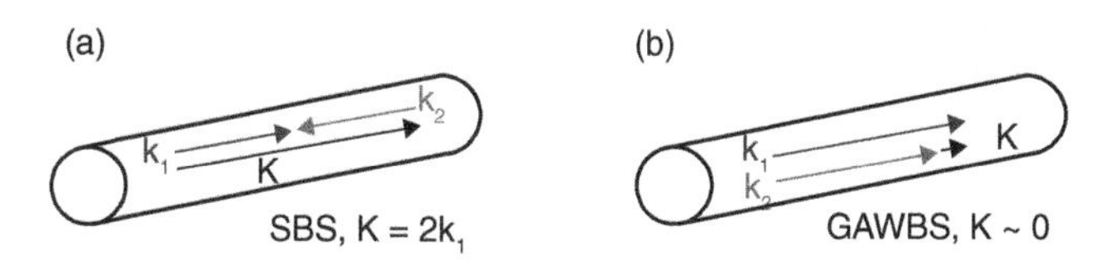

Figure 16.1 Phase-matching conditions for (a) backward and (b) forward Brillouin scattering.

scanning the detuning frequency Ω [31]. Different numerical methods could be used to solve the partial differential equation (16.3). Since the displacement fields $\bar{u}_i$ can be assumed to be continuous everywhere in the area of interest and the solid boundaries can be set arbitrary, one can implement a finite element method (FEM). Substituting Eqs. (16.1) and (16.2) into Eq. (16.3), one get

$$-\rho\omega^2 u_i e^{-iKz} - c_{ijkl}\frac{\partial}{\partial x_j}\left[\frac{\partial}{\partial x_l}u_k e^{-iKz}\right] = \epsilon_0 \chi_{klij}\frac{\partial}{\partial x_j}\left[E_k E_l^* e^{-iKz}\right]. \tag{16.4}$$

By left-multiplying by virtual displacement field $v_i^* e^{+iKz}$ and integrating over the cross-section of the waveguide, S, we get

$$-\omega^2\int_S \rho v_i^* u_i - \int_S c_{ijkl} v_i^* e^{iKz}\frac{\partial}{\partial x_j x_l}\left[u_k e^{-iKz}\right] = -\int_S \epsilon_0\chi_{lkij}\left[v_i^* e^{iKz}\right]\frac{\partial}{\partial x_j}\left[E_k E_l^* e^{-iKz}\right] \tag{16.5}$$

We then apply Green's theorem to transform integrals with a divergence operator into integrals with the gradient of the test function and free boundary conditions. To get more insight into the structure of the FEM equations, we can then rewrite the above equation by showing the K dependence as

$$\rho\omega^2\int_S v_i^* u_i + ik_z c_{i3kl}\int_S v_i^*\frac{\partial u_k}{\partial x_l} + c_{i3kl}\int_S\frac{\partial v_i^*}{\partial x_j}\frac{\partial u_k}{\partial u_l} + k_z^2 c_{i3k3}\int_S v_i^* u_k - ik_z c_{ijk3}\int_S u_k\frac{\partial v_i^*}{\partial x_j}$$

$$= -ik_z\int_S v_i^*\epsilon_0\chi_{lkiz}E_k E_l^* - \int_S \epsilon_0\chi_{lkij}E_k E_l^*, \tag{16.6}$$

This formulation of the elastodynamics equation can be readily encoded using a FEM method such as FreeFem++ or Comsol software (for details, see [31]). Solutions of Eq. (16.6) yield exact distributions of all displacements within waveguide's cross-section. The associated kinetic energy density is given by

$$E_c(\Omega) = \frac{1}{2}\Omega^2\int_S \rho|u_i|^2 = \frac{1}{2}\rho\Omega^2(u_x^2 + u_y^2 + u_z^2) \tag{16.7}$$

with ρ the material density and Ω the acoustic frequency. Note also that the transverse variations of the two pump and Stokes optical waves are taken into account. For the sake of simplicity, they are normalized to the z component of the Poynting vector as

$$P_z = \mathrm{Re}\left\{\int_\sigma E^* \times H dS\right\}. \tag{16.8}$$

We also include the phonon lifetime in the numerical model by taking into account the material elastic losses. This is done by assuming a complex elastic tensor of the form $c_{ijkl} + i\Omega\eta_{ijkl}$ where η_{ijkl} is the viscosity tensor [32]. This loss model is compatible with the usual assumption that the product of the quality factor Q and the acoustic frequency is constant for a given material (e.g., for silica, $Q \times \Omega/2\pi = 5$ THz [29]).

Although different from the stochastic dynamical model derived by Gaeta and Boyd in 1991 [32], the elastodynamics model provides a very good estimate of all acoustic waves and the displacements generated by electrostriction, in particular for small wavelength-scale optical waveguides. When solved as function of the acoustic frequency Ω, the kinetic energy density E_c can be directly compared to the experimental Brillouin spectrum. It also allows the Brillouin gain characteristics with arbitrary refractive index and materials compositions to be accurately calculated. In the following, we will apply this electrostriction-based model to the specific case of silica optical microwires and then of photonic crystal fibers.

16.4 Tapered Optical Fibers

Tapered optical fibers (TOFs), also known as optical microwires, microfibers, and nanofibers, are the tiny cousins of optical fibers [33, 34]. These hair-like slivers of glass, fabricated by tapering standard optical fibers down to the micro and nanoscale, enable enhanced nonlinear optical effects, large evanescent field, widely tailorable dispersion, and thus applications not currently possible with comparatively bulky optical fibers. Although TOFs have helped greatly to enhance the optical Kerr effect and stimulated Raman scattering (SRS) for supercontinuum generation [35, 36], Brillouin scattering in these tiny waveguides has not been explored yet. This section presents a complete experimental investigation and numerical modeling of Brillouin scattering in a sub-wavelength diameter optical fiber, revealing the full elastic wave distribution of such ultrathin optical fibers. Specifically, the generation of a new class of surface acoustic waves (SAW) is reported along with the observation of surface acoustic wave Brillouin scattering (SAWBS) in the backward direction. In addition to surface waves, experimental and theoretical investigations also show that silica microwires also support several widely-spaced Brillouin frequency peaks involving hybrid shear and longitudinal acoustic waves.

16.4.1 Principles

When a laser beam is coupled and guided into an uniform optical microwire as that shown schematically in Figure 16.2, the light both excites and feels several types of elastic waves with similar micrometer-scale wavelengths. In standard optical fibers, the light is guided into the core and is therefore solely sensitive to both shear and longitudinal bulk acoustic waves. In contrast, in subwavelength-diameter optical microwires, the guided light and particularly the evanescent field interact with the outer surface. Light can thus shake the wire and generate surface acoustic waves (SAW). The associated mechanical ripples will lead to small periodic changes of the effective refractive index along the optical microwire. When passing through this moving refractive index grating, light undergoes Bragg scattering in the backward direction according to the phase-matching condition, as in fiber-based Brillouin scattering from longitudinal acoustic waves (LAW). The backscattered Brillouin signal also undergoes a slight shift of its carrier frequency due to the Doppler effect according to the photon-phonon energy conservation law $\nu_B = 2n_{eff}V/\lambda$, with n_{eff} the effective refractive index of the microwire, λ the optical wavelength in vacuum, and V the acoustic phase velocity. The acoustic velocity, however, significantly differs for surface, shear, and longitudinal waves. Surface waves travel at a velocity between 0.87 and 0.95 of a shear wave (for fused silica, $V_S = 3400$ m.s^{-1}). This gives rise to new optical sidebands down-shifted from 6 GHz in the light spectrum. On top of surface waves, there are also bulk hybrid acoustic waves (HAW) involving both shear and longitudinal components because of the mechanical boundary

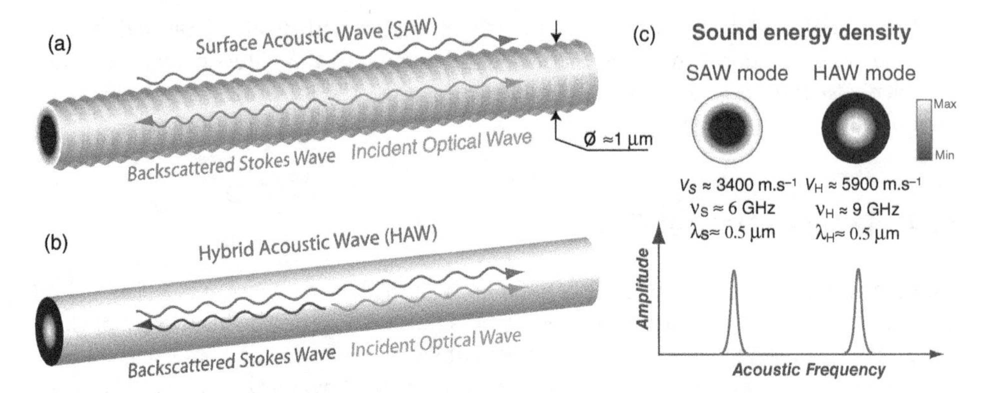

Figure 16.2 Illustration of surface and hybrid acoustic wave Brillouin scattering in silica microwire: (a) Schematic representation of the tapered silica fiber and of the wavevector interaction for surface acoustic waves (SAW). (b) Same for hybrid acoustic waves (HAW). (c) Comparison of sound energy densities, acoustic velocities and Brillouin frequency shifts between surface and hybrid waves. The surface wave is confined below the microwire surface and gives rise to small mechanical ripples, whereas the hybrid bulk wave remains confined within the core without altering the microwire surface. *Source*: Beugnot 2014 [15]. Reproduced with permission of Nature Publishing Group.

conditions [10] (Figure 16.2 (b)). Hybrid acoustic waves propagate at an intermediate speed between shear and longitudinal waves with acoustic frequencies around 9 GHz. Furthermore, the sound energy density of surface waves is confined at the air-silica interface of the microwire, leading to small mechanical ripples of a few picometer, whereas the HAW energy density remains trapped within the core without altering the wire shape (Figure 16.2 (c)).

16.4.2 Experiments

Experimentally, we investigated several sub-wavelength-diameter silica optical fibers, as those shown in Figure 16.3 (c). They were drawn from a commercial single-mode fiber (SMF) using the heat-brush technique [37,38] (for details, see Fabrication method). They have a waist diameter around 1 μm, a length of 8-cm and the input/output tapers are 15-mm long. Figure 16.3 (b) shows the experimental setup [39]. As a pump laser, we used a narrow-linewidth continuous-wave distributed-feedback laser running at a wavelength of 1,550 nm. The laser output was split into two beams using a fiber coupler. One beam was amplified and injected in the optical microwire through an optical circulator, while the other beam served for detection. We then implemented a heterodyne detection in which the backscattered light from the microwire was mixed with the input coming from a second fiber coupler. The resulting beat note was then detected using a fast photodiode and averaged Brillouin spectra were recorded using an electrical spectrum analyzer.

Figure 16.4 (a) shows the experimental Brillouin spectrum for an input power of 100 mW and for a wire diameter of 1 μm. We see the clear emergence of several frequency peaks with different weights and linewidths in a radio-frequency range from 6 GHz to 11 GHz. First, the high-frequency at 10.86 GHz originates from standard Brillouin scattering in the 2-m-long untapered fiber sections and can be disregarded. More importantly, three other peaks appear at 8.33 GHz, 9.3 GHz, and slightly above 10 GHz,

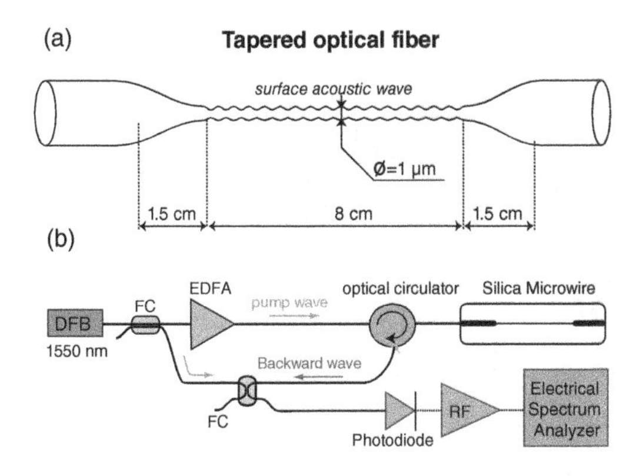

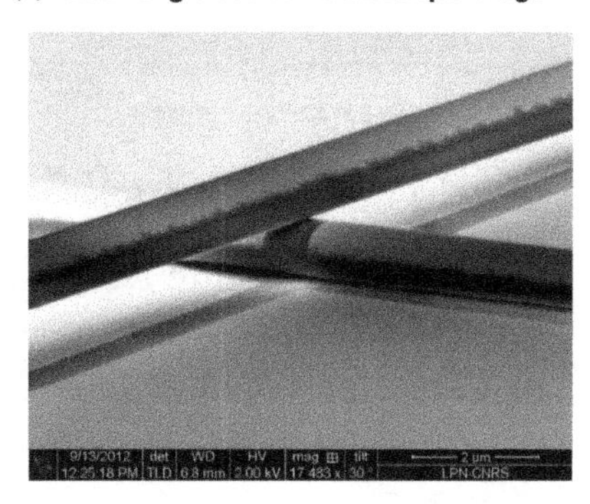

Figure 16.3 (a) Schematic of the silica tapered optical fiber. (b) Experimental setup. The backscattered light is mixed with the input and the Brillouin signal is detected using a fast photodiode and an electrical spectrum analyzer. DFB: Distributed feedback laser. FC: 10:90 fiber coupler. EDFA: Erbium-doped fiber amplifier. (c) Scanning electron microscope (SEM) image of typical silica wire. *Source*: Beugnot 2014 [15]. Reproduced with permission of Nature Publishing Group.

respectively. The two first peaks exhibit a linewidth around 25 MHz, in good agreement with the acoustic amplitude lifetime in fused silica (~ 10 ns). Based on numerical simulations described below, we identified them as resulting from Brillouin scattering from hybrid waves while the two other resonances around 6 GHz are clearly the signature of surface acoustic waves.

16.4.3 Numerical Simulations

To better understand the nature of these elastic modes, we performed numerical simulations based on the elastodynamics equation described in previous section [29, 31]. This new modeling provides an excellent estimate of the theoretical Brillouin gain spectrum by computing the kinetic energy density of acoustic phonons generated by electrostriction. For the sake of simplicity, we ignored the two conical taper sections of the microwire and we considered the silica microwire as a rod-type cylinder, taking into account all elastic and optical parameters of silica glass. The results of our numerical simulations are shown in Figures 16.4 (b–h). The elastic energy is plotted in Figure 16.4 (b) versus acoustic frequency and for a wire diameter varying from 1 µm to 1.35 µm. As can be seen, we retrieve most of the surface and hybrid acoustic resonances, as those observed experimentally in Figure 16.4 (a). There are, however, slight differences with experiment regarding the precise acoustic resonant frequencies. We attribute them to both the microwire uniformity and diameter uncertainty. Nevertheless, we can clearly identify two surface acoustic modes around 6 GHz, and three hybrid acoustic modes around 9 GHz. More information is provided by the modal distribution of the interacting waves. Figure 16.4 (c) shows the spatial distribution of the optical mode intensity in the waist region at a wavelength of 1,550 nm. The black circle marks out the interface between silica and air. The optical mode is guided in the microwire with a rather long evanescent tail extending outside silica. The tight confinement of the light into the microwire provides an electrostrictive stress up to 7 kPa, which is about 30 times larger than in usual optical fiber under same incident optical power. The electrostrictive stress

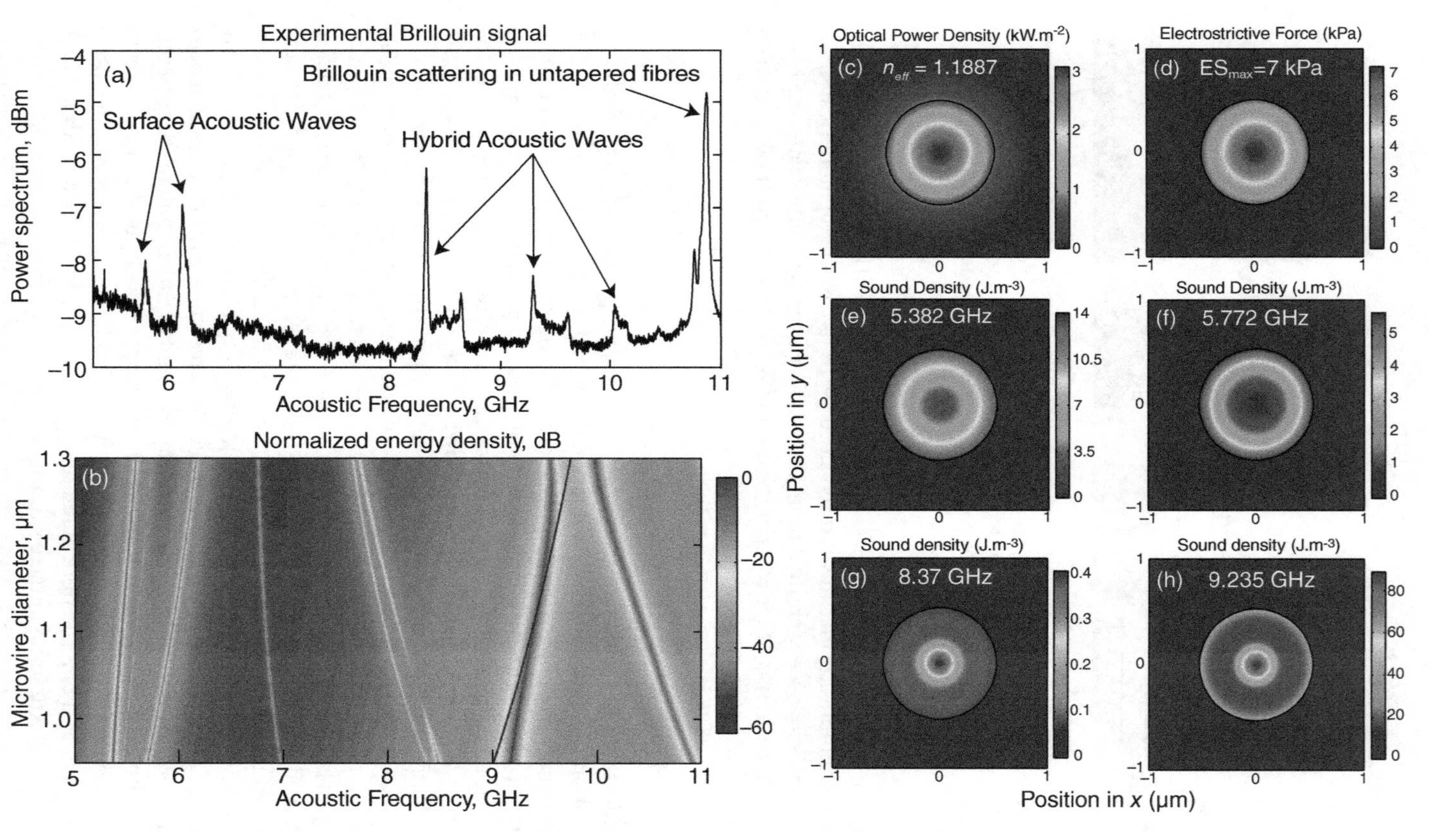

Figure 16.4 (a) Experimental Brillouin spectrum measured with a c.w. laser at 1,550 nm in the silica tapered optical fiber shown in Figure 16.3 (a). (b) Numerical simulations of Brillouin scattering spectrum for a wire waist diameter increasing from 1 μm to 1.35 μm. The black line shows the acoustic frequency given by the phase-matching condition for pure longitudinal waves ($K = 2k_1$). (c) Optical power density in the fundamental TE-like mode at a wavelength of 1,550 nm (n_{eff} = 1.1887). (d) Spatial distribution of the electrostrictive stress under 1 W optical power. (e,f) Sound density of surface acoustic waves at 5.382 GHz and 5.772 GHz. (g,h) Sound density of hybrid bulk acoustic modes at 8.37 GHz and 9.235 GHz. *Source*: Beugnot 2014 [15]. Reproduced with permission of Nature Publishing Group.

distribution, shown in Figure 16.4 (d), follows the optical mode distribution without extending outward the microwire. The sound energy density of the two surface acoustic waves at 5.382 GHz and 5.772 GHz are plotted in Figures 16.4 (e) and (f) for a waist of 1.05 μm. As can be seen, the elastic energy density is mainly localized below the wire surface. In contrast, the kinetic energy density of hybrid acoustic waves at 8.37 GHz and 9.235 GHz, shown in Figures 16.4 (g) and (h), are mostly confined within the core of the microwire. The role of confining elastic energy is due to the elastic waveguide properties of such microwires, mostly its small diameter and hard mechanical boundary conditions.

To go further into details, we plot in Figure 16.5 (a) the full elastic wave spectrum over a wider range of wire diameter till 3.6 μm. For every diameter, the fundamental optical mode is recomputed and normalized. The refractive index n_{eff} increases smoothly with the diameter as the optical mode becomes more and more localized within silica rather than air. The two SAWs that we identified before exist for every diameter with a slight shift in frequency. The sound density is significant for small core but strongly reduces as the diameter increases beyond the optical wavelength (1.5 μm). Figure 16.5 (a) also shows that, for a given microwire diameter, multiple HAW with widely-spaced frequencies can be simultaneously excited, as in small-core photonic crystal fibers [10, 13]. In such small sub-wavelength waveguides, the light-sound interaction is fundamentally different from standard optical fibers. Brillouin spectrum is not simply the signature of a single bulk longitudinal sound wave. Instead, waveguide boundary conditions induce a strong coupling of shear and longitudinal displacements, resulting in a much richer dynamics including hybrid and surface acoustic waves. We can also see the emergence of several avoided crossings in Figure 16.5 (a), due to the fact that HAWs are not orthogonal and thus strongly interact. In contrast, the two SAW branches cross without interacting for a diameter ~ 0.85 μm because they have orthogonal polarizations. Figure 16.5 (b) shows the associated transverse and axial displacements in the microwire for one of surface and hybrid modes, denoted A and B in Figure 16.5 (a). Transverse and axial displacements are simply defined as $\sqrt{(u_x^2 + u_y^2)}$ and $(|u_z|)$, respectively. R is defined as the ratio between spatial integrals of the shear and longitudinal displacements ($R = 1$ for a pure shear wave). As can be seen, the surface acoustic mode A at 5.382 GHz exhibits a transversal displacement of a few picometer and a weak axial displacement just below the microwire surface. Clearly, this is the signature of a surface Rayleigh wave that combines both a longitudinal and transverse motion to create mechanical ripples. In contrast, the displacements for the bulk hybrid mode B are smaller ($R = 0.589$) and mainly localized within the core of the wire.

16.4.4 Photonic Crystal Fibers

Photonic crystal fibers, also known as microstructured or holey fibers, have also generated great interest in the scientific community for their ability to guide and manipulate both photons and phonons [10–13, 40, 41]. The periodic wavelength-scale air-hole microstructure of PCFs indeed deeply changes the elastic wave distribution compared to what is commonly observed in conventional single-mode fibers. This leads to new interesting characteristics for both forward and backward Brillouin scattering, including the generation of high-frequency shear and hybrid acoustic waves, in a way similar to optical microwires.

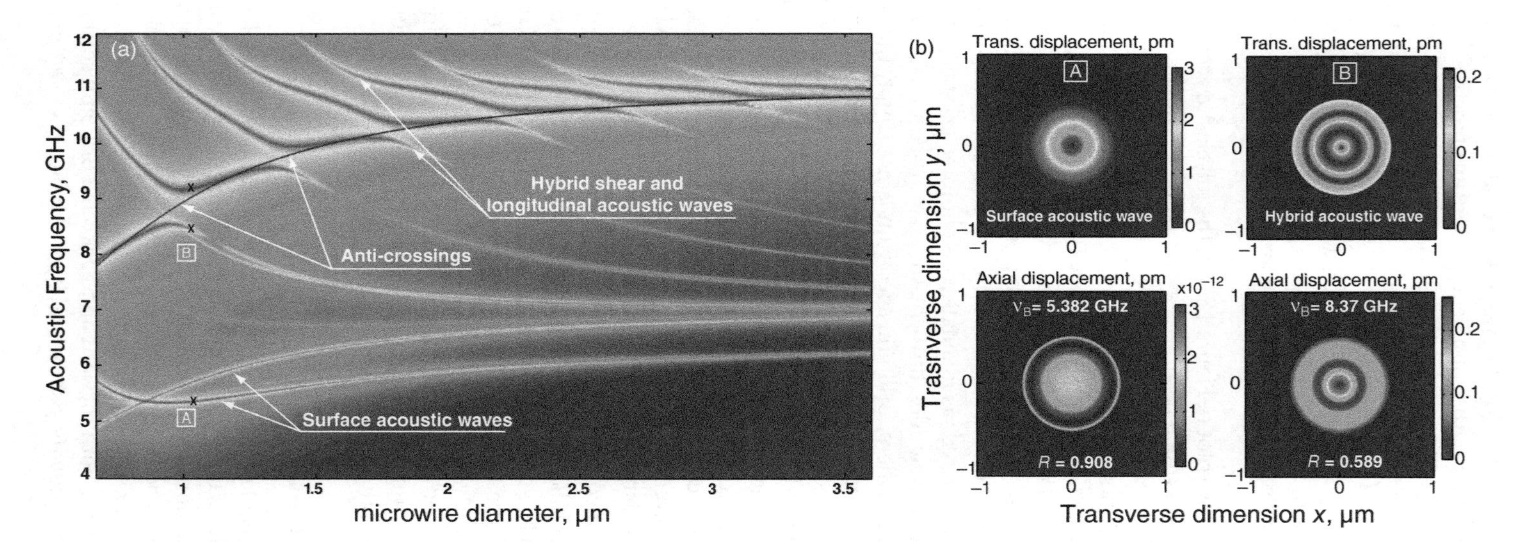

Figure 16.5 (a) Color plot of energy density of acoustic waves as a function of frequency and for a wire diameter varying from 0.6 μm to 3.5 μm. The black curve shows the acoustic frequency given by the standard phase-matching condition ($K = 2k_1$) for pure longitudinal waves. White arrows indicate the surface and hybrid acoustic waves, and the anti-crossing points due to the strong coupling regime. (b) Transverse and axial displacements in the microwire associated with surface and hybrid acoustic modes labelled A and B in (a). *Source*: Beugnot 2014 [15]. Reproduced with permission of Nature Publishing Group.

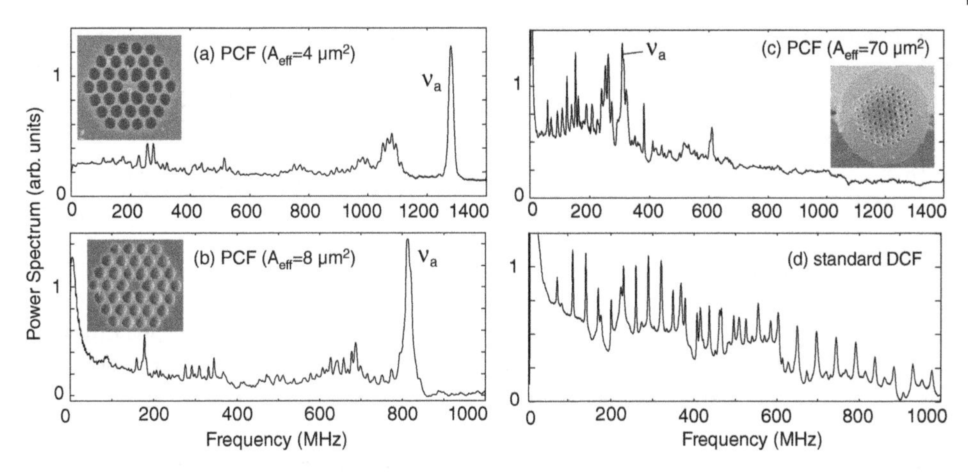

Figure 16.6 Forward Brillouin scattering spectra measured in (a-c) three solid-core PCFs with different effective mode area and (d) a standard all-solid dispersion-compensating fiber (DCF). The resolution is 300 kHz. The insets show scanning electron microscope images of the fiber cross-sections.
Source: Beugnot 2007 [12]. Reproduced with permission of OSA.

To observe these new characteristics, GAWBS was experimentally investigated in several solid-core photonic crystal fibers using the fiber loop mirror technique [12]. Polarized GAWBS in an optical fiber results from the radial dilatation or R_{0m} modes which mainly give rise to weak phase modulation [3]. Note that there also exist other torso-radial TR_{2m} modes that lead to polarization modulation [3]. To observe the R_{0m} modes, it is useful to set up an optical interferometer such as the fiber loop mirror, so as to convert the phase modulation to amplitude modulation [12]. The GAWBS spectrum can then be recorded by use of photodetector and an electrical spectrum analyzer [12]. Figure 16.6 shows the experimental spectra measured in three PCFs with different effective mode area. For comparison, that of a standard dispersion-compensating fiber (DCF) is also shown in Figure 16.6 (d). As can be seen, all three PCFs exhibit GAWBS spectra completely different from that of the DCF, which shows the generation of a set of radial acoustic modes over a broad frequency range up to 1000 MHz [3]. First, we can see a substantial suppression of GAWBS modes in small-core PCFs in the low frequency range. Second, Figures 16.6 (a) and (b) clearly show the efficient generation of shear elastic waves at 812 MHz and 1281 MHz, respectively. They have large Q factors, $\nu_a/\Delta\nu_a \approx 50$, where ν_a is the acoustic frequency and $\Delta\nu_a$ is the full width at half maximum (FWHM). The generation of these phonons in PCFs can be understood by considering the case of a conventional fiber. In SMFs, GAWBS mainly depends on the fiber cladding diameter (typically 125 μm) that allows a wide spatial domain of existence of acoustic modes to be generated [3]. However, the optical mode is guided within the small core and thus only partially overlaps with acoustic modes. In PCF, a strong dependence on boundary conditions is introduced by the air-hole microstructure leading to both the suppression of most cladding acoustic modes and to the significant enhancement inside the PCF core. The suppression of most of GAWBS modes in the lower frequency range results from the fact that the mechanical coupling between core and cladding is significantly reduced by the surrounded holes [41]. In the large-core PCF however, GAWBS resonances in this lower frequency range below 300 MHz are more closely spaced in PCFs as compared to the standard DCF. It is important to stress that the frequency ν_a of the main transverse acoustic modes observed in small-core PCFs, as those shown in Figures 16.6 (a) and (b), can readily be related to the core diameter, by $\nu_a = c_{R01}\frac{V_S}{D}$, where D is the fiber core diameter, $V_S = 3740$ m.s^{-1} is the shear acoustic velocity and c_{R01} is a factor close to 1 which depends on the type of acoustic waves, in this case the fundamental radial R_{01}-mode [13, 42].

We then performed numerical simulations based on the elastodynamics equation in a silica-based triangular lattice PCF, with a hole diameter of 2 μm, an air filling ratio $d/\Lambda = 0.6$, and a core size of 4.4 μm. Figures 16.7 (a) and (b) shows both the computed forward and backward Brillouin spectra from the elastodynamics model. As can be seen, the forward Brillouin spectrum exhibits a set of transverse acoustic modes including one in particular at 925 MHz that is very significant. The associated transverse displacements and sound energy density are plotted in Figures 16.7 (c)–(e). On the other hand, the backward Brillouin spectrum shown in Figure 16.7 (b) exhibits two main frequency-detuned peaks around 11 GHz. These peaks are in fact the signature of hybrid shear and longitudinal acoustic waves, as reported in the silica optical microwire in the previous section. Their associated axial displacements and sound energy density are plotted in Figures 16.7 (d)–(f). The longitudinal deformation of the phonon wave packet induced by electrostriction forces in the SBS case is indeed confined by a combination of the air-hole microstructure and the optical force. To go further into details, we compare in Figure 16.8 the experimental and computed Brillouin spectra in another PCF with a larger core diameter of 11 μm. The agreement between computed and experimental

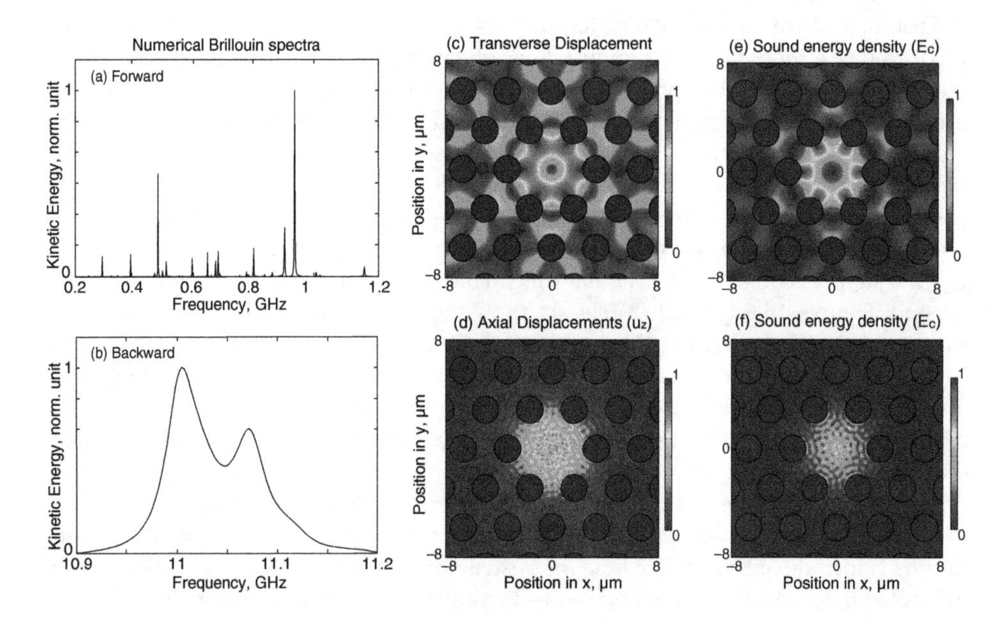

Figure 16.7 (a) Numerical forward and (b) backward Brillouin spectrum. (c) Spatial distribution of transverse displacements and (e) kinetic energy density of the elastic resonance at 925 MHz. (d) Axial displacements and (f) elastic energy of the longitudinal acoustic wave at 11 GHz.

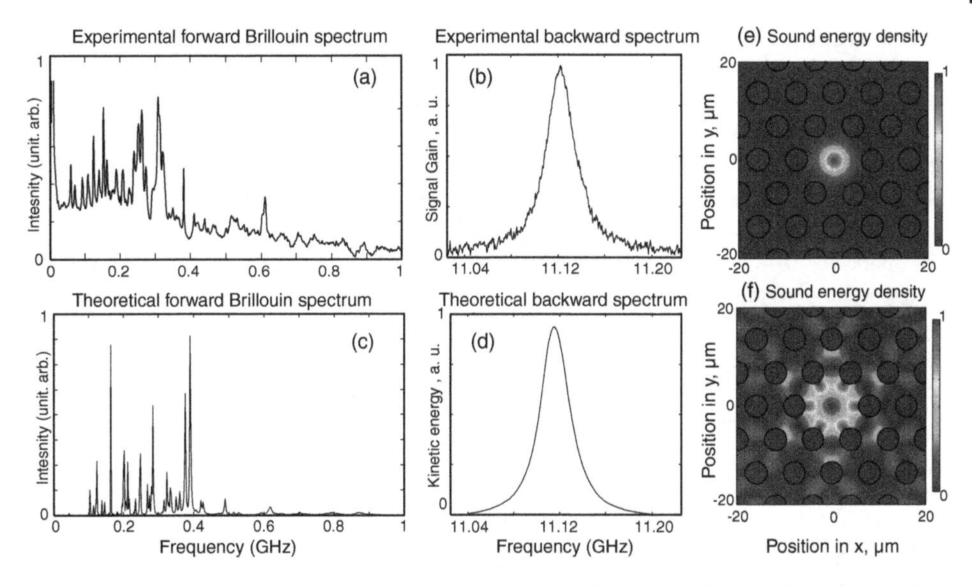

Figure 16.8 Comparison between (a–b) experimental and (c–d) theoretical forward and backward Brillouin spectra in a photonic crystal fiber. Computed sound energy density of (e) the longitudinal one at 11.12 GHz and (f) the shear elastic wave at 390 MHz.

spectra is excellent. We found in particular most of radial elastic waves up to 0.4 GHz and the main longitudinal elastic wave at 11.12 GHz with nearly the same linewidth.

16.5 Conclusion

This chapter was devoted to the complete description of Brillouin light scattering in specialty optical fibers including photonic crystal fibers and tapered optical fibers. It has been shown in particular that light-sound interactions in these tiny waveguides are very different from standard optical fibers. These include the generation of high-frequency shear, hybrid and surface acoustic waves that do exist in standard optical fibers. These results suggest that PCFs or tapered optical fibers can be advantageously used to enhance and harness Brillouin scattering in view of potential applications to fiber-optic sensors and all-optical processing for telecommunication industry. Specifically, surface acoustic wave Brillouin scattering can find strong potential applications for optical sensing and detection because these acoustic waves are inherently sensitive to surface features and defects and are already used in many sensing systems (e.g., MEMS). They could be seen as a new class of Brillouin optical sensors that rely on the modulation of surface acoustic waves to sense a physical phenomenon (e.g. temperature, stress, gas). Changes in amplitude and frequency of the Brillouin signal through the silica microwire could be advantageously used to measure the presence of the desired phenomenon. One can also imagine a strong interaction between surface waves and surface plasmons by depositing a thin layer of gold or silver [43,44]. Finally, this chapter contributes to the further understanding of the intriguing light-sound interactions in sub-wavelength optics and in nanophotonics.

References

1 Kobyakov, A., Sauer, M., and Chowdhury, D. (2010) Stimulated Brillouin scattering in optical fibers. *Advances in Optics and Photonics*, **2** (1), 1–59.
2 Agrawal, P. (2001) *NonLinear Fiber Optics*, Academic Press, 3rd edn.
3 Shelby, R.M., Levenson, M., and Bayer, P. (1985) Guided acoustic-wave Brillouin scattering. *Physical Review B*, **31**, 5244–5252.
4 Chiao, R.Y., Townes, C.H., and Stoicheff, B.P. (1964) Stimulated Brillouin scattering and coherent generation of intense hypersonic waves. *Physical Review Letters*, **12** (21), 592–595.
5 Smith, R.G. (1972) Optical power handling capacity of low loss optical fibers as determined by stimulated Raman and Brillouin scattering. *Applied Optics*, **11** (11), 2489.
6 Thévenaz, L. (2010) Brillouin distributed time-domain sensing in optical fibers: state of the art and perspectives. *Frontiers of Optoelectronics in China*, **3** (1), 13–21.
7 Lee, M.W., Stiller, B., Hauden, J., *et al.* (2012) Differential phase-shift-keying technique-based Brillouin echo-distributed sensing. *Photonics Technology Letters, IEEE*, **24** (1), 79 –81.
8 Carry, E., Beugnot, J.C., Stiller, B., *et al.* (2011) Temperature coefficient of the high-frequency guided acoustic mode in a photonic crystal fiber. *Applied Optics*, **50** (35), 6543–6547.
9 Kang, M.S., Butsch, A., and Russell, P.S.J. (2011) Reconfigurable light-driven opto-acoustic isolators in photonic crystal fibre. *Nat. Photonics*, pp. 1–5.
10 Dainese, P., Russell, P.S.J., Joly, N., *et al.* (2006) Stimulated Brillouin scattering from multi-GHz-guided acoustic phonons in nanostructured photonic crystal fibres. *Nature Physics*, **2** (6), 388–392.
11 Kang, M.S., Nazarkin, A., Brenn, A., and Russell, P.S.J. (2009) Tightly trapped acoustic phonons in photonic crystal fibres as highly nonlinear artificial Raman oscillators. *Nature Physics*, **5** (4), 276–280.
12 Beugnot, J.C., Sylvestre, T., Maillotte, H., Mélin, G., and Laude, V. (2007) Guided acoustic wave Brillouin scattering in photonic crystal fibers. *Optics letters*, **32** (1), 17–19.
13 Stiller, B., Delqué, M., *et al.* (2011) Frequency-selective excitation of guided acoustic modes in a photonic crystal fiber. *Optics Express*, **19** (8), 7689–7694.
14 Poulton, C.G., Pant, R., and Eggleton, B.J. (2013) Acoustic confinement and stimulated Brillouin scattering in integrated optical waveguides. *Journal of the Optical Society of America B*, **30** (10), 2657.
15 Beugnot, J.C., Lebrun, S., Pauliat, G., *et al.* (2014) Brillouin light scattering from surface acoustic waves in a subwavelength-diameter optical fibre. *Nature Communications*, **5**.
16 Wang, J., Zhu, Y., Zhang, R., and Gauthier, D.J. (2011) FSBS resonances observed in a standard highly nonlinear fiber. *Opt. Express*, **19** (6), 5339–5349.
17 Brillouin, L. (1922) Diffusion de la lumiere et des rayons x par un corps transparent homogène. *Ann. Physique*, **17**, 88.
18 Mandelstam, L.I. (1926) *Zh. Russ. Fiz.-Khim.* **58**, 381 (1926).
19 Ippen, E.P. and Stolen, R.H. (1972) Stimulated Brillouin scattering in optical fibers. *Applied Physics Letters*, **21** (11), 539–541.

20 Thévenaz, L. (2008) Slow and fast light in optical fibres. *Nature Photonics*, **2** (8), 474–481.
21 Marpaung, D., Morrison, B., Pagani, M., *et al.* (2015) Low-power, chip-based stimulated Brillouin scattering microwave photonic filter with ultrahigh selectivity. *Optica*, **2** (2), 76–83.
22 Tchahame, J.C., Beugnot, J.C., Kudlinski, A., and Sylvestre, T. (2015) Multimode Brillouin spectrum in a long tapered birefringent photonic crystal fiber. *Optics Letters*, **40** (18), 4281–4284.
23 Laude, V., Khelif, A., Benchabane, S., *et al.* (2005) Phononic band-gap guidance of acoustic modes in photonic crystal fibers. **71**, 045 107.
24 Beugnot, J.C., Ahmad, R., Rochette, *et al.* (2014) Reduction and control of stimulated brillouin scattering in polymer-coated chalcogenide optical microwires. *Optics Letters*, **39** (3), 482–485.
25 Bahl, G., Tomes, M., Marquardt, F., and Carmon, T. (2012) Observation of spontaneous Brillouin cooling. *Nature Physics*, **8** (3), 203–207.
26 Lin, G., Diallo, S., Saleh, K., *et al.* (2014) Cascaded Brillouin lasing in monolithic barium fluoride whispering gallery mode resonators. *Applied Physics Letters*, **105** (23), 231103.
27 Rakich, P.T., Reinke, C., Camacho, R., Davids, P., and Wang, Z. (2012) Giant enhancement of stimulated Brillouin scattering in the subwavelength limit. *Physical Review X*, **2** (1), 011 008.
28 Van Laer, R., Kuyken, B., Van Thourhout, D., and Baets, R. (2015) Interaction between light and highly confined hypersound in a silicon photonic nanowire. *Nat Photon*, **9** (3), 199–203.
29 Beugnot, J.C. and Laude, V. (2012) Electrostriction and guidance of acoustic phonons in optical fibers. *Physical Review B*, **86** (22), 224 304.
30 Royer, D. (2000) *Elastic Waves in Solids I: Free and Guided Propagation*, Springer, Berlin.
31 Laude, V. and Beugnot, J.C. (2013) Generation of phonons from electrostriction in small-core optical waveguides. *AIP Advances*, **3** (4), 042109.
32 Gaeta, A.L. and Boyd, R. (1991) Stochastic dynamics of stimulated Brillouin scattering in an optical fiber. *Optics Letters*, **44** (5), 3205–3208.
33 Wu, X. and Tong, L. (2013) Optical microfibers and nanofibers. *Nanophotonics*, **2**, 407–428.
34 Brambilla, G. (2010) Optical fibre nanowires and microwires: a review. *Journal of Optics*, **12** (4), 043 001.
35 Foster, M.A., Turner, A.C., Lipson, M., and Gaeta, A.L. (2008) Nonlinear optics in photonic nanowires. *Optics Express*, **16** (2), 1300–1320.
36 Leon-Saval, S., Birks, T., Wadsworth, W., Russell, P.S.J., and Mason, M. (2004) Supercontinuum generation in submicron fibre waveguides. *Optics Express*, **12** (13), 2864–2869.
37 Shan, L., Pauliat, G., Vienne, G., Tong, L., and Lebrun, S. (2013) Stimulated Raman scattering in the evanescent field of liquid immersed tapered nanofibers. *Applied Physics Letters*, **102** (20), 201110.
38 Baker, C. and Rochette, M. (2011) A generalized heat-brush approach for precise control of the waist profile in fiber tapers. *Opt. Mater. Express*, **1** (6), 1065–1076.

39 Beugnot, J.C., Sylvestre, T., Maillotte, H., *et al.* (2007) Complete experimental characterization of stimulated Brillouin scattering in photonic crystal fibers. **15**, 15 517–15 522.

40 Russell, P.S.J. (2006) Photonic-crystal fibers. *Optics Communications*, **24** (12), 4729–4749.

41 Elser, D., Andersen, U.L., Korn, A., *et al.* (2006) Reduction of guided acoustic wave Brillouin scattering in photonic crystal fibers. **97**, 133 901.

42 Kang, M.S., Brenn, A., Wiederhecker, G.S., and Russell, P.S. (2008) Optical excitation and characterization of gigahertz acoustic resonances in optical fiber tapers. *Applied Physics Letters*, **93** (13), 131110.

43 Kauranen, M. and Zayats, A.V. (2012) Nonlinear plasmonics. *Nature Photonics*, **6** (11), 737–748.

44 Eberle, R. and Pietralla, M. (2008) Plasmon mediated Brillouin scattering of surface acoustic waves. *Thin Solid Films*, **516** (15), 4803–4808.

Index

Shaping Light in Nonlinear Optical Fibers, First Edition. Edited by Sonia Boscolo and Christophe Finot.